Date Due

Disciplines in Combinational and Sequential Circuit Design

McGRAW-HILL ELECTRICAL AND ELECTRONIC ENGINEERING SERIES

Frederick Emmons Terman, *Consulting Editor*
W. W. Harman *and* **J. G. Truxal,** *Associate Consulting Editors*

ANGELAKOS AND EVERHART · *Microwave Communications*
ANGELO · *Electronic Circuits*
ANGELO · *Electronics: BJTs, FETs, and Microcircuits*
ASELTINE · *Transform Method in Linear System Analysis*
ATWATER · *Introduction to Microwave Theory*
BENNETT · *Introduction to Signal Transmission*
BERANEK · *Acoustics*
BRACEWELL · *The Fourier Transform and Its Application*
BRENNER AND JAVID · *Analysis of Electric Circuits*
BROWN · *Analysis of Linear Time-invariant Systems*
BRUNS AND SAUNDERS · *Analysis of Feedback Control Systems*
CARLSON · *Communication Systems: An Introduction to Signals and Noise in Electrical Communication*
CHEN · *The Analysis of Linear Systems*
CHEN · *Linear Network Design and Synthesis*
CHIRLIAN · *Analysis and Design of Electronic Circuits*
CHIRLIAN · *Basic Network Theory*
CHIRLIAN AND ZEMANIAN · *Electronics*
CLEMENT AND JOHNSON · *Electrical Engineering Science*
CUNNINGHAM · *Introduction to Nonlinear Analysis*
D'AZZO AND HOUPIS · *Feedback Control System Analysis and Synthesis*
ELGERD · *Control Systems Theory*
EVELEIGH · *Adaptive Control and Optimization Techniques*
FEINSTEIN · *Foundations of Information Theory*
FITZGERALD, HIGGINBOTHAM, AND GRABEL · *Basic Electrical Engineering*
FITZGERALD AND KINGSLEY · *Electric Machinery*
FRANK · *Electrical Measurement Analysis*
FRIEDLAND, WING, AND ASH · *Principles of Linear Networks*
GEMMLICH AND HAMMOND · *Electromechanical Systems*
GHAUSI · *Principles and Design of Linear Active Circuits*
GHOSE · *Microwave Circuit Theory and Analysis*
GLASFORD · *Fundamentals of Television Engineering*
GREINER · *Semiconductor Devices and Applications*
HAMMOND AND GEHMLICH · *Electrical Engineering*
HANCOCK · *An Introduction to the Principles of Communication Theory*
HARMAN · *Fundamentals of Electronic Motion*
HARMAN · *Principles of the Statistical Theory of Communication*
HARMAN AND LYTLE · *Electrical and Mechanical Networks*
HAYASHI · *Nonlinear Oscillations in Physical Systems*
HAYT · *Engineering Electromagnetics*
HAYT AND KEMMERLY · *Engineering Circuit Analysis*
HILL · *Electronics in Engineering*
JAVID AND BROWN · *Field Analysis and Electromagnetics*

Johnson · *Transmission Lines and Networks*
Koenig and Blackwell · *Electromechanical System Theory*
Koenig, Tokad, and Kesavan · *Analysis of Discrete Physical Systems*
Kraus · *Antennas*
Kraus · *Electromagnetics*
Kuh and Pederson · *Principles of Circuit Synthesis*
Kuo · *Linear Networks and Systems*
Ledley · *Digital Computer and Control Engineering*
LePage · *Complex Variables and the Laplace Transform for Engineering*
LePage and Seely · *General Network Analysis*
Levi and Panzer · *Electromechanical Power Conversion*
Ley, Lutz, and Rehberg · *Linear Circuit Analysis*
Linvill and Gibbons · *Transistors and Active Circuits*
Littauer · *Pulse Electronics*
Lynch and Truxal · *Introductory System Analysis*
Lynch and Truxal · *Principles of Electronic Instrumentation*
Lynch and Truxal · *Signals and Systems in Electrical Engineering*
McCluskey · *Introduction to the Theory of Switching Circuits*
Manning · *Electrical Circuits*
Meisel · *Principles of Electromechanical-energy Conversion*
Millman and Halkias · *Electronic Devices and Circuits*
Millman and Taub · *Pulse, Digital, and Switching Waveforms*
Minorsky · *Theory of Nonlinear Control Systems*
Mishkin and Braun · *Adaptive Control Systems*
Moore · *Traveling-wave Engineering*
Murdoch · *Network Theory*
Nanavati · *An Introduction to Semiconductor Electronics*
Oberman · *Disciplines in Combinational and Sequential Circuit Design*
Pettit · *Electronic Switching, Timing, and Pulse Circuits*
Pettit and McWhorter · *Electronic Amplifier Circuits*
Pfeiffer · *Concepts of Probability Theory*
Reza · *An Introduction to Information Theory*
Reza and Seely · *Modern Network Analysis*
Ruston and Bordogna · *Electric Networks: Functions, Filters, Analysis*
Ryder · *Engineering Electronics*
Schilling and Belove · *Electronic Circuits: Discrete and Integrated*
Schwartz · *Information Transmission, Modulation, and Noise*
Schwarz and Friedland · *Linear Systems*
Seely · *Electromechanical Energy Conversion*
Seely · *Electron-tube Circuits*
Seifert and Steeg · *Control Systems Engineering*
Shooman · *Probabilistic Reliability: An Engineering Approach*
Siskind · *Direct-current Machinery*
Skilling · *Electric Transmission Lines*
Stevenson · *Elements of Power System Analysis*
Stewart · *Fundamentals of Signal Theory*
Strauss · *Wave Generation and Shaping*
Su · *Active Network Synthesis*
Terman · *Electronic and Radio Engineering*
Terman and Pettit · *Electronic Measurements*
Thaler · *Elements of Servomechanism Theory*

Thaler and Brown · *Analysis and Design of Feedback Control Systems*
Thaler and Pastel · *Analysis and Design of Nonlinear Feedback Control Systems*
Tou · *Digital and Sampled-data Control Systems*
Tou · *Modern Control Theory*
Truxal · *Automatic Feedback Control System Synthesis*
Tuttle · *Electric Networks: Analysis and Synthesis*
Valdes · *The Physical Theory of Transistors*
Van Bladel · *Electromagnetic Fields*
Weeks · *Antenna Engineering*
Weinberg · *Network Analysis and Synthesis*

Disciplines in Combinational and Sequential Circuit Design

R. M. M. OBERMAN

Professor of Information Processing
Technical University
Delft, Netherlands

McGraw-Hill Book Company

New York St. Louis San Francisco Düsseldorf London Mexico
Panama Sydney Toronto

To my wife
and to my laboratory staff past and present,
whose invaluable cooperation was instrumental
in the preparation of this book.

Disciplines in Combinational and Sequential Circuit Design

Library of Congress Catalog Card Number 78-90017

47495

1234567890 MAMM 79876543210

This book was set Modern by Spottiswoode, Ballantyne and Co. Ltd., and printed on permanent paper and bound by The Maple Press Company. The designer was Paula Tuerk; the drawings were done by Philip Cohen. The editors were B. G. Dandison, Jr., and Peter R. Karsten. Robert R. Laffler supervised the production.

Preface

The most difficult part of a book from the author's point of view is the preface in which he is expected to state his intentions. In a rapidly evolving technique, it is hardly possible to write a preface before the text itself has been written. In a new field—and switching techniques are indeed still new in spite of the fact that automatic telephony has been with us for more than 70 years—things are moving so fast that important new aspects emerge even while a book on the subject is being written.

A "prefabricated" preface therefore requires a more or less fundamental readjustment as soon as the first draft of the text itself is completed. Moreover, in such a case, the danger exists that the preface might be transfigured into a kind of epilogue in which all the facts and findings of the book may be considered as part and parcel of the acquired knowledge of the reader. Of course, this enables the author to maintain a high level of discussion in his preface—but then the preliminary remarks mean nothing to the reader who looks at the preface to see what are the author's main intentions and whether the book meets—or fails to meet—his particular requirements.

The first and foremost aim of the author as he was writing this book about disciplines in combination and sequential circuit design was to provide a textbook for students at the graduate level. As the preface itself will indicate, the task was made all the more difficult because American and European university systems hold widely divergent views about what a textbook should be. As the author sees it, the American system generally prefers that the textbook provide a comprehensive course with problems and bibliography, from which a subject can be taught by following the textbook closely. The students, moreover, are not required to go further into the matter or to study other texts than those mentioned in the chosen book.

Generally speaking, in most European universities, students do not use books that are written specifically for a given course. Thus, without following any specific book, the courses aim at acquainting the student with the fundamentals of the basic technique and indicating the possible applications of these fundamentals to engineering. The student is also provided with a choice bibliography on the subject so that he can, if need be, find his own way about.

Part of this book can be used as a textbook for graduate student courses (that on the fundamentals of switching algebra, simplification of switching equations, the theory of memory elements, the Huffman-Moore-Mealy-methods of circuit design); the other part provides the student with all the information he needs on switching from the logical point of view (pulse counting, transfer of digital information, arithmetic operations on digital information, etc.) so that he will be able to tackle the design of more complicated information-processing circuits. No attempt was made to subdivide the book into two such parts, however; instead, the sequence of the chapters dealing with these various topics was handled in such a way that it would not interfere with the systematic development of the various subject items.

The study of switching circuits has been limited in this book to their logical design. It does not purport to discuss methods of dimensioning circuit components, such as resistors, capacitors, and inductors. The dimensioning methods for switching circuits are not typical for this class of circuits; indeed, they can be brought under the heading of general design methods of electronic circuits, which do not fall within the scope of this book.

Nevertheless, a number of useful hints are given regarding some dimensioning problems, and the book even includes some actual examples of the basic electronic unit circuit most frequently used. This circuit in a prefabricated form could provide students with an easy tool for practical exercises in switching and thus be a first step toward the building of a university switching laboratory.

The first chapters deal with the principles of Boolean algebra and their application to contact circuits, electronic switching elements, and similar kinds of components. Circuits in which such logical elements are incorporated can be described by means of switching algebra expressions. In the design of switching circuits, it is economically justified to use the smallest possible number of such logical elements. This minimum number of logical elements can be determined by means of various factoring methods of the switching functions. A number of these methods are discussed. However, it is the author's view that sometimes too much attention is paid to perfecting these simplification methods which, generally speaking, do not account for a major part of the actual design of an information-processing circuit. Moreover, the feasibility of a circuit, as will be shown in the later chapters of the book, is greatly dependent on the basic organization of the circuit to be designed; in which case, the factoring of the switching formulas is of secondary importance.

The introduction of amplification and feedback in switching circuits makes it possible to design memory elements and to apply state equations to these elements. The circuits of some types of electronic elements (trigger or flip-flop) are treated. The design of the master and slave memory element, which consists in principle of two triggers or a trigger in combination with one or two capacitors or inductors, is analyzed thoroughly, because this memory element plays an important role in all sequential circuits.

The fundamental properties of a number of switching components, such as relays, diodes, tubes, transistors, and magnetic cores, are discussed from the point of view of the switching theory so that their application in the later chapters can be fully understood. The author is not limiting himself to one type of component; he is endeavoring to show that the design of switching circuits is to a large extent independent of the specific type of switching component. Automatic devices with very different physical components, such as electronic triggers and pneumatic cylinders, are essentially designed in the same way.

Following the introductory chapter on the binary unit counter, there follow a number of chapters devoted to switching methods that have become more or less standard. The operating principle of these standard methods is explained by means of time charts in which the basic design of circuits are presented in a more readily accessible form than that of the state diagrams used in the later chapters. The simplicity of a circuit and, correspondingly, its ease of maintenance are largely determined by the soundness of its basic organization.

Various methods for counting pulses in relay and electronic circuits are given. The shifting of digital information through various types of shift registers is also treated. Reference is made to the similarities of these basic types of circuits, which appear to be so very much different in operation. Separate chapters are devoted to the start-stop transmission of information, to arithmetic operations such as addition, subtraction, multiplication, and division, and to some coding problems of digital information. Strictly speaking, a thorough knowledge of these standard operations is not a prerequisite for the study of the last chapters of the book, in which the design of sequential circuits is dealt with from a very general point of view. The methods known as the Huffman-Moore-Mealy methods lead to the simplest state diagram for a given set of switching requirements. It is shown that these methods must be handled with the utmost care, otherwise the simplest solution for the switching problem under consideration will not be found. The simplest state diagram does not necessarily result in the simplest or best circuit possible, for a more complicated state diagram may very well simplify the internal organization of the circuit concerned and, in general, result in a better circuit. The basic design features involved are mostly counting and shifting, two types of operations which can be realized by means of very simple circuits, as is shown in the chapters on standard switching methods.

It is of course possible to design a circuit which will operate correctly without going through the process of assimilating all the information in this book that is of a more or less mathematical and logical nature; but it is the author's view that without such basic knowledge it is virtually impossible to design a circuit that one can truthfully claim does not contain more than the minimum number of components.

Each chapter of this book has a selective bibliography, which does not mean, however, that the book is a survey of the available literature. Many chapters give an original view of the subject of switching circuitry. In some cases, the bibliography lists authors who are the originators of the basic idea underlying a given chapter or paragraph. In other cases, it includes authors who give other ways of dealing with the same subject. Finally, it also lists books that can be used as basic tools for a more fundamental or thorough study of the topic involved.

The author hopes that his book will be useful to many people other than the group of graduate students for whom it was originally written. It incorporates a lot of the author's experience in switching gained during some 30 years of work in this field and in which he is still actively engaged with a staff of long-time enthusiasts and a rather rapidly changing group of students in his laboratory at the Technical University, Delft.

The author gained much of his experience in the field of switching during his employment with the Netherlands General Post Office, to which he is all the more grateful because the work there was ultimately to form the basis of the present book. A special tribute must be paid to Mr. A. Snijders, a close associate for some 25 years, for his many valuable ideas and remarks which, in some way or the other, have been incorporated in this book; to Mr. C. H. Eversdijk and Dr. F. W. Hogesteeger for their mathematical assistance in the preparation of this manuscript; to Mr. Th. van Onzen for the final form of many state diagrams and circuits; to Messrs. G. G. Sipkes and M. Koppenberg, all of the Laboratory for Information Processing Machines of the Technical University at Delft; and, finally, to Mr. G. R. E. T. Hondequin. The author also wishes to express gratitude to a great number of his students, now full-fledged electrical engineers, who have contributed in some way or the other to the production of this book during their graduate training.

R. M. M. Oberman

Contents

6. Switching Elements 116

7. Magnetic Cores as Switching Elements 173

8. Code Converters 214

9. Static Memory Circuits 234

10. Dynamic Memory Circuits 260

11. Pulse-Frequency-halving Circuits 298

12. Transformation of Relay Circuits 320

13. Relay Counters 334

14. The Design of Electronic Counters and Distributors 361

15. Reflected Binary Codes 388

16. Shift Registers 412

17. The Addition and Subtraction of Binary Numbers 492

1
Boolean Algebra

1-1 INTRODUCTION

In order to acquire a proper understanding of switching techniques and switching algebra it is necessary to familiarize oneself with a kind of mathematics called Boolean algebra after its founder George Boole. Boolean algebra is a deductive theory just as is euclidean geometry and ordinary algebra, which means that Boolean algebra is governed by a set of basic concepts and a set of postulates that express relations between these concepts.

For basic concepts and postulates mentioned in this text, we are indebted to E. V. Huntington.

BASIC CONCEPTS

1. There is a class K of elements A, B, C,..., which may be thought of as items in a list K (possibly an infinite list).
2. There are two operations + and · (read "plus" and "times"), which may be thought of as rules by means of which any two items in the list determine another item (which may or may not be contained in this list).

Postulates

1a. *If a is an element of class K and b is also an element of class K, then $a + b$ is also an element of class K.*

1b. *If a is an element of class K and b is also an element of class K, then $a \cdot b$ is also an element of class K.*

2a. *Class K contains an element 0 such that $a + 0 = a$ for every element of class K.*

2b. *Class K contains an element 1 such that $a \cdot 1 = a$ for every element of class K.*

3a. *If a and b are elements of class K, then $a + b = b + a$.*

3b. *If a and b are elements of class K, then $a \cdot b = b \cdot a$.*

4a. *If a, b, and c are elements of class K, then $a + (b \cdot c) = (a + b) \cdot (a + c)$.*

4b. *If a, b, and c are elements of class K, then $a \cdot (b + c) = (a \cdot b) + (a \cdot c)$.*

5. *If elements 0 and 1 of Postulate 2 are unique, then for every element of class K there is an element a' (read "a prime") of class K such that $a + a' = 1$ and $a \cdot a' = 0$.*

6. *Class K contains at least two elements a and b such that $a \neq b$.*

The first eight postulates form four pairs, which thereby illustrates the perfect symmetry or duality of this algebra. If in any of these postulates 0 is replaced by 1, 1 by 0, + by ·, and · by +, one automatically finds the dual of the original postulate.

In the formulation of the postulates given above, use was made of certain principles of logical procedure. For example, the variables a, b, and c appear in the postulates instead of the constants A, B, and C. These variables correspond to elements of class K but without reference to a specific element. So variable a may correspond to element A or B or C, etc. When class K contains n different elements the variables $a, b, \ldots$ are said to be *n-valued*.

Furthermore, use was made of the = sign. This sign means that the quantities which it separates are identical. For example:

$A = B$ means that elements A and B are identical
$a = A$ means that variable a corresponds to element A
$a = b$ means that the variables a and b represent the same element but the element itself is unspecified

In a deductive theory, the postulates must have the following properties:

1. They must be consistent, i.e., they must not be self-contradictory.
2. They must be independent, i.e., it must be impossible to derive any one of the postulates from any of the others.

The proof of these two properties of the postulates is not given here because it does not fall within the scope of this book.

1-2 THEOREMS

By means of a logical process, it is possible to derive a number of theorems from the above postulates. These theorems express relationships between the concepts in much the same way as do the postulates.

Theorem 1a *Element 0 of Postulate 2a is unique.*

Proof Let us assume that there are two elements 0 which we will call 0_1 and 0_2. Then for every element a of class K

$$a + 0_1 = a \qquad \text{and} \qquad a + 0_2 = a \tag{1-1}$$

in accordance with Postulate 2a. Since 0_2 is an element of class K (Postulate 2a), we may substitute $a = 0_2$ in the left-hand Eq. (1-1); and for the same reason, $a = 0_1$ may be substituted in the right-hand Eq. (1-1). We thus obtain

$$0_2 + 0_1 = 0_2 \qquad \text{and} \qquad 0_1 + 0_2 = 0_1 \tag{1-2}$$

It follows from Postulate 3a that

$$0_1 + 0_2 = 0_2 + 0_1 \tag{1-3}$$

so that

$$0_2 = 0_1 \tag{1-4}$$

Theorem 1b *Element 1 of Postulate 2b is unique.*

Proof Let us assume that there are two elements 1, which we will call 1_1 and 1_2. Then for every element a of class K,

$$a \cdot 1_1 = a \qquad \text{and} \qquad a \cdot 1_2 = a \tag{1-5}$$

in accordance with Postulate 2b. Since 1_2 is an element of class K (Postulate 2b), we may substitute $a = 1_2$ in the left-hand Eq. (1-5); and for the same reason $a = 1_1$ may be substituted in the right-hand Eq. (1-5). We thus obtain

$$1_2 \cdot 1_1 = 1_2 \qquad \text{and} \qquad 1_1 \cdot 1_2 = 1_1 \tag{1-6}$$

It follows from Postulate 3b that

$$1_2 \cdot 1_1 = 1_1 \cdot 1_2 \tag{1-7}$$

so that

$$1_2 = 1_1 \tag{1-8}$$

Since the elements 0 and 1 are proved to be unique, the element a' of Postulate 5 does exist and the following theorem can now be demonstrated:

Theorem 2 *The element a' of Postulate 5 is unique.*

Proof Let us suppose that there are two elements a' which we will call a_1' and a_2'; then

$$a + a_1' = 1 \qquad \text{and} \qquad a + a_2' = 1 \tag{1-9}$$

$$a \cdot a_1' = 0 \qquad \text{and} \qquad a \cdot a_2' = 0 \tag{1-10}$$

in accordance with Postulate 5.

The further postulates used in this proof are given in brackets after each successive step:

$$\begin{aligned} a_2' &= a_2' \cdot 1 && \text{(P2}b\text{)} \\ &= a_2' \cdot (a + a_1') && \text{(P5)} \\ &= a_2' \cdot a + a_2' \cdot a_1' && \text{(P4}b\text{)} \\ &= a \cdot a_2' + a_1' \cdot a_2' && \text{(P3}b\text{)} \\ &= 0 + a_1' \cdot a_2' && \text{(P5)} \\ &= a_1' \cdot a_2' + 0 && \text{(P3}a\text{)} \\ &= a_1' \cdot a_2' && \text{(P2}a\text{)} \end{aligned} \tag{1-11}$$

Exactly in the same way it is found that

$$a_1' = a_1' \cdot a_2' \tag{1-12}$$

so that

$$a_1' = a_1' \cdot a_2' = a_2' \tag{1-13}$$

Theorem 3a *$a + 1 = 1$ for every element a of class K.*

Proof

$$\begin{aligned} a + 1 &= (a + 1) \cdot 1 && \text{(P2}b\text{)} \\ &= (a + 1) \cdot (a + a') && \text{(P5)} \\ &= a + 1 \cdot a' && \text{(P4}a\text{)} \\ &= a + a' \cdot 1 && \text{(P3}b\text{)} \\ &= a + a' && \text{(P2}b\text{)} \\ &= 1 && \text{(P5)} \end{aligned} \tag{1-14}$$

Theorem 3b $a \cdot 0 = 0$ *for every element a of class K.*

Proof

$$\begin{aligned} a \cdot 0 &= a \cdot 0 + 0 && \text{(P2}a\text{)} \\ &= a \cdot 0 + a \cdot a' && \text{(P5)} \\ &= a \cdot (0 + a') && \text{(P4}b\text{)} \\ &= a \cdot (a' + 0) && \text{(P3}a\text{)} \\ &= a \cdot a' && \text{(P2}a\text{)} \\ &= 0 && \text{(P5)} \end{aligned} \tag{1-15}$$

Theorem 4 *The elements 0 and 1 are not identical.*

Proof Let us suppose that the elements 0 and 1 are identical so that this element can be represented by symbol E. Then

$$0 = E = 1 \tag{1-16}$$

However, it follows from Postulate $2a$ and Theorem $3a$ that

$$a + 0 = a \tag{1-17}$$

and

$$a + 1 = 1 \tag{1-18}$$

The elements 0 and 1 in Eqs. (1-17) and (1-18) may be replaced by E as given by Eq. (1-16) so that

$$a + E = a \tag{1-19}$$

and

$$a + E = E \tag{1-20}$$

Therefore, $a = E$ for every element a of class K. This means that all elements of class K are identical with E. Thus class K would only contain identical elements. But this is in conflict with Postulate 6. Therefore, the assumption that the elements 0 and 1 are identical must be wrong.

Theorem 5a $a + a = a$ *for every element a of class K.*

Proof

$$\begin{aligned} a + a &= (a + a) \cdot 1 && \text{(P2}b\text{)} \\ &= (a + a) \cdot (a + a') && \text{(P5)} \\ &= a + a \cdot a' && \text{(P4}a\text{)} \\ &= a + 0 && \text{(P5)} \\ &= a && \text{(P2}a\text{)} \end{aligned} \tag{1-21}$$

Theorem 5b *$a \cdot a = a$ for every element a of class K.*

Proof

$$\begin{aligned} a \cdot a &= a \cdot a + 0 && \text{(P2}a\text{)} \\ &= a \cdot a + a \cdot a' && \text{(P5)} \\ &= a \cdot (a + a') && \text{(P4}b\text{)} \\ &= a \cdot 1 && \text{(P5)} \\ &= a && \text{(P2}b\text{)} \end{aligned} \tag{1-22}$$

It follows from Theorems $5a$ and $5b$ that expressions such as $4a$ or a^3 do not appear in Boolean algebra. Furthermore, the proofs of Theorems $1a$, $1b$, $3a$, $3b$, $5a$, and $5b$ demonstrate the symmetry or duality of this algebra. So if in the proof of Theorem $5a$ 0 is replaced by 1, 1 by 0, + by $\cdot$, and $\cdot$ by +, and if the corresponding dual forms of the postulates mentioned above are applied, what one obtains will be the proof of Theorem $5b$.

This property, which follows from the duality of the postulates, can generally be expressed by the following theorem.

Theorem 6 *The theorem of duality. If in any valid formula 0 is replaced by 1, 1 by 0, + by $\cdot$, and $\cdot$ by +, one finds another valid formula, which is called the dual of the original one.*

It is obvious that if the dual of the dual of a formula is formed, one obtains the original formula again; thus, duality is a mutual property.

There are other theorems in Boolean algebra, but they will be dealt with in the next chapter.

1-3 THE TWO-VALUED BOOLEAN ALGEBRA

It has already been mentioned that the variables a, b, c,... are said to be n-valued if class K contains n different elements. It follows from Postulate 6 that the smallest value of n is 2. Since the application of the Boolean algebra to the theory of switching is in most cases confined to binary or two-valued quantities, the two-valued Boolean algebra is of great importance with respect to switching algebra.

In two-valued Boolean algebra, class K contains two elements only. It follows from Postulate 2 that the elements 0 and 1 are among them; and from Theorem 4, it follows that these elements are not identical. Class K thus contains the elements 0 and 1 and no others.

This fact can be stated in the following way. In order to define the two-valued Boolean algebra (as compared with the n-valued case) one additional postulate has to be stated:

Postulates

7a. *If* $a \neq 0$, *then* $a = 1$.

7b. *If* $a \neq 1$, *then* $a = 0$.

The calculating rules can easily be derived from the postulates and theorems given above. They are:

$$\begin{array}{llll} 0+0=0 & (\text{P}2a) & 0\cdot 0=0 & (\text{T}3b) \\ 0+1=1 & (\text{T}3a) & 0\cdot 1=0 & (\text{P}2b) \\ 1+0=1 & (\text{P}2a) & 1\cdot 0=0 & (\text{T}3b) \\ 1+1=1 & (\text{T}3a) & 1\cdot 1=1 & (\text{P}2b) \end{array} \qquad (1\text{-}23)$$

and

$$0' = 1 \qquad\qquad 1' = 0 \qquad (1\text{-}24)$$

Since $0' = 1$ satisfies Postulate 5 and $0'$ is unique (Theorem 2), there is no other possibility but that $0' = 1$. It can be shown that $1' = 0$ in the same way.

All theorems of the two-valued algebra can be derived from the calculating rules and from Postulate 7 by means of the method of perfect induction (see Chap. 2). This same method can also be used to derive Postulates 1 to 6 from Postulate 7 and from the calculating rules. This means that the calculating rules together with Postulate 7 may be considered as a set of postulates which define the two-valued Boolean algebra in much the same way as Postulates 1 to 7.

By way of an example, Postulates 1a and 4b will be proved from the calculating rules and from Postulate 7.

Proof of Postulate 1a From Postulate 7 it follows that variables a and b can form four different combinations. These are shown in the first column of Table 1-1. The second column shows the corresponding sum calculated by means of rules (1-23) and (1-24). Since this sum is in all cases equal to 0 or 1, the proof is completed.

Table 1-1

a	b	$a+b$
0	0	0
1	0	1
0	1	1
1	1	1

Proof of Postulate 4b From Postulate 7 it follows that variables a, b, and c can form eight different combinations as shown in the first column of Table 1-2. In the next columns, the corresponding values of ab, ac, $b+c$, $a(b+c)$, and $ab + ac$ have been calculated. Since the last two columns show identical results, the proof is completed.

Table 1-2

a	b	c	ab	ac	$b+c$	$a\cdot(b+c)$	$a\cdot b + a\cdot c$
0	0	0	0	0	0	0	0
1	0	0	0	0	0	0	0
0	1	0	0	0	1	0	0
1	1	0	1	0	1	1	1
0	0	1	0	0	1	0	0
1	0	1	0	1	1	1	1
0	1	1	0	0	1	0	0
1	1	1	1	1	1	1	1

1-4 THE ALGEBRA OF CHARACTERISTIC FUNCTIONS OF CLASSES

The concepts and postulates of Boolean algebra can be interpreted in many different ways; or, in other words, the rules of Boolean algebra can be applied to many different aspects of real life. Some of these interpretations will be considered here. Let us take first of all the algebra of characteristic functions of classes.

A class is considered as defined when the things or elements belonging to it are specified and when it is known which elements do not belong to it. For example, if it is stated that "class A consists of all male citizens of the Netherlands who have reached the age of 30" or "class B contains all points in a plane with coordinates $0 \leqslant x \leqslant 1$ and $0 \leqslant y \leqslant 1$," classes A and B are completely defined. Two classes P and Q are said to be equivalent if every element of P is also an element of Q and if every element of Q is also an element of P; or in other words, if they contain the same elements.

There are two classes which require special attention:

1. *The empty class.* This class contains no elements at all. Whether or not this class has any practical significance, it will still be very convenient to treat this class as if it really exists.
2. *The universal class or the universe.* This is the class that groups all things contained in a given discussion. It changes from one discussion to another, but in a given discussion it remains the same throughout. For example, in a discussion about the points in a plane, the universal class will contain all points in that plane. The classes treated in a given discussion are all subclasses of the universe, which means that every element of these classes is also an element of the universe.

A function a of all elements of the universe can now be defined: it takes the value 1 for all elements which belong to class A and the value 0 for all elements which do not. This function is called the characteristic function of class A because it characterizes or defines that class. It can be shown that the characteristic functions of the classes treated in a given discussion may be considered as corresponding to the variables of the two-valued Boolean algebra. In order to prove this, one need only show that these functions satisfy the calculating rules (1-23) and (1-24) as well as Postulate 7.

Since an element either belongs to a given class or does not belong to it, the characteristic function of that class takes either the value 1 or the value 0. So Postulate 7 is satisfied.

The constant 0 may be considered as the characteristic function of the empty class and the constant 1 as that of the universe.

The equivalence $a = b$ of the characteristic functions of the classes A and B means that no elements exist for which a takes the value 0 and b the value 1 or for which a takes the value 1 and b the value 0. In other words, no elements exist which belong to class A and not to class B and no elements exist which belong to class B and not to class A. The equivalence $a = b$ therefore expresses the equivalence of classes A and B.

The expression $a = 0$ (or $a = 1$) may have two different meanings:

1. $a = 0$ means that a equals 0 for all elements of the universe, so class A is empty.
2. $a = 0$ means that a has the value 0, so "if $a = 0$, then ..." actually means "for all elements which do not belong to class A, the following holds ...."

Which of these two meanings is under discussion will be clear from the context.

It is easy to see that if a' is considered as the characteristic function of the class containing all elements of the universe which do not belong to class A, the calculating rules (1-24) are satisfied. So the sign $'$ has the meaning "not."

The sum $s = a + b$ of the characteristic functions a and b of classes A and B can be considered as the characteristic function of class S which contains all elements belonging to class A or to class B or to both. So $s = 1$ if either a or b or both are equal to 1.

The product $p = a \cdot b$ of the characteristic functions a and b of classes A and B can be considered as the characteristic function of class P which contains all elements belonging as well to class A as to class B. So $p = 1$ if and only if $a = b = 1$.

It will now be clear that if the $+$ sign has the meaning "or ... or both" (briefly "or") and the $\cdot$ sign has the meaning "as well ... as" (briefly "and"), the following relations are found to exist:

1. All elements which belong neither to class A nor to class B belong neither to class S nor to class P.

   So if $a = 0$ and $b = 0$, then $a + b = 0$ and $a \cdot b = 0$

2. All elements which belong to class B but not to class A belong to class S but not to class P.

   So if $a=0$ and $b=1$, then $a+b=1$ and $a \cdot b=0$

3. All elements which belong to class A but not to class B belong to class S but not to class P.

   So if $a=1$ and $b=0$, then $a+b=1$ and $a \cdot b=0$

4. All elements which belong as well to class A as to class B belong to class S and to class P.

   So if $a=1$ and $b=1$, then $a+b=1$ and $a \cdot b=1$

These four relations show that the calculating rules (1-23) are satisfied.

The relations between the signs $+$, $\cdot$, and $'$ and the concepts of or, and, and not, respectively, are not limited only to the interpretation of Boolean algebra as the algebra of characteristic functions of classes but are of primary importance in every application. It might even be stated that Boolean algebra is the algebra of or, and, and not.

Since the calculating rules and Postulate 7 are all proved to be satisfied, all theorems of the two-valued Boolean algebra may be applied to the characteristic functions of classes in the interpretation given above.

To illustrate how Boolean algebra can be applied to the characteristic functions of classes, the following simple example is given which requires no other theorems than those treated in this chapter.

A given club has the following statutory rules:

1. No one shall be a member of the financial committee unless he is also a member of the library committee.
2. Members of the general committee shall be those persons who are at the same time on the financial committee and on the library committee.

Simplify these rules.

Let the universe be the class grouping all the members of the club. Furthermore, let f be the characteristic function of the class which is formed by the members of the financial committee; let l be the function for the class consisting of the members of the library committee; and let g be the function for the class containing the members of the general committee. The two rules of the club can now be expressed as follows:

$$f \cdot l' = 0 \qquad (1\text{-}25)$$

and

$$g = f \cdot l \qquad (1\text{-}26)$$

But

$$\begin{aligned} g &= f\cdot l + 0 && \text{(P2}a\text{)} \\ &= f\cdot l + fl' && \text{(1-25)} \\ &= f\cdot(l + l') && \text{(P4}b\text{)} \\ &= f\cdot 1 && \text{(P5)} \\ &= f && \text{(P2}b\text{)} \end{aligned} \tag{1-27}$$

The simplified rules of the club can now be written as follows:

1. No one shall be a member of the financial committee unless he is also a member of the library committee.
2. The general committee has the same membership as the financial committee.

1-5 VENN DIAGRAMS

Classes and their relations can be expressed graphically by means of Venn diagrams. In such diagrams the universe is represented by a rectangle as shown in Fig. 1-1.

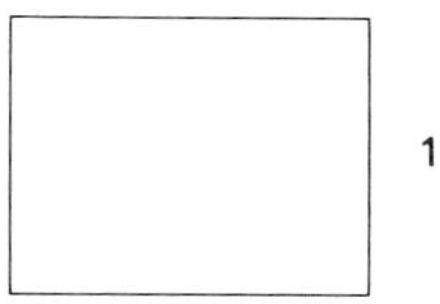

Fig. 1-1

A class A (of which all elements belong to the universe) is drawn as a closed area within the rectangle as indicated in Fig. 1-2a. The shaded area is the set of all points of the rectangle for which the characteristic function a has the value 1. In the same way, the shaded area in Fig. 1-2b is the set of all points of the rectangle for which the function a' has the value 1. So this shaded area corresponds to the class "not A" or the "inverse of A," that is, it contains all elements of the universe which do not belong to class A.

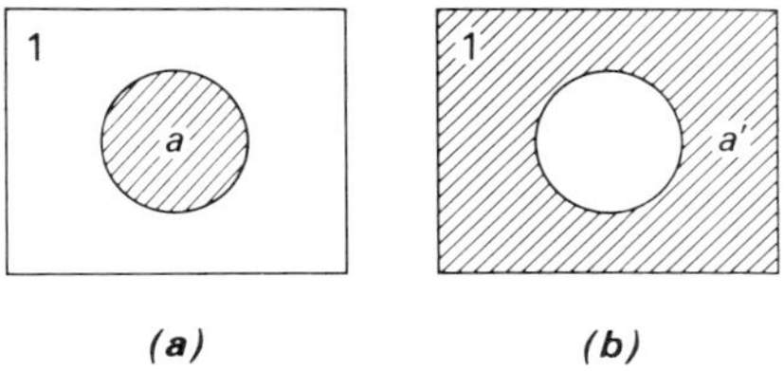

Fig. 1-2

Figure 1-3*a* shows two classes A and B, which may or may not overlap. The shaded area of Fig. 1-3*b* incorporates all points of the rectangle for which the function $a + b$ has the value 1, and the shaded area of Fig. 1-3*c* contains all points for which $a \cdot b$ has the value 1. So these areas are the representations of the classes A or B and A and B, respectively. The first one is called "the union of A and B" and the second "the intersection of A and B."

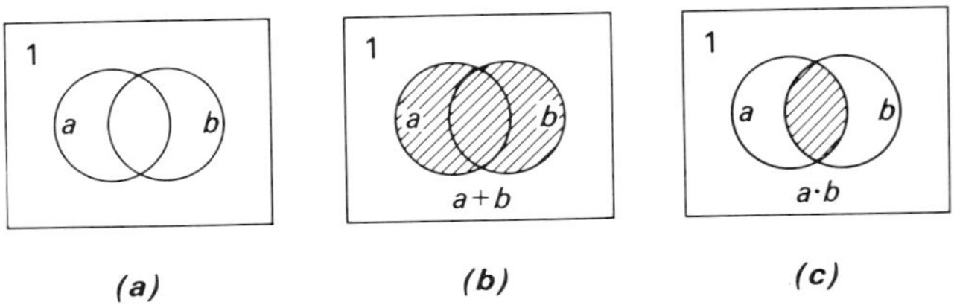

Fig. 1-3

A class which is known or proved to be empty is marked by blacking the corresponding area, and a class which is proved to contain at least one element is marked by placing an asterisk in the area.

Example 1 The simplification processes of the club rules mentioned in the preceding paragraph will now be illustrated graphically by means of Fig. 1-4. Since the first rule is expressed by Fig. 1-4*d*, it follows from Fig. 1-4*e* that class F and the intersection of class F and class L are equivalent. This fact leads directly to the given simplification of the second rule.

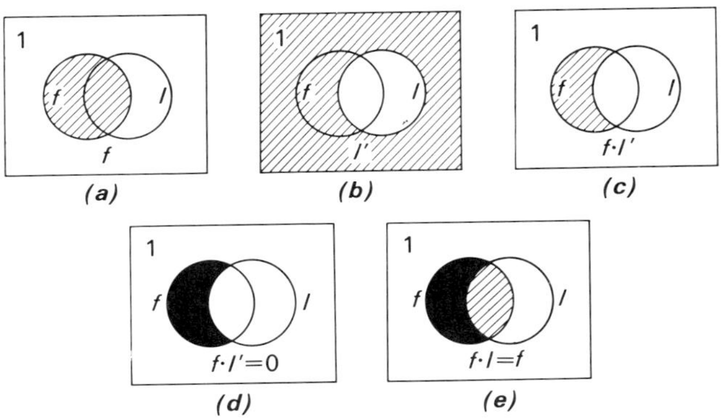

Fig. 1-4

Example 2 A given club has the following statutory rules:

1. The members of the financial committee shall be chosen among the members of the general committee.

2. No one shall be a member of both the general and the library committee unless he is also a member of the financial committee.
3. No member of the library committee shall be a member of the financial committee.[1]

Simplify these rules by means of a Venn diagram. Let the classes grouping the members of the general, the financial, and the library committee be represented by G, F, and L, respectively. Rules 1 to 3 can be expressed by means of the following formulas:

$$f \cdot g' = 0 \tag{1-28}$$

$$g \cdot l \cdot f' = 0 \tag{1-29}$$

$$l \cdot f = 0 \tag{1-30}$$

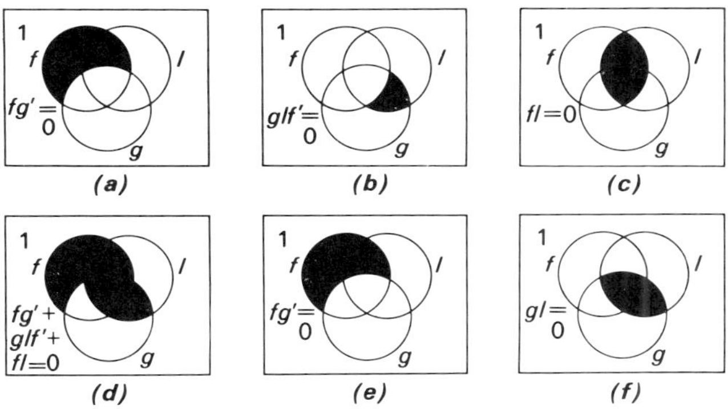

Fig. 1-5

These formulas are illustrated in Fig. 1-5a to c. By combining these, one obtains Fig. 1-5d. But this figure can also be derived from Fig. 1-5e and f, which correspond to the following formulas:

$$f \cdot g' = 0 \qquad \text{and} \qquad g \cdot l = 0 \tag{1-31}$$

So the simplified rules are as follows:

1. The members of the financial committee shall be chosen among the members of the general committee.
2. No member of the general committee shall be a member of the library committee.

1-6 THE ALGEBRA OF TRUTH VALUES AND PROPOSITIONS

We shall now consider a second interpretation of Boolean algebra—the algebra of truth values of propositions. This algebra is a branch of symbolic logic, or more precisely, of the proposition calculus.

[1] For this example, we are indebted to John Venn (1888).

In general, a proposition or statement is an expression which can be true or false: "Jack is ill"; "Delft is a university town between Rotterdam and the Hague"; "It is raining outside." To such a statement X, a truth value x can be assigned which equals 1 if the statement X is true and 0 if this same statement X is false. It can be shown that these truth values of propositions may be considered as variables in a two-valued Boolean algebra. In order to prove this, it is again sufficient to show that the calculating rules (1-23) and (1-24) and Postulate 7 are satisfied.

The validity of Postulate 7 follows directly from the fact that a proposition is either true or false. Furthermore, it is easy to see that the calculating rules (1-24) are satisfied if a' is the truth value of the proposition "not A" or, in other words, if here again the sign $'$ means "not."

Since the proposition "A or B (or both)" is true if A is true or B is true or both are true, the first four calculating rules (1-23) are satisfied if $a + b$ is considered as the truth value of "A or B." In the same way, the last four calculating rules (1-23) are satisfied if $a \cdot b$ is considered as the truth value of the proposition "A and B," since this proposition is true only if A and B are true (as is indicated in Table 1-3). So here again the signs $+$, $\cdot$, and $'$ correspond to or, and, and not, respectively.

Table 1-3

A	B	A or B	A and B	a	b	$a+b$	$a \cdot b$
False	False	False	False	0	0	0	0
False	True	True	False	0	1	1	0
True	False	True	False	1	0	1	0
True	True	True	True	1	1	1	1

In Boolean algebra, the word "or" is always used in the sense of "or ... or ... or both" (inclusive or) and not in the sense of "either ... or ..." (exclusive or). The truth value of the proposition "either A or B" can be represented be means of $a'b + ab'$, as shown in Table 1-4.

Table 1-4

a	b	$a'b$	ab'	$a'b + ab'$
0	0	0	0	0
0	1	1	0	1
1	0	0	1	1
1	1	0	0	0

A table in which a complex truth value is represented as a function of all combinations of the elementary truth values is called a truth table. Tables 1-3 and 1-4 are truth tables.

Let us now consider the following simple example:

We will go sailing if it does not rain and the sun is shining, or if it does not rain and the sun is not shining.

Let R represent the proposition "it is raining" and let S represent the proposition "the sun is shining." Then the condition for sailing is satisfied if

$$r' s + r' s' = 1 \tag{1-32}$$

But

$$\begin{aligned} r' s + r' s' &= r'(s + s') && \text{(P4}b\text{)} \\ &= r' \cdot 1 && \text{(P5)} \\ &= r' && \text{(P2}b\text{)} \end{aligned} \tag{1-33}$$

So

$$r' = 1$$

This statement can be simplified to

We will go sailing if it does not rain

The example is trivial but it gives an idea of the possible applications of Boolean algebra to symbolic logic.

Boolean algebra has still a large number of other uses—in the algebra of contact circuits, in the algebra of electronic circuits, etc. These will be dealt with in the following chapters.

PROBLEMS

1. Prove that $a + ab = a$, using Postulates $2b$, $4b$, and $3a$, Theorem $3a$, and Postulate $2b$.

2. Prove that $a \cdot (a + b) = a$, using Postulate $4b$, Theorem $5b$, and the formula given in Prob. 1.

3. Prove that $a + a'b = a + b$, using Postulates $4a$, 5, $3b$, and $2b$.

4. Prove in an analogous way that $a(a' + b) = ab$. Indicate all postulates and theorems used.

5. Prove that $(a + b) + c = a + (b + c)$. Please note that it is convenient to show first that $(a + b) + c = b + (a + c)$. Subsequently demonstrate that $a + (b + c) = b + (a + c)$. List all postulates and theorems used. The formulas given in the first three problems may be used to solve this problem.

6. Prove in an analogous way, using the duals of the postulates, theorems, and formulas of Prob. 5, that $(a \cdot b) \cdot c = a \cdot (b \cdot c)$.

7. Prove that $(a + b) + a'b' = 1$. Indicate all postulates and theorems used. The formulas given in the preceding problems may be used to solve this problem.

8. Prove that $(a + b) \cdot (a'b') = 0$. Indicate all postulates and theorems used. The formulas given in the preceding problems may be used to solve this problem.

9. Prove that $(a')' = a$, using Postulates $2b$, 5, $4b$, $3b$, 5, 5, $4b$, $3a$, 5, and $2b$, in this order.

10. Prove that $(a + b)' = a'b'$. Indicate all postulates and theorems used. The formulas given in the preceding problems may be used.

11. Prove in a way analogous to that used in Prob. 10 that $(ab)' = a' + b'$. List all postulates and theorems used. The formulas given in the preceding problems may be used in this demonstration.

12. Show by means of a Venn diagram that rules 2 and 3 of Example 2 in Sec. 1.5 cannot be simplified as mentioned in that section if the first rule is not valid.

13. Show by means of a Venn diagram that $(a + b)' = a'b'$.

14. Show by means of a Venn diagram that $(a' + b)(a + c)(b + c) = (a' + b)(a + c)$.

15. Show by means of a Venn diagram that $a'b + ac + bc = a'b + ac$.

16. Prove the consistency of Postulate 4a by means of a truth table and a Venn diagram.

17. Given are three classes: males, graduates, and students. Draw a Venn diagram for the following classes:

(*a*) Graduate students
(*b*) Male students
(*c*) Female nongraduate students

and shade the appropriate areas.

18. Given the three classes of Prob. 17, draw a Venn diagram for the class consisting of either female graduate students or male nongraduate students.

SELECTIVE BIBLIOGRAPHY

1. Berkeley, E. C.: "Symbolic Logic and Intelligent Machines," Reinhold Publishing Corporation, New York, 1959.
2. Boole, George: "The Mathematic Analysis of Logic," Cambridge, 1847.
3. Boole, George: "An Investigation of the Laws of Thought," Walton, London, 1854. Reprint as part of "Collected Logical Works," vol. 2, "Laws of Thought," Open Court Publishing Company, Chicago, 1916, and Dover Publications, Inc., New York, 1954.
4. Huntington, E. V.: The Algebra of Logic, *Trans. Am. Math. Soc.*, vol. 5, pp. 288–309, 1904.
5. Jevons, W. S.: "Pure Logic, or the Logic of Quality apart from Quantity," London, 1864.
6. Lewis, C. I., and C. H. Langford: "Symbolic Logic," Century Company, 1932. Reprint, Dover Publications, Inc., New York, 1954.
7. Martin, N. M.: On Completeness of Boolean Element Sets, *J. Computing Systems*, vol. 1, no. 3, pp. 150–154, 1953.
8. Phister, M., Jr.: "Logical Design of Digital Computers," chap. 3, John Wiley & Sons, Inc., New York, 1958.
9. Reichenback, H.: "Elements of Symbolic Logic," The Macmillan Company, New York, 1950.
10. Rosenbloom, P. C.: "The Elements of Mathematical Logic," Dover Publications, Inc., New York, 1950.
11. Schröder, E.: "Vorlesungen über die Algebra der Logik," Leipzig, 1890–1905.
12. Sheffer, H. M.: "A Set of Five Independent Postulates for Boolean Algebra," *Trans. Am. Math. Soc.*, vol. 14, pp. 481–488, 1913.
13. Venn, J.: Boole's Logical System, *Mind*, vol. 1, pp. 479–491, 1876.
14. Venn, J.: "Symbolic Logic," 2d ed., London, 1894.

2 Switching Algebra

2-1 POSTULATES AND THEOREMS

The postulates and theorems of Boolean algebra as given in Chap. 1 can be used as a basis for a switching algebra with two-valued switching components such as contacts (open or closed), diodes (low or high voltages), magnetic cores (clockwise or counterclockwise magnetization), pneumatic or hydraulic valves (open or closed), or any other type of two-valued switching component. This means that the design of an automatic circuit or of an information-processing device consisting of two-valued components will to a large extent be independent of the specific type of components used.

The modern setup of Boolean algebra with its own group of symbols does not make it quite so suitable for the design of switching circuits. Thus, in the opinion of the author, it is necessary to choose a group of symbols for the mathematical part of circuit design that does not differ too much from ordinary algebra. The switching algebra would otherwise be accessible to only the select group of circuit designers with a university background; it would not reach the large group of professional circuit designers who did not have the opportunity to receive such formal training. The efficiency of switching algebra in the design of complex circuits is such that the trial-and-error methods of practical circuit design have become obsolete.

With the purpose of developing a switching algebra for general use, we can use calculating rules (1-23) and (1-24) and Postulate 7 as a set of postulates on which to base this algebra. The theorems which will be derived from these postulates are valid for every interpretation of the algebra and may thus be used for the design of switching circuits with any type of two-valued switching elements. However, they will be explained in terms of contact circuits because the validity of the various theorems can be proved most easily by means of the method of "perfect induction," i.e., by checking all possible combinations for this type of circuit. The postulates and theorems will be arranged in pairs in order to stress the duality that exists between the operations of logical addition and logical multiplication and the state symbols 0 and 1 of the switching components.

The postulates are based on the following concepts: an open circuit is represented by 0 and a closed circuit by 1; there are two types of relations between these symbols, viz., a series connection for which the sign $\cdot$ is used and a parallel connection which is represented by $+$; the negation of these symbols is shown by means of $(\ )'$.

Postulates

1a. $0 \cdot 0 = 0$ *An open circuit in series with an open circuit is an open circuit.*

1b. $1 + 1 = 1$ *A closed circuit in parallel with a closed circuit is a closed circuit.*

2a. $0 \cdot 1 = 1 \cdot 0 = 0$ *An open circuit in series with a closed circuit (or the reverse) is an open circuit.*

2b. $1 + 0 = 0 + 1 = 1$ *A closed circuit in parallel with an open circuit (or the reverse) is a closed circuit.*

3a. $1 \cdot 1 = 1$ *A closed circuit in series with a closed circuit is a closed circuit.*

3b. $0 + 0 = 0$ *An open circuit in parallel with an open circuit is an open circuit.*

4a. $(0)' = 1$ *The negative of an open circuit is a closed circuit.*

4b. $(1)' = 0$ *The negative of a closed circuit is an open circuit.*

5. $z = 0$ *or* $z = 1$ *A circuit can be either open or closed.*

These postulates are all that is required to develop all the following theorems in switching algebra dealing with planar contact circuits with series and parallel connections only or with their equivalent in other types of switching techniques.

The postulates are given in terms of contact circuits. A zero has been chosen to show that a circuit is open and a one that it is closed. The · sign is used for series and the + sign for parallel connections. It should be noted that two contact circuits are called equal or equivalent (=) if for every possible combination of the contacts the two circuits are both open or both closed.

The assignment of this meaning to the signs · and + and the symbols 0 and 1 is one out of four possible choices, two of which can be discarded without further ado because they lead to numerical relations between zeros and ones which differ considerably from the corresponding relations between zeros and ones in ordinary numerical algebra. The postulates given above correspond to one of the remaining two possibilities. The other solution can be obtained by interchanging the numerical part of these postulates and by using the text of the corresponding dual forms. In theory, there is no reason for preferring either system. However, there are some practical features (see Chap. 12) which make the group of postulates as given slightly more attractive than the fully dual system.

Some users might try to attach a physical meaning to the symbols 0 and 1, the sole purpose of which is to describe in a symbolic way the state or the condition of a contact or even of a whole circuit. The chosen group of symbols makes it possible to represent a contact circuit by means of ideal conductances. These conductances can take the values 0 or ∞. Instead of ∞, the cardinal number 1 can be used so that Postulate 1*b* as given in this chapter becomes acceptable even for those who are used to working with ordinary numerical algebra exclusively.

In the fully dual group of postulates, it is possible to represent a contact by means of resistances which can be either ∞ or 0. It is also possible to use the concept of "hindrance." An open circuit is said to have a hindrance of 1 and a closed circuit a hindrance of 0.

These physical concepts are valid for contact circuits only. In circuits using diodes (which can be considered in certain respects as the electronic equivalent of contact circuits), both basic concepts of the contact circuits, viz., "open-closed" and "series-parallel," lose their meaning entirely. The basic concepts of diode circuits that lead to a set of postulates analogous to the ones given above for contact circuits are "low-high" voltages and the "and-or" relation of Boolean algebra.

The following theorems or laws of switching algebra will be illustrated by means of contact circuits. The diagrams of these contact circuits lend themselves admirably to the perfect-induction method for checking the corresponding theorem. However, any other type of switching circuit may be used for this purpose. To prove the theorems, one may also apply the postulates or the previous theorems. However, in view of its general applicability, the method of perfect induction is preferable for nonmathematicians. If the proof of one of the theorems of a dual pair has been given, the other theorem will be also true.

Theorem 1 *The commutative law*

a. *Of the logical multiplication*

$y \cdot z = z \cdot y$

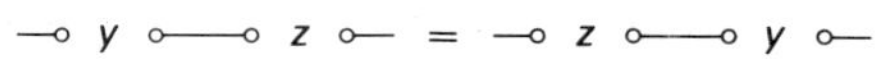

b. *Of the logical addition*

$y + z = z + y$

Theorem 2 *The associative law*

a. *Of the logical multiplication*

$x \cdot (y \cdot z) = (x \cdot y) \cdot z$

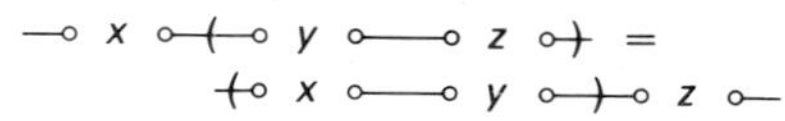

b. *Of the logical addition*

$x + (y + z) = (x + y) + z$

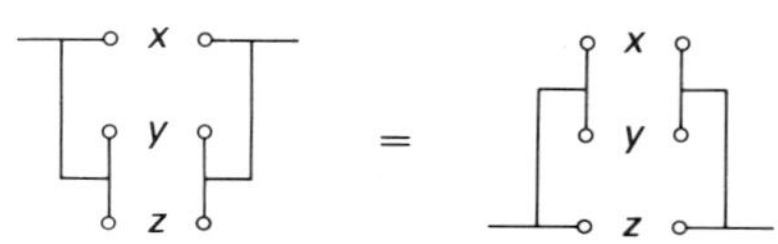

Theorem 3 *The distributive law*

a. *Of the logical multiplication*

$x \cdot (y + z) = xy + xz$

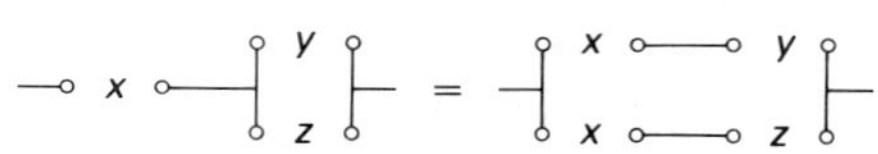

b. *Of the logical addition*

$x + y \cdot z = (x + y) \cdot (x + z)$

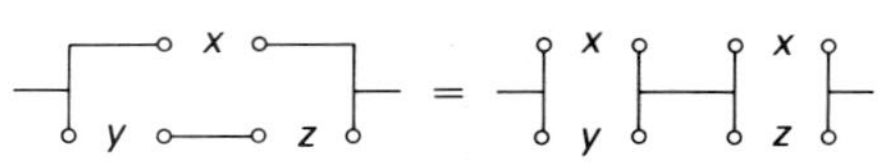

Laws 1a and b, 2a and b, and 3a are analogous to laws in ordinary algebra; law 3b, the second distributive law of the logical addition, has no equivalent in ordinary algebra. Both distributive laws, when used in the reverse direction, are reduction formulas by means of which switching formulas can be factored or simplified. The factoring of switching functions is an important process for finding the simplest form of the switching formula and, accordingly, the simplest form of the desired circuit. This is important for the sake of economy in the use of switching components.

The second distributive law is difficult to handle, whereas there is no problem with the first distributive law among workers in this field because it is the same as in ordinary algebra. It is desirable that a student acquire some skill in handling the second distributive law for the simplifying of switching functions. However, with the help of the dual pairs of laws, any difficulty that he has may be overcome by applying the first distributive law to the dual form of the switching function in question and by determining the dual form of the result.

Theorem 4 *Modulus*

a. Of the logical multiplication

$1 \cdot z = z$

b. Of the logical addition

$0 + z = z$

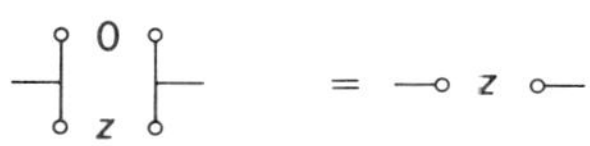

Theorem 5

a. $0 \cdot z = 0$

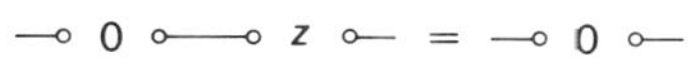

b. $1 + z = 1$

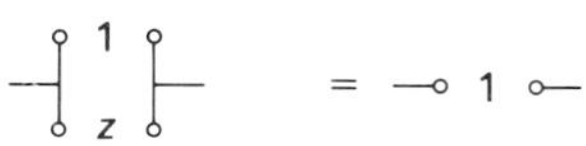

Theorem 6 *Negation*

a. $z \cdot z' = 0$

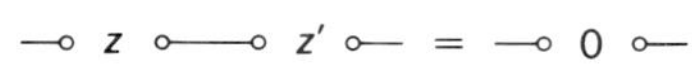

b. $z + z' = 1$

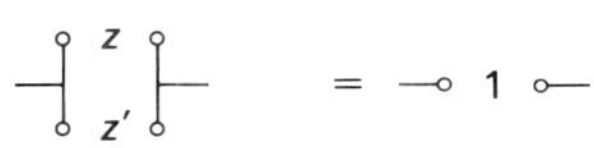

Theorem 7 *Involution*

$(z')' = z$

This last theorem states that the negative of a negative of a switching function corresponds to the original function. In terms of contact circuits, this means that the negative of a make contact is a break contact and that the negative of a break contact is the negative of the negative of a make contact and thus a make contact.

Theorem 7 leads to the following important theorem.

Theorem 8 *De Morgan's theorem*

a. $(x \cdot y \cdot z \cdots)' = x' + y' + z' + \cdots$

b. $(x + y + z + \cdots)' = x' \cdot y' \cdot z' \cdot \cdots$

De Morgan's theorem indicates that the negative of a function in the form of a logical product or a logical sum corresponds to the negative of the factors or the terms of this function. The proof of this theorem can be given for any number of variables starting, for example, with the proof by the method of perfect induction for two variables and then extending this proof by mathematical induction.

The next theorem states that any number of contacts with the same designation connected in either series or parallel is exactly equivalent to one contact.

Theorem 9 *Idempotent properties*

a. $z \cdot z \cdot z \cdot \cdots = z$

b. $z + z + z + \cdots = z$

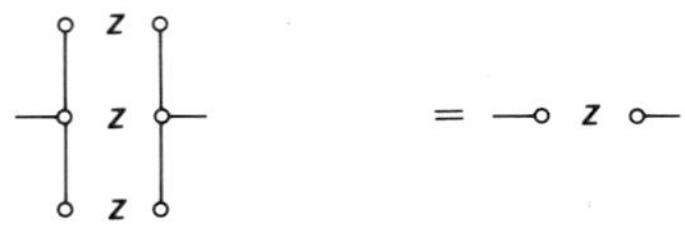

Theorem 10 *Absorption*

a. $z \cdot (y + z) = z$

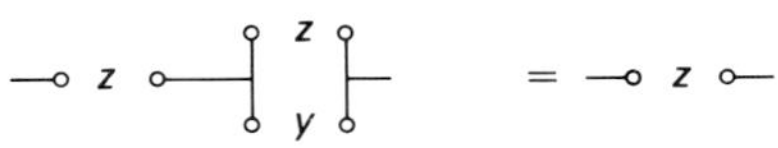

b. $z + y \cdot z = z$

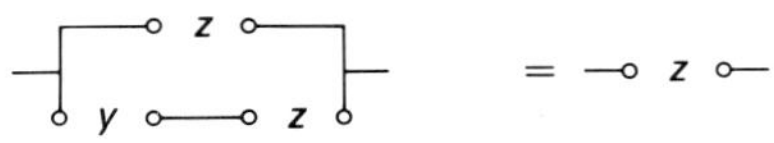

Theorem 10 is useful for simplifying switching functions. Its proof can easily be derived from the circuits as illustrated.

Theorem 11 is equally useful for simplifying switching functions.

Theorem 11

a. $y + y' \cdot z = y + z$

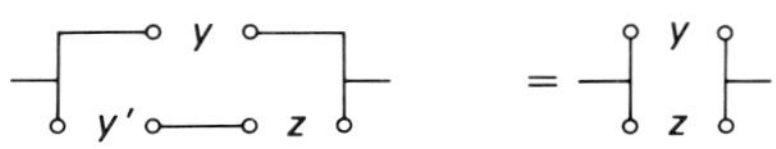

b. $y \cdot (y' + z) = y \cdot z$

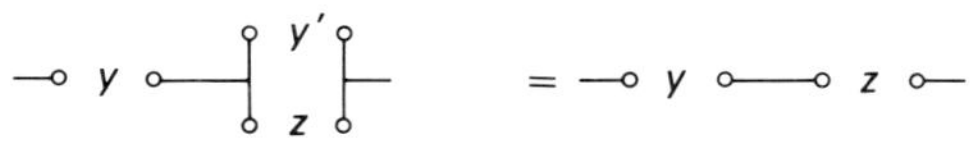

Theorem 11*b* can be proved by "multiplying out" this expression by means of the first distributive law (Theorem 3*a*) and applying Theorem 6*a*. The validity of Theorem 11*a* is thereby also established. The proof of Theorem 11*a* can also be derived directly from the second distributive law (Theorem 3*b*) ("adding out").

Finally, there is a group of three theorems which is very useful for circuit simplification.

Theorem 12

a. $(x + y) \cdot (x' + z) \cdot (y + z) = (x + y) \cdot (x' + z) =$

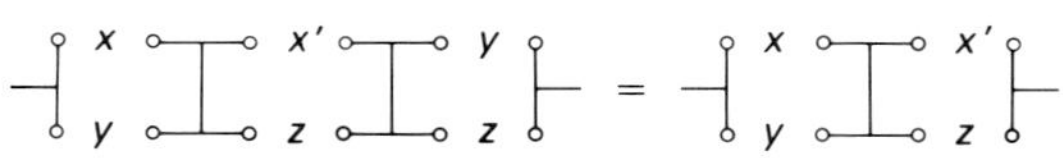

b. $x \cdot z + x' \cdot y + y \cdot z = x \cdot z + x' \cdot y =$

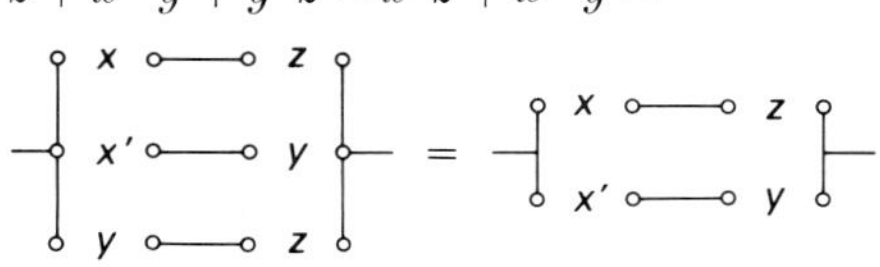

c. $(x + y)\cdot(x' + z) = x\cdot z + x'\cdot y =$

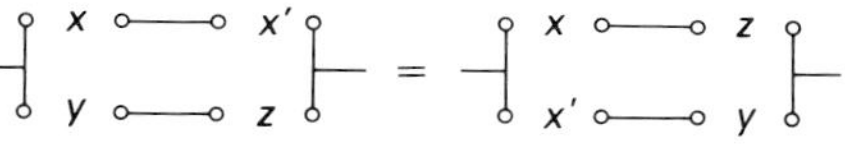

These theorems are easily proved by means of the perfect-induction method applied to the corresponding contact circuits. Proof can also be given in the following way. When adding the nonsignificant factor $(x + x')$ to Theorem 12*a* and applying the second distributive law, it follows that

$$\begin{aligned}(x + y)\cdot(x' + z)\cdot(y + z) &= (x + x')\cdot(x + y)\cdot(x' + z)\cdot(y + z)\\ &= (x' + xz)\cdot(y + xz)\\ &= x\cdot z + x'\cdot y \qquad (2\text{-}1)\end{aligned}$$

When adding the nonsignificant term $x\cdot x'$ to Theorem 12*b* and applying the first distributive law, it is found that

$$\begin{aligned}x\cdot z + x'\cdot y + y\cdot z &= x\cdot x' + x\cdot z + x'\cdot y + y\cdot z = x\cdot(x' + z) + y\cdot(x' + z)\\ &= (x + y)\cdot(x' + z) \qquad (2\text{-}2)\end{aligned}$$

If the first distributive law alone is applied to Theorem 12*a*, it is found that

$$(x + y)\cdot(x' + z)\cdot(y + z) = x\cdot z + x'\cdot y + y\cdot z \qquad (2\text{-}3)$$

To prove Theorem 12, all that is required is to combine Eqs. (2-1) to (2-3).

2-2 CANONICAL FORMS OF SWITCHING FUNCTIONS

Each group of variables $(\ldots,x,y,z)$ which is formed by means of the operations of logical addition and/or logical multiplication and/or negation is called a switching function and can be represented by $f(\ldots,x,y,z)$. The switching functions can be developed in a series expansion by means of the following equations:

$$f(\ldots,x,y,z) = f(\ldots,x,y,0)\cdot z' + f(\ldots,x,y,1)\cdot z \qquad (2\text{-}4a)$$

$$f(\ldots,x,y,z) = [f(\ldots,x,y,1) + z']\cdot[f(\ldots,x,y,0) + z] \qquad (2\text{-}4b)$$

These equations reduce to identities for $z = 0$ and $z = 1$. Equation (2-4*a*) shows the development of the function in the form of a logical sum of logical products, whereas Eq. (2-4*b*) gives the dual representation of that development as the logical product of logical sums. Equations (2-4) give the expansion of switching function $f(\ldots,x,y,z)$ in terms of variable z.

The series expansion of switching functions as given in Eqs. (2-4) is of primary importance in contact circuits because it shows that an arbitrarily chosen switching function can be converted into a layout consisting of no more than one make contact and one break contact of one specific variable. This may be illustrated by means of Fig. 2-1, which expresses Eqs. (2-4) in terms of contact circuits.

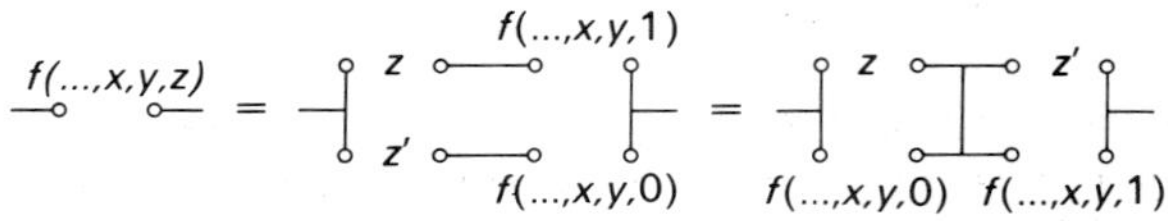

Fig. 2-1

If switching function $f(\ldots,x,y,z)$ has n variables, the expansion process must be repeated n times before arriving at a complete series expansion. The complete expansion of a function $f(\ldots,x,y,z)$ has the following form:

$$
\begin{aligned}
f(\ldots,x,y,z) = f(\ldots,0,0,0)\cdot x'y'z' \quad & + f(\ldots,0,0,1)\cdot x'y'z + \\
\cdots\cdots\cdots \quad & \cdots + \\
f(\ldots,1,1,0)\cdot xyz' \quad & + f(\ldots,1,1,1)\cdot xyz \qquad \text{(2-5}a\text{)}
\end{aligned}
$$

$$
\begin{aligned}
f(\ldots,x,y,z) + [f(\ldots,1,1,1) + x' + y' + z'] & \cdot [f(\ldots,1,1,0) + x' + y' + z]\cdot \\
\cdots\cdots\cdots & \cdots \\
[f(\ldots,0,0,1) + x + y + z'] & \cdot [f(\ldots,0,0,0) + x + y + z] \\
& \qquad \text{(2-5}b\text{)}
\end{aligned}
$$

In the complete expansion, the functions $f(\ldots)$ have become coefficients which are either 0 or 1. In the sum case, the 0's in the coefficients correspond to the primed variables and the 1's to the unprimed variables. This allows the numbering of these coefficients by means of their characterizing group of 0's and 1's taken as a binary number. The complete expansion, which is called the canonical form of the switching function, comprises 2^n terms or factors in which n represents the number of variables. Using as an index the decimal equivalent of the binary numbers represented by the f coefficients in Eqs. (2-5), the canonical form of a switching function with three variables can be written as follows:

$$
\begin{aligned}
f(x,y,z) &= f_0 x'y'z' + f_1 x'y'z + f_2 x'yz' + f_3 x'yz \\
&\quad + f_4 xy'z' + f_5 xy'z + f_6 xyz' + f_7 xyz \qquad \text{(2-6}a\text{)} \\
&= (f_7 + x' + y' + z')\cdot(f_6 + x' + y' + z)\cdot(f_5 + x' + y + z') \\
&\quad \cdot(f_4 + x' + y + z)\cdot(f_3 + x + y' + z')\cdot(f_2 + x + y' + z) \\
&\quad \cdot(f_1 + x + y + z')\cdot(f_0 + x + y + z) \qquad \text{(2-6}b\text{)}
\end{aligned}
$$

Equation (2-6a) is called the minterm canonical form and Eq. (2-6b), the maxterm canonical form. If two switching functions have identical canonical forms, then the two functions must be equal.

The f coefficient of a term of the minterm form or in a factor of the maxterm form of a switching function represents the value which this function will take for that combination of the values of the variables which correspond to this term or factor. Every possible switching function can therefore be defined by means of a truth table that shows the corresponding f coefficient (the desired value of the function) for every possible combination of the values of the variables.

The indices of the f coefficients of the minterms of a canonical form can easily be derived from the minterms themselves by adding the decimal values of the variables of these terms considered as a binary number. These values can be taken, for example, as

$$x = 4 \qquad y = 2 \qquad z = 1 \qquad \text{and} \qquad x' = y' = z' = 0 \tag{2-7a}$$

The indices of the f coefficients of the maxterms of a canonical form can also be derived from the maxterms themselves in a similar way but with the following group of values:

$$x' = 4 \qquad y' = 2 \qquad z' = 1 \qquad \text{and} \qquad x = y = z = 0 \tag{2-7b}$$

These indices of the maxterms are the decimal equivalent of the binary ones complement of the indices as obtained for the minterms.

The expressions minterm and maxterm canonical form of a switching function, which are used for the logical sum of logical products and for the logical product of logical sums, respectively, are so chosen because the terms that are logical products occupy a minimum area in a Venn diagram and the terms that are logical sums occupy a maximum area in a Venn diagram. This is illustrated in Figs. 2-2 and 2-3 for the functions

$$f(y, z) = y'z' = (y' + z')\cdot(y' + z)\cdot(y + z') \tag{2-8}$$

and

$$f(y, z) = y' + z' = y'z' + y'z + yz' \tag{2-9}$$

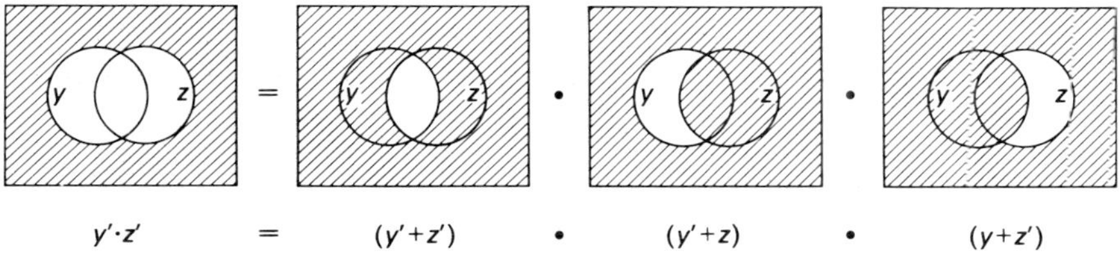

Fig. 2-2

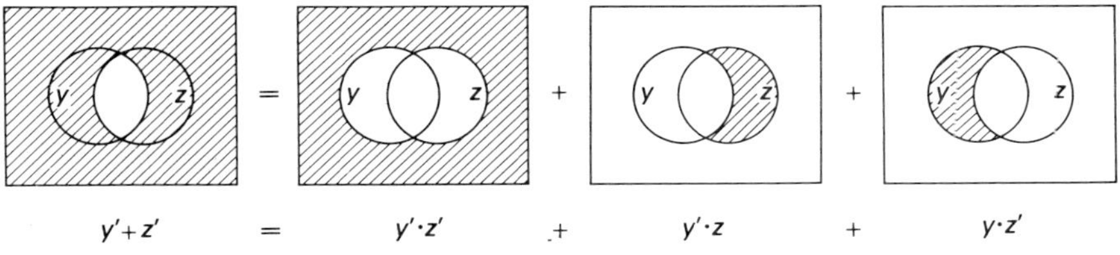

Fig. 2-3

Minterms and maxterms can now be indicated by some form of shorthand notation. A small m will be used to indicate a minterm and capital M to represent a maxterm. The index of a specific minterm or maxterm will be the same as that of the corresponding f coefficient, so that $m_i' = M_i$ and $M_i' = m_i$.

The following properties of canonical expressions (2-6a) and (2-6b) are sometimes useful.

1. $$m_0 + m_1 + m_2 + \cdots + m_{2^n-1} = 1 \tag{2-10a}$$

$$M_{2^n-1} \cdot \cdots \cdot M_2 \cdot M_1 \cdot M_0 = 0 \tag{2-10b}$$

The proof of these formulas can be given by means of an iterative process using Theorems 3a and 6b. This will be demonstrated for a minterm canonical form with three variables. Equation (2-10b) can be demonstrated in an analogous way by the iterative use of the dual forms of the theorems used in the proof of Eq. (2-10a).

$$\begin{aligned} f(x,y,z) &= x'y'z' + x'y'z + x'yz' + x'yz \\ &\quad + xy'z' + xy'z + xyz' + xyz \\ &= x'(y'z' + y'z + yz' + yz) + x(y'z' + y'z + yz' + yz) \\ &= (x + x')\cdot(y'z' + y'z + yz' + yz) \\ &= y'z' + y'z + yz' + yz \end{aligned} \tag{2-11}$$

This process must be repeated n times for a minterm canonical form with n variables; it ends with:

$$z + z' = 1 \tag{2-12}$$

2. The minterms and maxterms of the canonical expressions (2-10a) and (2-10b) can be divided into any two groups which are always the negative of each other. For example:

$$(m_0 + m_1 + m_5)' = m_2 + m_3 + m_4 + m_6 + m_7 \tag{2-13a}$$

and

$$(M_0 \cdot M_1 \cdot M_5)' = M_2 \cdot M_3 \cdot M_4 \cdot M_6 \cdot M_7 \tag{2-13b}$$

The validity of these equations is based on Eqs. (2-10) and Theorem 6.

Some auxiliary equations derived from the expansion equation (2-4) can also be useful; so are, for example, the following equations:

$$z \cdot f(\ldots, x, y, z) = z \cdot f(\ldots, x, y, 1) \tag{2-14a}$$

$$z' + f(\ldots, x, y, z) = z' + f(\ldots, x, y, 1) \tag{2-14b}$$

and

$$z + f(\ldots, x, y, z) = z + f(\ldots, x, y, 0) \tag{2-15a}$$

$$z' \cdot f(\ldots, x, y, z) = z' \cdot f(\ldots, x, y, 0) \tag{2-15b}$$

These equations can be proved by developing the function $f(\ldots, x, y, z)$ about z or z' and applying the idempotent theorem.

It follows from Eqs. (2-5a) and (2-5b) that the minterm form of a switching function and the maxterm form of the same switching function will have identical groups of f coefficients. The f coefficients of both forms that have the same sequence of 0's and 1's result in the same value, which is either 0 or 1. However, the effect of these two identical groups of 0's and 1's of the minterm and maxterm forms of the same function is different. The f coefficients that have the value 1 determine the significant minterms in the minterm canonical form, whereas the f coefficients that are 0 determine the significant maxterms in the maxterm canonical form. To illustrate this, the minterm and maxterm forms of all switching functions with two variables are listed in Table 2-1.

Table 2-1

Switching function	*f coefficient* f_3	f_2	f_1	f_0	*Minterm expansion*	*Maxterm expansion*
0	0	0	0	0	0	$M_0 \cdot M_1 \cdot M_2 \cdot M_3$
$y'z'$	0	0	0	1	m_0	$M_1 \cdot M_2 \cdot M_3$
$y'z$	0	0	1	0	m_1	$M_0 \cdot M_2 \cdot M_3$
y'	0	0	1	1	$m_0 + m_1$	$M_2 \cdot M_3$
yz'	0	1	0	0	m_2	$M_0 \cdot M_1 \cdot M_3$
z'	0	1	0	1	$m_0 + m_2$	$M_1 \cdot M_3$
$y'z + yz' = (y' + z')(y + z)$	0	1	1	0	$m_1 + m_2$	$M_0 \cdot M_3$
$y' + z'$	0	1	1	1	$m_0 + m_1 + m_2$	M_3
yz	1	0	0	0	m_3	$M_0 \cdot M_1 \cdot M_2$
$y'z' + yz = (y' + z)(y + z')$	1	0	0	1	$m_0 + m_3$	$M_1 \cdot M_2$
z	1	0	1	0	$m_1 + m_3$	$M_0 \cdot M_2$
$y' + z$	1	0	1	1	$m_0 + m_1 + m_3$	M_2
y	1	1	0	0	$m_2 + m_3$	$M_0 \cdot M_1$
$y + z'$	1	1	0	1	$m_0 + m_2 + m_3$	M_1
$y + z$	1	1	1	0	$m_1 + m_2 + m_3$	M_0
1	1	1	1	1	$m_0 + m_1 + m_2 + m_3$	1

The choice of the indices for the f coefficients of the minterm and maxterm forms as determined by Eqs. (2-5) greatly simplifies the conversion of one type of switching function into the other by determining the f coefficients of the canonical form.

It is also possible to transform the minterm canonical form of a switching function into the maxterm form (and inversely) by means of the following operation:

$$\begin{aligned} f(y,z) &= f_0 m_0 + f_1 m_1 + f_2 m_2 + f_3 m_3 \\ &= [(f_0 m_0 + f_1 m_1 + f_2 m_2 + f_3 m_3)']' \\ &= (f_0' m_0 + f_1' m_1 + f_2' m_2 + f_3' m_3)' \\ &= (f_0 + m_0') \cdot (f_1 + m_1') \cdot (f_2 + m_2') \cdot (f_3 + m_3') \\ &= (f_0 + M_0) \cdot (f_1 + M_1) \cdot (f_2 + M_2) \cdot (f_3 + M_3) \end{aligned} \tag{2-16}$$

To illustrate the use of transformation equation (2-16), the maxterm form of switching function $f(y,z) = m_0$ will be determined.

$$\begin{aligned} f(y,z) &= 1 \cdot m_0 + 0 \cdot m_1 + 0 \cdot m_2 + 0 \cdot m_3 \\ &= (1 + M_0) \cdot (0 + M_1) \cdot (0 + M_2) \cdot (0 + M_3) \\ &= M_1 \cdot M_2 \cdot M_3 \end{aligned} \tag{2-17}$$

Two useful laws with respect to the minterms and maxterms of the canonical form of a switching function should also be mentioned here:

$$m_i \cdot m_j = 0 \qquad \text{for } i \neq j \tag{2-18a}$$

and

$$M_i + M_j = 1 \qquad \text{for } i \neq j \tag{2-18b}$$

Law (2-18*a*) can be written as follows:

$$(f_i x_i y_i z_i) \cdot (f_j x_j y_j z_j) = 0 \tag{2-19}$$

Because $i \neq j$, at least one of the variables of the logical product of two minterms will appear in one of the minterms in the negative form, so that the logical product of Eq. (2-19), if z is taken to be that variable, will contain a logical product zz' which is 0; therefore, the complete product of Eq. (2-19) is 0 in accordance with Theorems 6*a* and 5*a*.

Equation (2-18*b*) can be proved easily by means of the dual theorems in a way analogous to the proof of Eq. (2-18*a*). The dual theorems to be used are 6*b* and 5*b*.

The expansion of an arbitrary switching function into the minterm or maxterm canonical form can be performed in one of the following ways.

EXPANSION IN MINTERM CANONICAL FORM

This expansion need not be performed by means of the rather elaborate process based on the development equation (2-4*a*). A possible shortcut is provided by Theorems 3*a*, 4*a*, and 6*b*. The use of these theorems in this expansion process will be demonstrated by means of the minterm expansion of the following function:

$$f(x, y, z) = (x + y')\cdot(y + z) \tag{2-20}$$

Application of Theorem 3*a* gives:

$$f(x, y, z) = x\cdot y + y'\cdot z \tag{2-21}$$

Application of Theorem 4*a* gives:

$$f(x, y, z) = xy\cdot 1 + 1\cdot y'z \tag{2-22}$$

Application of Theorem 6*b* gives:

$$f(x, y, z) = xy\cdot(z + z') + (x + x')\cdot y'z \tag{2-23}$$

Application of Theorem 3*a* gives:

$$\begin{aligned} f(x, y, z) &= xyz + xyz' + xy'z + x'y'z \\ &= m_7 + m_6 + m_5 + m_1 \end{aligned} \tag{2-24}$$

In each of the minterms of Eq. (2-21), a missing variable (and this is not necessarily limited to one variable) is marked with a 1 (Theorem 4*a*). In accordance with Theorem 6*b*, this 1 can be replaced by the logical sum of the missing variable and its negative. The desired minterm canonical expression will then result from application of the first distributive law (Theorem 3*a*) to Eq. (2-23).

With some experience, most of these transformation steps can be taken in one operation.

EXPANSION IN MAXTERM CANONICAL FORM

This operation can be performed in a way analogous to the minterm canonical expansion. If the function to be expanded is given as the logical sum of logical products, it is necessary to apply in succession the second distributive law (Theorem 3*b*) and Theorems 6*b*, 4*a*, 4*b*, 6*a*, 3*b*, 1*a*, and 9*a*. The same example as above may be used to illustrate this expansion.

$$f(x, y, z) = x\cdot y + y'z \tag{2-25}$$

$$= (xy + y')\cdot(xy + z) \qquad \text{(T3}b\text{)} \tag{2-26}$$

$$= (x + y')(y + y')(x + z)(y + z) \qquad \text{(T3}b\text{)} \tag{2-27}$$

$$= (x + y')(x + z)(y + z) \qquad \text{(T6}b\text{, T4}a\text{)} \tag{2-28}$$

$$= (x + y' + 0)(x + 0 + z)(0 + y + z) \qquad \text{(T4}b\text{)} \tag{2-29}$$

$$= (x + y' + zz')(x + yy' + z)(xx' + y + z) \qquad \text{(T6}a\text{)} \tag{2-30}$$

$$f(x, y, z) = (x + y + z)(x + y' + z)(x + y' + z')(x' + y + z) \quad \text{(T3}b\text{, T1}a\text{, T9}a\text{)}$$

$$= M_0 \cdot M_2 \cdot M_3 \cdot M_4 \tag{2-31}$$

The results found in Eqs. (2-24) and (2-31) agree because the maxterm indices are precisely those that are missing in the minterm expansion result, and vice versa.

2-3 TRUTH TABLES

It is often very useful to define or to express the desired operation of a logical circuit by means of a truth table. A truth table consists of two parts, viz., columns stating the cause and a column giving the result of the desired operation that corresponds to the f coefficients of the canonical form of the desired switching function. In most cases, all possible combinations of the causes or input variables are arranged systematically in the cause columns.

In the following chapters, the input variables of a circuit will be indicated by means of small letters and the output variables, which describe the result of a circuit, will be capitalized. Such notation can, in turn, be used again as the cause of the next operation, in which case the symbol will be changed into a small letter at the input of the circuit for the next operation.

The use of truth tables may be illustrated by means of some examples. Figures 2-4 and 2-5 show a series connection and a parallel connection, respectively, of two contacts. The operation of these two circuits is defined by the corresponding truth tables 2-2 and 2-3, respectively.

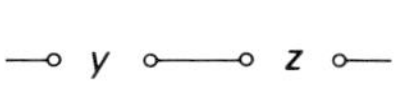

Fig. 2-4

Table 2-2

y	z	S	
0	0	0	f_0
0	1	0	f_1
1	0	0	f_2
1	1	1	f_3

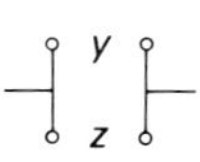

Fig. 2-5

Table 2-3

y	z	S	
0	0	0	f_0
0	1	1	f_1
1	0	1	f_2
1	1	1	f_3

The input variables are in both cases y and z. They represent the controlling contacts of the circuits. The state of the complete circuit is indicated by S. The truth table gives in a systematic way all the steps to be found in a proof by perfect induction. The desired switching function can now be expressed in either of its canonical forms. The f coefficients of the terms or factors of the canonical form are made to correspond in their order of numbering with the corresponding rows of the truth table, the input variables being arranged as increasing binary numbers.

The minterm canonical form of the switching function that results from truth table 2-2 is

$$\begin{aligned} S &= f_0 y'z' + f_1 y'z + f_2 yz' + f_3 yz \\ &= 0\cdot y'z' + 0\cdot y'z + 0\cdot yz' + 1\cdot yz = y\cdot z \end{aligned} \tag{2-32}$$

The resulting formula $S = y\cdot z$ corresponds to the already-known formula that describes the series connection of two contacts.

The minterm canonical form of the switching function resulting from truth table 2-3 is

$$\begin{aligned} S &= f_0 y'z' + f_1 y'z + f_2 yz' + f_3 yz \\ &= 0\cdot y'z' + 1\cdot y'z + 1\cdot yz' + 1\cdot yz \\ &= y'z + yz' + yz = y + z \end{aligned} \tag{2-33}$$

Equation (2-33), of course, corresponds to the already-known equation that describes the parallel connection of two contacts.

The desired switching equations (2-32) and (2-33) can also be obtained starting with the maxterm canonical form:

$$\begin{aligned} S &= (f_3 + y' + z')\cdot(f_2 + y' + z)\cdot(f_1 + y + z')\cdot(f_0 + y + z) \\ &= (1 + y' + z')\cdot(0 + y' + z)\cdot(0 + y + z')\cdot(0 + y + z) \\ &= (y' + z)\cdot(y + z')\cdot(y + z) = y\cdot z \end{aligned} \tag{2-34}$$

and

$$\begin{aligned} S &= (f_3 + y' + z')\cdot(f_2 + y' + z)\cdot(f_1 + y + z')\cdot(f_0 + y + z) \\ &= (1 + y' + z')\cdot(1 + y' + z)\cdot(1 + y + z')\cdot(0 + y + z) \\ &= y + z \end{aligned} \tag{2-35}$$

Finally, the concepts of the dual and the inverse of a switching function must be clarified. The dual of a function is obtained by replacing all operating signs (+ and $\cdot$) by their duals, all constants (0 and 1) by their negatives, and all variables by arbitrary corresponding variables. This will be illustrated by means of the following examples.

1. The given function is

$$f(y, z) = y + z + 0 \tag{2-36}$$

The dual function is

$$\phi(y, z) = y\cdot z\cdot 1 \tag{2-37}$$

2. The original function is

$$f(x, y, z) = x'y + xz \tag{2-38}$$

Then the dual form is

$$\phi(a, b, c) = (a + b) \cdot (a' + c') \tag{2-39}$$

In the first example, the variables have been left unchanged—but this is not necessary. In the second example, the variable a' corresponds to the variable x, b to y, and c' to z. Generally, two corresponding variables are independent. So variable a' may take the value 0 or 1 if x takes the value 0. The property of duality expresses only a relation between the structures of two functions and not a relation between the values of these functions.

The inverse of a function is obtained by replacing all operating signs by their duals, all constants by their negatives, and all variables by their negatives. The inverse of a function is thus the dual of the function in which the corresponding variables are the negatives of those of the original function.

From De Morgan's theorem (Theorem 8), it follows that the inverse of a function is the negative of that function. So inversion and negation are identical concepts. The concept of inversion or negation is a specialization of the concept of duality, because the property of inversion expresses not only a relation between the structures of two functions but also a relation between the values of these functions. As an illustration, Eqs. (2-40) and (2-41) show the inverses of the functions (2-36) and (2-38):

$$f'(y, z) = y' \cdot z' \cdot 1 \tag{2-40}$$

$$f'(x, y, z) = (x + y') \cdot (x' + z') \tag{2-41}$$

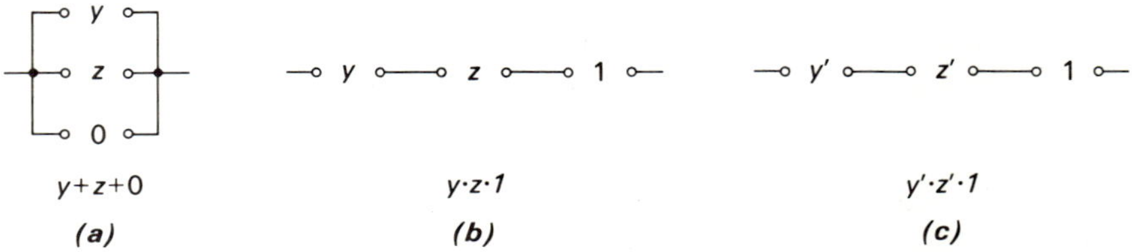

Fig. 2-6

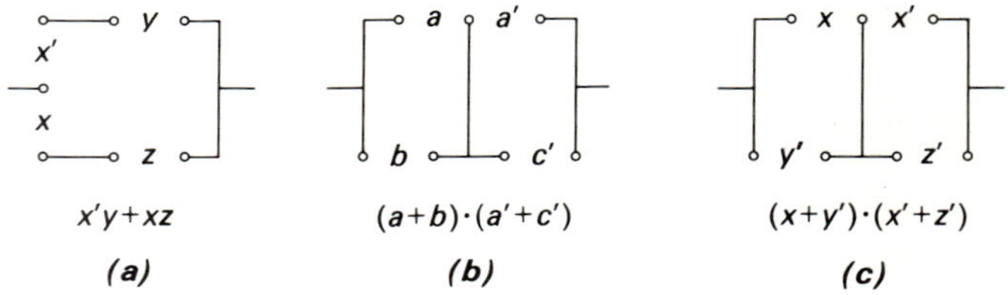

Fig. 2-7

Since the dual of a switching function is obtained by replacing all operating signs by their duals, the dual of a contact switching network is obtained by replacing all parallel connections by series connections and all series connections by parallel connections.

It is obvious that the duality of contact networks expresses only a relation between the network configurations and not one between the values of the switching functions. The inversion of contact networks, however, expresses not only a relation between the network configurations but also one between the switching functions. This will be illustrated by means of Figs. 2-6 and 2-7, which show the networks corresponding to Eqs. (2-36) to (2-41).

PROBLEMS

1. Prove by means of perfect induction that $1 \cdot z = z$. The proof should be based exclusively on the postulates. Indicate the postulates used for each successive step.

2. Prove by means of perfect induction that $z + z' = 1$, using the postulates only. Indicate the postulates used for each step.

3. Prove that $(xy + y'z)' = x'y + y'z'$. Indicate the theorems used for each successive step.

4. Prove that $[(x + y') \cdot (x + z')]' = x'(y + z)$. Indicate the theorems used for each successive step.

5. Prove that $(xz' + x'y' + yz')' = (x + y)z$, using Theorems 7*a*, 3*b*, 3*a*, and 6*a*. List all theorems used.

6. Prove that $(x'z')' \cdot (x' + z) \cdot (y + z) = z$, using Theorems 7*b*, 3*b*, 6*a*, 4*b*, and 10*a*. List all theorems used.

7. Prove that $(x + y) + (y + z')' = x + y + z$, using Theorems 11 and 7. List all theorems used.

8. Prove that $xz + x'y + yz = (x + y) \cdot (x' + z)$, using Theorems 4*b*, 6*a*, and 3*a*. List all theorems used.

9. Expand function $f(x,y,z) = xy + y'z$ in its minterm and maxterm canonical form using Eqs. (2-4*a*) and (2-4*b*).

10. Prove that $y \cdot (x + z) \cdot (x' + y + z') = y \cdot (x + z)$.

11. Prove that $(m_i)' = M_i$.

12. Expand function $w(x + y') + (w + y) \cdot (x' + z)$ in the minterm and maxterm canonical form.

13. Expand function $[w + (x + y')']' \cdot [(w + y)' \cdot (x' + z)']'$ in the minterm and maxterm canonical form.

14. Simplify the three-variable function characterized by the following f coefficients:

$$f_0 = f_2 = f_3 = f_4 = f_5 = f_6 = 1 \qquad \text{and} \qquad f_1 = f_7 = 0.$$

Draw this contact circuit with five contacts.

15. Simplify the three-variable function characterized by the following f coefficients:

$$f_0 = f_3 = f_5 = f_6 = 1 \qquad \text{and} \qquad f_1 = f_2 = f_4 = f_7 = 0$$

Draw this nonplanar circuit with eight contacts.

16. Write the truth table for the four-variable function describing a contact circuit that is closed if two, and only two, contacts are closed.

17. Write the truth table for the addition mod 2 of two variables y and z.

18. Write the truth table for the functions mentioned in Probs. 12 through 16.

SELECTIVE BIBLIOGRAPHY

1. Berlin, R. D.: Synthesis of N-valued Switching Circuits, *IRE Trans.*, vol. EC-7, pp. 52–56, 1958.
2. Buffery, G. H.: A Contribution to the Algebra of Relay and Switch Contacts, *Proc. Inst. Elec. Engrs. (London)*, pt. 1, vol. 97, pp. 357–363, 1950.
3. Gilbert, E. N.: N-terminal Switching Circuits, *Bell System Tech. J.*, vol. 30, pp. 668–688, 1951.
4. Keister, W., A. E. Ritchie, and S. H. Washburn: "The Design of Switching Circuits," D. Van Nostrand Company, Inc., New York, 1951.
5. Lee, C. Y., and W. H. Chen: Several Valued Combinational Switching Circuits, *Trans. AIEE*, pt. 1, vol. 75, pp. 278–283, 1956.
6. Marcus, M. P.: The Detection and Identification of Symmetric Switching Functions with the Use of Tables of Combinations, *IRE Trans.*, vol. EC-5, pp. 237–239, 1956.
7. Miller, R. E.: Formal Analysis and Synthesis of Bilateral Switching Networks, *IRE Trans.*, vol. EC-7, pp. 231–234, 1958.
8. Montgomerie, G. A.: Sketch for an Algebra of Relay and Contactor Circuits, *J. Inst. Elec. Engrs. (London)*, pt. 3, vol. 95, pp. 303–312, 1948.
9. Nakasima, A.: The Theory of Equivalent Transformation of Simple Partial Paths in the Relay Circuit, *J. Inst. Elec. Engrs. (Tokyo)*, no. 165, 1936, and no. 167, 1937 (in Japanese); condensed English translation by A. Nakasima and M. Hanzawa in *Nippon Electrical Communication*, pp. 32–39, 1938.
10. Oberman, R. M. M.: Switching Algebra (Schakelalgebra), *Het P.T.T.-Bedrijf*, vol. 4, pp. 125–175, 1952 (in Dutch).
11. Piesch, H.: Begriff der Algemeinen Schaltungstechnik, *Archiv für Elektrotech.*, vol. 33, pp. 672–686, 1939.
12. Serrel, R.: Elements of Boolean Algebra for the Study of Information Handling Systems, *Proc. IRE*, vol. 43, pp. 1366–1380, 1953.
13. Shannon, C. E.: Symbolic Analysis of Relay and Switching Circuits, *Trans. AIEE*, vol. 57, pp. 713–723, 1938.
14. Staehler, R.: An Application of Boolean Algebra to Switching Circuit Design, *Bell System Tech. J.*, vol. 31, pp. 125–175, 1952.
15. The Staff of the Computation Laboratory: "Synthesis of Electronic Computing and Control Circuits," Harvard University Press, Cambridge, Mass., 1951.
16. Washburn, S. H.: Application of Boolean Algebra to the Design of Electronic Switching Circuits, *Trans. AIEE Commun. Electron.*, vol. 8, pp. 380–388, 1953.

3
Simplification of Switching Functions

3-1 SIMPLIFICATION BY MEANS OF THE MANIPULATION OF LOGICAL LAWS

A large portion of the theorems or laws given in the previous chapter are in fact simplification formulas, because a different number of variables appear on either side of the sign of equality. Many types of switching functions can be manipulated into their simplest form by means of a correct use of these simplification formulas. Some experience in the handling of these formulas is required in order to arrive at the simplest form. However, a separate chapter must be devoted to this topic because there is no general rule for manipulating a switching function, nor is there any theorem or law which is more important than the other formulas for this particular purpose. Moreover, some of the theorems given in Chap. 2 have no equivalent in ordinary algebra, so that in the beginning it may be somewhat difficult to see in what possible ways they could be applied.

Many articles have been written about this subject that try to achieve minimal solutions in the representation of the switching formulas. However, in electronic switching circuits, it is seldom necessary to go further than two-level factoring. In contact circuits, a good circuit can be obtained from a switching formula in the two-level form. After an explication of simplification by means of the manipulation of logical laws and topological figures, we shall limit our discussion to the use of so-called "Karnaugh," or "Veitch," maps for this purpose. With up to six variables, no difficulties will be encountered in the use of these maps; and for switching functions with more variables, it is perhaps better to review the original problem in order to decrease the number of variables than to obtain the simplification of the formulas by more elaborate methods. The smaller the number of variables involved in a switching function, the greater are the possibilities of implementing the switching functions with building stones of standard construction.

Some of the possible methods for manipulating switching functions will be described below by means of simple examples. The first example, which is perhaps the simplest example which could be thought of, is still difficult from the point of view of simplification by means of trial-and-error methods. The switching function to be simplified is

$$f(a,b) = a' + ab \tag{3-1}$$

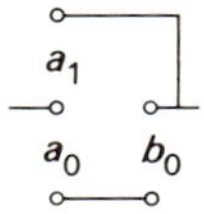

Fig. 3-1

This function has a contact-circuit configuration as given in Fig. 3-1. The variable a represents a make contact by a_0 in Fig. 3-1, and the variable a' corresponds to a break contact a_1. This convenient method of indicating contacts will be used throughout this book. In the chapters on relay circuits, this method will sometimes also be used for the variables in the switching functions themselves; but in this chapter, which deals exclusively with the most general simplification methods, these symbols will be used only in the drawings representing such contact circuits.

ELIMINATION OF REDUNDANT VARIABLES

Expansion equations The expansion equations (2-14) and (2-15) can be used to eliminate redundant variables from a minterm or maxterm of a switching function. The switching function (3-1) is similar to the expression given in rule (2-14b).

$$\begin{aligned} z' + f(\ldots,x,y,z) &= z' + f(\ldots,x,y,1) \\ a' + f(a,b) &= a' + f(1,b) \\ a' + a \cdot b &= a' + 1 \cdot b = a' + b \end{aligned} \tag{3-2}$$

In accordance with the right-hand portion of rule (2-14*b*), a 1 has to be substituted for every a in the function $f(a,b)$, and a 0 for every a' in that same function. The resulting contact circuit consists of a simple parallel connection of a break contact a_1 and a make contact b_0.

Distributive laws The application of the first distributive law (Theorem 3*a*) is fully analogous to the use of the same law in ordinary algebra. The example of Fig. 3-1 can be solved in an elegant fashion by means of the second distributive law (Theorem 3*b*). This theorem states that

$$x + y \cdot z = (x + y) \cdot (x + z)$$

By applying this theorem, Eq. (3-1) may now be converted as follows:

$$\begin{aligned} a' + a \cdot b &= (a' + a) \cdot (a' + b) = 1 \cdot (a' + b) \\ &= a' + b \end{aligned} \tag{3-3}$$

The right-hand side of Eq. (3-3) is derived from Theorems 6*b* and 4*a*.

Negation of a function It has been shown in Chap. 2 that all switching functions can be written as the logical sum of logical products (the minterm form) or as the logical product of logical sums (the maxterm form). This latter form is difficult to handle because one cannot see very well whether or not the second distributive law (Theorem 3*b*) can be applied to this particular form of the function. In the case of a maxterm form and in the absence of any experience in the handling of the second distributive law, it is advisable to take the dual or negative form of the given function. The first distributive law and all other possible simplification formulas can then be applied. After that, the dual or negative form of the resulting function must be taken. Switching function (3-1) can be used to illustrate this method.

$$\begin{aligned} a' + a \cdot b :: (a' + a \cdot b)' &= a \cdot (a' + b') = a \cdot a' + a \cdot b' \\ &= (a' + b)' :: a' + b \end{aligned} \tag{3-4}$$

Canonical form A general manipulation method for switching formulas involves using the canonical form of the switching function, as was already shown in Chap. 2. In this manipulation method, the problem is restated in its most general terms, which sometimes allow for the best possible manipulation—although this may not be apparent in the given example. This manipulation process, when applied to our example, can best be described as follows:

$$\begin{aligned} f(a,b) &= a' + ab = a' \cdot 1 + ab = a'(b + b') + ab = a'b + a'b' + ab \\ &= (a'b' + a'b) + (a'b + ab) = a'(b' + b) + b(a' + a) = a' + b \end{aligned} \tag{3-5}$$

This result is, of course, the same as the result obtained by the other methods.

ELIMINATION OF REDUNDANT TERMS OR FACTORS

The laws of Theorem 12 (Chap. 2) can be used to test a switching function for any redundant terms or factors. Since these laws do not exist in ordinary algebra, it makes elimination of redundant terms or factors difficult for the beginner. The following expression is a simple example of a switching function containing such a redundant term:

$$f(a, b, c) = ab + ac + bc' \tag{3-6}$$

The redundant term in this function can be detected by a systematic inspection of all pairs of terms containing a certain variable and its negative respectively. The function given above contains such a pair of terms, that is, ac and bc'. In accordance with Theorem 12*b*, the logical product of the other variables of this pair of terms corresponds to the redundant term—in this case the term ab.

An analogous process can also be used for a switching function consisting of a logical product of logical sums, such as

$$f(a, b, c) = (a + b) \cdot (a + c) \cdot (b + c') \tag{3-7}$$

It follows from Theorem 12*a* that $(a + b)$ is the redundant factor.

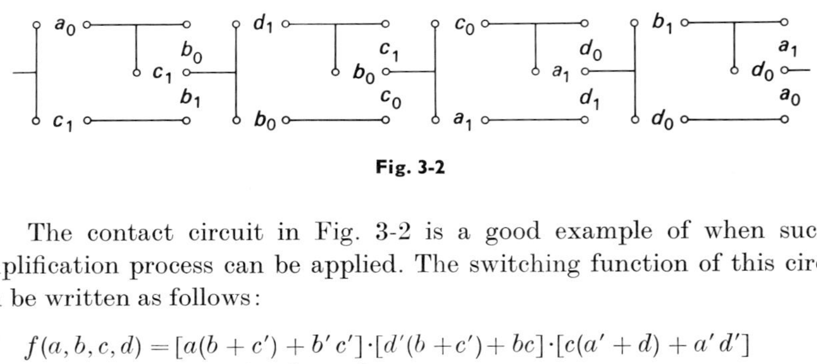

Fig. 3-2

The contact circuit in Fig. 3-2 is a good example of when such a simplification process can be applied. The switching function of this circuit can be written as follows:

$$\begin{aligned} f(a, b, c, d) &= [a(b + c') + b'c'] \cdot [d'(b + c') + bc] \cdot [c(a' + d) + a'd'] \\ &\quad \cdot [b'(a' + d) + ad] \\ &= (ab + ac' + b'c') \cdot (bd' + c'd' + bc) \cdot (a'c + cd + a'd') \\ &\quad \cdot (a'b' + b'd + ad) \end{aligned} \tag{3-8}$$

By Theorem 12*b*, this reduces to

$$f(a, b, c, d) = (ab + b'c') \cdot (bc + c'd') \cdot (a'd' + cd) \cdot (a'b' + ad)$$

Application of Theorem 12*c* gives

$$\begin{aligned} f(a, b, c, d) &= (a + b') \cdot (b + c') \cdot (b + c') \cdot (c + d') \cdot (a' + d) \cdot (c + d') \\ &\quad \cdot (a' + d) \cdot (a + b') \end{aligned}$$

Application of Theorem $9a$ results in

$$f(a,b,c,d) = (a+b')\cdot(b+c')\cdot(c+d')\cdot(d+a') = abcd + a'b'c'd' \tag{3-9}$$

The resulting circuit is shown in Fig. 3-3.

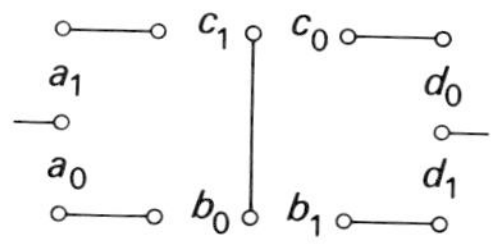

Fig. 3-3

INTRODUCTION OF REDUNDANT TERMS OR FACTORS

In the simplification process a switching function can sometimes be manipulated into one of a number of final forms that may not be its simplest form. Such a final form cannot be simplified further by means of the rules and methods given above. It is possible to test a switching function to see whether it has in fact reached such a form.

A switching function cannot be simplified by introducing redundant terms or factors in the following two cases:

1. When there are no pairs of terms or factors in the function containing a certain variable and its negative, respectively.
2. When the function does contain such pairs of terms (or factors) as mentioned under case 1, but the logical product (or logical sum) of the other variables of these terms (or factors) is 0 (or 1).

If the logical product (or the logical sum) of the variables as mentioned under 2 is not 0 (or 1), it is sometimes possible to reduce the number of terms (or factors) of the switching function under consideration. In order to find out whether this is possible, it is necessary to expand the function with all possible redundant terms (or factors) and to try all possible recombinations of terms (or factors) again.

This procedure will be illustrated by means of an example. The circuit to be reduced is shown in Fig. 3-4. The switching function of this circuit can be written as follows:

$$\begin{aligned} f(a,b,c,d) &= a'b'(c'+d') + (bc'+b'c)\,d \\ &= \underset{1}{a'b'c'} + \underset{2}{a'b'd'} + \underset{3}{bc'd} + \underset{4}{b'cd} \end{aligned} \tag{3-10}$$

Obviously, all means for reducing the number of minterms of this switching function would seem to be exhausted. The four minterms of the switching

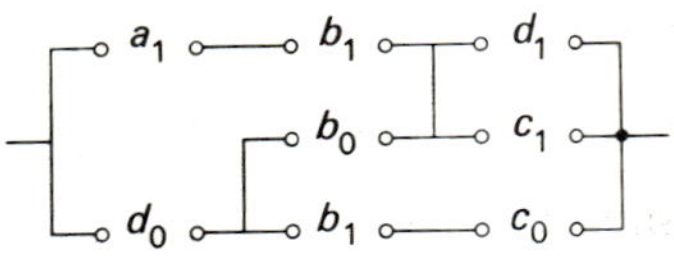

Fig. 3-4

function are numbered 1 to 4. Redundant minterms can be obtained from the minterms 1 and 3, 1 and 4, and 2 and 4:

$$\begin{array}{ccc} a'c'd & a'b'd & a'b'c \\ 1+3 & 1+4 & 2+4 \end{array} \tag{3-11}$$

The complete switching function which does not correspond to the minterm canonical form then becomes

$$\begin{aligned} & a'c'd + a'b'd + a'b'c + a'b'c' + a'b'd' + bc'd + b'cd \\ &= a'c'd + a'b' + bc'd + b'cd = a'b' + (bc' + b'c)\,d \end{aligned} \tag{3-12}$$

The additional minterm $a'c'd$ is still redundant after the recombination of terms. The resulting switching function contains seven variables. The corresponding circuit, however, can be designed as a bridge circuit with only six contacts as shown in Fig. 3-5.

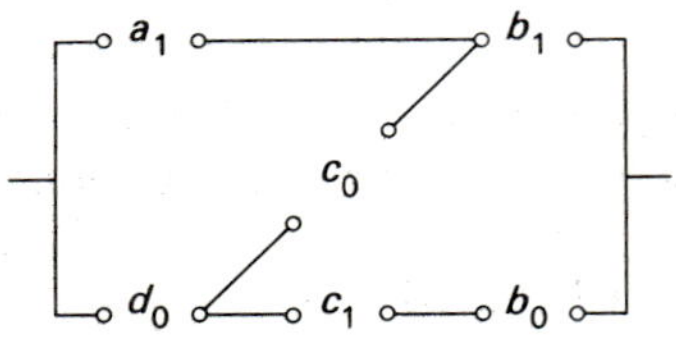

Fig. 3-5

The fundamental methods for simplifying switching functions can be summarized as follows, in the order in which they will normally be tried:

1. A switching function is first tested for terms (minterms or maxterms) which differ in one variable only. The simplification depends mainly on the application of Theorem 6 ($z + z' = 1$ or $z \cdot z' = 0$).
2. The second step is to investigate whether the switching formula is of a type analogous to the expressions $y + y'z = y + z$ or $y \cdot (y' + z) = y \cdot z$.
3. Sometimes a third step is possible, which is based on Theorem 12:

$$az' + bz + ab = az' + bz$$

or

$$(a + z')\,(b + z)\,(a + b) = (a + z')\,(b + z)$$

This step can be applied in both directions. In the reverse direction, it leads initially to a function with more terms than the original expression, but these can sometimes be grouped in a better way so that a further simplification can finally be achieved.

3-2 SIMPLIFICATION BY MEANS OF TOPOLOGICAL FIGURES

It is possible to represent switching functions by means of topological figures. When a limited number of variables is involved these figures give a good insight into the various methods—and their variants—for simplifying a given switching function, using the minterm or maxterm canonical form as a starting point. This method requires no special skill.

All possible switching functions with one variable can be represented by the two terminal points of a straight line, as shown in Fig. 3-6. One end represents the function z', the other end the function z. The line connecting these two points defines in terms of contacts a permanently closed circuit.

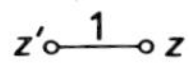

Fig. 3-6

A switching function with two variables y and z can be represented graphically by means of the four angular points of a square, as indicated in Fig. 3-7.

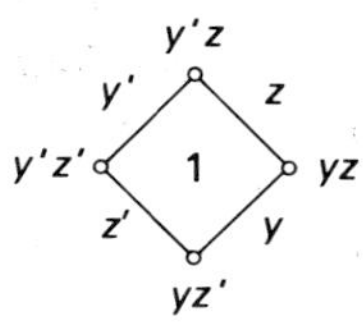

Fig. 3-7

On this figure we find:

1. The switching functions $y'z'$, $y'z$, yz', and yz are the four angular points of the square.
2. The switching functions z', z, y', and y, which appear as a simplification of the functions mentioned under 1, are in the form of the sides of the square.
3. The switching function 1, the permanently closed circuit, is the surface of the square.

If the canonical form of a switching function has two terms represented by adjacent angular points, these two terms can be combined into one term containing one variable less. This combined term is characterized by the side of the square connecting these points.

The switching function $a' + ab$ may be used again as an example. The canonical form of this function is:

$$f(a, b) = a'b' + a'b + ab$$

The square representing this switching function with two variables is shown in Fig. 3-8. The diagram shows two pairs of adjacent angular points

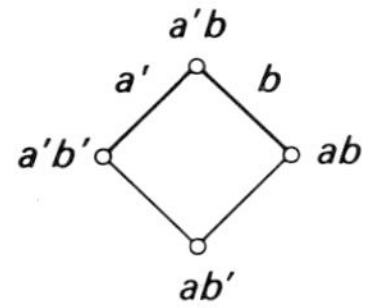

Fig. 3-8

that each have one variable in common. These two pairs of angular points may be replaced by the connecting sides of the square. The original switching function can thus be reduced to

$$f(a, b) = a' + b$$

This procedure can be extended to three variables. The resulting diagram is then the projection of a three-dimensional cube on a two-dimensional surface, as shown in Fig. 3-9*a*. The diagram is still simple enough for practical use. But the number of angular points, edges, and lateral faces increases with the number of variables. It is therefore more practical to indicate the angular

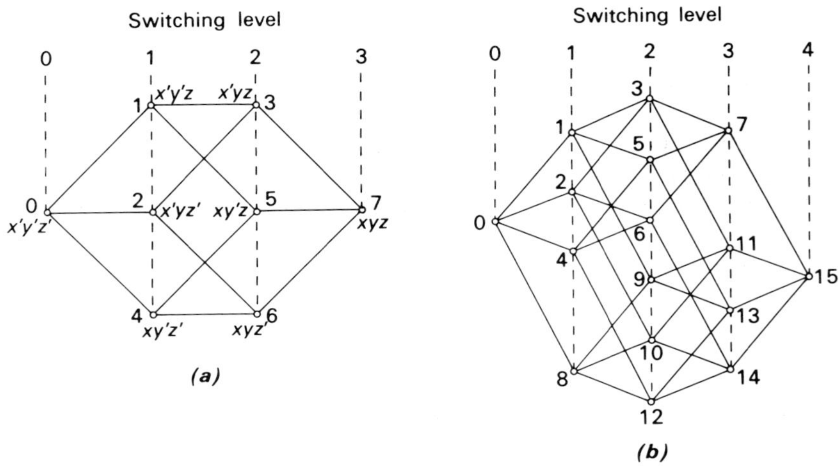

Fig. 3-9

points by means of the f-coefficient indices of the canonical form of the switching function. The canonical form of a switching function with three variables is as follows:

$$\begin{aligned} f(x,y,z) = & f_0 x'y'z' + f_1 x'y'z + f_2 x'yz' + f_3 x'yz \\ & + f_4 xy'z' + f_5 xy'z + f_6 xyz' + f_7 xyz \end{aligned} \tag{3-13}$$

The indices of the f coefficients are also shown in Fig. 3-9a. The procedure to be followed for this figure is in principle the same as that for Fig. 3-8. Moreover, if the canonical form comprises four terms forming the angular points of a square, these four terms can be combined into one term having two variables less. It is rather difficult to provide clear indications and symbols for all edges and all lateral faces of the cube of Fig. 3-9a.

It should be pointed out here that this topological representation of switching functions may also be used for maxterm canonical forms.

The diagram of Fig. 3-9a is arranged in such a way that all points which represent terms of the canonical form with the same number of complementary variables are shown on the same vertical line. These lines can be called switching levels.

Switching functions with four variables can be represented in the same way by means of the projection of a four-dimensional cube on a two-dimensional surface. This projection can be drawn in a rather symmetrical way, as shown in Fig. 3-9b. It comprises 16 angular points which correspond to the 16 terms of the canonical form of a 4-variable switching function. The possible combinations of two angular points into a line, of four angular points into a square, and of eight angular points into a cube are clearly seen in the figure.

This representation method for switching functions can be extended to switching functions with five variables, with adequate readability for discerning the various simplification possibilities. The corresponding diagram is shown in Fig. 3-10. It can be seen rather easily in this diagram that the method of choosing the prime implicants as far as possible with a minimum number of variables may lead to difficulties for some types of switching functions with five variables, as was pointed out by De Troye. The diagram of Fig. 3-10 corresponds to the following switching function:

$$\begin{aligned} f(v,w,x,y,z) = & m_0 + m_1 + m_3 + m_5 + m_7 + m_8 + m_{12} + m_{16} + m_{19} \\ & + m_{20} + m_{21} + m_{23} + m_{24} + m_{28} \end{aligned} \tag{3-14}$$

Simplification of this switching function in the way indicated above will yield the following result:

$$f(v,w,x,y,z) = v'w'z + w'xz + w'yz + vy'z' + wy'z' + x'y'z' \tag{3-15}$$

Each of the minterms of Eq. (3-15) describes a square, represented in the diagram of Fig. 3-10 by means of thick lines. The result in this pure sum form, however, is not the simplest to be obtained from the original switching

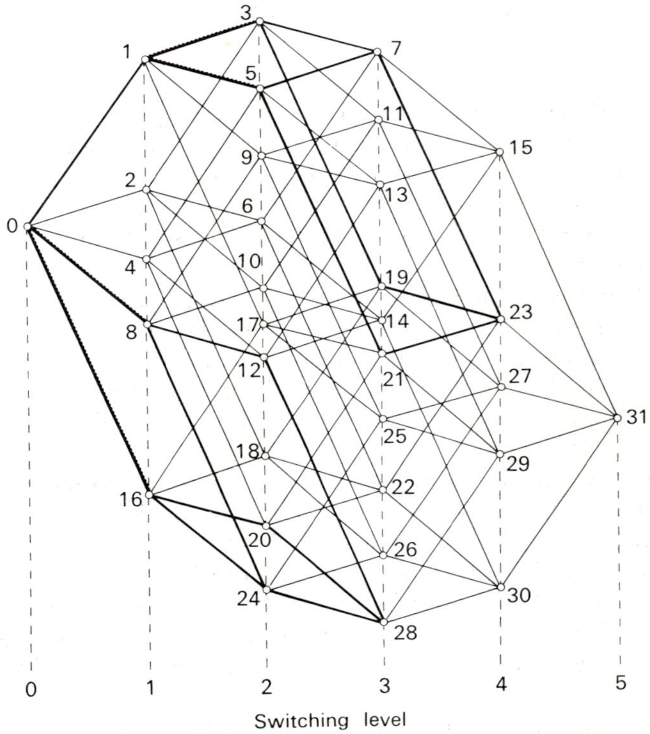

Fig. 3-10

function. The squares corresponding to the minterms $m_1 + m_3 + m_5 + m_7$ and $m_0 + m_8 + m_{16} + m_{24}$ can be replaced by a line which covers the minterms $m_0 + m_1$, as can easily be seen in the diagram of Fig. 3-10, because all other angular points of these squares are already part of other squares.

The resulting switching function is therefore

$$f(v, w, x, y, z) = v' w' x' y' + w' xz + w' yz + vy' z' + wy' z' \tag{3-16}$$

The sum form of Eq. (3-15) comprises six minterms, each with three variables; the sum form of Eq. (3-16) has five minterms, one of which has four variables; and the other four terms have three variables. As a sum form, the latter is more advantageous than the first because a smaller number of variables are involved.

However, if one factors the functions (3-15) and (3-16) into a mixed product-and-sum form, the result will be just the reverse:

$$\text{Equation (3-15)} = w'z(v' + x + y) + y'z'(v + w + x') \tag{3-17}$$

$$\text{Equation (3-16)} = v'w'x'y' + w'z(x + y) + y'z'(v + w) \tag{3-18}$$

Both equations are factored to the same level. The number of variables in Eq. (3-17) is slightly less than that of Eq. (3-18). It is possible to factor Eq. (3-18) one step further, but it does not affect the result.

3-3 SIMPLIFICATION BY MEANS OF KARNAUGH MAPS

The Venn diagram can be used as a graphical aid in the simplification or factoring of switching formulas. For example, a switching function in its canonical form can be represented easily on a Venn diagram by means of shaded areas. The function

$$f(y, z) = yz + y'z \tag{3-19}$$

occupies the shaded areas of a Venn diagram, as indicated in Fig. 3-11. The two shaded areas which represent both terms of the switching function (3-19)

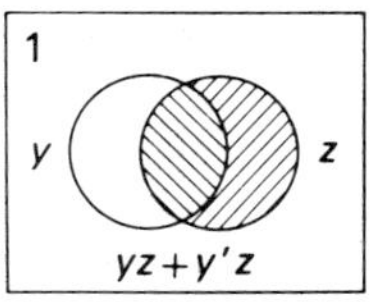

Fig. 3-11

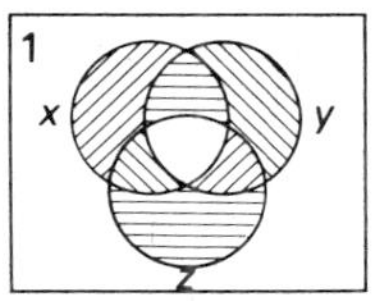

Fig. 3-12

together fill up the z area. This diagram illustrates graphically the obvious simplification of Eq. (3-19) to

$$f(y, z) = z \tag{3-20}$$

This graphical method of factoring is slightly more difficult in Venn diagrams for three variables, and it becomes impractical for switching functions with four variables.

The Venn diagram shown in Fig. 3-12 corresponds to the three-variable switching function

$$f(x, y, z) = x'y'z + x'yz' + x'yz + xy'z' + xy'z + xyz' \tag{3-21}$$

The diagram of Fig. 3-12 can be divided into three parts as shown in Fig. 3-13a to c. The shaded areas in these three diagrams taken together cover the same area as indicated in the diagram of Fig. 3-12.

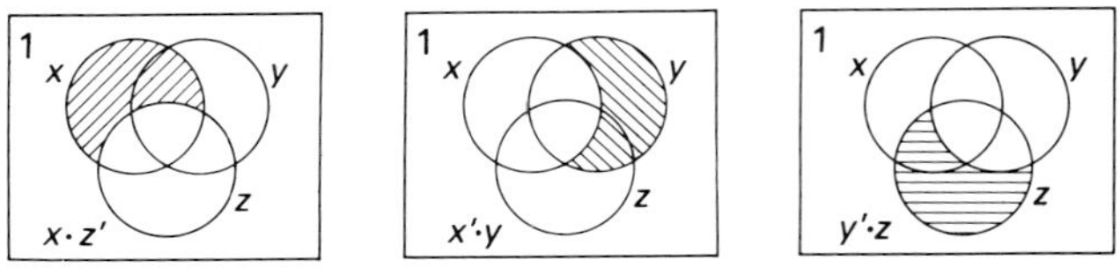

(a) (b) (c)

Fig. 3-13

The resulting simplified switching function corresponds to the logical sum of the three shaded areas as shown in Fig. 3-13a to c. This simplified function is

$$f(x, y, z) = x'y + y'z + xz' \tag{3-22}$$

A given area in a Venn diagram may be covered more than once, as will be illustrated by means of the two-variable switching function

$$f(y, z) = yz' + yz + y'z = y + z \tag{3-23}$$

and its graphical representation in the Venn diagram of Fig. 3-14. The area yz is in fact covered twice in this Venn diagram corresponding to the simplest

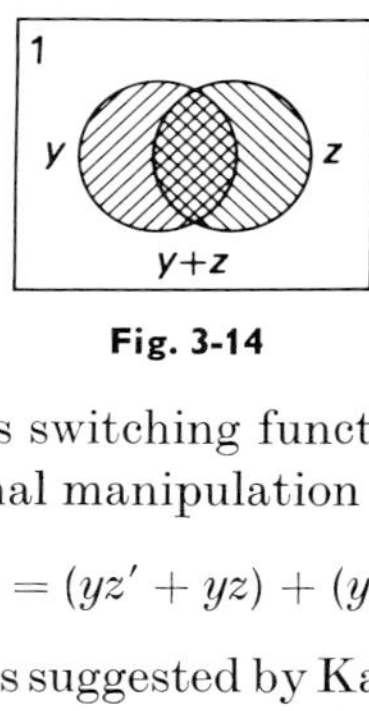

Fig. 3-14

	z'	z
y'	y'z'	y'z
y	yz'	yz

Fig. 3-15

form of this switching function. This is actually equivalent to what happens in the normal manipulation of Eq. (3-23) where the term yz is used twice:

$$f(y, z) = (yz' + yz) + (y'z + yz) = y + z \tag{3-24}$$

As was suggested by Karnaugh—and also by B. W. Veitch—the graphical representation of the Venn diagram as a handy tool for factoring switching formulas can be improved considerably. Instead of circles, rectangles are used for the various variables; these rectangles intersect each other and thus form a number of squares. Each of those squares represents a term of the canonical form of the switching function. These areas are arranged in such a systematical way that one can easily see the specific relations between the various areas, i.e., between the various terms of the switching function. The general form of a Karnaugh map for two variables is given in Fig. 3-15.

The variables y, y', z, and z' are each related to a pair of squares on the map in such a way that each area on this diagram represents one specific term of the canonical form of the two-variable switching function. A specific switching function can be indicated in this type of diagram by marking with a 1 the areas corresponding to the actual terms of the switching function in its canonical form and using a 0 to denote those areas representing the terms which are not found in the function under consideration. The switching function (3-23) is represented in this way in Fig. 3-16. All adjacent y areas

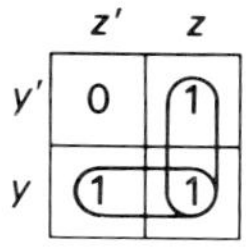

Fig. 3-16

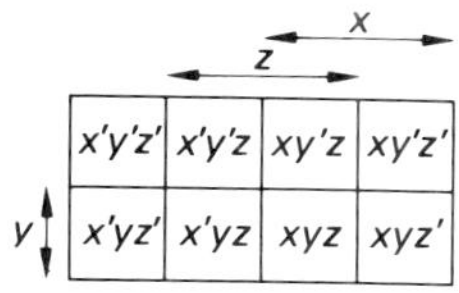

Fig. 3-17

and all adjacent z areas are marked with 1's. If in such diagrams two adjacent areas are marked with a 1, the corresponding terms of the switching function can be factored. Areas affording this possibility of simplification are enclosed on the map. This means that in the example given in Fig. 3-16, the switching function (3-22) can be simplified as was already shown in Eq. (3-24).

The basic map of a three-variable function is given in Fig. 3-17. The variables x, x', y, y', z, and z' are now related to groups of four adjacent squares. Two adjacent squares marked with 1's here again result in the canceling out of one term and one variable; and four such adjacent squares grouped on one line or in a square indicate that this group of three-variable terms of the canonical form of the switching function can be simplified into one term with a single variable.

The Karnaugh map for the switching function given by Eq. (3-21) is shown in Fig. 3-18. The associated 1 areas show that this switching function can be factored as already indicated in Figs. 3-12 and 3-13.

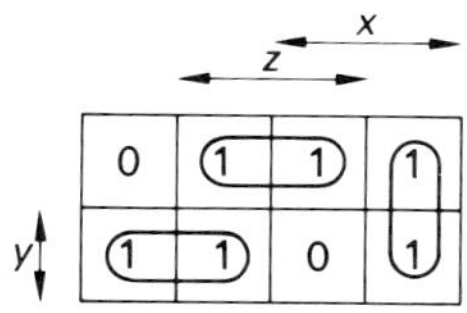

Fig. 3-18

The switching function given in Eq. (3-21) can be factored in still another way than that shown in Fig. 3-13a to c. The various areas of the Venn diagram of Fig. 3-12 can obviously also be grouped in such pairs as are shown in Fig. 3-19a to c, in which case the resulting switching function is

$$f(x, y, z) = xy' + yz' + x'z \tag{3-25}$$

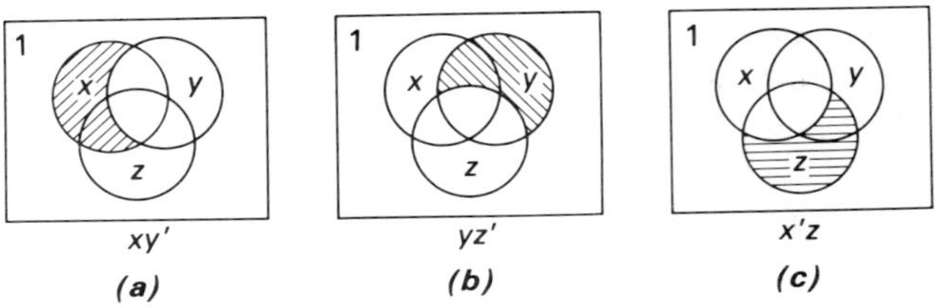

Fig. 3-19

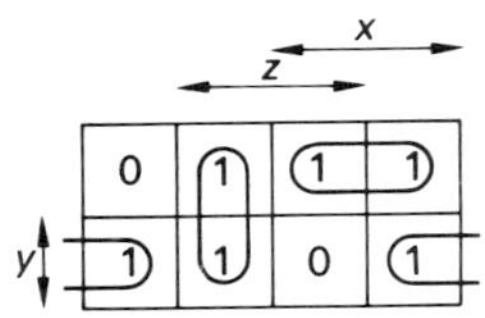

Fig. 3-20

This result means that the factoring enclosure in the corresponding Karnaugh map may also be drawn as indicated in Fig. 3-20 or, in other words, that this three-variable Karnaugh map must be considered as a flat representation of a toroidal surface.

The manipulation of a three-variable Karnaugh map is slightly easier than the processing of Venn diagrams for functions with three variables. The advantages of the Karnaugh-map method begin to show clearly in the factoring of switching functions with four variables. Of course, one can still draw a Venn diagram for functions with four variables, using ellipses instead of circles to represent the various variables.

The Karnaugh map for switching functions with four variables is a flat representation of a toroidal surface covered with 16 areas corresponding to all possible terms of the various switching functions. All terms of these functions which differ from each other in one variable only are found in adjacent areas; accordingly, where two adjacent areas are marked with a 1, the switching function in question contains two terms with four variables, each of which can be combined into one term with only three variables. Four adjacent areas marked with a 1 and grouped on one line or in a square correspond to four terms of the switching function with four variables that can be simplified into one term with two variables only. Finally, two adjacent squares each containing four areas marked with a 1 represent eight terms of the switching function with four variables that can be combined into one term with only one variable. These various possible simplifications are illustrated in Fig. 3-21*a* to *d*.

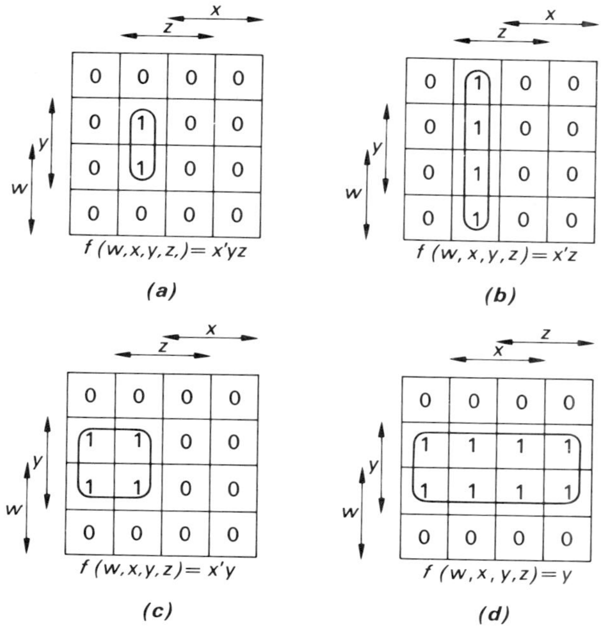

Fig. 3-21

Switching functions with more than four variables cannot be represented on a Karnaugh map in which all terms differing in one variable only are to be indicated by means of adjacent map areas. Before going into a more general method for extending the graphical representation of switching functions on a Karnaugh map to the many-variable case, it is necessary to deal with some useful applications of the Karnaugh maps, which are equally suitable for many-variable maps.

FACTORING VIA INVERTED FUNCTIONS

For a specific switching function, the 0's found on the Karnaugh map represent the inverse of that function, since these 0's correspond to the unused terms of that function in its canonical form. Thus, factoring of the 0's in a map leads to the simplified form of the inverted function, from which a simplified form of the original function can then be obtained by inversion. This second form of the simplified switching function will be a logical product of logical sums if the original function is given as a logical sum of logical products, or vice versa.

The two forms of the simplified switching functions obtained by means of these two different factoring methods will differ in appearance and sometimes in the number of components necessary for the actual circuit. It is important to know both forms of the function because either may be more advantageous from the point of view of actual circuit design.

The factoring of the inverted form of a switching function may be illustrated by means of the following example:

$$f(w,x,y,z) = m_0 + m_1 + m_2 + m_3 + m_6 + m_8 + m_9 + m_{12} + m_{13} \qquad (3\text{-}26)$$

In Eq. (3-26), the switching function is defined by means of a group of minterms the indices of which correspond to those of the f coefficients of the minterm canonical form. The Karnaugh map for normal factoring (i.e., of 1's) of Eq. (3-26) is shown in Fig. 3-22 and that for factoring the inverted function (i.e., of 0's) in Fig. 3-23.

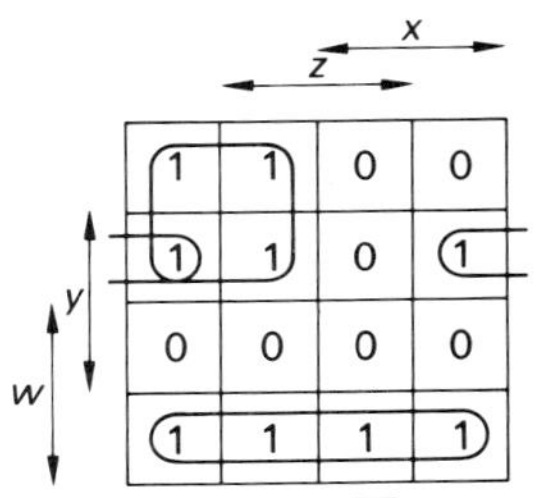

Fig. 3-22

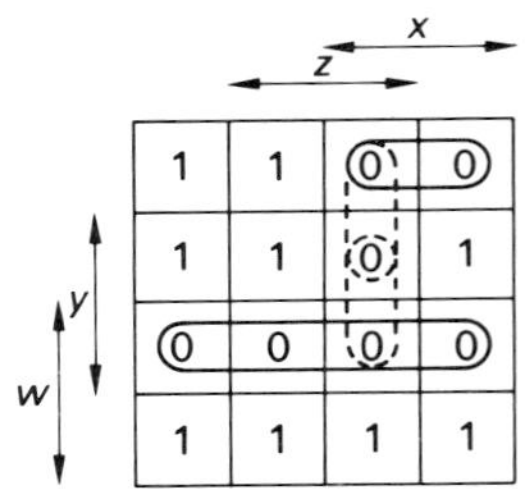

Fig. 3-23

If the best factoring result is desired, it is necessary to achieve the smallest possible number of enclosures on the map, grouping as many adjacent 1's as possible. In this example, the only suitable way of factoring the 1's is obviously that which is given in Fig. 3-22. The diagram of Fig. 3-22 leads to the following switching function:

$$f(w,x,y,z) = wy' + w'x' + w'yz' \qquad (3\text{-}27)$$

Zero factoring opens up two possibilities, i.e.,

$$w'xyz + w'xy'z$$

or

$$w'xyz + wxyz$$

Accordingly, the factored switching formula can be written in two different ways:

$$f'(w,x,y,z) = wy + w'xy' + w'xz \qquad (3\text{-}28)$$

$$f'(w,x,y,z) = wy + w'xy' + xyz \qquad (3\text{-}29)$$

Inversion of Eqs. (3-28) and (3-29) gives

$$f(w,x,y,z) = (w' + y')(w + x' + y)(w + x' + z') \tag{3-30}$$

$$f(w,x,y,z) = (w' + y')(w + x' + y)(x' + y' + z') \tag{3-31}$$

Applying the first distributive law to Eqs. (3-30) and (3-31) will yield Eq. (3-27). Inversely, when applying the second distributive law to Eq. (3-27), one will obtain either Eq. (3-30) or (3-31).

For those who are not accustomed to manipulating switching formulas by means of the second distributive law, the desired result can also be obtained by inversion of all + signs into · signs and all · signs into + signs in the original switching function, followed by application of the first distributive law (which is easy to handle). After that, all + and · signs have to be inverted again.

It may be concluded that zero factoring on a Karnaugh map followed by inversion of the result does not lead to anything which could not be obtained by manipulating the same switching function when factoring 1's. However, when the number of 0's on a Karnaugh map is considerably less than the number of 1's, zero factoring may lead more rapidly to the desired result.

MULTILEVEL FACTORING

The prescription that only adjacent pairs of 1's, groups of four on a line or in a square, or groups arranged in two adjacent squares of four areas, etc., can be used in the factoring process limits the result of this process to a pure logical sum of logical products or the reverse. This is called "two-level factoring."

It may be obvious, for example, when considering Eqs. (3-27), (3-28), or (3-29), that the two-level result will not always be the one containing the minimum number of variables. The equations mentioned above can be rewritten as follows:

(3-27): $$f(w,x,y,z) = wy' + w'(x' + yz') \tag{3-32}$$

(3-28): $$f(w,x,y,z) = wy + w'x(y' + z) \tag{3-33}$$

(3-29): $$f(w,x,y,z) = wy + x(w'y' + yz) \tag{3-34}$$

The level of the resulting function can be determined by numbering the consecutive operations found in the equation. For example, in Eq. (3-32),

$$f(w,x,y,z) = w\cdot y' \overset{4}{+} w' \overset{3}{\cdot} (x' \overset{2}{+} y \overset{1}{\cdot} z') \tag{3-35}$$

It is not necessary to award a level number to the · sign of the term wy'; this term may be taken to correspond to any of the numbers 1, 2, or 3. Equation (3-32) is found to be a four-level switching function. Equation (3-33) proves to be a three-level function, whereas Eq. (3-34) also represents a four-level function. Generally no difficulties will be encountered in the algebraic simplification of two-level functions into higher-level functions.

The Karnaugh map offers the possibility of factoring a suitable switching function into a three-level function. The zero factoring of Eq. (3-27) leading to Eq. (3-33) will be used to illustrate this extension of the factoring process. The corresponding map of this function is shown in Fig. 3-24. If the 1 in column

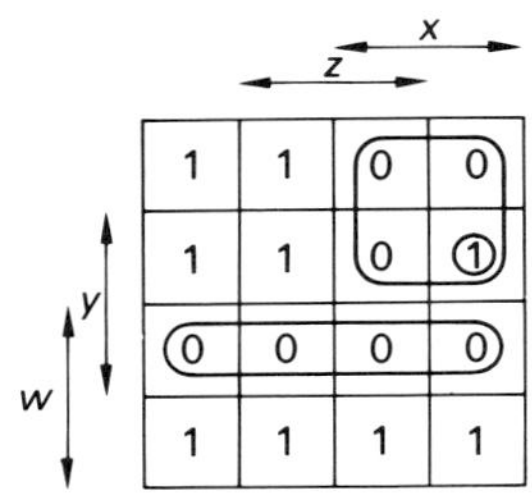

Fig. 3-24

xz' which corresponds to the term $w'xyz'$ had been a 0, the inverted form of the function represented by Fig. 3-24 would have been

$$f'(w,x,y,z) = wy + w'x \tag{3-36}$$

The term $w'x$ covers too many areas of the map. This term must therefore be complemented by means of an extra factor to exclude the term $w'xyz'$. In this case, the exclusion can be obtained by means of the formula

$$w'x(yz')' = w'x(y' + z) \tag{3-37}$$

This formula shows that the desired result can be obtained by inverting the part of the function that does not appear in the oversimplified form of the function.

The switching function resulting from this process is the same as the one given above in Eq. (3-33).

The same reasoning, of course, also holds for the 1 factoring of switching functions. In electronic switching circuits in general, no factoring higher than the three-level operation is required because in actual circuit design, switching functions with four or more levels lead to difficulties in the dimensioning of the components.

FACTORING WITH DON'T-CARE CONDITIONS

A specific circuit of an automatic device to be designed will be determined by a number of operating requirements as defined by the f coefficients 1 and 0 of the canonical form of the switching function. The unspecified f coefficients form the so-called "don't-care conditions." They correspond to such terms of the canonical form that are not determined in any way by the circuit requirements. The f coefficients of these terms may be chosen at will by the designer so that the largest possible simplification can be achieved.

These don't-care conditions may be taken into consideration in the factoring of switching functions inasmuch as they can help simplify the various circuits. Perhaps the most practical advantage of Karnaugh maps is that these don't-care conditions can be taken into account so that the best combination of the con't-care conditions with the given circuit requirements can be arrived at with practically no trial-and-error manipulations. This will be illustrated by means of some examples.

The switching function to be factored is

$$f(w,x,y,z) = f_3, f_4, f_5, f_6, f_{11}, f_{15} \tag{3-38}$$

The following don't-care conditions are given:

$$d(w,x,y,z) = f_1, f_7, f_{10} \tag{3-39}$$

The resulting Karnaugh map is shown in Fig. 3-25. The don't-care condition represented by coefficient f_7 can be used for simplifying two different groups

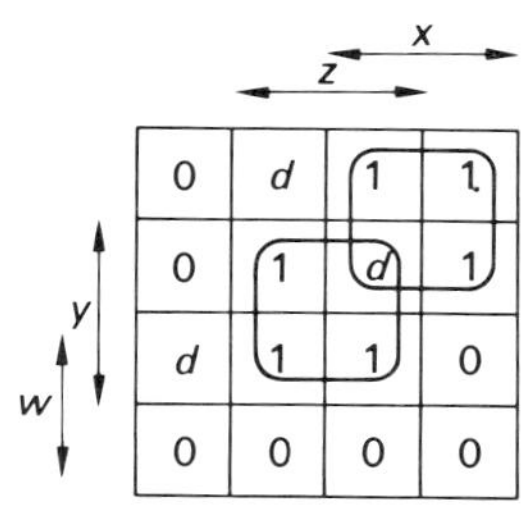

Fig. 3-25

of minterms f_4, f_5, f_6 and f_3, f_{11}, f_{15}. It is not difficult to see that the other don't-care conditions (f_1 and f_{10}) do not permit any further simplification of Eq. (3-38) because they cannot be used for "filling out" such unenclosed 1 areas so as to combine them into pairs or double pairs of 1's.

If the coefficient f_7 did not belong to the don't-care group, switching Eq. (3-38) could only be factored to

$$f(w,x,y,z) = w'x(y' + z') + yz(w + x') \tag{3-40}$$

The don't-care condition f_7 by itself leads to the following simpler result:

$$f(w,x,y,z) = w'x + yz \tag{3-41}$$

For cases in which a three-level factoring result can be obtained, it is sometimes useful to know that a Karnaugh map can be split up into two or more maps each of which represents a specific 1 configuration. In doing so, it is not necessary to cover the corresponding areas in the various maps more than once with a 1. If an area is already covered with a 1 in one of the sub-maps, that area may be considered as a don't-care condition in the others. This possibility is illustrated in the factoring of the following function:

$$f(w,x,y,z) = f_1, f_4, f_5, f_6, f_7, f_9, f_{12}, f_{13}, f_{14} \tag{3-42}$$

The corresponding Karnaugh map without enclosures is shown in Fig. 3-26. This map can be split up into two maps, that is, Fig. 3-27a and b. Since the term $f_{13} = wxy'z$ has already been used in the map of Fig. 3-27a, the corresponding area is marked with a d in Fig. 3-27b.

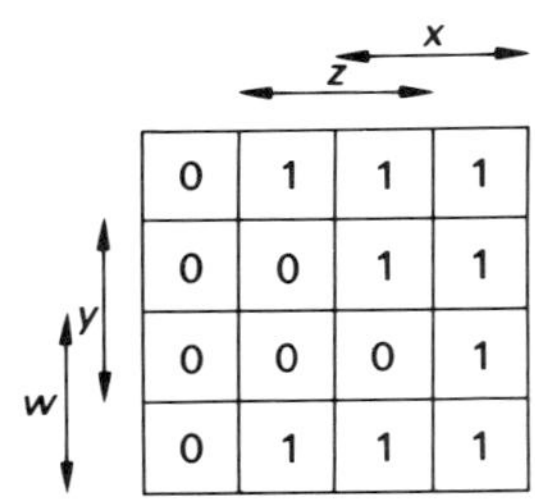

Fig. 3-26

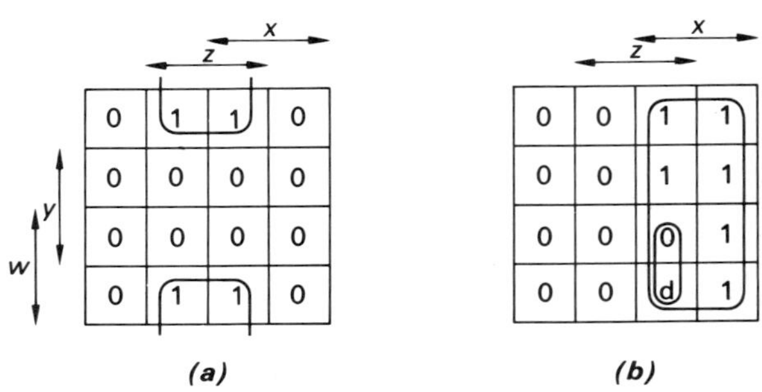

Fig. 3-27

If this area is considered as a 1 area in Fig. 3-27b, the factored result of the map will be

$$\begin{aligned} f(w,x,y,z) &= w'x + wx(yz)' = w'x + wx(y' + z') \\ &= x \cdot (w' + y' + z') \end{aligned} \tag{3-43}$$

If the f_{13} area in Fig. 3-27b is considered as a 0 area, the factored result will be

$$f(w,x,y,z) = w'x + wxz' = x \cdot (w' + z') \tag{3-44}$$

This result is more advantageous than Eq. (3-43) in spite of the fact that a smaller number of 1's is involved.

The total result of the factoring of Eq. (3-42), as shown in the maps of Fig. 3-27, will then be as follows:

$$f(w,x,y,z) = y'z + x(w' + z') \tag{3-45}$$

It remains to be seen whether this refinement of the Karnaugh map process is really worthwhile. The same result can also be obtained by means of a straightforward factoring process, followed by a simple application of the first distributive law. Both steps are just as easy or even easier than is the method given above. Of course, such rather mechanical minimizing aids can be very helpful, but one is too lightly inclined to overlook the basic laws of switching algebra when using these aids. To demonstrate this, let us consider again Eq. (3-42) as presented graphically in Fig. 3-28 with the necessary enclosures. The result, which is obtained in two steps, is

$$f(w,x,y,z) = w'x + y'z + xz' = y'z + x(w' + z') \tag{3-46}$$

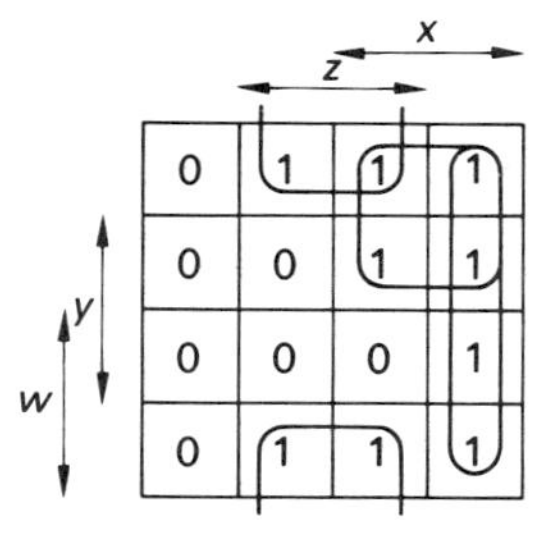

Fig. 3-28

MANY-VARIABLE KARNAUGH MAPS

It is possible to extend the use of Karnaugh maps to the factoring of switching functions with many variables. This extension will be explained by means of a map for six variables, showing the various steps to develop it from smaller maps. A map for six variables is by no means the outer limit of the extension process, but it becomes increasingly difficult to recognize the factoring patterns with the size of these maps. Moreover, practical cases are generally not so complicated as to require minimizing aids for more than six variables.

The principle of the Karnaugh map is that adjacent areas represent minterms or maxterms of the canonical form of a switching function which can be factored into a term with one variable less. Maps for switching functions with six variables can be built up of four maps for four variables on top of each other. This three-dimensional arrangement, which is difficult to handle, can be replaced by a two-dimensional setup in which the four maps are combined into a square.

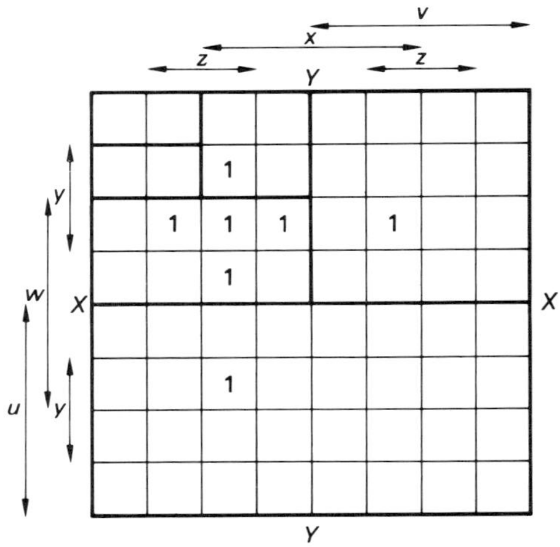

Fig. 3-29

The seven minterms (or maxterms) of the canonical form of a six-variable switching function which differ in one variable only and which form specific groups are arranged in Fig. 3-29 into a group of five on one map, as in the original four-variable map, whereas the other two terms are entered in two of the other maps symmetrically arranged with respect to X- and Y-coordinate axes separating the four 4-variable maps shown in Fig. 3-29. An arbitrary group of seven terms differing in one variable only is chosen as an example in Fig. 3-29.

In the diagram of Fig. 3-29 it is shown by means of thick lines how each Karnaugh map for one variable more can be obtained by reflection of the present diagram, this starting from a single-variable diagram. This process does not stop with a six-variable Karnaugh map. As a result, the variables covering groups of rows or columns appear alongside the map in a so-called reflected binary sequence. This process will be explained in more detail in Chap. 15.

Within each of the four squares representing a four-variable map, the rules given for the factoring process remain valid. In addition, each symmetric grouping of areas marked with a 1 in each adjacent pair of these four-variable maps, or in all four of these maps, corresponds to a factoring possibility. All that was said about the factoring of the negative of a function, three-level factoring, and don't-care conditions also holds.

In the example given in Fig. 3-30, some of the possible symmetric configurations in a six-variable map are shown. These will be sufficient to illustrate the use of this type of map for an arbitrary switching function.

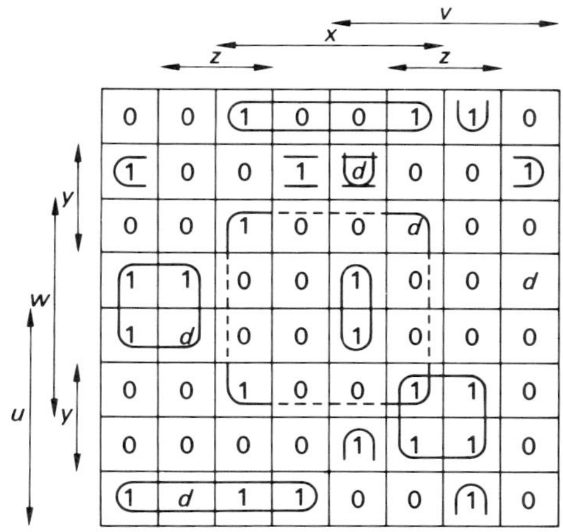

Fig. 3-30

MULTIOUTPUT CIRCUITS

In certain cases, the Karnaugh-map method can be helpful in combining a number of switching circuits into a multioutput circuit. Use of these maps for combining a number of two-terminal contact networks into a $n + 1$ terminal network is completely different from that for combining a number of logical diode circuits into a multioutput circuit.

The combination of contact networks by means of Karnaugh maps is possible under certain conditions, but in this process hard-to-find sneak paths are easily introduced. A more effective way of combining contact circuits without this hazard will be discussed in Chap. 5.

Two map configurations that enable an advantageous combination are shown in Fig. 3-31a and b. The diagrams in these maps represent the following switching functions:

$$f(w, x, y, z) = a = wz \tag{3-47}$$

$$f(w, x, y, z) = b = wz' \tag{3-48}$$

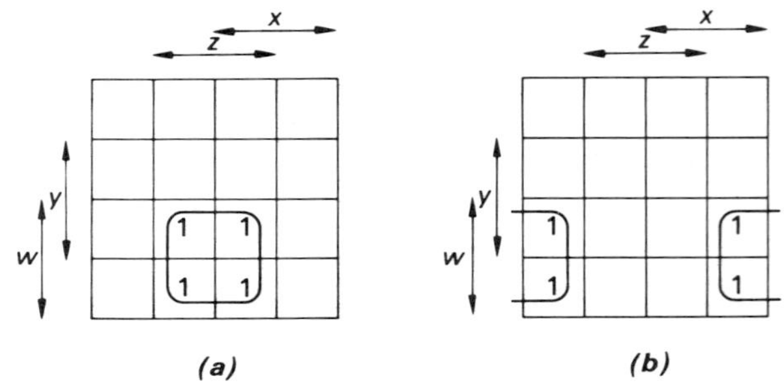

Fig. 3-31

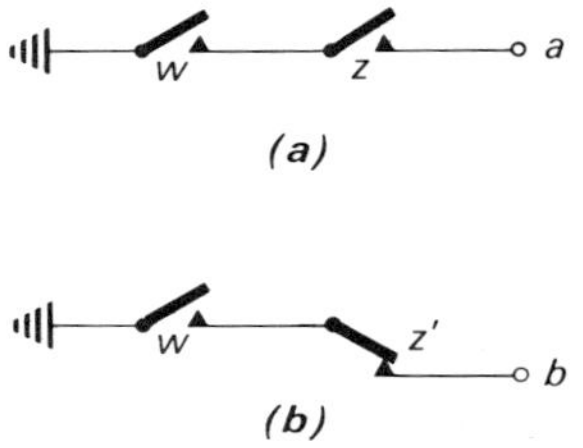

Fig. 3-32

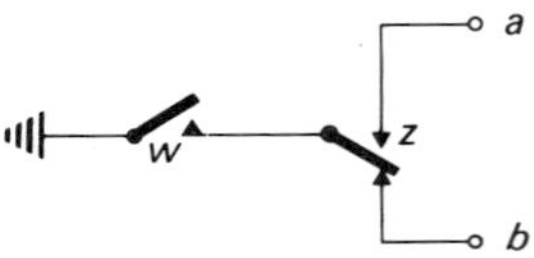

Fig. 3-33

It is not difficult in this example to conclude from Eqs. (3-47) and (3-48) that the corresponding circuits shown in Fig. 3-32*a* and *b* can be combined into one circuit as shown in Fig. 3-33.

In this case, it can easily be seen that in the combination of the circuits of Fig. 3-32*a* and *b*, no sneak paths have been introduced. In practical cases, however, such circumstances are usually not so simply obtained.

From the above it can be concluded that for the combination of contact circuits, complementary groups of 1's and not identical groups of 1's have to be traced in the maps. Identical groups of 1's lead to identical contact circuits, resulting in the connection of two different output terminals, which is not allowed when still other contact circuits to these output terminals are present. This may be illustrated by the following example.

$$f(w,x,y,z) = a = wz + (w')\,x'\,yz \tag{3-49}$$

$$f(w,x,y,z) = b = wz' + w'\,xy'\,z \tag{3-50}$$

The corresponding Karnaugh maps are shown in Fig. 3-34*a* and *b* and the resulting combined contact circuit in Fig. 3-35.

In the circuit configuration of Fig. 3-35, a sneak path $wx'y'z$ has been introduced for output terminal b and a sneak path $wx'y'z'$ for output terminal a. This is the result of the combination of two correctly designed circuits, as shown in Fig. 3-36*a* and *b*.

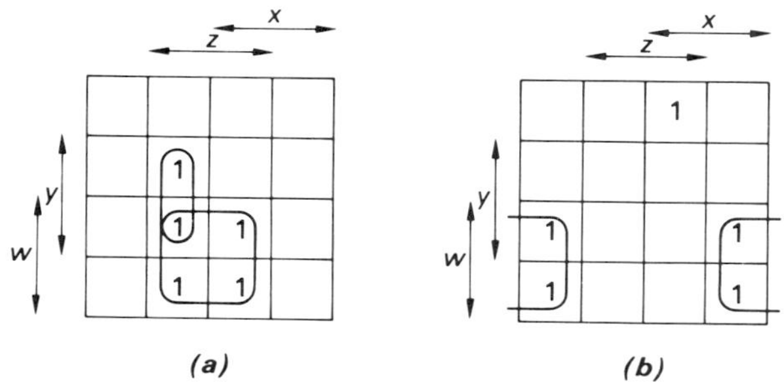

Fig. 3-34

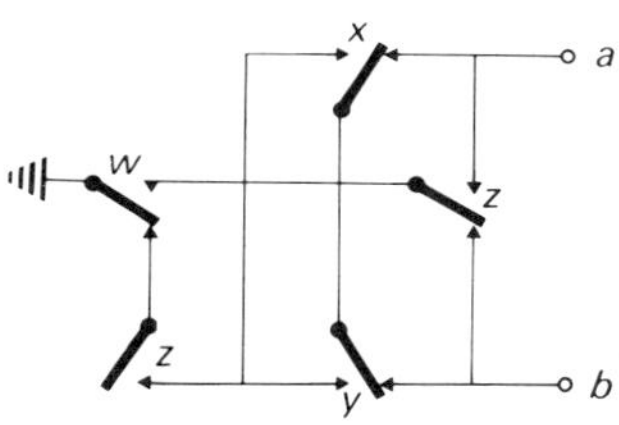

Fig. 3-35

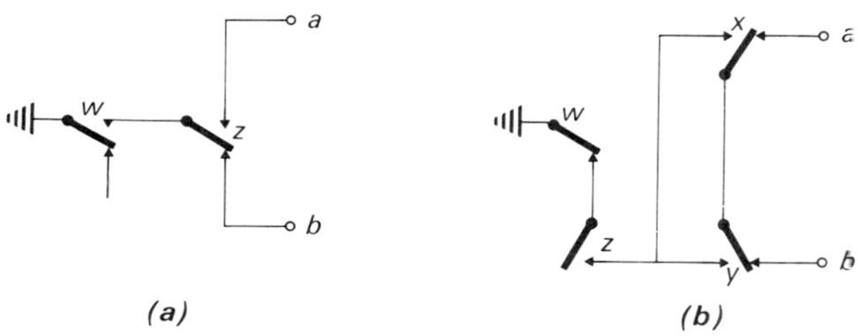

Fig. 3-36

There is no way of verifying by means of Karnaugh maps whether or not sneak paths have been introduced, and most sneak paths are not so easy to find as that shown in the example.

In the combination of diode circuits, Karnaugh maps can play a useful role. For this purpose, identical configurations of 1's (or 0's) have to be traced in a number of maps. These configurations imply gates that can be commonly used for a number of output terminals with no risk of sneak paths, as will be explained in Chap. 6.

PROBLEMS

1. Simplify the function $f(a,b,c,d,e) = (a + b')e + a'b(c + d')e$ by means of formula manipulations.

2. Simplify by means of a Karnaugh map the function given by $f_0 = f_1 = f_2 = f_3 = f_{12} = 1$.

3. Simplify the function given by the f coefficients $f_0 = f_1 = f_2 = f_{13} = 1$ into a three-level form containing only seven variables. Use only equations.

4. Simplify the function given by the f coefficients $f_0 = f_3 = f_{12} = f_{15} = 1$ into a three-level form comprising eight variables only. Use only equations.

5. Simplify by means of formula manipulation the function given by

$$f_0 = f_1 = f_2 = f_3 = f_4 = f_5 = f_{14} = f_{15} = 1$$

into a three-level form containing six variables.

6. Simplify the function $x'(w'y' + yz' + y'z) + xyz + w'y'z'$ by means of formula manipulation into a form with two variables less.

7. Simplify the function given in Prob. 6 by means of a single application of Theorem 11.

8. Simplify by means of a Karnaugh map the function given by

$$f_0 = f_1 = f_2 = f_4 = f_9 = f_{10} = f_{11} = 1$$

9. Simplify by means of formula manipulation the function $w'x' + y'z'(w' + x) + xyz$ into a form with two variables less.

10. Simplify by means of a Karnaugh map the function given by

$$f_0 = f_1 = f_2 = f_4 = f_7 = f_9 = f_{15} = 1$$

11. Simplify by means of a Karnaugh map the function given by

$$f_0 = f_1 = f_2 = f_3 = f_4 = f_7 = f_9 = f_{10} = 1$$

The resulting logical sum form contains 12 variables in the two-level form.

12. Simplify by means of a Karnaugh map the function given by

$$f_0 = f_1 = f_3 = f_4 = f_5 = f_{10} = f_{11} = 1$$

13. Simplify by means of a Karnaugh map the function given by

$$f_0 = f_1 = f_2 = f_3 = f_4 = f_5 = f_6 = f_{11} = 1$$

The resulting logical sum form contains seven variables.

14. Simplify the function given by $f_1 = f_2 = f_4 = f_7 = f_8 = f_{11} = f_{13} = f_{15} = 1$.

15. Simplify the function

$$f(a,b,c,d) = a'b'c'd + a'b'cd' + a'b'cd + a'bc'd + a'bcd' + ab'c'd + ab'cd + abcd'$$

by means of

(*a*) Formula manipulation

(*b*) A Karnaugh map

Give the simplest form of this function as logical product and sum.

16. Simplify the function $f(a,b,c,d) = ab'd + a'bc' + a'bcd' + b'd'$ into the logical product-and-sum forms by means of

(*a*) Formula manipulation

(*b*) A Karnaugh map

17. Simplify the function

$$f(a,b,c,d) = b'c'd' + abc'd' + a'b'cd' + ab'd + abc'd + a'b'c' + b'cd$$

by means of

(a) Formula manipulation

(b) A Karnaugh map

Give the simplest form of this function as logical product and sum.

SELECTIVE BIBLIOGRAPHY

1. Epstein, G.: Synthesis of Electronic Circuits for Symmetric Functions, *IRE Trans.*, vol. EC-7, pp. 57–60, 1958.
2. Ghazala, M. J.: Irredundant Disjunctive and Conjunctive Forms of a Boolean Function, *IBMJ Res. Develop.*, vol. 1, pp. 171–176, 1957.
3. Karnaugh, M.: The Map Method for Synthesis of Combinational Logic Circuits, *Trans. AIEE Commun. Electron.*, vol. 9, pp. 593–598, 1953.
4. McCluskey, E. J., Jr.: Minimization of Boolean Functions, *Bell System Tech. J.*, vol. 35, pp. 1417–1444, 1956.
5. Shreeram, Abhyankar: Minimal "Sum of Products of Sums" Expressions of Boolean Functions, *Proc. IRE*, vol. EC-7, pp. 268–276, 1958.
6. Shreeram, Abhyankar: Absolute Minimal Expressions of Boolean Functions, *IRE Trans.*, vol. EC-8, pp. 3–8, 1959.
7. The Staff of the Computation Laboratory: "Synthesis of Electronic Computing and Control Circuits," Harvard University Press, Cambridge, Mass., 1951.
8. Troye, N. C. de: Classification and Minimization of Switching Functions, Ph.D. Thesis, University of Amsterdam, 1958.
9. Urbano, R. H., and R. K. Mueller: A Topological Method for the Determination of the Minimal Forms of a Boolean Function, *IRE Trans.*, vol. EC-5, pp. 126–131, 1956.
10. Veitch, E. W.: A Chart Method for Simplifying Truth Functions, *Proc. Assoc. Computing Machinery Conf.*, pp. 127–133, 1952.

4
Contact Circuits

4-1 CLASSIFICATION OF SWITCHING FUNCTIONS

The first and perhaps the simplest step in the design of switching circuits is the realization of a switching function by means of a contact circuit. It is perhaps also the step about which most has been written even though in practice it is found to account for the least number of difficulties. This topic will therefore be dealt with only inasmuch as such a survey is useful for purposes of contact-circuit design.

Switching functions, finite as their number may be although rapidly increasing with the number of variables involved, can be subdivided systematically into classes and groups. The systematic classification of switching functions as laid down by the Computation Laboratory of Harvard University will be followed here. This classification uses as a basis the f coefficients of the canonical form of a switching function. The canonical form of a switching function with three variables is as follows:

$$f(x,y,z) = f_0x'y'z' + f_1x'y'z + f_2x'yz' + f_3x'yz + f_4xy'z' + f_5xy'z + f_6xyz' + f_7xyz \qquad (4\text{-}1)$$

The canonical form of a switching function consists of 2^n minterms or maxterms, where n is the number of variables involved. Because each f coefficient can be 0 or 1 and all functions with n variables are obtained by awarding the f coefficients the values 0 or 1 in every possible combination, the number of different switching functions will be

$$N = 2^{2^n} \tag{4-2}$$

Some typical values for N are shown in Table 4-1. These numbers contain degenerated functions with less than n variables. The largest possible degenerations are the ever-open and the ever-closed circuits 0 and 1. They correspond in fact to forms of a switching function with n variables where all f coefficients are equal to 0 or 1, respectively.

Table 4-1

n	N	DN	F	DF
1	4	2	3	2
2	16	6	6	3
3	256	38	22	6
4	65,536	942	402	22
5	4,294,967,296	325,262	1,228,158	402

Switching functions can be subdivided into classes according to the number of terms found in the canonical form. These various classes cannot contain equal switching functions. The number of classes for a switching function with n variables is $2^n + 1$. The number of switching functions in each of these $2^n + 1$ classes can be found by means of the binomial formula for 2^n or the corresponding Pascal triangle as shown in Table 4-2.

Table 4-2

n	*Number of switching functions per class*								
1				1	2	1			
2			1	4	6	4	1		
3	1	8	28	56	70	56	28	8	1
					etc.				

Switching functions belonging to the same class of functions with n variables can be subdivided into groups of functions using the same type of basic circuit. All the functions of such a class can be derived from one another by inverting and/or permuting the variables. The number F of these fundamental groups of functions or circuits is not excessive up to four variables;

but from five variables onwards, this number F becomes prohibitive in that it no longer allows the drawing up of a systematic list. The F numbers are also shown in Table 4-1.

In addition, Table 4-1 gives the total number of degenerations DN of the various possible functions; this number can be calculated, as can the number of degenerations DF of the fundamental functions, for which there is no basic formula to the author's knowledge.

The various switching functions will be represented by means of the indices of the f coefficients which have the value 1 in the canonical form. For example,

$$f(x,y,z) = f_0 x'y'z' + f_3 x'yz + f_5 xy'z = f(0,3,5) \tag{4-3}$$

It is also possible to describe the function by means of the sum of the powers base 2 of the indices of the f coefficients. In this system, the function of Eq. (4-3) is described as follows:

$$f(x,y,z) = 2^0 + 2^3 + 2^5 = 1 + 8 + 32 = 41 \tag{4-4}$$

With the first classification system it is generally possible to find the index groups of the fundamental type of a class of switching functions without having to draw up a complete reference list. The fundamental type of a class of functions will be indicated by means of the group with the lowest possible f coefficient indices. This group always starts with a 0.

The fundamental type of a function can be derived from an arbitrary function by inverting the variables to a maximum and permuting the variables so that the indices of the f coefficients yield a minimum. This process does not always lead directly to the desired result, e.g., when the number of the inverted and noninverted variables of a certain type is equal or when there is no f coefficient f_0 after the inversion. This process may be illustrated by means of the following examples.

$$\begin{aligned} f(x,y,z) &= f_3 x'yz + f_4 xy'z' + f_5 xy'z = f(3,4,5) \\ f(x,y,z) &\equiv f_6 xyz' + f_1 x'y'z + f_0 x'y'z' = f(0,1,6) \end{aligned} \tag{4-5}$$

The resulting classification $f(0,1,6)$ is obtained by inverting the variables x and z of the original function. It is not possible to reduce these numbers further.

The following function does not lead directly to a result:

$$\begin{aligned} f(x,y,z) &= f_1 x'y'z + f_4 xy'z' + f_7 xyz = f(1,4,7) \\ f(x,y,z) &\equiv f_4 xy'z' + f_1 x'y'z + f_2 x'yz' = f(1,2,4) \end{aligned} \tag{4-6}$$

Inversion of the variables x and z leads to an improvement, but no index group starting with a 0 has been obtained. Since all the terms have one noninverted variable, a term with an f_0 coefficient will be obtained by inverting one of the variables. The inversion of the z variable leads to the desired result without permutation of any of the variables:

$$f(x, y, z) \equiv f_5 xy'z + f_0 x'y'z' + f_3 x'yz = f(0, 3, 5) \tag{4-7}$$

It is pointed out here that the sum of the indices of the f coefficients in Eq. (4-7) is higher than that of Eq. (4-6).

The following function requires a permutation of the variables:

$$\begin{aligned} f(x, y, z) &= f_1 x'y'z + f_3 x'yz + f_4 xy'z' = f(1, 3, 4) \\ f(x, y, z) &\equiv f_0 x'y'z' + f_2 x'yz' + f_5 xy'z = f(0, 2, 5) \end{aligned} \tag{4-8}$$

Inversion of variable z leads to the second line of Eq. (4-8). The second term of this line has one noninverted variable, that is, y. The f_2 coefficient of this minterm can be changed into f_1 by permuting the variables y and z. This leads to the fundamental indication:

$$f(x, y, z) \equiv f_0 x'y'z' + f_1 x'y'z + f_6 xyz' = f(0, 1, 6) \tag{4-9}$$

In actual practice, a more complicated manipulation than the one used in the above examples is sometimes required. Functions with an even number of terms and variables both inverted and noninverted sometimes require an inversion of these variables in order to arrive at the desired solution.

$$\begin{aligned} f(w, x, y, z) &= f_2 w'x'yz' + f_3 w'x'yz + f_8 wx'y'z' + f_{11} wx'yz \\ f(w, x, y, z) &\equiv f_0 w'x'y'z' + f_1 w'x'y'z + f_{10} wx'yz' + f_9 wx'y'z \\ &= f(0, 1, 9, 10) \end{aligned} \tag{4-10}$$

Equation (4-10) is obtained by inverting the y variable as required to find the desired minimum index combination. Inverting the variable z, of which two are inverted and two are not, leads to a form of the function with two terms with one noninverted variable. This function form is preferable from the point of view of the lowest possible f-coefficient indices:

$$\begin{aligned} f(w, x, y, z) &\equiv f_1 w'x'y'z + f_0 w'x'y'z' + f_{11} wx'yz + f_8 wx'y'z' \\ &= f(0, 1, 8, 11) \end{aligned} \tag{4-11}$$

Permutation of the variables w (8) and x (4) leads to a more favorable index combination because two variables w change into x and no variables x change into w.

$$\begin{aligned} f(w, x, y, z) &\equiv f_1 w'x'y'z + f_0 w'x'y'z' + f_7 w'xyz + f_4 w'xy'z' \\ &= f(0, 1, 4, 7) \end{aligned} \tag{4-12}$$

Finally, permutation of the variables x (4) and y (2) leads to the desired minimum index combination because two variables x change into y and one variable y changes into x, whereas no further reductions of this type are possible:

$$f(w,x,y,z) \equiv f_1 w'x'y'z + f_0 w'x'y'z' + f_7 w'xyz + f_2 w'x'yz' \\ = f(0,1,2,7) \tag{4-13}$$

By means of this method, three undegenerated fundamental functions with two variables can be determined. These are shown in Fig. 4-1 with their

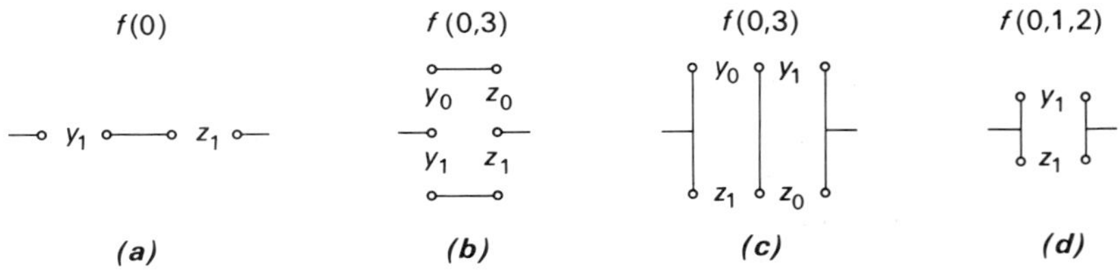

Fig. 4-1

corresponding contact circuits. The fundamental circuits for switching functions with two variables are a series connection of two (break) contacts (Fig. 4-1*a*), a series connection of two transfer contacts (Fig. 4-1*b*), or the equivalent parallel connection of two transfer contacts (Fig. 4-1*c*), and a parallel connection of two (break) contacts (Fig. 4-1*d*). The form of the last network is the dual of the first type of network. The combination shown in Fig. 4-1*b* is the dual of that of Fig. 4-1*c*. The operation of both circuits, however, is identical.

This duality of pairs of fundamental circuits exists in principle for switching functions with any number of variables. For each class of functions with a number p of canonical terms $(0 < p < 2^{n-1})$, there is a complemental class of functions with $2^n - p$ terms. These pairs of classes correspond to dual forms of the fundamental contact networks, because the network form of the inverse of a switching function is the dual of the original form.

All fundamental functions of the classes consisting of 2^{n-1} canonical terms can be represented by means of a network of which the dual form performs the same operation (see Fig. 4-1*b* and *c*). Some of these networks even have the same topological form.

The systematic classification of all functions with three variables is shown in Table 4-3. The corresponding circuits are given in Figs. 4-2 to 4-21. The columns marked m and $\overline{m}$ of Table 4-3 give the numbers of minterms of the canonical function. The column marked f_i gives the characteristic index groups of the f coefficients of the fundamental form. Columns s and $\bar{s}$ contain the corresponding circuit number of Figs. 4-2 to 4-21. Columns a and $\bar{a}$ give the numbers of functions in the class.

Table 4-3

m	$\bar{m}$	f_i	s	$\bar{s}$	a	$\bar{a}$
0	8				1	1
1	7	0	4–2	4–21	8	8
2	6	0, 1	4–3	4–20	12	12
2	6	0, 3	4–4	4–19	12	12
2	6	0, 7	4–5	4–18	4	4
3	5	0, 1, 2	4–6	4–17	24	24
3	5	0, 1, 6	4–7	4–16	24	24
3	5	0, 3, 5	4–8	4–15	8	8
4		0, 1, 2, 3	4–9		6	
4		0, 1, 2, 4	4–10		8	
4	4	0, 1, 2, 5	4–11*a*	4–11*b*	24	
4	4	0, 1, 2, 7	4–12*a*	4–12*b*	24	
4	4	0, 3, 4, 7	4–13*a*	4–13*b*	2	
4	4	0, 3, 5, 6	4–14*a*	4–14*b*	6	

Wherever possible, the fundamental contact circuits of the various classes have been drawn in Figs. 4-2 to 4-21 in such a way as to allow a maximum number of make and break contacts to be combined into transfer contacts. This type of circuit will generally be most economical because it contains the smallest number of contact springs. The design of the fundamental circuits that correspond to switching functions with four variables and make-and-break contacts combined to the maximum into transfer contacts is not always as simple as in the case of the functions with three variables given below.

4-2 THE DESIGN OF A CONTACT CIRCUIT

To the author's knowledge, there is no simple systematic method of circuit design by means of which a switching function of the simplest form can be converted into the best possible contact circuit. Moreover, there is no unambiguous definition of what is to be considered as the best contact circuit; and the simplest form of the switching function does not always correspond to the simplest form of the contact circuit.

Contact circuits can be designed in two different ways:

1. With the smallest possible number of springs
2. With the smallest possible number of contacts

The construction of most types of relays makes it possible to combine a two-spring make contact and a two-spring break contact into a three-spring transfer contact. This represents a saving of one spring and of the space that

Fig. 4-2

Fig. 4-3

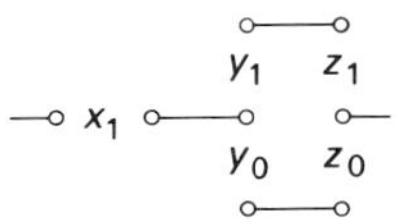

Fig. 4-4

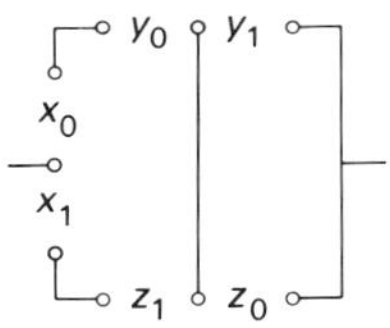

Fig. 4-5

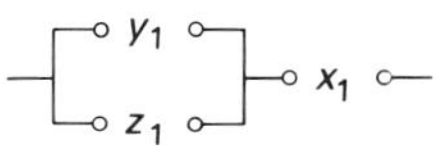

Fig. 4-6

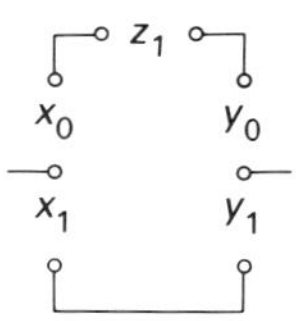

Fig. 4-7

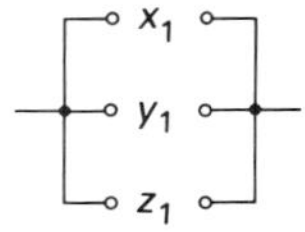

Fig. 4-21

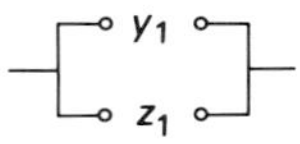

Fig. 4-20

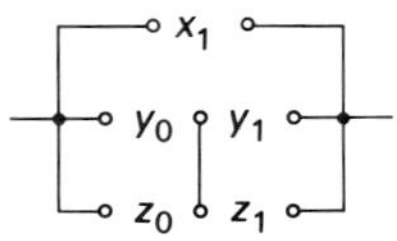

Fig. 4-19

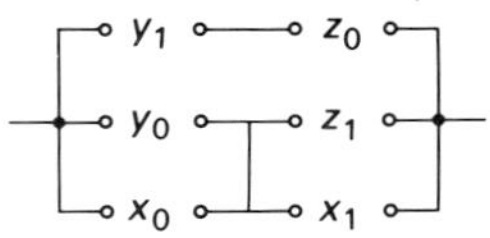

Fig. 4-18

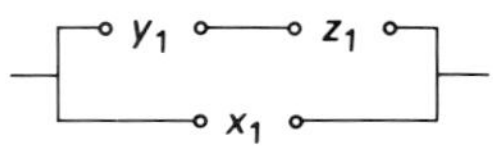

Fig. 4-17

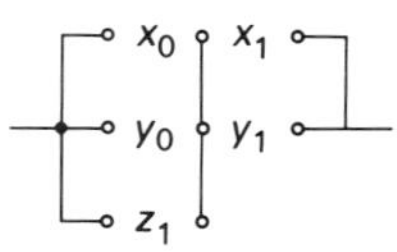

Fig. 4-16

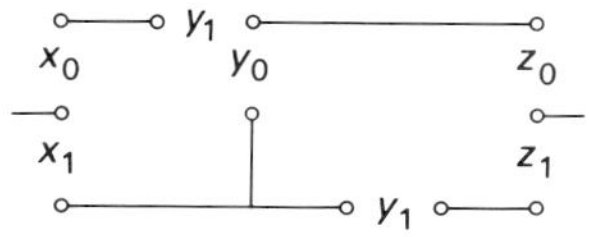

Fig. 4-8

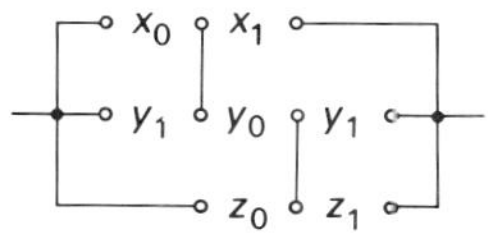

Fig. 4-15

Fig. 4-9

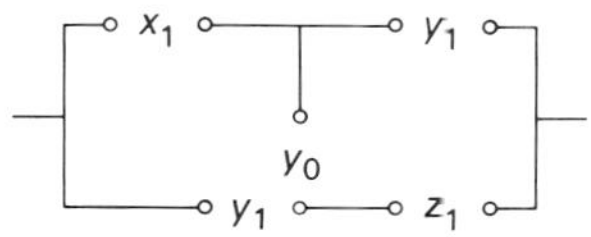

Fig. 4-10

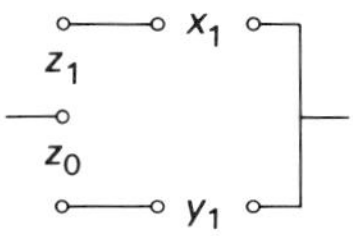

(a)

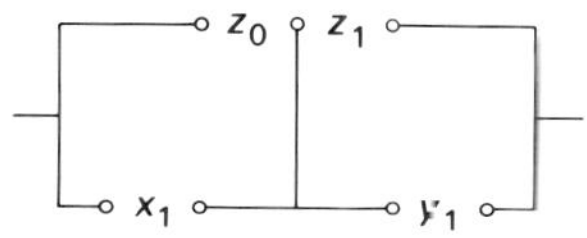

(b)

Fig. 4-11

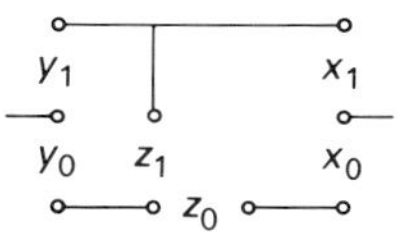

(a)

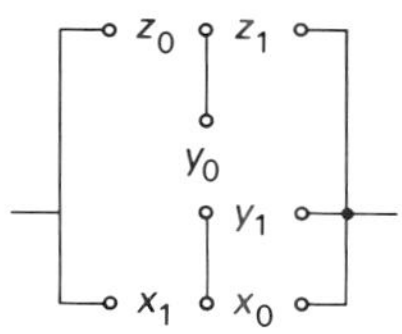

(b)

Fig. 4-12

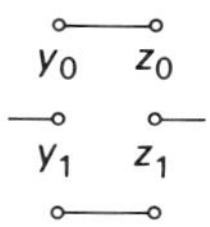

(a)

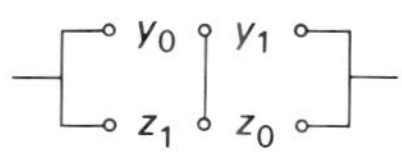

(b)

Fig. 4-13

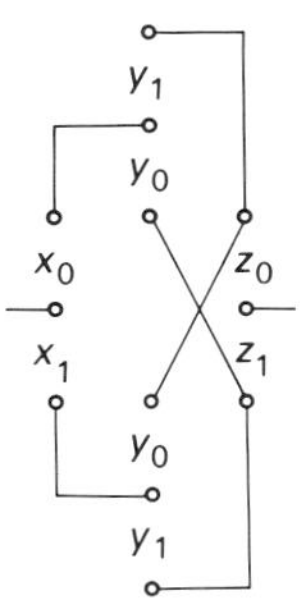

(a)

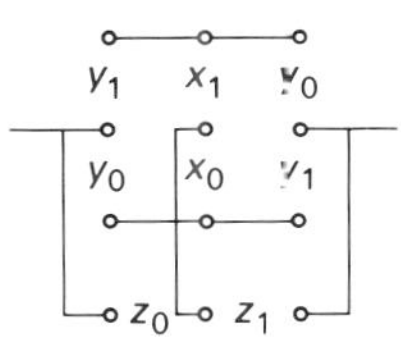

(b)

Fig. 4-14

spring would occupy as well as a reduction in the number of amp-turns necessary for the operation of that relay.

Circuits with the smallest number of contacts do not always allow for a combination of springs into transfer contacts, whereas a well-chosen circuit form with more contacts does provide for such a combination, so that ultimately the total number of springs is considerably smaller.

A contact circuit that allows a maximum number of springs to be combined into transfer contacts can be designed in a more or less systematic way. The switching function has to be written in the form of a logical sum of products, because in this form, which corresponds to a parallel connection of a number of series-connected contacts, the contacts can be combined in the easiest way.

SYMMETRICAL FUNCTIONS

Each canonical switching function with n variables corresponds to 2^n individual circuits with n variables. According to the switching levels of these individual circuits a switching function with, say, three variables can be written as follows:

$$\begin{aligned} f(x,y,z) = {} & f_0 x'y'z' \\ & + f_1 x'y'z + f_2 x'yz' + f_4 xy'z' \\ & + f_3 x'yz + f_5 xy'z + f_6 xyz' \\ & + f_7 xyz \end{aligned} \tag{4-14}$$

These four lines represent the so-called symmetrical functions. These functions consist of all terms located on the same switching level or all canonical terms with the same number of inverted variables. The most elaborate way of representing all terms of a canonical switching function by means of contact circuits is that consisting of 2^n individual circuits, as shown in Fig. 4-22. All

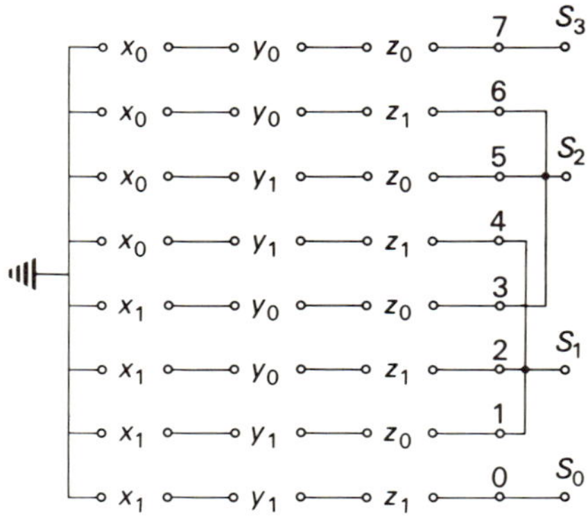

Fig. 4-22

desired switching functions can be obtained by means of a parallel connection of the output terminals of the individual circuits for which the f coefficient must be 1.

The output terminals of the circuit shown in Fig. 4-22 are connected to give the four symmetrical switching functions with three variables, which are indicated by S_0, S_1, S_2, and S_3. The indices are chosen in accordance with the switching level. The circuit shown in Fig. 4-22 is, of course, too elaborate. It can be transformed into a circuit consisting of transfer contacts only, as is shown in Fig. 4-23. This circuit is a complete "contact tree," as this combination is called. The circuit of Fig. 4-22 consists of $8 \times 3 = 24$ contacts or 48 springs; the one shown in Fig. 4-23 has 14 contacts or 21 springs. This

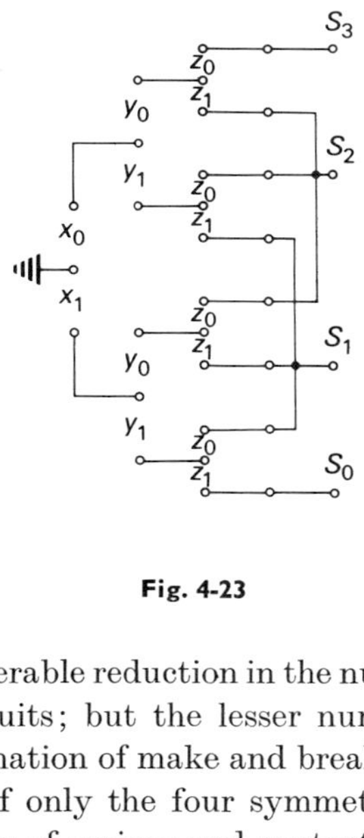

Fig. 4-23

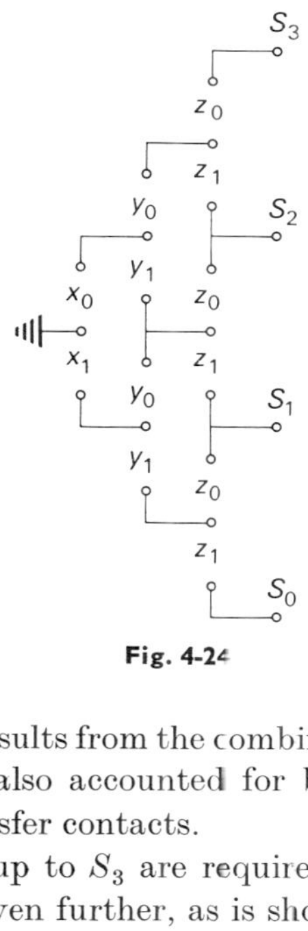

Fig. 4-24

considerable reduction in the number of contacts results from the combination of circuits; but the lesser number of springs is also accounted for by the combination of make and break contacts into transfer contacts.

If only the four symmetrical functions S_0 up to S_3 are required, the number of springs and contacts can be reduced even further, as is shown in Fig. 4-24. This circuit contains 12 contacts and 18 springs, i.e., the smallest possible number of contacts and springs for this type of circuit.

A circuit corresponding to one or more of the symmetrical functions can be obtained by omitting the undesired output terminals and associated circuit elements. In some cases, a further combination of contacts into transfer contacts is then possible. A circuit derived from Fig. 4-24 representing more than one symmetrical function can sometimes be simplified even further by means of shifting down the output terminals. This process converts the original planar network into a nonplanar network.

In the process of shifting down the output terminals of the higher-order symmetrical functions, part of the circuit corresponding to that function can be combined with part of the circuit of a lower-order symmetrical function. This process is illustrated by means of Fig. 4-25*a* and *b*.

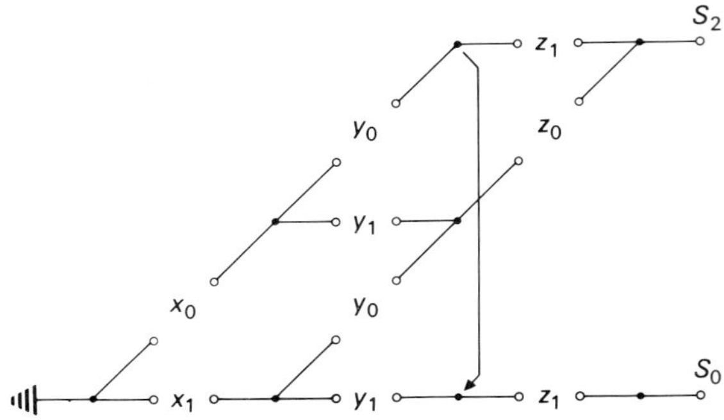

(a)

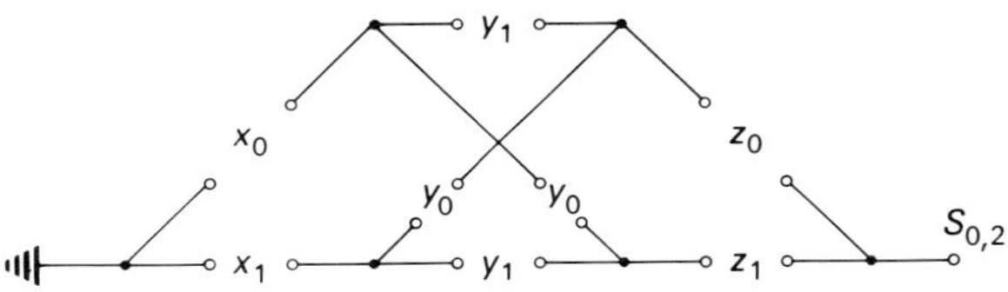

(b)

Fig. 4-25

These figures show the shifting-down process for the combination of the functions $S_0 + S_2$. One contact is saved and all springs can be combined into transfer contacts.

ARBITRARY FUNCTIONS

The best form of a contact circuit representing an arbitrary switching function can be obtained by means of the manipulation of matrices. Boolean matrices representing n-terminal networks are treated in detail in Chap. 5. One of the rules of matrix manipulation, however, will be used here, because it is possible to obtain a good circuit representing an arbitrary switching function by means of this rule. But this is possibly not the best circuit. This single rule can be applied in such a way that, when possible, make and break contacts of the same relay can be combined into transfer contacts. This can lead to a circuit with a minimum number of contact springs. From a practical point of view, this type of circuit will most often be preferred.

The manipulation rule will be explained by means of the following example. Design a contact circuit with a minimum of contact springs operating in accordance with the following switching function:

$$f(x, y, z) = x'yz' + xy'z \tag{4-15}$$

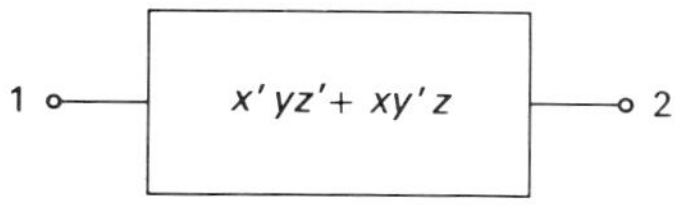

Fig. 4-26

This formula represents a two-terminal network, as shown in the block diagram of Fig. 4-26. Such a circuit can be represented by the following 2×2 matrix:

$$f(x, y, z) = \begin{bmatrix} 1 & x'yz' + xy'z \\ x'yz' + xy'z & 1 \end{bmatrix} \tag{4-16}$$

$$= \begin{bmatrix} c_{11} & c_{12} \\ c_{21} & c_{22} \end{bmatrix} \tag{4-17}$$

The entries c_{12} and c_{21} are the switching function of the circuit between the two external nodes 1 and 2. In contact circuits, entries $c_{ij} = c_{ji}$ because a contact is a bidirectional element. Entries $c_{ii} = 1$ because any circuit node is permanently connected to itself. In contact circuits, only one-half of the matrix need be shown.

A matrix can be extended by means of a process of internal-node insertion in the switching network. For that purpose the switching function has to be factorized in one or the other way. The possible ways of factorization determine the kind of circuit that will finally appear. For each extra internal node of the switching network, the matrix grows with one column and one row. The node-insertion process in the example of Eq. (4-15) can be handled in three different ways, viz., by means of the minterm formula, the factorized formula, or the maxterm formula:

$$f(x, y, z) = x'yz' + xy'z = (x' + y')(yz' + xz) = (x' + y')(x + z')(y + z) \tag{4-18}$$

The node-insertion process will first be tried starting with the original formula. The complete transformation is shown in Eq. (4-19).

$$\begin{bmatrix} 1 & x'yz' + xy'z \\ & 1 \end{bmatrix} = \begin{bmatrix} 1 & 0 & yz' & y'z \\ & 1 & x' & x \\ & & 1 & 0 \\ & & & 1 \end{bmatrix} = \begin{bmatrix} 1 & 0 & 0 & 0 & z' & z \\ & 1 & x' & x & 0 & 0 \\ & & 1 & 0 & y & 0 \\ & & & 1 & 0 & y' \\ & & & & 1 & 0 \\ & & & & & 1 \end{bmatrix} \tag{4-19}$$

In the first step of the node-insertion process, columns 3 and 4 and rows 3 and 4 are added, so that the second matrix of Eq. (4-19) represents a switching network with two external and two internal nodes. Node 3 corresponds to column 3 and row 3, and node 4 to column 4 and row 4. Node 3 is placed in the series-contact network $x'yz'$ between contact x' and the series connection of contacts y and z'. Node 4 has been placed in the same way in the series circuit $xy'z$.

The term $x'yz'$ of entry c_{12} has been shifted to column 3 to the entry places c_{13} and c_{23}, both entries together forming the required series circuit:

$$c_{13} \cdot c_{23} = x' \cdot yz' \tag{4-20}$$

The shifting of term $xy'z$ of entry c_{12} to column 4 is represented by

$$c_{14} \cdot c_{24} = x \cdot y'z \tag{4-21}$$

so that in this case

$$c_{12} = c_{13} \cdot c_{23} + c_{14} \cdot c_{24} \tag{4-22}$$

This process of node insertion is a manipulation of indices, as can be seen readily in the above. It can be repeated on entries $c_{13} = yz'$ and $c_{14} = y'z$, respectively, leading to entries c_{15}, c_{35} and c_{16}, c_{46}, as shown in the last matrix of Eq. (4-19).

All three matrices of (4-19) represent equivalent contact circuits—and in this case for the drawing of the final contact circuit, none of the node-insertion steps was necessary—but the last matrix shows without any ambiguity the sequence of the contacts in the various paths of the circuit.

This can be seen as follows. The rows and the columns of a matrix representing a switching function can be considered as contact strips. The various horizontal and vertical contact strips are interconnected at the places of the 1 entries. The entries have to be considered as contact circuits between the strips crossing at the place of entry. The last matrix of Eq. (4-19) can be transformed into the contact circuit shown in Fig. 4-27. The final circuit shown here could of course have been drawn directly from the given switching equation (4-15). However, this is not the case when using the middle part of Eq. (4-18) as the first step for the insertion of one node.

$$\begin{bmatrix} 1 & (x'+y')(xz+yz') \\ & 1 \end{bmatrix} = \begin{bmatrix} 1 & 0 & xz+yz' \\ & 1 & x'+y' \\ & & 1 \end{bmatrix} = \begin{bmatrix} 1 & 0 & 0 & z & z' \\ & 1 & x'+y' & 0 & 0 \\ & & 1 & x & y \\ & & & 1 & 0 \\ & & & & 1 \end{bmatrix} \tag{4-23}$$

Because in the second matrix of (4-23) entry c_{13} consists of two terms, two extra nodes will have to be introduced in this case in such a way that entries

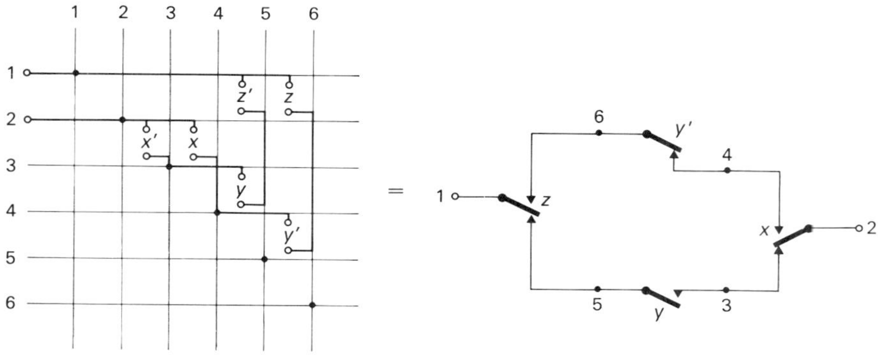

Fig. 4-27

$f_{14} = z$ and $f_{15} = z'$ appear on one row so that contacts z' and z together can form a transfer contact. The right-hand matrix need not be extended further because it already represents a circuit with three transfer contacts, as can be seen in the circuit of Fig. 4-28. The second means of matrix manipulation leads to a circuit with nine springs instead of ten. It should be noted here that this is not a virtue of the matrix manipulation but stems from the fact that the possibility of factorization as given by Eq. (4-18) was recognized. The maxterm form as a starting point leads in this case to a more complicated matrix manipulation.

The following switching function is a more complicated example of the recognition of possible ways of factorization:

$$\begin{aligned} f(w,x,y,z) &= wx'y' + w'yz' + xy'z' + w'x'z \\ &= (w'+y')(wx' + yz' + xz' + x'z) \end{aligned} \tag{4-24}$$

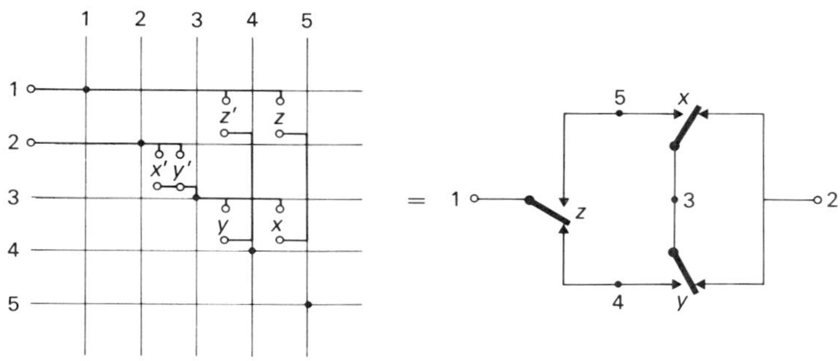

Fig. 4-28

In this factorization, two redundant terms are introduced, namely, $w'xz'$ and $x'y'z$, as can be easily proved.

The corresponding matrices are given below.

$$\begin{bmatrix} 1 & wx'y' + w'yz' + xy'z' + w'x'z \\ & 1 \end{bmatrix} = \begin{bmatrix} 1 & 0 & wx' + yz' + xz' + x'z \\ & 1 & w' + y' \\ & & 1 \end{bmatrix} \tag{4-25}$$

$$\text{Entry } c_{13} = wx' + yz' + xz' + x'z = (x' + z')(w + x + y + z) \tag{4-26}$$

In this factorization, redundant terms $x'y$ and wz' appear, as can be easily proved. By the introduction of a fourth node, using Eq. (4-26), the following matrix will appear:

$$\begin{bmatrix} 1 & 0 & 0 & x' + z' \\ & 1 & w' + y' & 0 \\ & & 1 & w + x + y + z \\ & & & 1 \end{bmatrix} \tag{4-27}$$

Further factorization is not necessary because all contacts can now form transfer contacts, as can be seen in the circuit of Fig. 4-29.

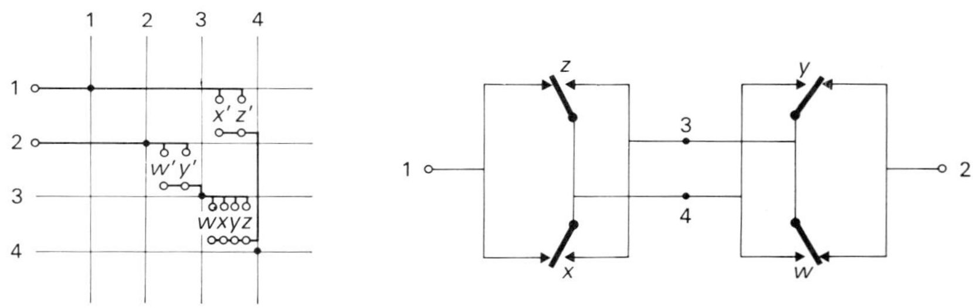

Fig. 4-29

Another interesting example is found in the following switching function:

$$f(w, x, y, z) = wx'y'z + w'x'yz' + w'xy'z + wxy \tag{4-28}$$

When trying to find the circuit with the minimum number of contact springs, the above method will yield the best result; however, when this formula has to be implemented by the circuit with the minimum number of contacts, more complicated matrix methods (to be described in the next chapter) are required. In order to obtain a good possibility of combining make and break contacts to transfer contacts, Eq. (4-28) can be written as follows:

$$\begin{aligned} f(w, x, y, z) &= (y + z)(x + wy' + w'z')(x' + wy + w'y') \\ &= (y + z)(w + x + z')(w + x' + y')(w' + x + y')(w' + x' + y) \end{aligned} \tag{4-29}$$

This equation leads directly to the following matrix representation:

$$\begin{bmatrix} 1 & (y+z)\,(x+wy'+w'z')\,(x'+wy+w'y') \\ & 1 \end{bmatrix}$$

$$= \begin{bmatrix} 1 & 0 & y+z & 0 \\ & 1 & 0 & x'+wy+w'y' \\ & & 1 & x+wy'+w'z' \\ & & & 1 \end{bmatrix} \qquad (4\text{-}30)$$

Terms wy of entry c_{24} and wy' of entry c_{34} can be used to form a fifth internal node with entries $c_{45}=w$, $c_{25}=y$, and $c_{35}=y'$, provided that the introduced sneak path $c_{23}=0$—which is the case because $c_{25}\cdot c_{35}=y\cdot y'=0$. Terms $w'y'$ of entry c_{24} and $w'z'$ of entry c_{34} can be used to form a sixth internal node with entries $c_{46}=w'$, $c_{26}=y'$, and $c_{36}=z'$, provided that the introduced sneak path $c_{23}=c_{26}\cdot c_{36}=y'z'$ is of no influence. This happens to be because $c_{12}=c_{13}\cdot c_{23}=(y+z)\,y'z'=0$.

Using these entries, the following matrix will be obtained from matrix (4-30).

$$\begin{bmatrix} 1 & 0 & y+z & 0 & 0 & 0 \\ & 1 & 0 & x' & y & y' \\ & & 1 & x & y' & z' \\ & & & 1 & w & w' \\ & & & & 1 & 0 \\ & & & & & 1 \end{bmatrix} \qquad (4\text{-}31)$$

Matrix (4-31) is implemented with contacts in the circuit of Fig. 4-30. The resulting circuit contains five transfer contacts involving fifteen springs.

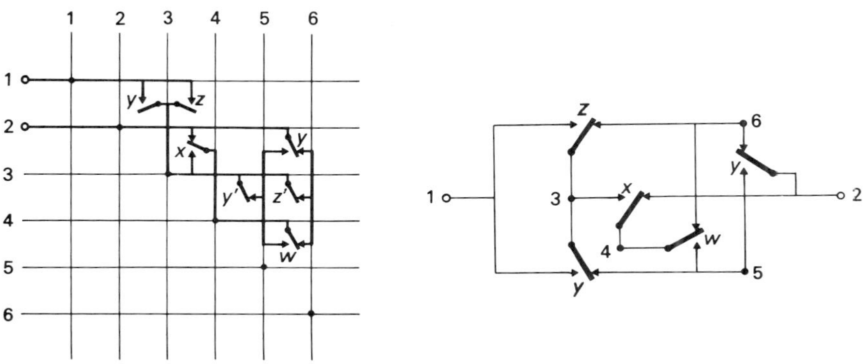

Fig. 4-30

As Shannon and Moore have shown, function (4-28) can be realized by means of a double-bridge circuit with only nine contacts. This circuit is shown in Fig. 4-31. Two of the contacts of this circuit can be combined into a transfer contact, which brings the total number of relay springs of this circuit to 17, that is, two more than in the circuit of Fig. 4-30.

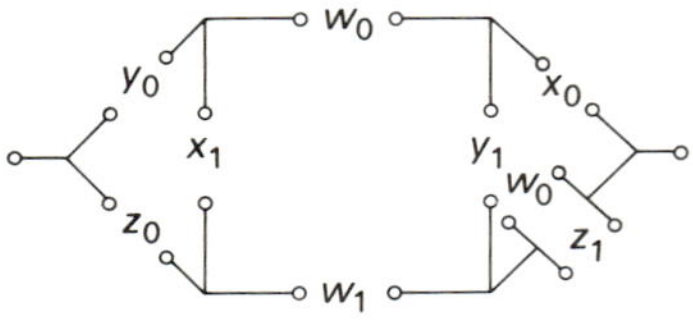

Fig. 4-31

The design of circuits such as the one shown in Fig. 4-31 is a time-consuming process. For all practical purposes, it does not generally pay to try to achieve this special type of circuit, which probably has the minimum number of contacts. The time spent on the actual investigation will generally cost many times more than the saving of one contact. And it sometimes results in no saving at all, especially if a solution is available with more contacts but a smaller number of relay springs.

4-3 SPECIAL TYPES OF CONTACT NETWORKS

TWO-TERMINAL NETWORKS

Bridge-type networks A special type of two-terminal network is the bridge circuit, which is shown in Fig. 4-32. The switching function of this network can be determined by following all possible paths connecting the two terminals:

$$f(a, b, c, d, e) = ab + aed + ceb + cd \tag{4-32}$$

In the first instance, Eq. (4-32) might be realized by means of the circuit shown in Fig. 4-33 if the designer had no knowledge whatsoever about bridge circuits. This circuit consists of four parallel-connected contact chains with a total of ten contacts, whereas only five are used in the bridge circuit.

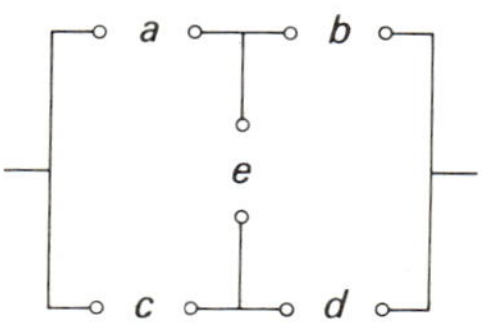

Fig. 4-32

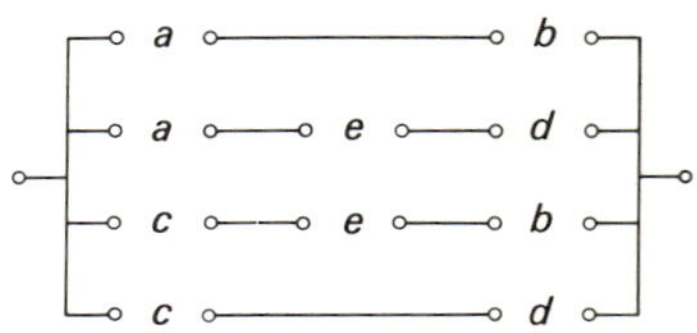

Fig. 4-33

Using the diagram of Fig. 4-33 as a starting point one might, after some manipulation, arrive at the bridge circuit of Fig. 4-32. If, however, the switching function of this circuit had been given in the form of a logical product of logical sums (the maxterm form) [Eq. (4-33)], it would be rather difficult to start from the corresponding circuit (Fig. 4-34) and arrive by some manipulation process at the type of circuit given in Fig. 4-32.

$$f(a, b, c, d, e) = (a + c)(a + e + d)(c + e + b)(b + d) \tag{4-33}$$

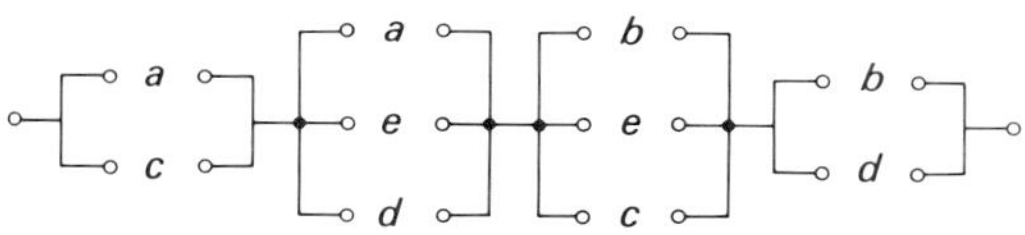

Fig. 4-34

The bridge circuit is a powerful tool for simplifying a number of contact circuits, but it is rather difficult to detect and recognize a bridge circuit among the various switching formulas. This is even harder or practically impossible for more complicated bridge-type networks.

Symmetrical functions These have already been discussed in the preceding section. This type of two-terminal network is of importance in information-processing systems that use so-called error-detecting codes consisting of a constant number of n elements with the value 1 within a group of m binary information elements forming the signal. If, for example, a seven-unit code is used in which three of the elements are bound to be "marks" and four must be "spaces," the contact circuit of Fig. 4-35 detects whether or not this condition is fulfilled. This figure represents the network for the symmetrical switching function ${}^7S_3$ (third switching level of a circuit with seven relays)

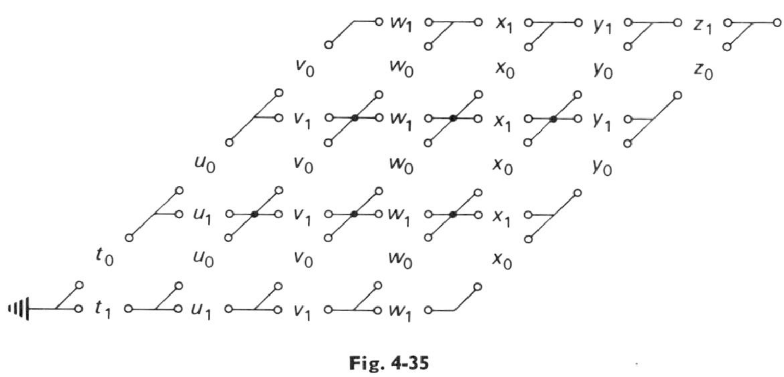

Fig. 4-35

consisting of 31 contacts. This circuit is closed as long as the 3:4 ratio between mark and space information elements is preserved.

All other kinds of n-out-of-m networks can be designed on the same principle.

A common type of circuit of this kind is represented by the switching formula $f(x,y,z) = {}^nS_{0,2}$. This circuit, which is shown in Fig. 4-36, is used to switch a lamp on and off from different control points.

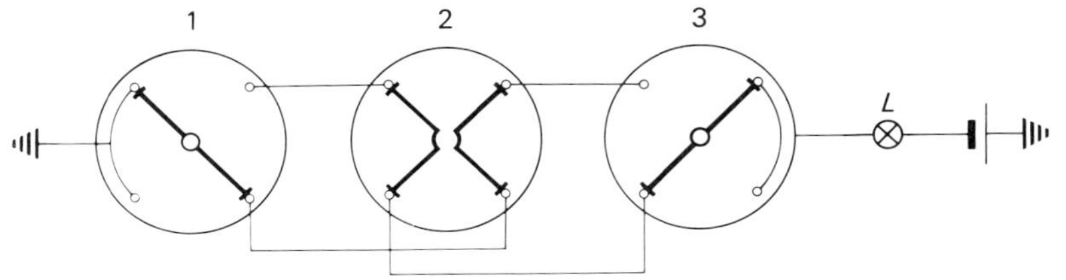

Fig. 4-36

Circuits for comparing This type of circuit is required in some information-processing applications such as automatic-telephone systems. It is sometimes necessary to compare the number of pulses of a train of pulses with a predetermined stored number and to generate a control pulse when this number of pulses has been received or transmitted. This type of control can be achieved by means of a series connection of transfer contacts as shown in Fig. 4-37. This figure shows a pulse receiver consisting of four pairs of relays (see Chap. 13) which store the pulse series to be controlled in a binary parallel form on the elements Z, Y, X, and W. The predetermined number is stored, using the same code, in the store with the four relays A, B, C, and D. The control circuit consists of a series connection of transfer contacts of Z and A, Y and B, X and C, and W and D, as shown in Fig. 4-37. This chain is closed only if the contacts of each pair are in the same position, which is the case for all identical code combinations in the pulse receiver and in the store. A control relay CR can then be operated to stop transmission of the pulse series or to stop the movement of a selector.

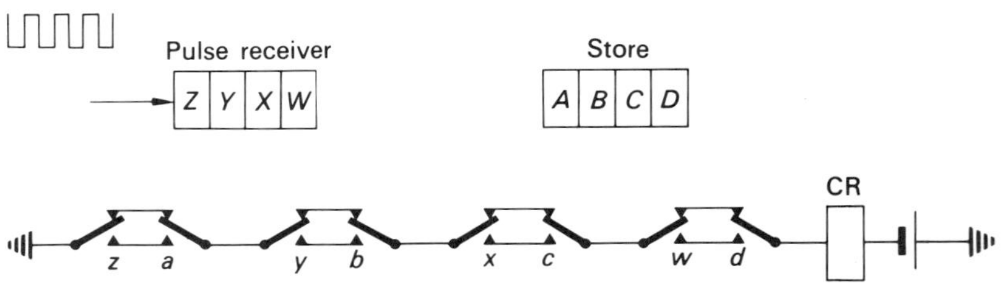

Fig. 4-37

Another application of this type of circuit is the stopping of a tape reader on a predetermined code combination. A tape reader consists, for example, of five transfer contacts that sense the perforations of a paper tape. The five moving springs of the transfer contacts are very often electrically connected. By means of a parallel connection of five control circuits consisting of a transfer contact of the tape reader and a transfer contact of the control device, a circuit can be made that is open for corresponding code combinations. In such a case, the transfer contacts of each pair have different positions. This control circuit is shown in Fig. 4-38.

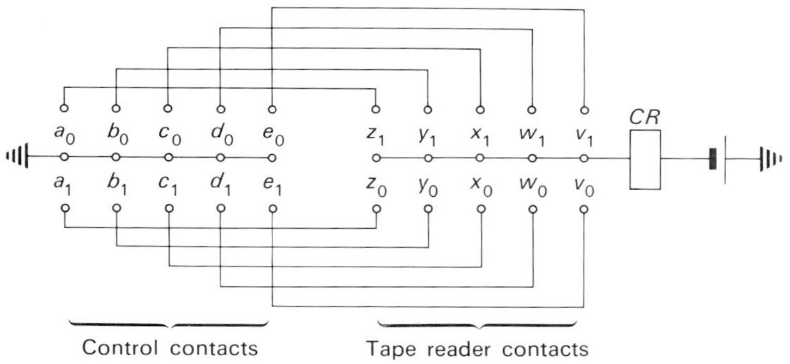

Fig. 4-38

Switching functions with two minterms or maxterms and complementary inverted variables This type of function allows an easy implementation. A switching function of this type can be written as follows:

$$f(x, y, z) = (x' + y' + z') \cdot (x + y + z) \tag{4-34}$$

This function can be represented by a two-terminal network consisting exclusively of transfer contacts, as shown in Fig. 4-39. The circuit representing the switching function

$$f(x, y, z) = x'y'z' + xyz \tag{4-35}$$

can be found by means of the normal process of inverting the network of Fig. 4-39. The resulting circuit, which is shown in Fig. 4-40, consists of two transfer

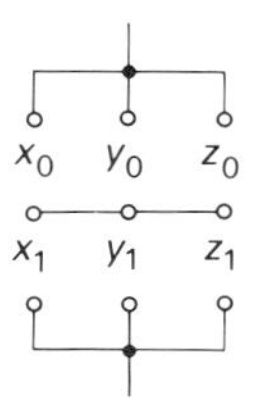

Fig. 4-39

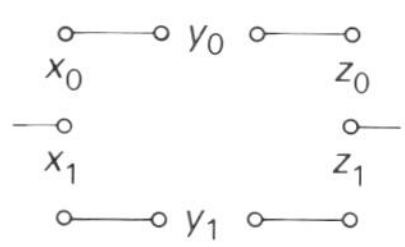

Fig. 4-40

contacts, one make and one break contact. But when drawn in still another way—as in Fig. 4-41—the contacts of all three variables can be combined again into transfer contacts. The dual of the network of Fig. 4-41 is shown in Fig. 4-42. The circuits of Figs. 4-39 and 4-42, however, perform the same

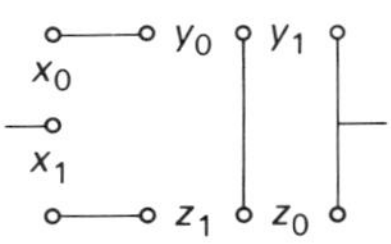

Fig. 4-41

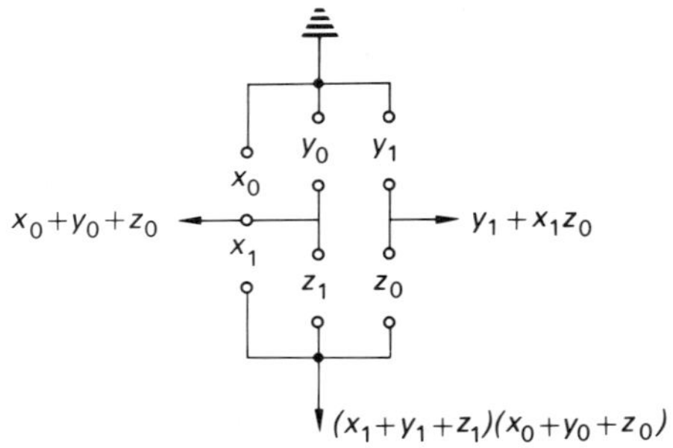

Fig. 4-42

operation, but they differ in composition. The switching function corresponding to the circuit of Fig. 4-42 is as follows:

$$
\begin{aligned}
f(x,y,z) &= (x+y)(x'+z')+y'z \\
&= xz'+x'y+yz'+y'z \\
&= xy'+xz'+x'y+yz'+x'z+y'z \\
&= (x'+y'+z')(x+y+z)
\end{aligned}
\tag{4-36}
$$

In the third line of proof (4-36), the redundant terms xy' and $x'z$ are introduced.

The importance of this type of change of configuration of the network lies in the fact that the partial series-parallel layout of the contact circuits offers more possibilities for combination with other circuits. So, in Fig. 4-42, the circuit offers the partial functions $y'+x'z$ and $x+y+z$.

The circuit form of Fig. 4-41 can be extended indefinitely, as is shown in Fig. 4-43.

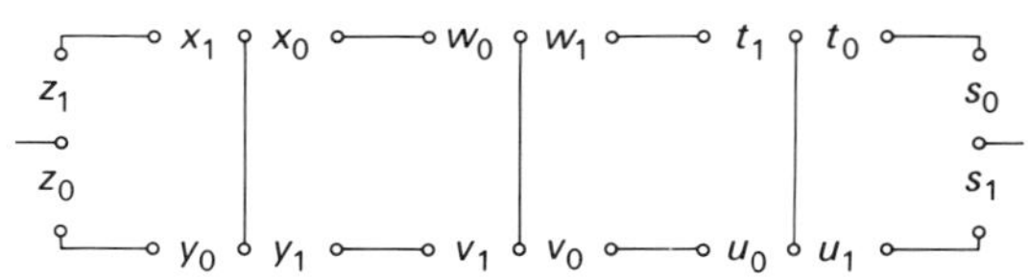

Fig. 4-43

THREE- AND FOUR-TERMINAL NETWORKS

It does not happen very often that the circuit to be designed consists of a contact circuit and one relay only. The desired automatic operation is normally obtained by a more or less sophisticated contact network controlling a number of relays or other switching elements. This raises the problem of the common use of contacts by different circuits and consequently of the combination of circuits into three- and more-terminal networks. Some of the important types of multiterminal networks will be given below.

Δ-type and Y-type networks Δ-type networks can be transformed into Y-type networks, and vice versa, by means of the normal laws of network theory. The circuit of Fig. 4-44*a* shows a Δ-type network with three contacts. The equivalent Y-type network consisting of six contacts is shown in Fig. 4-44*b*. This transformation applied in the reverse direction would have resulted in a circuit simplification. The same also holds for the Y- and Δ-type networks of Fig. 4-45*a* and *b*.

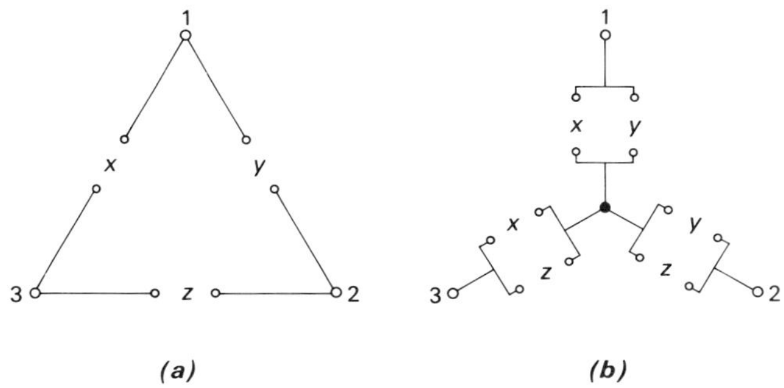

Fig. 4-44

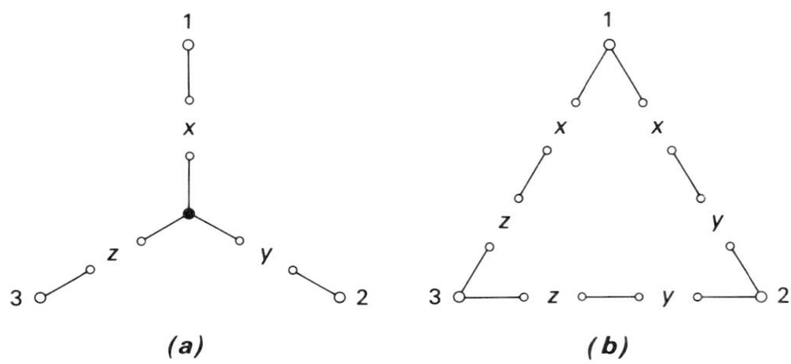

Fig. 4-45

Various forms of three- and four-terminal contact networks It is useful to know some special forms of three- and four-terminal contact networks. These networks are shown in Figs. 4-46 to 4-51.

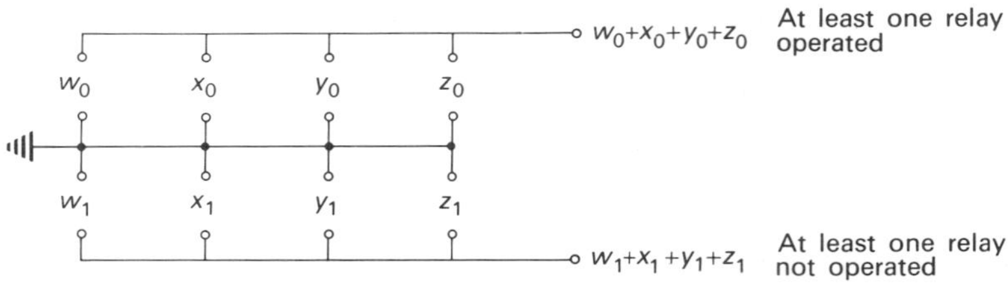

Fig. 4-46

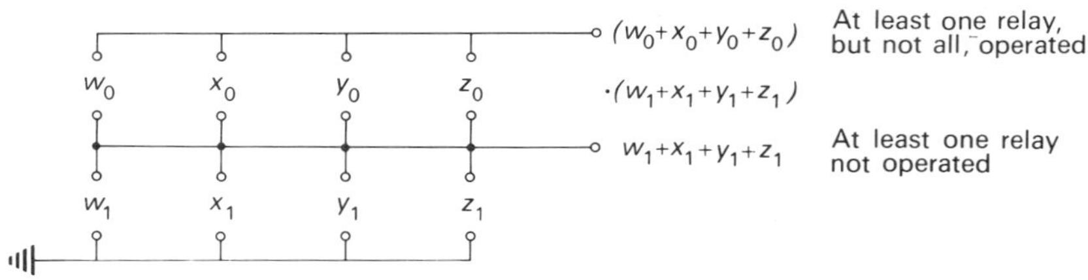

Fig. 4-47

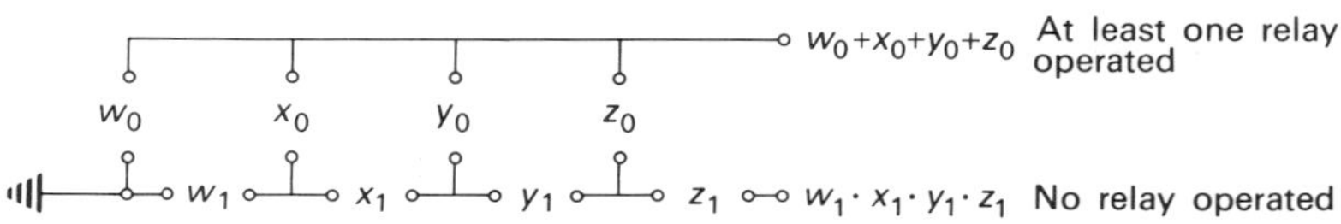

Fig. 4-48

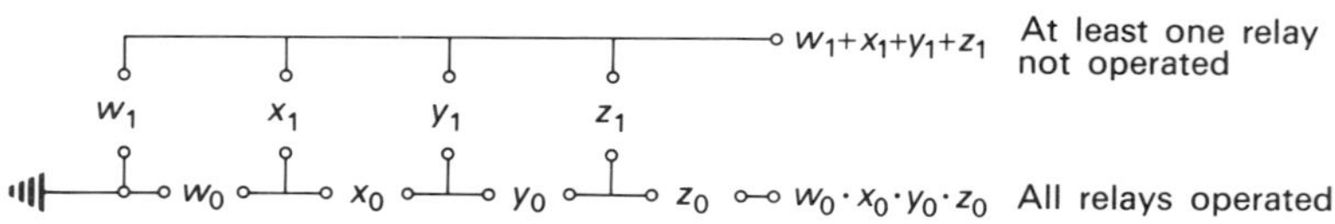

Fig. 4-49

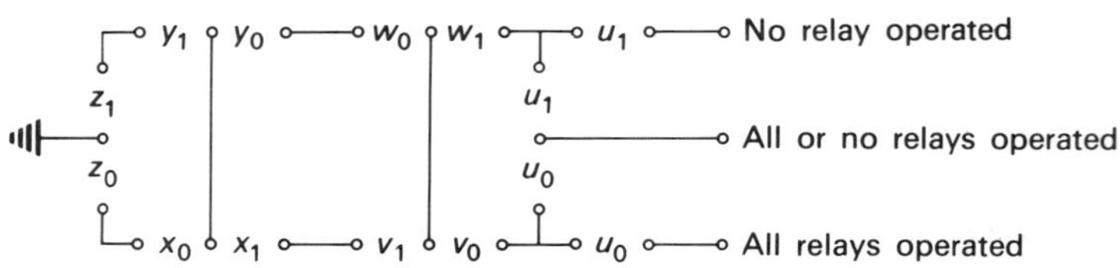

Fig. 4-50

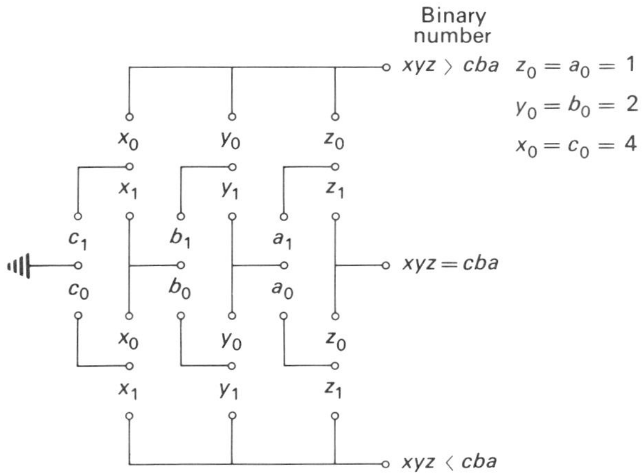

Fig. 4-51

$(n+1)$**-terminal contact networks** A number of relay counting circuits conform to the following types of switching functions for the control of the successive sections:

$$Z = \alpha \qquad Y = \alpha' a \qquad X = \alpha' a' b \qquad W = \alpha' a' b' c \tag{4-37}$$

These switching formulas can be realized by means of an $(n+1)$-terminal chain of transfer contacts, as shown in Fig. 4-52.

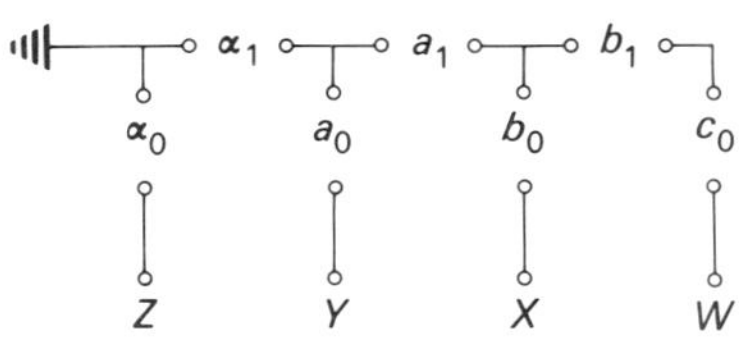

Fig. 4-52

Sometimes a more complicated set of switching functions will be encountered, such as

$$\begin{aligned} Z &= a\,[b'(cd + c'd') + b\,(cd' + c'd)] + \\ &\quad\; a'[b\,(cd + c'd') + b'(cd' + c'd)] \\ Y &= \quad b\,(cd + c'd') + b'(cd' + c'd) \\ X &= \qquad\qquad\qquad cd' + c'd \\ W &= \qquad\qquad\qquad\quad d \end{aligned} \tag{4-38}$$

or

$$\begin{aligned}
Z' &= a'[b'(cd + c'd') + b\,(cd' + c'd)] + \\
&\quad\; a\,[b\,(cd + c'd') + b'(cd' + c'd)] \\
Y' &= b'(cd + c'd') + b\,(cd' + c'd) \\
X' &= cd + c'd' \\
W' &= d'
\end{aligned} \tag{4-39}$$

These two sets of switching formulas can be realized by means of the circuit shown in Fig. 4-53.

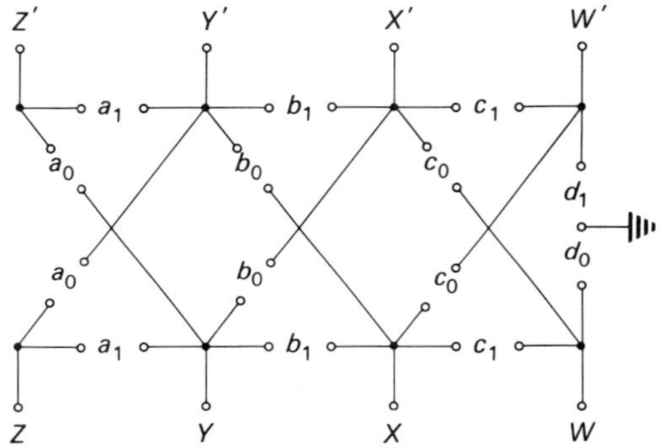

Fig. 4-53

These serial types of $(n + 1)$-terminal contact circuits are typical for operating circuits of relay counters. The hold or release circuits of the relays in counting circuits can consist of parallel connections of contacts, the number of which increases with the number of sections concerned. The following set of switching formulas may be considered as representative for such a situation:

$$\begin{aligned}
r_z &= z\alpha' \\
r_y &= y(z' + \alpha) = y(z' + z\alpha) \\
r_x &= x(y' + z' + \alpha) = x[y' + y(z' + \alpha)] \\
&= x[y' + y(z' + z\alpha)]
\end{aligned} \tag{4-40}$$

Equations (4-40) show the hold and release circuits of a sectional circuit. The combination of the various parallel-connected contacts is possible through the introduction of redundant variables by means of Theorem 11. The combination is made in such a way that the switching formula of section n comprises the switching formula of section $n - 1$, and so on. The resulting contact chain is shown in Fig. 4-54. The number of contacts involved in this type of contact chain is smaller than that of the individual parallel-connected contact circuits for four and more sections.

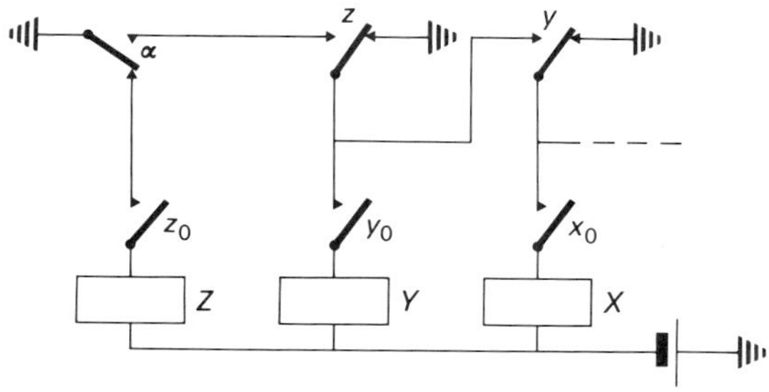

Fig. 4-54

CONTACT TREES

A contact tree is a special form of code converter that translates an n-element binary code into a 2^n-element decimal code. Such types of circuits are used in automatic-telephone registers. A pulse series transmitted by the dial of a subscriber's set can be received by a so-called binary relay counting chain, which translates the maximum 10-element serial-decimal code into a four-element parallel binary code. The decimal equivalent of the registered binary code combination can always be found by means of a so-called tree of relay contacts. An example of such a relay tree for a four-element binary code is given in Fig. 4-55.

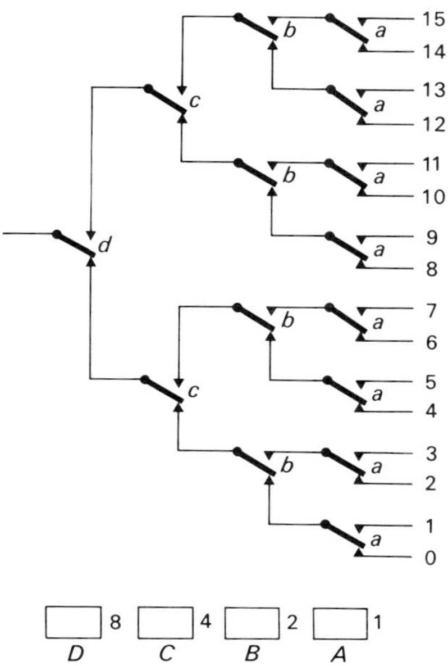

Fig. 4-55

The number of transfer contacts of the four register relays involved is 1, 2, 4, and 8, a total of 15. The physical layout of contacts and relays is entirely arbitrary. A complete tree can always be formed by choosing a number of transfer contacts that is equal to the maximum number of output terminals required. It is not necessary, however, to distribute the transfer contacts in the ratio 1:2:4:8 over the various relays. Except for the transfer contact which forms the top of the tree, and which appears necessarily in single, the other transfer contacts can be distributed rather evenly over the remaining relays. The best possible distribution over four relays is the 1:4:5:5 ratio shown in Fig. 4-56.

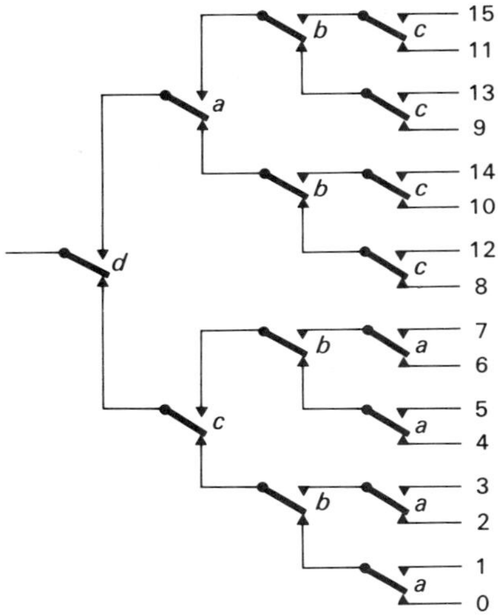

Fig. 4-56

It should be pointed out in this connection that no sneak paths exist in the contact tree between the various output terminals because the contact tree consists exclusively of transfer contacts, the make and break springs being connected only to a moving spring of another transfer contact.

In a number of applications, such as automatic telephony and relay computers, contact trees must sometimes provide an output terminal for the rest condition and 10 other output terminals corresponding to the decimal digits 1 to 10. Such a tree can be achieved by means of ten transfer contacts, which may be distributed in a 1:3:3:3 ratio over the four relays concerned. This type of contact tree is shown in Fig. 4-57.

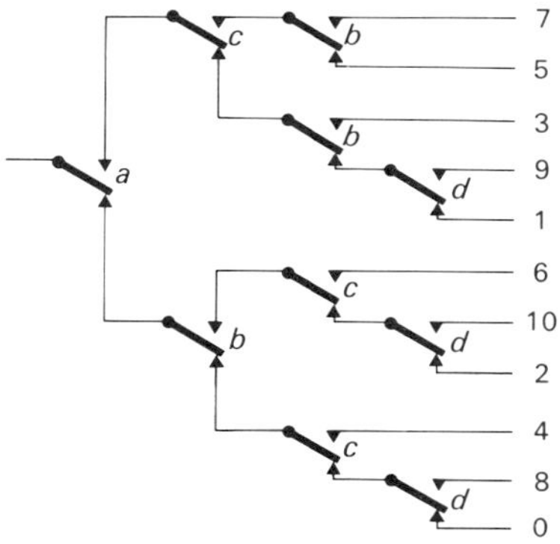

Fig. 4-57

PROBLEMS

1. Determine the classification of the following functions:

(a) $w'xz' + w'x'y'z$

(b) $w'x'yz + w'xyz' + wx'y'z$

(c) $wx'y' + w'x'y'z' + wxyz + w'xyz'$

(d) $wx'yz + wxyz' + w'x'yz' + wx'y'z + wx'y'$

(e) $(w'x' + wx)\cdot(y' + z') + yz\cdot(w'x + wx')$

2. Determine the classification of the following functions:

(a) $w'x'y'z + w'x'yz' + w'xyz' + wx'y'z'$

(b) $w'x'y'z' + w'x'yz + wx(y'z + yz')$

3. Determine the correct classification of the following functions:

(a) $f(0,1,2,8,10,11,12)$

(b) $f(0,1,4,7,11,12,15)$

(c) $f(0,1,2,4,5,11,13,14)$

(d) $f(0,3,4,6,7,8,10,11)$

(e) $f(0,4,6,8,10,11,13,15)$

4. Design the dual form of the contact circuit of Fig. 4-14*a*, and explain why this circuit contains one contact more than does the circuit given in Fig. 4-14*b*.

5. Design the contact circuit corresponding to the symmetrical function $S_{0,2}$ with three variables. Apply the shifting-down process.

6. Draw the contact circuits that correspond to the following switching functions:

(a) $f(0,1,6,11,14)$ (four transfer contacts)

(b) $f(0,3,5,9,14)$ (six transfer contacts)

(c) $f(0,1,2,3,4,11,13)$ (four transfer contacts)

(d) $f(0,1,2,3,4,5,8,15)$ (five transfer contacts)

(e) $f(0,1,2,7,11,12)$ (six transfer contacts)

7. Draw up the table of a five-unit code in which two elements are always 1, and design the corresponding contact circuit on a parallel basis.

8. Design a contact circuit consisting only of transfer contacts for the function $f(w,x,y,z) = w'xy'z + wx'yz'$.

9. Design the dual form of the contact circuit of Prob. 8. What is its switching function? Prove that this function is the negative of the function mentioned in Prob. 8.

10. Design a contact tree for the code symbols 0 up to 10 of the 6421 code with, respectively, one, three, three, three transfer contacts on the relays.

11. Simplify the two-terminal contact circuit of Fig. 4-58.

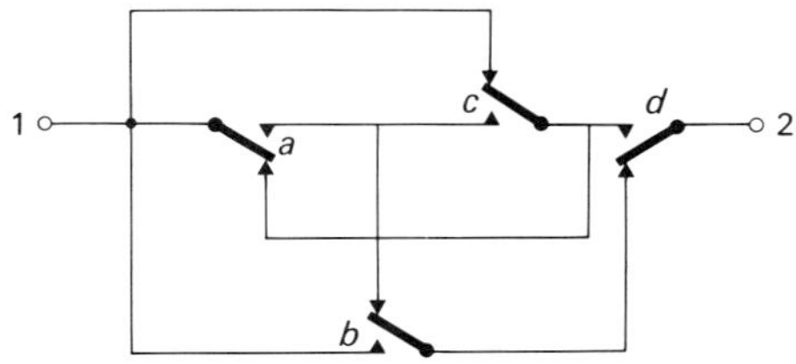

Fig. 4-58

12. Determine the simplest form of the contact network given in Fig. 4-59.

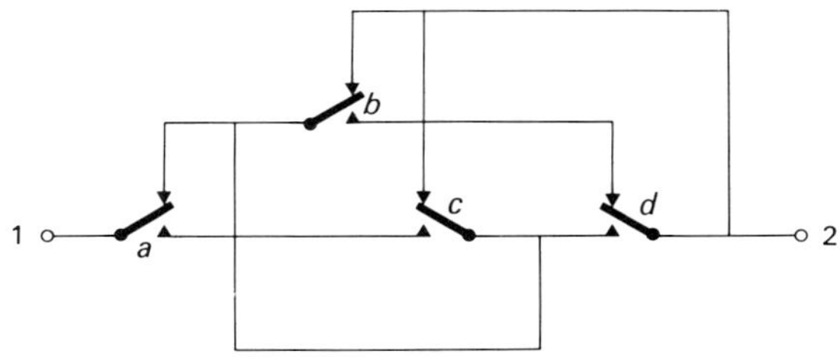

Fig. 4-59

13. Determine the simplest form of the contact network given in Fig. 4-60.

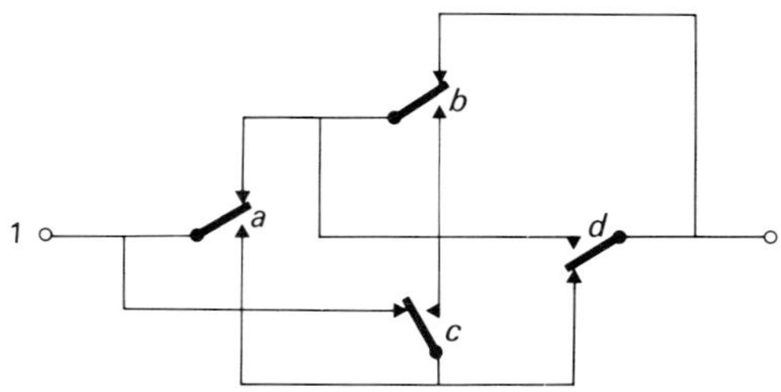

Fig. 4-60

14. Determine the simplest form of the contact network of Fig. 4-61.

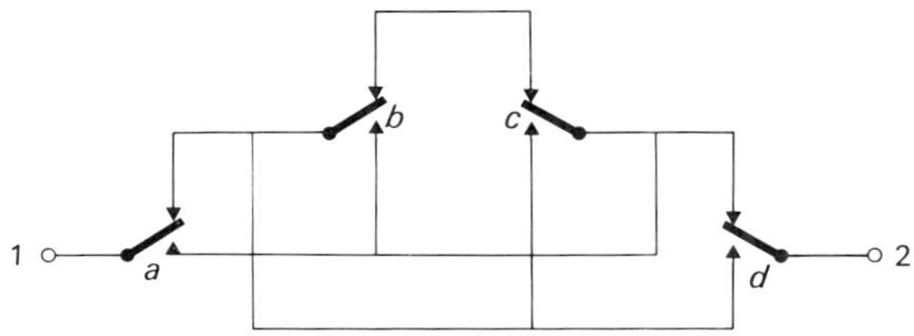

Fig. 4-61

15. Determine the simplest form of the contact network of Fig. 4-62.

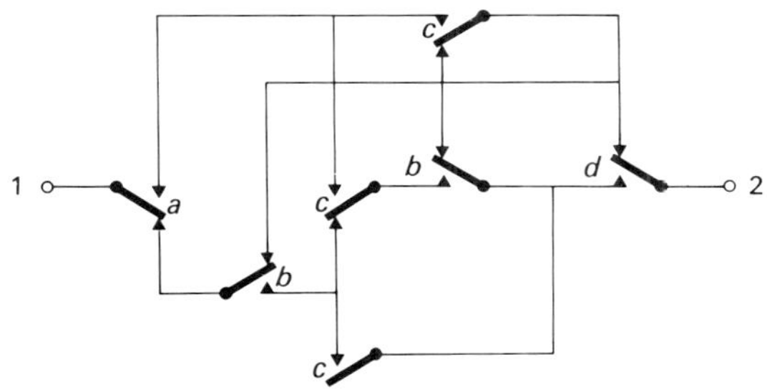

Fig. 4-62

SELECTIVE BIBLIOGRAPHY

1. Caldwell, S. H.: The Recognition and Identification of Symmetric Switching Circuits, *Trans. AIEE*, vol. 73, pp. 142–146, 1954.
2. Gawrilow, M. A.: Relaisschalttechnik für Stark- und Schwachstromanlagen, VEB Verlag Technik, Berlin, 1953.
3. Gould, R.: A Note on Contact Networks for Switching Functions of Four Variables, *IRE Trans.*, vol. EC-7, pp. 196–198, 1958.
4. Higgonet, R., and R. Gréa: "Étude Logique des Circuits Électriques et des Systèmes Binaires," pp. 404–428, Éditions Berger-Levrault, Paris, 1955.
5. Scheimann, A. H.: A Numerical-graphical Method for Synthesizing Switching Circuits, *Trans. AIEE*, pt. 1, vol. 76, pp. 687–689, 1957.
6. Shannon, C. E.: The Synthesis of Two-terminal Switching Circuits, *Bell System Tech. J.*, vol. 28, pp. 59–98, 1949.
7. Shannon, C. E., and E. F. Moore: Machine Aid for Switching Circuit Design, *Proc. IRE*, vol. 41, pp. 1348–1351, 1953; Bell Monograph 2153.
8. The Staff of the Computation Laboratory: "Synthesis of Electronic Computing and Control Circuits," app. I, Harvard University Press, Cambridge, Mass., 1951.
9. Troye, N. C. de: Classification and Minimization of Switching Functions, Ph.D. thesis, University of Amsterdam, 1958.

5
Matrix Methods in the Design of Contact Circuits

5-1 MATRIX REPRESENTATION OF SWITCHING FORMULAS

The design method of a contact circuit described in Sec. 4-2 is quite sound from the engineering point of view, but the example given at the end of this same section shows clearly that this method does not offer enough possibilities to find the variants with the minimum number of contacts. In this discussion the examples of multiterminal circuits were left unexplained, which may have given the impression that in this very logical field, there is no logical way to manipulate a set of switching formulas.

It is possible, however, to represent switching formulas corresponding to two- and multiterminal networks by means of matrices, to render the topology of the network by means of a process of nonterminal-node insertion or nonterminal-node removal, and, finally, to eliminate redundant elements. The use of matrices will be explained hereafter without overemphasizing mathematical considerations.

A contact network generally consists of contacts, terminals, and nodes. The terminals and the nodes may also be subdivided into terminal nodes and nonterminal nodes. Both classes have the same properties. The terminals and nodes are numbered. The switching functions f_{ik} of the circuit elements between any combination of terminals and nodes can be arranged in the form of a matrix because in a network with p terminals and nodes, there are p^2 such switching functions. If the network is a combinational circuit controlled by n variables, the p^2 switching functions have the form $f_{ik}(a,b,c,\ldots,n)$, this function being equal to 1 for all f_{ii}, since a terminal is invariably connected with itself. Furthermore, $f_{ik} = f_{ki}$ because relay contacts have the same features in both directions. It is therefore possible to introduce diodes as circuit elements in the matrix representation of switching functions. If the only switching elements used are contacts, the matrix will be symmetrical with respect to the main diagonal.

There are three types of switching matrices as given below.

1. Output matrix This is the $p \times p$ symmetrical matrix, which describes the switching functions corresponding to all possible paths between the p terminal nodes. The following is an example of such an output matrix:

$$F = \begin{bmatrix} f_{11} & f_{12} & f_{13} \\ f_{21} & f_{22} & f_{23} \\ f_{31} & f_{32} & f_{33} \end{bmatrix} = \begin{bmatrix} 1 & x_1 + y_0 + z_0 & x_0 + y_1 z_0 \\ x_1 + y_0 + z_0 & 1 & y_1 z_0 + x_0 y_0 \\ x_0 + y_1 z_0 & y_1 z_0 + x_0 y_0 & 1 \end{bmatrix} \tag{5-1}$$

This matrix corresponds to the circuit shown in Fig. 5-1.

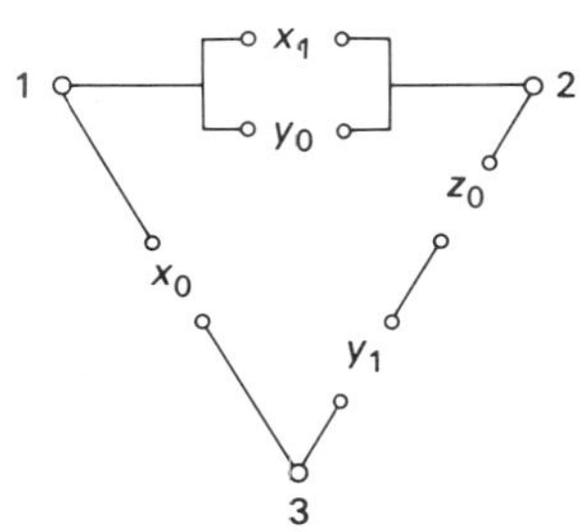

Fig. 5-1

2. Primitive connection matrix This is the $(p + k) \times (p + k)$ symmetrical matrix containing the switching functions of the contact circuits between p terminal nodes and k nonterminal nodes. The nonterminal nodes are chosen in such a way that one finds between any two nodes, terminal or nonterminal, either a single contact, a number of contacts in parallel, or no circuit at all.

The terminal nodes will be indicated as small circles and the nonterminal nodes as solid dots. An example of a primitive connection matrix is

$$P = \begin{bmatrix} 1 & 0 & z_0 & y_1 \\ 0 & 1 & 0 & y_0 + z_1 \\ z_0 & 0 & 1 & x_1 \\ y_1 & y_0 + z_1 & x_1 & 1 \end{bmatrix} \tag{5-2}$$

The corresponding circuit is shown in Fig. 5-2. Where two nodes are not connected, the corresponding switching function is 0.

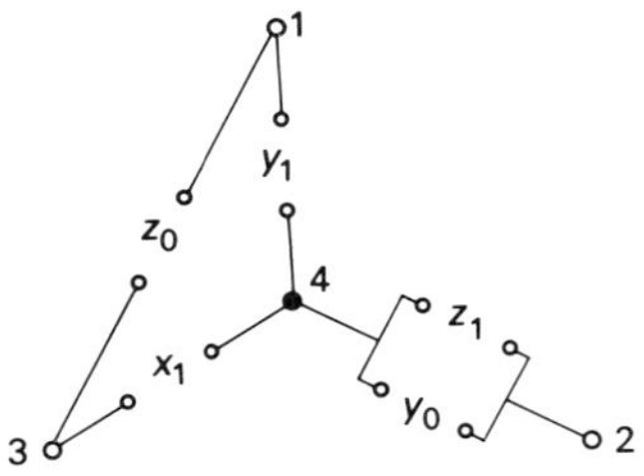

Fig. 5-2

3. Connection matrix This is a type of matrix that falls between the two extremes mentioned under 1 and 2. This matrix has so many nonterminal nodes that all circuits connecting the nodes (terminal or nonterminal) are two-terminal networks. An example of such a matrix is given in (5-3) and the corresponding circuit is shown in Fig. 5-3.

$$C = \begin{bmatrix} 1 & x_0 & x_1 z_0 & y_1 + z_0 \\ x_0 & 1 & x_1 + z_1 & y_0 \\ x_1 z_0 & x_1 + z_1 & 1 & y_1 z_1 \\ y_1 + z_0 & y_0 & y_1 z_1 & 1 \end{bmatrix} \tag{5-3}$$

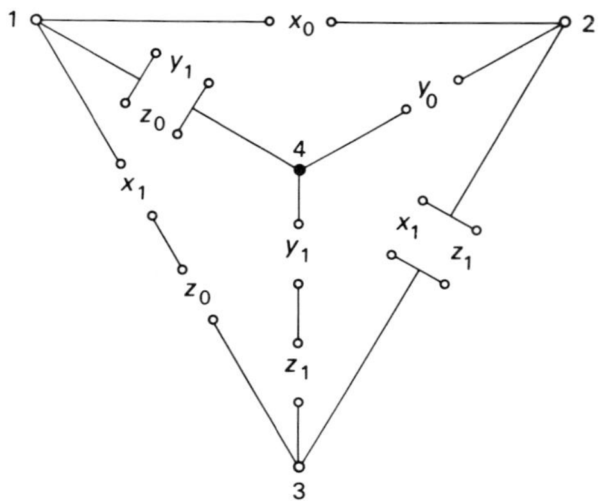

Fig. 5-3

The output matrix specifies the connections between output terminals over all existing paths. By means of this type of matrix, one can easily determine whether or not there is a connection for a certain combination of input variables. This can be of importance for checking whether or not in the circuit to be designed two or more relays release their armature in parallel. If they do, the release time of such relays can increase considerably. The requirements for a certain circuit converted into switching formulas lead, in the first instance, to the drawing up of an output matrix. The simplest type of output matrix is the two-terminal case:

$$\begin{bmatrix} 1 & f_{12} \\ f_{12} & 1 \end{bmatrix} \tag{5-4}$$

The connection matrix shows all connections over existing direct paths between the nodes, terminal or nonterminal. By means of a process of insertion of nodes and of elimination of redundant elements, the output matrix F can be transformed into a primitive connection matrix C_0.

5-2 DEFINITIONS

Switching matrices are generally rectangular arrays consisting of m columns and n rows, the elements of which are switching functions. Some of the rules for manipulating these matrices are analogous to those given for the manipulation of switching functions. The matrices used in this chapter will consist of relay contacts only and describe such contact circuits as are shown in Sec. 5-1. These matrices are symmetric with respect to the main diagonal. Since the matrices used here present a network as a combination of paths between the nodes, terminal or nonterminal, they are all of the $m = n$ type.

The following definitions constitute these switching matrices as a Boolean algebra. Matrix A contains the elements a_{ij}, matrix B the elements b_{ij}.

1. *Equality* $A = B$ if and only if $a_{ij} = b_{ij}$ for all i and j.
2. *Logical sum* $A + B = [a_{ij} + b_{ij}]$. This sum is obtained by adding the corresponding elements of both matrices. The elements of the resulting switching matrix are formed by connecting the elements a_{ij} and b_{ij} in parallel between the nodes i and j.
3. *Logical product* $A \cdot B = [a_{ij} \cdot b_{ij}]$. This product is formed by multiplying all corresponding elements. The elements of the resulting switching matrix are formed by connecting the elements a_{ij} and b_{ij} in series between the nodes i and j.
4. *Complement* $A' = [a'_{ij}]$ for $i = j$; $a'_{ij} = 1$ because a node is invariable to itself. The complement A' of a matrix is obtained by replacing all two-terminal circuits between nodes by their complement.

5. *Inclusion* $A \leqslant B$. Matrix A is contained in B if and only if $a_{ij} \leqslant b_{ij}$ for all i and j. The expression $A \leqslant B$ means that for any combination of the variables resulting in a closed path between the nodes i and j in the circuit described by matrix A, there is also a closed path between the nodes i and j in the circuit described by matrix B.

Some other types of operations can be performed with matrices. The matrix product $A \times B$ follows the same rules as in ordinary algebra. In the transposition process, all column elements are replaced by the symmetrical—i.e., inversely numbered—row elements.

It is worthwhile to remember these definitions and to be aware of the fact that matrices defined in this way are a fundamental part of the switching algebra; however, the way in which they will be used in the following discussion of the manipulation of switching formulas and contact circuits will be more or less a systematic bookkeeping method.

5-3 METHODS FOR SIMPLIFYING MATRICES

The purpose of manipulating matrices is to find simpler forms of circuits. This manipulation can consist of translating the operating requirements of the circuit to be designed into an output matrix that has to be simplified into a primitive or near-primitive connection matrix in order to find the circuit with the smallest number of switching elements. This simplification or change in circuit topology can be obtained by means of three methods that can either be used separately or be combined, depending on the character of entries to be found in the matrix.

REMOVAL OF NONTERMINAL NODES

The process of node removal will be demonstrated by means of the following general example. Figure 5-4 shows part of a circuit with five nodes numbered 1 to 5, which may be part of a larger circuit. All nodes which are linked to node 5, that is, the node to be removed from the circuit, are shown in Fig. 5-4. The associated connection matrix C is given as the first part of Eq. (5-5). Node 5 can be removed from the network by means of a simultaneous operation on the terms of row 5 and column 5. The logical products of all pairs of entries in this row and column which represent the possible paths from one node to another over the node to be removed have to be added to the entry representing the existing connection between the same pair of nodes. The resulting connection matrix is shown in the right-hand side of Eq. (5-5). This procedure is generally applicable and can be pursued as a maximum until one arrives at connection matrix ${}^{p}C_0$ (with no nonterminal nodes). This matrix differs from the output matrix F.

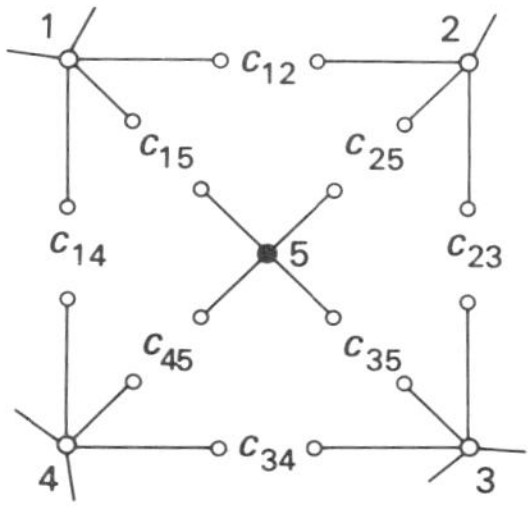

Fig. 5-4

$$\begin{bmatrix} 1 & c_{12} & 0 & c_{14} & c_{15} \\ c_{12} & 1 & c_{23} & 0 & c_{25} \\ 0 & c_{23} & 1 & c_{34} & c_{35} \\ c_{14} & 0 & c_{34} & 1 & c_{45} \\ c_{15} & c_{25} & c_{35} & c_{45} & 1 \end{bmatrix} \sim$$

$$\begin{bmatrix} 1 & c_{12} + c_{15}\cdot c_{25} & c_{15}\cdot c_{35} & c_{14} + c_{15}\cdot c_{45} \\ c_{12} + c_{15}\cdot c_{25} & 1 & c_{23} + c_{25}\cdot c_{35} & c_{25}\cdot c_{45} \\ c_{15}\cdot c_{35} & c_{23} + c_{25}\cdot c_{35} & 1 & c_{34} + c_{35}\cdot c_{45} \\ c_{14} + c_{15}\cdot c_{45} & c_{25}\cdot c_{45} & c_{34} + c_{35}\cdot c_{45} & 1 \end{bmatrix} \tag{5-5}$$

Application of Eq. (5-5) will be illustrated by means of the star-mesh transformation of the circuit shown in Fig. 5-5. The matrix of Eq. (5-6) is the connection matrix representing the Y circuit shown in Fig. 5-5; the second

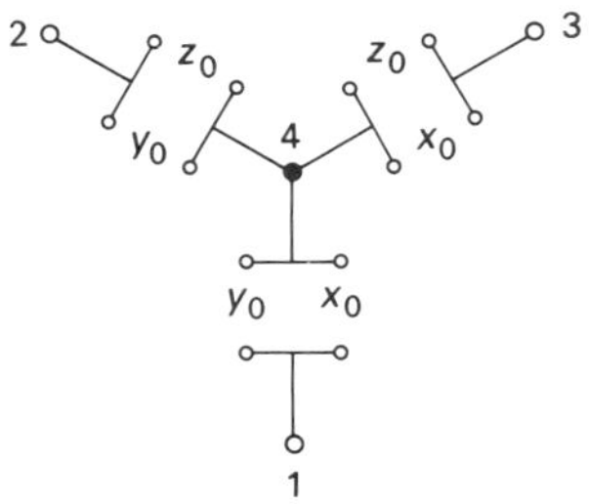

Fig. 5-5

matrix gives the result after removal of node 4. Since entries c_{13} and c_{23} comprise a term x_0 and z_0, respectively, the term x_0z_0 in entry c_{12} is redundant. This means that node 1 and node 2 are interconnected not only directly by a circuit x_0z_0 but also over a path of the series connection of the two links passing through node 3. The direct path x_0z_0 is therefore redundant.

$$
{}^3C_1 = \begin{bmatrix} 1 & 0 & 0 & x_0+y_0 \\ 0 & 1 & 0 & y_0+z_0 \\ 0 & 0 & 1 & z_0+x_0 \\ x_0+y_0 & y_0+z_0 & z_0+x_0 & 1 \end{bmatrix}
$$

$$
\sim \begin{bmatrix} 1 & y_0+x_0z_0 & x_0+y_0z_0 \\ y_0+x_0z_0 & 1 & z_0+x_0y_0 \\ x_0+y_0z_0 & z_0+x_0y_0 & 1 \end{bmatrix} = \begin{bmatrix} 1 & y_0 & x_0 \\ y_0 & 1 & z_0 \\ x_0 & z_0 & 1 \end{bmatrix} \tag{5-6}
$$

It can be shown that the terms y_0z_0 of entry c_{13} and x_0y_0 of entry c_{23} are also redundant. The resulting circuit is shown in Fig. 5-6.

Node removal from a Y-type circuit does not necessarily always lead to a reduction in the number of switching elements, as can be illustrated by means of the circuit shown in Fig. 5-7.

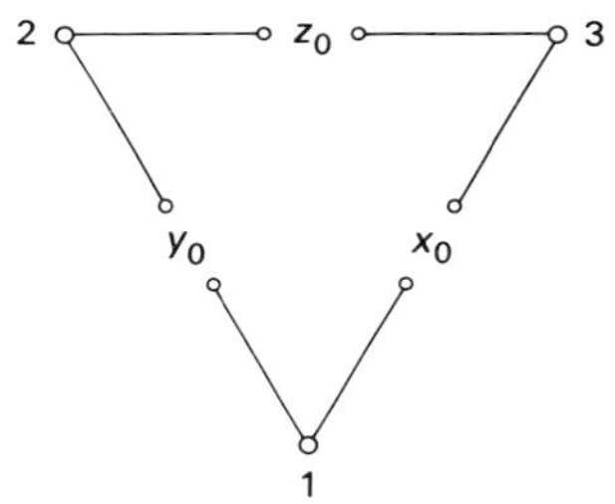

Fig. 5-6

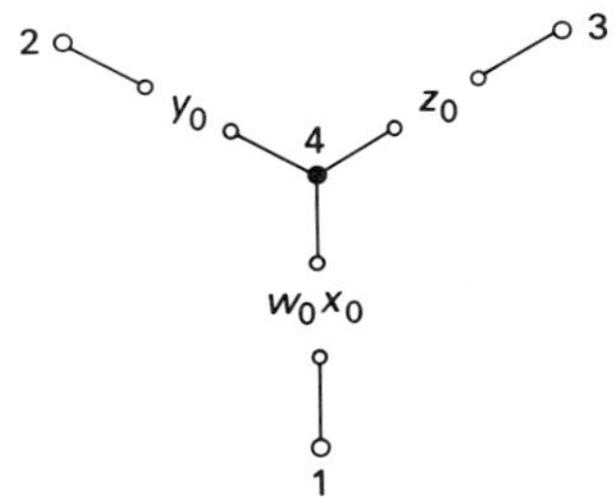

Fig. 5-7

$$
{}^3C_1 = \begin{bmatrix} 1 & 0 & 0 & w_0x_0 \\ 0 & 1 & 0 & y_0 \\ 0 & 0 & 1 & z_0 \\ w_0x_0 & y_0 & z_0 & 1 \end{bmatrix}
$$

$$
\sim \begin{bmatrix} 1 & w_0x_0y_0 & w_0x_0z_0 \\ w_0x_0y_0 & 1 & y_0z_0 \\ w_0x_0z_0 & y_0z_0 & 1 \end{bmatrix} \tag{5-7}
$$

Transformation equation (5-5) is applied to the left-hand matrix of (5-7), which represents the circuit shown in Fig. 5-7. The right-hand matrix of (5-7) gives the result of the transformation, which is represented by the circuit shown in Fig. 5-8.

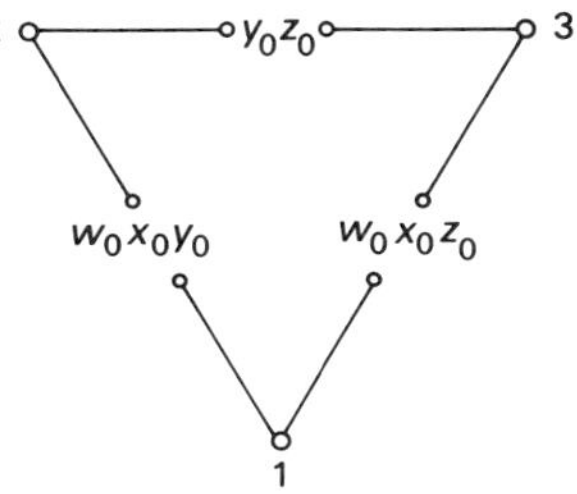

Fig. 5-8

INSERTION OF NONTERMINAL NODES

Instead of node removal, one may resort to the reverse process, i.e., node insertion. It can be used as a possible way of changing the topology of a network in a predetermined way. The node-insertion process can lead to a simplification or to a complication of the circuit, here again depending on the nature of the entries. This process can be applied to an output matrix or to a connection matrix until a primitive connection matrix is obtained. The principle of the node-insertion process is described by Eq. (5-5), if read in the reverse direction, or by Eq. (5-8):

$$\begin{bmatrix} 1 & a+xy & b+xz \\ a+xy & 1 & c+yz \\ b+xz & c+yz & 1 \end{bmatrix} \sim \begin{bmatrix} 1 & a & b & x \\ a & 1 & c & y \\ b & c & 1 & z \\ x & y & z & 1 \end{bmatrix} \tag{5-8}$$

The node-insertion process may be illustrated by means of the following example of a two-terminal network. As was shown in the preceding chapter, a bridge-type contact circuit may help save an important number of contacts, provided, of course, that such topology is possible with the switching formula on hand. However, it is sometimes difficult to determine through systematic manipulation whether or not a bridge circuit is incorporated in the formula.

The general switching formula of a bridge circuit is given in Eq. (4-32):

$$f(a,b,c,d,e) = ab + aed + ceb + cd$$

The possibility of transforming a two-terminal circuit into a four-node bridge circuit can be ascertained as follows:

$$F = \begin{bmatrix} 1 & ab+aed+ceb+cd \\ ab+aed+ceb+cd & 1 \end{bmatrix} \tag{5-9}$$

The output matrix of the desired circuit is given by Eq. (5-9). The topology of the two-terminal network has to be changed as shown in Fig. 5-9. As a first step in the transformation process, one has to decide which of the variables of matrix (5-9) are going to be used as entry c_{13} and entry c_{14} of the four-node connection scheme of Fig. 5-9*b*. In the present example, it is rather obvious that this choice is limited to the common variable of the first pair of terms and that of the second pair of terms, or to the common variable of the first

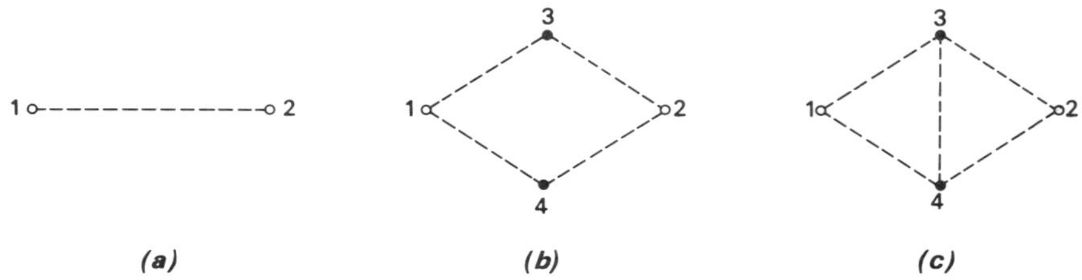

Fig. 5-9

and third terms and that of the second and fourth terms. These pairs of variables are, respectively,

$$a \quad \text{and} \quad c \tag{5-10a}$$

$$b \quad \text{and} \quad d \tag{5-10b}$$

When using $c_{13} = a$ and $c_{14} = c$, output matrix (5-9) can be transformed into the following connection matrix ${}^2C_2$:

$${}^2C_2 = \begin{bmatrix} 1 & 0 & a & c \\ 0 & 1 & b + ed & d + eb \\ a & b + ed & 1 & 0 \\ c & d + eb & 0 & 1 \end{bmatrix} \tag{5-11}$$

The second step is the introduction of a suitable entry c_{34} as required by the topology of the network shown in Fig. 5-9*c*. The term *ed* of entry c_{23} may be replaced by a term *d* in entry c_{24} (which is already present) and a term *e* in entry c_{34}, whereas the term *eb* of entry c_{24} may be replaced by a term *b* in entry c_{23} (which is already present) and a term *e* in entry c_{34}. This leads to the fundamental form (5-12) of the matrix representation of a bridge network. The resulting circuit is shown in Fig. 5-10.

$${}^2C_2 = \begin{bmatrix} 1 & 0 & a & c \\ 0 & 1 & b & d \\ a & b & 1 & e \\ c & d & e & 1 \end{bmatrix} \tag{5-12}$$

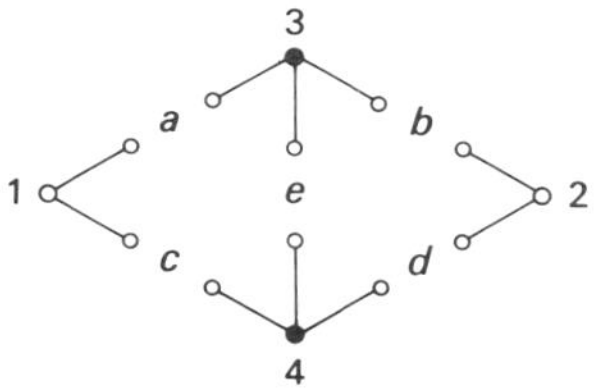

Fig. 5-10

INSERTION AND REMOVAL OF REDUNDANT ELEMENTS

The entries of a matrix are sometimes of such a form that a node can be reached over different paths corresponding to the same switching functions. In the star-mesh transformation shown in Eq. (5-6), use was already made of this fact, whereby part of an entry proved to be redundant and could be omitted from the matrix. The removal of redundant elements will be illustrated by means of some examples. The examples refer to three-node networks; one should bear in mind, however, that these networks can be part of much larger networks.

One redundant element One of the x elements is redundant (see Fig. 5-11).

$$\begin{bmatrix} 1 & \alpha+x & \gamma+x \\ \alpha+x & 1 & \beta+x \\ \gamma+x & \beta+x & 1 \end{bmatrix} \tag{5-13}$$

One may arbitrarily choose the one to be omitted from this matrix.

Two redundant elements The initial circuit is given in Fig. 5-12. Since

$$\begin{bmatrix} 1 & \alpha+\beta\gamma & \beta \\ \alpha+\beta\gamma & 1 & \gamma \\ \beta & \gamma & 1 \end{bmatrix} \tag{5-14}$$

$c_{13} \cdot c_{23} = \beta\gamma$, the term $\beta\gamma$ of entry c_{12} is redundant and may be omitted from matrix (5-14) without affecting the operation of the network.

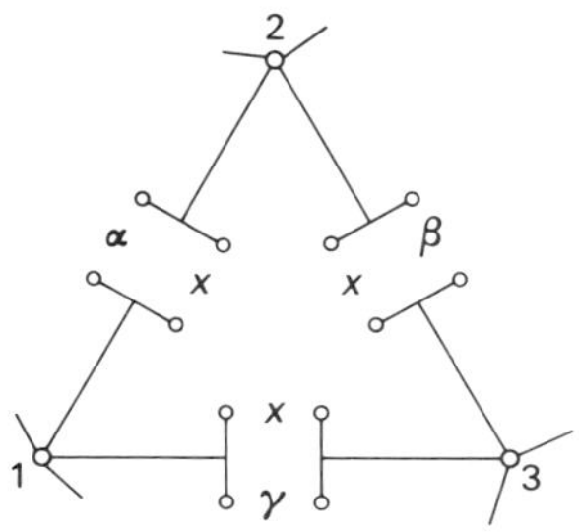

Fig. 5-11

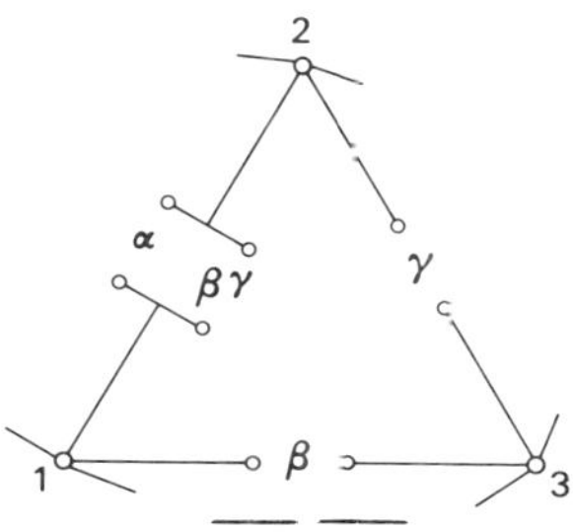

Fig. 5-12

Sometimes redundant elements have to be introduced in order to make it possible to simplify the matrix by manipulation. This may be illustrated by the following example.

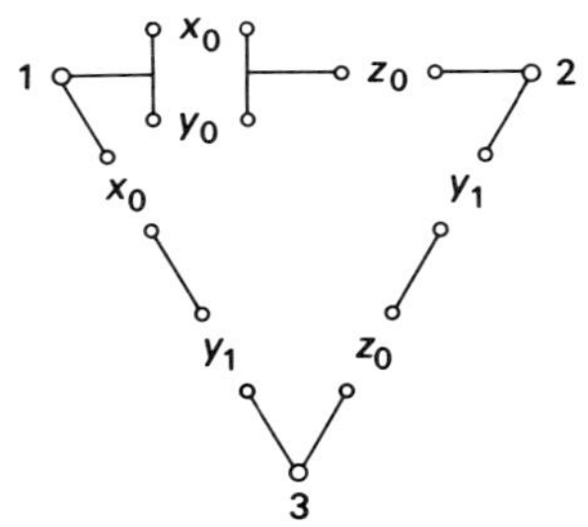

Fig. 5-13

The initial circuit is given in Fig. 5-13. The corresponding connection matrix [first matrix of Eq. (5-15)] contains entry $c_{13} = x_0y_1$, which may be replaced by $(x_0 + y_0)\,y_1$. It then becomes possible to insert a fourth node and thus achieve a reduction of elements, as is shown by the last matrix of Eq. (5-15).

$$\begin{bmatrix} 1 & (x_0+y_0)\,z_0 & x_0y_1 \\ (x_0+y_0)\,z_0 & 1 & y_1z_0 \\ x_0y_1 & y_1z_0 & 1 \end{bmatrix}$$

$$= \begin{bmatrix} 1 & (x_0+y_0)\,z_0 & (x_0+y_0)\,y_1 \\ (x_0+y_0)\,z_0 & 1 & y_1z_0 \\ (x_0+y_0)\,y_1 & y_1z_0 & 1 \end{bmatrix}$$

$$\sim \begin{bmatrix} 1 & 0 & 0 & x_0+y_0 \\ 0 & 1 & 0 & z_0 \\ 0 & 0 & 1 & y_1 \\ x_0+y_0 & z_0 & y_1 & 1 \end{bmatrix} \tag{5-15}$$

The resulting circuit is shown in Fig. 5-14. This process is generally more complicated in practice.

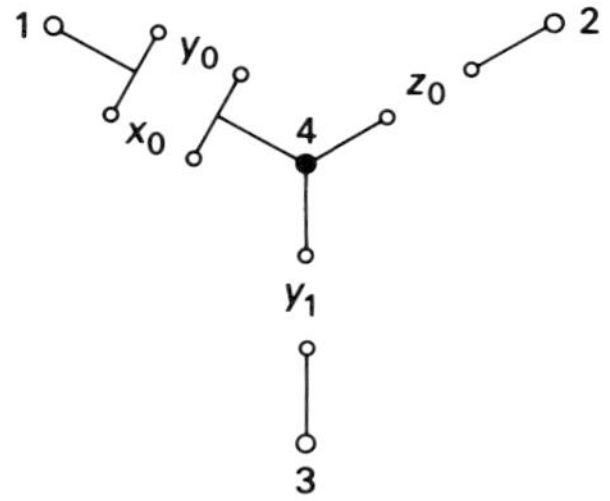

Fig. 5-14

INSERTION OF NONTERMINAL NODES UNDER DON'T-CARE CONDITIONS

If the rules governing the introduction of nodes are applied correctly, no sneak paths will occur. As a general rule, sneak paths can be allowed to occur provided that they appear only during don't-care conditions, which do not affect the original conditions of the circuit to be designed. These don't-care conditions can serve to simplify a circuit even further. The resulting sneak paths can be easily checked. The following examples of Y-Δ transformations can be used as an illustration. The network shown in Fig. 5-15 can be represented by the connection matrix ${}^3C_0$ given in the left-hand side of Eq. (5-16):

$$ {}^3C_0 = \begin{bmatrix} 1 & x_0y_0 & x_0y_1 \\ x_0y_0 & 1 & 0 \\ x_1y_1 & 0 & 1 \end{bmatrix} \sim \begin{bmatrix} 1 & 0 & 0 & x_0 \\ 0 & 1 & 0 & y_0 \\ 0 & 0 & 1 & y_1 \\ x_0 & y_0 & y_1 & 1 \end{bmatrix} = {}^3C_1 \qquad (5\text{-}16) $$

Since $y_0y_1 = 0$, a fourth node may be introduced as shown in the connection matrix ${}^3C_1$ of the right-hand side of Eq. (5-16). The products of the pairs of entries in column 4 correspond to the switching functions between the three original nodes 1, 2, and 3 (see Fig. 5-16).

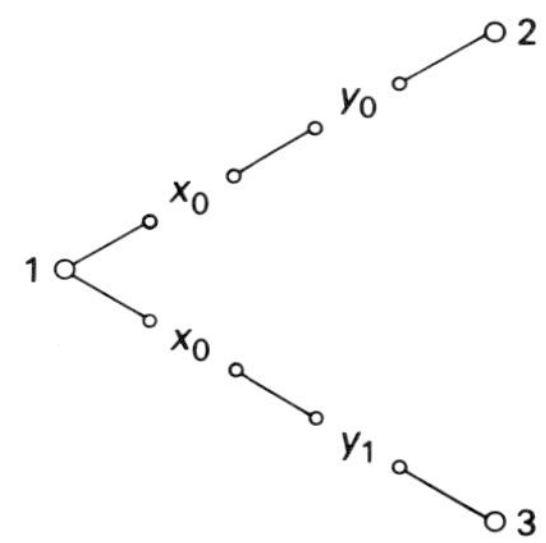

Fig. 5-15

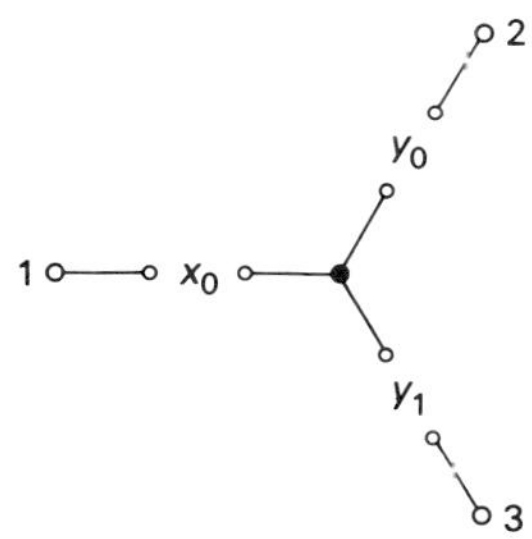

Fig. 5-16

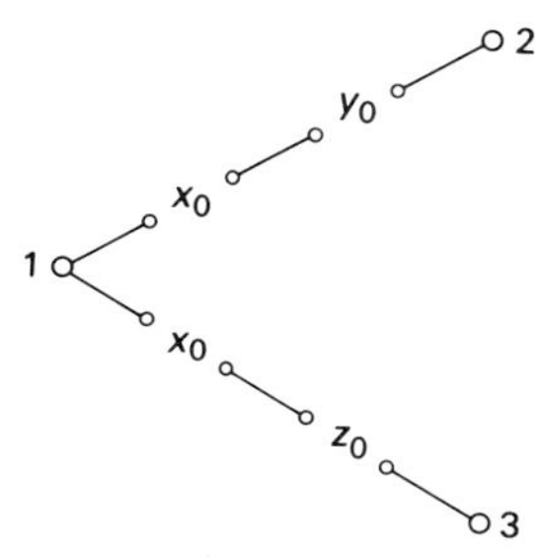

Fig. 5-17

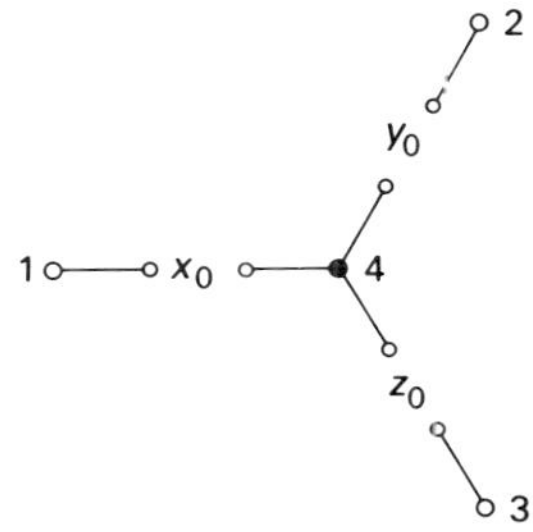

Fig. 5-18

If, however, the switching function x_0y_1 of the path 1-3 is changed into x_0z_0, the star-mesh transformation cannot be performed in the manner shown in Figs. 5-17 and 5-18 without altering the properties of the network. The switching function $x_0y_0z_0$ of the circuit between nodes 2 and 3 has been changed into y_0z_0.

In circuits where such a change is immaterial, this type of transformation can be used as a conditional circuit simplification that can be described by the matrices of Eq. (5-17).

$$^3C_0 = \begin{bmatrix} 1 & x_0y_0 & x_0z_0 \\ x_0y_0 & 1 & 0 \\ x_0z_0 & 0 & 1 \end{bmatrix} \sim \begin{bmatrix} 1 & 0 & 0 & x_0 \\ 0 & 1 & [y_0z_0] & y_0 \\ 0 & [y_0z_0] & 1 & z_0 \\ x_0 & y_0 & z_0 & 1 \end{bmatrix} = {}^3C_1 \qquad (5\text{-}17)$$

This type of node insertion and matrix simplification under restrictive conditions can be shown in the resulting matrix by indicating the sneak-path functions between brackets. In the present case, the introduced sneak path is $c_{24} \cdot c_{34} = [c_{23}] = [y_0z_0]$, which is indicated in Eq. (5-17) as a reminder of the difference between the left-hand and right-hand side of this equation.

5-4 THE SHAPING OF A TWO-TERMINAL CONTACT NETWORK

The problem of converting the two-terminal circuit represented by Eq. (4-11) into a double-bridge circuit may now be analyzed.

$$f(w, x, y, z) = wx'y'z + w'x'yz' + w'xy'z + wxy \qquad (5\text{-}18)$$

The topology of the desired circuit is shown in Fig. 5-19. The possible paths in the type of circuit shown in Fig. 5-19 may consist of three, four, or five links. There are two 3-link paths, four 4-link paths, and two 5-link paths, that is, a total of eight. Since switching formula (5-18) contains four terms only (five terms in its canonical form), some of these paths have to be blocked by a suitable sequence of complemental variables.

The first step in this node-insertion process is the transformation of the two-terminal network into a four-node network of the type given in Fig. 5-20.

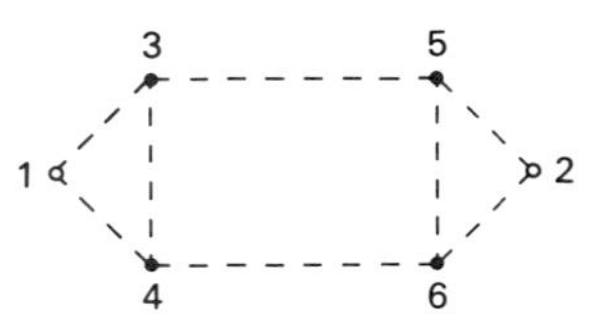

Fig. 5-19

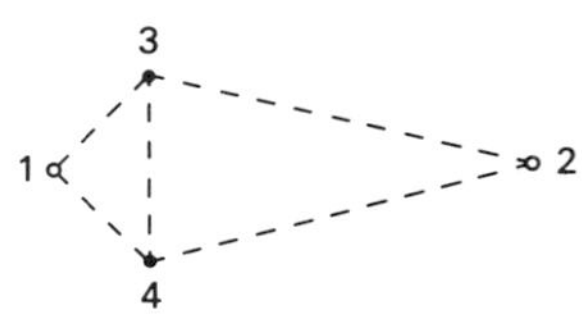

Fig. 5-20

Two variables have to be chosen from Eq. (5-18) for use as entries c_{13} and c_{14} in the corresponding connection matrix $^2C_2$. It is obvious that one of these variables must be w_0, x_0, or y_0, since the last term of Eq. (5-18) will probably have to be realized by means of a three-link circuit. Another normal requirement for the choice of the variables for entries c_{13} and c_{14} is that they appear in different groups, i.e., pairs of terms. A last requirement, since switching function (5-18) is not a symmetrical function, is that

$$c_{13} \neq (c_{14})' \tag{5-19}$$

If this were not the case, a number of blocking possibilities of paths would not be available for reducing the eight possible paths of this network to the required number of four.

These requirements make it mandatory to choose variables y_0, and therefore z_0, as entries c_{13} and c_{14}. The insertion of nodes 3 and 4 leads to matrix (5-20):

$$^2C_2 = \begin{bmatrix} 1 & 0 & y_0 & z_0 \\ 0 & 1 & w_1 x_1 z_1 + w_0 x_0 & w_0 x_1 y_1 + w_1 x_0 y_1 \\ y_0 & w_1 x_1 z_1 + w_0 x_0 & 1 & 0 \\ z_0 & w_0 x_1 y_1 + w_1 x_0 y_1 & 0 & 1 \end{bmatrix} \tag{5-20}$$

The second step in this circuit transformation is the detection of a suitable entry c_{34}; in the present case, this is not so easy as it is for a normal-bridge circuit. The variables w_0 and x_0 of the $w_0 x_0 y_0$ term are excluded because it is very likely that this term will have to be obtained by means of a three-link path without any diagonal connections. The choice of y_1 and z_1 is unlikely because their complements are already used as entries c_{13} and c_{14}. Such a choice would impede the proper distribution of the various terms over different paths.

A choice has to be made between w_1 and x_1. Permutation of the variables w and x does not change entries c_{23} and c_{24} of matrix (5-20). This choice will in principle lead to the same result.

If $c_{34} = w_1$, the term $w_1 x_1 z_1$ of entry c_{23} may be replaced by entry $c_{34} = w_1$ and a term $x_1 z_1$ in entry c_{24}, whereas the term $w_1 x_0 y_1$ of entry c_{24} may be replaced by entry $c_{34} = w_1$ and a term $x_0 y_1$ in entry c_{23}. This results in the matrix given by Eq. (5-21):

$$^2C_2 = \begin{bmatrix} 1 & 0 & y_0 & z_0 \\ 0 & 1 & x_0(w_0 + y_1) & x_1(w_0 y_1 + z_1) \\ y_0 & x_0(w_0 + y_1) & 1 & w_1 \\ z_0 & x_1(w_0 y_1 + z_1) & w_1 & 1 \end{bmatrix} \tag{5-21}$$

The third step provides for the conversion of the connection matrix $^2C_2$ [Eq. (5-21)] into a connection matrix $^2C_4$ by the introduction of nodes 5 and 6, as shown in the desired network configuration of Fig. 5-19. Looking closely at matrix (5-21), one can see that the variables x_0 and x_1 are the only possible choices for entries c_{35} and c_{46}, respectively. The resulting matrix (5-22) is

$$
{}^2C_4 = \begin{bmatrix} 1 & 0 & y_0 & z_0 & 0 & 0 \\ 0 & 1 & 0 & 0 & w_0 + y_1 & w_0 y_1 + z_1 \\ y_0 & 0 & 1 & w_1 & x_0 & 0 \\ z_0 & 1 & w_1 & 1 & 0 & x_1 \\ 0 & w_0 + y_1 & x_0 & 0 & 1 & 0 \\ 0 & w_0 y_1 + z_1 & 0 & x_1 & 0 & 1 \end{bmatrix} \tag{5-22}
$$

The fourth step in this process is the detection of entry c_{56} and the final form of entries c_{25} and c_{26}. The final form of entry c_{25} is determined by the fact that $c_{13} \cdot c_{35} = x_0 y_0$, so that for $c_{25} = w_0$ the term $w_0 x_0 y_0$ of Eq. (5-18) is satisfied.

The variable y_1 seems to be the most suitable choice for entry c_{56}, w_0 having been chosen for entry c_{25}. The term $w_0 y_1$ of entry c_{26} is then redundant; however, term y_1 of entry c_{25} cannot be eliminated if sneak paths are to be avoided. This term y_1 can be made redundant by means of the introduction of a suitable redundant factor. Entry c_{23} in matrix (5-21) may be written as follows:

$$
x_0(w_0 + y_1) = x_0(w_0 + x_0 y_1) \tag{5-23}
$$

This leads to entry $c_{25} = w_0 + x_0 y_1$ in matrix (5-22), which is rewritten below as matrix (5-24).

$$
{}^2C_4 = \begin{bmatrix} 1 & 0 & y_0 & z_0 & 0 & 0 \\ 0 & 1 & 0 & 0 & w_0 + x_0 y_1 & w_0 y_1 + z_1 \\ y_0 & 0 & 1 & w_1 & x_0 & 0 \\ z_0 & 0 & w_1 & 1 & 0 & x_1 \\ 0 & w_0 + x_0 y_1 & x_0 & 0 & 1 & 0 \\ 0 & w_0 y_1 + z_1 & 0 & x_1 & 0 & 1 \end{bmatrix} \tag{5-24}
$$

Matrix (5-24) allows the introduction of an entry $c_{56} = y_1$, so that the term $w_0 y_1$ of entry c_{26} becomes redundant, and the term $x_0 y_1$ of entry c_{25} can be replaced by entry $c_{56} = y_1$ and an extra term x_0 in entry c_{26}. The final form of the connection matrix ${}^2C_4$ is given in Eq. (5-25).

$$
{}^2C_4 = \begin{bmatrix} 1 & 0 & y_0 & z_0 & 0 & 0 \\ 0 & 1 & 0 & 0 & w_0 & x_0 + z_1 \\ y_0 & 0 & 1 & w_1 & x_0 & 0 \\ z_0 & 0 & w_1 & 1 & 0 & x_1 \\ 0 & w_0 & x_0 & 0 & 1 & y_1 \\ 0 & x_0 + z_1 & 0 & x_1 & y_1 & 1 \end{bmatrix} \tag{5-25}
$$

The redundant term x_0 in entry c_{26} does not lead to the occurrence of unwanted paths, because $x_0 \cdot c_{46} = x_0 \cdot x_1 = 0$.

The resulting circuit is shown in Fig. 5-21. This circuit is a variant of the circuit already given in Fig. 4-29.

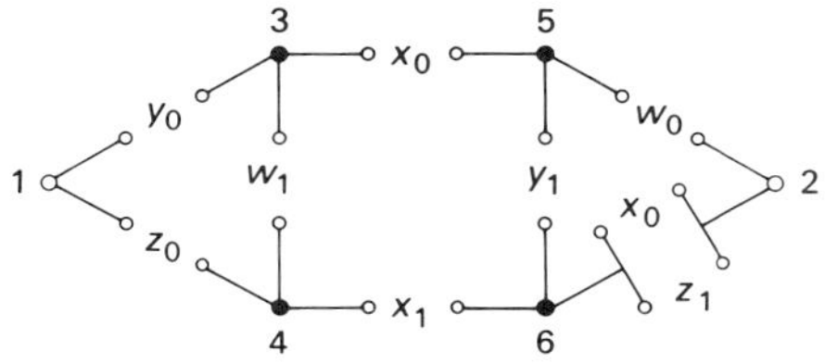

Fig. 5-21

5-5 MULTITERMINAL CIRCUITS

The manipulation of matrices in case of multiterminal circuits will be explained by means of a useful example. The translation of a two-out-of-five code into the binary 8421 code, as given in Table 5-1, is required. This translation has to be implemented by means of a contact circuit. No other operating conditions of relays V to Z than those listed in Table 5-1 occur. The 21 combinations not listed may be used as don't-care conditions. The problem of a multiterminal contact circuit is to find the solution with a maximum combination of the various contact paths without the introduction of undesired sneak paths. By means of a careful choice of the possible switching formulas for the various outputs, a number of contact paths can partially be combined and the switching formulas of these combined paths can be shaped in such a way that undesired sneak paths are prevented.

Table 5-1

No.	z	y	x	w	v	8	4	2	1
0	0	0	0	0	0	0	0	0	0
1	0	0	0	1	1	0	0	0	1
2	0	0	1	0	1	0	0	1	0
3	0	0	1	1	0	0	0	1	1
4	0	1	0	0	1	0	1	0	0
5	0	1	0	1	0	0	1	0	1
6	0	1	1	0	0	0	1	1	0
7	1	0	0	0	1	0	1	1	1
8	1	0	0	1	0	1	0	0	0
9	1	0	1	0	0	1	0	0	1
10	1	1	0	0	0	1	0	1	0

It is also desirable to avoid parallel connection of open output circuits when releasing relays because this increases the release time—but it is rather difficult to do so. In many cases circuit simplifications can be obtained in not doing so.

Using don't-care conditions, the following group of four switching formulas can be derived from Table 5-1. In three of these formulas, a possible choice is introduced.

$$1 = wz' + \begin{matrix} vz + xz \\ w'\,y'\,z \end{matrix} \qquad 3 = yz' + \begin{matrix} vz \\ vw'\,x' \end{matrix}$$

$$2 = xz' + \begin{matrix} vz + yz \\ w'\,x'\,z \end{matrix} \qquad 4 = v'\,z \tag{5-26}$$

The numbers in Eqs. (5-26) indicate the output terminal to which the corresponding circuit must lead, beginning with a common terminal 5. When using Karnaugh maps, the variations of Eqs. (5-26) given above follow easily. The problem is which variations to choose. The simplest form of an individual switching formula among a group of formulas of a multiterminal circuit is not always the most suitable form for use in the combined circuit. Matrix (5-27) can be derived from Eqs. (5-26), thereby introducing a number of paths not specified between the output terminals 1, 2, 3, and 4. These paths can be allowed.

$$\begin{bmatrix} 1 & 0 & 0 & 0 & 0 & w & w'y' \\ & 1 & 0 & 0 & 0 & x & w'x' \\ & & 1 & 0 & 0 & y & v(y') \\ & & & 1 & 0 & 0 & v' \\ & & & & 1 & z' & z \\ & & & & & 1 & 0 \\ & & & & & & 1 \end{bmatrix} \tag{5-27}$$

Matrix (5-27) is of a very suitable form for perfecting by changing entry c_{37} in vy'. This change can easily be checked by means of a Karnaugh map according to the original specification. It is noted that in this code translator, ground is connected to terminal 5.

Matrix (5-27) shows that transfer contact z connects ground-terminal 5 to the contacts shown in columns 6 and 7, respectively. These contacts, in turn, form the various ways to the output terminals 1 to 4. Output terminal 1 can be connected to columns 6 and 7, respectively, via transfer contact w, output terminal 2 via transfer contact x, and output terminal 3 via transfer contact y. In this type of contact circuit, sneak paths between output terminals are basically avoided. The contact circuit representing matrix (5-27) is shown in Fig. 5-22. It contains 12 contacts and only 19 springs.

The circuit shown in Fig. 5-22 is easily found even with slight experience in matrix manipulation, but one has to know the fundamental type of combinational circuit suitable for multioutput purposes.

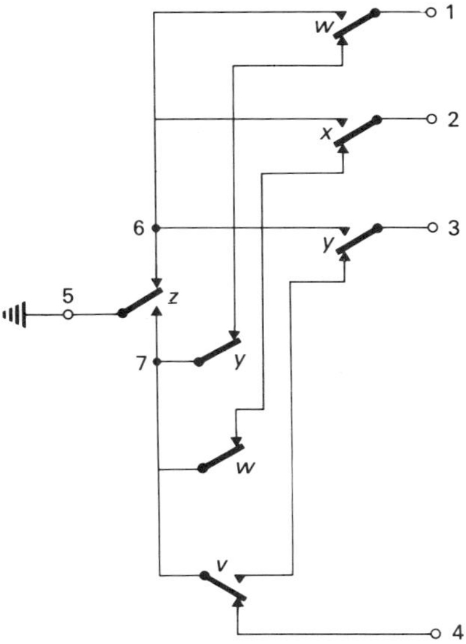

Fig. 5-22

In this example, one of the variations of Eqs. (5-26) leads to a slightly better result with respect to the number of springs. The following group of formulas leads to this result:

$$\begin{aligned} 1 &= wz' + w'y'z & \qquad 3 &= yz' + vy'z \\ 2 &= xz' + x'z(v+y) & \qquad 4 &= v'z \end{aligned} \tag{5-28}$$

From these formulas, the following near-primitive connection matrix (5-29) follows without difficulties.

$$\begin{bmatrix} 1 & 0 & 0 & 0 & 0 & w & w'y' & 0 \\ & 1 & 0 & 0 & 0 & x & 0 & x' \\ & & 1 & 0 & 0 & y & 0 & y' \\ & & & 1 & 0 & 0 & v' & 0 \\ & & & & 1 & z' & z & 0 \\ & & & & & 1 & 0 & 0 \\ & & & & & & 1 & v+y \\ & & & & & & & 1 \end{bmatrix} \tag{5-29}$$

Inspection of matrix (5-29) shows that all contacts can be grouped into transfer contacts. The diagram of Fig. 5-23 shows this result in the form of a circuit. The result is obtained at the cost of a path between output terminals 2 and 3 when relays X and Y are released, but this is not objectionable. The number of springs involved in this circuit (Fig. 5-23) is 18.

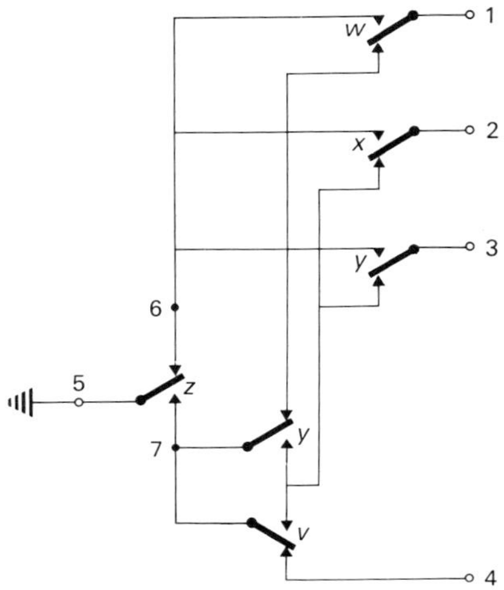

Fig. 5-23

Which of the circuits (Fig. 5-22 or Fig. 5-23) is preferable is hard to say without a further knowledge of possible other functions of relays V to Z.

In multiterminal circuits, sneak paths disturbing the desired output requirements can easily be introduced. The previous example can be used to illustrate this point. Starting from matrix ${}^5C$ [Eq. (5-30)] matrix ${}^5C_2$ [Eq. (5-31)] is readily obtained.

$$
{}^5C = \begin{bmatrix} 1 & 0 & 0 & 0 & wz' + w'y'z \\ & 1 & 0 & 0 & xz' + w'x'z \\ & & 1 & 0 & yz' + vz \\ & & & 1 & v'z \\ & & & & 1 \end{bmatrix} \tag{5-30}
$$

$$
\approx \begin{bmatrix} 1 & 0 & 0 & 0 & 0 & w & w'y' \\ & 1 & 0 & 0 & 0 & x & w'x' \\ & & 1 & 0 & 0 & y & v \\ & & & 1 & 0 & 0 & v' \\ & & & & 1 & z' & z \\ & & & & & 1 & 0 \\ & & & & & & 1 \end{bmatrix} = {}^5C_2 \tag{5-31}
$$

Matrix ${}^5C_2$ is a connection matrix, not an output matrix. This means that there are no entries in this matrix indicating sneak paths between output terminals. However, with some experience in manipulating matrices, one will be careful about entries c_{36} and c_{37}, which form a path between nodes 6 and 7 possibly leading to a sneak path between output terminals.

Checking a matrix for sneak paths between output terminals can be handled in two different ways. The first contains two steps: elimination of all internal nodes from the connection matrix; and multiplication of this reduced connection matrix with itself until $C^n = C^{n+1}$. The output matrix is then obtained that shows in its entries the complete switching formulas of all paths between all pairs of output terminals.

In a connection matrix, each entry shows the switching formula of the direct path between two nodes indicated by the corresponding indices. A connection matrix C^2 originating from the multiplication of connection matrix C with itself has entries showing the switching formulas of the direct path between two nodes indicated by their entries and all indirect paths leading via one other node, and so on. Finally, matrix $C^{p-1} = F$ has entries taking all possible paths from one node to another into account because in a network with p nodes, a connection from one node to another cannot pass more than $p - 2$ nodes. In actual cases, output matrix F can already result with a lower power than $p - 2$, but that depends on the special properties of the network.

Notwithstanding the fact that any off-main-diagonal entry of matrix C^{p-2} can be determined by means of a systematic and not-too-complicated formula, it is in the opinion of the author more practical to check the primitive or near-primitive connection matrix resulting from a node-insertion process for sneak paths by means of a node-elimination process, thereby considering the multiterminal circuit as a two-terminal circuit between the pair of terminals to be checked.

In not too many steps, matrix ${}^5C_2$ [Eq. (5-31)] can be reduced to matrix (5-32), checking path 2-5 of matrix (5-30):

$$\begin{bmatrix} 1 & xz' + w'x'z + vw'x'yz' + vxyz \\ & 1 \end{bmatrix} \tag{5-32}$$

From this matrix, it follows that there are two possible sneak paths between earth (terminal 5) and terminal 2, namely, $vw'x'yz'$ and $vxyz$. These sneak paths can be found in the circuit of Fig. 5-24, which corresponds to matrix (5-31). Sneak path $vxyz$ is unimportant because it represents one of the don't-care conditions. The other, $vw'x'yz'$, is fatal because it does not belong to the original set of requirements. It is the result of a not-too-careful choice of entry c_{37} in matrix (5-31), as could be expected from study of the preceding correctly treated examples.

With regard to engineering, it should be noted in this example that any one of the five input relays V to Z is in fact superfluous and that a rather

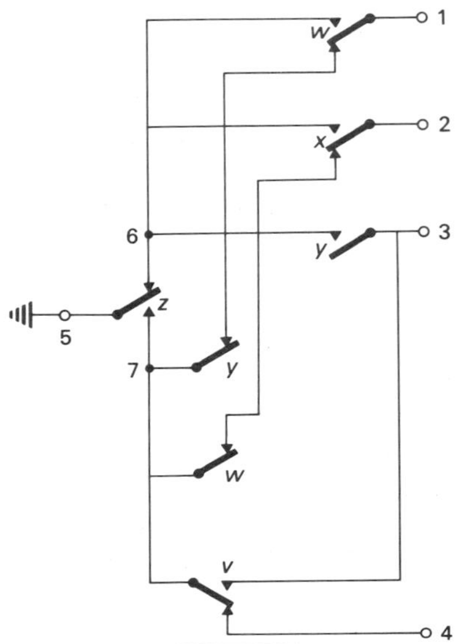

Fig. 5-24

good correspondence between the two codes is obtained when omitting relay V. This leads to the following group of switching formulas:

$$\begin{aligned} 1 &= wz' + w'y'z \qquad & 3 &= yz' + w'x'y'z \\ 2 &= xz' + w'x'z \qquad & 4 &= wz + (w')xz + yz \end{aligned} \tag{5-33a}$$

The five-node matrix representing Eqs. (5-33a) can be expanded into a ten-node circuit. The corresponding near-primitive switching matrix is

$$\begin{bmatrix} 1 & 0 & 0 & 0 & 0 & w & 0 & 0 & w' & 0 \\ & 1 & 0 & 0 & 0 & x & 0 & x' & 0 & 0 \\ & & 1 & 0 & 0 & y & 0 & y' & 0 & 0 \\ & & & 1 & 0 & 0 & w+y & 0 & 0 & x \\ & & & & 1 & z' & z & 0 & 0 & 0 \\ & & & & & 1 & 0 & 0 & 0 & 0 \\ & & & & & & 1 & 0 & y' & w' \\ & & & & & & & 1 & 0 & x' \\ & & & & & & & & 1 & 0 \\ & & & & & & & & & 1 \end{bmatrix} \tag{5-33b}$$

It is not difficult to detect that the contact circuit representing matrix (5-33) consists of transfer contacts only. In this type of circuit, sneak paths cannot exist. It should be noted that when relays X and Y are released, a path between output terminals 2 and 3 will be present but will not affect the logical operation

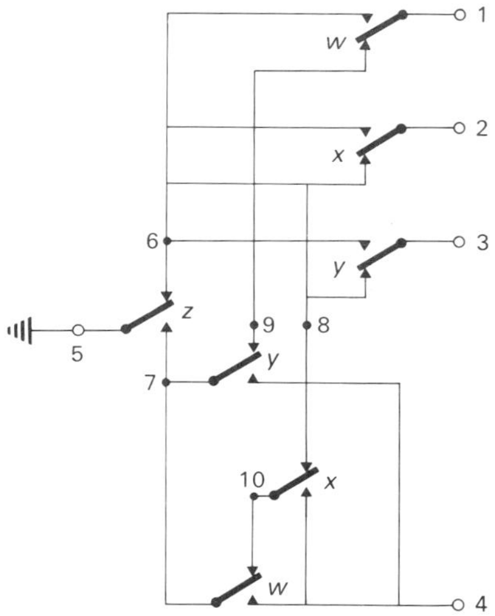

Fig. 5-25

of the circuit. The resulting circuit is shown in Fig. 5-25. It contains 14 contacts and 21 springs in a good distribution over the various relays.

From a practical point of view, this result is better than that obtained in the circuit of Fig. 5-23, because the cost of a relay is always higher than that of three springs.

PROBLEMS

1. Simplify by means of matrix methods the circuit in Fig. 5-26.

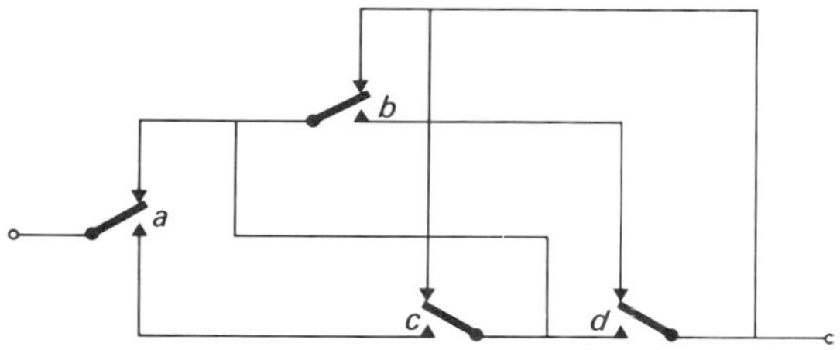

Fig. 5-26

The final circuit contains four springs.

2. Simplify by means of matrix methods the circuit in Fig. 5-27.

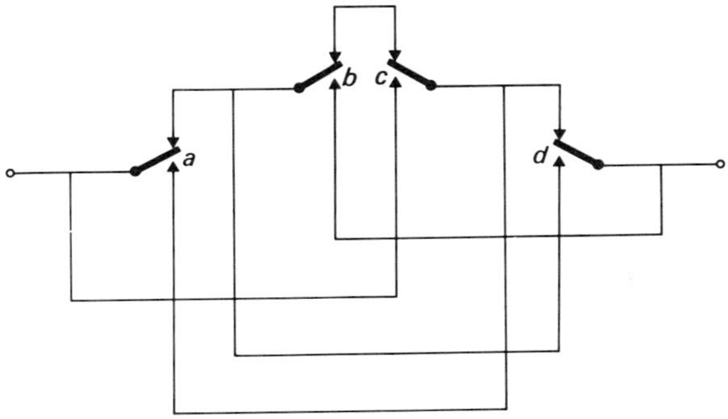

Fig. 5-27

The diagram given can be simplified to a circuit with eight springs.

3. Design a contact circuit for the following switching function:

$$S = ad + abc + b'cd + a'bc'd'$$

This switching function can be implemented with a contact circuit with 12 springs.

4. Design a contact circuit for the following switching function:

$$S = ace + ade' + cd'e + bc'de'$$

Thirteen contact springs are sufficient to implement this switching function.

5. Design a contact circuit implementing the following two switching functions:

$$S_1 = a'c' + ac + b'c$$

and

$$S_2 = ab$$

The combined circuit can be implemented with a contact circuit with nine springs.

6. Design a contact circuit implementing the following two switching functions:

$$S_1 = a' + c$$

and

$$S_2 = ad' + bd + cd'$$

The final circuit contains 10 springs.

7. Design a contact circuit implementing the following three switching functions:

$$S_1 = ad' + b'd' + cd'$$
$$S_2 = be + ce$$
$$S_3 = ad + b'd + be' + cd + ce'$$

The combined final circuit will contain 13 springs.

8. Design a contact circuit implementing the following three switching functions:

$$S_1 = a' + b'cd'$$
$$S_2 = b'cd$$
$$S_3 = a'bcd' + abc'$$

The final circuit will contain 12 springs.

SELECTIVE BIBLIOGRAPHY

1. Campeau, J. O.: The Synthesis and Analysis of Digital Systems by Boolean Matrices, *IRE Trans.*, vol. EC-6, pp. 231–241, 1957.
2. Lunts, A. G.: The Application of Boolean Matrix Algebra to the Analysis and Synthesis of Relay and Contact Networks, *Dokl. Akad. Nauk SSSR*, vol. 70, pp. 421–423, 1950.
3. Hohn, F. E.: A Matrix Method for the Design of Relay Circuits, *IRE Trans.*, vol. CT2, pp. 154–161, 1955.
4. Hohn, F. E., and L. R. Schissler: Boolean Matrices and the Design of Combinational Relay Switching Circuits, *Bell System Tech. J.*, vol. 34, pp. 177–202, 1955.
5. Semon, W. L.: "The Application of Matrix Methods in the Theory of Switching," Ph.D. thesis, Harvard University, April, 1954.

6
Switching Elements

6-1 CONTACTS AND RELAYS

The simplification of switching functions has been illustrated by means of a number of examples taken from the field of contact circuits. The simplification methods of switching functions, however, are entirely general and independent of any specific kind of circuit, but they can best be explained by means of contact or equivalent types of circuits. It is therefore necessary to say something about this commonly used switching element.

Contacts in switching circuits are mostly used as electromechanically operated contacts on relays. There are many different types of relays, and there exists an extensive literature on this subject. A push-button-operated electromechanical switch used to close the power circuit of a motor is in fact a relay. Most existing relays are applied in automatic-telephone exchanges. These relays, however, are of a completely different structure, but they are basically the same. A brief description of the design and operation of a relay will therefore be sufficient to understand the following chapters of this book.

A sketch of a commonly used type of relay is given in Fig. 6-1. This relay is extensively used in telephone exchanges and business machines. There are so many different types of relays in this field alone that a general survey would fall outside the scope of this textbook; accordingly the example given in Fig. 6-1 will be adequate for our purposes.

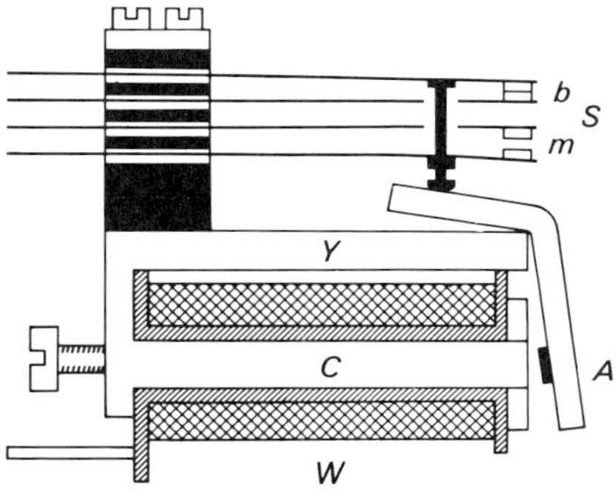

Fig. 6-1

The main components of this relay are the core C and the yoke Y in which a magnetic field can be induced by means of the winding W, which is made of insulated copper wire. The armature A hinges on the upper edge of the yoke. A set of contact springs S is mounted on top of the yoke, so that the moving springs of the spring set force the armature to stay in the rest position shown in Fig. 6-1 when no current flows through the winding. The moving springs are provided with thin insulating pins or studs, which eventually pass through a hole in the other contact springs. The pins of the contact combinations of the spring set rest upon each other and ultimately on the upper-left portion of the armature.

When a current of sufficient strength flows through the winding, the magnetized core attracts the armature and brings it in a second position. The moving springs are thereby mechanically operated, so that contacts which were closed in the rest position are opened and contacts which were open in the rest position are closed. The first type are referred to as break contacts and the second as make contacts. In Fig. 6-1 a make contact (m) and a break contact (b) are shown.

When the current through the winding is interrupted, the magnetic field disappears and the armature is released. The moving springs of the various contacts force the armature back in the rest position, as shown in Fig. 6-1.

If Fig. 6-1 were to be used in circuit diagrams as the symbol for a relay and its contacts, the drawings which would be obtained in this way would be far too complicated. Relays will be indicated by means of schematic symbols as indicated in Figs. 6-2 and 6-3. Figure 6-2 shows the winding of a relay in the form of a rectangle and a make contact. The circuit wires are always connected to the long sides of this rectangle. The symbol of a relay with a break contact is given in Fig. 6-3.

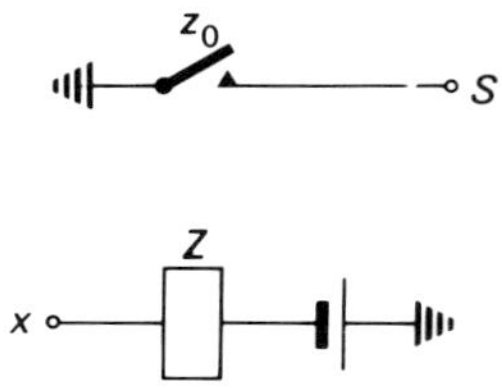

Fig. 6-2

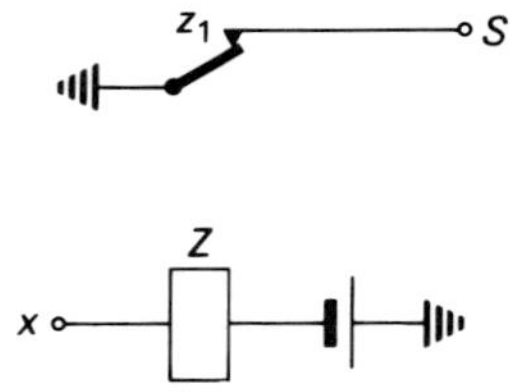

Fig. 6-3

The windings of a relay will always be indicated by means of capitals and the corresponding contacts, by means of small letters; the winding of relay Z will always be indicated in the diagrams with Z, the make contacts of this relay by z_0, and the break contacts by z_1.

There are two methods of drawing contact circuits. In the first method, all contacts of a relay are drawn on one line on both sides of the relay winding. This leads, especially in complicated circuits, to diagrams with many long and parallel lines and a large number of line crossings, which make a diagram unreadable. In the second method, the so-called detached-contact method, the contacts are spread all over the diagram so that the various circuits of the relays are drawn with a minimum number of line crossings. This method is particularly useful for complicated circuits if the various relay circuits are drawn in the diagram from left to right in their sequential order of operation. The author has seen examples of very complicated telephone-register circuits which could be read practically at a glance.

If a make contact and a break contact of the same relay are directly connected to each other in a circuit, an example of which is shown in Fig. 6-4*a*, the four springs of the separate make and break contacts can be reduced to three within one spring set, as is shown in Fig. 6-5*a* and symbolically indicated in Fig. 6-4*b*. This type of contact is called a changeover or transfer contact.

The idealization of contact circuits implied in the previous chapters appears quite clearly on looking more closely at the transfer contact and its

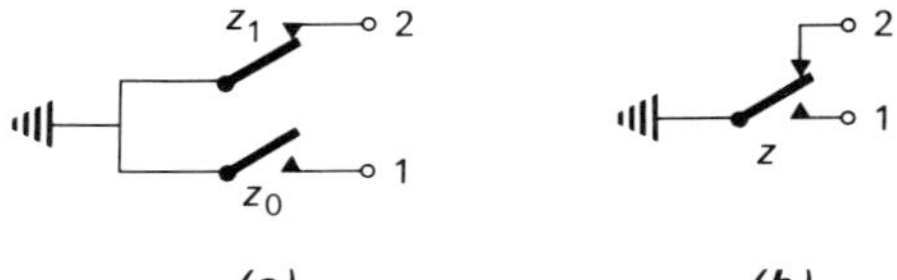

(a) (b)

Fig. 6-4

features. Ideally, the earth in Fig. 6-4 is to be transferred from one terminal to the other without interruption and without contact between the two output terminals. It is hardly possible in practice to have transfer contacts perform in this way. In a number of cases, it is even undesirable to have a through connection even for a short moment between the output terminals of a circuit of the type shown in Fig. 6-4. In other cases, it may be necessary to provide such a through connection.

Specially constructed transfer contacts, called make-before-break contacts, perform this type of operation. These types of contacts can be designed

(a) (b)

Fig. 6-5

by means of separate make and break contacts and special three-spring transfer contacts as shown in Fig. 6-5*b*.

It should be pointed out in this connection that Theorems 12*a* and *b* given in Chap. 2 apply to ideal contact circuits. If in a circuit as shown in Fig. 6-4*b* an uninterrupted transfer of the earth between the two output terminals is required, then Theorem 12*b* is no longer valid. It is then not possible to convert the left-hand side of that theorem into the simpler right-hand side.

In the case of Theorem 12*a*, a make-before-break feature in the operation of contacts x and x' would result in a closed circuit during the transition time of these contacts for the right-hand side of the formula. If in the circuit shown as illustration of the right-hand side of Theorem 12*a* no through connection is desired during the transition of the x contacts, then it is not possible to convert the left-hand side of that rule into the simpler right-hand side.

Theorems 12*a* and 12*b* show that there is a difference in static and dynamic behavior with respect to the types of circuit represented by the right-hand portion of these formulas.

The formulas given so far describe in principle the static behavior of a contact circuit. During the operating and release time of the various relays of the circuit, there will be a transition period in which the circuit tries to adapt itself to the new set of external control signals. The final condition of the circuit will generally be reached through a number of race conditions, because the operating and release times of the various relays of a circuit can vary widely.

For normal types of telephone relays with not-too-heavy spring sets, the operating and release time is about 10 msec. Special high-speed relays with only one transfer contact have been developed which have operating times of about 1 msec.

The release times of relays can be controlled to a large extent by means of the air gap between the armature and the core in the operating condition, and also by means of a copper sleeve surrounding the core. Release times up to 500 msec can be obtained with not-too-heavy spring sets; however, a tolerance of about 50 percent in the release time should not be considered exceptional.

For quite a number of applications, the simple definition of operating and release time of a relay will not be adequate. In a normally adjusted transfer contact there will be a short time lag between the opening of the break contact and the closing of the make contact. A more precise definition of the operation of a relay should therefore take the following time intervals into account:

1. The operating break time, i.e., the interval between the closing of the circuit of the operating winding and the opening of a break contact on that relay. The symbol for this interval of time will be t_{ob}.
2. The operating make time, i.e., the interval between the closing of the circuit of the operating winding and the closing of a make contact on that relay. The symbol for this interval will be t_{om}.
3. The transfer time t_{ot}, which is the interval between the opening of a break contact and the closing of a make contact of a relay.

The relation between these various time intervals can be written as follows:

$$t_{ot} = t_{om} - t_{ob} \tag{6-1}$$

An analogous group of time intervals can be defined for the release of a relay:

4. The release break time t_{rb}, that is, the interval between the opening of the energizing winding of the relay and the moment that the closed make contact is opened.
5. The release time t_{rm}, that is, the interval between the opening of the energizing winding and the closing of a break contact of the relay.
6. The release transfer time t_{rt}, that is, the interval between the opening of a make contact and the closing of a break contact during the release of a relay.

The relation between this second group of time intervals is as follows:

$$t_{rt} = t_{rm} - t_{rb} \tag{6-2}$$

The transfer times t_{ot} and t_{rt} are of importance because they account for a difference in the static and dynamic behavior of a circuit. Theorem 12 (Chap. 2) is no longer valid if these transfer times become too large; and they indeed become too large for slow release relays because the copper sleeve on the core not only causes a long release time but also causes the armature to move slowly, so that the release transfer time t_{rt} in particular can become rather long.

Normally when operating or releasing a relay, the contacts do not close at once. Sometimes they do not even open at once. Contacts that close normally chatter or bounce, which means that an intermittent contact interruption of a rather high frequency will occur. This intermittent contact makes it difficult to combine contact circuits and electronic circuits, so that auxiliary circuits (triggers) have to be provided to eliminate this effect.

The symbolic representation of contact circuits by means of switching functions can be applied with practically no changes to two other kinds of switching components of a completely different physical nature and structure. The first is the so-called cryotron, the second the pneumatic (or hydraulic) cylinder and valve.

THE CRYOTRON

The cryotron as devised by D. A. Buck is a switching element which operates on the principle of the disappearance of the superconductive state of a metal in a magnetic field. This component can be made by using as a core a small piece of metal wire, e.g., of tantalum—a superconductive metal below 4.4°K—covered by a winding (i.e., one layer) of another metal, e.g., niobium, which is superconductive below 8°K, as indicated by D. A. Buck.

A current through the winding induces a magnetic field that can neutralize the superconductive state of the core wire, depending on the current through this core and on its temperature. The winding itself retains its superconductive state, so that the magnetic field of the winding is maintained without power consumption and, consequently, without dissipation of heat.

The very low temperature required for the operation of the cryotron can be obtained by immersion in liquid helium, which has a boiling point of 4.2°K under normal atmospheric-pressure conditions.

The superconductive properties of tantalum and niobium as a function of the magnetic field are shown graphically in Fig. 6-6. The dotted line indicates the boiling point of liquid helium (4.2°K), that is, the operating temperature of the circuit. The graph shows that the superconductive state of tantalum will vanish in a magnetic field of a couple of hundred oersteds. The superconductive state of niobium, however, remains unchanged at that temperature and in a magnetic field of that strength.

A sketch of a cryotron is given in Fig. 6-7. The actual length of the core wire can be about 1 in. The function of the winding of a cryotron is

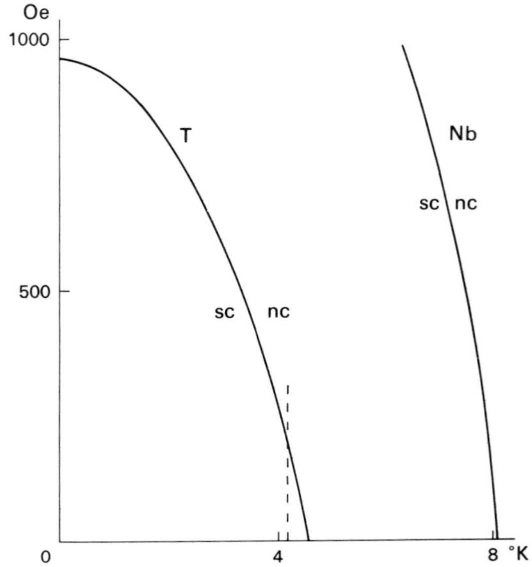

Fig. 6-6

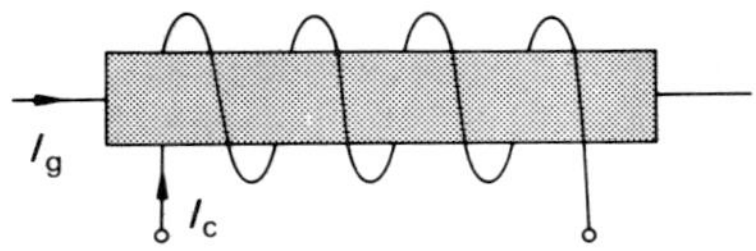

Fig. 6-7

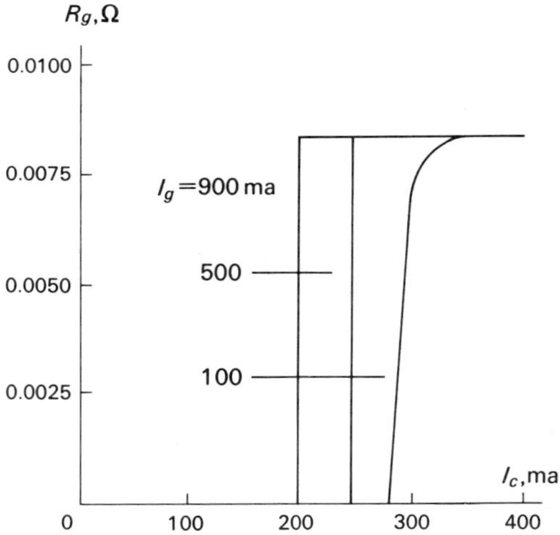

Fig. 6-8

analogous to the winding of a relay, whereas the core wire performs as a break contact. The switching effect of a cryotron results from a change in the conductivity of the core wire.

The change in resistance of the core wire of a specific cryotron as a function of the current passing through the winding is shown in Fig. 6-8. The resistance of the core wire is also dependent on the current passing through the core wire itself, because it generates a magnetic field that sustains the magnetic field of the winding. Large currents I_g in the core wire and small currents I_c through the winding cause a rather sudden increase in resistance of the core wire.

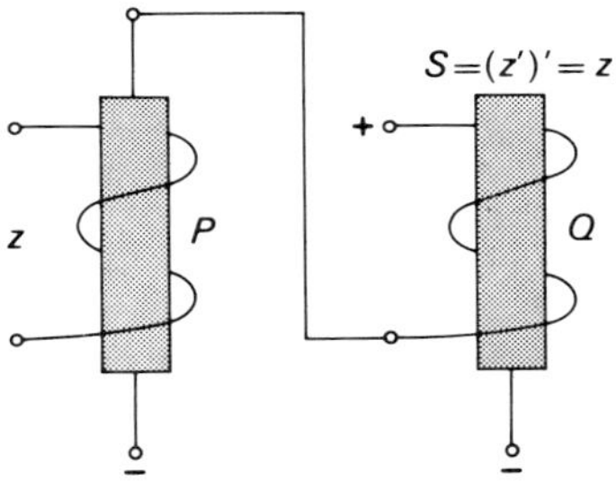

Fig. 6-9

The current gain of the cryotron is shown in Fig. 6-8. Power is supplied to a cryotron circuit by means of a current source; this source can be derived from a voltage source via a series resistor. The terminal voltage of a cryotron circuit with one or more superconductive elements is 0. There can be no current flow in a nonsuperconductive wire of such a circuit because a current source with an emf of 0 volts cannot send a current through a resistor of finite resistance.

The switching effect of a cryotron is such that a magnetizing current passing through the winding causes an interruption in a current passing through the core wire. Its operation is therefore the same as that of a relay with a break contact and can be described by means of the switching function $S = z'$. The switching function $S = z$ can be realized by means of two cryotrons connected in series, as shown in Fig. 6-9. The input signal z is applied to the winding of cryotron P. The core wire of P is connected in series with the winding of cryotron Q. The core wire of cryotron Q is connected to the output terminal of the circuit. When the winding of cryotron P is energized, the core wire of P becomes nonsuperconductive, so that the current through the winding of cryotron Q is interrupted. The core wire of cryotron Q then becomes superconductive, so that an output current can be produced. The switching function at the output terminal S of this circuit layout will be $S = (z')' = z$.

Except for the transient conditions during which the magnetic fields are being formed, a cryotron circuit does not consume power because the resistance of all paths carrying current is 0. The heat dissipation of these types of circuits is therefore very low, so that the helium does not evaporate too soon.

The operating speed of a cryotron is in the order of 1 msec, but it can be further reduced. Complicated circuits with cryotrons require very little space. The cryotron is still under development. It is mentioned here merely as an example of a rapidly evolving technique based on entirely different technological means, but one that can nevertheless be described by means of the same laws as those for relay techniques.

HYDRAULIC OR PNEUMATIC SWITCHING ELEMENTS

Automatic relay circuits cannot be used in coal mines because of explosion hazards. Pneumatic devices are preferable under such circumstances. For most applications, relay circuits can also be realized by pneumatic (or hydraulic) cylinders and valves which may even be of miniature construction. The cylinder contains a piston held in one position by a spring and brought in the operating position by means of pneumatic or hydraulic pressure. The piston controls a valve that it can open or close, so that the pneumatic or hydraulic stream is put into action or interrupted. If the valve is opened, the cylinder-and-valve combination is analogous to a relay with a make contact; if the valve is closed, the cylinder-and-valve combination is analogous to a relay with a break contact. These two types of devices are illustrated in Fig. 6-10*a* and *b*.

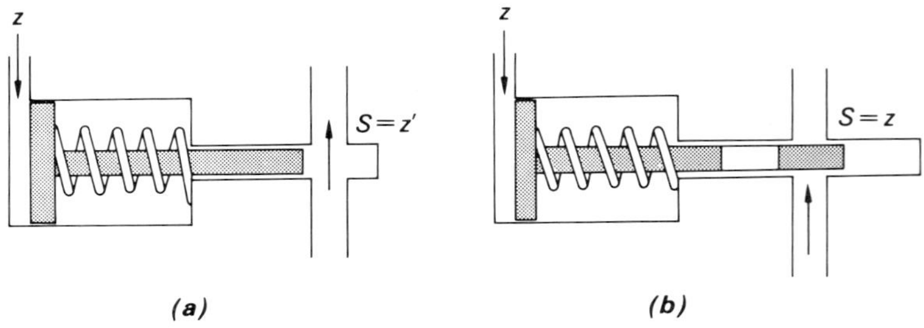

Fig. 6-10

6-2 *AND* AND *OR* GATES

In the preceding chapters, contact circuits were used to illustrate the validity of switching theorems and formulas; this discussion was followed by application of these rules to logical contact circuits. There are numerous types of relays and switches, and there is an even larger number of different types of contacts, contact materials, and contact combinations on relays; however, all contact circuits using any type of relay or contact are such that they can be treated and explained in the same way. The physical properties of contacts and relays are so analogous that it is self-evident that all contact circuits can be dealt with as one and the same category.

This is not the case with electronic switching components. There is such a tremendous difference between the various types of electronic switching components that it is necessary to deal with the principles of a number of more or less commonly used types, inasmuch as their features may affect the behavior of the component when incorporated in a digital-information-processing circuit.

In spite of the tremendous difference in physical structure and features among contact circuits and circuits with electronic components, the switching algebra as given in the preceding chapters can be applied to all types of electronic circuits.

As the following chapters will show, the switching algebra is a general means of describing circuit requirements independent of the type of component used. Once these requirements have been converted into switching functions and once the required simplifications have been performed, the switching functions can be translated, by means of operators that are specific for a given type of component, into a form from which one can derive the wanted circuit type.

In spite of the fact that semiconductor diodes and transistors have become the most widely used electronic switching components, some attention will be given to high-vacuum diodes, triodes, and pentodes and to tubes of the cold-cathode type, because their features differ rather considerably in some respects from the properties of the former class of elements. These differences could lead to difficulties should the circuit designer be unaware of them.

RESISTANCE GATES

First we will investigate to what extent a resistance circuit can perform the logical AND and OR operation on two variables y and z. Take a circuit such as the one shown in Fig. 6-11 with $R = R_y = R_z$ and signal voltages of -10 and $+10$ volts, respectively, corresponding to 0 and 1 (positive logic). The operation of the circuit on all possible combinations of the input variables is given in terms of voltages in the first three columns of Table 6-1. Depending on the voltage reference level of the switching circuit connected with output terminal S, the resistance-gate circuit of Fig. 6-11 can act in two different ways as is shown in terms of 0 and 1 in the last two columns of Table 6-1.

With a reference level of 0 volts, the following output function will be obtained:

$$S = y + z \qquad \text{OR gate} \tag{6-3a}$$

Table 6-1

y	z	S	*Reference level* 0 *volts*	+ 5 *volts*
−10	−10	$-3\frac{1}{3}$	0	0
−10	+10	$+3\frac{1}{3}$	1	0
+10	−10	$+3\frac{1}{3}$	1	0
+10	+10	+10	1	1

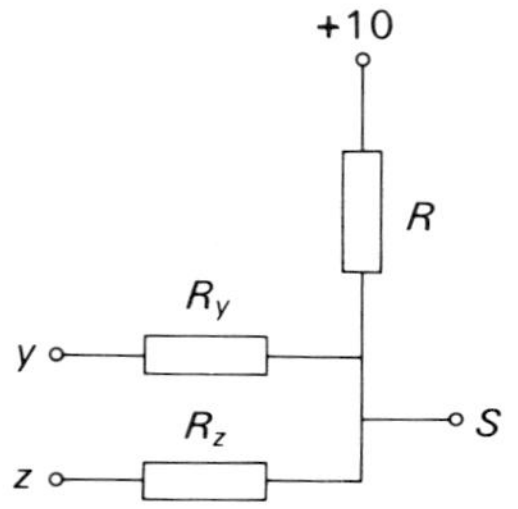

Fig. 6-11

With a reference level of +5 volts, the output function will be as follows:

$$S = y \cdot z \qquad \text{AND gate} \tag{6-3b}$$

Circuits of this type will be used in the Kirchhoff adder, as described in Chap. 17. They have a marginal operation, which is why these gates will be avoided as much as possible in digital switching systems. The marginal operation of resistance gates results in a tremendous loss in signal level in each gate, this depending on the combination of input signals. Gates of this type can be used successfully only with direct amplification of their output signals.

The advantage of this type of gate is that the same circuit generates the OR function as well as the AND function of its input signals, this depending on the reference level of the output amplifier.

DIODE GATES

The operation of the resistance gate can be improved by replacing resistors R_y and R_z (Fig. 6-11) by diodes, as is shown in the circuit of Fig. 6-12. Let the signal voltages be, for example, −10 volts (0) and +10 volts (1). Then it is not difficult to see that if $y = z = -10$ volts, $S = -10$ volts; and if $y = z = +10$ volts, $S = +10$ volts. In both cases, diodes D_y and D_z are conducting. If the signal voltages are, for example, $y = -10$ volts and $z = +10$ volts, then diode D_y is conducting and diode D_z is blocking. An output signal of

-10 volts (0) is then generated. For reasons of symmetry, this will also be the case if $y = +10$ volts and $z = -10$ volts. The operation of the gate of Fig. 6-12 is shown in Table 6-2. This table shows the AND operation $S = y \cdot z$.

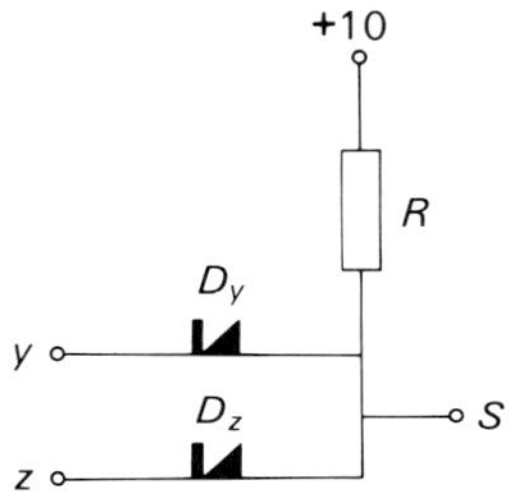

Fig. 6-12

Table 6-2

y	z	S
0	0	0
0	1	0
1	0	0
1	1	1

The OR operation of a diode-gate circuit can be obtained by reversing the polarity of the power supply and that of the diodes, as is shown in the circuit of Fig. 6-13. The symbolic operation of this gate circuit is given in Table 6-3 and by the equation $S = y + z$.

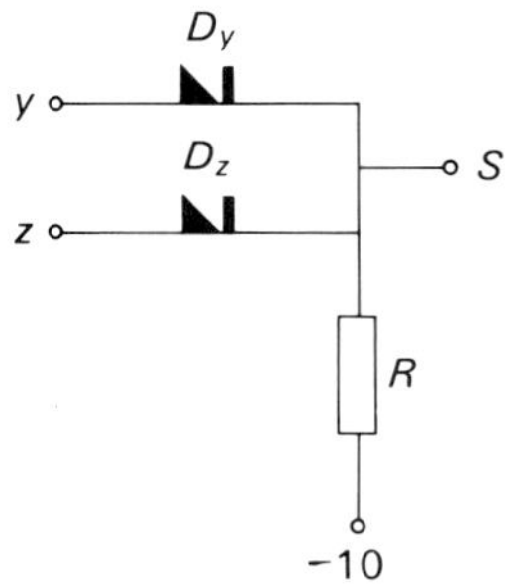

Fig. 6-13

Table 6-3

y	z	S
0	0	0
0	1	1
1	0	1
1	1	1

The operation of the OR and AND diode gates can be explained in somewhat plainer language than a truth table. One might say that the output voltage of an OR gate follows the most positive voltage of any one of the input terminals and that the output voltage of an AND gate follows the most negative of any one of the input terminals.

It should be noted that Eqs. (6-3a) and (6-3b) are analogous to the expressions describing the series and parallel connections of two make contacts. Since any of the variables in Eqs. (6-3a) and (6-3b) may be replaced by its negative, one might even generalize and state that these formulas are analogous to the series and parallel connections of two contacts. The OR gate as shown in Fig. 6-13 may be somewhat similar to a parallel connection of two contacts; however, the AND gate as shown in Fig. 6-12 has no resemblance whatsoever to a series connection of two contacts. It is appropriate to forget about parallel and series in electronic switching circuits and to think exclusively in terms of OR and AND.

Resistance gates and diode gates as shown are not limited to two inputs. The switching functions that result from a minimizing process are two-level or even multilevel functions that can be realized by means of diode gates with the same number of levels as the function itself. Diode gates with two levels (whether of the minterm or maxterm type), which can be formed by means of an AND gate followed by an OR gate, or vice versa, are very reliable. Three-level gates can be realized without too much difficulty. However, it is necessary to resort to amplification in order to restore the voltage level of the output signals of such groups of diode gates.

An example of a diode-gate circuit with two levels is shown in Fig. 6-14, and another one with two levels is given in Fig. 6-15.

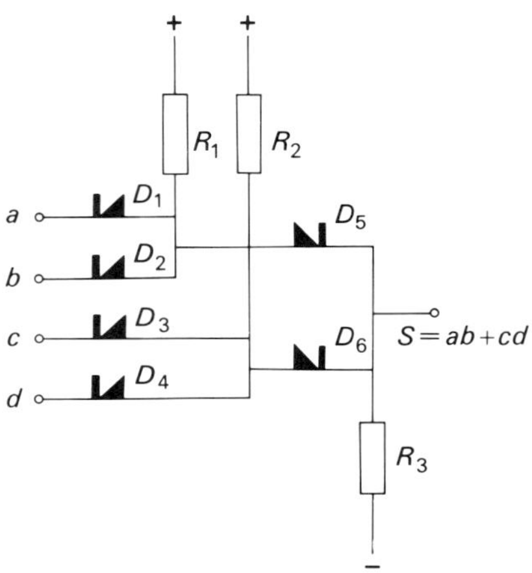

Fig. 6-14

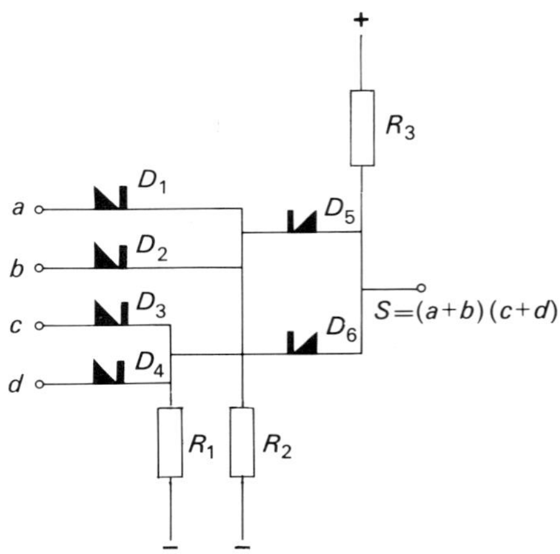

Fig. 6-15

The drawing of gate circuits such as those shown in Figs. 6-14 and 6-15 becomes too involved in the case of complex circuits. The same circuits can also be presented by means of simpler, commonly used symbols, as shown in Figs. 6-16 and 6-17.

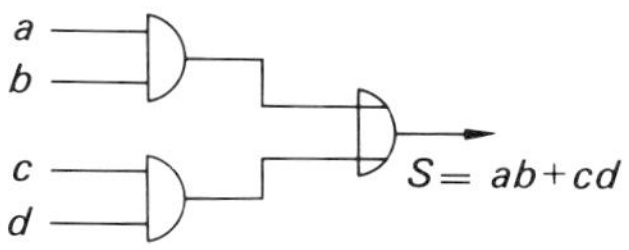

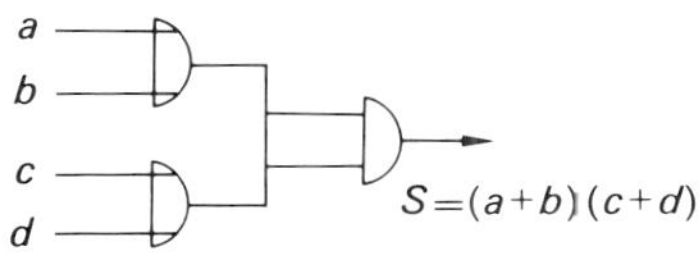

MULTIINPUT AND MULTIOUTPUT DIODE GATES

In practice, it will be found that there is indeed a good diode-gate circuit for most types of contact circuits, except for a number of multioutput circuits in which the outputs appear as taps on series connections of contacts. Contacts with near-zero resistance are extremely well suited for connection in series,

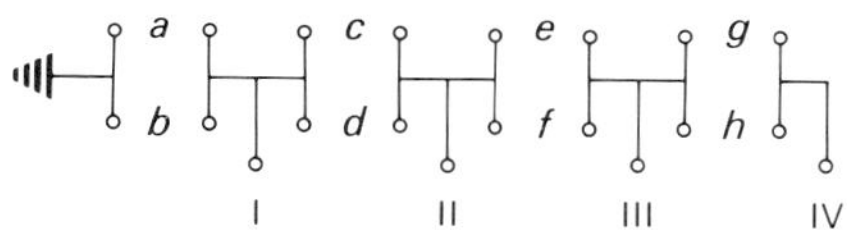

Fig. 6-18

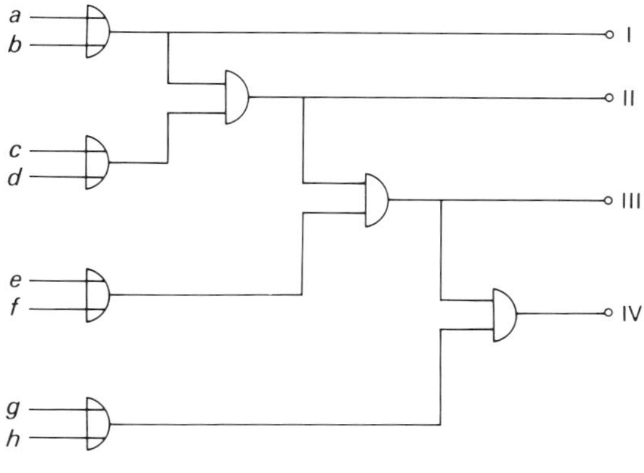

Fig. 6-19

especially when taps are needed on that series connection of contacts. The diode-gate equivalent of such types of circuit would generally be multilevel diode-gate circuits with an impractical number of levels. Figure 6-18 gives an example of a four-level tapped contact circuit, Fig. 6-19 being its diode-gate equivalent. The required number of taps equals the required number of diode-gate levels. The circuits of Figs. 6-18 and 6-19 correspond to the following expressions:

$$\text{I} = (a + b) \tag{6-4}$$

$$\text{II} = (a + b) \cdot (c + d) \tag{6-5}$$

$$\text{III} = (a + b) \cdot (c + d) \cdot (e + f) \tag{6-6}$$

$$\text{IV} = (a + b) \cdot (c + d) \cdot (e + f) \cdot (g + h) \tag{6-7}$$

From the point of view of dimensioning, the sequence of three gates of the same kind that is found in the circuit of Fig. 6-19 is not so bad as an alternating sequence of gates.

The two-level diode-gate circuits are not limited to the type of combinations shown in Figs. 6-14 and 6-15, in which a number of inputs control a single output. In principle, this is also the case with the tapped circuit of Fig. 6-19. The most general type of diode-gate circuit embodies a number of inputs which control a number of outputs, as is the case in the following set of switching functions:

$$S_1 = (a + b) \cdot (c + d) \tag{6-8}$$

$$S_2 = (c + d) \cdot (e + f) \tag{6-9}$$

$$S_3 = (a + b) \cdot (g + h) \tag{6-10}$$

$$S_4 = (e + f) \cdot (g + h) \tag{6-11}$$

The corresponding diode-gate circuit is shown in Fig. 6-20.

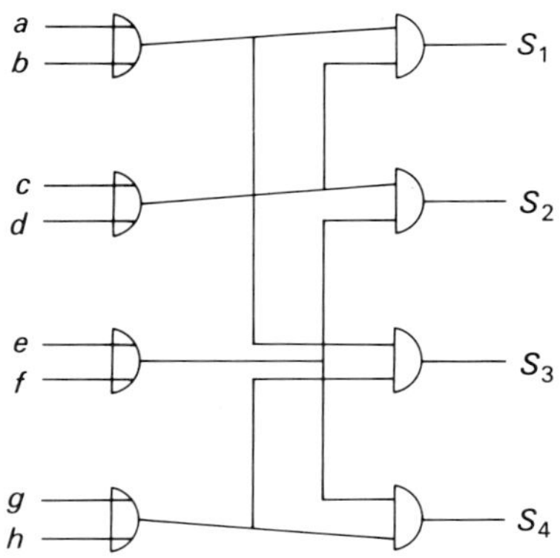

Fig. 6-20

The most fundamental forms of multiinput and multioutput diode gates are presented in Figs. 6-21 and 6-22. These figures show that a gate with an

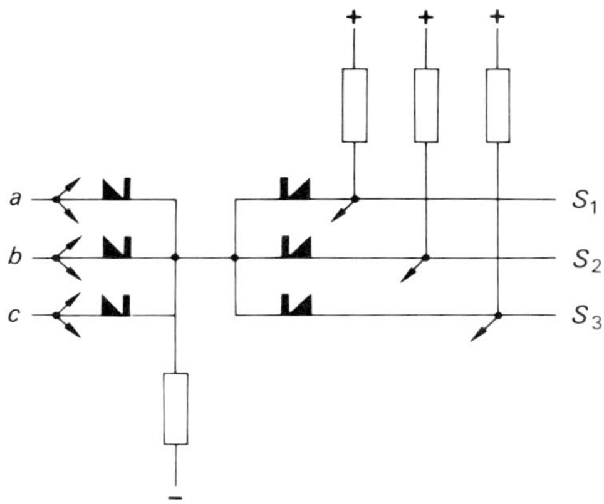

Fig. 6-21

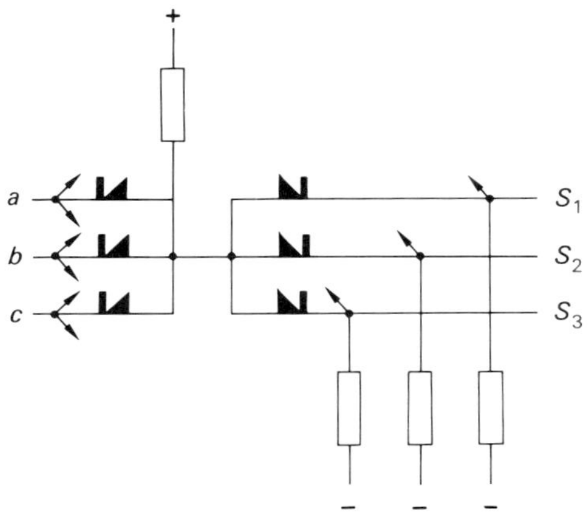

Fig. 6-22

arbitrary number of inputs can control any one of the inputs of an arbitrary number of gates in the next level. The problem involved with designing gate circuits of this kind is to recognize in the switching functions as given which parts of these expressions can be represented by means of a common part of the gate circuit. This is generally not possible if the group of switching functions for which such a gate circuit must be designed is presented in its simplified form. In such a case, it is necessary to rewrite the switching functions in the minterm (or maxterm) canonical form, so that common minterms (or maxterms) can be traced. Such terms can be represented by a common diode-gate circuit. The complete complex circuit will then be obtained by adding diode-gate circuits representing the simplest form of the remaining minterms (or maxterms), whereby the simplest form of the complete function has to be taken into account. It is very convenient to use Karnaugh maps, because this form of graphical representation of the canonical forms of the various switching functions makes it rather easy to recognize the common minterms (or maxterms). Another workable method for finding these common terms would be a systematic arrangement of the f coefficients that describe the canonical forms of the functions.

This may be illustrated by means of an example. A three-output diode-gate circuit is to be designed for the following switching functions given in their simplest two-level form:

$$\begin{aligned} F_1 &= wx'y + wz' + w'x'y'z + yz' \\ F_2 &= w'x + w'y'z + wx'yz \\ F_3 &= w'x'y' + w'xyz + w'z' + y'z' \end{aligned} \tag{6-12}$$

The canonical form of these switching functions can be written as follows:

$$\begin{aligned} F_1 &= m_1 + m_2 + m_6 + m_8 + m_{10} + m_{11} + m_{12} + m_{14} \\ F_2 &= m_1 + m_4 + m_5 + m_6 + m_7 + m_{11} \\ F_3 &= m_0 + m_1 + m_2 + m_4 + m_6 + m_8 + m_{11} + m_{12} \end{aligned} \tag{6-13}$$

The corresponding Karnaugh maps are given in Fig. 6-23a to c.

The common m terms are m_1, m_6, and m_{11}. The auxiliary function common to all three functions is as follows:

$$F_a = w'x'y'z + w'xyz' + wxyz' \tag{6-14}$$

The designer of multioutput gate circuits aims at representing the common function F_a by a separate set of diodes, so that the terms contained in F_a can be disregarded or considered as don't-care conditions when sorting out the individual Karnaugh maps.

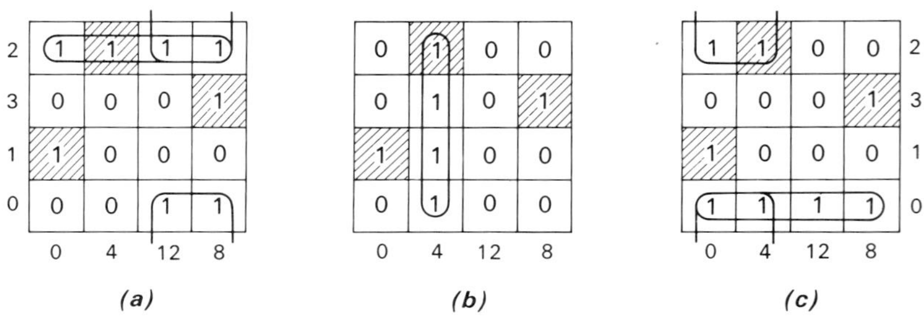

Fig. 6-23

Looking at the maps shown in Fig. 6-23a to c, one finds that the minterm m_6 will be part of the normal simplification process in each of the three maps; the minterm $w'xyz'$ is therefore no longer needed in the common function F_a. The necessary common part of all three functions (6-12) will then be

$$F_b = w'x'y'z + wxyz' \tag{6-15}$$

The extension of Eq. (6-15) with the expressions corresponding to the three Karnaugh maps of Fig. 6-23 leads to the following results:

$$\begin{aligned} F_1 &= F_b + wz' + yz' \\ F_2 &= F_b + w'x \\ F_3 &= F_b + w'z' + y'z' \end{aligned} \tag{6-16}$$

The corresponding circuit with 28 diodes is shown in Fig. 6-24. This circuit is no longer a true two-level circuit, nor is it a true three-level circuit, because two diode levels of the same kind are connected in series. If a three-level gate circuit is acceptable, the circuit of Fig. 6-24 can be simplified further, the result of which is shown in Fig. 6-25. This circuit requires two diodes less than the preceding one.

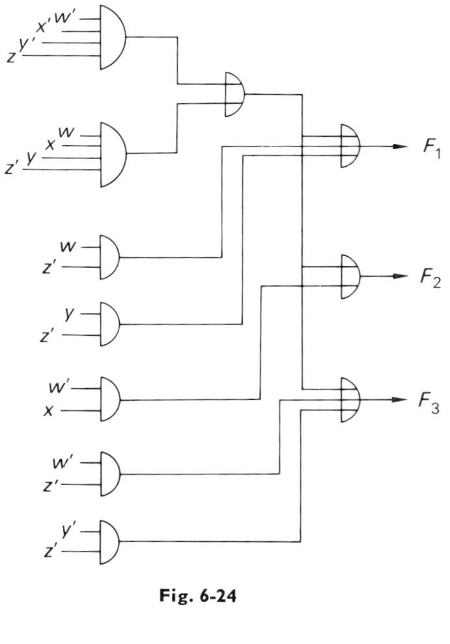

Fig. 6-24

6-3 NEGATION ELEMENTS

The various variables that are required for designing a circuit corresponding to a given switching function are not always directly available in electronic circuits. Some are available in the ordinary form, others in the negative or complemental form. For that reason, one must have a means for generating the negative or complemental form of a variable, or even of a whole switching function. Moreover, as was already explained, once switching signals have

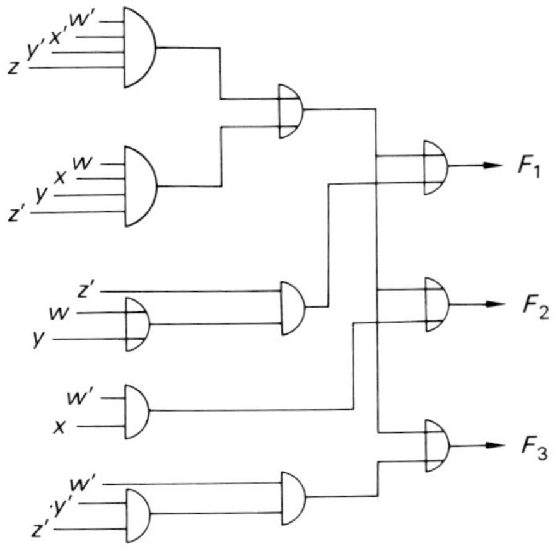

Fig. 6-25

passed through one or a number of gate levels, they must be amplified. Negation and amplification are simultaneously performed by means of the same fundamental type of active element, which will be referred to here as a triode.

The function of the triode need not be limited only to the negation and amplification. It is also possible to realize complete logical circuits exclusively by means of triodes as was the case in the beginning of the art of electronic switching, because at that time reliable diodes were not available, whereas reliable high-vacuum triodes were. The triode is in certain respects analogous to a relay with one break contact; therefore, just as it proved possible to realize all logical circuits by means of this type of relay exclusively, similarly one could conceivably design all logical circuits by means of triodes.

A triode is an electronic circuit element with three electrodes, which will be referred to here as collector, emitter, and base. The latter controls the state of the emitter-collector path in the triode. The characteristic of an ideal triode is shown in Fig. 6-26. This graph shows the emitter-collector current I_c as a function of the voltage V_b of the base with respect to the emitter. It is basically the same type of characteristic as that of the ideal diode, except that here on the abscissa, the anode voltage V_a is replaced by the base voltage V_b.

The emitter-collector path is blocked ($I_c = 0$) for all voltages V_b, which are negative with respect to the emitter (for example, *npn* transistor). This is shown in the left-hand portion of the graph of Fig. 6-26.

With other types of physical layout of the triode, it is possible to block the emitter-collector path by voltages V_b, which are positive with respect to the emitter (for example, *pnp* transistor). This is shown in the right-hand portion of Fig. 6-26.

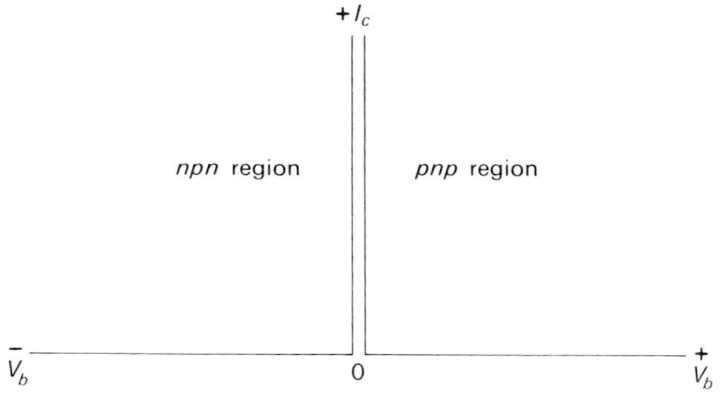

Fig. 6-26

The emitter-collector path of an ideal triode is conducting when the voltage difference between the base and emitter is 0. The lines in Fig. 6-26 show both possible ideal-triode characteristics. In this ideal case, the emitter-collector path of both types of triodes has 0 resistance in its conducting state.

The relation between base and emitter in a triode can be considered as equivalent to the relation between the two electrodes of a diode of the same physical structure. The blocking condition of the path between base and emitter coincides with the blocking condition in the emitter-collector path. It is therefore required, in this condition of the triode, that the voltage of the base be of opposite polarity with respect to the emitter as the collector.

The conducting conditions of both paths also coincide. In the blocking region of its operation, the ideal triode is a voltage-controlled element; whereas in its conducting range of operation, it is a current-controlled element. The amplification of the ideal triode defined as dI_c/dV_b is infinite in its conducting range.

By means of Fig. 6-27*a* and *b* it will be shown that both types of triodes perform the logical operation of the negation. The circuits of Fig. 6-27*a* and *b* correspond, respectively, to the left-hand and right-hand characteristics of Fig. 6-26. In these circuits, the triodes are shown as a more or less self-explaining symbol similar to the diode symbol already used.

In the circuit of Fig. 6-27*b*, the emitter-base path is blocked by a base voltage V_b, which is "high" or positive (1) with respect to the emitter, so that the collector voltage will equal that of the power supply (low or 0). The emitter-base path is conducting when earth potential (low or 0) is applied to the base. Earth potential (high or 1) is then fed to the output terminal of the circuit via the conducting emitter-collector path.

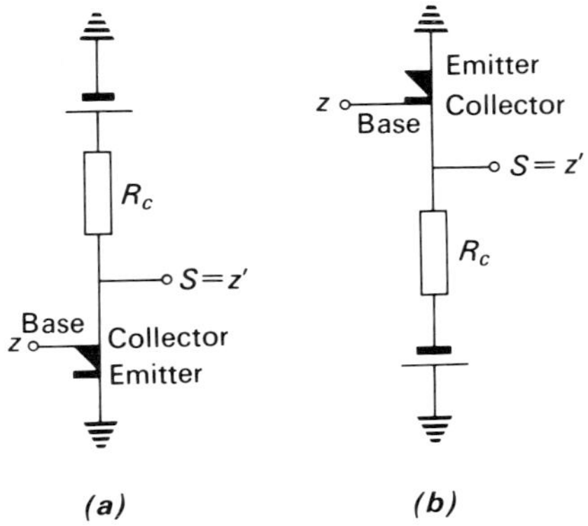

Fig. 6-27

The logical operation of this triode circuit, which does not show its amplification, can be described by the following formula:

$$S = z' \tag{6-17}$$

It should be noted that the voltage levels of the controlling signals and the output signals do not correspond but, as will be shown later, can be made so by means of an extra power supply and a voltage divider.

The operation of the triode circuit of Fig. 6-27*a* can be described in dual terms, leading to the fact that its logical operation can also be described by means of Eq. (6-17).

In the following sections, various electronic components will be described that perform the logical operation $S = z'$.

6-4 TUBES

Tubes used as switching components can be subdivided into two classes:

1. High-vacuum tubes
2. Cold-cathode tubes

HIGH-VACUUM TUBES

A high-vacuum tube consists at least of a cathode heated by an internal filament, so that it emits electrons by thermal agitation, mostly via a special surface material, and an anode normally surrounding the cathode, both elements being sealed in a glass or steel envelope in which a high vacuum has been created.

This type of tube is called diode (two active electrodes). If a positive potential is applied to the anode and a negative potential to the cathode, an electron flow is created from cathode to anode. The diode is conducting in the forward direction. Because electrons are emitted from the cathode with a certain velocity, a current will already flow in the diode at 0 anode-cathode potential. This starting current of the diode decreases practically to 0 for negative anode potentials higher than about −1 volt. The anode-cathode gap of the diode is then a rather good nonconductor. The usefulness of the diode for switching purposes, however, is greatly reduced by the rather-limited resistance between cathode and filament. The typical V_a-I_a characteristic of a diode is shown in Fig. 6-28. The symbol normally used for high-vacuum diodes is shown in Fig. 6-29.

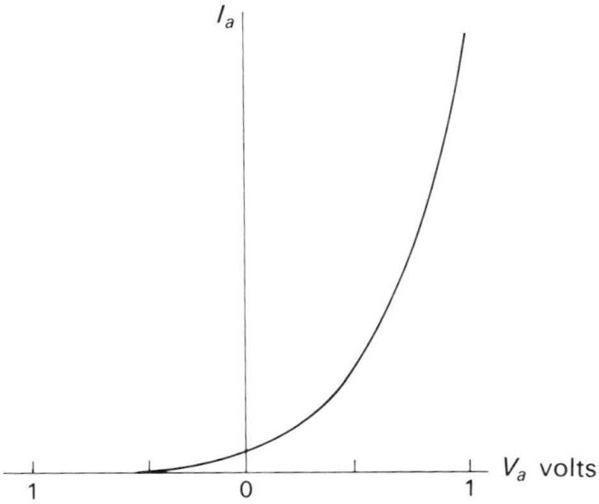

Fig. 6-28

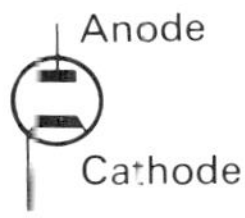

Fig. 6-29

The heat dissipation and the required space limit the use of high-vacuum diodes to those switching applications where an extremely high blocking resistance is required and the cathode-filament resistance is of no importance.

HIGH-VACUUM TRIODE

The electron flow from the cathode to the positive anode can be controlled by one or more grids. These grids increase the usefulness of the high-vacuum tube where switching applications are concerned. Before the advent of the transistor, the high-vacuum triode tube (one control grid) and the high-vacuum pentode tube (two control grids) were the only practical means available for electronic switching.

In a high-vacuum triode tube, a sufficiently negative control grid blocks the electron flow from cathode to positive anode. A typical group of I_a-I_g-triode characteristics ($V_a =$ constant) is shown in Fig. 6-30. The high-vacuum triode symbol is given in Fig. 6-31.

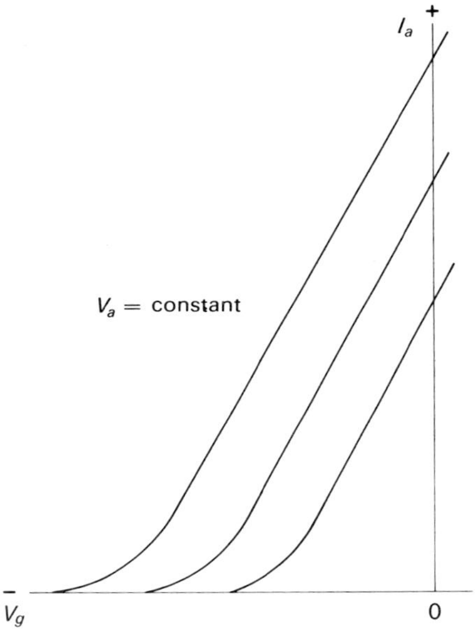

Fig. 6-30

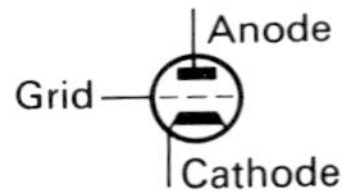

Fig. 6-31

The high-vacuum triode tube when used as a switching device normally operates with a low (negative) control-grid voltage V_g (symbolic value 0) or with a control-grid voltage which is about 0 volts (symbolic value 1). The anode current I_a of the high-vacuum triode is then 0 or a certain value—or in other words, the triode is then nonconducting or conducting. In these two conditions, the anode voltages V_a are, respectively, high (1) or low (0), so that function $S = z'$ represents the switching operation of the high-vacuum triode tube. The symbolic grid-control signal is represented by z and the output signal produced by the anode by S.

A basic remark must be introduced here about grid-controlled high-vacuum tubes. These tubes are essentially voltage-controlled switching elements that consume almost no control current. This may have its effect on the use of high-vacuum tubes in the design of practical switching circuits.

Since it is a well-known requirement in switching circuitry that a large number of these tube switching elements must sometimes be connected in successive switching orders, the anode of the high-vacuum triode tube is normally connected to a level-shifting voltage divider as shown in Fig. 6-32*a*. The voltage divider of Fig. 6-32*a* is dimensioned in such a way that the mid-levels of input and output voltages of the circuit tally.

The switching function $S = z$ can be obtained by means of a high-vacuum triode tube in a cathode-follower circuit as shown in Fig. 6-32*b*.

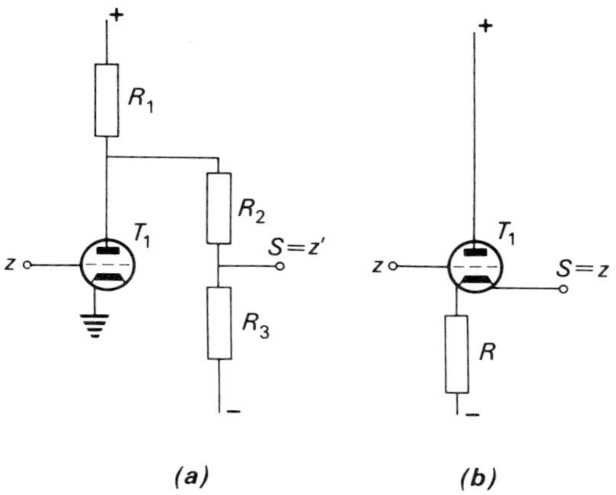

Fig. 6-32

This type of circuit is a very suitable arrangement for providing the power to operate all kinds of electromechanical relays, e.g., at the output terminal of a more or less complex automatic circuit. On the debit side, however, it should be noted that the actual voltage difference between the two output switching levels is in principle smaller than the difference between the two input switching levels. This means that the cathode-follower circuit cannot be used as a general-purpose building block in large automatic circuits.

The switching function $S = z$ can also be obtained by means of the double negation of this formula to $S = (z')'$. The elementary circuits of Fig. 6-32*a* connected in series are shown in Fig. 6-33 to illustrate the commonly used realization of the formula $S = z$.

All switching formulas can be realized by means of inverting circuits such as the high-vacuum-tube triode circuit and one type of the logical gate circuits.

One of the logical operations can be performed easily by means of a double triode, as is shown in Fig. 6-34. The operation of this circuit is illustrated in the truth table presented as Table 6-4. From this table it can be seen that

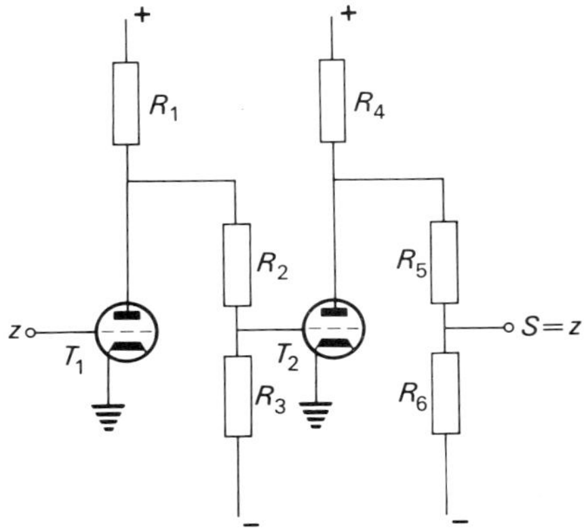

Fig. 6-33

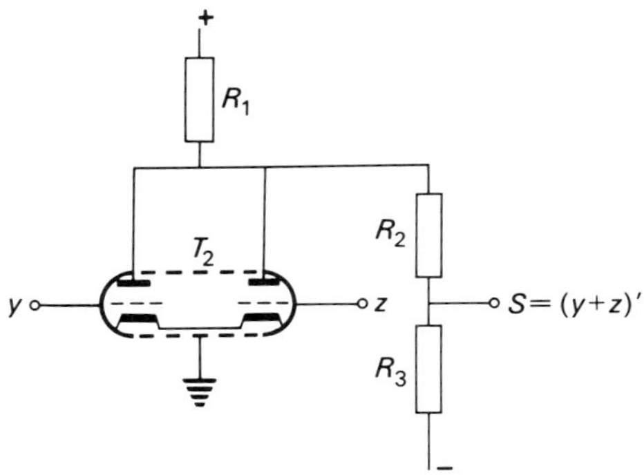

Fig. 6-34

Table 6-4

y	z	S
0	0	1
0	1	0
1	0	0
1	1	0

the operation of this circuit can be written as follows:

$$S = (y + z)' = y' \cdot z' \tag{6-18}$$

The function $S = y + z$ can be obtained by connecting a circuit of the type shown in Fig. 6-32a in series with the output terminal of the circuit shown in Fig. 6-34. The function $S = y \cdot z$ can be obtained by connecting a circuit similar to Fig. 6-32a to each of the output terminals of the circuit of Fig. 6-34. Even more complex switching functions can easily be converted into these types of switching layouts.

HIGH-VACUUM PENTODE

A high-vacuum pentode when used as a switching device can be controlled by means of the first and third grids. The second or screen grid is left out

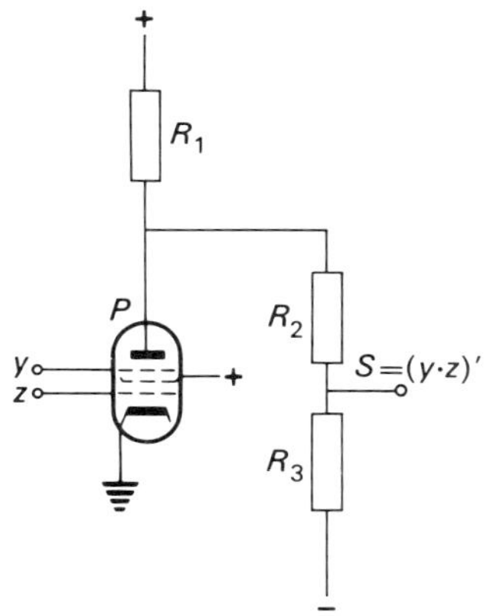

Fig. 6-35

of consideration in this connection. The basic pentode switching circuit is shown in Fig. 6-35. The operation of this circuit is defined by truth table 6-5.

Table 6-5

y	z	S
0	0	1
0	1	1
1	0	1
1	1	0

This table shows that the operation of this circuit can be described by the following expression:

$$S = (y \cdot z)' = y' + z' \tag{6-19}$$

With some groups of switching functions, it is sometimes easier to resort to pentodes than it is to use triodes.

Switching techniques based on the exclusive use of high-vacuum tubes were common practice in the initial stages of development of electronic switching because reliable semiconductor diodes were not available at that time. This type of technique is now obsolete, since it is now possible to find good diodes for logical circuits from many different sources and in many different types.

The high-vacuum tube has a number of important practical disadvantages with respect to its switching applications:

1. The filament of a tube dissipates heat (about 2 watts for a double triode for switching purposes)
2. The anode-cathode path in the conducting condition and the anode resistor also dissipate heat (about 2 watts for an ordinary computer triode)

As a result, the problem of ventilation of the switching equipment already arises for circuits with more than 10 to 20 tubes. There is also an additional disadvantage in that the resistance of the anode-cathode path is so high that a series connection of tubes in the same circuit, though it could be achieved with contacts, is impractical when using tubes.

COLD-CATHODE TUBES

The hot-cathode gas tubes will not be discussed here. There are some possible applications in switching (thyratrons), but these tubes have not gained general applicability in this field. They combine some of the disadvantages of the high-vacuum tubes and of the cold-cathode gas tubes which will be dealt with now.

The cold-cathode gas tube is a gaseous discharge tube with two or more electrodes. The cathode can consist of a small nickel plate covered with a thin layer of activated barium oxide that is partly reduced to barium. The anode is also made of nickel. The tube can be controlled by means of an auxiliary or starter anode located at a distance from the cathode, which is substantially shorter than the anode-cathode gap. The electrodes are sealed in a glass envelope, which is filled with an ideal gas, e.g., argon or neon under a pressure of about 15 mm.

The cold-cathode diode is often filled with neon. The I_a-V_a characteristic of such a diode is shown in Fig. 6-36. It consists of four parts. If the anode voltage is raised from about 65 volts to a potential of 70 volts a small current (the "dark" current) starts to flow in the anode-cathode gap. This is phase 1 of the characteristic. At about 70 volts, the ionizing value is reached and the gas suddenly becomes conducting. The tube is fired and runs through phase 2 of the characteristic, where its dynamic resistance is negative. For switching purposes, the cold-cathode diode will normally be operated by means of pulses of sufficient potential and length to ionize the gas.

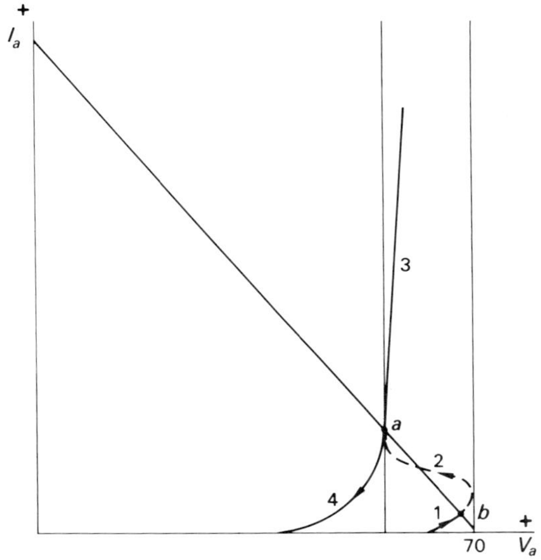

Fig. 6-36

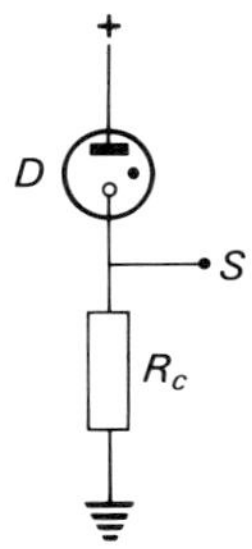

Fig. 6-37

Once the tube is fired, the voltage drop in the anode-cathode gap is reduced to about 60 volts and tends to be constant with increasing anode current, as shown in phase 3 of the characteristic curve. This implies that a current-limiting resistor must always be connected in series with these types of tubes in order to prevent a damaging overload of the cathode. This simple circuit is shown in Fig. 6-37. Phase 3 of the curve corresponds to the characteristics of a stabilizing tube, and in fact a cold-cathode diode is a tube of this kind. The output voltage S of the circuit shown in Fig. 6-37 is nearly constant once the tube has been fired.

The cold-cathode diode can be deionized by interruption of the anode voltage or by reduction of that voltage, the diode passing then through phase 4 of its characteristic curve of Fig. 6-36.

The cold-cathode diode has two stable operating conditions, condition a (tube conducting) and condition b (tube nonconducting). This feature makes the cold-cathode diode well suited for use as a memory element. As can be seen in Fig. 6-36, for a voltage source of 70 volts, the load line crosses the curve in points a and b.

The firing voltage of the anode-cathode path of a gaseous discharge tube can be increased to 140 volts and even more by the proper choice of gas, pressure, and anode-cathode distance. A starter anode can be introduced in the anode-cathode path, whereby a cold-cathode triode tube is obtained. The starter path will fire the tube, for example, at 70 volts and ionize the gas, so that the main anode-cathode path is also fired. A potential of about 75 volts is then sustained in the main gap. The firing of the starter gap corresponds to phases 1, 2, and 3 of the characteristic curve of Fig. 6-36. The cold-cathode triode can be deionized in accordance with phase 4 of the characteristic curve only if the voltage of the starter and main anode is decreased for a sufficiently long time and well below the minimum sustaining voltage. The deionizing time depends on the cathode material.

When a mixed direct-current and alternating-current anode voltage is used, the cold-cathode triode tube can be controlled by means of the starter anode only. The tube then fires if the potential of the starter anode is raised above a certain level, and deionization occurs only when that voltage drops below the minimum sustaining voltage of the starter gap, because the anode voltage drops periodically below the anode-sustaining voltage.

There is a wide range of special types of cold-cathode tubes. There even exist complete decimal counting tubes with a direct-reading feature. Such special tubes, however, fall outside the scope of this book.

The cold-cathode tubes have the advantage that their operating condition is automatically visible. They have no filament. Their life is long if the cathode is not overloaded. If the cathode is overloaded, their lifetime decreases exponentially. However, these attractive features are compensated for by a number of serious disadvantages, such as the unstable operating characteristics which are dependent on the light conditions under which the tube is used—to which one should add the fact that the tube cannot be deionized by the sole starter voltage if nothing but direct current is used.

The first disadvantage requires either that the tubes be mounted in specially lighted closed cabinets or that they be provided with special features with a view to maintaining a low degree of ionization in the tube. This can be achieved by means of a continuously operated ionizing anode or small particles of radioactive material. The latter method, however, results in variable firing times.

Cold-cathode triodes and diodes have been sometimes used in large numbers, but this type of tube has not gained the confidence of the designer of switching circuits.

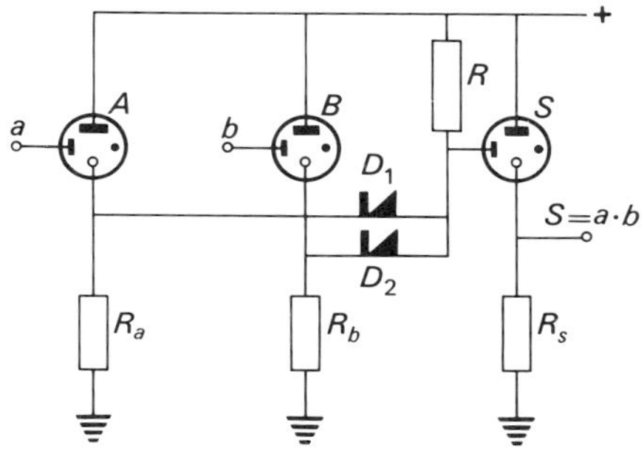

Fig. 6-38

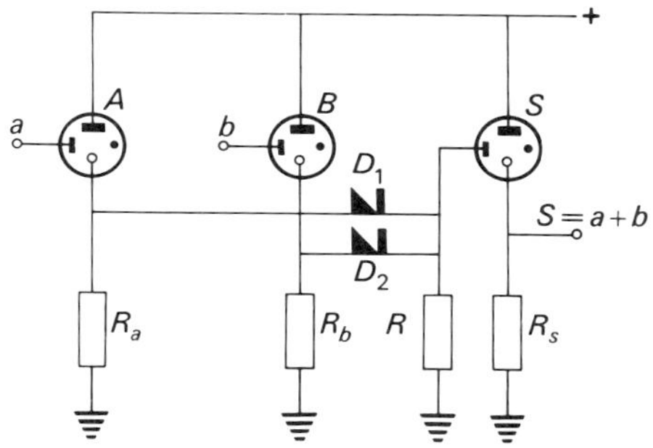

Fig. 6-39

The practical realization of some of the simpler switching functions will now be shown. The functions $S = a \cdot b$ and $S = a + b$ can be realized by means of the circuits shown in Figs. 6-38 and 6-39. The circuit element with the diodes D_1 and D_2 in Fig. 6-38 allows the potential of the starter anode of tube S to reach the firing voltage only after both tubes A and B have been fired, whereas the diodes in Fig. 6-39 allow the starter anode to fire the tube S if tube A or tube B has been fired.

The logical functions obtained in this way are in fact not inherent in the cold-cathode-tube logic. Functions $S = a \cdot b$ and $S = a + b$ are already determined by the circuits containing diodes D_1 and D_2.

Deionization of a gas triode tube, if not performed by means of auxiliary contact circuits or of a mixed dc-ac power supply, can be achieved by inserting a capacitive coupling between the anodes or the cathodes of two inversely operating tubes, as is shown in Fig. 6-40.

If, for example, tube B is conducting, tube A being nonconducting, the firing of tube A will cause a sudden voltage rise on the cathode of tube A. This rise is transmitted over capacitor C to the cathode of tube B. The resulting positive pulse on the cathode of tube B causes the voltage in the main gap to drop below the sustaining voltage. If cathode resistors R_a and R_b and capacitor C are correctly dimensioned, the voltage drop in the main gap will be kept below the sustaining voltage for a sufficiently long time to deionize the tube. The operation of the circuit is reversible.

This circuit can be used, for example, for switching functions with primed variables. If such fundamental circuits are applied systematically, they lead to the use of pairs of cold-cathode tubes, as shown in Fig. 6-40. One of

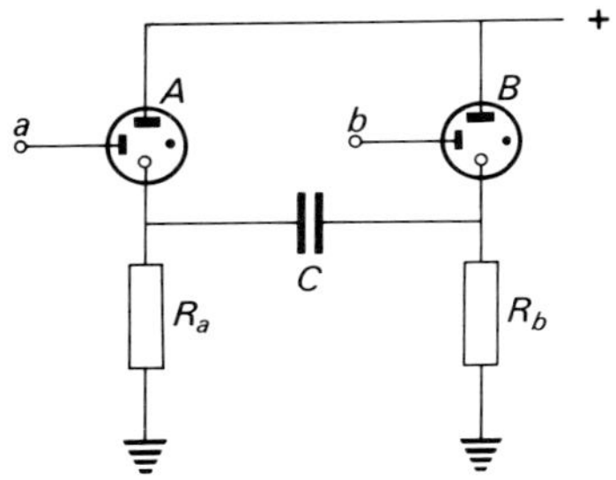

Fig. 6-40

the disadvantages of this type of circuit, attractive as it may appear at first sight, is that if one tube of a pair fails to deionize, the circuit will be locked in an erroneous condition.

Ingenious circuits have been proposed to overcome the various difficulties inherent in the use of cold-cathode tubes. The advent of semiconductor diode and triode switching devices, however, has brought so many advantages, with so few disadvantages, that the use of cold-cathode-tube switching circuits will be limited, in the author's opinion, to some special applications for which no semiconductor devices exist.

6-5 SEMICONDUCTOR DEVICES

The majority of digital switching circuits operate with diodes and triodes with silicon or germanium as semiconductor materials. In this text, the solid-state physics of semiconductor materials will not be discussed. The properties of semiconductor switching elements will be treated here in so far as they are of importance for the design of a switching circuit. The simple semiconductor diode having a high resistance on one polarity of the signal voltage and a low resistance on the other polarity need not to be discussed in this respect.

The *npn* and *pnp* transistors need some further explanation after the aforesaid about triodes and their operation. The drawing symbols of these transistors to be used further in this text are shown in Fig. 6-41.

One of the most important characteristics of transistors used as switching components is the V_c-I_c characteristic (collector voltage–collector current for constant base current and grounded emitter). A group of such characteristic curves is shown in Fig. 6-42. The $I_b = 0$ line is in fact the reverse diode

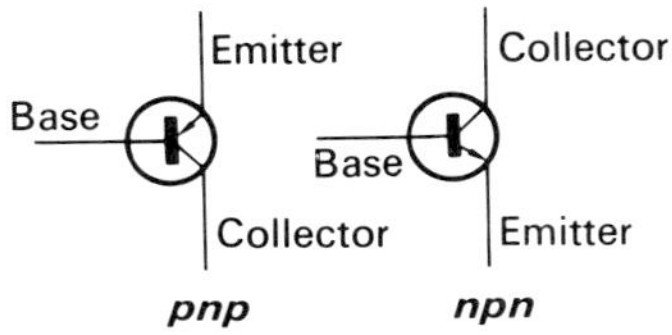

Fig. 6-41

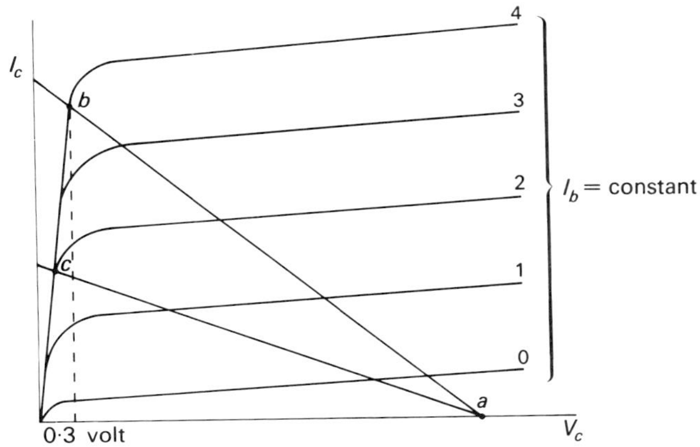

Fig. 6-42

characteristic of the base-collector junction. This line corresponds to the diode characteristic shown in Fig. 6-28. The load line for a collector resistance R_c yields the characteristic points *a* and *b*. The emitter-collector path of the transistor is blocked in point *a*, whereas this path is fully conducting if the transistor is operated in point *b*. In such a case, the transistor is "bottomed."

There is practically no heat dissipation in a blocked transistor operating in point *a* of Fig. 6-42, whereas the heat developed in a bottomed transistor operating in point *b* of its characteristic is very small because the emitter-collector voltage drop is very small (0.05 to 0.3 volt). Switching transistors are operated preferably in these two conditions only. All points of the V_c-I_c

characteristic between the origin and point *b* can be used as bottomed conditions. The collector voltage is practically constant for these rather varying operating conditions.

Line *ac* is another load line of this transistor for a smaller collector load. The characteristic of Fig. 6-42 shows that for a certain collector load a minimum base current is required to bring the transistor in its bottomed condition. With a varying collector load, the transistor can stay in its bottomed condition when a high-enough current is fed into the base. This involves saturation of the base-emitter path of the conducting transistor, which causes a delay in the switching off and thus a difference in fall and rise times of the collector current of the transistor. This delay depends on the particular features of the transistor and the circuit in which it is used.

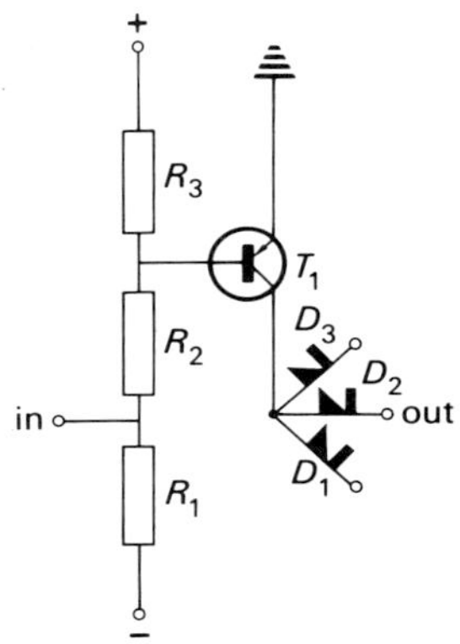

Fig. 6-43

A switching transistor is normally used in connection with a voltage-level-shifting voltage divider, so that the various switching levels can simply be connected in succession, essentially in the same way as with high-vacuum triodes (see Figs. 6-32*a* and 6-33). However, the semiconductor triode is a current-controlled switching element, whereas the high-vacuum triode is voltage-controlled. A transistor that controls a number of transistors in a next switching level via a number of diode gates will carry a varying load, this depending on the logical properties of these gates. The signal voltages can be kept constant by avoiding connecting a driving transistor exclusively to one output voltage divider, as is shown schematically in Fig. 6-43 for a *pnp* transistor. Resistor R_1 forms part of the collector load of the driving transistor, which is not shown. The driving transistor can be connected via an appropriate diode gate to a number of circuits, as shown in Fig. 6-43. All resistors R_1 in parallel together form the collector load of the single driving transistor. The input voltage of the basic transistor circuit of Fig. 6-43 will be either a 0 potential via the conducting collector-emitter path of the driving transistor or, in the blocking state of the driving transistor, a potential determined by the resistors of the voltage divider $R_{1/2/3}$.

Depending on the situation of the gate circuits between successive transistor switching levels, a driving transistor will have to provide the base current for a varying number of transistors in the next switching level.

In practice, this does not affect the input voltage of the transistors in the next transistor switching level as long as the driving transistor load does not lead to a working point on the right of point b on the maximum-base-current line (line $I_b = 4$) in the graph of Fig. 6-42.

In case of a varying collector load of a driving transistor—this being the general rule with switching techniques based on the fundamental circuit of Fig. 6-43—the base current of that transistor must be rendered so high that the required maximum collector current, which is determined by the current amplification factor, can flow through the transistor. A safety factor to allow for tolerances in this amplification factor and for its decrease in time must be taken into account. In those cases where this maximum base current is not used up, the balance between the collector and base current of the transistor is more or less disturbed. The base-current surplus drives the transistor into a situation of saturation that causes a delay to occur in the fall of the collector current, as already mentioned.

Another delay in the response of a transistor in the basic circuit of Fig. 6-43 is due to the capacitance of the base in combination with resistor R_2.

A number of steps can be taken to improve the switching response of a transistor. These measures can either be applied separately or be combined.

ACCELERATING CAPACITOR

A good improvement in the fall and rise time of the collector current of a transistor can be obtained by means of an accelerating capacitor C_1 shunted on resistor R_2 of the input voltage divider, as shown in the circuit of Fig. 6-44. The accelerating capacitor has a rather small capacitance. It operates mainly as a short circuit of resistor R_2 on the very steep leading and trailing edges of the input pulses. This means that the full jump of the input signal is passed on to the base of the transistor, whereby the transistor base at the leading edge of the pulse becomes positive with respect to the emitter (*pnp* transistor). Without accelerating capacitor, this results in a rather delayed blocking of the collector current (line 2 of Fig. 6-45); with accelerating capacitor, this delay is much smaller. With accelerating capacitor, an improvement is also obtained in the rise of the collector current at the end of the controlling pulse. With smaller and heavier loads, the improvement of rise and fall times, which can be obtained by means of an accelerating capacitor, is not very significant.

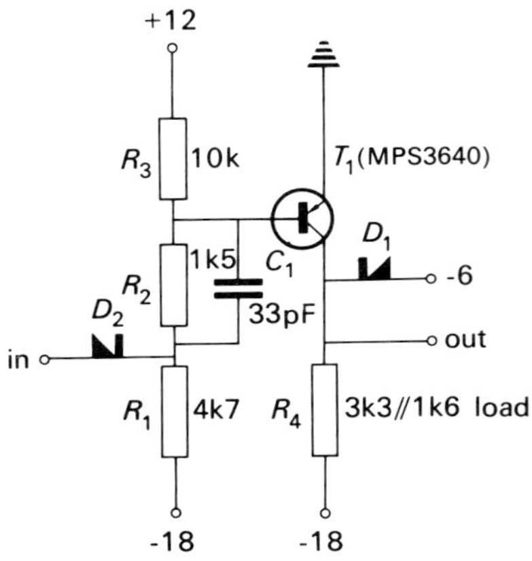

Fig. 6-44

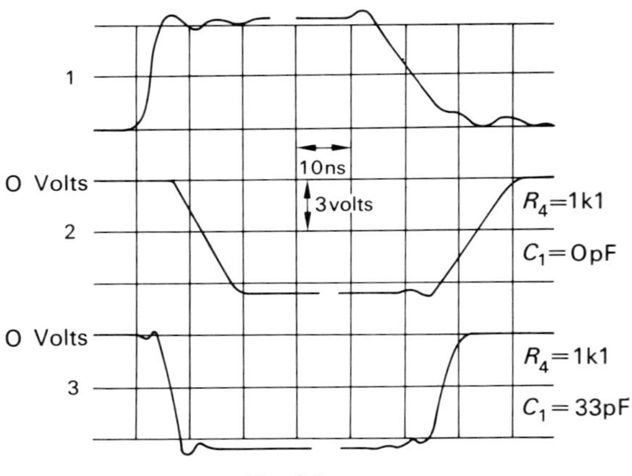

Fig. 6-45

COLLECTOR-CLIPPING DIODE

The circuit of Fig. 6-44 shows a collector-voltage-clipping diode D_1. The collector voltage in this circuit was clipped off at –6 volts, whereas the collector power supply provided a voltage of –18 volts. This has the following effect on the rise and fall of the collector voltage of the transistor.

If the collector load of a transistor has a capacitive character, the collector voltage will clip off rather sharply at –6 volts. In circuits with a collector power supply of –6 volts only and a capacitive load on the collector, the collector will follow a rather slowly evolving e curve. However, in circuits with an increased collector power-supply voltage and a clipping diode, the collector voltage may initially follow a much steeper e curve pointing to that increased voltage. This curve suddenly reaches its final value at the predetermined clipping voltage.

This is an effect which has nothing to do with transistor features; it is normally resorted to in electrical engineering and may be very useful in transistor circuits.

The clipping diode in the collector circuit of a transistor has some other important effects.

1. In both phases of the output signal voltage the output has a low internal resistance. Without a clipping diode in the blocked condition of the transistor, the output circuit, being at the same time the input circuit of the next switching stage, is more sensitive to interference.
2. The input voltages of the next switching stage are always very close to the predetermined levels when using clipping diodes. This means that there are no varying base currents of the transistors in the next switching stage with varying logical conditions in the intermediate gate circuits.
3. Finding faults in transistor switching circuits with collector-voltage-clipping diodes is facilitated by the fact that except for certain component faults both phases of the signal voltage are always well determined and close to the predetermined levels.

6-6 DIODE-TRANSISTOR LOGIC (DTL)

SINGLE-LEVEL DIODE GATES

In the transistor control circuit described in the preceding section, each level-shifting voltage divider was individually connected to a transistor base. In this type of circuit, the power-supply voltage may, if so required, be set to a much higher level than the admissible collector-emitter voltages. The operating margins of this type of circuit in the presence of variations in the voltage of the power supply can be very high.

However, a difficulty arises in this type of control circuit since resistor R_1 (Fig. 6-43) is in fact part of the collector circuit of the preceding transistor, as is indicated in Fig. 6-46. This figure shows that the collector circuit of transistor T_2 in fact consists of a combination of several resistors (R_7, R_{10}, ...). The inputs of transistors T_3 and T_4, however, have to be separated by means of diodes D_2 and D_3, because each of these transistors can be controlled by means of a number of other transistors as the multiple arrows suggest.

The number of resistors ($R_7, R_{10}, \ldots$) forming the load of transistor T_2 in Fig. 6-46 is immaterial as long as the transistor is used in a bottomed condition. In that case, the output voltage is practically constant, and the heat dissipation of the transistor is of no practical importance.

The control of a transistor by a number of other transistors is illustrated further in the circuit of Fig. 6-46, in which transistors T_1 and T_2 control transistor T_3. Both inputs to transistor T_3 have to be separated by means of the OR-gate-forming diodes D_1 and D_2 in order to avoid sneak paths. It should be borne in mind that resistor R_7 in Fig. 6-46 has in fact a two-fold function. The first, as was already explained, is related to it being part

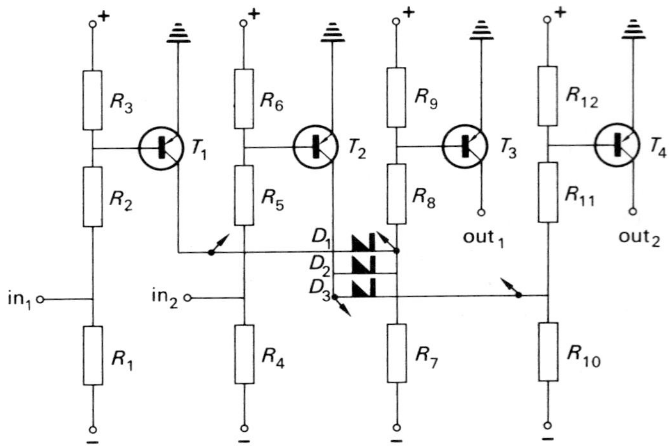

Fig. 6-46

of the collector circuit of the preceding transistor; the second is that it operates as a gate resistor for the OR gate formed by diodes D_1 and D_2. It then follows that one cannot simply resort to reversing the polarity of diodes D_1 and D_2 in order to obtain an AND gate at the input of a transistor circuit of the type described above. This would result in an interruption of the collector current of the controlling transistor. It is possible to add extra individual collector resistors to transistors T_1 and T_2 in the circuit shown in Fig. 6-46 to prevent this difficulty.

It should be pointed out that if the transistor collectors are not clipped, these collectors will get full battery voltage in certain switching combinations.

A *pnp* transistor in a switching system with individual voltage dividers can be preceded by an AND gate if an auxiliary nonsignificant OR gate is added. An example of a circuit in which this process is used, and an over-loading of the base of the controlled *pnp*-transistor (T_3) is avoided, is shown in Fig. 6-47.

The configuration of the diode gates (D_1, R_{11}; D_2, R_{12}) in this circuit is such that if all transistors (T_1, T_2) are nonconducting, resistors R_{11} and R_{12} are connected in parallel (via diodes D_3 and D_4) to resistor R_8 of the input voltage divider of the controlled transistor T_3. If transistor T_1 becomes conducting, point z is raised to earth potential and point y remains at, say, –6 volts, so that diode D_3 is blocking.

If no clipping diodes D_6 and D_7 are provided, the potential of point x will vary with the number of conducting diodes of the group D_3, D_4. However, since each of the collectors of the controlling transistors T_1, T_2 is clipped at -6 volts, points y, z also have that clipping voltage, and, consequently, also point x, irrespective of the number of conducting diodes (D_3,

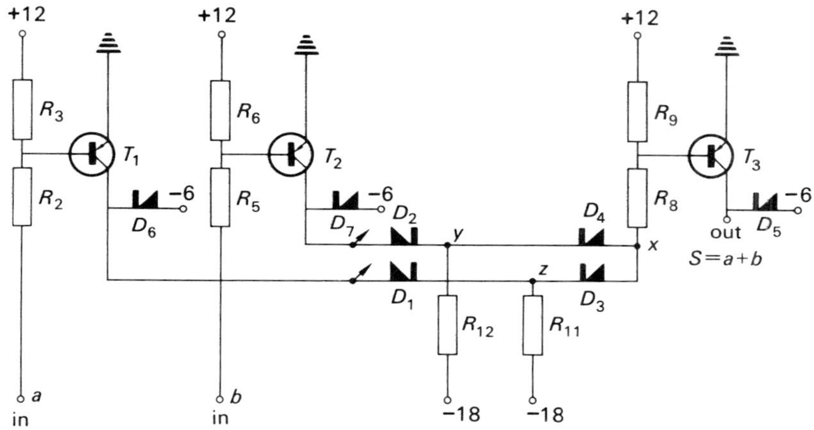

Fig. 6-47

D_4), until all points y, z are connected to earth potential via one or the other conducting transistor of the group T_1, T_2. In this case, point x gets earth potential, so that transistor T_3 is made nonconducting. Voltage divider $R_{8/9}$ has to be so dimensioned that for $x = -6$ volts, the base current of transistor T_3 is just allowed to flow. If no clipping of the collector voltage of the controlling transistors is provided, the base of the controlled transistor can be heavily overloaded.

TWO-LEVEL DIODE GATES

The diode-gate structure, as shown in Fig. 6-47, is directly applicable to two-level diode gates between successive transistor switching levels. It is in fact a simplified form of the most-general two-level diode-gate structure already shown in Fig. 6-15. In Fig. 6-48, this most-general form of the two-level diode gate is shown between successive transistor switching levels.

Transistor T_1 drives one input of the OR gates with resistors R_1, R_2, and R_3. To illustrate this point, one of the gates is shown in a dotted enclosure. The diodes on the right-hand side of these OR gates are part of the various AND gates, one of which is also surrounded by a dotted line. This AND gate controls transistor T_2. In Fig. 6-48, the input of this transistor is controlled by the function

$$S = (a + b + c)\,(d + e + f)\,(g + h + i)$$

This function is a maxterm form. When using *npn* transistors, the two-level diode gates between successive transistor switching levels will be analogous, the types of gates being used in the reverse order.

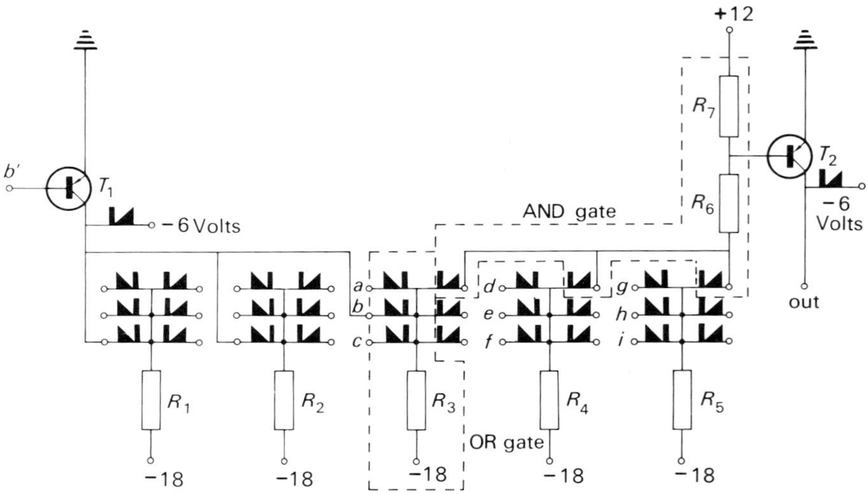

Fig. 6-48

It should be borne in mind that a single-level diode gate between successive transistor switching levels can in fact be considered more or less as a degenerated form of the general two-level gate circuit (Fig. 6-48). Furthermore, it should be pointed out that the very way in which Fig. 6-48 is drawn would tend to suggest that all diodes connected to one resistor should form one building block.

The general form of the two-level diode-gate circuit of Fig. 6-48 can be simplified in the special case where design requires something corresponding to a switching function of the type $(a + b)\,(a + c) = a + bc$ (second distributive law). This simplification is shown in the diagrams of Figs. 6-49 and 6-50.

6-7 BUILDING BLOCKS

Since all switching functions can be implemented by means of an OR gate followed by an inverter or an AND gate followed by an inverter, many types of building blocks with separate components or with integrated circuits incorporating these unit circuits have been manufactured. The first type

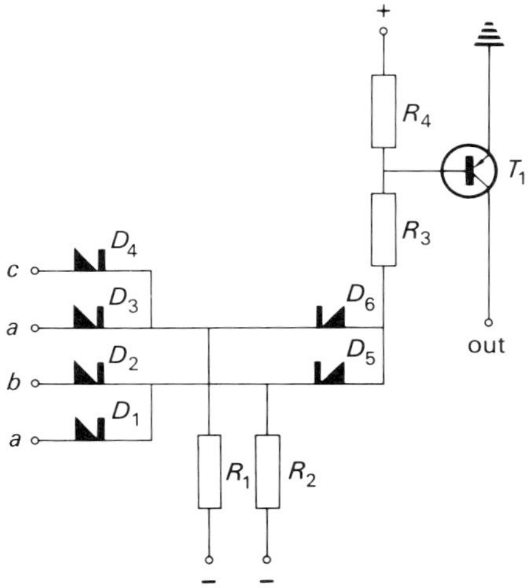

Fig. 6-49

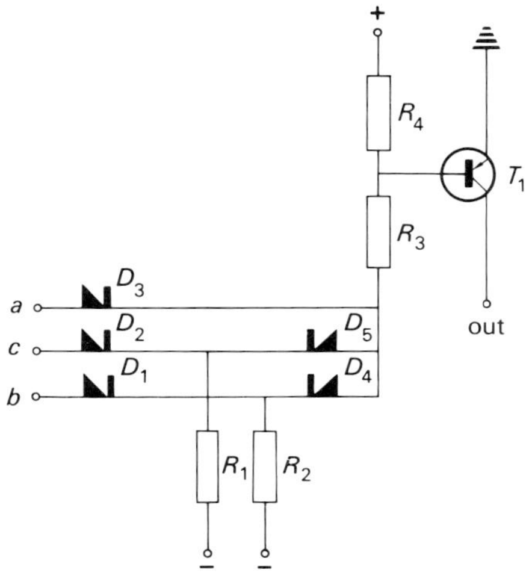

Fig. 6-50

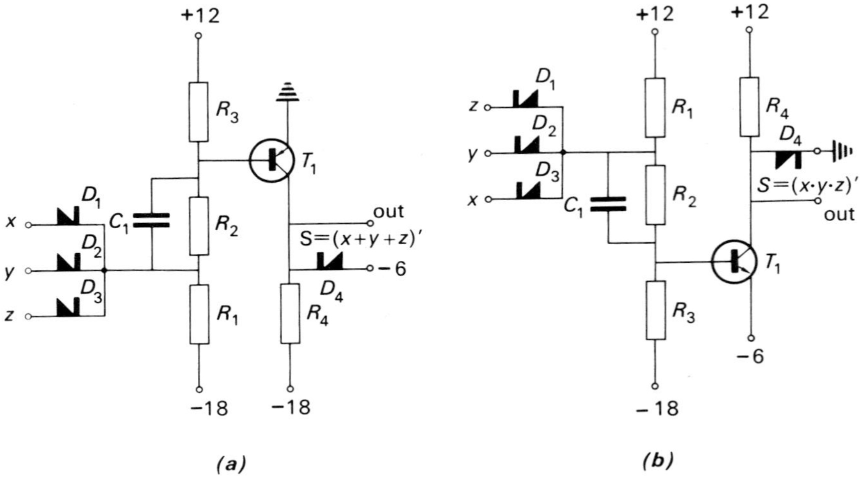

Fig. 6-51

containing a diode OR gate and a *pnp* transistor inverter is called a NOR gate; the complementary type containing a diode AND gate and an *npn* transistor inverter is called a NAND gate. These names are self-explanatory.

Both these DTL-circuit types are shown in Fig. 6-51. In these circuits, the same power supply voltages and the same signal voltage levels (0 volts, or 1, and −6 volts, or 0) have been used. However, in most commercially available building blocks the useful clipping diode is missing.

The symbols for these building blocks to be used further in this text are shown in Fig. 6-52.

The realization on this basis of some of the most frequently encountered switching functions with two variables is illustrated in the diagrams of Figs. 6-53 to 6-57.

Using ordinary components, such mixed diode-transistor gate circuits built with the unit circuit of Fig. 6-51 are not always economical because the

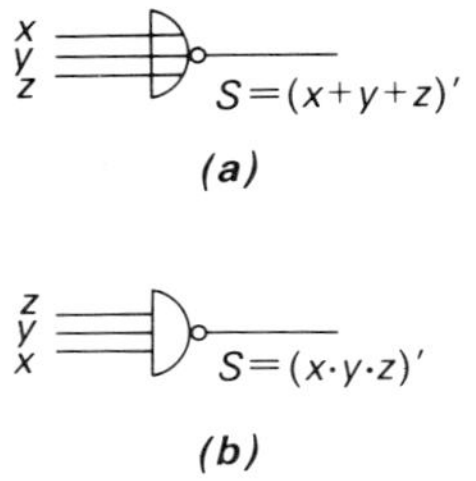

Fig. 6-52

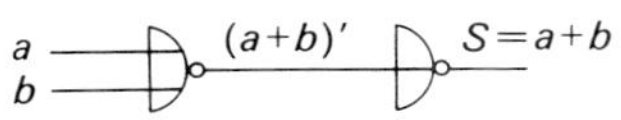

Fig. 6-53

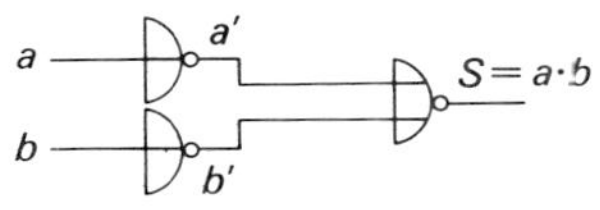

Fig. 6-54

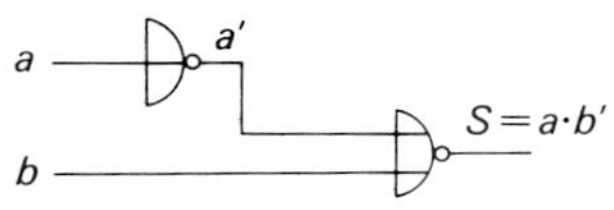

Fig. 6-55

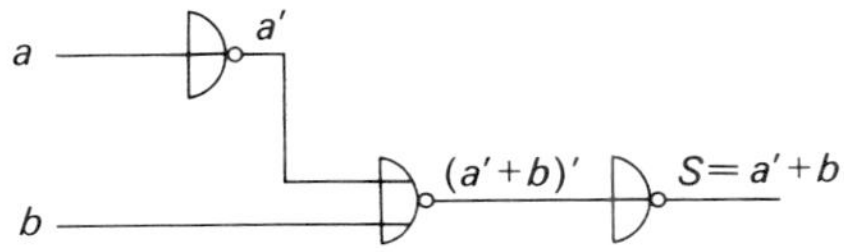

Fig. 6-56

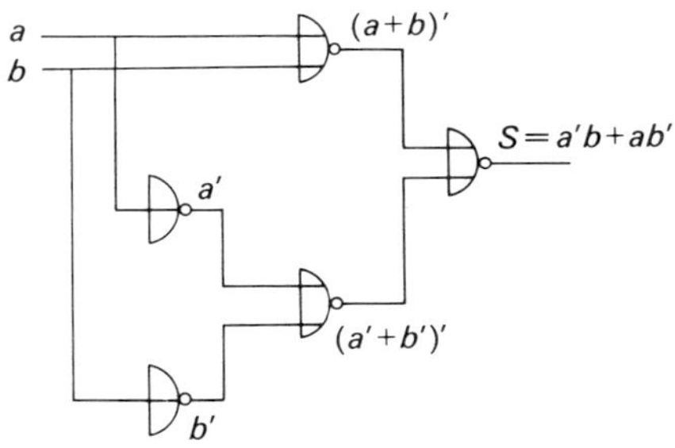

Fig. 6-57

number of diodes and triodes involved is too high. Some computer manufacturers exclusively use these NAND or NOR circuits in integrated form. However, the saving in space is not very important.

The one advantage in using NAND or NOR circuits is that no difficulties will ever arise with the switching power level, this level being automatically restored by the transistor within each unit circuit. Building blocks of this type are now available from many different sources. Nearly every manufacturer uses another type of basic circuit, especially in integrated circuits.

The simplification of switching functions for implementation with NAND or NOR gates can be performed in the same way by means of Karnaugh maps, as shown in Chap. 4.

If a circuit has to be implemented by means of NAND gates, the optimal two-level form of the logical sum of logical products has to be used as a starting point. Then, for example,

$$f(x, y, z) = \sum m_p$$

in which formula m_p is a logical product not necessarily of canonical form.

Using De Morgan's theorem, this equation may be written as follows:

$$f(x,y,z) = [f(x,y,z)']' = [(\sum m_p)']' = (\prod m'_p)'$$

Example

$$\begin{aligned} f(x,y,z) &= x'yz + xz' + y'z' \\ &= [(x'yz + xz' + y'z')']' \\ &= [(x'yz)' \cdot (xz')' \cdot (y'z')']' \end{aligned}$$

The resulting circuit implemented with NAND gates only is shown in Fig. 6-58*a*. If this circuit has to be implemented by means of NOR gates, then the optimal two-level form of the logical product of logical sums will be the starting point.

$$\begin{aligned} f(x,y,z) &= (x' + z') \cdot (y + z') \cdot (x + y' + z) \\ &= [(x' + z')' + (y + z')' + (x + y' + z)']' \end{aligned}$$

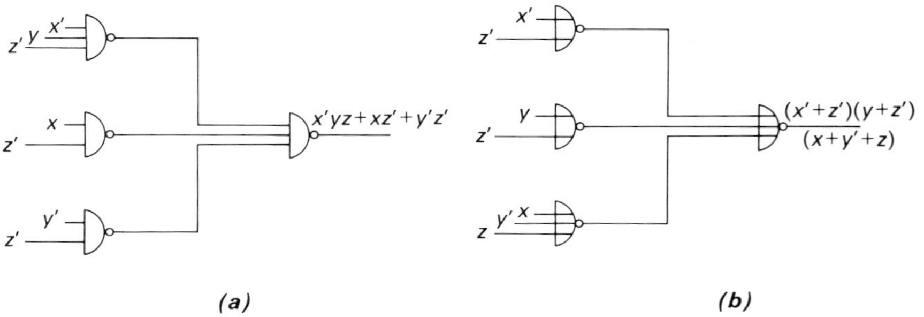

Fig. 6-58

The resulting circuit is shown in Fig. 6.58*b*. It should be noted that in both examples the switching formula can be manipulated into a three-level form with one less diode but one extra inverter.

DIRECT-COUPLED TRANSISTOR LOGIC (DCTL)

In the circuit of Fig. 6-59, transistors T_1, T_2, and T_3 in parallel with a common collector resistor form a NOR gate when using positive logic. All three transistors are blocked on a low input signal voltage (0), this resulting in a high output signal voltage (1). If any one of the transistors is made conducting on a high input signal voltage (1) the output-signal voltage drops to low (0) so that the switching formula of the circuit of Fig. 6-59 will be as follows:

$$S = (x + y + z)' \tag{6-20}$$

Hence the DCTL circuit of Fig. 6-59 is a NOR gate. There are some nonbasic variations of this type of circuit that depend on the base circuit of the transistors used.

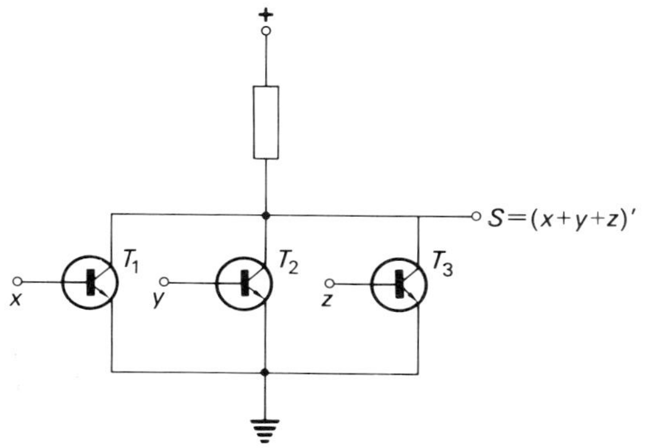

Fig. 6-59

CURRENT-MODE LOGIC (CML)

The circuit of Fig. 6-60 is dimensioned in such a way by means of emitter resistor R_E that transistor T_4 can be conducting only when neither one of transistors T_1 to T_3 is conducting. This means that the operation of the output of transistor T_4 will be represented by the following formula:

$$S = x + y + z \tag{6-21}$$

The circuit of Fig. 6-60 operates on this output terminal as an amplified OR gate. On the other output terminal, i.e., the collectors of transistors T_1 to T_3, the circuit operates as a NOR gate.

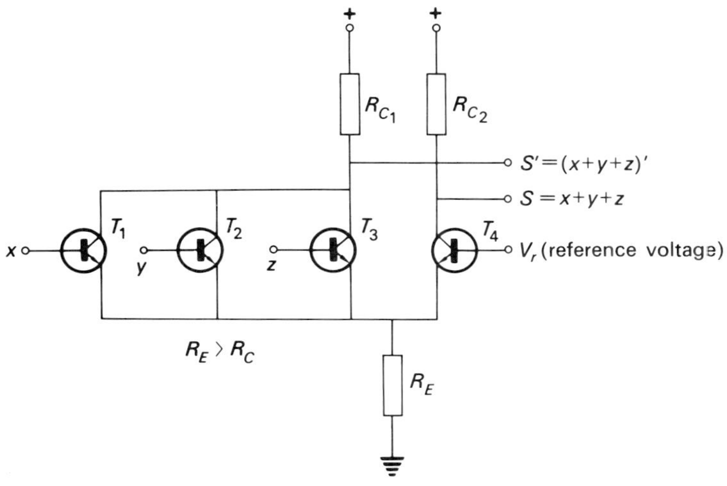

Fig. 6-60

TRANSISTOR-TRANSISTOR LOGIC (TTL)

A variation of the DTL NAND gate of Fig. 6-51*b* is shown in Fig. 6-61*a*. The diode input gate is replaced by multiple-emitter transistor T_1, which is easier to produce in integrated circuits than are diodes. The base of transistor T_1 is connected to + power supply via resistor R_1, whereas its collector is

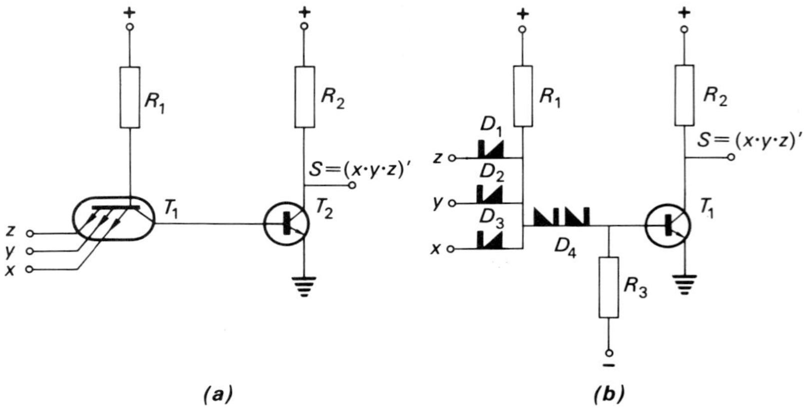

Fig. 6-61

connected to the base of the inverting output transistor T_2. The emitters of transistor T_1 form an AND gate, so that the switching function performed by the circuit element of Fig. 6-61*a* is as follows:

$$S = (x \cdot y \cdot z)' \qquad (6\text{-}22)$$

DIODE-TRANSISTOR LOGIC (DTL)

Another variation of the DTL NAND gate of Fig. 6-51*b* is shown in Fig. 6-61*b*. In this last circuit, resistor R_2 of Fig. 6-51*b* has been replaced by a couple of diodes D_4. This provides for a constant voltage shift between signal voltage and base voltage of the inverting transistor. This circuit is better suited for discrete component design than for integrated circuits. This circuit type is sometimes called low-level logic (LLL).

RESISTOR-TRANSISTOR LOGIC (RTL)

The resistor gate of Fig. 6-11 can be connected to a transistor inverter, as is shown in the circuit of Fig. 6-62. This type of gate circuit is not suited for high-speed work. Because resistors are not easy to produce in integrated circuits, this type of circuit seems better suited for discrete component design.

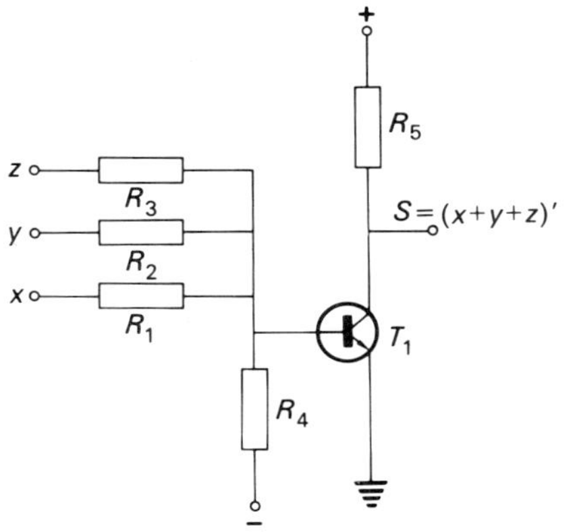

Fig. 6-62

In integrated circuitry the design philosophy is that transistors are cheap. This is the reason that most integrated NAND and NOR circuits are provided with a special output stage with two or three transistors, as is shown in the circuit of Fig. 6-63. In this circuit, transistor T_2 has been provided with a collector resistor and an emitter resistor (R_3 and R_2, respectively). The collector of transistor T_2 controls transistor T_4 via emitter follower T_3, whereas the emitter of transistor T_2 controls transistor T_5. This is performed in such a way that either transistor T_4 or transistor T_5 is conducting. Resistor R_5 has a low value, so that the output signal voltage levels are practically +

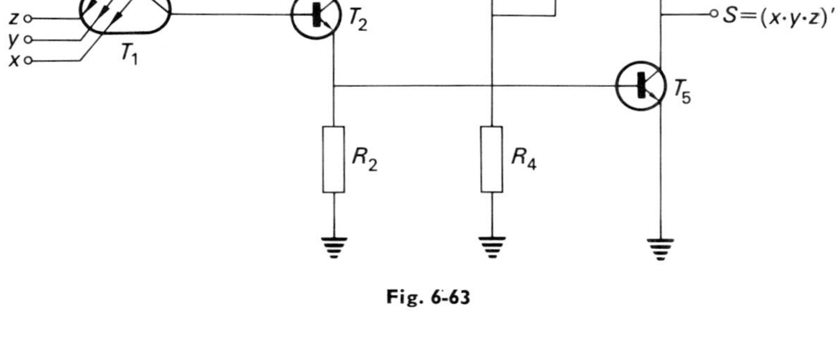

Fig. 6-63

power supply or earth potential. Both phases of the output circuit have a low internal resistance. In this respect, the output circuit of Fig. 6-63 is equivalent to that of Fig. 6-44 with the collector-voltage-clipping diode.

6-8 TRANSISTOR-GATE CIRCUITS

THE TRANSISTOR AS A SINGLE-INPUT SWITCHING ELEMENT

In switching techniques, there is a tendency to replace diode gates with transistor gates. The diode gate can be credited with the disadvantage that the number of switching levels which can be connected in succession is rather limited. Series-parallel connections, which can be achieved in contact circuits, have no equivalent in diode circuitry. This imposes serious limitations on diode-circuit techniques. In certain types of electronic counters, this results in an increasing number of diodes in successive counter orders.

The low collector-emitter resistance of transistors makes it possible to a certain extent to use a transistor in the same way as one would use a contact. The two fundamental types of circuit using *pnp* transistors are shown in Figs. 6-64 and 6-65; their operation is defined by truth tables 6-6 and 6-7, respectively.

The operation of the gate circuit of Fig. 6-64 is described by the following formula:

$$S = (x + y + z)' = x' \cdot y' \cdot z' \tag{6-23}$$

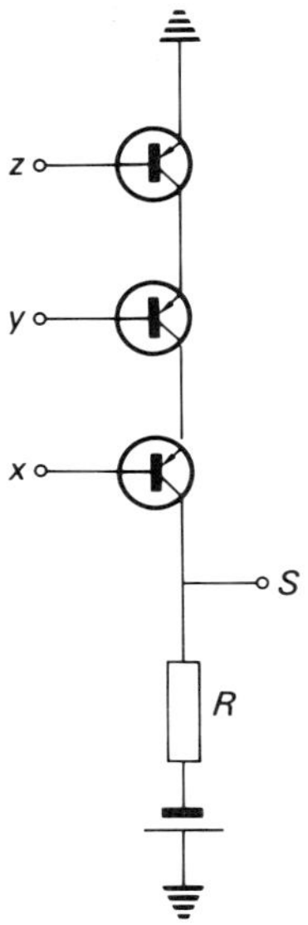

Fig. 6-64

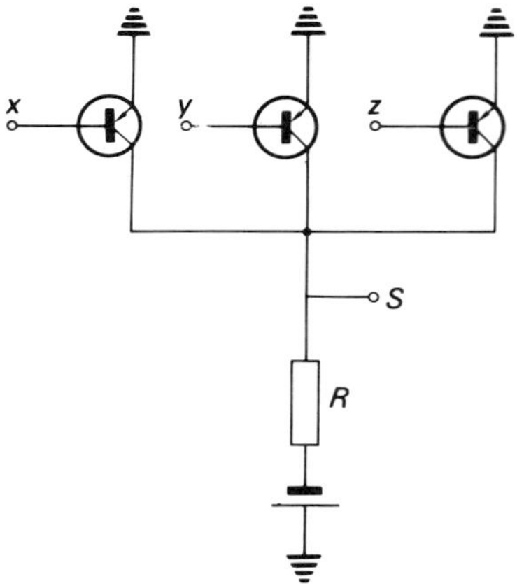

Fig. 6-65

Table 6-6

x	y	z	S
0	0	0	1
0	0	1	0
0	1	0	0
0	1	1	0
1	0	0	0
1	0	1	0
1	1	0	0
1	1	1	0

Table 6-7

x	y	z	S
0	0	0	1
0	0	1	1
0	1	0	1
0	1	1	1
1	0	0	1
1	0	1	1
1	1	0	1
1	1	1	0

The gate circuit of Fig. 6-65 can be described by

$$S = (x \cdot y \cdot z)' = x' + y' + z' \tag{6-24}$$

The gate circuit of the type shown in Fig. 6-64 may be defined as a *pnp* inverter AND gate, whereas circuits of the type shown in Fig. 6-65 could be referred to as *pnp*-inverter OR gates.

The *npn*-inverter OR and AND gates are shown in Fig. 6-66 and 6-67; their operation is defined by the corresponding truth tables 6-8 and 6-9.

Table 6-8

x	y	z	S
0	0	0	1
0	0	1	1
0	1	0	1
0	1	1	1
1	0	0	1
1	0	1	1
1	1	0	1
1	1	1	0

Table 6-9

x	y	z	S
0	0	0	1
0	0	1	0
0	1	0	0
0	1	1	0
1	0	0	0
1	0	1	0
1	1	0	0
1	1	1	0

A comparison of the truth tables shows that the operation of the gate circuit of Fig. 6-66 is analogous to that of Fig. 6-65, whereas the gate circuit of Fig. 6-67 operates in a way analogous to that of Fig. 6-64.

The operation of the circuit in Fig. 6-66 can be described by Eq. (6-24), whereas the gate circuit of Fig. 6-67 is defined by Eq. (6-23). The question of whether the series and parallel concepts can be used in these types of transistor gates will depend on the type of transistor used. Nevertheless, it is best to forget about these concepts in electronic switching.

Some other possible ways of interconnecting three transistors may be of importance. The gate circuits shown in Figs. 6-68 and 6-69 are equivalent. Figure 6-68 incorporates a group of three *pnp* transistors, whereas the other gate circuit shown in Fig. 6-69 is its equivalent with *npn* transistors. The operation of both circuits is defined by the same truth table, i.e., Table 6-10.

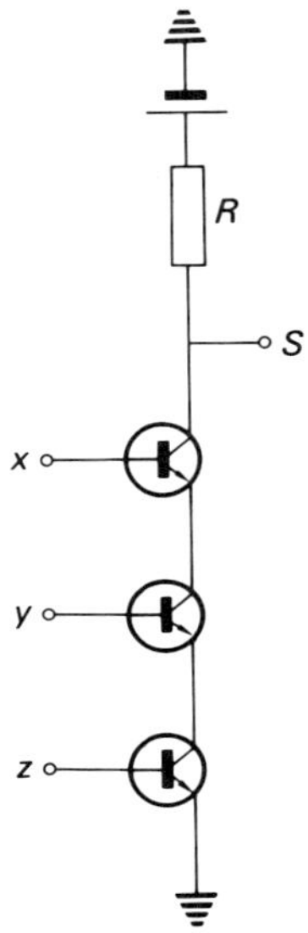

Fig. 6-66

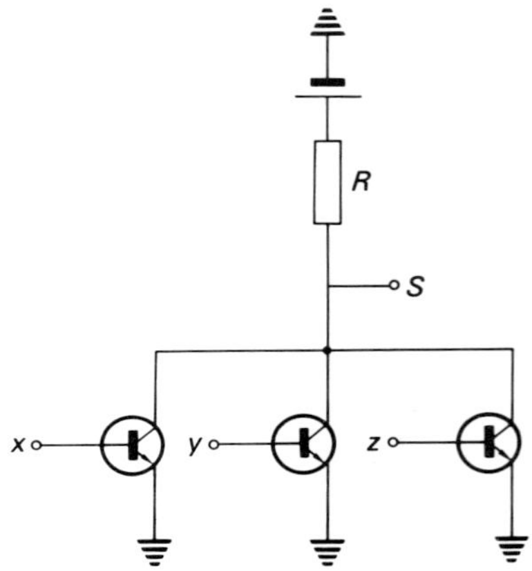

Fig. 6-67

This truth table defines the operation of the circuits of Figs. 6-68 and 6-69 as follows:

$$S = (x + y \cdot z)' = x'(y' + z') \tag{6-25}$$

Table 6-10

x	y	z	S
0	0	0	1
0	0	1	1
0	1	0	1
0	0	1	1
1	0	0	0
1	0	1	0
1	1	0	0
1	1	1	0

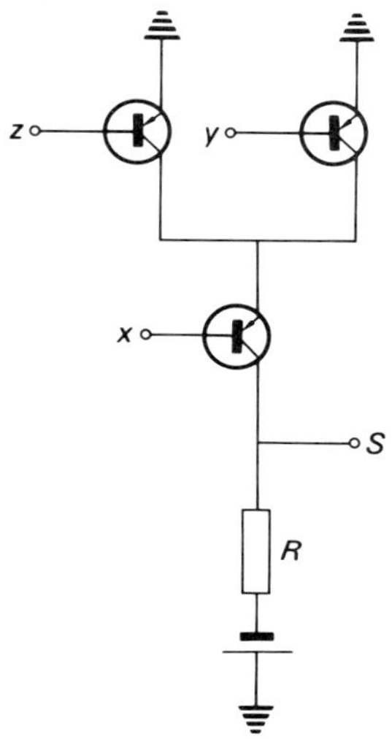

Fig. 6-68

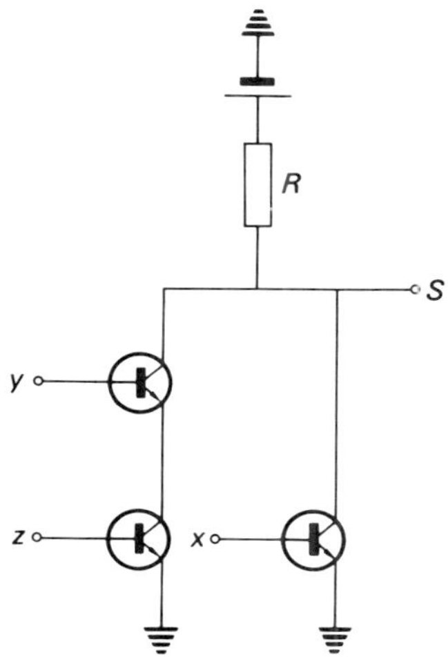

Fig. 6-69

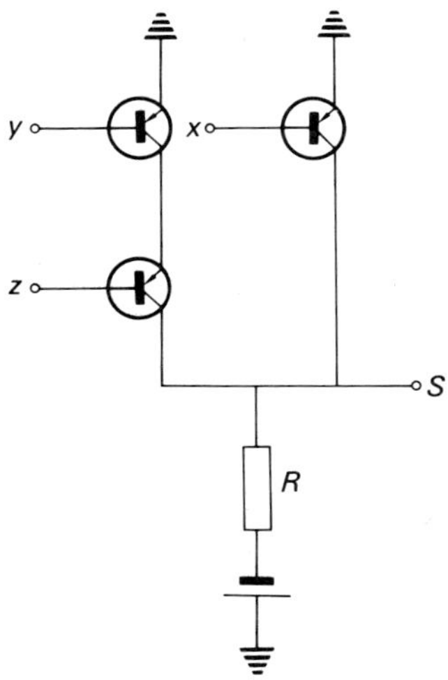

Fig. 6-70

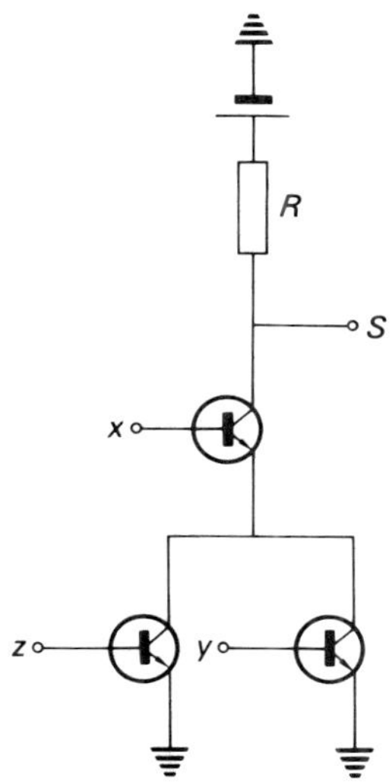

Fig. 6-71

Three transistors of the same type can be grouped in still another way, as shown in Figs. 6-70 and 6-71. The circuit of Fig. 6-70 uses *pnp* transistors, whereas the circuit of Fig. 6-71 is its equivalent with *npn* transistors. The operation of both circuits is again defined by the same truth table (Table 6-11).

This truth table describes the operation of the circuits shown in Figs. 6-70 and 6-71 in terms of the following formula:

$$S = [x(y + z)]' = x' + y'z' \tag{6-26}$$

Table 6-11

x	y	z	S
0	0	0	1
0	0	1	1
0	1	0	1
0	1	1	1
1	0	0	1
1	0	1	0
1	1	0	0
1	1	1	0

An example might be helpful to illustrate the use of transistors in logical circuits. A circuit is to be designed which produces an output for every combination of a two-out-of-four input, i.e., the required circuit corresponds to the following switching function:

$$f(w, x, y, z) = w'x'yz + w'xy'z + w'xyz' + wxy'z' + wx'yz' + wx'y'z \tag{6-27}$$

Since there are no don't-care conditions in this case, the function as given is the simplest possible expression. It corresponds to the symmetrical function S_2. When using inverted input variables and *pnp* transistors, the circuit can be made to correspond to the analogous contact circuit. The resulting transistor circuit is shown in Fig. 6-72.

It should be borne in mind that the size of the transistor-gate structures is not unlimited. It depends a good deal on the type of transistor used. Even so, five transistors in series and five in parallel give a good practical circuit.

Transistors similar to the high-vacuum cathode followers (as shown in Fig. 6-32*b*) can be used as an emitter follower. Both types of gate circuits have the same disadvantages. The voltage level of the control signals decreases when passing through an emitter-follower gate. However, there are some applications where an emitter follower can be very useful. The operation speed of emitter-follower gates is very high because no build-up of the conducting mechanism between emitter and base is required.

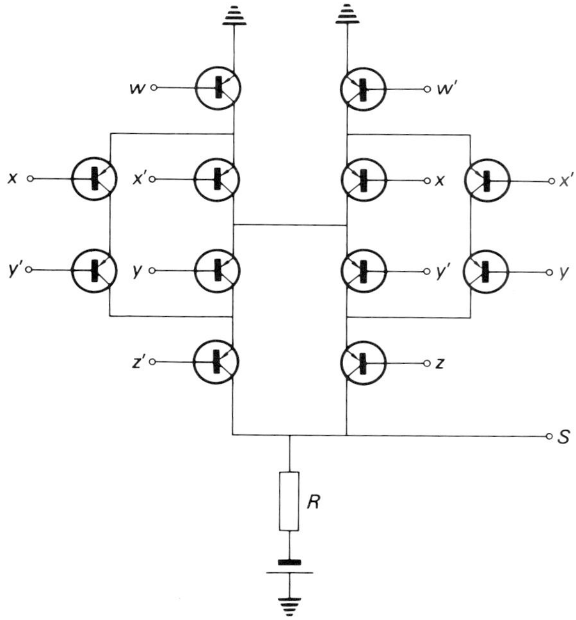

Fig. 6-72

The basic circuits for two-input *pnp* and *npn* emitter-follower gates are given in Figs. 6-73 and 6-74. The corresponding truth tables are shown in Tables 6-12 and 6-13, respectively. The truth tables show that the operation

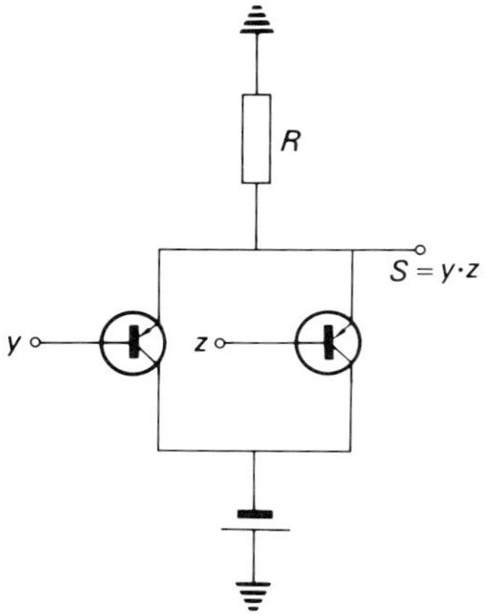

Fig. 6-73

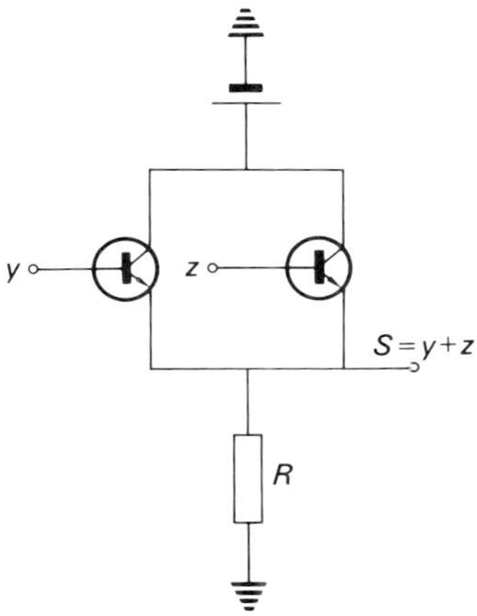

Fig. 6-74

Table 6-12

y	z	S
0	0	0
0	1	0
1	0	0
1	1	1

Table 6-13

y	z	S
0	0	0
0	1	1
1	0	1
1	1	1

of the circuit of Fig. 6-73 can be described by the following formula:

$$S = y \cdot z \tag{6-28}$$

whereas the circuit of Fig. 6-74 corresponds to the expression

$$S = y + z \tag{6-29}$$

TRANSISTORS USED AS DOUBLE-INPUT SWITCHING ELEMENTS

For certain applications, a triode can be considered as a two-input switching element. Because of the rather limited resistance between cathode and filament in a high-vacuum triode, this application is not particularly appropriate with respect to hot high-vacuum tubes. This difficulty does not arise with transistors, since they have no filament. The basic circuits for a two-input *pnp* transistor gate and its *pnp* equivalent are given in Figs. 6-75 and 6-76; the corresponding truth tables are given as Tables 6-14 and 6-15.

The switching formula for the basic circuit of Fig. 6-75 is as follows:

$$S = y \cdot z' \tag{6-30}$$

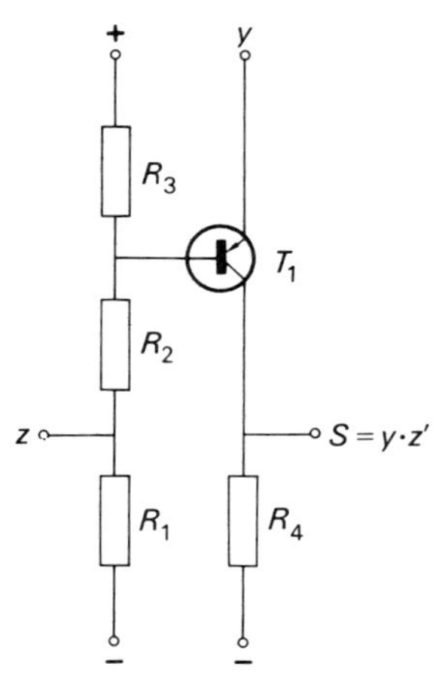

Fig. 6-75

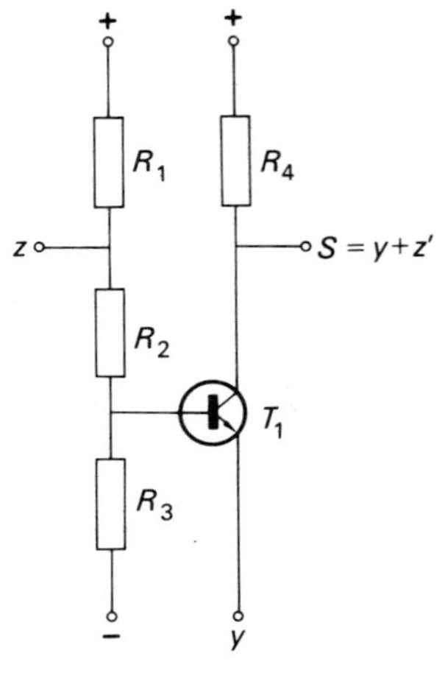

Fig. 6-76

On the other hand, the basic circuit of Fig. 6-76 can be described by the following expression:

$$S = y + z' \tag{6-31}$$

Table 6-14

y	z	S
0	0	0
0	1	0
1	0	1
1	1	0

Table 6-15

y	z	S
0	0	1
0	1	0
1	0	1
1	1	1

The two inputs in both circuits are not equivalent. The z input is formed by a transistor base and requires the normal base current as control current. The y input, however, consists of an emitter, so that the control current in this case has to equal the collector current of the transistor. The base input inverts and amplifies its control signal, whereas the emitter input does not.

The circuits of Figs. 6-75 and 6-76 are particularly well suited for the implementation of the function

$$f(y, z) = y'z + yz' \tag{6-32}$$

The corresponding circuit is given in Fig. 6-77. It is provided with a clipping diode D_1 to secure two output voltage levels.

Special transistor-gate circuits can be designed by means of a combination of *pnp* and *npn* transistors. For most practical purposes, however,

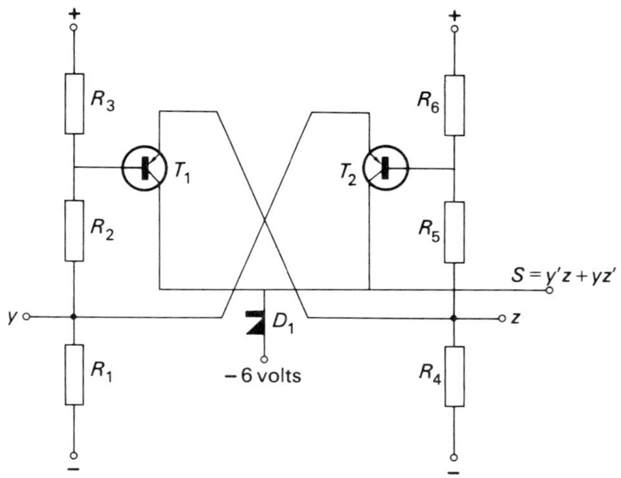

Fig. 6-77

the techniques given above will be adequate. The larger the number of different components involved, the more difficult it will be to develop a building-block system for use as a basis for complicated circuits. This is true for circuits with lumped components, but not for integrated circuits.

PROBLEMS

1. Explain how the logical OR operation can be performed by means of the resistor circuit shown in Fig. 6-78. What is the name and what are the characteristics of the switching element that can perform the function of resistors R_1 and R_2 in a more satisfactory manner?

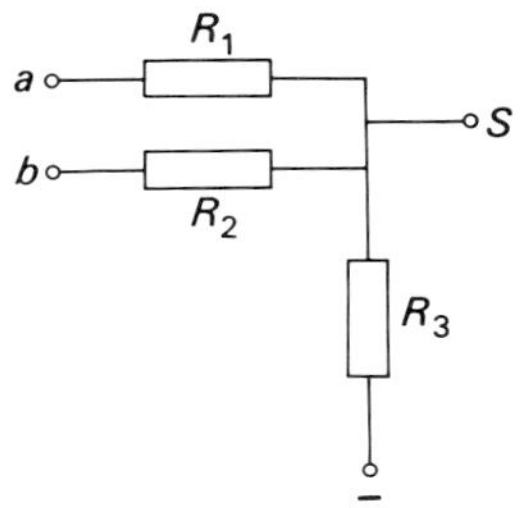

Fig. 6-78

2. Explain why the number of diode gates that can be connected in series is limited. Under what circumstances can this number of series-connected gate orders be fairly large?

3. Draw the multiple-output gate circuit corresponding to the following group of switching functions:

$$X = ab + cd \qquad Y = cd + ef \qquad Z = ef + ab$$

In addition, draw this circuit by means of symbols representing a complete gate circuit.

4. By means of Karnaugh maps, detect the common parts of the following functions and draw the corresponding diode-gate circuits in symbolic form.

$$F_1 = w'y'z + wy(x + z) + wx'y'z'$$
$$F_2 = w'x'(y' + z) + w'y'z + wx'y'z'$$

The resulting circuit will be a three-level circuit with 23 diodes.

5. Determine the number of diodes to be used in the two-level gate circuit represented in the Karnaugh map of Fig. 6-79.

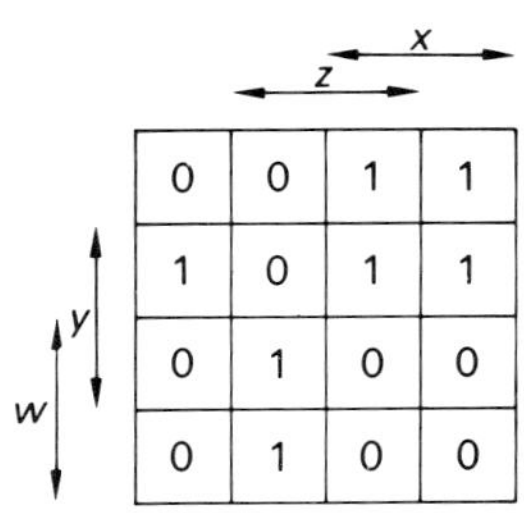

0	0	1	1
1	0	1	1
0	1	0	0
0	1	0	0

Fig. 6-79

6. Explain how a triode switching element can be used as an inverter.

7. What are the advantages and disadvantages of a high-vacuum diode versus a semiconductor diode?

8. How does a high-vacuum diode operate?

9. Show by means of simple circuits that the basic AND and OR operations can be performed with triodes only.

10. Describe the characteristic of the anode-cathode gap of a cold-cathode tube and state the advantages and disadvantages of this type of tube.

11. Draw the circuit of a switching transistor with its corresponding input signal level-shifting voltage divider. Explain the effect of the base capacity with respect to the operating speed, and indicate possible improvements.

12. What effect does the clipping of the collector voltage of a switching transistor have? Explain this by means of a diagram.

13. What is the "bottoming" of a transistor? Explain how a bottomed transistor can control a number of other transistors so that the signal level remains constant.

14. Show that all basic switching functions can be realized by means of NOR or NAND gates.

15. Draw the most-general form of a two-level diode-gate circuit used as an interconnection between several transistor switching stages.

16. Show that the diode OR gates used in Prob. 14 may be replaced by resistors (Kirchhoff networks).

17. Explain the use of transistors in logical gates.

18. Show that transistors can be used as two-input logical gates.

19. Draw the circuit corresponding to the following expression, using transistors only:

$$S = (a'b + ab')c + (ab + a'b')c'$$

20. Draw the circuit representing the switching function of Prob. 19 using NOR gates only.

21. Draw the circuit representing the switching function of Prob. 19 using NAND gates only.

SELECTIVE BIBLIOGRAPHY

1. Beter, R. H., W. E. Bradley, R. B. Brown, and M. Rubinoff: Directly Coupled Transistor Circuits, *Electronics*, vol. 28, pp. 132–136, 1955.
2. Booth, G. W., and T. P. Bothwell: Logic Circuits for a Transistor Digital Computer, *IRE Trans.*, vol. EC-5, pp. 132–136, 1955.
3. Booth, G. W., and T. P. Bothwell: Basic Logic Circuits for Computer Applications, *Electronics*, vol. 30, pp. 196–200, 1957.
4. Brown, D. R., and N. Rochester: Rectifier Networks for Multiposition Switching, *Proc. IRE*, vol. 37, pp. 139–147, 1949.
5. Chao, S. C.: A Generalized Resistor-transistor Logic Circuit and Some Applications, *IRE Trans.*, vol. EC-8, pp. 8–12, 1959.
6. Cook, C. R., Jr., and B. M. Martin: New Semiconductor Networks Reduce System Complexity, *Electronics*, vol. 37, pp. 25–29, 1964.
7. Flood, J. E., and J. B. Warmen: The Design of Cold-cathode Valve Circuits, *Electronic Eng.*, vol. 28, pp. 416, 489, and 528, 1956.
8. Gluck, S. E., H. J. Gray, C. T. Leondes, and M. Rubinoff: The Design of Logical OR-AND-OR Pyramids for Digital Computers, *Proc. IRE*, vol. 41, pp. 1388–1392, 1953.

9. Harris, J. R.: Direct Coupled Transistor Logic Circuitry, *IRE Trans.*, vol. EC-7, pp. 1–6, 1958.

10. Henle, R. A., and J. L. Walsh: The Application of Transistors to Computers, *Proc. IRE*, vol. 46, pp. 1240–1254, 1958.

11. Heron, K. M., H. Baker, and D. L. Benson: An Experimental Electronic Director: Basic Principles and Circuit Elements, *Post Office Elec. Eng. J.*, vol. 44, pp. 97–102 and 169–175, 1951; Field Trial Results, vol. 47, pp. 197–202, 1955.

12. Hussey, L. W.: Semiconductor Diode Gates, *Bell System Tech. J.*, vol. 32, pp. 1137–1154, 1953.

13. Ingram, S. B.: Cold-cathode Gas-filled Tubes as Circuit Elements, *Electron. Eng.*, vol. 58, pp. 342–246, 1939.

14. McCluskey, E. J., Jr.: Iterative Combinational Switching Networks—General Design Considerations, *IRE Trans.*, vol. EC-7, pp. 285–291, 1958.

15. Marcovitz, M. W., and E. Seif: Analytical Design of Resistor Coupled Transistor Logical Circuits, *IRE Trans.*, vol. EC-7, pp. 109–119, 1958.

16. Page, C. H.: Digital Switching Circuits, *Electronics*, vol. 21, pp. 110–118, 1948.

17. Prom, G. J., and R. L. Crosby: Junction Transistor Switching Circuits for High Speed Digital Computer Applications, *IRE Trans.*, vol. EC-5, pp. 192–196, 1956.

18. Richards, R. K.: Diode Switching Circuits, chap. 2 of "Digital Computer Components and Circuits," D. Van Nostrand Company, Inc., New York, 1957.

19. Rowe, W. D., and G. H. Royer: Transistor NOR-Circuit Design, *Winter Gen. Meeting AIEE Paper* 57–196, 1957.

20. Snijders, A.: Code Converter, U.S. Patent 2,620,395, filed in the Netherlands, June 30, 1947.

21. The Staff of the Computation Laboratory, Rectifiers, chap. 10 of "Synthesis of Electronic Computing and Control Circuits," Harvard University Press, Cambridge, Mass., 1951.

22. Unraveling the Maze in Semiconductor Integrated Circuits, *Intern. Electron.*, vol. II, pp. 24–29, 1966.

23. Wanlass, C. L.: Transistor Circuits for Digital Computers, *IRE Trans.*, vol. EC-4, pp. 11–15, 1955.

24. Yokelson, B. J., and W. Ulrich: Engineering Multistage Diode Logic Circuits, *Trans. AIEE*, vol. 74, pp. 466–475, 1955.

7

Magnetic Cores as Switching Elements

7-1 MAGNETIC MATERIALS WITH A NEAR-RECTANGULAR HYSTERESIS LOOP

The magnetic core is a switching element falling more or less outside the range of components dealt with in the preceding chapter. The theoretically possible uses as switching elements are numerous; however, there are some practical difficulties that somewhat restrict the use of this component as a switching element. The physical properties of magnetic cores will be described here only inasmuch as they are essential with respect to their use as switching elements.

A magnetic core is basically a memory element which retains its information in the form of one of two possible states. It does not provide a direct indication of its information state, for example, as a voltage level that can be used immediately for further switching. Nevertheless, the magnetic core has proved to be a useful switching device for a limited group of applications.

There are two types of magnetic cores:

1. Cores consisting of a torus of ferrite material
2. Cores consisting of a very thin tape of a suitable magnetic material wound on a nonmagnetic cylindrical core

The materials in these types of cores have in common the fact that they both have a near-rectangular magnetic hysteresis loop. The first type of cores has found large-scale application in the so-called immediate-access matrix core memories by reason of their small diameter and comparatively low price. The second type is more expensive because these cores are not pressed as are ferrite cores, but they are individually wound with ultrathin metallic tape, a rather more complex process than is the case with ferrite cores. The use of wound magnetic cores is more or less limited to a number of special applications in the switching field itself. A substantially lower number of ampere-turns is required for their operation than is the case with ferrite cores.

As was already said, this book is not the proper place for dealing exhaustively with the characteristics of magnetic materials of both types of cores. The use of both these types in switching is based on the near-rectangular hysteresis loop of the magnetic material.

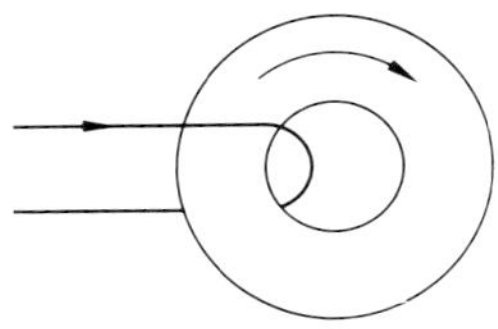

Fig. 7-1

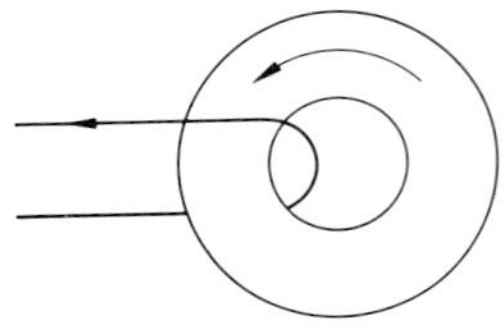

Fig. 7-2

The principle of the magnetic core is shown in Figs. 7-1 and 7-2. In Fig. 7-1 the core is magnetized clockwise; the core shown in Fig. 7-2 has counterclockwise magnetization. These figures also illustrate the normal correspondence between the direction of the magnetizing current I through the winding and the resulting magnetic induction B in the core. The magnetic state of a core can be reversed rather easily.

Figure 7-3 shows the near-rectangular hysteresis loop of a commercially available type of ferrite core. This loop is given as a B-I characteristic.

The two possible stable magnetic states ($I = 0$) of a magnetic core are indicated in Fig. 7-3 as $+B_r$ (1) and $-B_r$ (0). The magnetic core can be used as a switching element because it has two stable states.

The hysteresis loop of a magnetic material will be called near-rectangular if a magnetizing force of a certain strength will reverse the magnetic state of the magnetic material, but this same magnetic state will remain practically unchanged if exposed to a magnetic force of half this strength.

This special property of magnetic cores with a near-rectangular hysteresis loop is in some way fundamental to their use for the implementation of switching functions. In its simplest form, a core is provided with two windings, each of which can be energized by a current $I_m/2$. A reversal of the magnetic state of the core will be achieved only if both windings are energized in coincidence with a current $I_m/2$, so that the core receives in total a magnetizing current I_m. This is illustrated in the B-I-characteristic of Fig. 7-3.

In switching circuits, each core is provided with at least one read-in or write winding for setting the core in the 1 state, one driving or clock-pulse winding, and one read-out winding. Because the state of a core can be detected

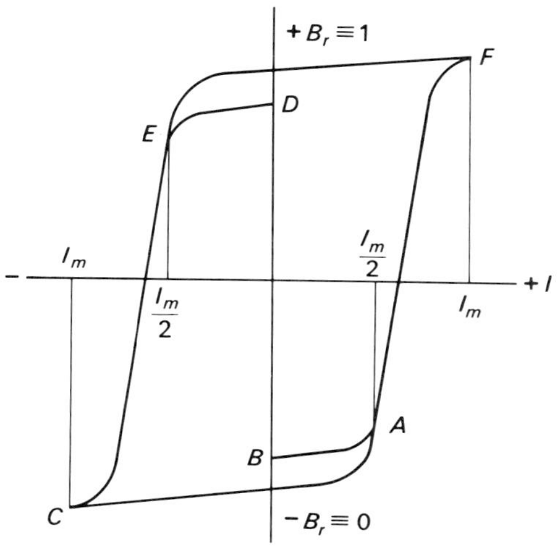

Fig. 7-3

only by trying to reverse it, a driving or clock-pulse winding is provided for this purpose. As a result, a voltage will be induced in the read-out winding. This leads to the following cases:

1. A current of, say, $-(I_m/2)$ (driving or read current) or a current of $+(I_m/2)$ (set or write current) brings the core, respectively, from $+B_r$ (1) to point E of the loop, and from $-B_r$ (0) to point A. When the read current reverts to 0, the state of the core changes, respectively, to points D and B on the B ordinate. One of the basic properties of ferrite cores is that these values on the B ordinate do not decrease very much further, even after applying a number of "half-read" and "half-write" currents at a fast rate. A core can thus remain in two slightly different 0 states, viz., point $-B_r$, which is the "undisturbed-zero" point, and point B, which is the "disturbed-zero" point. A read current of $-I_m$ brings the core in state C of the loop of Fig. 7-3. This causes a flux change in the core consisting of of a nonreversible part caused by the difference between points B and $-B_r$ and a reversible part resulting from the gradient of the loop between points $-B_r$ and C. A core remaining in the 0 state thus induces a voltage

in the read-out winding under the influence of a half-read current through the driving or clock-pulse winding.

2. It will also be immediately apparent that a core can also remain in two slightly different 1 states, namely, $+B_r$, the undisturbed-one point, and D, the disturbed-one point. A read current of $-I_m$ brings the core in state C of the loop of Fig. 7-3, so that the magnetic state of the core is reversed. This induces a relatively high voltage in the read-out winding, the peak value of which will depend on the starting point ($+B_r$ or D) on the hysteresis loop.

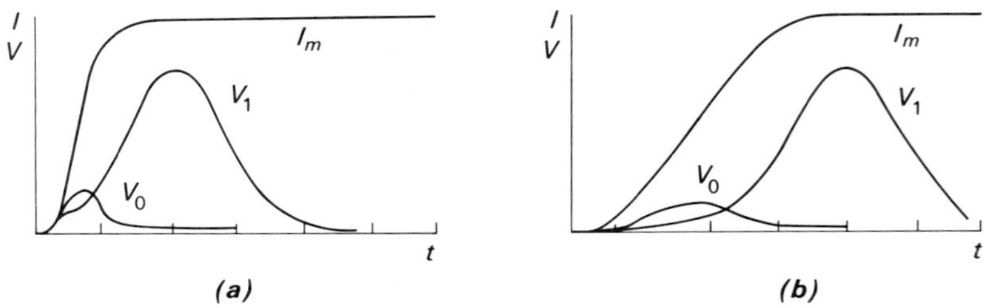

Fig. 7-4

It should be pointed out here that a half-read current causes a magnetic core in the undisturbed-one state to induce a small voltage in the read-out winding of this core.

The graphs of Fig. 7-4*a* and *b* show the output voltages V_1 and V_0 induced in the read-out winding of a ferrite core, as they might be found to occur in an actual case. The read current I_m is also indicated in these graphs. The voltage V_0 is the disturbed-zero read-out voltage, this being the most unfavorable case.

These graphs show a delay in the occurrence of the V_1 pulse, which depends on the gradient of the read current I_m. This delay is due to the irreversible processes in the ferrite material. The unwanted V_0 voltage pulse occurs earlier in time than the wanted V_1 pulse. This makes it possible to discriminate between read-out pulses, wanted and unwanted, not only on the basis of amplitude but also on the basis of time.

The peak values of the V_1 pulses in Fig. 7-4*a* and *b* do not differ very much. The higher the gradient of the peak current I_m, the earlier the V_1 pulse will occur.

There is a rather large difference between the peak value of the V_0 pulses in both graphs, so that the worst case has to be taken into account.

The winding of a core shows very different impedances depending on the magnetic state of the core. During the complete reversal of the magnetic state of a core from the 0 state ($-B_r$) to the 1 state ($+B_r$), or vice versa, a winding on a magnetic core shows a high impedance. However, a core winding shows a very low impedance when the state of the core is changed from point $-B_r$ to A and B as well as from points B or $-B_r$ to C and back to $-B_r$. The same holds for the analogous state changes starting from the 1 state.

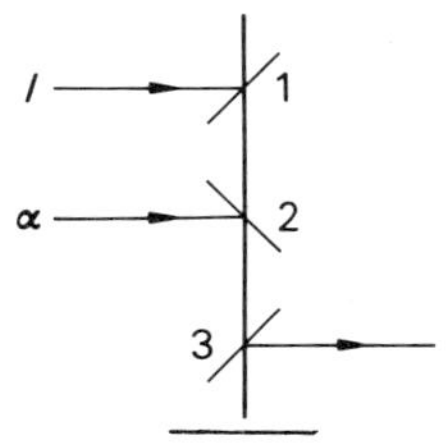

Fig. 7-5

An important operating condition where magnetic cores are concerned is the temperature, in that the properties of a magnetic core change with temperature. A core shows a decreasing coercive force with increasing temperature. This feature can be compensated by automatically decreasing the magnetizing current I_m with increasing temperature. If this automatic current compensation is not applied, a loss will occur in the operating margin of the circuit. This effect can easily be ascertained because an increase in temperature with constant magnetizing current has the same effect on a ferrite core as an increase in current under constant-temperature conditions.

From the point of view of the practical use of ferrite cores, it is desirable that these cores have the smallest possible diameter, because magnetizing currents decrease with core size, and so does heat dissipation. As a consequence of this decreased heat dissipation, higher operating speeds can be achieved.

The graphic representation of a magnetic core used in Fig. 7-1 is not very practical. The most commonly used symbolic notation for magnetic cores and windings is the mirror method of Karnaugh. A core will be indicated as a straight vertical line, as shown in Fig. 7-5. The windings will be shown in this book as short lines crossing the core at 45° angles, and the input and output wires will be drawn perpendicularly to the core. In symbolic diagrams, these wires terminate in or originate from the cross point of a winding and the core. The direction of the current is indicated by an arrow on the input or output wire. The symbolic notation for a three-winding core is shown in Fig. 7-5. The signal current I through the read-in winding 1 brings the core in the 1 state. The input current I "reflects against the read-in mirror" so that the 1 state of the magnetic core follows the direction indicated by the arrow on the core. The inquiry or read pulse α, generally a clock pulse, is reflected downward on winding 2. This resets the core from the 1 state into the 0 state and induces a current in the read-out winding. This induced output current is opposite to an input current in the same winding causing the same magnetic reversal of the core.

If for the sake of clearness in the diagram a core has to be drawn as a horizontal line, it is advisable to turn the whole symbol 90° clockwise. The direction of the currents is always chosen from left to right or from top to bottom.

The design of switching circuits by means of magnetic cores used as switching elements is more complex than the corresponding design work with contacts, diodes, transistors, or tubes. There are several reasons for this. In relay circuits, the switching function $S = z'$ can be obtained simply by means of a break contact. It is possible, however, to use a make contact to generate the switching function $S = z'$; all one need do is take a differential relay and magnetize one winding of that relay permanently. The desired switching function is then generated in the magnetic medium and not in an electric circuit. If in core circuits the switching functions are generated in the magnetic medium, some complications will arise.

1. The magnetic core has the same type of input and output media, i.e., windings. A switching signal can therefore pass through them in both directions. Relays, tubes, transistors, and diodes, on the other hand, are unidirectional switching elements.
2. The resulting switching effect of all windings on a magnetic core is determined by the physical addition of the cooperating and counteracting magnetomotive forces. This is a marginal process that imposes severe limitations on the stability of switching circuits using magnetic cores. Moreover, core characteristics more or less vary with temperature.
3. Some switching functions can be generated rather easily; others need marginal methods and timing effects, which reduce the stability of the switching circuit.

It should be pointed out here that logical circuits with magnetic cores can in principle be set up in the following two different ways:

1. When writing in a core, by addition of the magnetic fields originating from the write currents in the various windings, each of which represents a variable in the wanted switching function.
2. When reading a group of cores, each of which represents a variable or a part of the desired switching function, by algebraic addition of the voltages induced by a clock pulse in a series connection of the read windings on these cores.

Special effects can be achieved by means of cores with several apertures. The stability of operation can be increased by such switching devices. However, the decision about whether the use of rather expensive multiaperture cores with a simple system of core interconnection would be more satisfactory than single-aperture cores with a more-or-less complicated interconnection system will largely be based on economic considerations.

Some of the many possible ways of generating switching functions by means of magnetic cores will be discussed in the following sections. In principle, all switching functions can be implemented by means of magnetic cores, but some simple functions are not easy to generate, whereas some more complicated functions can be realized in a very simple way.

7-2 THE REALIZATION OF SWITCHING FUNCTIONS BY ADDITION OF MAGNETIC FIELDS ON A CORE

CLOCK-PULSE CIRCUITS

Functions of one variable The switching function $S = z$ can be implemented by means of the basic circuit shown in Fig. 7-6. The input signal

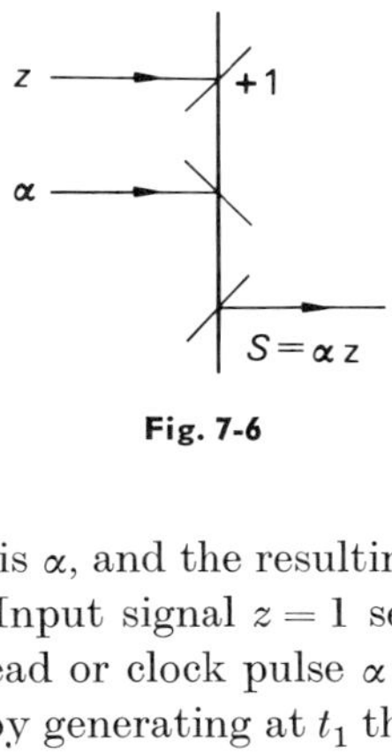

Fig. 7-6

is z, the read or clock pulse is α, and the resulting read-out pulse symbolizing the state of the core is S. Input signal $z = 1$ sets the core in the 1 state at a certain moment t_0; the read or clock pulse α resets the core in the 0 state at a later moment t_1, thereby generating at t_1 the read-out signal $S = 1$.

A core must always be reset in the 0 state before a new input signal is applied. Input signal $z = 0$ (no pulse) leaves the core in the 0 state at moment t_0 so that the read or clock pulse α only generates a noise pulse in the read-out winding at moment t_1.

The second function with one variable is $S = z'$. This function is not quite as simple to generate in the magnetic-core technique. This function can be implemented by using the function $S = y'z$ as a starting point, where z is always 1. The layout corresponding to this function will be shown later.

Functions of two variables The easiest two-variable switching function when using core techniques is $S = y + z$. The diagram corresponding to this switching function is shown in Fig. 7-7. A signal $y = 1$, or a signal $z = 1$, or both simultaneously, bring the core in the 1 state. As a result, the read or clock pulse α applied at a later moment will induce a pulse $S = \alpha(y + z)$ in the read-out winding, which will be the same as $S = y + z$, since $\alpha = 1$.

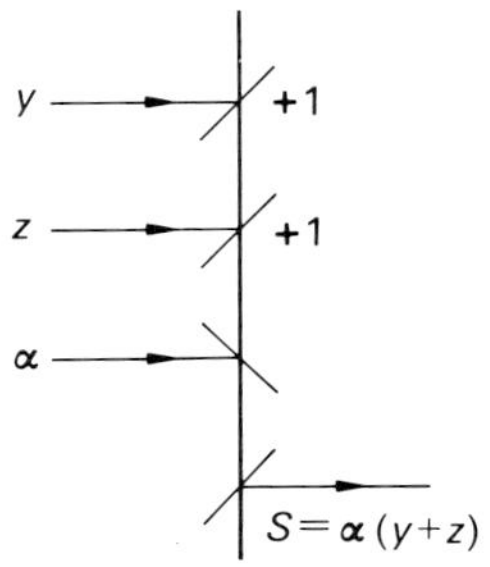

Fig. 7-7

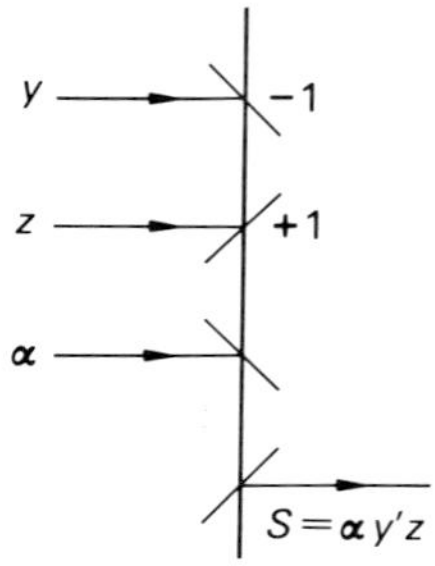

Fig. 7-8

The second switching function with two variables that would seem to be suited for a one-core network is $S = y'z$. The truth table corresponding to this switching function is given in Table 7-1, and its core diagram is given in Fig. 7-8.

Table 7-1

y'	z	S
0	0	0
0	1	0
1	0	0
1	1	1

The switching function $S = y'z$ can be implemented by means of two opposite input windings on a single core. The noninverted variable is connected to the input winding that brings the core in the 1 state, whereas the inverted variable is connected to an input winding that brings the core in the 0 state.

The input-signal combinations $y = 1$, $z = 0$ and $y = 0$, $z = 0$ leave the core in the agreed rest condition.

The input combination $y = 0$ and $z = 1$ brings the core in its 1 state by magnetization of the z signal.

The input-signal combination $y = 1$ and $z = 1$ results in the wanted output signal $S = 0$ *only if* the y signal is larger in ampere-turns and duration than the z signal. The first effect can easily be obtained by providing the Y winding with a larger number of turns. However, the time requirement is more difficult to meet. The y pulse has to start before the z pulse and has to end after the z pulse. It is of course possible to fulfil these requirements at the cost of some complexity. However, if the switching function to be generated is $S = y'z + yz'$, the requirements for the y and z signals become contradictory.

It is advisable to avoid the principle of the diagram of Fig. 7-8 in actual core circuits.

The switching function $S = y \cdot z$ will require at least three cores when using the elementary-core switching method described in this section. This may be considered as too complicated. Moreover, there exist other more practical core switching methods to generate this function, so that function $S = y \cdot z$ may be left out of consideration in this connection.

The implementation of $S = z'$ can be derived from the function $S = y'z$ where $z = 1$. The signal $z = 1$ must be a pulse with the same features with respect to the y signal as in the case of $S = y'z$. The $z = 1$ pulse can be generated by a special core.

CORE CIRCUITS WITH PREMAGNETIZED WINDINGS

The core circuits described in the preceding subsection operate under the control of an inquiry or clock pulse. The core circuits described in this subsection will be controlled directly by the input signals without using a clock pulse. The principle of these types of circuits can be described as follows: the core is always reset in the 0 state by the premagnetized winding, and it can attain magnetization into the 1 state only if the input-signal combination satisfies the required switching function. In these core circuits, there is no marginality in time requirements and current level as was found to occur in the circuits described in the preceding subsection. The various types of circuits developed along these lines have some advantages in that the required switching functions, complex as they may be, can be obtained in one clock-pulse period. The result of the operation is then available at the very moment the next operation must be performed. The memory function of the core is lost in this type of circuit.

This switching method is particularly well suited for polar relay windings.

The function $S = z$ can be implemented by means of the single-core circuit shown in Fig. 7-9. The winding in Fig. 7-9, which is indicated by $=$, magnetizes the core permanently with a magnetizing force with a relative value $H = -1$ that holds the core in the agreed 0 state. This is shown in the B-H hysteresis loop of Fig. 7-10. The z signal is applied to a second winding of the core, which causes a magnetizing force with a relative value $H = +2$, so that the total relative magnetic force is $+1$. This brings the core in the

1 state, so that a pulse $S = 1$ is induced in the read-out winding. It should be borne in mind that the resulting magnetizing force $H = +1$ must be high enough for the core to reach safely the saturated region, as shown in Fig. 7-10. After the end of the pulse $z = 1$, the core reverts to its initial 0 state under the influence of the permanent magnetizing force $H = -1$ of the = winding. This also induces a pulse in the read-out winding, but this pulse is of opposite polarity.

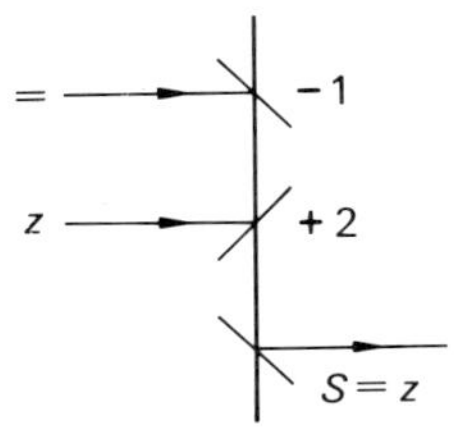

Fig. 7-9

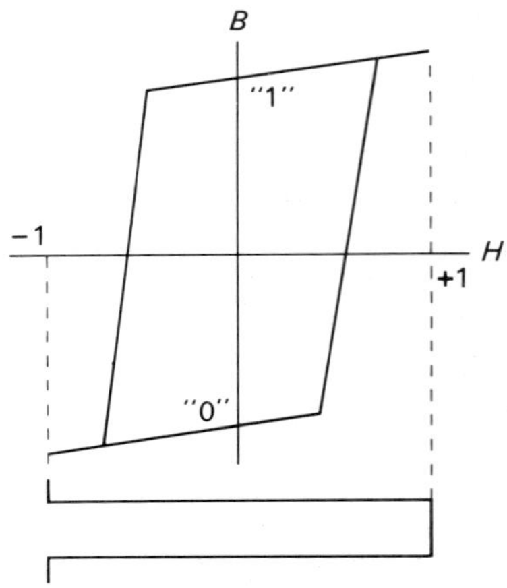

Fig. 7-10

Accordingly, a pulse is generated on the leading and trailing edges of the input-signal pulse $z = 1$. The transfer circuit to other cores will be the deciding factor as to which of the two generated signals will be used as read-out signal $S = 1$.

The switching function $S = z'$ is not easily generated in core circuits with premagnetized windings. The realization of this function will accordingly not be shown here.

The switching function $S = y + z$ is easily obtained by means of a core with two input or read-in windings. Both windings have to exercise a relative magnetizing force $H = +2$ against the constant magnetizing force $H = -1$. A core generating this function is shown in Fig. 7-11. The input signal $y = 1$, or $z = 1$, or both input signals simultaneously, reverse the 0 state of the core into the 1 state.

The function $S = y \cdot z$ can also be easily implemented with a premagnetized core. The required core circuit (Fig. 7-12) is analogous to the layout of Fig. 7-11 except that the relative magnetizing force exercised by each of the read-in windings is $+1$; so that it is only when the y and z signal are applied

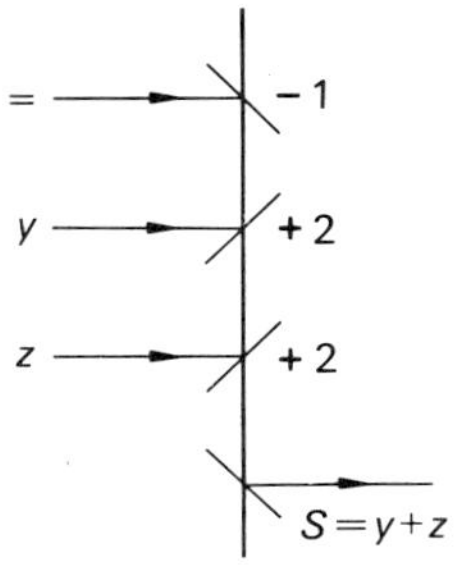

Fig. 7-11

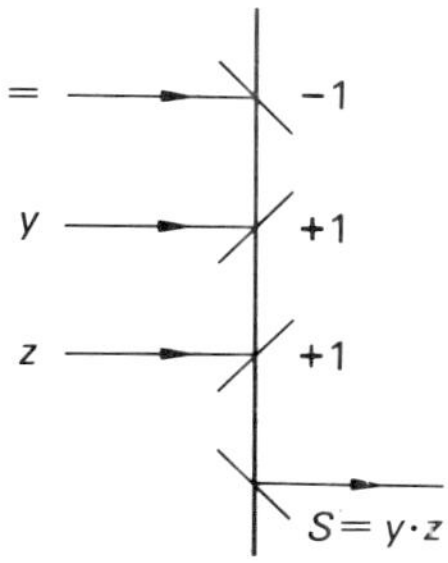

Fig. 7-12

simultaneously that the magnetizing force on the core can reach the relative value of +1, which results in the reversal of the core from the 0 state into the 1 state. This process could be referred to as a marginal switching method.

A single core can generate OR functions for an arbitrary number of variables. There is no practical limit to the number of variables, because one of the variables of the maxterm function is sufficient to reverse the magnetic condition of the core. Figure 7-13 shows the core circuit corresponding to the switching function $S = x + y + z$.

The minterms with several variables can also be implemented by means of single-core circuits, but the number of variables is stringently limited. The resulting magnetizing force $H = +1$ corresponds to the difference between a relatively large permanent magnetizing force $-H_p$ and the sum of a number of small magnetizing forces H_v, so that for n variables,

$$n \cdot H_v - H_p = +1 \tag{7-1}$$

The core circuit representing the function $S = x \cdot y \cdot z$ is shown in Fig. 7-14. The tolerances of a single-core circuit corresponding to a minterm with more than three variables become too small in general.

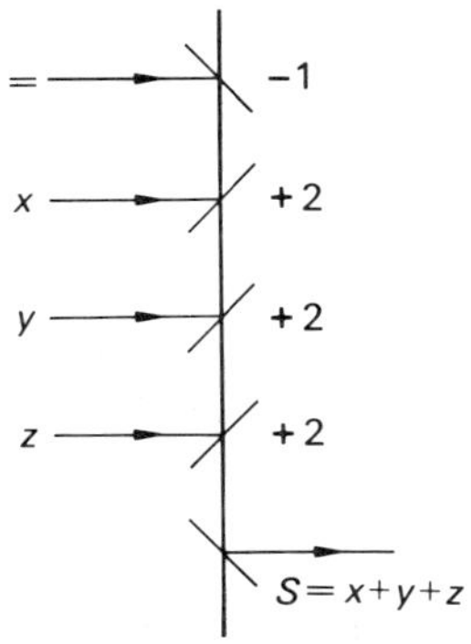

Fig. 7-13

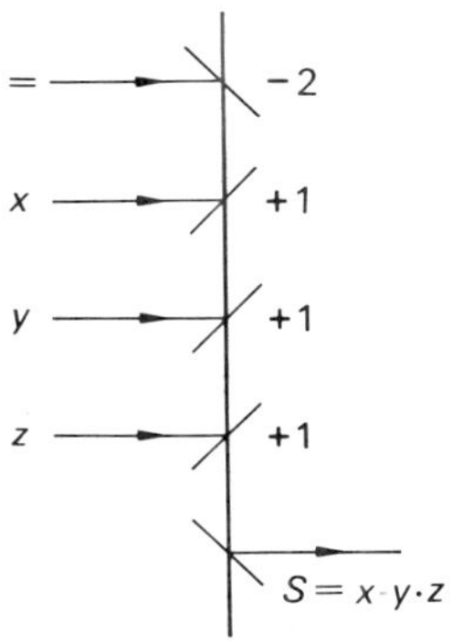

Fig. 7-14

The switching function $S = xy + yz + xz$ can also be implemented by means of a single-core circuit, as shown in Fig. 7-15. The relative value of the permanent magnetizing force must be $H_p = -1$, whereas the input windings for the variables x, y, and z have to exercise a relative magnetizing force

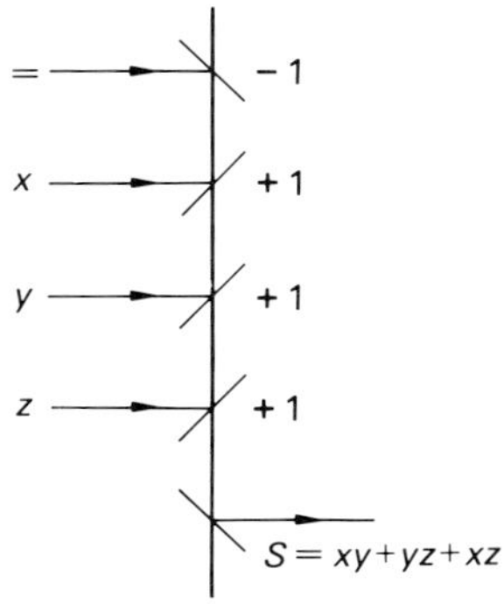

Fig. 7-15

of $H_v = +1$. The coincidence of each pair of variables reverses the magnetic condition of the core from the 0 state into the 1 state.

The switching function $S = y'z$ (inhibit function) does not lend itself readily to the use of this switching method based on the use of cores.

CORE CIRCUITS WITH PREMAGNETIZED WINDINGS AND READ-OUT BY CLOCK PULSES

A large group of switching circuits for computers operate on a system of clock pulses, i.e., pulses produced by a central pulse source, on which all operations of the switching system are synchronized. No change of state of any of the switching elements is possible unless it is triggered by a clock pulse $\alpha = 1$. Through the various windings of a magnetic core, the clock pulses generate the desired switching functions, which are then transferred in the form of a pulse to the next switching stage. Sometimes the clock-pulse source generates three such pulses with a difference in phase to arrive at complicated switching functions in a number of steps.

There are two different ways in which a switching function can be implemented in a system with premagnetized windings and clock pulses. These two different ways can best be illustrated by means of the switching function $S = y \cdot z$.

In the circuit of Fig. 7-16, for $y = 0$ and $z = 0$, the core is held permanently in the agreed 0 state by a relative magnetizing force of $H = -2$. Only the coincidence of $y = 1$ and $z = 1$ makes the relative magnetizing force $H = 0$ on the core. A clock pulse $\alpha = 1$ exercises a relative magnetizing force of

$H = +1$ on the core and then reverses the magnetic flux, so that the core changes from the 0 state into the 1 state. This induces a pulse $S = \alpha \cdot y \cdot z$ in the read-out winding of the core. In this system, all signal pulses have to start before the clock pulse begins and must end after the end of the clock pulse. This type of core logic is not very practical in a system where practically all signals representing switching functions are generated as pulses.

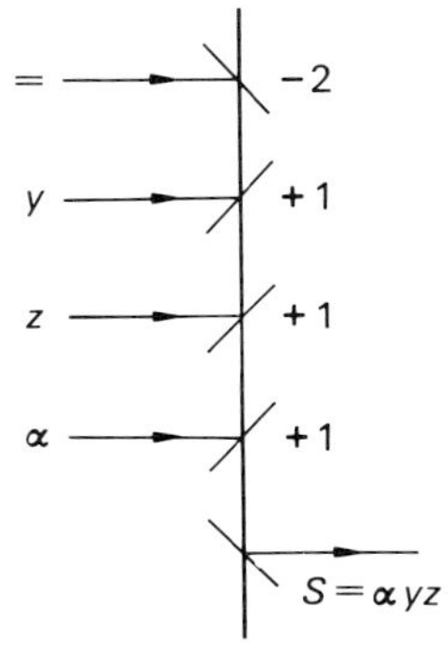

Fig. 7-16

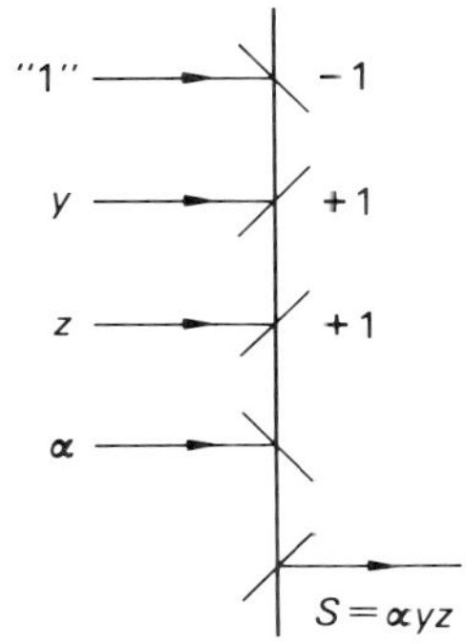

Fig. 7-17

The core circuit of Fig. 7-17 shows a more practical method for obtaining the switching function $S = y \cdot z$. The input signals y, z, and 1 are supposed to be applied as simultaneous signals generated by other core switching stages. The y and z pulses can produce a relative magnetizing force on the core of $H = +1$, whereas the 1 pulse applies a relative magnetizing force $H = -1$ on the core. If the core is normally in the 0 state, only the coincidence of $y = 1$ and $z = 1$ can put the core in the 1 state. As will be shown in Sec. 7.4, a pulse is then generated in the read-out winding of such a polarity as to be ineffective. The clock pulse $\alpha = 1$ exercising a relative magnetizing force of at least $H = -1$ rapidly reverses the state of the core from 1 to 0. This induces a pulse of the desired polarity in the read-out winding, which pulse corresponds to the switching function $S = \alpha \cdot y \cdot z$.

The advantage of this method is that it fits in with other core circuits of the same type and that the clock-pulse current may be given several times its minimum value, with a corresponding increase in the speed of the flux reversal in the core. Another advantage of the core circuit of Fig. 7-17 is that it fulfils a memory function for the switching function which is stored in it.

The realization of the switching functions $S = \alpha(y + z)$ and $S = \alpha y' z$ for which no premagnetized auxiliary winding is required was already dealt with above in this section.

A number of other switching functions can be implemented on one core by means of the various polarity combinations of the read-in windings and the degree of magnetization of the winding that is premagnetized with a 1 pulse. Some of the possible layouts for switching functions with three variables are shown in Figs. 7-18 to 7-21.

In these AND-type core circuits, the relative value of the magnetizing force of the permanently magnetized winding is equal to the number of noninverted input variables minus 1. The tolerances of these circuits decrease

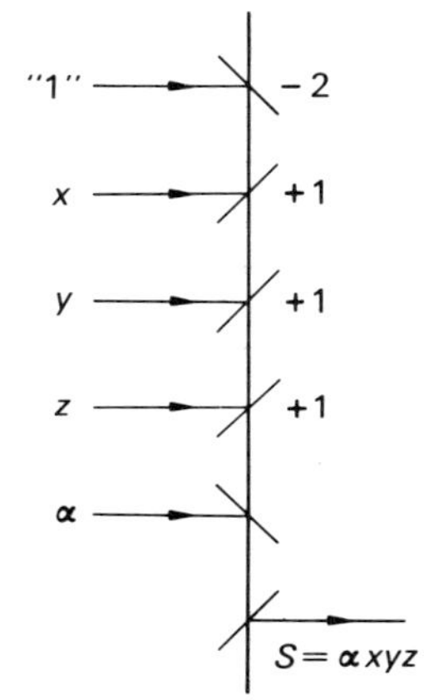

Fig. 7-18

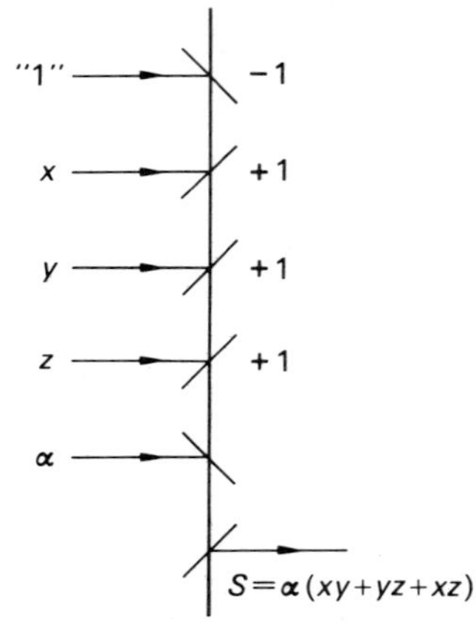

Fig. 7-19

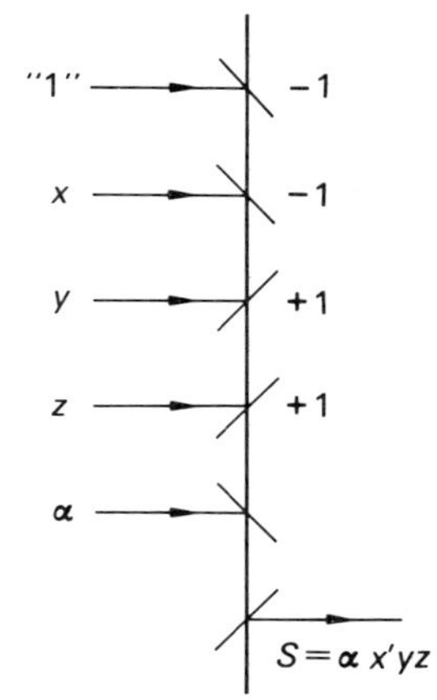

Fig. 7-20

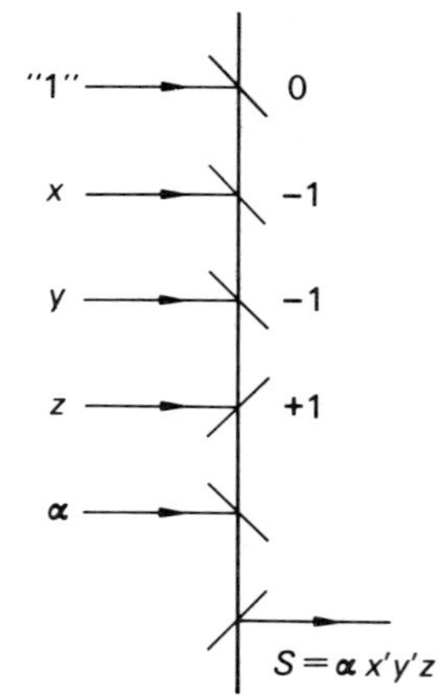

Fig. 7-21

very rapidly with increasing numbers of input variables. They depend not only on the properties of the cores themselves but also on the stability of the driving currents passing through the various windings.

7-3 THE REALIZATION OF SWITCHING FUNCTIONS BY ADDITION OF VOLTAGES INDUCED IN SERIES-CONNECTED READ-OUT WINDINGS

A group of switching functions can be implemented by a core circuit by connecting in series, with the correct polarity, the read-out windings of a number of cores. This method may be illustrated by some simple examples.

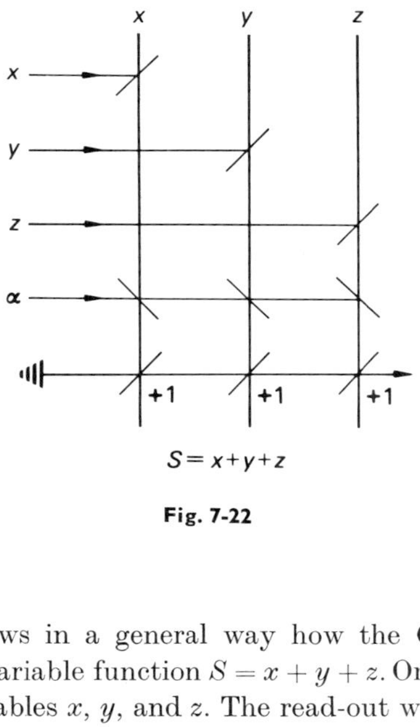

Fig. 7-22

Figure 7-22 shows in a general way how the OR operation can be obtained for a three-variable function $S = x + y + z$. One core is provided for each of the three variables x, y, and z. The read-out windings are connected in series, using the same polarity, so that the output voltages will be added. A read amplifier connected to the read-out circuit must generate a pulse when triggered by a V_1 read-out pulse generated by any one of the cores, by all possible pairs of cores, or by all three cores simultaneously.

The V_0 read-out pulse generated simultaneously by all three cores must not operate the read amplifier. Taking into account the limitations inherent in core circuits, this type of circuit can be made to operate reliably.

Figure 7-23*a* shows a possible AND-gate circuit for the switching function $S = x \cdot y \cdot z$. In this circuit, a positive read-out pulse to operate the amplifier will be generated only if all three input variables x, y, and z have been given, or "written" into the cores. In all other cases, no pulse, or a negative

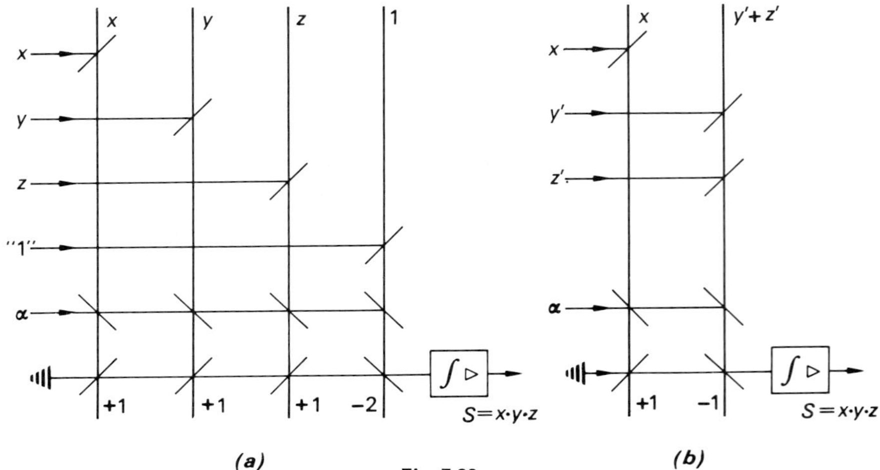

Fig. 7-23

pulse, will be generated in the read-out circuit. The operation of this type of circuit can also be reliable because all pulses can be of a standard length.

A variant of this circuit is given in Fig. 7-23*b*, in which only two cores are used. This variant is only possible if the variables y and z are available in their negated form.

The read amplifier must operate only on receiving a positive read-out pulse.

With ferrite cores being commercially available, there is one fact which must be taken into account, namely, that the V_1 read-out pulses are generated with a different delay even though they are generated under the control of the very same clock pulse. The resulting effect of two V_1 pulses of opposite polarity, as may be found in the read-out circuit of Fig. 7-23, is shown in Fig. 7-24 (line S). This fact by itself would make it impossible to provide for

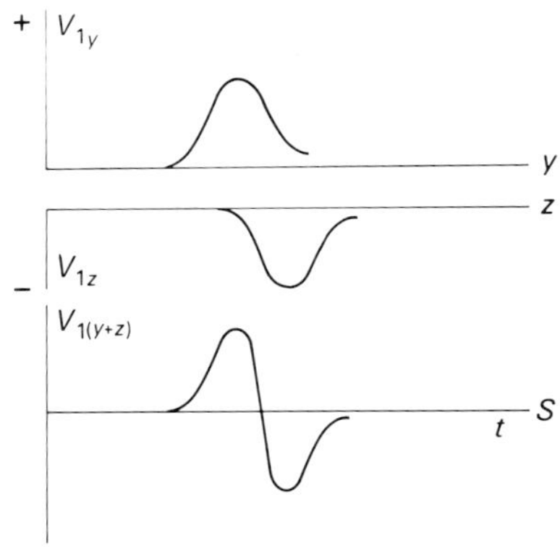

Fig. 7-24

the AND operation by means of the circuit of Fig. 7-23. However, it appears that for different cores of the same type, $\int V_1 dt$ is sufficiently zero after an integrating network of a very simple type for a 0 output of the core read-out circuit.

The negation of a variable or a function can be performed as follows, by means of the AND operation:

$$z' = 1 \cdot z' = 1 \cdot (z)'$$

This AND relation is indicated in the circuit of Fig. 7-25.

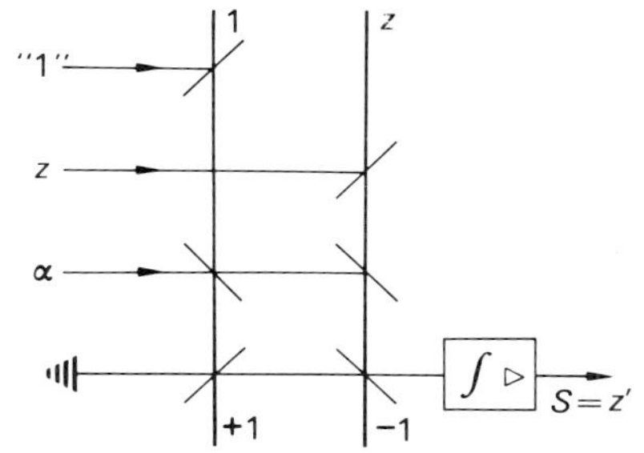

Fig. 7-25

In Sec. 7-7 a method will be reviewed in which this switching principle will be used to solve design problems for more complicated switching functions.

7-4 INFORMATION TRANSFER BETWEEN MAGNETIC CORES

If a magnetic core is part of a large core switching circuit, information has to be transferred from one core switching stage to the next stage. As was already indicated, the reversible character of a core may lead to some difficulties which may be solved in various ways.

DIODE TRANSFER LOOP

The circuit shown in Fig. 7-26 consists of two cores A and B. It is assumed that the 1 state of core A has to be transferred to core B, which is in the agreed 0 state. A clock pulse α resets core A in the 0 state so that a current is induced (polarity as shown in Fig. 7-26) into the read-out winding RO of core A. Diode D_1 is conducting for that polarity of the read-out current, while diode D_2 is blocking. The diodes do not hinder the normal information transfer from core A to core B.

The difficulties start to occur for core B when setting core A, and for core A when the information is transferred from core B to a core in the next switching stage.

1. Setting of core A. The signal pulse $z = 1$ tries to induce a current in the read-out winding of core A with a polarity opposite to that indicated in Fig. 7-26. Diode D_1 blocks this polarity, so that practically no forward kick is generated at the read-in moment of core A. Moreover, the read-in winding of core B does not form an undesired load for the set pulse $z = 1$ of core A.

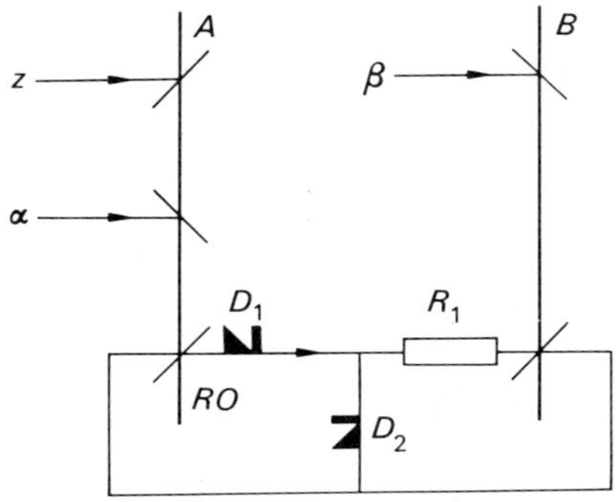

Fig. 7-26

2. Read-out of core B. During the read-out of core B, its 1 state is changed into the 0 state. This induces the desired output pulse in the read-out winding, but a voltage is also induced in the read-in winding of core B. This undesired voltage tries to set the A core in the 1 state. Diode D_2 forms a short circuit on the read-out winding RO of core A, so that an undesired effect on the A core is eliminated. In practice, a small resistance (R_1) will be inserted in the transfer loop between cores A and B in order to eliminate a short circuit of the read-in winding of core B during the read-out period, since this would mean an extra load for the clock pulse α. The A core remains in its 0 state for an information transfer $z = 0$ to the B core. The clock or inquiry pulse α induces a noise signal V_0 in the read-out winding of core A as is shown in Fig. 7-4.

During the flux reversal of a core, e.g., from the 1 state into the 0 state, a magnetic core and its windings act like a transformer. This means that if the current of the driving pulse α is set at several times the minimum required to reverse the state of the core with a sufficiently fast rise time, a rather large energy can be made available in the read-out winding to switch one or more other cores. This is somewhat comparable to an amplification effect for the information signals to be transferred.

It is necessary, however, to note another important factor in this process. The flux reversal in the driven cores has to be safely completed before the flux reversal in the driving core has come to an end. Otherwise the information-transfer process between cores cannot be extended arbitrarily as would be required in a practical switching system. This condition can be fulfilled by making the number of turns on the read-out winding of the driving core greater by an adequate margin than the number of turns on the read-in winding of the driven core.

The noise signal generated on transferring a 0 has practically no effect on the receiving core because its effect on that core will be substantially smaller than $I_m/2$ (Fig. 7-3).

DIODE TRANSFER LOOP WITH NOISE-SIGNAL COMPENSATING CORES

In the transfer loop as shown in Fig. 7-26, a noise signal V_0 is generated for transferring a 0 from core A to core B because the hysteresis loop is not ideally rectangular. It is theoretically possible to generate in two cores equal but opposite noise voltages V_0 which compensate each other. A transfer loop in which this compensation is performed is shown in Fig. 7-27. Core A_1 is

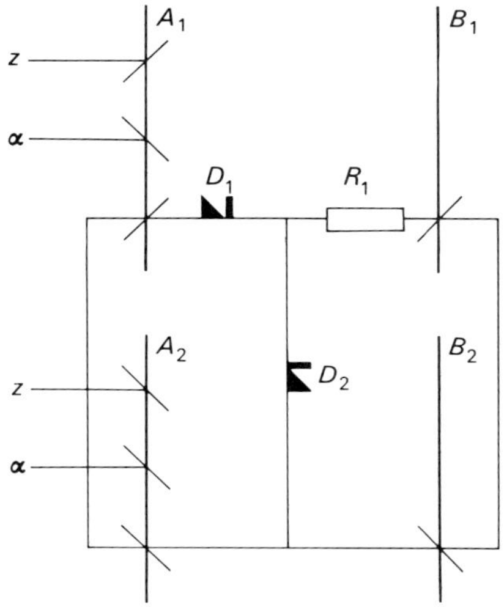

Fig. 7-27

set into the 1 state by an incoming signal $z = 1$, whereas the auxiliary core A_2 is brought in the 0 state by that same signal. The voltages induced in the read-out windings of cores A_1 and A_2 have opposite polarity. The α-pulse windings on both cores have the same polarity. During the transfer of a 1 from core pair A to core pair B, core A_1 produces the desired signal pulse, whereas core A_2 produces a noise pulse which is subtracted from the desired signal pulse, because the read-out windings of cores A_1 and A_2, being of opposite polarity, are connected in series. This results in a small loss in amplitude. For the transfer of a 0, both cores A_1 and A_2 stay in the 0 state. Both cores generate a noise signal V_0 which is of opposite polarity, so that no current will flow in the transfer loop if both cores have the same hysteresis loop.

The circuit of Fig. 7-27 makes it possible theoretically to use cores with poorly shaped hysteresis loops, but the differences between these types of cores will be so large that the effect will generally be worse than if a transfer loop were used of the type shown in Fig. 7-26 with only one core with a good hysteresis loop. The wiring complexity and the use of two cores instead of one increases the cost of the circuit of Fig. 7-27 by such a margin that the price of cores good enough for the transfer loop of Fig. 7-26 may be at least three times as high as that of cores just adequate for the circuit of Fig. 7-27.

TRANSISTOR TRANSFER LOOP

A complete separation of the cores in the successive switching stages can be obtained by means of a transistor, as is shown in Fig. 7-28. The read-out winding of core A is connected to the base of transistor T_1.

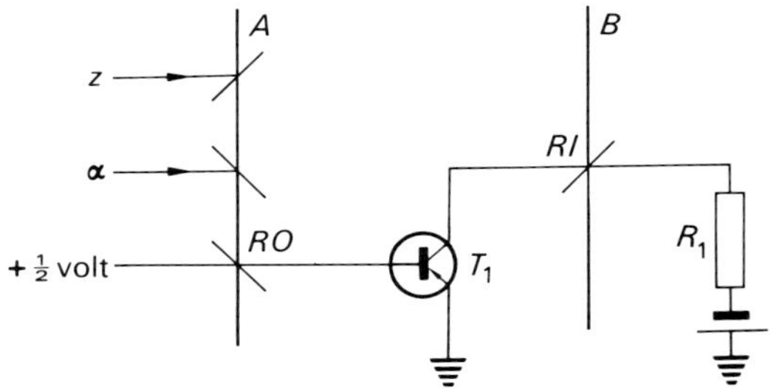

Fig. 7-28

The transistor is blocked by a small positive voltage applied to the base. The read-in winding of core B is connected to the collector circuit of transistor T_1. A read-in signal $z = 1$ in core A induces a voltage in the read-out winding of this core, its polarity being such that the base voltage of transistor T_1 is made more positive; accordingly, the transistor remains in the blocking condition. No through kick is generated and no extra load of the inquiry or clock pulse α can delay the setting of core A from the 0 state into the 1 state.

A clock pulse α which resets the A core from its 1 state into the 0 state induces a negative voltage in the read-out winding RO of core A, so that the collector of transistor T_1 is made conducting. The collector current sets core B into the 1 state. It is possible to set a relatively large number of cores (say ten) with this type of transfer loop, whereas the number of cores to set would be rather limited (two or three) with the diode transfer loop.

It should be pointed out that a core cannot kick back through this type of transfer loop during the clock pulse, because a transistor is a unidirectional element.

The noise signal which is generated in the read-out winding of core A for the transfer of a 0 has so small an amplitude that it cannot overcome the small positive base voltage of $\frac{1}{2}$ volt. This bias voltage can be chosen in relation to the squareness ratio of the hysteresis loop of the cores.

The transistor transfer loop has the advantage that the cores in the various switching stages operate independently. Furthermore, the requirements as regards rectangular features of the hysteresis loop, amplitude and duration of the inquiry or clock pulses, are not so very stringent.

TRANSISTOR TRANSFER LOOP WITH FEEDBACK

The advantages of the latter method can be further increased by applying the collector current to a winding on core A, as is shown in Fig. 7-29. The

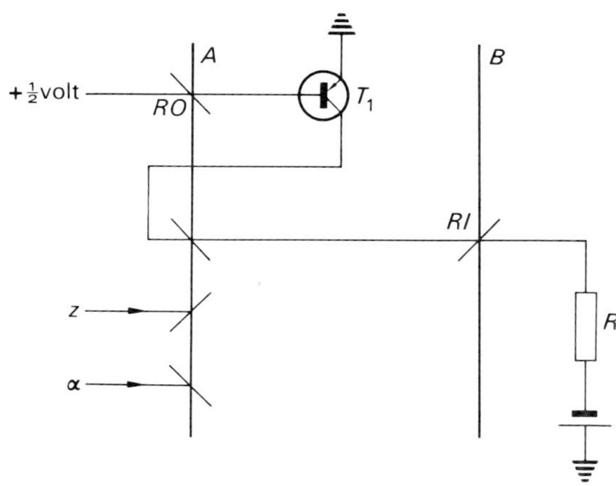

Fig. 7-29

feedback current increases the reset effect of the inquiry or clock pulse α and compensates for the already small load of the base current on the clock pulse α. A kind of blocking oscillator effect is thus obtained. This regenerating effect will last until the core has been completely reset to 0. A low-powered α pulse of short duration is sufficient to start the core-reset process from 1 to 0. The collector circuit of the transistor produces most of the reset power for the A core. The low positive bias on the base of transistor T has to be made slightly higher relative to the previous case where there was no feedback, because an interference of very short duration, which results in a very small negative bias on the base, will open the transistor completely. This is a disadvantage of this type of transistor transfer loop.

THE SPLIT-WINDING TRANSFER LOOP

The fundamental circuit of the split-winding transfer loop is shown in Fig. 7-30. Core A is the transmitting core and core B the receiving core. The inquiry

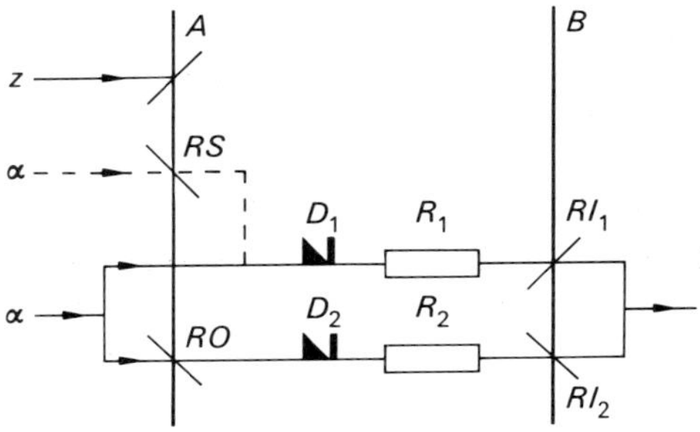

Fig. 7-30

or clock pulse α is sent directly through the read-out winding of core A. The receiving core B is provided with two read-in windings RI_1 and RI_2 which have an equal but opposite number of ampere-turns. The winding RI_2 is connected in series to the read-out winding RO or core A; the winding RI_1 is directly connected to the α-pulse source.

If core A is in its 0 state, the read-out winding of core A will have a very low impedance during the α pulse. In that case, the information transfer current will be divided practically equally over both branches of the transfer loop, so that core B receives equal but opposite magnetizations on both windings.

If, however, core A is in the 1 state, it will be reset by the α pulse through winding RO. During this pulse, the impedance of winding RO will be rather high. The currents in both branches of the transfer loop differ so much that the difference in magnetization on both windings of the receiving core B will set this core in the 1 state.

The split-winding transfer loop does not produce a through kick to the next core, nor a back kick to the preceding core in the switching logic, because all currents induced in the various windings of a transfer loop other than the α-pulse current are blocked by one direction or other of diodes D_1 and D_2. Resistors R_1 and R_2 have a balancing effect for compensating small differences in the diode characteristics. This results in an efficient type of transfer loop. A transmitting core can control about five receiving cores.

It should be pointed out that the driving currents through both windings of core B differ only during the time core A has not been completely reset. Since the reset current through the read-out winding RO of core A is relatively small, the current in the other branch of the transfer loop being consequently rather high, the setting of core B will be assured. Core B cannot be reset once the current in both branches of the transfer loop have become equal after the resetting of core A.

An important advantage of this type of transfer loop is that a core can be switched from its 1 state into the 0 state, and vice versa, by means of other windings on the core without generating undesired pulses in either direction in the split-winding transfer loops.

A variant of this split-winding transfer loop is shown in dotted lines in Fig. 7-30. An additional winding on core A, which is connected in series to both branches of the transfer loop, resets the A core in the 0 state on receiving an α pulse. This induces a voltage in winding RO, its polarity and magnitude being such that diode D_2 is made to block. This increases the current ratio between the two branches of the split-winding transfer loop.

THE SPLIT-PATH TRANSFER LOOP

The effect of the complete blocking of one of the transfer paths can also be obtained by means of a split-path transfer loop, as shown in Fig. 7-31. This

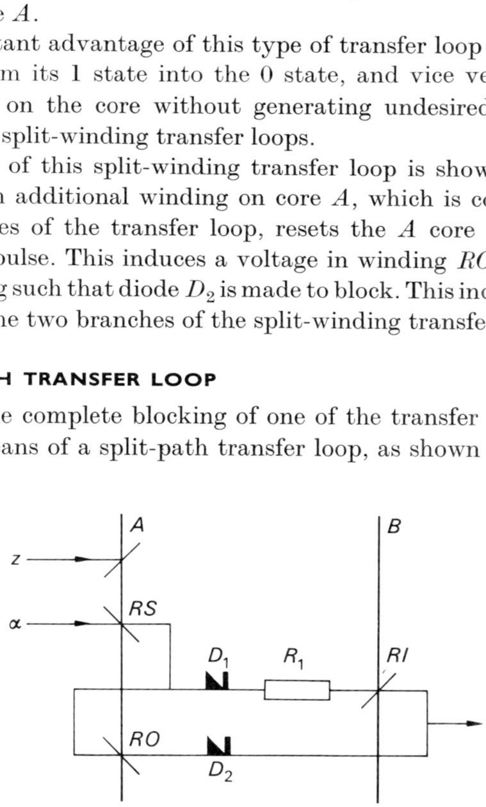

Fig. 7-31

circuit is in fact a variant of the split-winding transfer loop. For the transfer of a 1, the reset winding RS of core A again induces a voltage in the read-out winding RO of such magnitude and polarity that diode D_2 is completely blocking. The transfer pulse α then flows through diode D_1, resistor R_1, and the read-in winding RI of core B, and this pulse then sets core B in the 1 state.

In order to ensure the safe transfer of a 1 from core A to core B, it is necessary that the reversal of the flux in core B be completed before this same process is completed in core A. This can be done by providing the RI winding of core B with a higher number of turns than the RS winding of core A. Furthermore, it should be pointed out that a large circulating current will be generated in this type of transfer loop if diode D_2 is not put in a complete blocking condition during the transfer of a 1.

Other variations of this type of transfer loop can, of course, be devised, e.g., by means of extra cores. It is also possible to use the high-impedance effect of a read-out winding more or less as a break contact of a relay in a single-wire transfer loop.

THE BALANCED SPLIT-WINDING TRANSFER LOOP

The information transfer loops between magnetic cores as given above all have the disadvantage that nothing is transferred from one core stage to the next if the transmitting core is in the 0 state. This means that the load imposed on pulse sources and power supplies varies with the information to be transferred. There are switching applications where this is undesirable and where it is required to have a constant load on all pulse sources and power supplies. This requirement can be met in core circuits by means of the balanced split-winding transfer loop, as shown in Fig. 7-32.

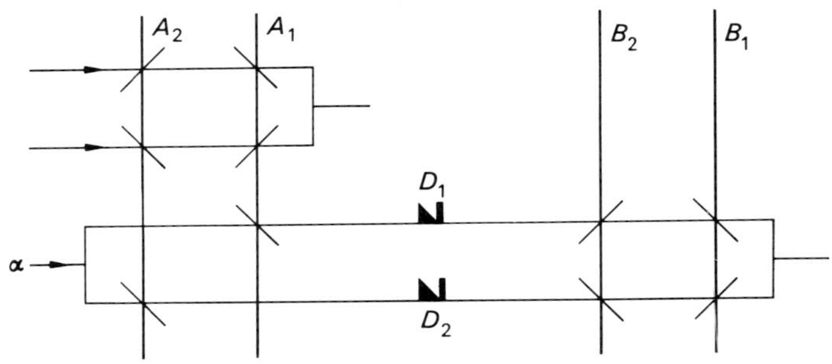

Fig. 7-32

All core stages consist of two cores, one of which is in the 0 state and the other in the 1 state, except after the read-out of a stage when both cores are in the 0 state. This situation is shown in Table 7-2.

The information contained in the core pair $A_{1/2}$ is transferred to the $B_{1/2}$ pair by the clock pulse α. The situation in the transfer loop is now such that one of the cores of a pair will always be reset to 0, whereas the other will remain in its initial 0 state. In this way, there will always be one high-impedance winding at the transmitting end of the information transfer loop, so that the core pair $B_{1/2}$ will always be set positively, either in the 0 combination or in the 1 combination. The transfer current is thus independent of the information to be transferred. The operating margin is better than with the split-winding transfer loop, since in the latter case there is no balance of currents to represent the 0 state.

Table 7-2

A	A_2	A_1
0	0	1
1	1	0
After read-out	0	0

Once a pair of cores has been read out, the pair remains in the 0 state until the next information is read in.

The design considerations for the windings of cores in a balanced split-winding transfer loop are the same as in the case of the normal split-winding transfer loop. It is possible, however, to use cores with slightly inferior hysteresis loops, since the transfer of a 1 or of a 0 will depend on the difference between a signal pulse generated by the flux reversal of a core and a noise pulse generated by the nonrectangularity of a core (see Figs. 7-3 and 7-4).

It should be borne in mind that the variant of the split-winding transfer loop, shown in dotted lines in Fig. 7-30, can also be applied to cores A_1 and A_2 of the balanced split-winding transfer loop of Fig. 7-32. This also holds for all further circuits in which the balanced split-winding transfer loop is used.

7-5 LOGICAL FUNCTIONS IN TIME-DIVISION OR CLOCK-PULSE CORE CIRCUITS

Section 7-2 was concerned with the way in which logical functions can be formed with a single core by means of marginal interaction of the various windings. These procedures proved to be applicable for a limited number of switching functions consisting of a limited number of variables. The implementation of some rather complicated switching functions proved to be very simple, whereas some very simple functions proved to be impossible to implement by means of a single core.

The latter type of switching functions can be implemented by means of a number of core stages controlled by different clock pulses in a time-division system. The transfer of information between these various core stages can be performed by means of transfer loops, as described in the previous section.

The clock-pulse logic described below is to a certain extent independent of the type of information transfer loop used, but some types of transfer loops may lead to a simpler logic than others. For this reason the various logical possibilities of the different types of transfer loops will be related to the switching function involved.

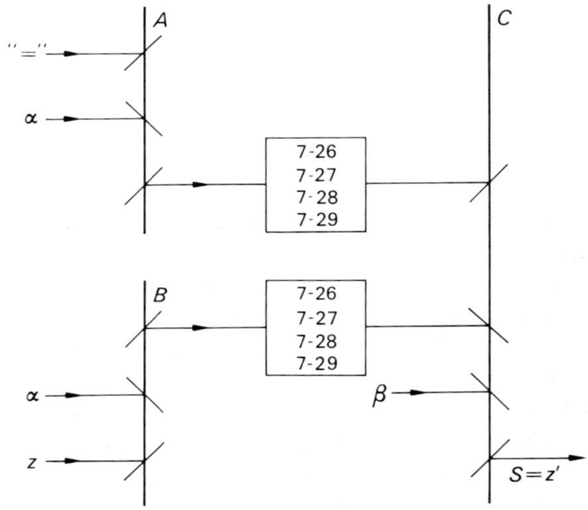

Fig. 7-33

FUNCTION $S = z'$

The function $S = z'$ can be implemented by means of three cores, as shown in Fig. 7-33. This circuit corresponds in fact to the function $S = yz'$ with $y = 1$. The function $y = 1$ is simulated by a dc current input signal on core A. A 1 signal will therefore always be transferred from core A to core C. The effect of this 1 signal on core C will be such that it will tend to set core C in the 1 state. If $z = 1$, a 1 signal will be transferred from core B to core C, but its effect on core C will be such that it will prevent the setting of core C in the 1 state by reason of the continuous presence of the 1 signal originating from core A. The state of core C can accordingly be represented by the function $S = z'$.

The information contents of core C can be transferred to further switching stages by a β clock pulse, which should necessarily not coincide with the α clock pulse used for the information transfer from cores A and B to core C.

The coupling circuit between cores A or B and C can be set up in conformity with the circuits given in Figs. 7-26 to 7-29. For the function $S = z'$, the split-winding transfer loop (Fig. 7-30) leads to the type of circuit of Fig. 7-34. No further explanation of these types of circuits is necessary.

The circuit implementing the function $S = z'$ could be limited to two cores if the split-winding transfer loop were used in connection with a receiving core with a rest state of 1 instead of the normal 0. This circuit is shown in Fig. 7-35. In this figure, the normal split-winding loop of Fig. 7-30 is used;

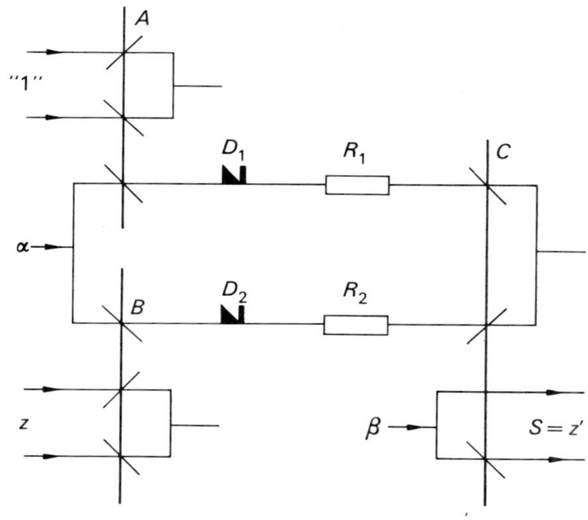

Fig. 7-34

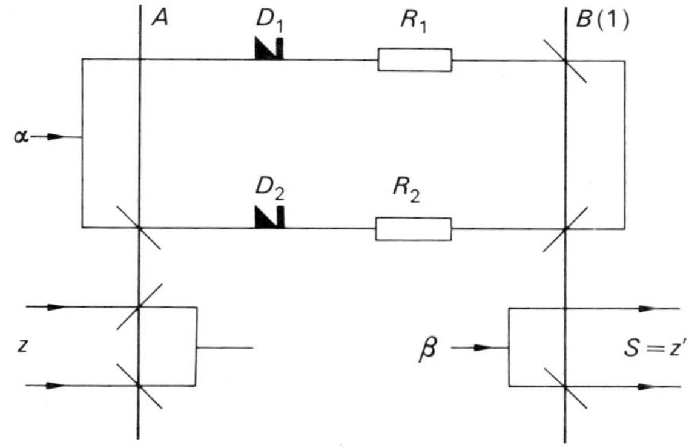

Fig. 7-35

in this version, core B will be reset to 0 on the transfer of a 1 from core A to core B, whereas core B will remain in its normal 1 rest state in the case of a transfer of a 0 from core A to core B. The read-out of core B can again be performed in a β clock-pulse phase.

The false rest condition of core B is normally immaterial because core B can be reset in its 1 state during a third γ clock-pulse phase.

The balanced split-winding transfer loop is actually the most elegant way of solving the problem of designing a circuit for the function $S = z'$. It can be done by means of a four-core circuit which is hardly more complex than the transfer loop shown in Fig. 7-32. The two wires of the transfer loop of Fig. 7-32 have to be crossed in order to obtain the output function $S = z'$. It is also possible to arrive at $S = z'$ by interchanging the read-out windings on cores A_1 and A_2. The circuit in which the function $S = z'$ is realized by means of a crossed transfer loop is shown in Fig. 7-36.

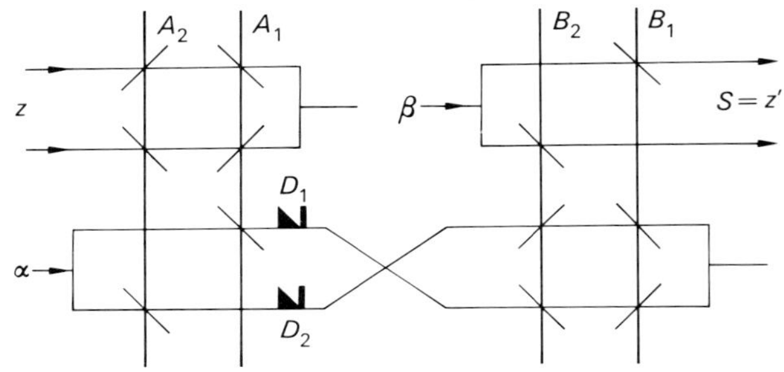

Fig. 7-36

The rest condition of each core pair after the read-out is 0-0. In this type of transfer loop, there is no need for a reset in either condition with a view to the next switching cycle because an incoming 0 signal as well as an incoming 1 signal automatically lead to the desired condition of a core pair.

A number of cores can be connected in series in a balanced split-winding transfer loop, as is demonstrated in the circuit of Fig. 7-37 in which the functions $S = z$ and $S = z'$ are generated. The effect of crossing both paths of a transfer loop can also be obtained by a reversal of the polarity of the read-in windings of a core pair.

In order to generate the function $S = z'$, there is no need to use two cores in the balanced split-winding transfer-loop system. The inversion of a function stored in a core is a matter of location of the read-out windings on the transmitting cores or of polarity of the read-in windings on the receiving cores in the transfer loop.

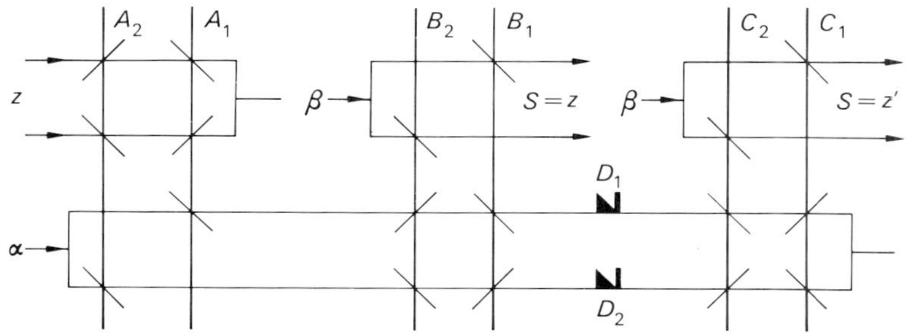

Fig. 7-37

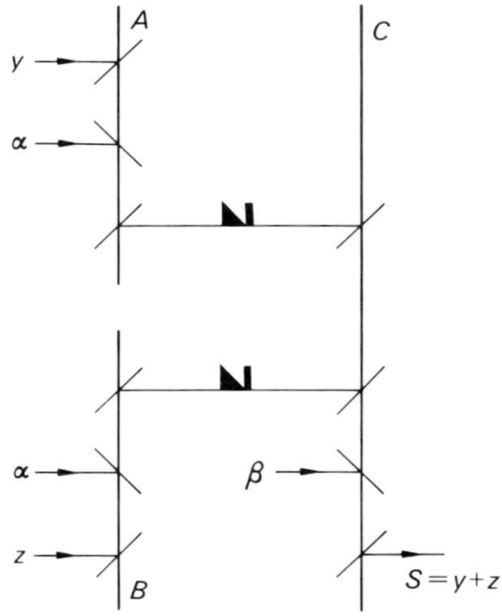

Fig. 7-38

SOME FUNCTIONS WITH TWO VARIABLES

The functions $S = y + z$ and $S = yz'$ can be implemented in about the same way by means of various types of transfer loops. The circuit of Fig. 7-38 represents the switching function $S = y + z$. The read-out windings of cores A and B can be connected separately to two different read-in windings of core C; they may also be connected in series, for which one transfer loop would be required; or else they may each be connected via a diode transfer loop to the same read-in winding of core C. Core C will be brought in the 1 state by the transfer of a 1 either from core A, core B, or both simultaneously.

The function $S = yz'$ can be obtained by means of the circuit of Fig. 7-39. It uses two transfer loops and two read-in windings on core C. The polarity of the read-in winding of the inverted variable is such that a 1 signal

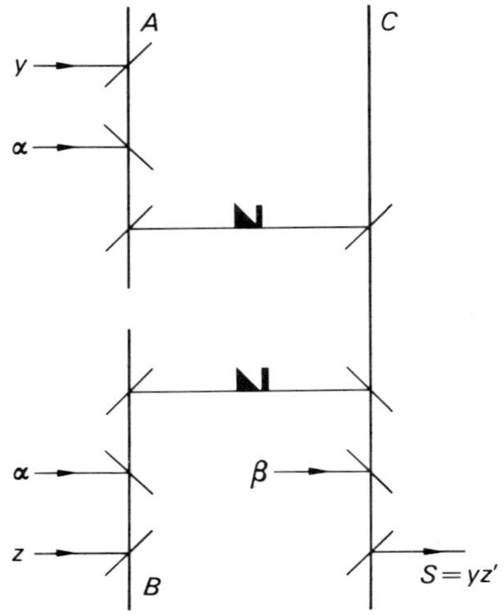

Fig. 7-39

puts core C into the 0 state. Timed by the same clock pulse α, core C will be set only if a 1 signal is received from core A and not from core B.

The transfer loops used may be any one of the first four types described in Sec. 7-4.

The functions $S = y + z$ and $S = yz'$ can be obtained in about the same way with all types of split transfer loops. Figures 7-40 and 7-41 show the basic circuit for both types of functions.

The function $S = y \cdot z$ is not so easy to realize when using this type of core logic. This function can be derived as follows from the functions $S = y + z$ and $S = yz'$:

$$S = z \cdot (y' \cdot z)' = z \cdot (y + z') = y \cdot z \tag{7-2}$$

The core circuit corresponding to Eq. (7-2) is shown in Fig. 7-42.

A rather large number of cores is already required to realize this rather simple switching function. The number of cores required for more complex switching functions in this type of switching logic will largely depend on the skill of the designer in the factoring process of switching formulas.

It is possible to devise a generally applicable core logic based on the use of the basic core element of the balanced split-winding transfer loop even though it may require a larger number of cores. The balanced split-winding transfer loop shown in Figs. 7-32 and 7-37 has two general features to show which allow a simple and systematic implementation of simple and complicated switching functions:

1. The inverse of the variable stored in a core pair can be obtained by interchanging the cores of that pair. In this way, a core pair can always

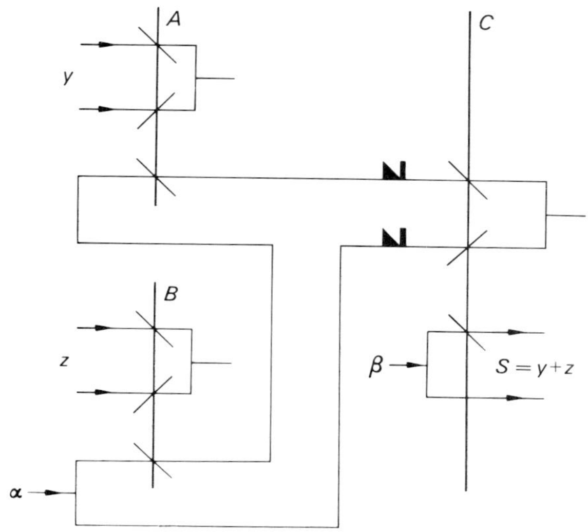

Fig. 7-40

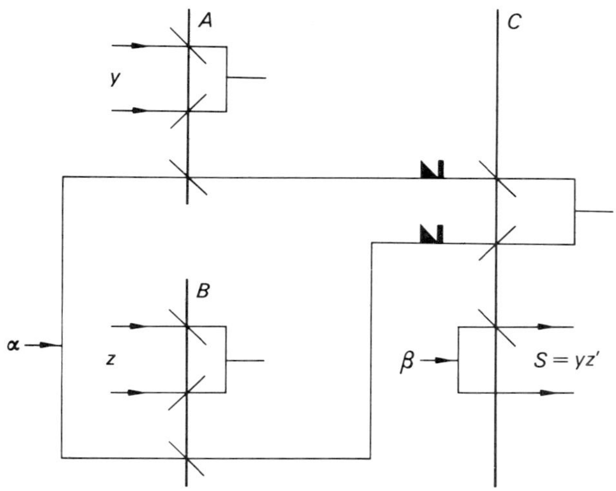

Fig. 7-41

be made to generate its inverse, without any special provision being required.

2. The switching function generated by a number of cores in a certain core stage can be inverted as a whole by interchanging two paths of the transfer loop leading to the next core stage.

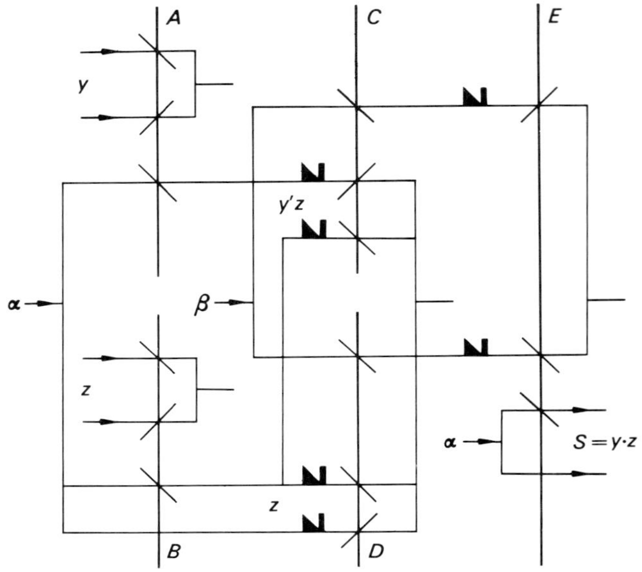

Fig. 7-42

This means that the switching function $S = y + z$ can easily be implemented and a number of functions can be obtained in the same way simply by interchanging the cores in one or both of the core pairs A and B, whereas the functions $S = (y + z)' = y'z'$, $S = (y' + z)' = yz'$, $S = (y + z')' = y'z$, and $S = (y' + z')' = yz$ will be arrived at, using the circuit of the first function as a starting point, by an interchange of the paths in the transfer loop.

By way of illustration, the circuits for the functions $S = yz$ and $S = y + z$ are shown in Figs. 7-43 and 7-44.

These circuits need no special core reset to 0 or to 1 once a function has been transferred to the next core stage. It should be pointed out that a similar type of core logic can in principle also be achieved with other types of split transfer loops; however, in many cases a special reset of some cores is required, which is avoided completely in the balanced split-winding transfer loop logic.

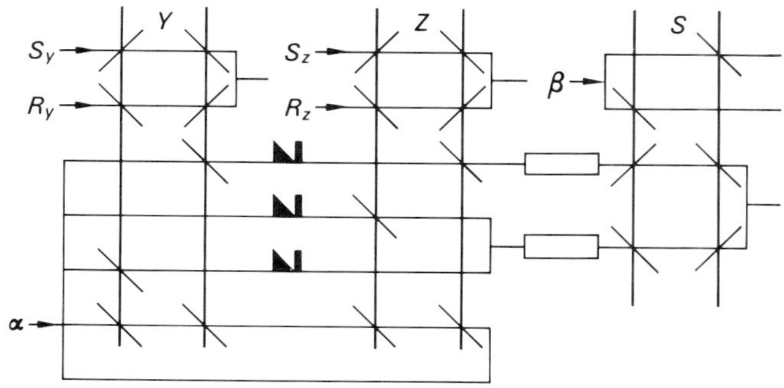

Fig. 7-43

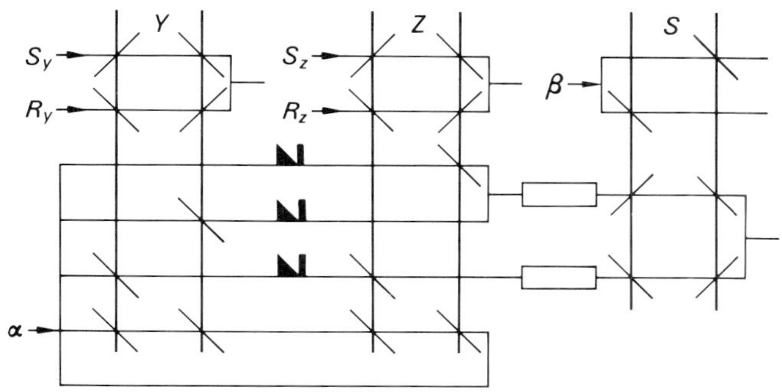

Fig. 7-44

7-6 THE BALANCED SPLIT-WINDING TRANSFER LOOP AS A BASIC ELEMENT FOR A STANDARDIZED CORE LOGIC

Relays with make and break contacts allow the implementation of all possible switching functions by means of rather simple, generally applicable methods. The same is true for diodes and/or transistor circuits. Magnetic cores, however, differ in certain respects from relay, diode, and transistor circuits, so that a generally applicable switching method does not seem to be feasible for these components.

In the early days of electronic switching, tube circuits were designed as one integrated system with many components. Gradually, the trigger consisting of a symmetrical two-phase switching unit became a commonly used building stone which, when combined with diodes or transistors, could be considered more or less as the electronic equivalent of a relay and its contacts.

The technique of the balanced split-winding transfer loop is the core-circuit equivalent of a relay and its contacts. A core pair as, for example, the pairs A, B, and C in Figs. 7-43 and 7-44 can be considered as relays with their associated contacts. The group of receiving windings on each core pair is analogous to the relay winding. The windings used in the transmitting end of a transfer loop can be assimilated to the contacts of a relay.

A core that is in the 1 state and is reversed into the 0 state by a clock pulse shows a high impedance in its transmitting winding, which more of less prevents the clock-pulse current from flowing through this path of the transfer loop. Accordingly, a transmitting winding on a core operates somewhat as a break contact.

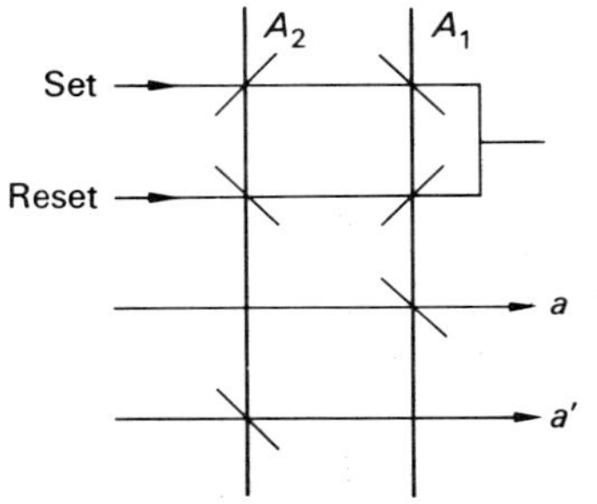

Fig. 7-45

For example, if the core pair of Fig. 7-45 operates as shown in Table 7-3, the winding of core A_1, which is in the 1 state, for the 0 state of the core pair acts as a contact a, whereas a winding of core A_2, which is in the 1 state if the core pair is in the 1 state, acts as a contact a'.

Table 7-3

A	A_2	A_1
0	0	1
1	1	0

The transfer loop between core pairs has to be considered as a set and a reset path. The windings of the transmitting core pairs, which have to be connected in the set path, are determined by the required switching function, and they are connected to each other in the same way as the equivalent contact circuits. Parallel paths have to be blocked against circulating currents by diodes. The interconnection of the windings of the transmitting cores in the reset path is determined by the inverse of the required switching function. This method can be checked easily in the circuits shown in Figs. 7-43 and 7-44. The set path of the circuit shown in Fig. 7-44 can be described by $S = y + z$; the reset path is described by $S' = (y + z)' = y'z'$. It should be noted that there is no simultaneous blocking of the set and reset path because the switching functions for these two paths are always one another's inverse.

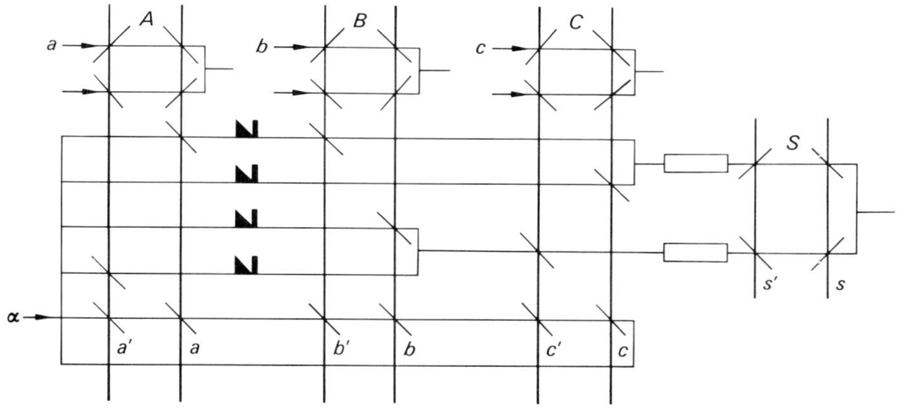

Fig. 7-46

Switching functions leading to multilevel core circuits by means of design methods other than the split-winding transfer loop can be implemented in a one-level circuit with this last method. This may be illustrated by the following example. The switching function $S = ab' + c$ is shown in Fig. 7-46. The reset path is described by $S' = (ab' + c)' = (a' + b) \cdot c'$.

The correct interconnection of the transmitting windings in the circuit of Fig. 7-46 can be checked easily by means of the required switching function and its inverse.

It is not necessary to limit the number of transmitting windings on a core to one. The effect of the complete blocking of the diode in a transfer path by the voltage induced in the corresponding read-out winding if special reset windings are provided on the transmitting cores does not overload the clock pulse α because only leak currents of the diode are involved. An example of such a circuit based on this consideration is shown in Fig. 7-47. This figure corresponds to the switching function $S = ab' + a'b$. The formula $S' = ab + a'b'$ describes the interconnection of the read-out windings which control the reset function of the core pair S. As is shown in Fig. 7-47, this function can be implemented by means of two core pairs. The design of this type of circuit is hardly more sophisticated than its equivalent in relay circuitry, but it requires more "contacts."

If required, the read-out windings on cores a and a' can be combined into one on each core.

A number of core pairs can be controlled simultaneously with this design method. This may be illustrated by means of the switching function $S = ab' + a'b$ and $D = ab$. The reset function for the core pair D will be $D' = ab' + a'b + a'b'$. All the terms necessary for implementing function D are already generated for functions S and S'. The resulting circuit is shown in Fig. 7-48.

Cores D and S can be read out by other clock pulses not indicated in the diagram.

As will be shown in more detail in Chap. 16 (on shift registers), the use of different clock pulses can be avoided by using auxiliary memories, e.g., capacitors. These auxiliary memories can also be used in the different types of core logic described in the preceding sections of this chapter without affecting the basic logic.

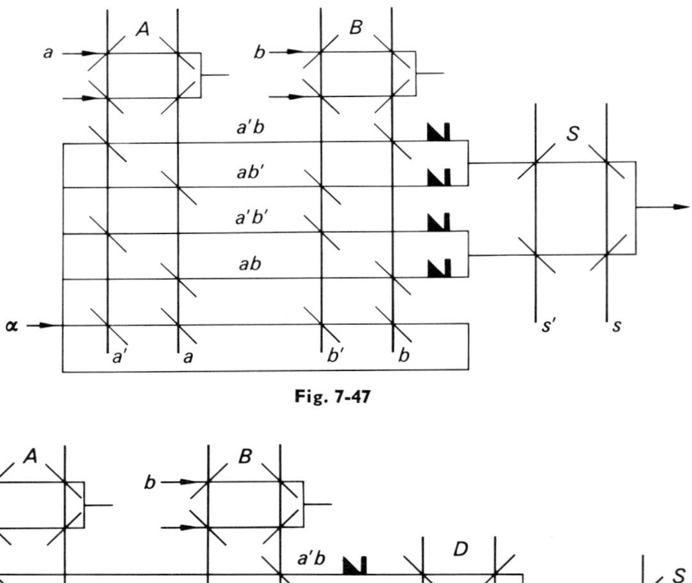

Fig. 7-47

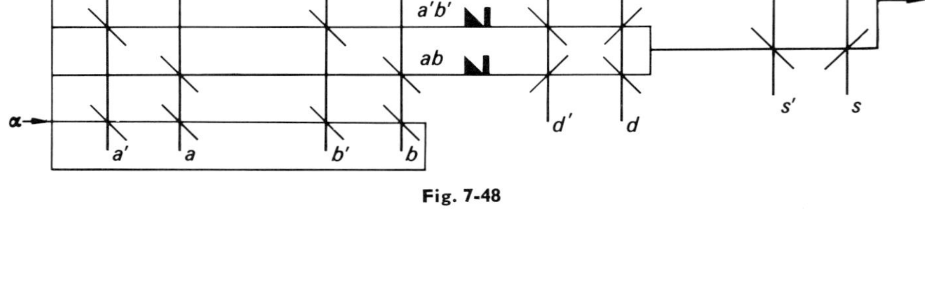

Fig. 7-48

7-7 MIXED WRITE AND READ LOGICAL CORE CIRCUITS AS BASIC ELEMENTS FOR A STANDARDIZED CORE LOGIC

A generally applicable core logic can also be obtained by the serial use of write and read logical core circuits. The OR operation can be performed in a reliable way by means of write logical core circuits; the read logical core circuits provide a better means of accomplishing the AND operation.

THE MAXTERM REPRESENTATION

Each Boolean expression can be written in a maxterm form, i.e., a logical product of logical sums. The logical sums can be formed by reading in the variables of one factor of a Boolean product into one core, whereas the logical product of the factors involved can be obtained, as shown diagrammatically in the core circuit of Fig. 7-23a, by a simultaneous read operation on the various cores corresponding to the factors of the Boolean product wanted.

This may be illustrated by means of a switching expression representing a logical product of logical sums.

$$f(u, v, w, x, y, z) = (u + v)\,(w + x)\,(y + z) \tag{7-3}$$

The circuit corresponding to Eq. (7-3) is shown in Fig. 7-49.

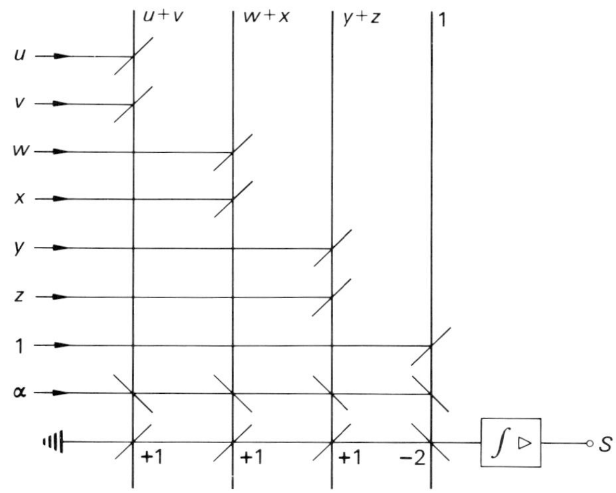

Fig. 7-49

THE MIXED ALGEBRAIC-BOOLEAN REPRESENTATION

The mixed algebraic-Boolean representation of a minterm switching function, a method originally devised by Heyn, is a variant of the maxterm representation. With this method, it is now possible to avoid using negated variables as input signals, for this is not always feasible in the maxterm representation; alternately, it opens the way for any desired combination of input variables and/or their negations. The algebraic part of the mixed algebraic-Boolean representation of a switching function is based on the switching principle of Fig. 7-22.

The mixed algebraic-Boolean representation of a given switching function is derived from its Boolean expression by (1) replacing all negated

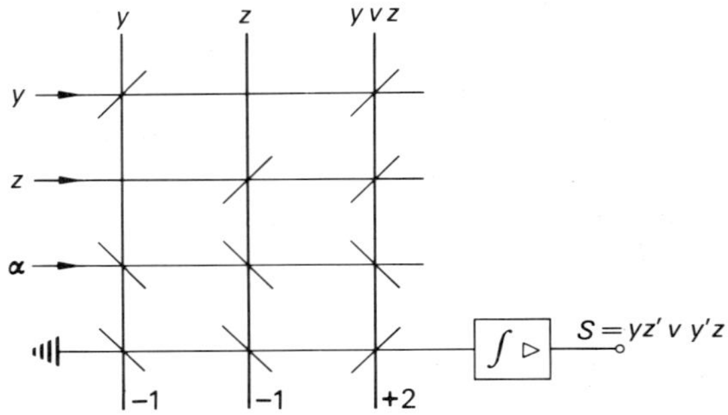

Fig. 7-50

variables by a 1− variable or the unwanted (input) negated variables by a 1− negated variable and all logical additions by algebraic additions, (2) multiplying and adding out the obtained expression as an ordinary algebraic expression, (3) replacing each product yz by $y + z - (y \vee z)$ (where $\vee$ is the symbol for the Boolean addition), and (4) multiplying and adding out. The result will be the desired representation.

It should be pointed out that the variables y and z mentioned above could conceivably be products that are to be developed further.

The process will be demonstrated by means of the following example:

$$\begin{aligned} f(y,z) &= y'z \vee yz' = (1-y)z + y(1-z) \\ &= z - yz + y - yz = y + z - 2yz \\ &= y + z - 2[y + z - (y \vee z)] \\ &= y + z - 2y - 2z + 2(y \vee z) \\ &= -y - z + 2(y \vee z) \qquad (7\text{-}4) \end{aligned}$$

This equation can easily be checked by means of a truth table. The resulting circuit is shown in Fig. 7-50.

In the author's opinion, a core logic based on the mixed algebraic-Boolean representation of switching functions is well suited for use as a standardized tool, provided that the operating speeds of the cores do not differ too much. The simplicity of a core circuit is striking, even for complicated switching functions, because no complicated information transfer loops are required in the circuit itself.

The transformation of a Boolean function into a mixed algebraic-Boolean expression as given above can easily be programmed on a computer so that elaborate manipulation of complicated functions can be avoided.

The above considerations on a standardized core logic call for some general comments on the use of magnetic cores for switching purposes. The development of core switching circuits as described in the preceding paragraphs has been somewhat analogous to the past evolution of electronic circuits. In the early days of electronic switching, many of the circuits proposed were one integrated assembly of tubes. A great deal of thought was then given to the possible use that might be made of the special features of some components, such as cold-cathode tubes and point-contact transistors. The subdivision of electronic switching circuits into a number of rather simple standardized building stones somewhat analogous to relays is now a well-established basic concept. These building blocks have been designed in such a way that they do not depend on the special features of characteristics of the components. These special features will alter during the lifetime of the component, thereby decreasing the reliability of the circuit. This fact is reflected in the extensive use of two-transistor triggers, highly independent as they are of the transistor characteristics, whereas one-transistor triggers do depend on a special characteristic of the particular transistor used.

The development of switching circuits with magnetic cores followed an analogous trend. Magnetic cores can be used in a single-core logic based on the marginal cooperation between a number of windings with different polarities and a core material with a nearly rectangular hysteresis loop. This type of core logic, which is described in Sec. 7-2, is dependent on the stability of the hysteresis loop of the core and on the stability of the magnetizing currents through the windings. However, the features of the hysteresis loop of available magnetic materials vary with temperature, so that the use of switching logic is severely limited unless stabilizing measures are provided.

The use of the switching measures described in Sec. 7-3 makes it possible to overcome some of the difficulties described in this section. Another perhaps less-economical way is the use of clock-pulse systems and balanced split-winding information transfer loops between a number of successive core stages, as described in Sec. 7-6. These methods can lead to a generally applicable core logic comparable to the relay and contact logic.

The standardized core logic given in Sec. 7-6 is based on the use of pairs of cores and of a number of read-out windings on each of these cores, by means of which core circuits can be designed in the same way as contact circuits. The characteristics of the rectangular hysteresis loop are then practically no longer a decisive factor for the reliability of this type of circuits.

PROBLEMS

1. Explain by means of a B-H diagram the behavior of a magnetic core with a nearly rectangular hysteresis loop. Draw a diagram of the output voltages that can appear in a read-out winding.

2. Implement the switching function $S = z'$ by means of a single-core circuit and explain the difficulties with respect to operating margins for this type of circuit.

3. Realize the switching function $S = xy + yz + xz$ by means of a single-core circuit with a premagnetized winding.

4. Implement the switching function $S = x \cdot y \cdot z$ by means of a premagnetized single-core circuit and elaborate on the difficulties encountered with respect to tolerances for this type of core circuit.

5. Show the implementation of the switching function $S = z'$ in a premagnetized single-core circuit controlled by a clock pulse.

6. Explain the operation of the double-diode information transfer loop between two magnetic cores and discuss the limitations of this type of circuit.

7. Show two types of transistor information transfer loops between two magnetic cores.

8. Explain the operation of the split-winding information transfer loop between two magnetic cores.

9. Explain the operation of the balanced split-winding information transfer loop between two magnetic cores and discuss its operating margins.

10. Draw the circuit corresponding to the switching function $S = z'$ in a system of core logic controlled by two interleaved clock pulses

(*a*) With a diode transfer loop

(*b*) With a balanced split-winding transfer loop.

11. Implement the switching function $S = y' + z$ in a system of core logic controlled by two interleaved clock pulses. All variables are given unprimed.

12. Draw the circuit for the switching function $S = y' \cdot z'$ in a magnetic core logic controlled by a system of two interleaved clock pulses. All variables are given unprimed.

13. Show the implementation of the switching function $S = (a' + b) \cdot c$ in a system of core logic with balanced split-winding information transfer loops.

14. Draw the circuit for the switching function $S = y \cdot z$ by means of a system of core logic with balanced split-winding information transfer loops.

15. Implement the switching function $S = ab + a'b'$ and $D = a'b'$ in a system of core logic with balanced split-winding information transfer loops.

16. Implement the switching functions of Probs. 11 to 15 in a system of core logic using the maxterm representation.

17. Draw the circuits for the switching functions of Probs. 11 to 15 in a system of core logic using the mixed algebraic-Boolean representation. State the advantages of this type of circuit.

SELECTIVE BIBLIOGRAPHY

1. Abbott, H. W., and J. J. Turan: Multihole Ferrite Core Configuration and Applications, *Proc. IRE*, vol. 45, pp. 1081–1093, 1957.
2. Blois, M. S.: Preparation of Thin Magnetic Films and their Properties, *J. Appl. Phys.*, vol. 26, pp. 975–980, 1955.
3. Bobeck, H.: A New Storage Element Suitable for Large-sized Memory Arrays. The Twistor, *Bell System Tech. J.*, vol. 36, pp. 1319–1340, 1957.

4. Crane, H. D.: A High Speed Logic System Using Magnetic Elements and Connecting Wire Only, *Proc. IRE*, vol. 47, pp. 63–73, 1959.
5. Guterman, S., R. D. Kodis, and S. Ruhman: Logical and Control Functions Performed with Magnetic Cores, *Proc. IRE*, vol. 42-1, pp. 291–298, 1955.
6. Hegyi, I. J.: Ferromagnetic Spinels with Rectangular Hysteresis Loops, *J. Appl. Phys.*, vol. 25, pp. 176–178, 1954.
7. Heyn, H. J.: Representations of Switching Functions and Their Application to Computers, Ph.D. thesis, Technical University, Delft, 1960.
8. Holt, A. D.: The Application of Square Hysteresis Loop Materials in Digital Computer Circuits, *Electron. Eng.*, vol. 30, pp. 196–199, 1958.
9. Karnaugh, M.: Pulse Switching Circuits Using Magnetic Cores, *Proc. IRE*, vol. 42-1, pp. 570–584, 1955.
10. Littman, M. F.: Ultrathin Tapes of Magnetic Alloys with Rectangular Hysteresis Loops, *Trans. AIEE, Comm. Electron.*, vol. 71, pp. 220–223, 1952.
11. Menyuk, N.: Magnetic Characteristics of Ultrathin Molybdenum Permalloy Cores, *J. Appl. Phys.*, vol. 26, pp. 692–697, 1955.
12. Menyuk, N., and J. B. Goodenough: Magnetic Materials for Digital Computer Components: A Theory of Flux Reversal in Polycrystalline Ferromagnetics, *J. Appl. Phys.*, vol. 26, pp. 8–18, 1955.
13. Minnick, R. C.: The Use of Magnetic Cores as Switching Devices, Ph.D. thesis, Harvard University, 1953.
14. Minnick, R. C.: Magnetic Switching Circuits, *J. Appl. Phys.*, vol. 25, pp. 479–485, 1954.
15. Parkin, B. G.: Magnetic Materials with Rectangular Hysteresis Loops, *Post Office Elec. Eng. J.*, vol. 48, pp. 1–6, 1955 (has detailed bibliography).
16. Perry, G. H., G. R. Hoffman, and E. W. Shallow: A New and Simple Type of Digital Circuit Technique Using Junction Transistors and Magnetic Cores, *Proc. Inst. Elec. Engrs. (London)*, vol. 103, pp. 412–417, 1956.
17. Rajchman, Jan. A.: Magnetics for Computers—A Survey of the State of the Art, *RCA Rev.*, vol. 20, pp. 92–135, 1959.
18. Sands, E. A.: The Behavior of Rectangular Hysteresis Loop Magnetic Materials under Current Pulse Conditions, *Proc. IRE*, vol. 40, pp. 1246–1250, 1952.
19. Sant, O. J. van: Considerations for the Selection of Magnetic Core Materials for Digital Computer Elements, *Conv. Record IRE*, vol. 2, pt. 4, pp. 109–113, 1954.
20. Scarrot, C. G., W. J. Harwood, and K. C. Johnson: Logical Devices Using Saturable Magnetic Cores, *Proc. Inst. Elec. Engrs. (London)*, vol. 103, pp. 302–312, 1956.
21. Wang, A.: Magnetic Delay Line Storage, *Proc. IRE*, vol. 39, pp. 401–407, 1951.
22. Wang, A., and W. D. Woo: Static Magnetic Storage and Delay Line, *J. Appl. Phys.*, vol. 21, pp. 49–54, 1950.
23. Westmijze, W. K.: Properties and Applications of Magnetic Cores in Data Handling, *Elektrotech. Z.*, vol. A 81, pp. 779–783, 1960.

8
Code Converters

8-1 INTRODUCTION

A logical circuit can in principle be considered as a conversion device, a certain code of received control signals being generally converted into a different code of resulting control signals. In general, such code conversion can be broken down into two categories:

1. A binary-decimal code conversion which may be incomplete
2. A decimal-binary code conversion which may also be incomplete

In many types of code-conversion circuits, only one of these two categories is used, whereas other types of code conversions do sometimes take the form of a hardly recognizable combination of the two categories mentioned under 1 and 2.

Code converters can be designed on the sole basis of the laws dealt with in the preceding chapters. These circuits contain neither memory elements nor time sequences. They consist exclusively of contact chains or their equivalents based on other switching techniques. Code-converting circuits are essentially multiterminal circuits.

The switching formulas for a certain type of code conversion are practically independent of the switching elements used for their implementation. It is sometimes possible, however, to simplify the final code-converting circuit by taking advantage of the special features of the particular components used.

The analogy between circuits using different types of switching components is even so close that all code converters with binary switching elements, even those with binary switching elements of a yet undefined construction, will fit into the same general frame of reference.

8-2 BINARY-DECIMAL CODE CONVERSION

The general form of a binary-to-decimal code converter as given here provides for the conversion of a three-bit binary code to an eight-element decimal code. The canonical form of a switching function with three variables given in the form of a logical sum of logical products, i.e., in its minterm form, describes in principle all possible circuits with that number of variables:

$$f(x, y, z) = f_0 x' y' z' + f_1 x' y' z + f_2 x' y z' + f_3 x' y z + f_4 x y' z' + f_5 x y' z + f_6 x y z' + f_7 x y z \tag{8-1}$$

The desired code converter will generally comprise 3×2 input wires, i.e., one for each of the variables x, x', y, y', z, and z'. The result of the code conversion will be the marking of one of the 2^n output terminals, n being the number of variables (in this case $2^n = 8$). Each term of Eq. (8-1) corresponds to one of the output terminals. In general, a complete code conversion from a three-digit binary code to the equivalent decimal code can be implemented by means of $2^n = 8$ AND gates with three input terminals. The symbolic diagram of this code converter is shown in Fig. 8-1.

If relay contacts are used for the eight AND gates and if no advantage is taken of the normal simplification methods for contact circuits, the resulting layout will consist of eight series circuits representing all possible combinations of make and break contacts of the three controlling relays X, Y, and Z. The output terminal of one of these eight contact circuits is connected to earth; the terminal to be earthed will depend on the particular combination of the input variables. The seven other contact circuits are open (Fig. 8-2).

A contact circuit consisting of a series connection of contacts connected to a relay winding can be replaced by a circuit consisting of a parallel connection of the inverse contacts linked in parallel to the relay winding to be controlled, the operation of this array (Fig. 8-3) being unaffected by the change. In this case, the output terminal is marked by the opening of a short circuit, and no longer by the closing of a circuit.

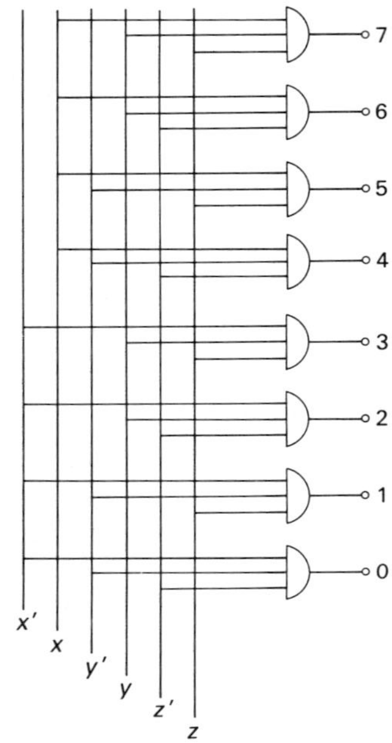

Fig. 8-1

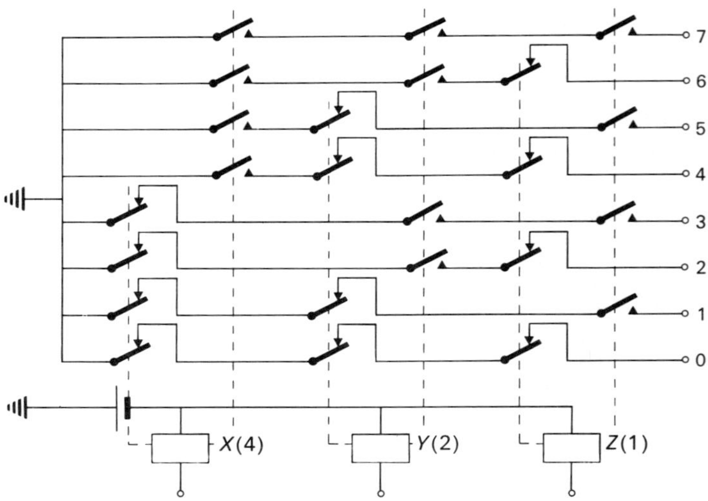

Fig. 8-2

It should be borne in mind that the contact circuits of the code converter of Fig. 8-2 can be simplified by the combined use of contacts, since these are part of a series circuit. The resulting circuit is the normal contact tree. The parallel control circuit of Fig. 8-3 cannot be simplified further. Resistors R_0 to R_7 are necessary in order not to short-circuit the battery.

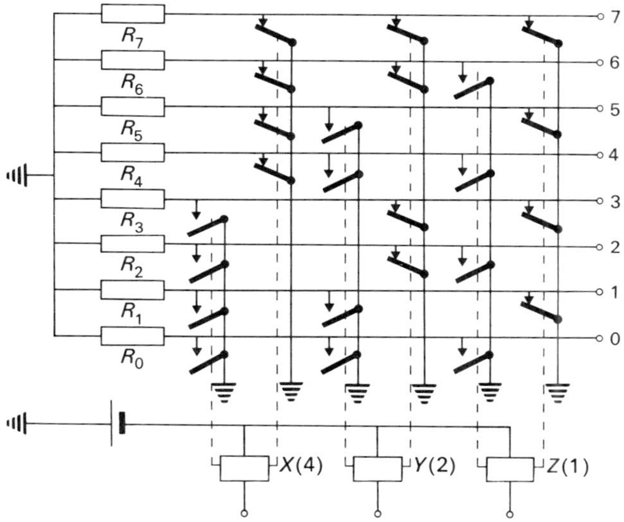

Fig. 8-3

The contacts of Fig. 8-3 can be replaced by diodes to provide the required AND gates, as is shown in the code converter of Fig. 8-4. For the sake of uniformity of presentation, the eight diode AND gates have been shown as being controlled by means of contacts x, y, and z. In general, such a type of diode circuit will be controlled electronically. In the code converter of Fig. 8-4, the selected output terminal is energized by a voltage with the symbolic value 1, whereas a voltage with the symbolic value 0 is fed to all other terminals.

Since transistors have a very small voltage drop between collector and emitter, a number of these components can be connected in series. The basic diagram of a code converter implementing Eq. (8-1) by means of *pnp* transistors is shown in Fig. 8-5. This circuit can be simplified in the same way as the corresponding circuit of Fig. 8-2. The controlling contacts x, y, and z are connected in such a way to the bases of the various transistors that only the group of transistors defining the desired output terminal experiences no blocking voltage, whereas at least one transistor of each of the other groups receives a blocking voltage on its base. This closed circuit to earth is found to occur at the selected output terminal.

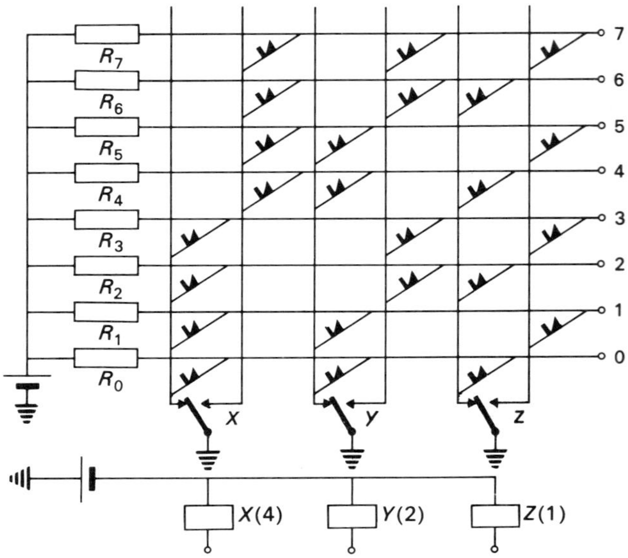

Fig. 8-4

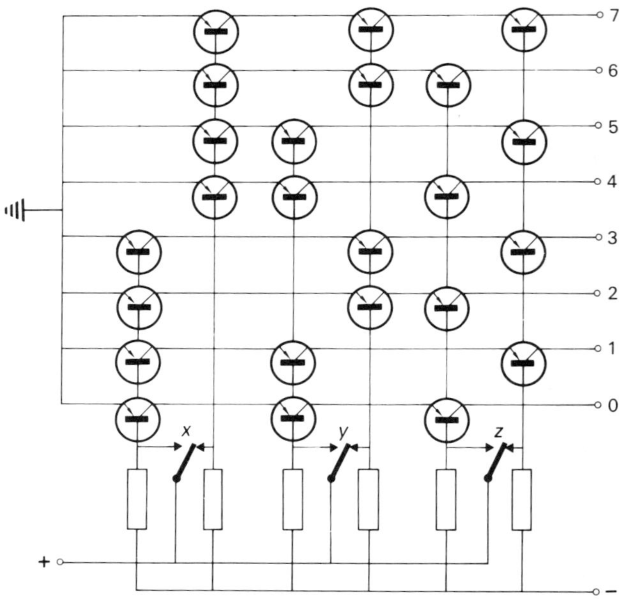

Fig. 8-5

The code converters described above are all controlled by means of a make and a break contact of each of the controlling relays. If advantage is taken of the magnetic condition of the core of a polar relay to implement one of the various terms of Eq. (8-1), it is possible to limit the controlling contacts of relays X, Y, and Z to one make contact, in view of the fact that

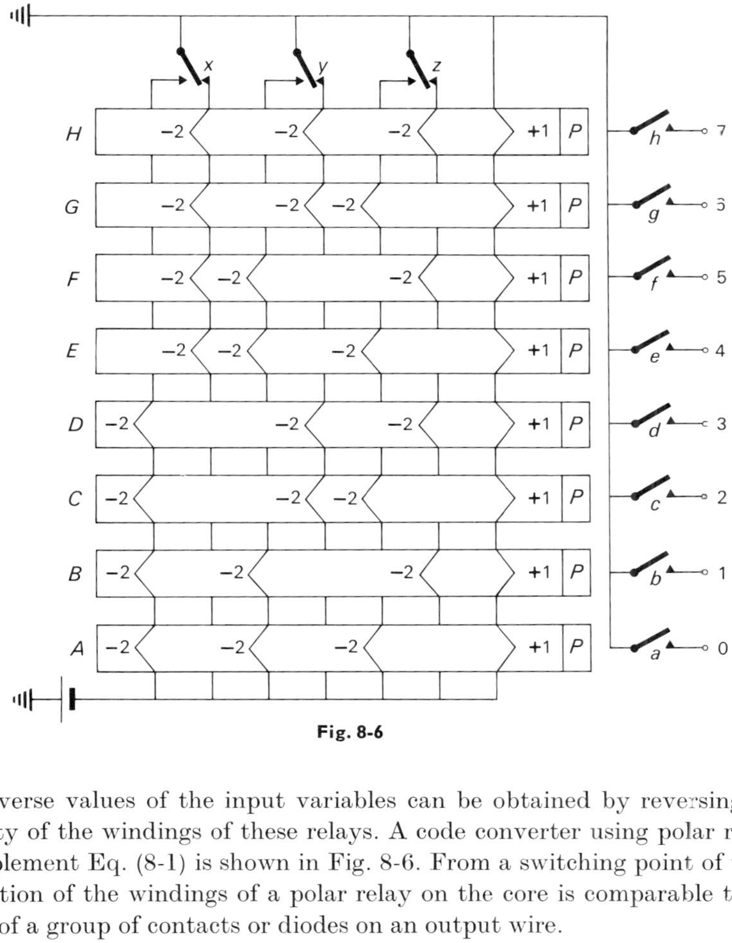

Fig. 8-6

the inverse values of the input variables can be obtained by reversing the polarity of the windings of these relays. A code converter using polar relays to implement Eq. (8-1) is shown in Fig. 8-6. From a switching point of view, the action of the windings of a polar relay on the core is comparable to the effect of a group of contacts or diodes on an output wire.

All windings of the polar relays controlled by the variables x, y, and z magnetize the core with a magnetic field with a relative value of -2. Each of the relays is furthermore provided with a fourth winding which produces a permanent magnetic field with the relative value of $+1$. Each of the possible input combinations of the variables x, y, and z results in the operation of one of the relays A to H. The operated relay receives a total magnetization of

+1, the magnetic field applied to all the other relays having a value of at least −1; so that in this type of circuit, one and only one relay can operate at any one time. This results in the closing of the corresponding contact the number of which is the decimal equivalent of the binary input combination.

In this type of code converter, the windings of the polar relays act in a way analogous to a series-contact circuit. A number of variants of this type of circuit are, of course, possible; for example, the single-current control of the polar relay windings can be replaced by a double-current control.

The switching principles of magnetic cores are basically the same as those of polar relays. Magnetic cores and polar relays differ in principle only in the way in which the condition of the magnetic core is made known to the outside world. Polar relays have a contact which is positioned by the flux resulting from the interaction of the various windings. This type of relay permanently shows the state it is in. The magnetic core, however, has its condition concealed in itself. With single-hole cores, this condition can only be made known empirically by means of two windings, the first for feeding in an inquiry pulse and the second which acts as an output winding, as was already explained in Chap. 7.

The control windings of the cores in Fig. 8-7 produce a magnetization of

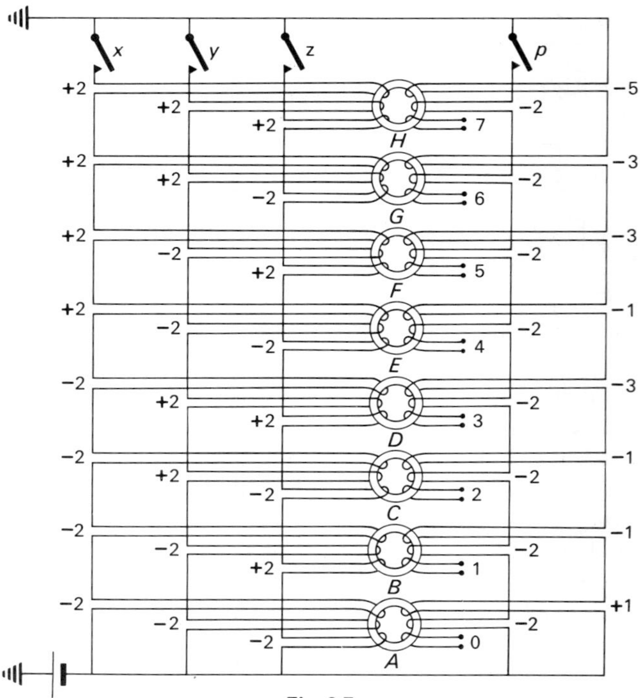

Fig. 8-7

the cores as shown in the diagram, with a relative value of + or − 2. These control windings interact with a group of permanently magnetized windings with different relative values. These values are so chosen that only one core will have a magnetization with a relative value of +1, the magnetic field applied to the other cores having at least a value of −1. An inquiry pulse with a relative value of −2, which can be passed via contact p through the series-connected inquiry windings, reverses the magnetization of the former core from +1 to −1. The relative magnetization of all other cores which was initially at least −1 changes during the inquiry pulse from −1 to at least −3. No output pulse is thereby generated. The principles of binary-decimal code conversion described above can be applied to the various types of binary codes given in Chap. 20.

A group of binary adding circuits can be considered as a special form of binary-decimal code converter. These circuits will be discussed in Chap. 17.

8-3 DECIMAL-BINARY CODE CONVERSION

The decimal-binary code conversion can best be explained by means of the example of Table 8-1. This table shows the ordinary binary code which must

Table 8-1

	Decimal number	8 *W*	4 *X*	2 *Y*	1 *Z*
A	1	0	0	0	1
B	2	0	0	1	0
C	3	0	0	1	1
D	4	0	1	0	0
E	5	0	1	0	1
F	6	0	1	1	0
G	7	0	1	1	1
H	8	1	0	0	0
I	9	1	0	0	1
J	10	1	0	1	0

be generated using the decimal code as a starting point. The switching equations describing this code conversion are as follows:

$$W = h + i + j \tag{8-2}$$

$$X = d + e + f + g \tag{8-3}$$

$$Y = b + c + f + g + j \tag{8-4}$$

$$Z = a + c + e + g + i \tag{8-5}$$

This code converter can be realized by means of four OR gates. The general form of circuit for this code converter is shown in Fig. 8-8. Any type of switching element can be used in such a converter.

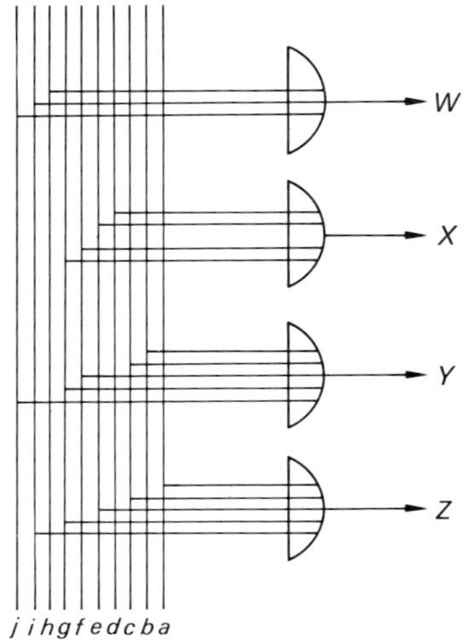

Fig. 8-8

The decimal-binary type of code conversion leads to an interesting circuit form for the conversion of the decimal code into the 74210 two-out-of-five code, as given in Table 8-2.

Table 8-2

	Decimal number	7 *V*	4 *W*	2 *X*	1 *Y*	0 *Z*
A	1	0	0	0	1	1
B	2	0	0	1	0	1
C	3	0	0	1	1	0
D	4	0	1	0	0	1
E	5	0	1	0	1	0
F	6	0	1	1	0	0
G	7	1	0	0	0	1
H	8	1	0	0	1	0
I	9	1	0	1	0	0
J	10	1	1	0	0	0

This code conversion can be described by the following equations

$$V = g + h + i + j \tag{8-6}$$

$$W = d + e + f + j \tag{8-7}$$

$$X = b + e + f + i \tag{8-8}$$

$$Y = a + c + e + h \tag{8-9}$$

$$Z = a + b + d + g \tag{8-10}$$

The resulting circuit will contain five OR gates with the same number of input terminals. Each of the variables is used twice, so that in the case of a relay code converter, all spring sets of the relays are equally loaded.

If there is no special relation at all between the original code and the code into which this code must be converted, a good circuit can be obtained easily by interconnecting the decimal output terminals of a binary-decimal converter to the input terminals of a decimal-binary converter. The best circuit may then be selected for each part of the particular code converter. One might, of course, design a circuit with even less components by using as a starting point all the switching formulas defining this special code converter; this circuit, however, would be much less flexible as regards arbitrary changes of the conversion system used as the converter consisting of a combination of a binary-decimal and a decimal-binary code converter.

To conclude, it may be stated that any code can be converted into another code provided that the conversion is unambiguous.

8-4 BINARY-BINARY CODE CONVERSION

The matrix representation of multiterminal circuits can be applied successfully to the design of code converters. This type of representation and handling of switching formulas has some advantage in that when combining circuits between the various output terminals, i.e., when performing in fact a Δ-Y transformation, sneak paths can be avoided or introduced intentionally. Where the manipulation of the various entries leads to simpler circuit forms by making use of don't-care conditions, harmless sneak paths are introduced. Sometimes a network simplification can be achieved by introducing a sneak path the effect of which on the original requirements cannot be directly ascertained. In this case, it may be necessary to write the matrix in full, putting all the introduced sneak paths in brackets. A fictitious reversed process of node elimination is then applied by means of which the effects of the sneak paths on the entries of the original matrix can be calculated.

The use of matrices for the design of code converters may be illustrated by the conversion of the first 11 combinations of the binary 1248 code into the 1245 code ($10 = 1 + 4$). This type of problem is better stated in the form of a table (see Table 8-3) giving the decimal number of the code symbol, the binary 1248 code symbol, and the corresponding 1245 code symbol.

Table 8-3

	8	4	2	1	5	4	2	1
No.	*w*	*x*	*y*	*z*	*d*	*c*	*b*	*a*
0	0	0	0	0	0	0	0	0
1	0	0	0	1	0	0	0	1
2	0	0	1	0	0	0	1	0
3	0	0	1	1	0	0	1	1
4	0	1	0	0	0	1	0	0
5	0	1	0	1	1	0	0	0
6	0	1	1	0	1	0	0	1
7	0	1	1	1	1	0	1	0
8	1	0	0	0	1	0	1	1
9	1	0	0	1	1	1	0	0
10	1	0	1	0	0	1	0	1
11	1	0	1	1				
12	1	1	0	0				
13	1	1	0	1				
14	1	1	1	0				
15	1	1	1	1				

It is assumed that in the rest condition (0000) of the code converter no through connection exists between the top (node 1) of the circuit and the output terminals (nodes 2 to 5). In the 1245 code the tenth combination, which is often referred to as the 0 in automatic telephony, is represented as $1 + 4$. The don't-care combinations 11 to 15 of the binary code are also given in Table 8-3 because they may be useful for simplifying the switching formulas which describe the code conversion. The simplest form of the switching functions given below can be obtained easily by means of Karnaugh maps:

$$a = w'x'z + wz' + xyz' \tag{8-11}$$

$$b = w'x'y + wy'z' + yz \tag{8-12}$$

$$c = wy + wz + xy'z' \tag{8-13}$$

$$d = wy' + xy + xz \tag{8-14}$$

The switching functions (8-11) to (8-14) are given in their simplest form, but some of the conversion steps of the matrix representing these equations make it necessary to amplify some of the terms with redundant factors in order to make room, if required, for further node insertion.

Before starting the conversion of connection matrix ${}^5C_0$ representing Eqs. (8-11) to (8-14), it should be noted that equations c and d can be written as follows:

$$c = w(y + z) + xy'z' = w(y + y'z) + xy'z' \tag{8-15}$$

$$d = x(y + z) + wy' = x(y + y'z) + wy' \tag{8-16}$$

The advantage of this form is that starting from the circuit $y + y'z$, variable y' can be obtained without extra springs and that with one extra spring the function $y'z'$ can also be derived. All three functions can be implemented by means of two transfer contacts.

The connection matrix ${}^5C_0$ describing the complete code converter can be written as follows:

$$ {}^5C_0 = \begin{bmatrix} 1 & \begin{matrix} w'x'z + wz' \\ + xyz \end{matrix} & \begin{matrix} w'x'y + wy'z' \\ + yz \end{matrix} & \begin{matrix} w(y + y'z) \\ + xy'z' \end{matrix} & \begin{matrix} x(y + y'z) \\ + wy' \end{matrix} \\ & 1 & 0 & 0 & 0 \\ & & 1 & 0 & 0 \\ & & & 1 & 0 \\ & & & & 1 \end{bmatrix} \tag{8-17}$$

Only the upper-right half of the full matrix is shown. Since the code combination wx belongs to the don't-care group of the binary code to be converted, a redundant entry $c_{45} = wx = 0$ may be introduced. This makes it possible to insert a sixth node using entries $c_{16} = y + y'z$, $c_{46} = w$, and $c_{56} = x$ to replace functions $w(y + y'z)$ of entry c_{14} and $x(y + y'z)$ of entry c_{15}.

The resulting matrix ${}^5C_1$ is then:

$$ {}^5C_1 = \begin{bmatrix} 1 & \begin{matrix} w'x'z + wz' \\ + xyz' \end{matrix} & \begin{matrix} w'x'y + wy'z' \\ + yz \end{matrix} & xy'z' & xy' & y + y'z \\ & 1 & 0 & 0 & 0 & 0 \\ & & 1 & 0 & 0 & 0 \\ & & & 1 & [wx] = 0 & w \\ & & & & 1 & x \\ & & & & & 1 \end{bmatrix} \tag{8-18}$$

Similarly, a redundant entry $c_{34} = wx = 0$ can be introduced in matrix (8-18) so that a seventh node can be inserted by using entries $c_{17} = y'z'$,

$c_{37} = w$, and $c_{47} = x$ to replace the term $wy'z'$ of entry c_{13} and entry $xy'z'$. The resulting matrix is as follows:

$$
{}^5C_2 = \begin{bmatrix}
1 & \begin{matrix} w'x'z + wz' \\ + xyz' \end{matrix} & w'x'y + yz & 0 & wy' & y + y'z & y'z' \\
 & 1 & 0 & 0 & 0 & 0 & 0 \\
 & & 1 & [wx] = 0 & 0 & 0 & w \\
 & & & 1 & 0 & w & x \\
 & & & & 1 & x & 0 \\
 & & & & & 1 & 0 \\
 & & & & & & 1
\end{bmatrix} \tag{8-19}
$$

A redundant entry $c_{67} = z \cdot z'$ can be used to form a new node, No. 8. New entries $c_{18} = y'$, $c_{68} = z$, and $c_{78} = z'$ can be introduced to replace the term $y'z$ of entry c_{16} and entry $c_{17} = y'z'$. The resulting matrix is now

$$
{}^5C_3 = \begin{bmatrix}
1 & \begin{matrix} w'x'z + wz' \\ + xyz' \end{matrix} & w'x'y + yz & 0 & wy' & y & 0 & y' \\
 & 1 & 0 & 0 & 0 & 0 & 0 & 0 \\
 & & 1 & 0 & 0 & 0 & w & 0 \\
 & & & 1 & 0 & w & x & 0 \\
 & & & & 1 & x & 0 & 0 \\
 & & & & & 1 & [z \cdot z'] = 0 & z \\
 & & & & & & 1 & z' \\
 & & & & & & & 1
\end{bmatrix} \tag{8-20}
$$

Entries c_{12} and c_{13} of matrix ${}^5C_3$ can best be written in the following form in order to enable further combination of circuits by taking advantage of the node-insertion process. The transformation of entry c_{13} follows from the original requirements given in Table 8-3. This transformation illustrates the fact that the simplest form of a switching function does not always lead to the simplest form of circuit.

$$c_{12} = w'x'z + wz' + w'y \cdot xz' \tag{8-21}$$

$$c_{13} = w'x'y + w'yz = w'y(x' + z) \tag{8-22}$$

Since $xz'(x' + z) = 0$, a ninth node can be inserted now with the new entries $c_{19} = w'y$, $c_{29} = xz'$, and $c_{39} = x' + z$ to replace the term $xyz' = w'xyz'$ of entry c_{12} and the term $w'y(x' + z)$ of entry c_{13} in matrix (8-20). The resulting matrix is given in Eq. (8-23).

$$
{}^5C_4 = \begin{bmatrix}
1 & w'x'z + wz' & 0 & 0 & wy' & y & 0 & y' & w'y \\
 & 1 & [xz'(x'+z)] = 0 & 0 & 0 & 0 & 0 & 0 & xz' \\
 & & 1 & 0 & 0 & 0 & w & 0 & x'+z \\
 & & & 1 & 0 & w & x & 0 & 0 \\
 & & & & 1 & x & 0 & 0 & 0 \\
 & & & & & 1 & 0 & z & 0 \\
 & & & & & & 1 & z' & 0 \\
 & & & & & & & 1 & 0 \\
 & & & & & & & & 1
\end{bmatrix} \tag{8-23}
$$

Furthermore, the terms wz' of entry c_{12} and xz' of entry c_{29}, combined with the fact that wx is redundant, lead to the introduction of a tenth node, these terms being then replaced by the corresponding entries. The new entries are $c_{1\cdot 10} = w$, $c_{2\cdot 10} = z'$, $c_{9\cdot 10} = x$, and $c_{19} = wx = 0$.

The circuit corresponding to matrix (8-24) will operate in accordance with the requirements described in Table 8-3. Further simplification along these lines is not possible.

The variables of entries $c_{12} = w'x'z$ and $c_{19} = w'y$ could be used in common, with the result that an eleventh node would be introduced with

$$
{}^5C_5 = \begin{bmatrix}
1 & w'x'z & 0 & 0 & wy' & y & 0 & y' & w'y + [wx = 0] & w \\
 & 1 & 0 & 0 & 0 & 0 & 0 & 0 & 0 & z' \\
 & & 1 & 0 & 0 & 0 & w & 0 & x'+z & 0 \\
 & & & 1 & 0 & w & x & 0 & 0 & 0 \\
 & & & & 1 & x & 0 & 0 & 0 & 0 \\
 & & & & & 1 & 0 & z & 0 & 0 \\
 & & & & & & 1 & z' & 0 & 0 \\
 & & & & & & & 1 & 0 & 0 \\
 & & & & & & & & 1 & x \\
 & & & & & & & & & 1
\end{bmatrix} \tag{8-24}
$$

the new entries $c_{1.11} = w'$, $c_{2.11} = x'z$, and $c_{9.11} = y$ provided that the necessary "sneak" entry $c_{29} = x'yz$ does not lead to allowing intolerable circuits to occur.

Furthermore, variable y' of entries $c_{15} = wy'$ and $c_{18} = y'$ can be used in a common circuit involving entries $c_{18} = y'$ and $c_{58} = w$, provided that the resulting sneak entries $c_{56} = wz$ and $c_{57} = wz'$ do not lead to a change in the required condition between the input circuit terminal 1 and the output circuit terminals 2, 3, 4, and 5.

The resulting matrix with the sneak entries is given in Eq. (8-25).

$$
{}^5C_6 = \begin{array}{c} \\ \end{array}
\begin{array}{ccccccccccc}
1 & 2 & 3 & 4 & 5 & 6 & 7 & 8 & 9 & 10 & 11 \\
\end{array}
$$

$$
{}^5C_6 = \left[\begin{array}{ccccccccccc}
1 & 0 & 0 & 0 & 0 & y & 0 & y' & 0 & w & w' \\
 & 1 & 0 & 0 & 0 & 0 & 0 & 0 & [x'yz] & z' & x'z \\
 & & 1 & 0 & 0 & 0 & w & 0 & x'+z & 0 & 0 \\
 & & & 1 & 0 & w & x & 0 & 0 & 0 & 0 \\
 & & & & 1 & x+[wz] & [wz'] & w & 0 & 0 & 0 \\
 & & & & & 1 & 0 & z & 0 & 0 & 0 \\
 & & & & & & 1 & z' & 0 & 0 & 0 \\
 & & & & & & & 1 & 0 & 0 & 0 \\
 & & & & & & & & 1 & x & y \\
 & & & & & & & & & 1 & 0 \\
 & & & & & & & & & & 1
\end{array}\right] \tag{8-25}
$$

We now have to check the effects of these sneak entries on the circuits between terminals 1-2, 1-3, 1-4, and 1-5. A complete check can be performed by canceling out all inserted nonterminal nodes 6 to 11. The formula describing the complete circuit layout, inclusive of sneak paths, is then obtained. It is possible, however, to avoid this rather intricate reverse process by limiting the operation to the sneak terms of these switching functions. For this purpose, the matrix given as Eq. (8-25) can be used as a starting point.

The sneak entry $c_{29} = x'yz$ can have an effect on path 1-2 if an entry c_{19} can be obtained from a fictitious process of node elimination. This entry can be found in the connection matrix ${}^5C_4$:

$$c_{19} = w'y$$

The effect on path 1-2 can be represented by the following expression:

$$c_{19} \cdot c_{29} = w'y \cdot x'yz = w'x'yz \tag{8-26}$$

As this term $w'x'yz$ is combination 3 of Table 8-3 and as path 1-2 must be closed for that combination of variables, no undesired sneak path is found to occur.

It should be pointed out in this connection that entry $c_{29} = x'yz$ also leads to an allowable sneak path between terminals 2 and 3. The other sneak paths introduced between output terminals are also acceptable. These are found between nodes 3 and 5 (wz') and between 4 and 5 ($wxz' + wz$).

Entry $c_{57} = wz'$ must be checked to see how it affects path 1-5. Entry $c_{17} = y'z'$ can be found in matrix ${}^5C_2$, so that that part of path 1-5 can be represented by the following expression:

$$c_{17} \cdot c_{57} = y'z' \cdot wz' = wy'z' \tag{8-27}$$

that is, an allowable path, as can be seen from Eq. (8-14).

The wz part of entry c_{56} must also be checked to determine its influence on path 1-5. Entry $c_{16} = y + y'z = y + z$ can be found in matrix ${}^5C_1$, so that the pertinent part of path 1-5 can be written as follows:

$$c_{16} \cdot c_{56} = (y + z) \cdot wz = wz \tag{8-28}$$

This equation corresponds to a path between nodes 1 and 5 that is closed for the binary combination 9 as required and the combinations 11, 13, and 15 which belong to the group of don't-care conditions, so that this path can also be tolerated.

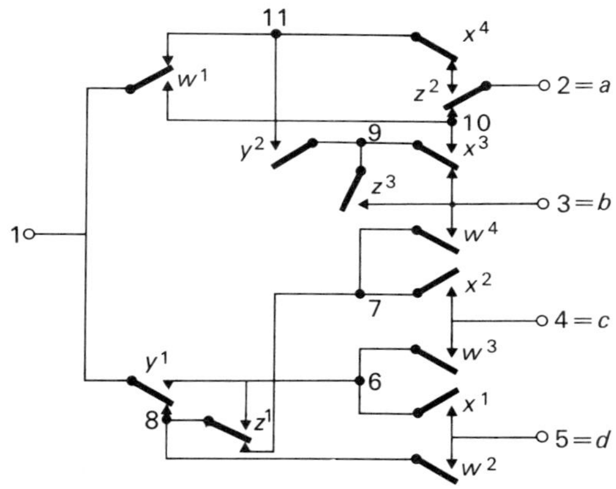

Fig. 8-9

Accordingly, matrix ${}^5C_6$ will lead to the correct circuit, which is shown in Fig. 8-9. The total number of springs involved is 31. Some experience in the handling of matrices is required before this type of problem can be tackled successfully.

To conclude this chapter, let us now look at an entirely different type of code converter. A code converter is to be designed to convert by means of magnetic cores the teleprinter code of the "International Telegraph Alphabet No. 2" into a three-out-of-seven error-detecting code used in radio transmission. The merits of this code are discussed in Chap. 20.

The desired code translation corresponds in fact to the conversion of a five-digit binary code into a seven-digit binary code, the latter code being limited to thirty-two of the thirty-five possible three-out-of-seven code symbols. The three extra symbols provided by the three-out-of-seven code are used in radio transmission. These three extra symbols can be generated in the designed code converter itself. Their use will not be discussed here. It is only by positioning switch S as shown in Fig. 8-10 that the 5/7 code translation can be performed.

The required code translation is defined by Table 8-4.

Table 8-4

Symbol	*Teleprinter code*	*Three-out-of-seven code*
—	0 0 0 0 0	0 0 0 0 1 1 1
e	1 0 0 0 0	0 1 1 1 0 0 0
l.f.	0 1 0 0 0	1 0 1 1 0 0 0
a	1 1 0 0 0	0 0 1 1 0 1 0
sp	0 0 1 0 0	1 1 0 1 0 0 0
s	1 0 1 0 0	0 1 0 1 0 1 0
i	0 1 1 0 0	1 1 1 0 0 0 0
u	1 1 1 0 0	0 1 1 0 0 1 0
c.r.	0 0 0 1 0	1 0 0 0 0 1 1
d	1 0 0 1 0	0 0 1 1 1 0 0
r	0 1 0 1 0	1 1 0 0 1 0 0
j	1 1 0 1 0	0 1 0 0 0 1 1
n	0 0 1 1 0	1 0 1 0 1 0 0
f	1 0 1 1 0	0 0 1 0 0 1 1
c	0 1 1 1 0	1 0 0 1 1 0 0
k	1 1 1 1 0	0 0 0 1 0 1 1
t	0 0 0 0 1	1 0 0 0 1 0 1
z	1 0 0 0 1	0 1 1 0 0 0 1
l	0 1 0 0 1	1 1 0 0 0 1 0
w	1 1 0 0 1	0 1 0 0 1 0 1
h	0 0 1 0 1	1 0 1 0 0 1 0
y	1 0 1 0 1	0 0 1 0 1 0 1
p	0 1 1 0 1	1 0 0 1 0 1 0
q	1 1 1 0 1	0 0 0 1 1 0 1
o	0 0 0 1 1	1 0 0 0 1 1 0
b	1 0 0 1 1	0 0 1 1 0 0 1
g	0 1 0 1 1	1 1 0 0 0 0 1
fig	1 1 0 1 1	0 1 0 0 1 1 0
m	0 0 1 1 1	1 0 1 0 0 0 1
x	1 0 1 1 1	0 0 1 0 1 1 0
v	0 1 1 1 1	1 0 0 1 0 0 1
ltt	1 1 1 1 1	0 0 0 1 1 1 0
rq		0 1 1 0 1 0 0
α		0 1 0 1 0 0 1
β		0 1 0 1 1 0 0

The code converter that can be derived from Table 8-4 is provided with a magnetic core for each line of the table. The corresponding circuit is shown in Fig. 8-10. This circuit is basically the one already shown in Fig. 8-6, since this type of circuit is more reliable than the one shown in Fig. 8-7.

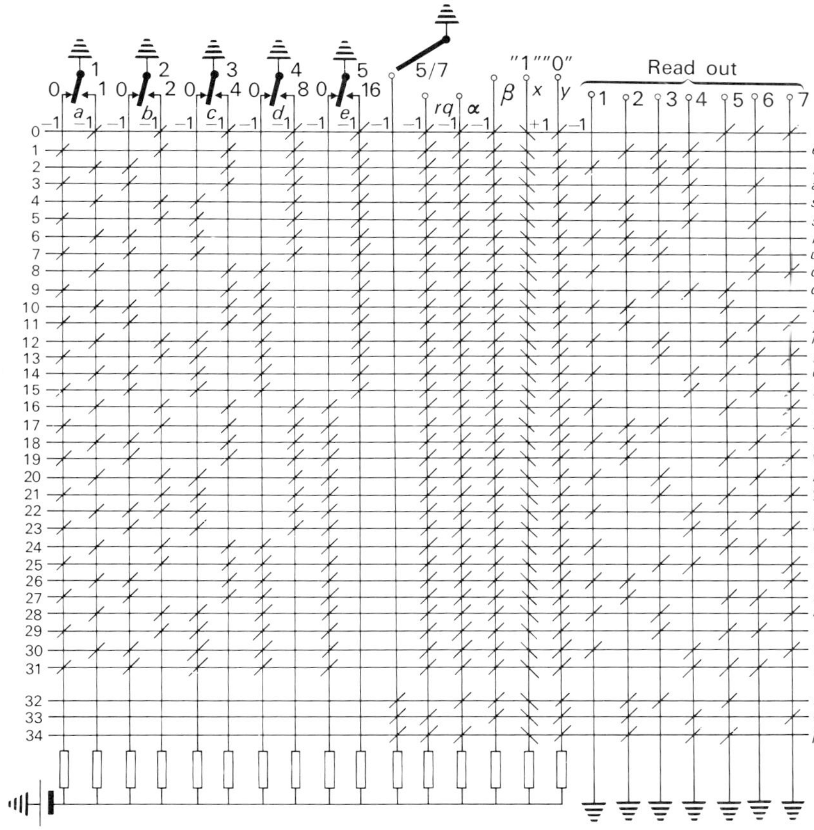

Fig. 8-10

The read-in windings of the magnetic cores in Fig. 8-10 are controlled by a group of five transfer contacts positioned in accordance with the five-unit code combination to be translated. With respect to the basic principles, the code converter of Fig. 8-10 is analogous to the code converter with polar relays shown in Fig. 8-6. In actual code converters of the type shown in Fig. 8-10, the read-in function will be performed by fully electronic means; however, for the sake of simplicity, this function is being entrusted to contacts in this diagram.

Before the translation of a code symbol is performed, all the cores will be preset, say, by a y pulse (-1), so that they remain in the symbolic 0 state. Contacts a to e are so connected to the 32 cores marked 0 to 31 that for every possible code symbol one core will always remain unmagnetized. All other cores will have a relative magnetization of at least -1.

With the read-in contacts a to e in the rest condition as shown, the core 0 receives no magnetization, core 1 has a magnetization of –1, and, finally, a magnetic field of –5 is induced in core 31.

The read-out function of the code converter is performed by an inquiry pulse which gives a relative magnetization of +1 to all cores. This pulse is fed over wire x. Core 0 changes its state only on this pulse, so that a read-out pulse is induced on wires 5, 6, and 7. The inquiry pulse on the x wire changes the magnetization of the other cores to 0 at most, so that the state of all these other cores is left unchanged.

The code converter shown in Fig. 8-10 is in fact presented in its entirely general form. It will entirely depend on the wiring itself—which five-unit code symbol is translated into which seven-unit symbol.

This example and that of Fig. 8-9 are sharp contrasts in that the circuit of Fig. 8-9 can be obtained only as a result of a rather elaborate design method, while the code converter of Fig. 8-10 can be arrived at without any further recourse to switching algebra, once the desired code translation has been defined.

PROBLEMS

1. Design a contact circuit that translates in parallel the 74210 two-out-of-five code (7 & 4 = 0) into a decimal one-out-of-ten code in such a way that the only decimal output circuit to be closed will be the one corresponding to a correctly translated code combination.

2. Design a magnetic-core circuit that translates in parallel the first ten combinations of the binary code (8421) into the 74210 two-out-of-five code (7 & 4 = 0).

3. Design a diode-gate circuit that translates in parallel the first ten combinations of the binary code (8421) into the biquinary code given below:

No.	5	00	4	3	2	1	0	*No.*	5	00	4	3	2	1	0
0	0	1	0	0	0	0	1	5	1	0	0	0	0	0	1
1	0	1	0	0	0	1	0	6	1	0	0	0	0	1	0
2	0	1	0	0	1	0	0	7	1	0	0	0	1	0	0
3	0	1	0	1	0	0	0	8	1	0	0	1	0	0	0
4	0	1	1	0	0	0	0	9	1	0	1	0	0	0	0

4. Design by means of matrix methods a code-translating contact circuit that translates in parallel the 5421 code into the 74210 two-out-of-five code.

5. Design a magnetic-core code translator which converts in parallel the four-digit binary code into the four-digit reflected binary code given below.

Reflected binary code

0	0	0	0	0	8	1	1	0	0
1	0	0	0	1	9	1	1	0	1
2	0	0	1	1	10	1	1	1	1
3	0	0	1	0	11	1	1	1	0
4	0	1	1	0	12	1	0	1	0
5	0	1	1	1	13	1	0	1	1
6	0	1	0	1	14	1	0	0	1
7	0	1	0	0	15	1	0	0	0

6. Design a diode-gate circuit that converts in parallel the 74210 two-out-of-five code into a decimal one-out-of-ten code.

SELECTED BIBLIOGRAPHY

1. Brown, D. R., and N. Rochester: Rectifier Networks for Multi-position Switching, *Proc. IRE*, vol. 37, pp. 139–147, 1949.
2. Oberman, R. M. M.: Some Analogies Between Contact, Resistor, Diode, Transistor, Polar Relay and Core Switching Circuits, *De Ingenieur*, pt. E, vol. 68, pp. 81–88, 1956.
3. Rajchman, Jan A.: Static Magnetic Matrix Memory and Switching Circuits, *RCA Rev.*, vol. 12, pp. 183–201, 1951.
4. Snijders, A.: Code converter, Netherlands Patent 68716, June 30, 1947; U.S. Patent 2,620,395, 1952.

9
Static Memory Circuits

9-1 THE DESIGN OF A SWITCHING CIRCUIT

Before actually designing a switching circuit on the basis of switching algebra, it is essential to draw up a complete set of operational requirements. These requirements may be stated in any of the following forms:

1. In a descriptive form
2. In the form of a state table
3. In the form of a time chart

It will be shown subsequently how requirements given in a descriptive form can be converted into a state table wherein all the operating states of the wanted circuit are listed systematically and how redundant states can be canceled out. The required states of the circuit to be designed must be realized in some fashion or other by means of switching components or combinations thereof which are so designed that they stay in a certain state after the controlling signals have disappeared.

The switching elements dealt with in the preceding chapters do not have ability to retain any information about the state of the circuit. These elements can be used to form logical circuits by means of which desired circuit states can be indicated; however, as soon as the control signals at the inputs of these logical circuits disappear, the output indication also disappears. In practically all cases, the next state of the circuit will depend on the present state. It is therefore necessary that this present state of the circuit be remembered by means of a certain type of switching or "memory" element. Circuits comprising memory elements are called sequential circuits.

The manipulation methods of state tables leading to state diagrams with a minimum number of memory elements will not raise any particular difficulties provided that some thought be given first to a number of standard switching-circuit layouts. The operation of these standard circuits, however, is better explained by means of time charts or sequence charts that present the operation or organization of the circuit graphically. Sometimes a time chart with a systematic regular operation that is easily recognizable will be preferable to another arrangement with less memory elements, operating as a rule in a more or less irregular sequence. Once this analysis of the time chart has been completed, the circuit requirements are now available in the complete form desired.

The switching algebra is then called upon to translate the operating stages of the memory elements as they are described in the time chart into a set of switching functions. These switching functions are not of the same type as those used in the previous chapters. The equations that describe the state of a combinational circuit without any time dimension will be used now to describe the state of a memory element that may adapt itself at a later stage to the state of the controlling combinational circuit. In the resulting state equation of a memory element, the variables that describe the present state of the circuit, i.e., those variables which describe the logical paths leading to the next state of the circuit, are indicated by means of small letters grouped on the right-hand side of the equation, and the resulting state of the operated memory element will be capitalized on the left-hand side of the state equation. There will normally be a time lag between the establishment of the logical paths described by the right-hand side of the equation and the moment in time when the corresponding memory element will be operated (or released).

It should be pointed out here that no circuit can be set up on the basis of switching algebra alone. The formulation of the circuit requirements and the construction of a state table or time chart in fact do determine the wanted circuit. Some experience is required to find the best and most practical solution. Switching algebra only comes into the picture once all circuit requirements have been established. This is somewhat analogous to a translating system by means of which a repertory of requirements (state table or time chart) is converted into actual circuit paths.

These transformation methods are to a large extent independent of the switching elements used in the final circuit.

9-2 THE STATE EQUATION OF A MEMORY ELEMENT

The truth table 9-1 describes the operation of a circuit that can function as

Table 9-1

r	s	z	Z
0	0	0	0
0	0	1	1
0	1	0	1
0	1	1	1
1	0	0	0
1	0	1	0
1	1	0	0
1	1	1	1

a memory element. The external input signals of this memory circuit are a set signal s and a reset signal r. The z signal is a feedback signal from output to input. These signals are indicated by small letters and represent present-state signals. The required next state is indicated in Table 9-1 by Z (capital letter). A number of lines in Table 9-1 show differing values of Z and z, which indicate that the memory circuit must adapt itself to a new combination of external input signals. Corresponding values of Z and z indicate that the memory circuit is in a stable state.

Truth table 9-1 shows six of these stable states of this binary memory element and two dynamic states. It is characteristic for a circuit with a memory for a past combination of input signals that different combinations of these signals can generate the same combination of output signals and also that the same combinations of input signals can generate different combinations of output signals. External input signals r, s (00, 10, and 11) generate an output signal 0 and external input signals r, s (00, 01, and 11) generate an output signal 1.

The switching requirements given in truth table 9-1 show that the memory element will be brought in state $Z = 1$ or set when a set signal $s = 1$ and no reset signal ($r = 0$) are applied, whereas the memory element will be brought in state $Z = 0$ or reset when a reset signal $r = 1$ and no set signal ($s = 0$) are applied. On the other hand, the combinations of input signals $r = 0$, $s = 0$ and $r = 1$, $s = 1$ leave the memory element in the state already adopted. On both these combinations, memory element Z can be in any one of both states, depending on its history.

In truth table 9-2, a number of variations are shown in the reactions of a memory element on the combination of external input signals $r = 1$ and $s = 1$. The column indicated by Z_0 shows the operation of a memory element when, in this case, a dominant reset signal is required. Then $Z_0 = 0$ when

Table 9-2

r	s	$z_{0,1,z}$	Z_0	Z_1	Z_z
0	0	0	0	0	0
0	0	1	1	1	1
0	1	0	1	1	1
0	1	1	1	1	1
1	0	0	0	0	0
1	0	1	0	0	0
1	1	0	0	1	0
1	1	1	0	1	1

$r = 1$ and $s = 1$. Column Z_1 relates to a memory element with a dominant set signal. In this case, $Z_1 = 1$ when $r = 1$ and $s = 1$. Column Z_z shows the same operation as already given in Table 9-1. This last type of memory element does not change its state under the influence of the input-signal combination $r = 1$ and $s = 1$. All three types of memory elements defined by Table 9-2 remain in their adopted state under the influence of the input-signal combination $r = 0$ and $s = 0$. This can be seen as a memory function.

The state Z of a memory element can be described in terms of present-state input signals, external or internal. Truth table 9-2 leads to the following three switching formulas:

$$Z_0 = r'(z_0 + s) \tag{9-1}$$

$$Z_1 = r' z_1 + s \tag{9-2}$$

$$Z_z = r' z_z + r' s + z_z s \tag{9-3}$$

These formulas follow easily from Table 9-2. Furthermore, it follows from Table 9-2 that there are some simple relations between output signals z_0, z_1, and z_z of the three types of memory elements and the set and reset signals s and r.

$$Z_0 = z_1 r' \tag{9-4}$$

$$Z_0 = z_z r' \tag{9-5}$$

$$Z_1 = z_0 + s \tag{9-6}$$

$$Z_1 = z_z + s \tag{9-7}$$

It should be noted that the generated output signals z always correspond in value to the state of the memory element obtained.

Equations (9-1) to (9-3) can easily be implemented by means of relay and contact circuits, as is shown in Figs. 9-1 to 9-3.

The implementation of Eqs. (9-1) to (9-3) by means of diode gates is not possible, only because in such a type of hardware the internal feedback circuit providing the memory function cannot generate a signal amplitude equal to or larger than that of the input signals. This is a necessary condition for a stable memory function.

Amplification of internal feedback signals can be obtained by means of inverters, as described in Chap. 6. In order to obtain the correct phase of the

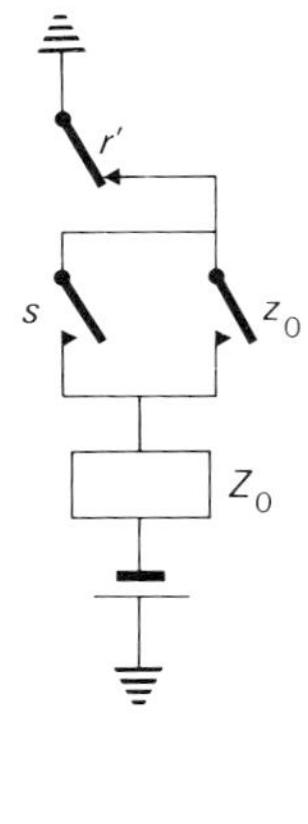

Fig. 9-1

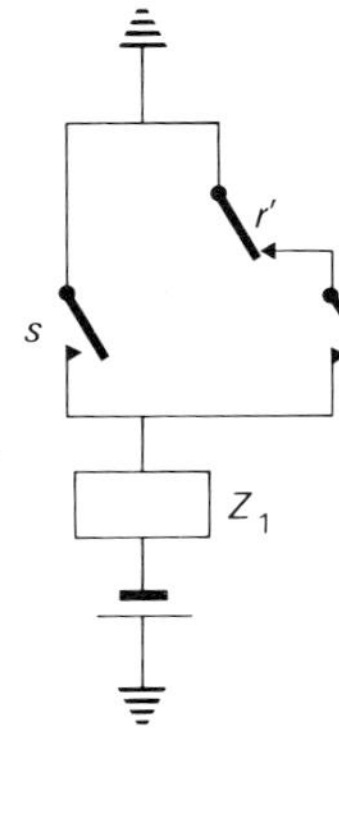

Fig. 9-2

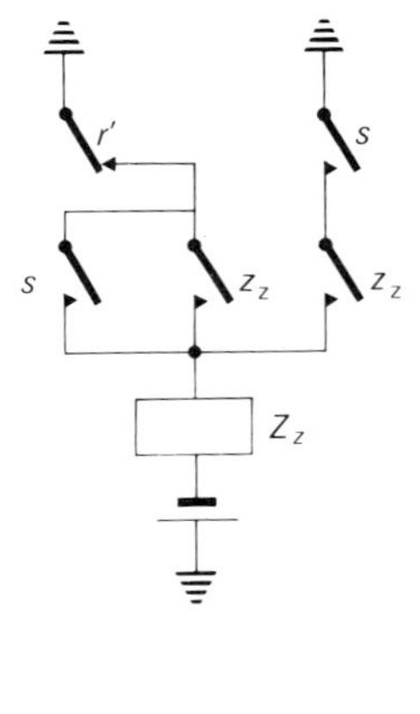

Fig. 9-3

feedback signals, two inverters have to be introduced. The switching formulas for electronic memory elements can be derived from Eqs. (9-1) to (9-3) by introducing a double negation of the z variable or the complete formula.

$$Z_0 = r'(z_0 + s) = \{[r'(z_0 + s)]'\}' = [r + (z_0 + s)']' = r'(z_0' s')' \tag{9-8}$$

Equation (9-8) gives three different ways of implementation of the circuit of this type of memory element. These variations are shown in Fig. 9-4.

It follows from Eq. (9-6), $Z_1 = z_0 + s$, that the memory circuit of Fig. 9-4a operates not only in the Z_0 mode but also in the Z_1 mode. The output of the z_1 signal has been indicated in the circuit of Fig. 9-4a.

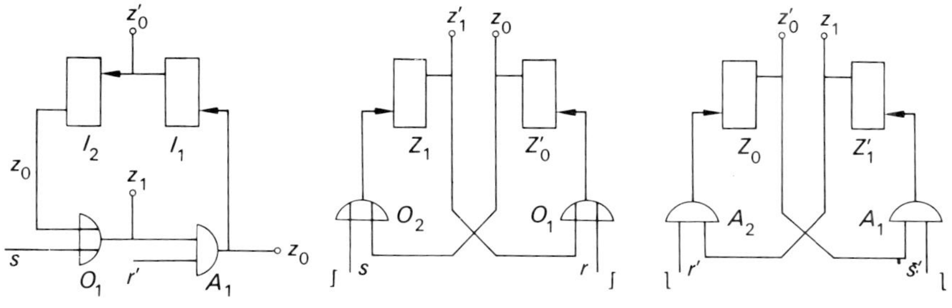

(a) (b) (c)

Figure 9-4*b* and *c* shows two symmetrical solutions of the elementary memory circuit, corresponding with the third and fourth portion of Eq. (9-8), respectively. The memory circuit shown in Fig. 9-4*b* operates on positive-going set and reset signals s and r, whereas the circuit shown in Fig. 9-4*c* operates on negative going set and reset signals s' and r'.

In the circuit of Fig. 9-4*b*, one output terminal generates a z_0 signal and the other a z_1' signal, as can easily be verified by means of Eq. (9-6). The circuit of Fig. 9-4*c* generates these output signals in the complementary form.

The symmetrical circuits of Fig. 9-4*b* and *c* will be called trigger in order to avoid ambiguity with the more complicated memory elements to be treated in Chap. 10. The drawing symbols for the circuits of Fig. 9-4*b* and *c* are shown in Fig. 9-5*a* and *b*, respectively.

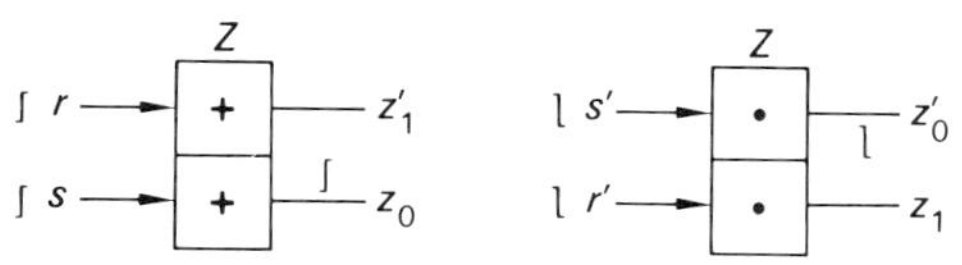

(a) (b)

Fig. 9-5

It should be noted that the memory circuits of Fig. 9-4*b* and *c* show in their stable states complementary output signals on both output terminals, except when $s = 1$ and $r = 1$.

Equation (9-2) can be transformed in the same way as Eq. (9-1):

$$Z_1 = r' z_1 + s = [(r' z_1 + s)']' = [(r' \cdot z_1)' \cdot s']' = (r + z_1')' + s \tag{9-9}$$

The third and fourth portions of Eq. (9-9) correspond to the circuits of Fig. 9-4*c* and *b*, respectively.

Equation (9-3) can be implemented in a number of ways. Some of the possible equation transformations follow below as Eq. (9-10):

$$Z_z = [(r' z_z + r' s + z_z s)']' = \{[r'(z_z + s) + z_z s]'\}' \qquad (9\text{-}10)$$

The corresponding memory circuits are shown in Fig. 9-6*a* and *b*, respectively. The last circuit generates not only z_z and z_z' but also signals z_0 and z_1. The Z_z memory element is not often used in practical switching.

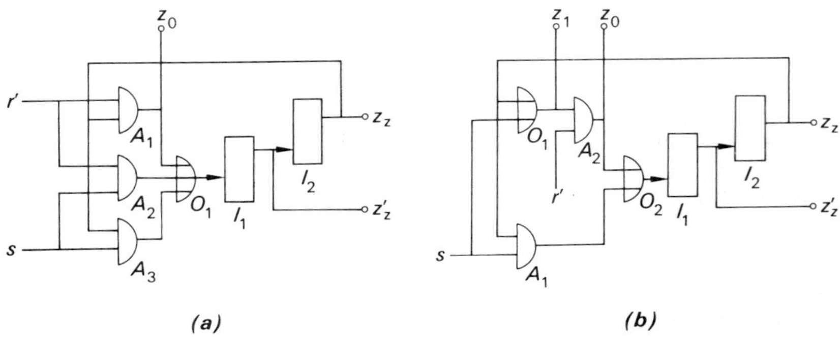

Fig. 9-6

The symmetric symbolic circuits of Fig. 9-4*b* and *c* can be implemented by means of relays and contacts, electronic components, or any other type of switching element with suitable logical properties. In Fig. 9-7*a* and *b*, the two-relay trigger circuits corresponding to these symbolic memory circuits are shown.

The individual circuits of both relays in both two-relay triggers have no memory character. However, the circuits of both relays of one trigger together form a memory circuit. This can be seen as follows:

$$Z_1 = s + z_0 \qquad \text{and} \qquad Z_0' = r + z_1'$$

Hence

$$Z_1 = s + (r + z_1')' = s + z_1 r'$$

and

$$Z_0 = r' z_1 \qquad \text{and} \qquad Z_1' = s' z_0'$$

Hence

$$Z_0 = r' \cdot (s' z_0')' = r'(s + z_0)$$

The time charts of both trigger relay pairs are presented in Fig. 9-8*a* and *b*. They show the two relays Z_1, Z_0' and Z_1', Z_0 almost inversely operating

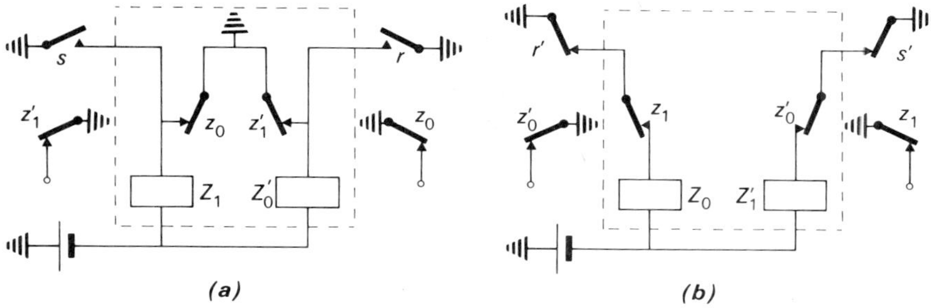

Fig. 9-7

on a set signal s and a reset signal r, respectively, a set signal s', and a reset signal r'. In the circuit of Fig. 9-7a, relay Z_1 is not operated in the agreed rest condition, while relay Z_0' is. The set signal s operates relay Z_1 at moment t_1. With a short delay, relay Z_1 interrupts the hold circuit of relay Z_0', so that this latter relay releases its armature also with a short delay after the operation of Z_1.

Reset signal r, which has the same polarity as the set signal, cannot directly release relay Z_1. Since the circuit has to be symmetrical, it is necessary that the r signal operates relay Z_0', which in its turn interrupts the hold circuit of relay Z_1. Reset signal r, which is in fact the set signal of relay Z_0', is found to occur in the time chart of Fig. 9-8a at moment t_3.

It should be noted that in the time chart of Fig. 9-8a, Z_0' and Z_1 are not simultaneously 0, whereas in the time chart of Fig. 9-8b, Z_0 and Z_1' are not simultaneously 1. This last time chart needs no further explanation.

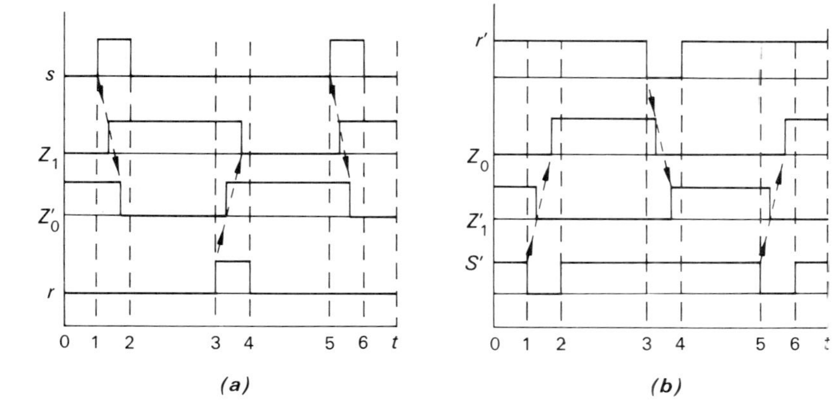

Fig. 9-8

It should also be noted that both types of symmetrical trigger circuits can be derived from each other by turning the circuit round its axis and inverting all indications (including AND gates in OR gates and v.v.). This can be seen in Figs. 9-4*b* and *c*, 9-5*a* and *b*, and 9-7*a* and *b*.

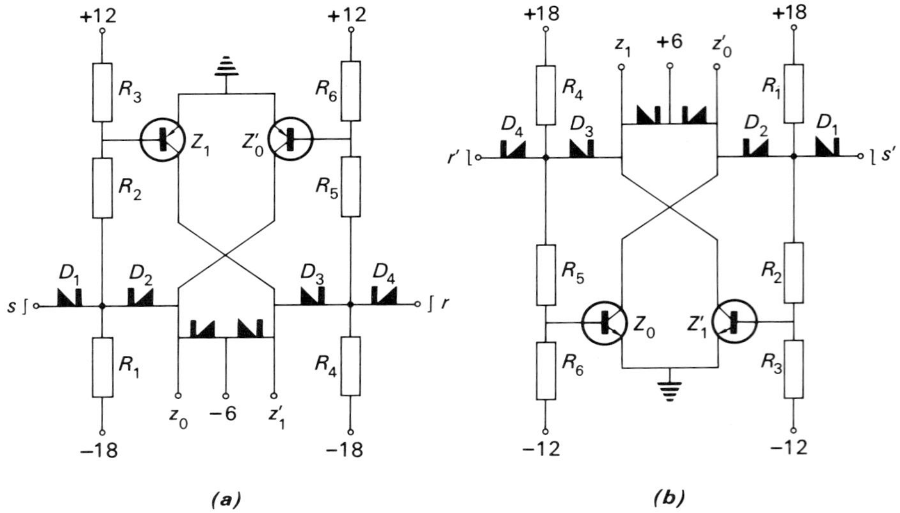

Fig. 9-9

The symbolic circuits of Fig. 9-4*b* and *c* can also be implemented by means of diodes and transistors, as is shown in Fig. 9-9*a* and *b*. In the circuit of Fig. 9-9*b*, *pnp* transistors are used in the inverters, whereas in the circuit of Fig. 9-9*b*, *npn* transistors are used for this purpose. The operation of the trigger circuit of Fig. 9-9*a* can be explained as follows. Let us assume, for instance, that transistor Z_1 is conducting and transistor Z'_0 is nonconducting in the rest condition of this trigger, in which case the z_0 signal, i.e., the collector voltage of transistor Z'_0, is 0 or negative with respect to earth potential. In this condition of the trigger, signals s and r are 0, so that both input signals of the OR gate consisting of diodes D_1 and D_2 are 0. This results in a 0-controlling signal on the base of transistor Z_1. In this situation, the base of transistor Z_1 is negative with respect to the emitter. This voltage causes transistor Z_1 to remain in conducting condition. Inversely, the z'_1 signal on the collector of transistor Z_1 is 1, which corresponds to earth potential in a circuit with grounded emitters. Via diode D_3, the 1 signal is passed to the base of transistor Z'_0. Voltage divider $R_{5/6}$ changes the voltage level of this signal from earth potential to a voltage that is positive, so that transistor Z'_0 is kept nonconducting.

A positive-going s signal applied over diode D_1 converts the base voltage of transistor Z_1 into a positive voltage that locks the base-emitter path and consequently also the collector-emitter path. This transistor then becomes nonconducting. The output signal of the collector of transistor Z_1 changes from 1 to 0, so that both inputs of the OR gate consisting of diodes D_3 and D_4 receive a 0-input signal. Consequently, the base voltage of transistor Z_0' is changed from positive to negative with respect to the emitter. Transistor Z_0' becomes conducting. The collector voltage of this transistor accordingly changes from 0 to 1, so that a feedback signal of correct phase and amplitude is produced to hold transistor Z_1 in its nonconducting condition.

A positive-going reset signal $r = 1$ triggers the circuit back in its agreed rest condition as described above. Simultaneously applied set and reset signals $s = 1$ and $r = 1$ bring both transistors of this trigger in their nonconducting condition, so that $z_0 = 0$ and $z_1' = 0$. The resulting condition of the trigger will depend on which of the two control signals ends last.

It should be noted that in the trigger circuits of Fig. 9-9, clipping diodes D_5, D_6 have been introduced so that both states of an output signal have a low internal resistance and are always rather well defined. In the given system of power supply, this leads to increased operating speed and decreased sensibility to noise. Moreover, in a switching system with well-defined signal voltages, overload of transistor bases can never occur.

The trigger circuit shown in Fig. 9-9*a* operates on signal voltages of 0 volts and -6 volts, represented by a symbolic 1 and 0, respectively. In the trigger circuit of Fig. 9-9*b* using *npn* transistors, the signal voltages are $+6$ volts and 0 volts represented by a symbolic 1 and 0, respectively. Shifting down all voltages of the power supply of this last type of trigger over 6 volts makes both types of triggers shown in Fig. 9-9 compatible from the point of view of the levels of the signal voltages.

It is also possible to design a trigger circuit operating on positive-going set and reset signals as well as on negative going set and reset signals. Such a circuit is shown in Fig. 9-10. It has four input terminals. The s' and r' input signals have to be applied to input terminals 1 and 2, whereas the s and r input signals are fed to input terminals 3 and 4.

The symmetrical trigger circuits of Fig. 9-9*a* and *b* show the diode-transistor logic. These circuits can also be implemented with transistors only. The all-*npn*-transistor trigger circuit corresponding to Eq. (9-8), $Z_0 = [r + (z_0 + s)']'$, is shown in Fig. 9-11*a*. The all-*pnp*-transistor trigger corresponding to Eq. (9-8), $Z_0 = r'(z_0's')'$, is shown in Fig. 9-11*b*.

Using series-connected transistors, Eq. (9-8) can be implemented in two other ways with *pnp* and *npn* transistors, as follows easily from this formula. Depending on the manufacturing process, a number of variations in triggers is possible in integrated circuit techniques. These variations will not be discussed in this text because they do not lead to a better insight in the logical operation of a memory element.

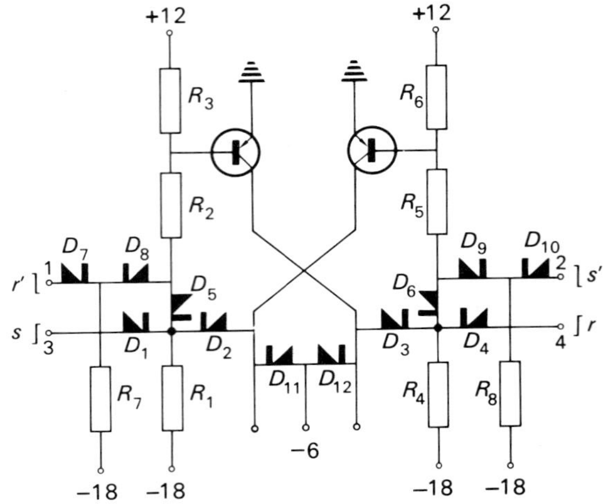

Fig. 9-10

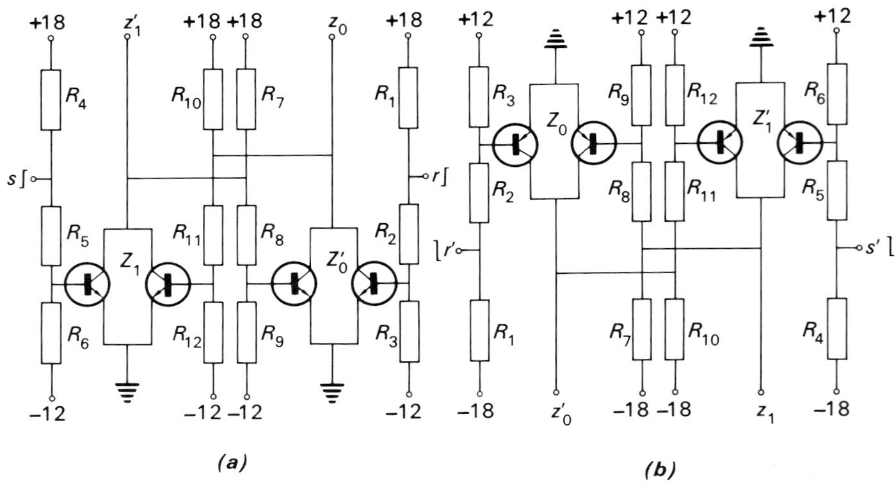

Fig. 9-11

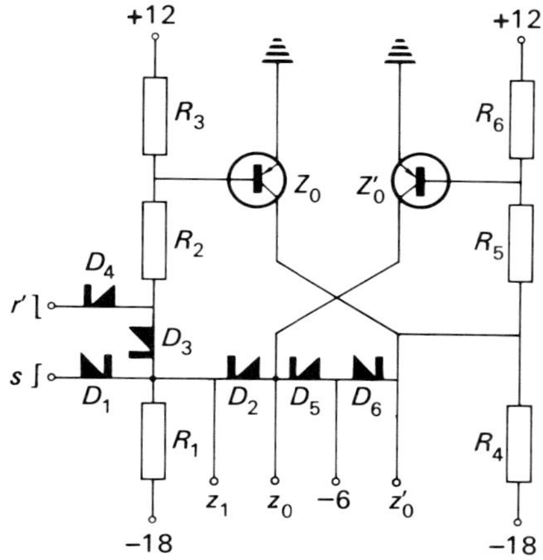

Fig. 9-12

Finally, a realization of the asymmetrical symbolic trigger circuit of Fig. 9-4*a* will be given (Fig. 9-12) in diode-transistor logic. The controlling signals are a set signal s and a reset signal r'. The input OR gate O_1 and the input AND gate A_1, as shown in Fig. 9-4*a*, can be formed in the base control circuit of one of the transistors (Z_0).

9-3 MULTIPOSITION TRIGGERS

The trigger discussed above was a two-state memory element. However, the two-state trigger is, in fact, no more than a special form of a circuit with a number of different states. These states can be formed by different combinations of operated and unoperated conditions of the switching elements. For a trigger circuit, it is necessary that one, or a constant number, of these switching elements remains always operated. It is not possible to describe the state of a multiposition trigger by means of one state equation because such equations describe binary events only. As is the case with the two-state triggers, these state equations degenerate to nonmemory formulas, but all the switching elements together form the desired memory circuit.

The time chart of Fig. 9-13 shows the operation of a three-position trigger. This time chart is analogous to the two-position trigger chart. Each of the elements is directly set by its corresponding control signal (α, β, or γ). A switching element can be reset by either of the other switching elements. It should be pointed out that the time chart of Fig. 9-13 shows the α, β, and γ pulses in this order; however, any arbitrary sequence may be used, so that conceivably the A element could be reset either by the B element or the C

element. In the time chart of Fig. 9-13, the set pulses are dominant over the reset signals produced by the other switching elements. Hence, state equation (9-2) has to be applied. The following state equations are thus derived for the elements A, B, and C:

$$A = \alpha + a(b + c)' = \alpha + ab'c' \tag{9-11}$$

$$B = \beta + b(a + c)' = \beta + a'bc' \tag{9-12}$$

$$C = \gamma + c(a + b)' = \gamma + a'b'c \tag{9-13}$$

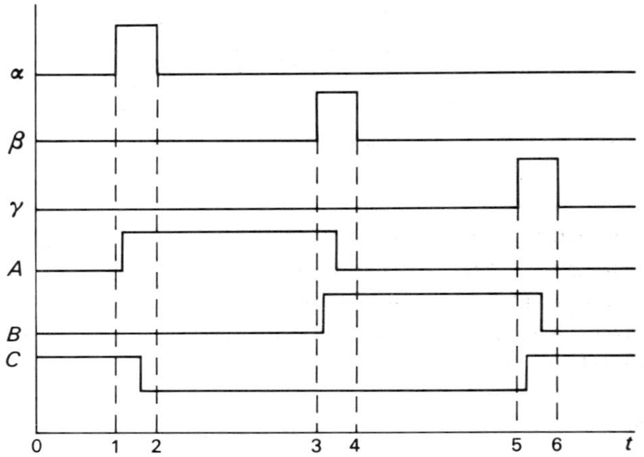

Fig. 9-13

In the time chart of Fig. 9-13, there are no periods during which all switching elements remain in the unoperated state; therefore $a'b'c' = 0$. Accordingly state equations (9-11) to (9-13) can be simplified as follows:

$$A = \alpha + b'c' \tag{9-14}$$

$$B = \beta + a'c' \tag{9-15}$$

$$C = \gamma + a'b' \tag{9-16}$$

The relay circuit corresponding to state equations (9-14) to (9-16) is given in Fig. 9-14.

The time chart of Fig. 9-15 illustrates the operation of a trigger in which the control signals perform the reset of the switching elements, whereas the set of a switching element is performed by the other switching elements of the trigger circuit. Each of the control pulses, which may be applied in an

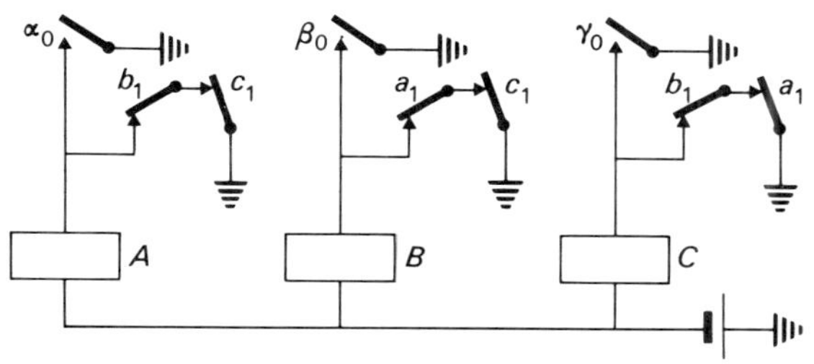

Fig. 9-14

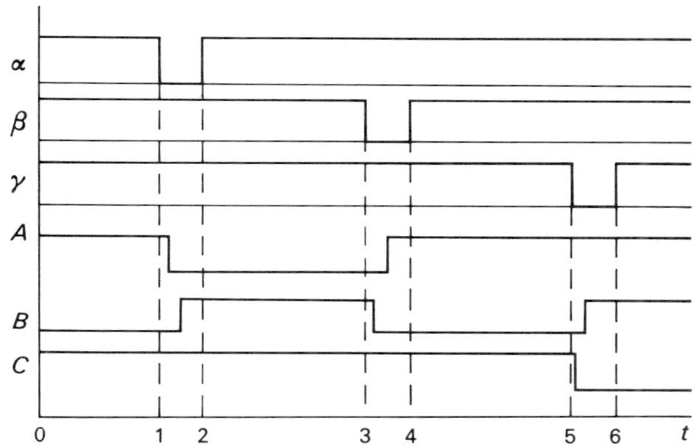

Fig. 9-15

arbitrary sequence, resets the corresponding switching element into its unoperated condition, after which the only other unoperated switching element is brought into its operated condition. In the steady state, two-out-of-three switching elements are always found to be in the operated condition in this type of time chart. The three-position trigger conforming to the time chart of Fig. 9-15 can be described by the following state equations:

$$A = r_a' \cdot (s_a + a) = \alpha'(b' + c' + a) \tag{9-17}$$

$$B = r_b' \cdot (s_b + b) = \beta'(c' + a' + b) \tag{9-18}$$

$$C = r_c' \cdot (s_c + c) = \gamma'(a' + b' + c) \tag{9-19}$$

In this type of trigger, the three switching elements are never found to be simultaneously operated; hence $abc = 0$. The introduction of this redundant term in state equation (9-17) makes it possible to prove that the term $a\alpha'$ is also redundant. This can be done once the term $a\alpha'$ has been converted to $ac\alpha'$. Ultimately the redundant term $abc = 0$ may be removed again.

$$A = \alpha'(b' + c' + a) = \alpha'(b' + c' + ac) + abc$$
$$= \alpha' b' + \alpha' c' + \alpha' ac + abc = \alpha' \cdot (b' + c') \tag{9-20}$$

$$B = \beta' \cdot (a' + c') \tag{9-21}$$

$$C = \gamma' \cdot (a' + b') \tag{9-22}$$

The state equations corresponding to the B and C elements can be manipulated in the same way. The resulting circuit is shown in Fig. 9-16.

State equations (9-20) to (9-22) can be implemented by means of any type of switching element, basically in the same way as was the case with the two-position trigger. By way of an example of such a three-position trigger, a mixed diode-transistor circuit is illustrated in Fig. 9-17.

The input circuits of this three-position trigger have been provided with diodes so that they can be connected immediately to a driving transistor collector.

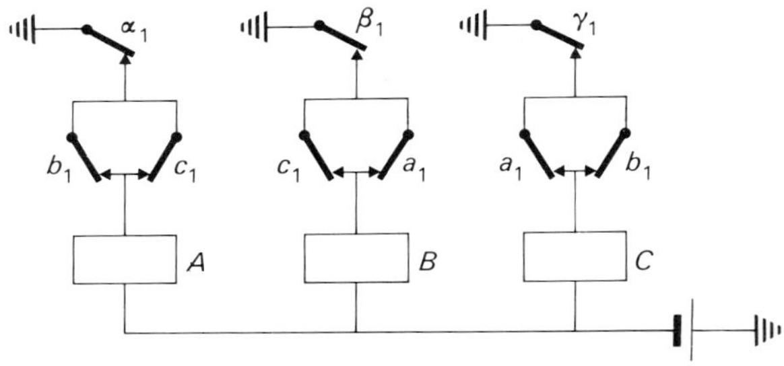

Fig. 9-16

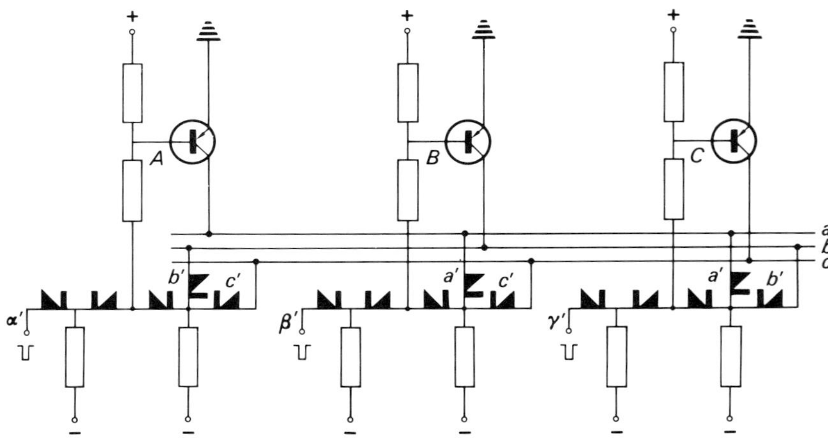

Fig. 9-17

It should be stressed that multiposition trigger circuits derived from either type of state equation are each other's dual. Accordingly, the choice of the set of state equations to be used is immaterial.

By way of extension, the multiposition trigger circuit of Fig. 9-17 can be made into a trigger circuit controlled by a central clock pulse. The setting indication of the trigger is derived from signals on three input wires. This circuit is presented symbolically in Fig. 9-18.

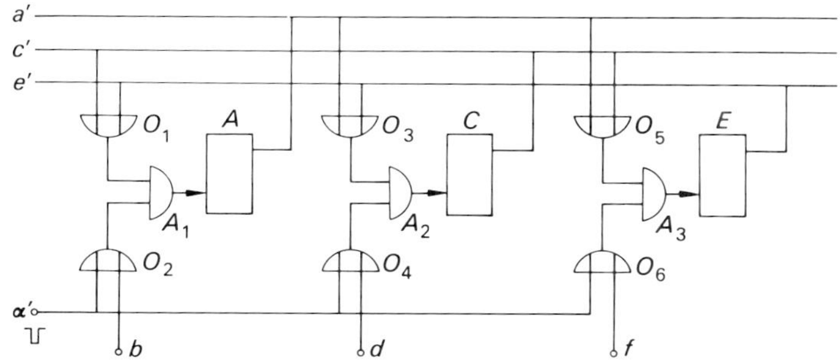

Fig. 9-18

For example, if the input signals of the A, C, and E inverters are, respectively, 1, 1, and 0, then $a' = 0$, $c' = 0$, and $e' = 1$. The e'-locking wire marks the OR gates O_1 and O_3 with 1, whereas both inputs of the OR gate O_5 receive a 0-level signal so that the trigger is in a stable state. If the next state $A = 1$, $C = 0$, and $E = 1$ is desired, the input wires have to be marked $b = 1$, $d = 0$, and $f = 1$. When feeding in a clock-pulse α', only the output of the OR gate O_4 will change from 1 to 0; consequently, the state of inverter C will also change from 1 to 0. The c'-locking wire is now marked with $c' = 1$, so that a 1 is passed on to inverters A and E. This does not alter the state of the A inverter, but the E inverter passes from 0 to 1.

The state equations for the inverters of this multiposition trigger are as follows:

$$A = (\alpha' + b) \cdot (c' + e') \tag{9-23}$$

$$C = (\alpha' + d) \cdot (a' + e') \tag{9-24}$$

$$D = (\alpha' + f) \cdot (a' + c') \tag{9-25}$$

All the multiposition triggers discussed so far are of the type in which the external control influenced the state of one of the inverters only, whereas the necessary change of state of other inverters was performed through the internal locking network. This external control signal determines which inverter has to be set (or reset).

It is possible to design another group of multiposition triggers whereby the external control acts on all the inverters but one. Hence, in the group of one-out-of-n triggers (one inverter operated, $n-1$ unoperated), the external control signals are applied to $n-1$ inverters and these signals will cause only one inverter to react, whereas $n-2$ inverters remain unaffected because these inverters are already in the desired state.

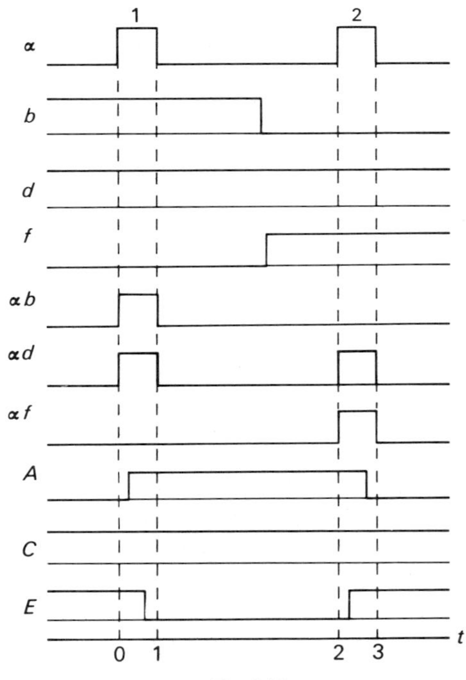

Fig. 9-19

The time chart picturing this operation of the inverters of a one-out-of-three trigger is shown in Fig. 9-19. Initially, the A, C, and E inverters are, respectively, in the 0, 1, and 1 state. On the first clock pulse, this combination has to change under the influence of $b=1$, $d=1$, and $f=0$ into $A=1$, $C=1$, and $E=0$. On the second clock pulse α, the trigger is restored to its initial state.

The lines b, d, and f in the time chart indicate what the next state of the multiposition trigger will be. With the α control pulse, they combine to form set pulses for the various inverters. In this example, pulses αb and αd are generated on the first α clock pulse. Since the A inverter was unoperated then, the pulse αb provides the set pulse for this inverter. The pulse αd does not affect the circuit because the C inverter is already set. As a consequence of the setting of the A inverter, the E inverter is reset via the locking network.

On the second clock pulse in the time chart of Fig. 9-19, the E inverter is set again, whereas the A inverter is reset over the locking network. This reset operation of the A inverter by means of the locking network must always be performed while the C and E inverters are in the operated state. The general state equation for the A inverter then becomes

$$A = s_a + ar_a' = \alpha b + a \cdot (ce)' = \alpha b + a(c' + e') \qquad (9\text{-}26)$$

As one can see merely by looking at the time chart of Fig. 9-19, at any one moment no two inverters are simultaneously unoperated. This is owing to the fact that the state equation used is of the type where the set signals are dominant. Hence,

$$a'c' = 0 \qquad a'e' = 0 \qquad c'e' = 0 \qquad (9\text{-}27)$$

Equations (9-27) can be used for simplifying state equation (9-26):

$$A = \alpha b + ac' + ae' + a'c' + a'e' = \alpha b + c' + e' \qquad (9\text{-}28)$$

The state equations of the other inverters can be arrived at in the same way:

$$C = \alpha d + a' + e' \qquad (9\text{-}29)$$

$$E = \alpha f + a' + c' \qquad (9\text{-}30)$$

If Eq. (9-1) instead of (9-2) had been used as a starting point (and interchanging the operated and unoperated states in the time chart of Fig. 9-19), an analogous set of state equations would have been obtained:

$$A = (\alpha' + b) \cdot c'e' \qquad (9\text{-}31)$$

$$C = (\alpha' + d) \cdot a'e' \qquad (9\text{-}32)$$

$$E = (\alpha' + f) \cdot a'c' \qquad (9\text{-}33)$$

The multiposition trigger circuit derived from Eqs. (9-31) to (9-33) is shown symbolically in Fig. 9-20.

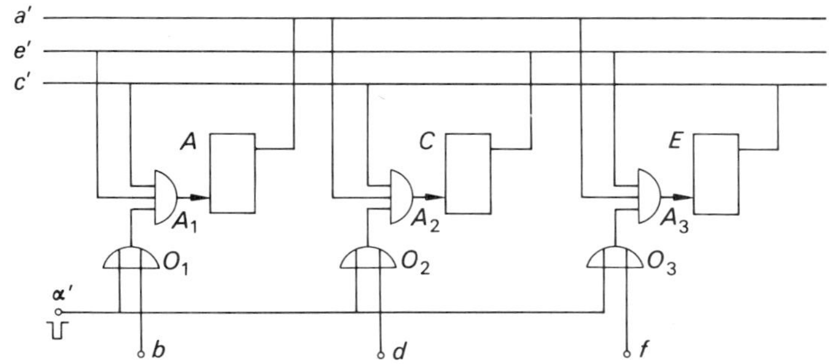

Fig. 9-20

The operation of the circuit of Fig. 9-20 can easily be checked by means of the time chart of Fig. 9-19 even though this chart would in fact correspond to the circuit to be derived from state equations (9-28) to (9-30). In the multiposition trigger shown in Fig. 9-20, there is one diode-gate level less in the locking network.

It is not necessary to limit the design of multiposition triggers to such triggers where one-out-of-n or $(n-1)$-out-of-n inverters must always be in the operated state. The number of diodes in the locking network of a one-out-of-n trigger increases with the square of the number of inverters. However, if the number of simultaneously operated inverters is increased, the number of diodes in the locking network decreases accordingly.

LOCKING NETWORKS

The design problem of multiposition triggers with more than one adjacent operated inverter (input signal is 1) lies mainly in the locking network. Some general laws help solve this problem. These laws can be shown easily by means of a few examples. The locking circuit of a ten-position trigger with three adjacent operated inverters is shown schematically in Fig. 9-21. The operated

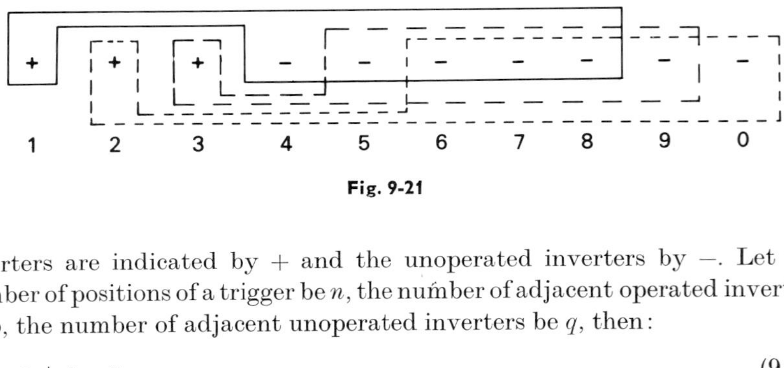

Fig. 9-21

inverters are indicated by + and the unoperated inverters by −. Let the number of positions of a trigger be n, the number of adjacent operated inverters be p, the number of adjacent unoperated inverters be q, then:

$$p + q = n \tag{9-34}$$

Let the number of operated inverters that give a combined locking signal be a, the number of inverters locked by a locking signal coming from a group of a operated inverters be b, then:

$$q - p = b - a \tag{9-35}$$

Equation (9-35) holds for all multiposition triggers of this type. All it says is that the complete locking circuit will consist of n identical locking circuits between n groups of a operated inverters and n groups of b unoperated inverters. This formula does not guarantee the adequacy of the locking network's preventing errors in the positioning of the trigger. In a ten-position trigger with a locking circuit of the type shown in Fig. 9-21, such errors cannot occur. The locking circuit is perfect. However, the locking circuit for the 15-position trigger shown in Fig. 9-22 is the first one in this group to be found lacking in this respect. The A inverter locks inverters G, H, I, and J. Inverter B locks inverters H, I, J, and K, and so on. In the present case, $q = 9$, $p = 6$, $b = 4$, and $a = 1$, which fulfils Eq. (9-35). However, as was pointed out already, this does not imply that the locking is perfect; it does mean that the locking circuit is identical for all inverters, this being also a requirement for a good circuit.

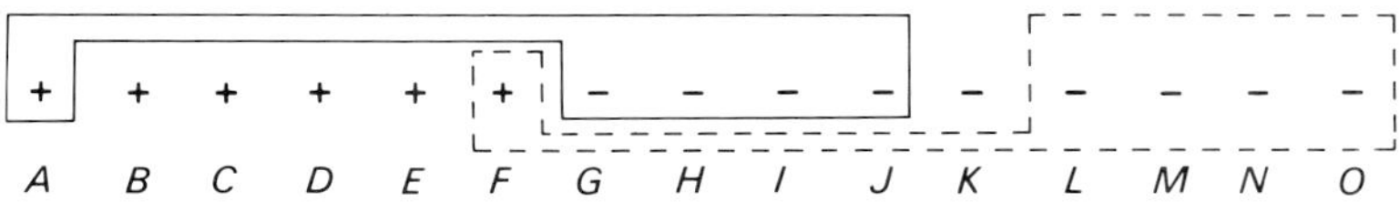

Fig. 9-22

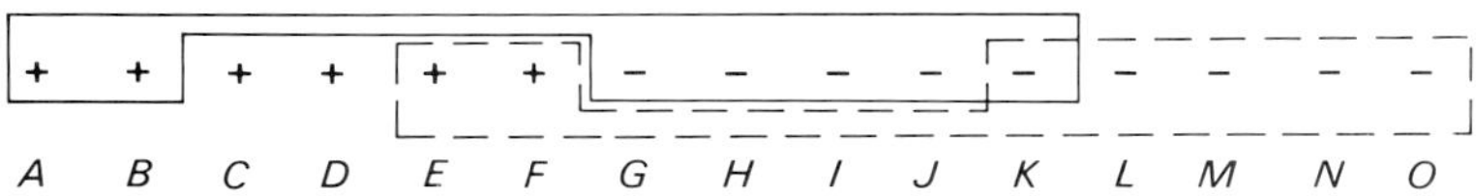

Fig. 9-23

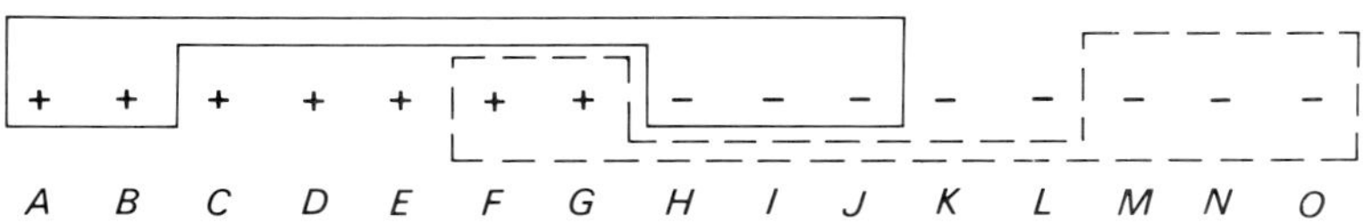

Fig. 9-24

If inverters A and F are operated, then inverters G, H, I, J and L, M, N, O are locked in the locking scheme of Fig. 9-22. Inverters B, C, D, and E can lock analogous groups of unoperated inverters, but it can also happen that the operated inverter K causes inverters B, C, D, and E to be locked in the unoperated state, thereby disturbing the perfect locking scheme.

As can be seen in Fig. 9-23, a perfect locking circuit for a 15-position trigger containing a group of 6 adjacent operated inverters can be arrived at for $a = 2$ and $b = 5$. Equation (9-35) also holds for these numbers. However, if the number of operated inverters in a 15-position trigger is increased to 7, the locking with $a = 2$ is no longer perfect, as shown in Fig. 9-24. The locking pattern given in Fig. 9-24 meets the requirements of Eq. (9-38), and this can easily be checked. However, the locking is not perfect. Indeed, if inverters A and B cause inverters H, I, and J to be locked and if inverters F and G do the same with inverters M, N, and O, then it is possible that inverters

K and L are operated, thereby locking inverters C, D, and E. The perfect-locking scheme is accordingly disturbed.

There is a general law for the design of perfect locking circuits for this type of multiposition trigger. The locking of an n-position trigger is no longer perfect if there are a or more inverters between the first and the last group of a number of b locked inverters, because these a inverters can lock a group of b inverters between the first and last group by means of these a operated inverters. Accordingly, the locking is not perfect for:

$$q \geqslant a + 2b \tag{9-36}$$

This general law can also be stated as follows in terms of the inverters of the operated group: the locking of an n-position trigger is not perfect if there are b inverters or more between the first and last group of locking inverters, because these b inverters can be locked by means of a group of a inverters between the first and last group of locked inverters. Thus, the locking is also not perfect for

$$p \geqslant 2a + b \tag{9-37}$$

Substitution of Eq. (9-35) in Eq. (9-36) gives Eq. (9-37), and vice versa.

If it is required to design a multiposition trigger circuit with a minimum number of diodes in the locking circuit, the number of operated inverters being immaterial, the type with $a = 1$ is to be preferred up to the limit of a perfect locking. The number of diodes per locking group is determined by $a + b$, which gives the following number of diodes for an n-position trigger

$$N = n(a + b) \tag{9-38}$$

For $a = 1$, this number is reduced in all cases to

$$N = nb \tag{9-39}$$

The type of multiposition trigger with $a = 1$ may be considered as the most economical.

This most economical type of n-position trigger with $a = 1$ can be readily designed by means of the following equations:

$$p = \frac{n + 1 - (2/\sqrt{3}) \sin (2n/3) (n + 1)}{3} \tag{9-40}$$

and

$$b = \frac{n + 1 + (4/\sqrt{3}) \sin (2n/3) (n + 1)}{3} \tag{9-41}$$

With these equations, the number of operated inverters p and the number b of inverters in the locked group can be determined immediately.

It is practically always necessary to have n outlets for an n-position trigger circuit. Multiposition triggers with $a = 1$ are in this respect the simplest examples. The only operated inverter always indicates the position of the trigger, so that only one diode per position is required. Multiposition triggers with an adjacent group of operated inverters always need two diodes per position. These two diodes are connected to the first and last operated inverter of the characteristic group.

By way of an example, let us look at the design of the locking circuit of a 15-position trigger with $p = 7$ operated and $q = 8$ unoperated adjacent inverters. Equation (9-35) states that in this case, for a symmetrical locking circuit, it is required that

$$q - p = b - a = 8 - 7 = 1$$

Hence,

$$b - a = 1 \tag{9-42}$$

Equation (9-36) shows that when $q \geqslant a + 2b$, locking is not perfect. Substitution of Eq. (9-42) in Eq. (9-36) shows that for

$$8 \geqslant 3a + 2 \tag{9-43}$$

locking will not be perfect. Hence, a perfect locking circuit will be obtained for

$$a > 2 \tag{9-44}$$

the minimum values for a and b being $a = 3$ and $b = 4$.

One section of the resulting locking network of a 15-position trigger with $a = 3$ and $b = 4$ is shown symbolically in Fig. 9-25. Each of the fifteen locking circuits comprises an OR gate with three inputs and one input to each of four AND gates with four input terminals.

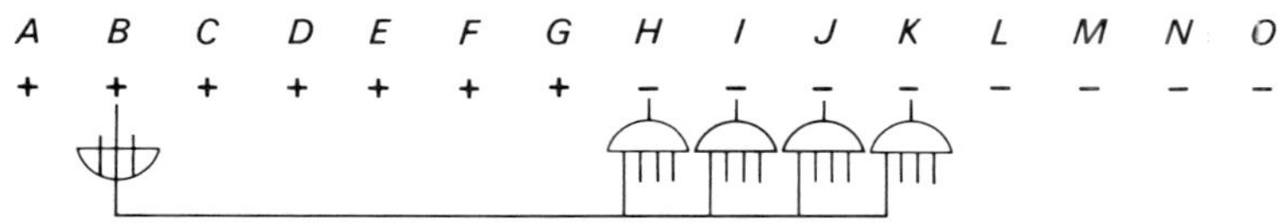

Fig. 9-25

MULTIPOSITION TRIGGERS WITH n-OUT-OF-m STATES

It is not necessary to limit the locking networks of a multiposition trigger to those circuits having a constant number of adjacent operated inverters. A more general type of multiposition trigger will be obtained if the restriction no longer applies that the operated inverters have to be adjacent. A trigger with n inverters, m of which have to be operated, can be put in N different states:

$$N = \frac{n!}{(m-n)!\,m!} \tag{9-45}$$

The locking network of such a trigger is basically analogous to the circuit shown in Fig. 9-25. By way of an example, a two-out-of-four trigger will be designed of the type shown in Fig. 9-20 [state equations (9-28) to (9-31)]. Let the four inverters of this trigger be called B, D, F, and H. The setting of each of these inverters will be controlled, respectively, by an a, c, e, and g signal, which will be applied externally under the control of a clock-pulse α. The six possible combinations of two operated and two unoperated inverters are given in Table 9-3. Although they are not directly relevant to the design of the multiposition trigger, the combinations of the externally applied a, c, e, and g signals have also been indicated in Table 9-3.

Table 9-3

External signal combination	*Operated inverters*	*Unoperated inverters*
$ace'g'$	$B\ D$	$F\ H$
$ac'eg'$	$B\ F$	$D\ H$
$a'ceg'$	$D\ F$	$B\ H$
$a'ce'g$	$D\ H$	$B\ F$
$a'c'eg$	$F\ H$	$B\ D$
$ac'e'g$	$B\ H$	$D\ F$

The state equations for these four inverters, which form together a six-state memory, then become

$$\begin{aligned} B = s_b + br_b' &= \alpha a + b(df + dh + fh)' \\ &= \alpha a + b(d' + f')\cdot(d' + h')\cdot(f' + h') \\ &= \alpha a + bd'f' + bd'h' + bf'h' \end{aligned}$$

Since no three inverters will be simultaneously in the unoperated state, $b'd'f'$, $b'd'h'$, and $b'f'h'$ are 0. Thus

$$\begin{aligned} B &= \alpha a + bd'f' + b'd'f' + bd'h' + b'd'h' + bf'h' + b'f'h' \\ &= \alpha a + d'f' + d'h' + f'h' \end{aligned} \tag{9-46}$$

The state equations of the other inverters follow in the same way:

$$D = \alpha c + b'f' + b'h' + f'h' \tag{9-47}$$

$$F = \alpha e + b'd' + b'h' + d'h' \tag{9-48}$$

$$H = \alpha g + b'd' + b'f' + d'f' \tag{9-49}$$

From these state equations, the memory circuit shown in Fig. 9-26 can be derived.

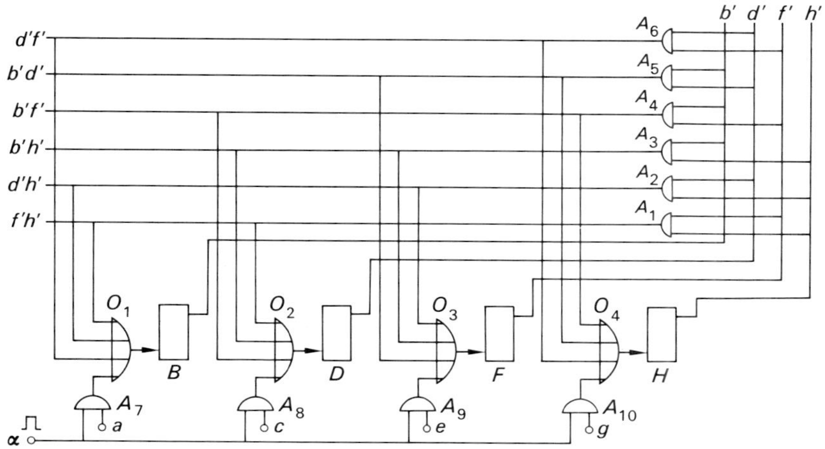

Fig. 9-26

In principle, the multiposition trigger circuit of Fig. 9-26 is similar to the circuit shown in Fig. 9-20, the difference being that in the trigger circuit of Fig. 9-20, there are as many locking bars as inverters. In both cases, the number of locking bars is equal to the number of positions of the trigger circuit. A group of six AND gates (A_1 to A_6) derives the locking indication from the state of the controlling inverters.

It should be pointed out that it is also possible to design an n-out-of-m trigger using the principle of the trigger circuit shown in Fig. 9-18. This circuit will not be discussed here because its gates are of a more complicated nature.

The n-out-of-m trigger is in fact the most fundamental type of trigger. The ordinary two-position trigger is derived from this circuit for $n = 1$ and $m = 2$.

PROBLEMS

1. Prove that the state equation of a complete trigger $A = \alpha' \cdot (\beta' a')'$ conforms to the fundamental state equation of a memory element $Z = s + zr'$.

2. Design an all-npn-transistor trigger operating on negative-going set and reset signals. Write the state equation of this memory circuit.

3. Design a four-transistor trigger circuit that sets on positive-going and resets on negative-going control signals.

4. A diode gate and inverter network is given in Fig. 9-27. The circuit is controlled by signals y and z. The output terminals are a and b. The output signals on terminals a and b are not each other's inverse on certain combinations of the input signals. Determine these combinations.

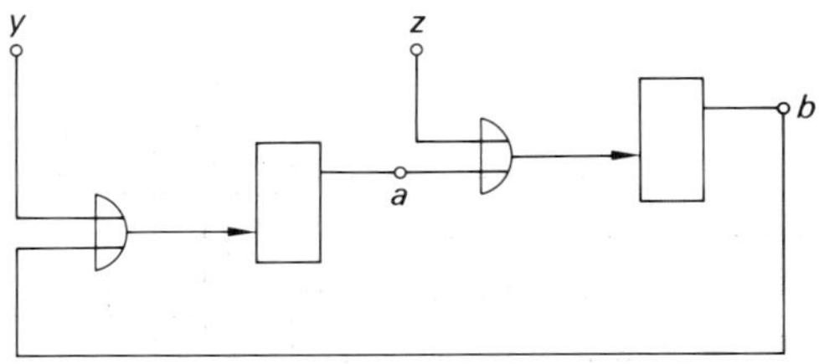

Fig. 9-27

5. Design the symbolic circuit of a trigger operating as follows on a set signal $s = 1$ and a simultaneous reset signal $r = 1$. If $s = 1$ and $r = 1$ follow the input combination $s = 1$ and $r = 0$, then z remains 1; and if $s = 1$ and $r = 1$ follow $s = 0$ and $r = 1$, then z remains 0. Note that no other input variables besides s' and r are available.

6. Determine the formula of output signal z of a memory element controlled by input signals x and y in the following way:

(*a*) Output signal z changes from 0 to 1 only if $x = y = 1$.

(*b*) Output signal z changes from 1 to 0 only if $x = y = 0$.

Draw the symmetrical circuit.

7. Draw two identical symmetrical triggers Y and Z using gate and inverter symbols. The gates are OR gates. Design a locking circuit excluding the possibility that both triggers are in the 1 state when their set and reset signals are 0. Leaving the locking circuit intact, design the input circuit for signals a, b, and c, one at a time being 1, so that signals a and b set triggers Y and Z, respectively, whereas signal c resets both triggers.

8. In the circuit of Fig. 9-28, three inverters and three OR gates are used. The input signals are α, β, and γ; the output signals are a, b, and c. Determine the combination of output signals for every possible combination of input signals. Determine a formula for output signal a not containing signals b and c.

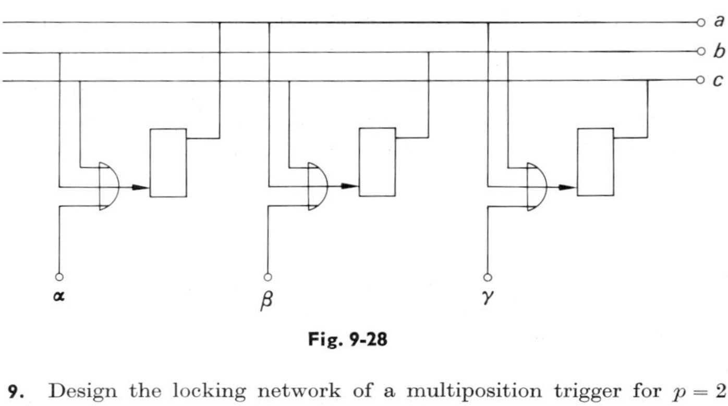

Fig. 9-28

9. Design the locking network of a multiposition trigger for $p = 2$ and $q = 2$ (two adjacent inverters operated out of four).

10. Design the locking network of a multiposition trigger for $n = 10$, $p = 2$, and $q = 8$. What is the number of diodes involved in this locking network? Does this 10-position trigger have the most economical locking network?

11. Design by means of Eqs. (9-40) and (9-41) the locking network of a 10-position trigger for $a = 1$. What is the number of diodes involved in this locking network?

12. Draw up the state equations and design the circuit of a two-out-of-four trigger that operates basically as the three-position trigger of Fig. 9-18.

SELECTIVE BIBLIOGRAPHY

1. Bay, Z., and N. T. Grisamore: High Speed Flip-flops for the Millimicrosecond Region, *IRE Trans.*, vol. EC-5, pp. 114–121, 1956.
2. Florida, C. D.: A New Bistable Element Suitable for Use in Digital Computers, *Electron. Eng.*, vol. 30, pp. 71–77 and pp. 148–153, 1958.
3. Moody, N. F.: Controlled Saturation in Transistors and Its Application in Trigger Circuit Design, *Electron. Eng.*, vol. 30, pp. 121–127 and pp. 200–204, 1958.
4. Oberman, R. M. M.: The Design of Triggers, Counters and Distributors, *Het. P.T.T.-Bedrijf*, vol. 6, pp. 58–74, 1955.
5. Oberman, R. M. M.: Construction Elements in the Electronic Switching Technique, *Tijdschrift Nederlands Radiogenootschap*, vol. 22, pp. 349–358, 1957.
6. Paivinen, J. O., and I. L. Auerbach: Design of Triode Flip-flop for Long Term Stability, *IRE Trans.*, vol. EC-2, pp. 14–26, 1953.
7. Prom, G. J., and R. L. Crosby: Junction Transistor Circuits for High Speed Digital Computer Applications, *IRE Trans.*, vol. EC-5, pp. 192–196, 1956.
8. Snijders, A.: Electronic Switching Unit, British Patent 780,003, Priority July 12, 1951.
9. The Staff of the Computation Laboratory: Triggers, Rings, and Digit Counters, chap. 8 of "The Synthesis of Electronic Computing and Control Circuits," Harvard University Press, Cambridge, Mass., 1951.
10. Wang, A.: Magnetic Triggers, *Proc. IRE*, vol. 38, pp. 626–629, 1950.
11. Wanlass, C. L.: Transistor Circuits for Digital Computers, *IRE Trans.*, vol. EC-4, pp. 11–15, 1955.

10
Dynamic Memory Circuits

10-1 PRINCIPLES

For the temporary storage of a binary information signal, there exist a number of dynamic memory circuits that are very useful in the design of switching circuits. The capacitor-resistor circuits are the most suitable of these single-bit memory circuits. However, inductor-resistor circuits can also be used for this purpose.

If a battery with a voltage E is connected via a switch a to an RC network as shown in Fig. 10-1, after the closing of switch a, the potential difference between points 1 and 2 will increase along the e curve shown schematically in Fig. 10-2. The potential difference between points 2 and 3 then decreases as shown in Fig. 10-3. The sum of both e curves must always be E.

The current in the circuit starts with $I = E/R$ and decreases to 0, thereby following the same e curve as the potential difference $V_{2/3}$ across resistor R. This current is shown as a dotted line in Fig. 10-2.

The v curve shown in Fig. 10-2 demonstrates the integrating effect of a capacitor, whereas the v curve of Fig. 10-3 shows its differentiating effect. The difference between these two effects is merely found in the way in which the charging process of the capacitor is considered to occur, viz., across the capacitor or across the resistor.

Each of these effects can be taken advantage of as a short-time dynamic memory for a control signal or even for a number of control signals.

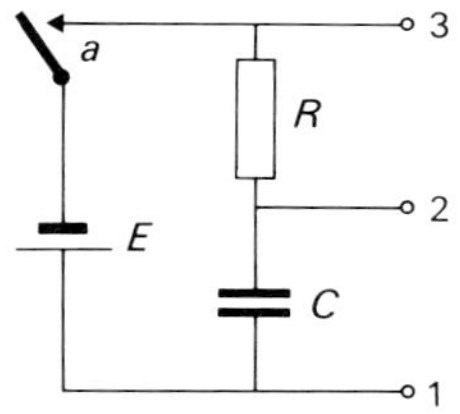

Fig. 10-1

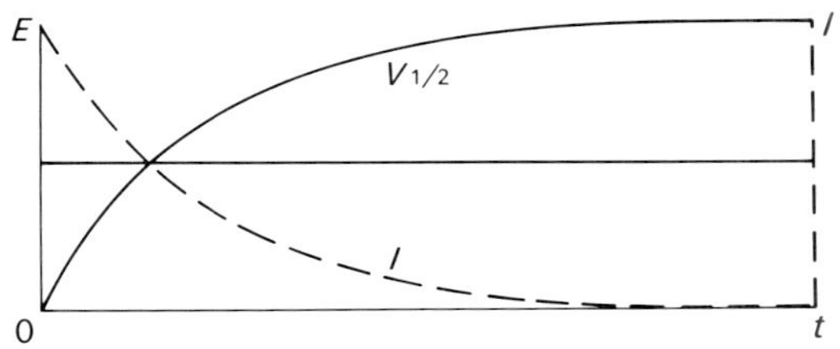

Fig. 10-2

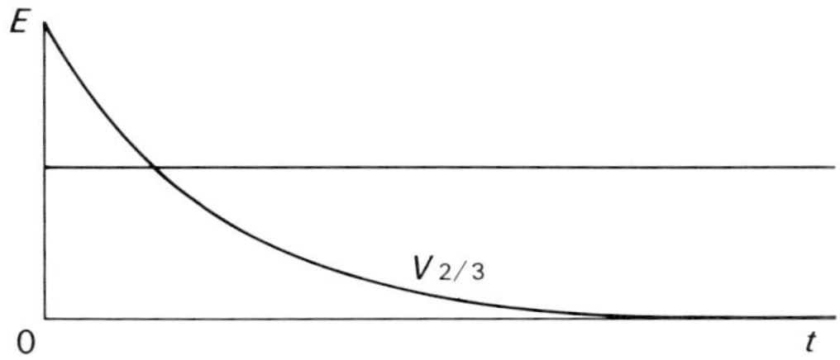

Fig. 10-3

A control signal with two levels, for example, $+E$ and $-E$, is now connected via an RC network as shown in Fig. 10-4 to the set input terminal of a trigger T. Contact a is shown in the rest position in which the capacitor voltage is 0. When the position of contact a is changed, the capacitor will be charged. The potential of the output terminal of the RC network will change from $-E$ to $+E$, as indicated in Fig. 10-5. After a short delay t_d, this potential will pass through the operating level of trigger T (earth potential in this example), so that the trigger is set.

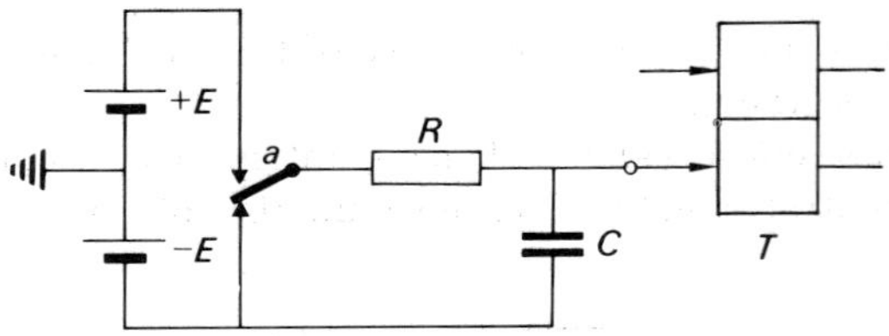

Fig. 10-4

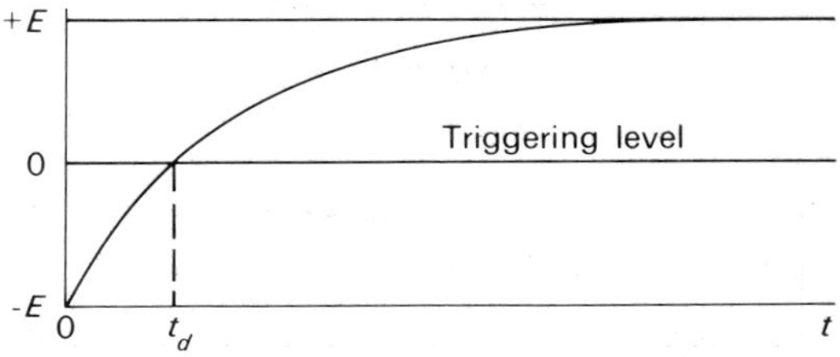

Fig. 10-5

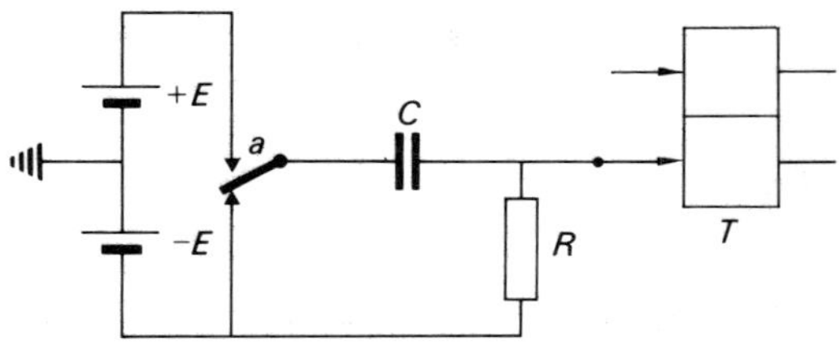

Fig. 10-6

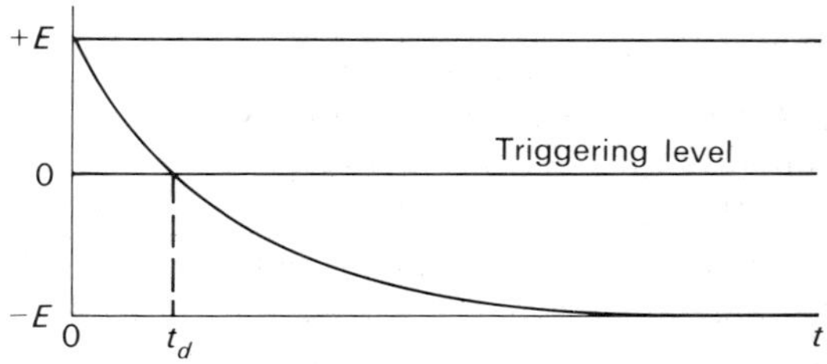

Fig. 10-7

Trigger T is operated after a time interval t_d measured from the moment the position of the controlling contact a is changed. It can also be said that the output of the integrating RC delay network memorizes, during an interval t_d, the previous state of the controlling contact a.

Trigger T stays in this new state, even after the controlling contact a of the circuit of Fig. 10-4 has been reset to its rest position; but this fact is of no importance for the time being.

The RC network can also be connected so as to control a trigger, as is indicated in Fig. 10-6. The control voltage of the output terminal of this differentiating RC network immediately increases to $+E$ once the condition of contact a is changed and decreases subsequently to $-E$, as shown by the curve of Fig. 10-7. During the interval t_d, this output potential of the RC network is above the operating level of trigger T. The trigger is thereby set. A safe change of state of this trigger has to be completed within the interval t_d, as is shown in Fig. 10-7. It can be said that during the interval t_d this network memorizes the fact that the position of contact a has been changed. As was pointed out already, a trigger controlled by this network must operate within this time interval if a more permanent storage of the state of the input signal prior to t_0 is required.

With the same dimensioning of resistor R and capacitor C, equal memory times t_d will be obtained in both types of circuits (Figs. 10-4 and 10-6).

The memory periods of both types of circuits can generally be used either to allow a change in the state of a trigger to occur or to block such a change. In the example of Fig. 10-4, the change of the state of trigger T is blocked during the interval t_d, whereas in the circuit of Fig. 10-6, the state of trigger T is allowed to change during this same interval t_d.

Before going into the operation of both types of RC networks, let us consider a practical way to use these RC networks in some relatively high-speed applications.

The intervals after which the position of the controlling contact a in the circuit of Fig. 10-6 can be changed are rather long. When contact a reaches its rest condition, as shown in Fig. 10-6, the potential of the output terminal of this circuit will change, as shown in Fig. 10-8, between moments t_0 and t_3 (full line). Once the potential of the output terminal has been restored to $-E$, that is, when the charging process of the capacitor is completed, then and only then can contact a be set in its operated position (moment t_3), so that an output pulse is produced between t_3 and t_4, as shown in Fig. 10-8.

If the position of contact a is changed too early, say at t_1, the output pulse produced will have too low a potential and too short a duration (t_{1-2}), so that the operating margins of the circuit are unfavorably affected.

The repetition rate of the operation of the a contact in Fig. 10-6 can be increased considerably by modifying this circuit as shown in Fig. 10-9. In modifying this circuit, the control function is performed by a transistor instead of a contact. In this example, the levels of the control signal have been chosen as 0 volts (1) and –6 volts (0), the battery voltages being +12 and –18 volts.

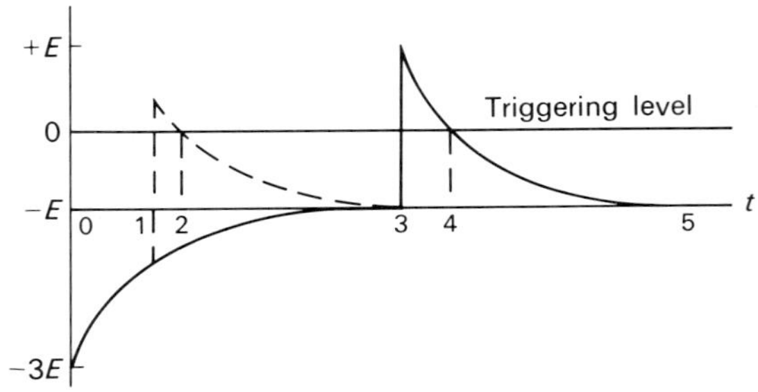

Fig. 10-8

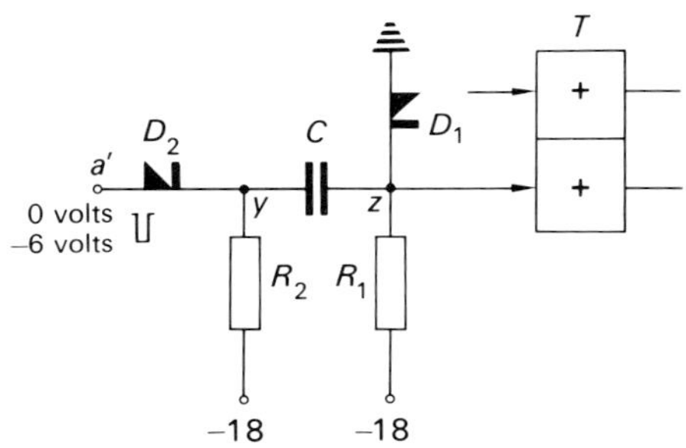

Fig. 10-9

In the agreed rest condition of the circuit, the a'-signal level is 0 volts (1). Diode D_2 is then conducting, and point y of the circuit has 0 potential. Point z of the circuit, i.e., the output, also has 0 potential because diode D_1 is also conducting in the rest condition of the circuit. Furthermore, the capacitor voltage V_c is 0 volts.

At t_0 an a' pulse is fed, with a duration t_{0-2}. The signal level of that pulse is –6 volts (0). The potential of point z of the circuit cannot change because diode D_1 remains conducting, so that point z remains connected to earth.

The a' signal given at moment t_0 initially blocks diode D_2 so that the potential of point y will initially vary in accordance with the e curve describing the charging of capacitor C to –18 volts. However, when the capacitor voltage $V_c = -6$ volts is reached, diode D_2 again becomes conducting. The potential of point y is clipped at t_1 at –6 volts, as shown by line y in the graph of Fig. 10-10.

The potential of point y remains at -6 volts until t_2, at which moment the a' pulse ends. The potential of point y is then raised from -6 volts to 0 volts because diode D_2 remains conducting. Consequently, the potential of point z of the circuit is suddenly raised from 0 volts to $+6$ volts, as is shown by line z in Fig. 10-10. Diode D_1 is now blocking until the potential of point z of the circuit has decreased to 0 volts, i.e., until t_3. In its decreasing portion, the potential of point z follows an e curve to -18 volts, but this curve is clipped off at earth potential where diode D_1 again becomes conducting.

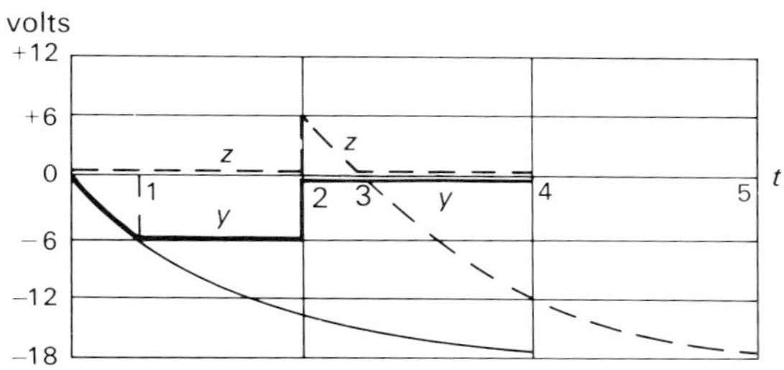

Fig. 10-10

The way in which the potential of points y and z to -18 volts would decrease if no clipping had been introduced is shown in thin lines in Fig. 10-10. In this graph, the discharge time of the capacitors has been so chosen that the potential decrease of points y and z reaches in the intervals t_{0-2} and t_{2-4} three-quarters of its final value.

If no clipping had been applied and if the power supply had been a battery of -6 volts only, curve y would have decreased to $-4\frac{1}{2}$ volts by moment t_2 and curve z to -3 volts by t_4 if the dimensioning of the RC networks were not changed. The circuit of Fig. 10-9 would have been unsuitable for use under these conditions.

If a clipping voltage is used, the a' signal may already end at moment t_1, whereas a new a' signal may be applied as early as t_3. This results in an important increase of the repetition rate of the control pulse.

It should be added that the repetition rate of the circuit of Fig. 10-9 without clipping could also be increased by using lower values for the capacitor or resistors, but this would impose rather high loads on the pulse source, and the delay times would not be well defined.

In this context, the discussion of the effects of clipping the capacitor voltages will be limited to the differentiating capacitor, since this circuit is the RC delay network most frequently used.

10-2 SWITCHING FORMULAS FOR DYNAMIC MEMORY CIRCUITS

It is fairly easy to describe the operation of the integrating RC network in a way suitable for the switching algebra used in this textbook. If a signal a is applied to the input terminal 1 of such a circuit (Fig. 10-11), the potential of output terminal 2 will, during the delay period t_d, remain below the operating level of a trigger T to which it is connected. At t_d, output signal

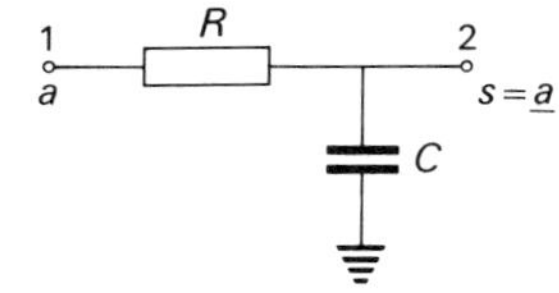

Fig. 10-11

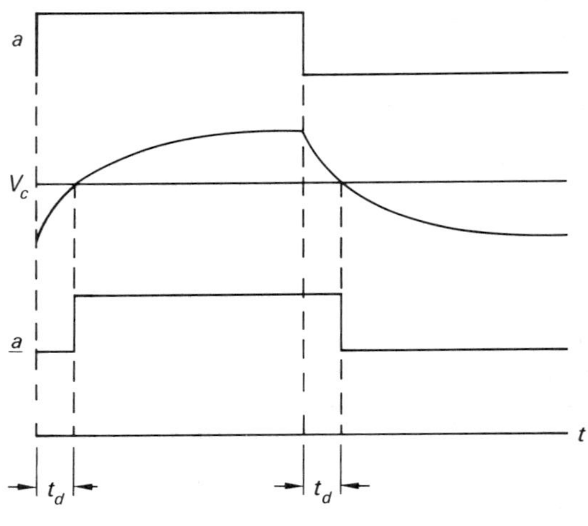

Fig. 10-12

s of the network will reach the same symbolic value as input signal a. This relation can be represented by

$$s = \underline{a} \tag{10-1}$$

Variable a is underlined to indicate that its effect starts with a delay t_d after applying signal a. The operation of an a signal and an $\underline{a}$ signal is shown in an idealized form in Fig. 10-12.

It is not quite so simple to describe by means of a formula the operation of a differentiating RC network. The principle of this network is shown in Fig. 10-13. A signal a' is applied to the input of this circuit as shown on line 1 of Fig. 10-14. The potential of point y changes as indicated on line 2 of this figure, whereas the potential of point z will change as shown by line 3. A fictitious asymmetrical trigger could be set by the potential of point y of the circuit at moment t_1 ($t_1 - t_0 = t_d$), whereas the same trigger could be reset by the potential of point z at moment t_3 ($t_3 - t_2 = t_d$). The output signal of this trigger is generated with a delay t_d with respect to the a' signal. The output signal of this imaginary trigger is shown as line 4 in the diagram of

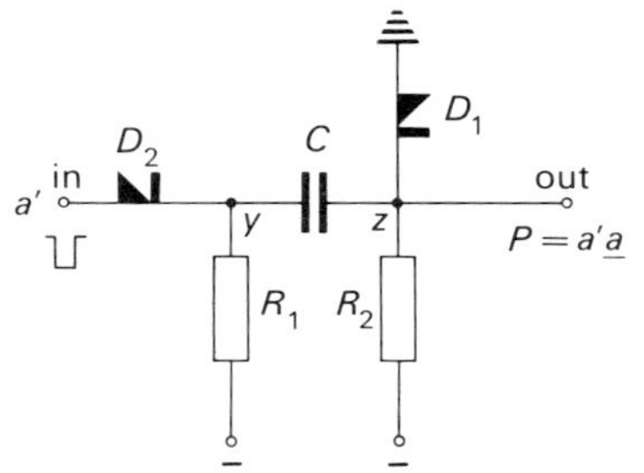

Fig. 10-13

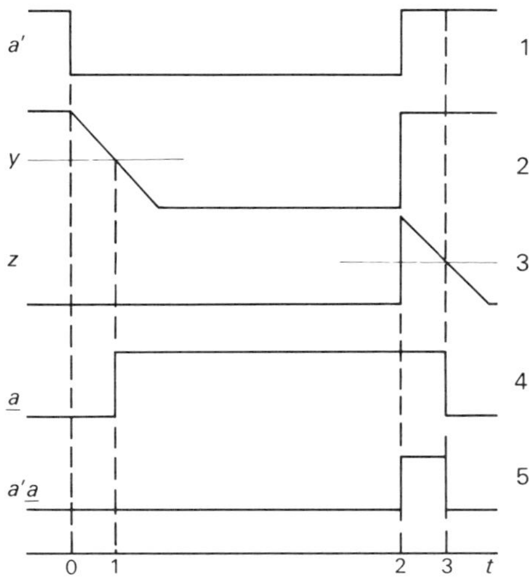

Fig. 10-14

Fig. 10-14. The signal of this line can be represented by $\underline{a}$. The resulting output signal s of the circuit of Fig. 10-13 is represented by line 5 of Fig. 10-14. This output signal can be derived as follows from input signal a' and output signal $\underline{a}$ of the trigger:

$$P = a' \cdot \underline{a} \tag{10-2}$$

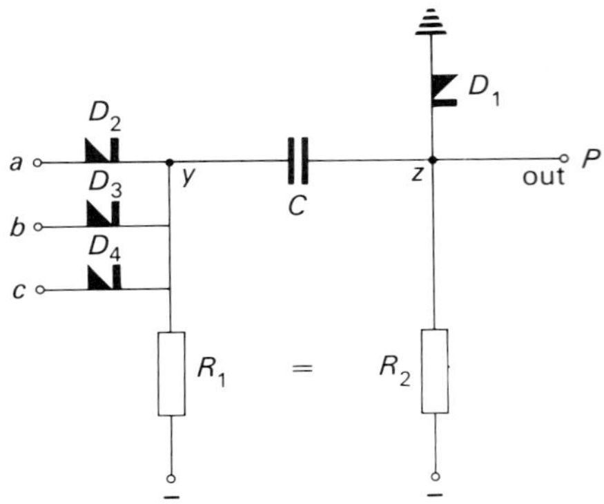

Fig. 10-15

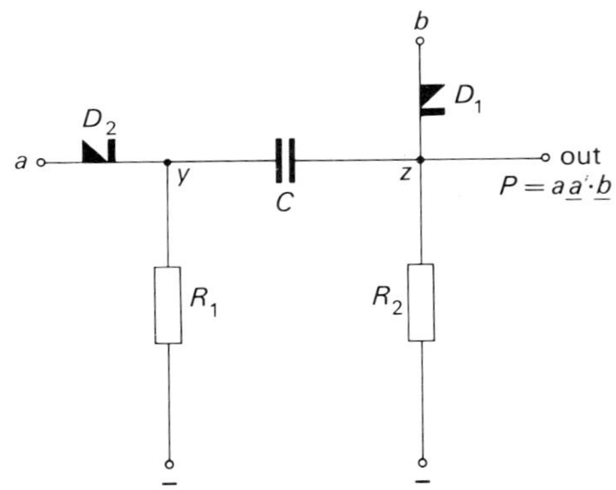

Fig. 10-16

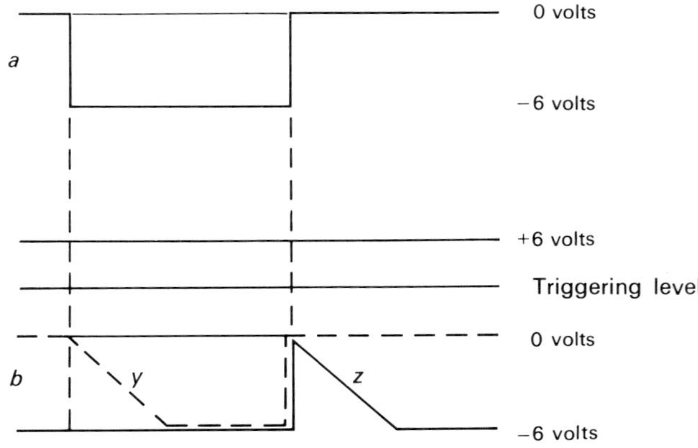

Fig. 10-17

The memory circuit of Fig. 10-13 can be extended in a number of ways.

1. A number of input signals can be applied to point y of the circuit in the way shown in Fig. 10-15. The input signals have to be applied via an OR gate, so that the resulting signal I is

$$I = a + b + c \tag{10-3}$$

This input signal can be substituted in Eq. (10-2), so that the output S then becomes

$$P = (a + b + c) \cdot (\underline{a + b + c})' \tag{10-4}$$

Any one of the input signals a, b, c, . . . may form the leading edge of the operating pulse of the network of Fig. 10-15 by means of which the potential of point y is decreased from 0 volts to -6 volts, just as any one of the input signals may be used as the trailing edge of that control pulse, provided that an interval is formed after the leading edge. Leading and trailing edges of the formed control pulse need not be part of the same signal.

2. A signal can be applied to both points y and z. This is shown in the circuit of Fig. 10-16, where a signal a is applied to point y and a signal b to point z. Both signals have, by definition, voltage levels of -6 volts (0) and 0 volts (1).

For $b = 1$, diode D_1 is kept at earth potential. The circuit then functions in a way completely analogous to the operation of the circuit of Fig. 10-13.

For $b = 0$, that is, if a signal voltage of -6 volts is fed to diode D_1, the potential of the output terminal z will change from -6 volts to 0 volts at the trailing edge of input signal a, as is shown in part b of Fig. 10-17. The potential of the output terminal is at most 0 volts, i.e., well below the operating level of the trigger. The signal $b = 0$ blocks the output terminal for a possible a signal applied to point y of the circuit.

It should be pointed out in addition that if the b signal varies from 1 to 0 in the interval between t_0 and t_d in Fig. 10-7, that is, within the interval t_d, the operation of the network cannot be changed anymore until the next input signal a. Accordingly, the circuit with the differentiating capacitor also performs as a memory element during the delay period for the b signal applied to point z.

The circuit of Fig. 10-16 will produce an output pulse only if a signal $b = 1$ and a pulse $a \cdot \underline{a}' = 1$ are connected to point z. The output signal P will then be

$$P = a \cdot \underline{a}' \cdot \underline{b} \tag{10-5}$$

Both variables underlined in Eq. (10-5) relate to the same delay period t_d.

The a signal applied to the front of the network of Fig. 10-15 and the b signal applied to the back of that network are not equivalent in all respects.

A change in the b signal from 0 to 1, or vice versa, occurring at any moment before t_d prior to the change of the a signal from 0 to 1, has no effect on a trigger connected to the output of the circuit; the a signal may then be either 0 or 1. However, for $b = 1$, a change of the a signal from 0 to 1 will produce an output pulse.

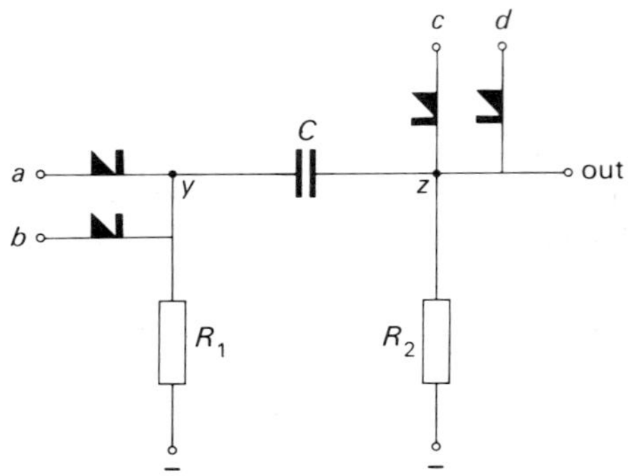

Fig. 10-18

3. A number of input signals can be applied to both points y and z, as is shown in the circuit of Fig. 10-18. The input signals a and b are connected to the "front" OR gate and the signals c and d to the back OR gate. All signals combine to form the following output signal:

$$P = (a + b)\ (\underline{a + b})' \cdot (\underline{c + d}) \tag{10-6}$$

The validity of this equation can be easily checked.

It should be stated that it is hard to distinguish front and back in a circuit with a differentiating capacitor when input signals are applied to both sides. This circuit has been considered so far as a unidirectional element; but a set or reset input terminal of a trigger may be connected either to point z or to point y of the circuit. A change of variable at the output-terminal side does not affect the state of the trigger connected to that output terminal. Assuming that the polarity is correct, this effect will only result from a change of a variable at the other side of the circuit.

Accordingly, depending on the particular variables that are changed, a useful signal can originate either from point y or from point z in this type of circuit. This means in fact that the type of circuit shown in Fig. 10-18 is a bidirectional element.

The RC circuit treated above, with signals applied both to the front and to the back of the circuit, can in fact no longer be considered as a genuine differentiating circuit. The signals applied to the front are submitted to the "differentiating" effect of the circuit; however, the signals applied to the back can be considered as being submitted to the "integrating" effect of the RC network, which is found to occur simultaneously. Both effects are related by the AND operation [see Eq. (10-5)], with signal polarities, voltages, and diodes being as chosen in the various examples. It is, of course, possible to reverse all these polarities and thereby obtain the dual forms of the formulas given above.

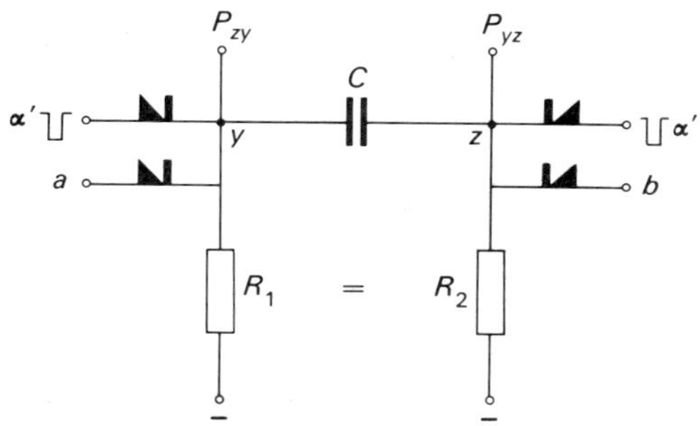

Fig. 10-19

A special case of this type of circuit is found in Fig. 10-19, where the input signals are a clock pulse α', a signal a, and a signal b. Both signals remain unchanged during clock pulse α'. The equations describing this circuit are as follows.

$$P_{yz} = (\alpha' + a)\,(\underline{\alpha' + a})' \cdot (\underline{\alpha' + b}) \tag{10-7}$$

$$P_{zy} = (\alpha' + b)\,(\underline{\alpha' + b})' \cdot (\underline{\alpha' + a}) \tag{10-8}$$

Equations (10-7) and (10-8) can be simplified on account of the following considerations.

If two equally delayed signals pass through an OR gate or an AND gate, the resulting signal can be considered as being generated by a circuit with the same delay and preceded by an OR gate or an AND gate, as is shown in Figs. 10-20 and 10-21.

These circuits can be represented by definition as follows:

$$\underline{a} + \underline{b} = \underline{a + b} \tag{10-9}$$

and

$$\underline{a} \cdot \underline{b} = \underline{a \cdot b} \tag{10-10}$$

Furthermore, if a signal is not changed during the clock-pulse period $\alpha' = 0$, then

$$\begin{aligned} &\alpha a \cdot \underline{a}' = 0 \qquad \alpha a' \cdot \underline{a} = 0 \\ &\alpha b \cdot \underline{b}' = 0 \qquad \alpha b' \cdot \underline{b} = 0 \end{aligned} \tag{10-11}$$

will be always true.

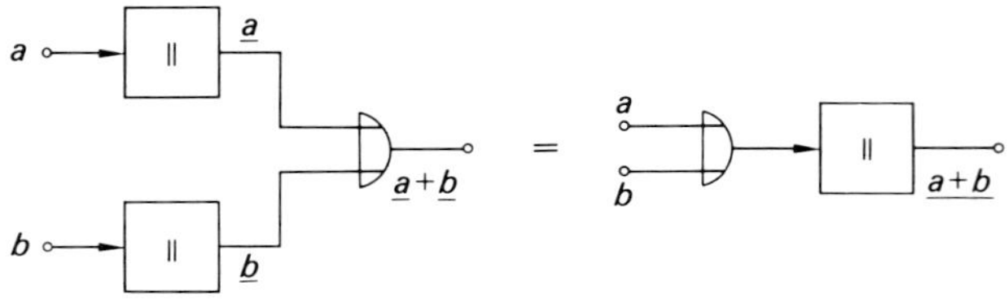

Fig. 10-20

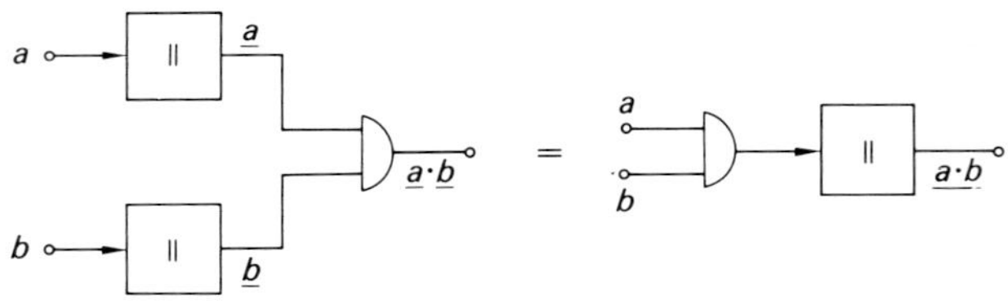

Fig. 10-21

Equations (10-7) and (10-8) can now be developed as follows:

$$\begin{aligned} P_{yz} &= (\alpha' + a)\,(\underline{\alpha' + a})'\,(\underline{\alpha' + b}) = (\alpha' + a)\,(\underline{\alpha}' + \underline{a})'\,(\underline{\alpha}' + \underline{b}) \\ &= (\alpha' + a)\,\underline{\alpha} \cdot \underline{a}'(\underline{\alpha}' + \underline{b}) = (\alpha' + a)\,\underline{\alpha}\underline{a}'\,\underline{b} \\ &= (\alpha' + \alpha a)\,\underline{\alpha}\underline{a}'\,b = \alpha'\,\underline{\alpha}\underline{a}'\,\underline{b} + \alpha\underline{\alpha}a\underline{a}'\,\underline{b} = \alpha'\,\underline{\alpha}\underline{a}'\,\underline{b} \end{aligned} \tag{10-12}$$

By means of an analogous simplification process, one obtains

$$P_{zy} = \alpha'\,\underline{\alpha}\underline{a}\underline{b}' \tag{10-13}$$

Equations (10-12) and (10-13) show that when $a = b = 0$ and $a = b = 1$, no output pulse P will be generated on either of the output terminals of the circuit of Fig. 10-19. But when $a = 0$ and $b = 1$, pulse P_{yz} will be formed; and when $a = 1$ and $b = 0$, pulse P_{zy} in the reverse direction will be generated. The capacitor is used here as a bidirectional element.

It should be noted that analogous reasoning can be used for dynamic memory elements with AND input gates.

An entirely different physical type of delay network also meets the basic requirements stated in Eq. (10-4). This network is shown in Fig. 10-22. It uses a self-inductor as a delay memory element. In the rest condition of the circuit, the output of the delay network is such that it does not disturb the conducting state of a transistor that is connected with its base to the output terminal.

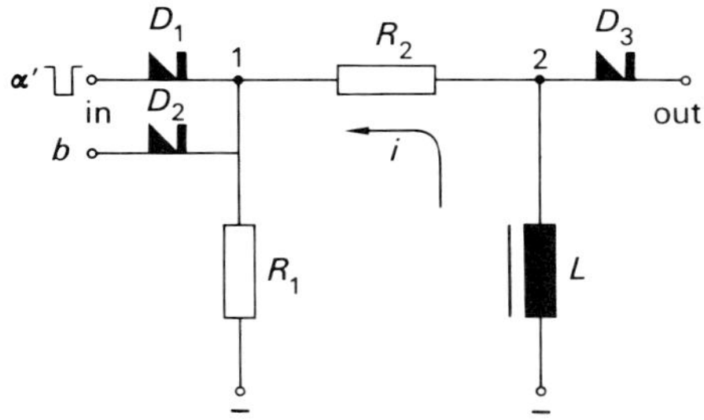

Fig. 10-22

If $b = 0$ (signal level –6 volts) and the control pulse $\alpha' = 0$, a current flows through self-inductor L as indicated in Fig. 10-22. At the end of the α' pulse, $\alpha' = 1$ (signal level 0 volts), so that the current through the self-inductance is suddenly interrupted. This will induce a voltage across the self-inductance with a polarity opposite to the driving voltage of the current before the interruption.

This voltage makes point 2 of the circuit of Fig. 10-22 positive with respect to earth potential, so that a positive pulse of short duration is generated. It can easily be checked that the switching function generated at the output terminal of this type of delay circuit is as follows:

$$P = (\alpha' + b) \cdot (\underline{\alpha' + b})' \qquad (10\text{-}14)$$

It is somewhat difficult to provide this type of delay network with operating margins as large as those of the RC networks of the type described above.

10-3 MASTER AND SLAVE MEMORY ELEMENTS

In the following chapters, the dynamic memory elements described in the preceding paragraph will generally be used in connection with a trigger circuit. With the chosen signal voltage levels, the output terminal of a dynamic memory element as shown in Fig. 10-15 has to be connected directly to the base of a transistor in a trigger, as, for example, the one in Fig. 9-9a.

A circuit of this type is shown in Fig. 10-23. In the rest condition of the circuit, there is no interaction of the RC circuit and the trigger. When transistor A is nonconducting, the base voltage is slightly positive, so that diode D_4 is nonconducting; when transistor A is conducting, both terminals of diode D_4 are kept at earth potential, so that the RC circuit does not influence the state of transistor A.

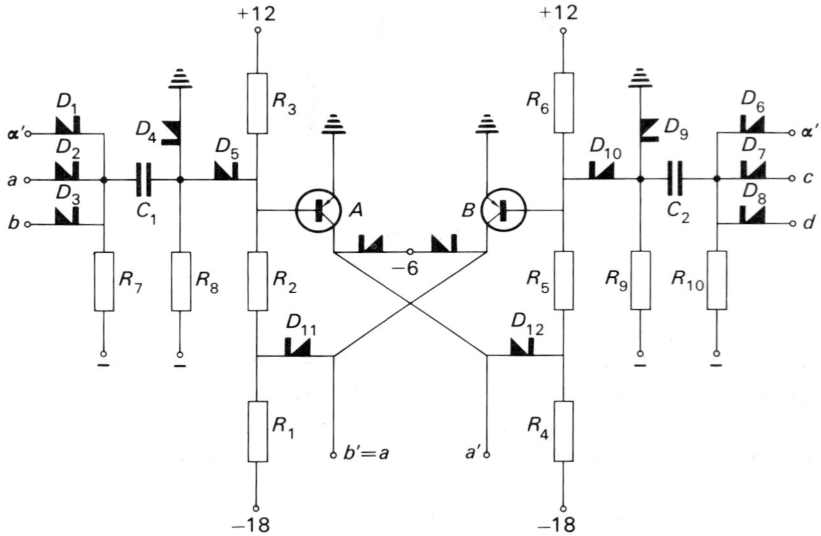

Fig. 10-23

A memory circuit of this type has the property that it changes its state on the trailing edge of the clock pulse. The input gates of the RC circuit are then closed, so that even feedback signals of the controlled trigger may be connected to these input gates without endangering the safe operation of the circuit. The delayed operation of the trigger controlled by the RC circuit is graphically shown in the time chart of Fig. 10-24.

In the following chapters, such a combination of a dynamic memory element in the form of an RC circuit followed by a static memory element in the form of a trigger will be referred to as a "master-and-slave," or MS, memory element because of their close interrelation.

This type of memory element will be the preferred basic building stone for many types of electronic circuits to be discussed in the following chapters. The terms "trigger" and "flip-flop" have been used in the technical literature to indicate memory elements of this type, but these names have also been earmarked for other types of circuits. In this textbook, the author has chosen the name MS memory element to indicate the combination of a dynamic and a static memory element—combining to form a memory element with delayed action, as is indicated in Fig. 10-24.

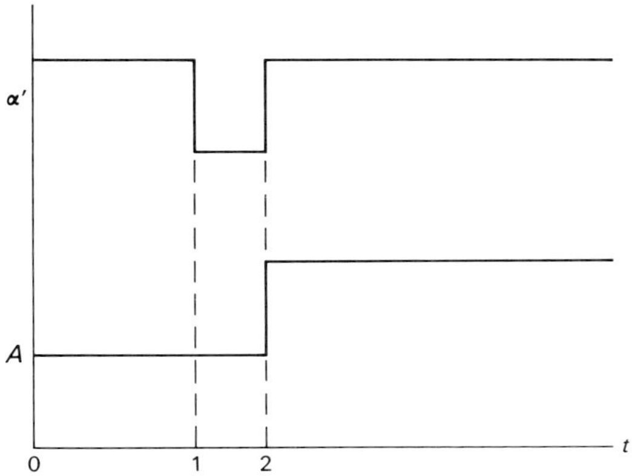

Fig. 10-24

MS memory elements can be formed by a trigger connected to two *RC* circuits (each if required with front and back control), as shown in detail in Fig. 10-23; however, in a number of cases, MS memory elements with single *RC* circuits, as symbolically shown in Fig. 10-25, will prove economical.

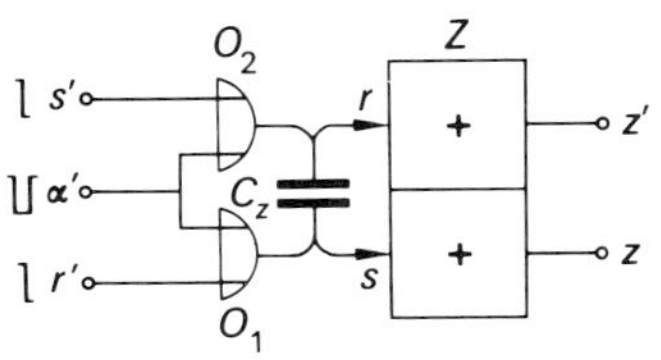

Fig. 10-25

Table 10-1

s	r	Z
0	0	z
1	0	1
0	1	0
1	1	z

For high-speed applications, the double-capacitor circuit is preferable because the value of the resistor connected to the input gate can be adapted to its function independently of the resistor of the *RC* circuit connected to its output terminal.

The signals represented by Eqs. (10-12) and (10-13) can be used as set and reset signals (s and r) of the trigger of the MS memory element shown in Fig. 10-25. Since these signals cannot be 1 simultaneously, the state of trigger Z controlled by these signals can be represented by the following pair of equations, as already discussed in Chap. 9:

$$\begin{aligned} Z &= s + \underline{z}r' = (s + \underline{z})\, r' = \alpha' \, \underline{\alpha}\underline{a}' \underline{b} + \underline{z}(\alpha + \underline{\alpha}' + \underline{a}' + \underline{b}) \\ &= \alpha' \, \underline{\alpha}[\underline{a}' \underline{b} + \underline{z}(\underline{a}' + \underline{b})] + \underline{z}(\alpha' \underline{\alpha})' \end{aligned} \tag{10-15}$$

It should be noted that the z variable has to be treated as a delayed variable because its place in time agrees with that of the other delayed variables. Now let $a = s'$ and $b = r'$ be the input signals of the MS memory element, then Eq. (10-15) shows the following form:

$$Z = \alpha' \, \underline{\alpha}[\underline{sr}' + \underline{z}(\underline{s} + \underline{r}')] + \underline{z}(\alpha' \, \underline{\alpha})' \tag{10-16}$$

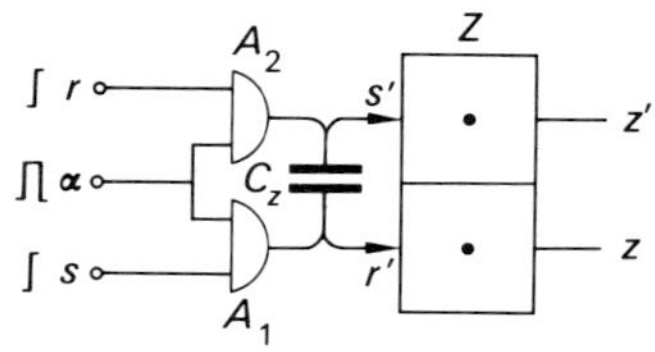

Fig. 10-26

This equation states that during pulse $\alpha' \, \underline{\alpha}$ the state of trigger Z will change in accordance with formula $sr' + z(s + r')$. The part of Eq. (10-16), $\underline{z}(\alpha' \, \underline{\alpha})'$, states that if outside pulse $\alpha' \underline{\alpha} = 1$, the state of trigger Z will be maintained.

The use of Eq. (10-16) can be facilitated by writing:

$$Z = D_{\alpha'}[sr' + z(s + r')] \tag{10-17}$$

In this equation the $D_{\alpha'}$ operator states that after clock pulse α' a change in the state of trigger Z will follow in accordance with the formula between the brackets. Equation (10-17) shows that the MS memory element of Fig. 10-25 represents the most general form of a memory element. It operates in accordance with truth table 10-1.

In Fig. 10-26, a symbol is shown that represents an MS memory element with AND input gates containing a trigger Z with AND input gates. Both symbols of Figs. 10-25 and 10-26 can be obtained from each other by inverting the gates and the signals and drawing the symbol upside down.

The operation of the MS memory element of Fig. 10-26 is in accordance with Table 10-1 and Eq. (10-17). This is not difficult to prove.

In the following, only MS memory elements with OR input gates will be used as examples. The resulting formulas will also hold for the equivalent type of MS memory element with AND input gates. The equations describing the operation of the following group of possibilities with MS memory elements can be derived in the same way as Eq. (10-17). Only the resulting equations will be given.

With correctly connected input terminals, the double-capacitor MS memory element can also operate in accordance with truth table 10-1. This case is shown in Fig. 10-27.

By the introduction of a feedback from the outputs of an MS memory element to its inputs, a number of well-known circuits can be obtained that are indicated in most literature as one or the other kind of flip-flop. The various names of the flip-flop circuits will be given here; however, the use of these names will be avoided in the text following because all these circuits are fundamentally MS memory elements.

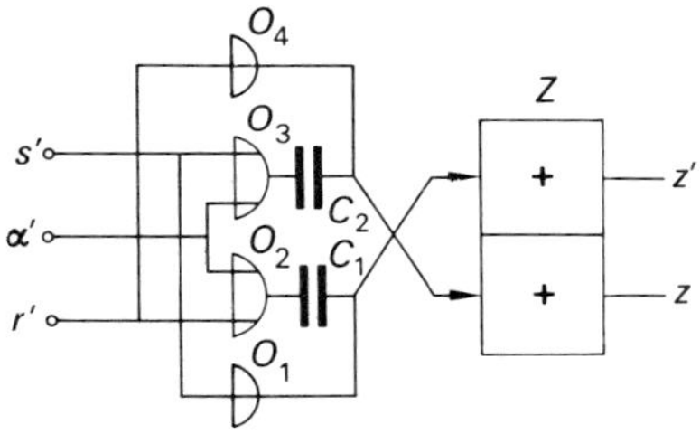

Fig. 10-27

All of the following types of flip-flops can be obtained from both types of MS memory elements. For each pair of flip-flop circuits, the corresponding truth table and formula are given. In a number of cases, various combinations of input and feedback signals can give the same operation of the circuit. This holds for the double-capacitor MS memory elements. The input variables indicated in the upper half of the columns giving the various possibilities with these MS memory elements can be combined with any pair of variables given in the lower half of these columns.

THE *C*(OMPLEMENTARY) FLIP-FLOP

An MS memory element connected as a *C* flip-flop or pulse halver has no other outside connections to the input gates than the pulse to be counted. The operation of this type of circuit is represented by the very simple truth table 10-2. Both types of circuits with their various combinations of input and feedback connections are shown in Fig. 10-28*a* and *b*.

With the single-capacitor MS memory element, there is only one possible feedback connection to the input terminals resulting in a pulse-halving circuit. The input terminals of the double-capacitor MS memory element can, in this case, be connected in 12 different ways.

The formula representing the operation of this element is in accordance with truth table 10-2:

$$Z = D_{\alpha'}(z') \tag{10-18}$$

Table 10-2

Z
z'

Equation (10-18) is easily obtained from Eq. (10-17) by substitution of $s' = z$ and $r' = z'$.

$$Z = D_{\alpha'}[sr' + z(s + r')] = D_{\alpha'}[z' \cdot z' + z(z' + z')] = D_{\alpha'}(z')$$

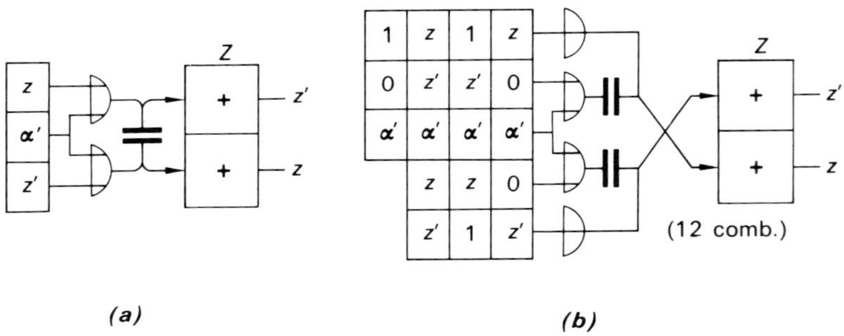

Fig. 10-28

THE D(ELAY) FLIP-FLOP

This type of circuit is mostly used as a section of a so-called shift register. Chapter 17 deals exclusively with shift registers. The operation of this type of circuit is defined by truth table 10-3. This table shows that the information

Table 10-3

y	Z
0	0
1	1

fed to the input of this switching element is registered at the end of clock pulse α' in the Z trigger of the MS memory element. With a number of these elements connected in series, the information runs through a delay line with delay steps equal to the period of the clock pulse. The various possibilities of connection of the input terminals of both types of MS memory elements are shown in Fig. 10-29*a* and *b*.

The operating formula of this type of MS memory element following from Table 10-3, $Z = D_{\alpha'}(y)$, follows also from Eq. (10-17) by substitution of one of the possible combinations of input signals given in Fig. 10-29*a* for s' and r'. With some complication in the calculation, the same operating formula can also be obtained for the circuit given in Fig. 10-29*b*.

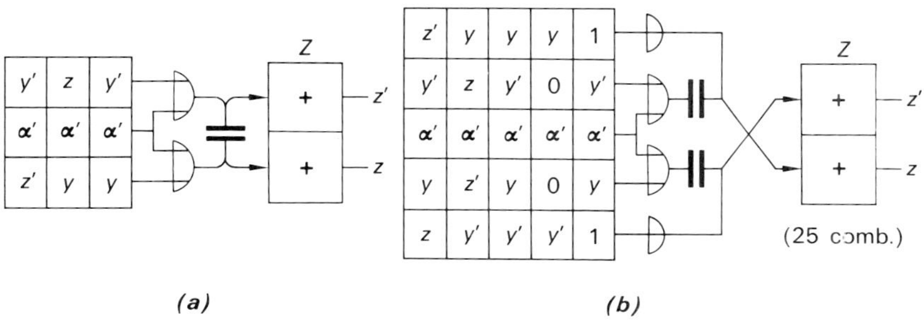

Fig. 10-29

For example, $s' = z$ and $r' = y$:

$$Z = D_{\alpha'}[sr' + z(s + r')] = D_{\alpha'}[yz' + z(z' + y)] = D_{\alpha'}(y) \tag{10-19}$$

THE R(ESET)-S(ET) FLIP-FLOP

In this category, three different circuit variations of the MS memory element may be grouped. In most of the literature, only one type of memory circuit is mentioned under this name. Just as we have seen with respect to relay circuits, MS memory circuits in the RS flip-flop group can be made with a dominant set signal so that when $s = 1$ and $r = 1$, $Z = 1$ also. It is also possible to make the reset signal dominant when $s = 1$ and $r = 1$ so that $Z = 0$; whereas it is also possible to connect the input terminals of the MS memory element in such a way that when $s = 1$ and $r = 1$, the state of the trigger Z will not be affected by the last given signal. Truth table 10-4 defines the operation of these three types of MS memory elements. Table 10-4 leads to the following group of operating equations:

$$Z_s = D_{\alpha'}(s + zr') \tag{10-20}$$

$$Z_r = D_{\alpha'}[(s + z)\, r'] \tag{10-21}$$

$$Z = D_{\alpha'}[sr' + z(s + r')] \tag{10-22}$$

Table 10-4

s	r	Z_s	Z_r	Z
0	0	z	z	z
0	1	0	0	0
1	0	1	1	1
1	1	1	0	z

In Fig. 10-30a and b, the various possible connections of the input terminals of the *RS flip-flop with dominant set signal* are given.

Operating equation (10-20) can easily be derived from Eq. (10-17) by using $r' + z'$ instead of r', as follows from Fig. 10-30a.

In Fig. 10-31a and b, the various possible connections of the input terminals of the *RS flip-flop with dominant reset signal* are given.

In Fig. 10-32a and b, the variations of the input connections of the *general type of RS flip-flop* are shown.

The JK flip-flop This is a special form of the RS flip-flop. The operation of this type of circuit is defined by truth table 10-5. The first three lines of this table equal the corresponding lines of Table 10-4. The last line of Table 10-5 shows that for $s = 1$ and $r = 1$, independent of which signal was applied first, the state of trigger Z is complemented. The variations of the input connections of the JK flip-flop are shown in Fig. 10-33a and b. The operating equation resulting from truth table 10-5 becomes:

Table 10-5

s	r	Z
0	0	z
0	1	0
1	0	1
1	1	z'

$$Z = D_{\alpha'}(zs'r' + sr' + z'sr) = D_{\alpha'}(sz' + r'z) \tag{10-23}$$

The same equation will be obtained when starting with Eq. (10-17) and using $s' + z$ instead of s' and $r' + z'$ instead of r'.

THE T(RIGGER) FLIP-FLOP

In this type of use of the MS memory element, the state of trigger Z is complemented only when receiving a signal $y = 1$ on the input terminals. On

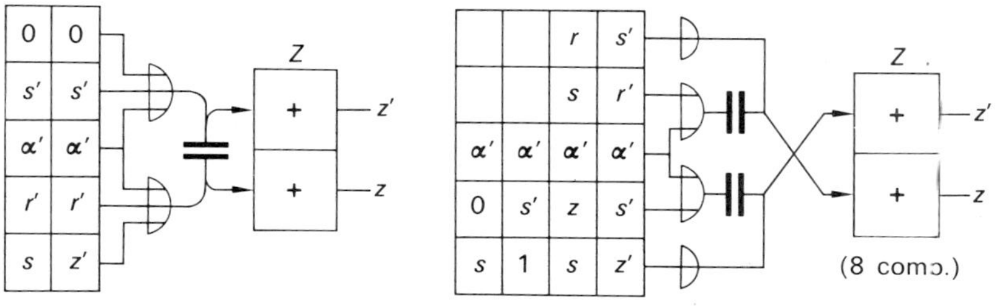

(a) Fig. 10-30 (b)

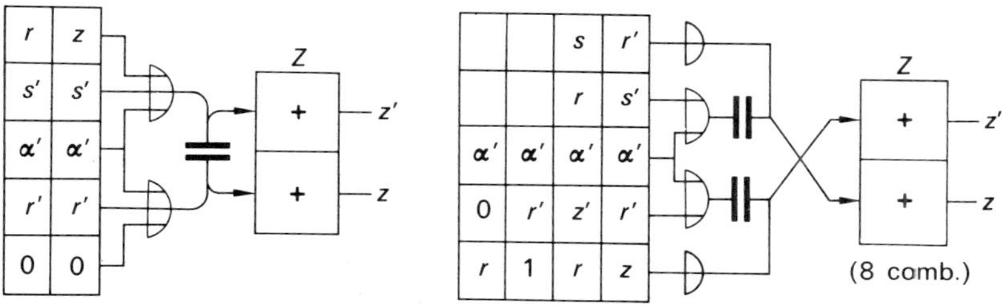

(a) Fig. 10-31 (b)

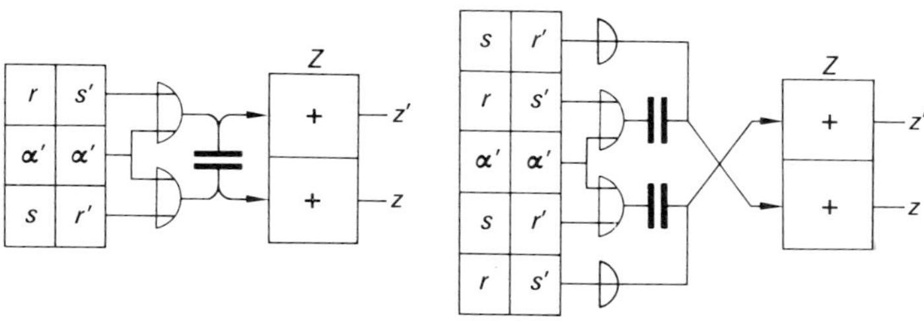

(a) Fig. 10-32 (b)

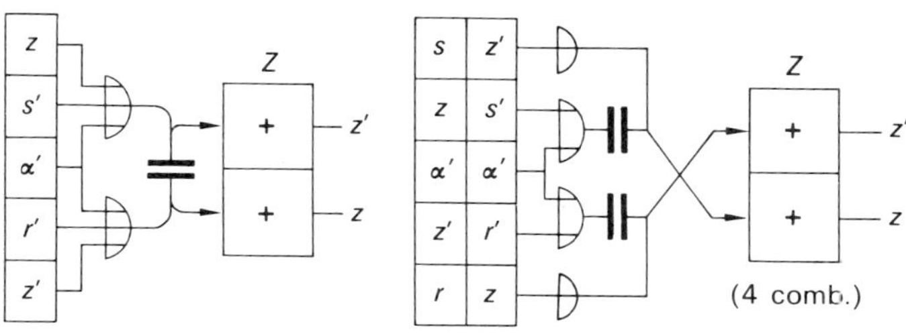

(a) Fig. 10-33 (b)

signal $y = 0$, trigger Z of this element remains in its state as is defined in truth table 10-6. The operating equation

$$Z = D_{\alpha'}(yz' + y'z) \tag{10-24}$$

Table 10-6

y	Z
0	z
1	z'

can also be obtained from Eq. (10-17) by substitution of $s' = y' + z$ and $r' = y' + z'$. The variations of the input connections of the T flip-flop are shown in Fig. 10-34*a* and *b*.

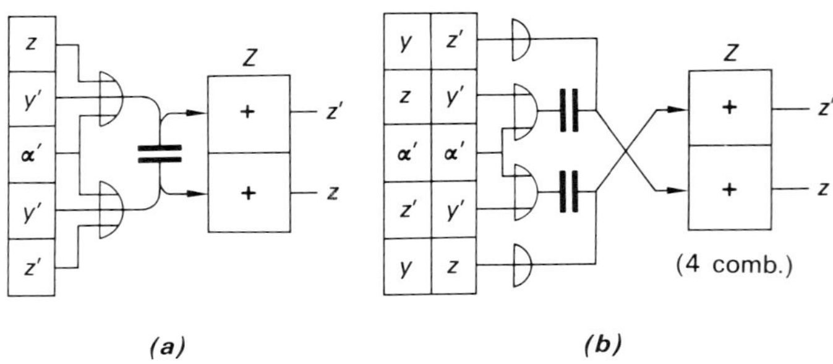

Fig. 10-34

Equation (10-24) may also be written as follows:

$$Z = D_{\alpha'}(y \oplus z) \tag{10-25}$$

in which the $\oplus$ sign indicates the addition mod 2.

Other types of flip-flops may also be thought of. The examples given above should be sufficient to show the general applicability of the MS memory element as developed in this chapter. It should be noted that it is possible to form analogous types of circuits by means of two triggers; however, this form of standard circuit will not be considered here.

10-4 MS ELEMENTS AS FUNDAMENTAL MEMORY ELEMENTS

The various types of flip-flops described in the preceding section are in fact nothing else than MS memory elements with variations in the input connections of the controlling signals. Thus, the problem of designing a circuit in a synchronous clock-pulse technique is not to take a certain type of flip-flop and determine the corresponding input-gate circuit from the given switching requirements but to determine from these requirements the simplest set of input signals and gates for an MS memory element and nothing else. In a given case, this will automatically lead to the simplest type of flip-flop.

A circuit can thus contain various types of flip-flops but only one type of MS memory element. This is of no difficulty or importance in the MS technique.

There are four different ways of connecting the input terminals of an MS memory element depending on the feedback connections, as are shown in Figs. 10-35 to 10-38.

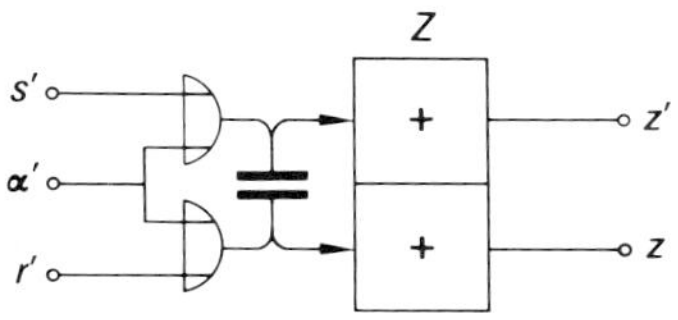

$Z = D_{\alpha'}[sr' + (s + r')z]$

Fig. 10-35

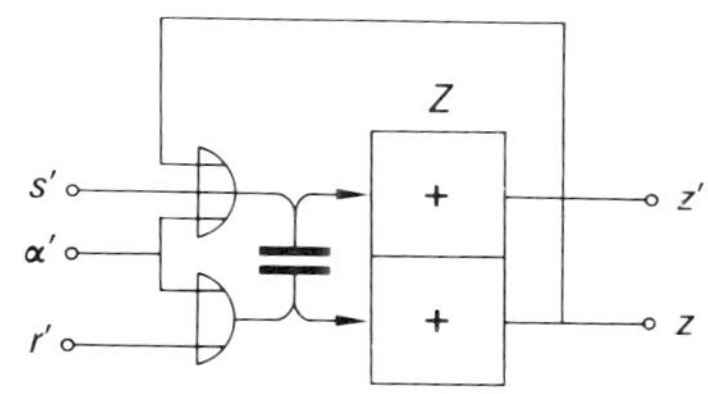

$Z = D_{\alpha'}(r's + r'z)$

Fig. 10-36

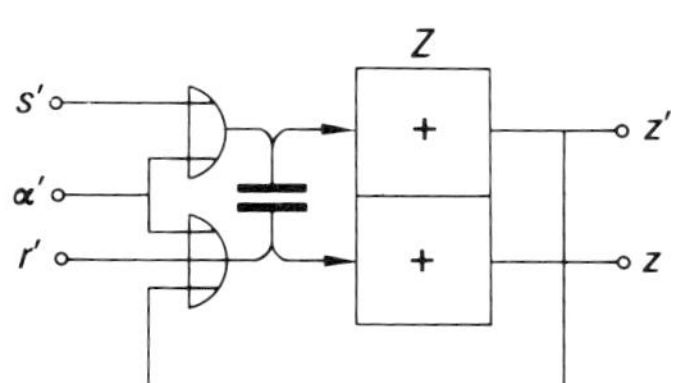

$Z = D_{\alpha'}(s + r'z)$

Fig. 10-37

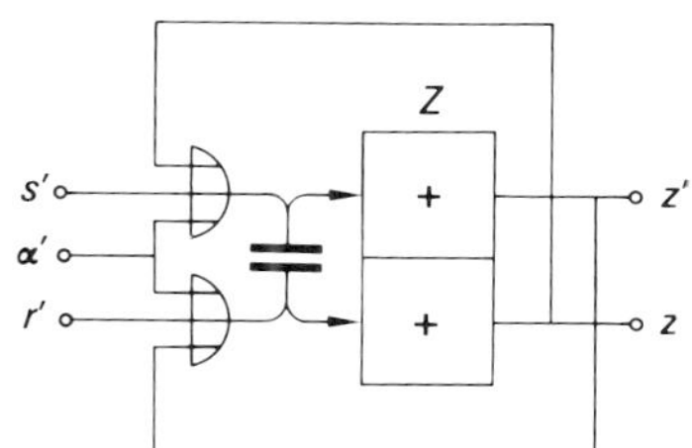

$Z = D_{\alpha'}(sz' + r'z)$

Fig. 10-38

In the equations accompanying these figures, only some types of flip-flops given in the preceding section can be recognized. However, the other types mentioned there can be obtained from Figs. 10-35 to 10-38 by making special relations between the s and r signals. It should be noted that the s and r signals need not be "pure" signals from a single other MS memory element but may be formed in a multiple input-gate circuit from the signals of a number of other MS memory elements. In such a case, there is a possibility for special relations between the s and r signals, eventually using don't-care conditions.

The D flip-flop, for example, will be obtained from the circuit of Fig. 10-35 when $s = r'$; the C flip-flop originates from the circuit of Fig. 10-38 when $s' = 0$ and $r' = 0$; and the T flip-flop is obtained from the circuit of Fig. 10-38 for $s = r$. Other special cases are possible.

The following discussion will be limited to MS memory elements with their own feedback signals connected to their actual input gates, as is shown in Figs. 10-35 to 10-38. It should be noted that it is also possible to connect these feedback signals to the s- and r-signal-forming gates, but in most cases this will not lead to a worthwhile economy in the number of diodes.

The input connections of an MS memory element can be derived from a given set of switching requirements by simplifying the switching formulas (e.g., by means of Karnaugh maps) and finding which of the four fundamental cases of Figs. 10-35 to 10-38 is applicable. By means of the corre-

Table 10-7

No.	*A*	*B*	*C*	*D*
0	0	0	0	0
1	1	0	1	0
2	1	0	0	1
3	1	0	0	0
4	0	1	0	0
5	1	1	1	0
6	1	1	1	1
7	0	0	0	1
8	0	1	1	1
9	1	1	0	1
10	0	1	0	1
11	1	1	0	0
12	0	1	1	0
13	0	0	1	0
14	0	0	1	1
15	1	0	1	1

sponding input tables given in Figs. 10-28 to 10-34, it can be checked whether there is another variation in connecting the inputs that is more suitable than the one found directly as a result of the simplification process.

The design procedure indicated above will be elucidated by means of an example also used by Phister. In Table 10-7, the operation of a 16-step sequence switch consisting of MS memory elements is given. This switch steps through the sequence of states of MS memory elements A, B, C, and D under the control of clock pulse α in the order given in the table. In each present state of the memory elements of the switch, the next states of these elements are determined. On the next clock pulse, the memory elements of the circuit will be set or reset accordingly. The circuit operation defined by Table 10-7 can be represented by means of so-called *application equations*, for each MS memory element 1. An application equation describes *all* the next 1 states of a memory element as a function of the present states of *all* the memory elements involved in the operation of the circuit. The following group of application equations corresponds with Table 10-7.

$$\begin{aligned} A = D_{\alpha'}[&(b'c'd' + bc'd' + bcd + bc'd + b'cd)\,a' \\ &+ (b'cd' + b'c'd + bcd')\,a] \end{aligned} \tag{10-26}$$

$$\begin{aligned} B = D_{\alpha'}[&(ac'd' + a'c'd)\,b' \\ &+ (a'c'd' + acd' + ac'd + a'cd + a'c'd + ac'd')\,b] \end{aligned} \tag{10-27}$$

$$\begin{aligned} C = D_{\alpha'}[&(a'b'd' + a'bd' + ab'd + abd')\,c' \\ &+ (abd' + a'bd' + a'b'd' + a'b'd)\,c] \end{aligned} \tag{10-28}$$

$$D = D_{\alpha'}[(ab'c + abc + a'b'c)\,d' + (abc + a'b'c' + a'bc + abc' + a'b'c)\,d] \tag{10-29}$$

From state equations (10-26) to (10-29), the following four 3-variable Karnaugh maps can be derived. Each three-variable minterm of a state equation is indicated in the corresponding Karnaugh map by means of the variable outside the brackets. Minterms belonging to both groups are indicated by a 1 and the nonpresent minterms by a 0. The resulting Karnaugh maps and the corresponding state equations are given in Figs. 10-39 to 10-42.

It can easily be found by inspection that the equation given in Fig. 10-39 corresponds in principle to that given in Fig. 10-38, so that

$$s' = r' = cd' + b'c'd \tag{10-30}$$

The resulting circuit of MS memory element A is given in Fig. 10-43.

The equation given in Fig. 10-40 corresponds to the type of MS memory element given in Fig. 10-35. It can be derived from the corresponding

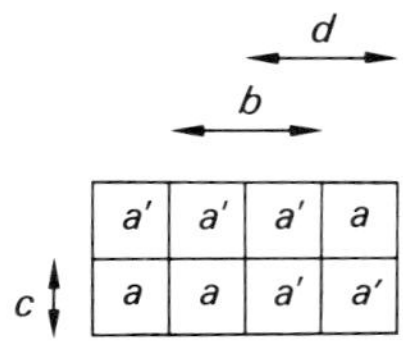

$A = D_{\alpha'}\left[(cd' + b'c'd)a + (cd' + b'c'd)'a'\right]$

Fig. 10-39

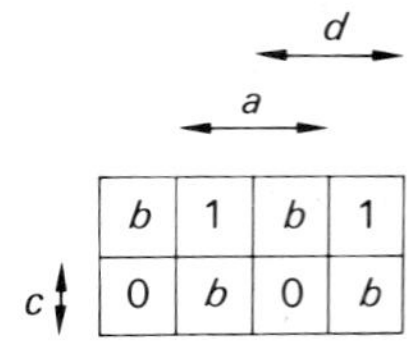

$B = D_{\alpha'}\left[(ad' + a'd + c')b + (ad' + a'd)c'\right]$

Fig. 10-40

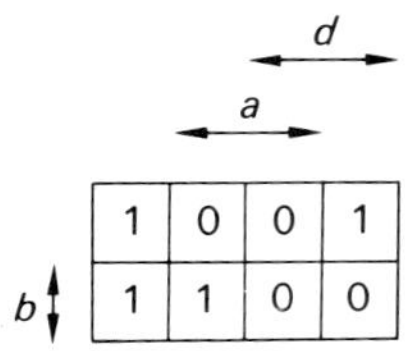

$C = D_{\alpha'}(a'b' + bd')$

Fig. 10-41

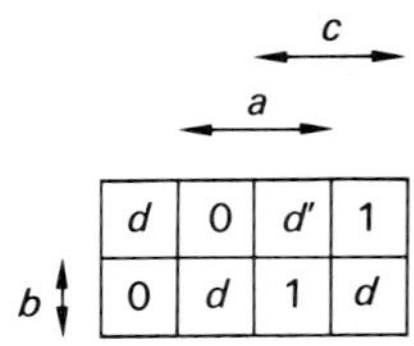

$D = D_{\alpha'}\left[(a'b' + ab + \substack{bc \\ a'c})\, d + (a + b')cd'\right]$

Fig. 10-42

fundamental state equation that in this case gives $sr' = (ad' + a'd)c'$ and $s + r' = ad' + a'd + c'$, which give the following solutions for s and r:

$$s = ad' + a'd \qquad r = c \tag{10-31}$$

The two possible solutions shown in Fig. 10-44 follow directly from Fig. 10-32a.

It should be noted that one of the two possible solutions might be more advantageous than the other from the point of view of combining one or more of the input AND gates of the MS memory elements.

The state equation for the C MS memory element given in Fig. 10-41 is independent of the c variable. In that case, a solution can be obtained with the circuits of Figs. 10-36 to 10-38; however, the solution obtained when using Fig. 10-38 is in this case the most complicated.

The case of Fig. 10-37 applies to that of Fig. 10-41 if $r' = 0$, so that

$$C = D_{\alpha'}(s + r'z) = D_{\alpha'}(s) \tag{10-32}$$

and

$$s = a'b' + bd' \tag{10-33}$$

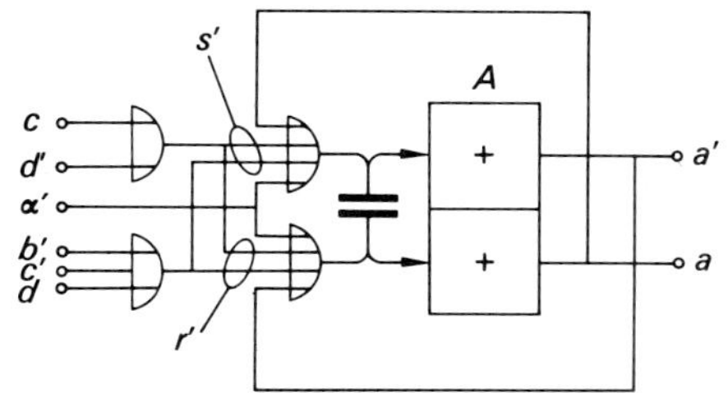

Fig. 10-43

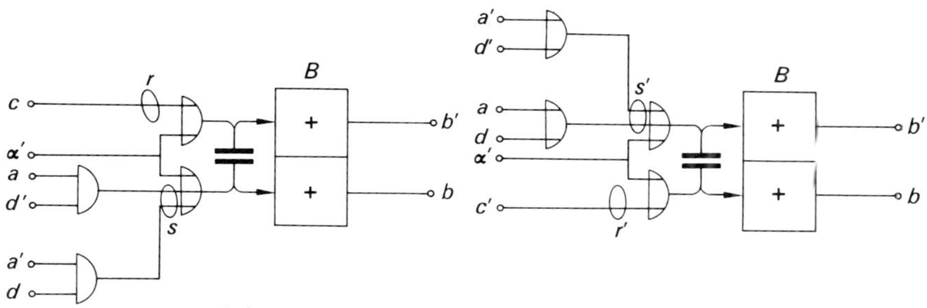

(a) (b)

Fig. 10-44

The case of Fig. 10-36 applies to that of Fig. 10-41 if $s = 1$, so that

$$C = D_{\alpha'}(r' s + r' z) = D_{\alpha'}(r') \tag{10-34}$$

and

$$r' = a' b' + bd' \tag{10-35}$$

The resulting circuits are shown in Fig. 10-45a and b.

It should be noted that the example of the C element shows that the cases of Figs. 10-36 and 10-37 lead to equivalent results if set and reset formulas consist of an equal number of minterms with the same number of variables.

The application or state equation of the D element is of a more or less arbitrary composition. The following expressions for set and reset can be derived directly from the equation given in Fig. 10-38.

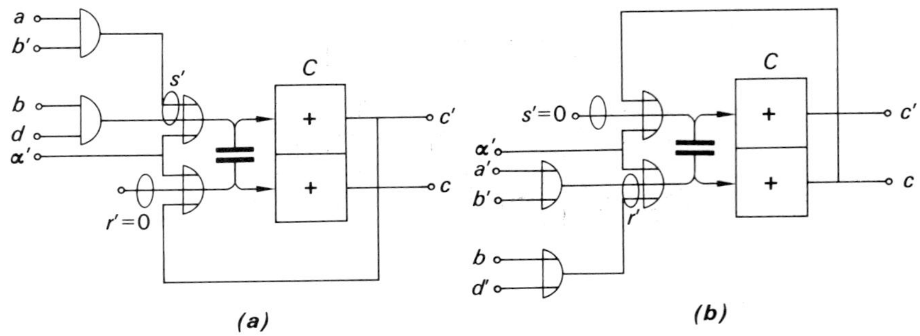

Fig. 10-45

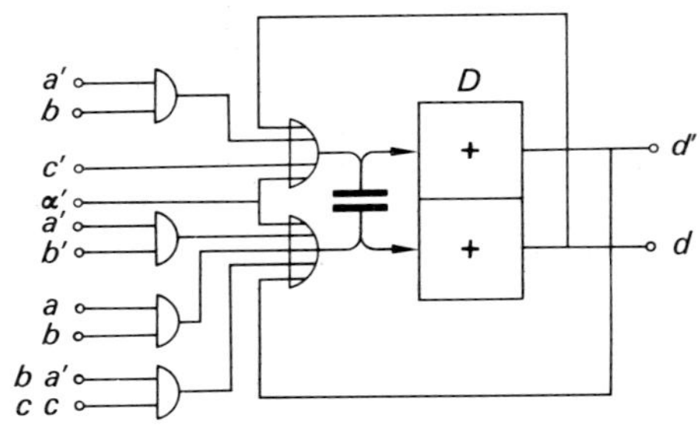

Fig. 10-46

$$s' = a'b + c' \qquad r' = a'b' + ab + \begin{matrix} bc \\ a'c \end{matrix} \tag{10-36}$$

The resulting diagram of the input connections is shown in Fig. 10-46.

It is possible to eliminate one $a'b'$ gate when for the C element the circuit of Fig. 10-45*b* is chosen.

The calculation of the number of input-gate diodes in this type of circuit is rather difficult because the input gates forming the set and reset signals are in most cases part of the actual OR input gates of the MS memory elements. In the most simple standard form, each of the OR input gates of an MS memory element requires three diodes. In the following count of diodes involved in the circuit of the 16-step-sequence switch designed above, the diodes of the AND input gates are counted as well as the diodes in excess of three in the OR input gates. This seems to be a good measure for the complexity of the circuit. The result is shown in Table 10-8.

Table 10-8

	A	B	C	D	Total
Number of diodes	7	4	4	10	25
	7	4	12		23

It is also possible to arrive at the correct formulas for set and reset signals of an MS memory element via an algebraic way that ends in a multiple

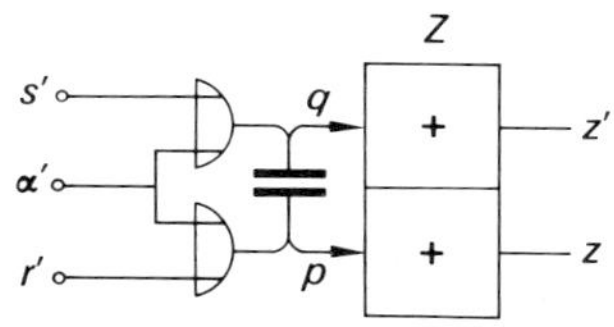

Fig. 10-47

choice. Leaving nonessential formula elements out of consideration, the following equations describe the operation of the trigger in the MS memory element of Fig. 10-47:

$$Z = g_0 z + g_1 z' \tag{10-37}$$

$$p = r' s \tag{10-38}$$

$$q = rs' \tag{10-39}$$

$$Z = p + q' z \tag{10-40}$$

The problem is to find s' and r' as a function of g_1 and g_0, which are the groups of minterms in the application equation indicating, respectively, the cases in which Z has to be set and in which Z has to be kept set.

Substitution of Eqs. (10-38) and (10-39) in Eq. (10-40) eliminates p and q:

$$Z = r' s + (r' + s) z = r' s + r' z + sz \tag{10-41}$$

Equations (10-37) and (10-41) describing the same state Z together give

$$r' s + r' z + sz = g_0 z + g_1 z' \tag{10-42}$$

which is equivalent to the following pair of relations:

$$r' s + r' z + sz \leqslant g_0 z + g_1 z' \tag{10-43}$$

and

$$g_0 z + g_1 z' \leqslant r' s + r' z + sz \tag{10-44}$$

In these formulas, the sign $\leqslant$ has the following meaning: it relates two groups of minterms (canonical form); moreover, the group on the left-hand side of this sign is included in the group of minterms on the right-hand side, or the reverse, when this sign is pointing in the other direction. For example,

$$xyz' + x'yz \leqslant x'y'z + xyz' + xy'z + x'yz$$

In the following discussion, three laws giving the rules of the use of the $\leqslant$ sign are required. These laws can be easily proved when one keeps in mind the fact that each variable may represent a sum of minterms and is a variable in a Boolean algebra. The three laws are as follows:

1. $$x + y \leqslant z \Leftrightarrow \begin{cases} x \leqslant z \\ y \leqslant z \end{cases} \tag{10-45}$$

$$xy \geqslant z \Leftrightarrow \begin{cases} x \geqslant z \\ y \geqslant z \end{cases} \tag{10-46}$$

2. $$x + y \geqslant z \Leftrightarrow x \geqslant y'z \qquad \text{or} \qquad y \geqslant x'z \tag{10-47}$$

$$xy \leqslant z \Leftrightarrow x \leqslant y' + z \qquad \text{or} \qquad y \leqslant x' + z \tag{10-48}$$

3. $$y \leqslant z \Leftrightarrow y' \geqslant z' \tag{10-49}$$

Applying law (10-45) to the relation given by Eq. (10-43), the following pair of relations is obtained:

$$s(r' + z) \leqslant g_0 z + g_1 z' \tag{10-50}$$

and

$$r'z \leqslant g_0 z + g_1 z' \tag{10-51}$$

Applying law (10-48), Eq. (10-50) changes as follows:

$$s \leqslant g_0 z + g_1 z' + rz' \tag{10-52}$$

Using law (10-49), Eq. (10-52) can be written in the following form:

$$(g_0' + z')(g_1' r' + z) \leqslant s'$$

or

$$g_0' z + g_1' z' r' \leqslant s' \tag{10-53}$$

Applying law (10-48), Eq. (10-51) changes as follows:

$$r' \leqslant g_0 z + g_1 z' + z' = g_0 + z' \tag{10-54}$$

Using law (10-49), relation (10-44) can be transformed as follows:

$$(g_0 z + g_1 z')' \geqslant (r's + r'z + sz)'$$

or

$$g_0' z + g_1' z' \geqslant rs' + rz' + s'z' \tag{10-55}$$

which is equivalent to the following pair of relations:

$$g_0' z + g_1' z' \geqslant rz' \tag{10-56}$$

and

$$g_0' z + g_1' z' \geqslant rs' + s' z' \tag{10-57}$$

Applying law (10-47) to Eq. (10-56), another relation can be obtained for the reset r:

$$z + g_0' z + g_1' z' = z + g_1' \geqslant r \tag{10-58}$$

Equation (10-58) can be changed as follows by application of law (10-49):

$$g_1 z' \leqslant r' \tag{10-59}$$

Applying law (10-47) to relation (10-57), the following equation can be obtained for s':

$$s' \leqslant r' z + g_0' z + g_1' z' \tag{10-60}$$

Equations (10-53) and (10-60) give the complete relation for the set signal s':

$$g_0' z + g_1' z' r' \leqslant s' \leqslant r' z + g_0' z + g_1' z' \tag{10-61}$$

And Eqs. (10-54) and (10-59) together give the complete relation for the reset signal r':

$$g_1 z' \leqslant r' \leqslant g_0 + z' \tag{10-62}$$

The left-hand sides of relations (10-61) and (10-62) show the minimum values of s' and r', respectively, that are the "must" conditions, whereas the right-hand sides of these relations show the maximum values, which include the don't-care conditions. Relations (10-61) and (10-62) offer many solutions for s' and r'; but when there are already special relations between g_0 and g_1, only a few will be practical. In the example of the 16-step-sequence switch, $g_0 = g_0'$ for MS memory element A, $g_0 = y + z$ and $g_1 = yz$ for MS memory element B, and $g_0 = g_1$ for MS memory element C. In these cases, the circuit design can best be performed via the simplification of the application equation.

The use of relations (10-61) and (10-62) will be explained by designing the D MS memory element of the 16-step-sequence switch as defined in Table 10-7.

MS MEMORY ELEMENT D

$$g_0 = a' b' + ab + a' c \qquad g_1 = (a + b') c$$
$$g_0' = ab' + a' bc' \qquad g_1' = a' b + c'$$

With these values of g_0 and g_1, relation (10-61) becomes

$$ab'd + a'bc'd + (a'bd' + c'd')r' \leqslant s' \leqslant ab'd + a'bc'd + a'bd' + c'd' + dr' \quad (10\text{-}63)$$

and relation (10-62) becomes

$$acd' + b'cd' \leqslant r' \leqslant a'b' + ab + a'c + d' \quad (10\text{-}64)$$

The left-hand side of Eq. (10-63) is shown in Fig. 10-48 in the form of a Karnaugh map. Minterms $ab'd$ and $a'bc'd$ are shown in the ordinary way by means of a 1. Minterms $a'bd'r'$ and $c'd'r'$ are indicated by $\geqslant\underline{r}'$ which means that these entries in the s' map must be greater than or equal to the corresponding entries in the r' map.

Figure 10-49 shows that map corresponding to the right-hand side of relation (10-63). This map shows the maximum values of s'.

The minterms without r' are shown as 1 entries. Minterm dr' fills all the open places in the d field with $\leqslant\underline{r}'$, indicating that these entries in the s' map must be less than or equal to the corresponding entries in the r' map.

The maps of Figs. 10-48 and 10-49 can be united into one map, as shown in Fig. 10-50. Because map 10-48 shows the musts, entries $\geqslant\underline{r}'$ replace 1's in the diagram of Fig. 10-50. The places in the diagram of Fig. 10-50 not covered by a 1, $\geqslant\underline{r}'$ or $\leqslant\underline{r}'$, are filled with 0.

The map diagram representing Eq. (10-64) is slightly easier to draw. It is shown in Fig. 10-51 in its final form. The left-hand-side minterms of Eq. (10-64) are entered first as 1's. The rest of the diagram is filled with 0's. The right-hand-side minterms, insofar as they are not already present in the diagram as a 1, are indicated by $\emptyset$, which means a don't care.

It follows from relation (10-62) that $r' = 1$ if $g_1 z' = 1$. Hence in this case $g_1' = 0$ and $z = 0$, so that the right-hand part of relation (10-61) is 0, and as a consequence, $s' = 0$. Moreover it follows from relation (10-62) that $r' = 0$ if $g_0 + z' = 0$. In this case $g_0 = 0$, $z' = 0$, and $g_0' z = 1$; hence the left-hand part of relation (10-61) is 1, and as a consequence, $s' = 1$. The result is that 0 and 1 entries in any r' map coincide with 1 and 0 entries, respectively, in the corresponding s' map, as can be seen in Figs. 10-50 and 10-51. The $\geqslant\underline{r}'$ and $\leqslant\underline{r}'$ entries in any r' map coincide with $\emptyset$ entries in the corresponding s' map.

Finding the simplest solution for s' and r' involves manipulating the limiting entries $\leqslant\underline{r}'$ and $\geqslant\underline{r}'$ in the map of Fig. 10-50 and the don't cares in the map of Fig. 10-51. The simplest constellation is not easy to detect among the many possibilities. A computer program can be helpful. The diagrams of Figs. 10-52 and 10-53 give a possible not-too-complicated solution.

The resulting circuit of the D MS memory element is shown in Fig. 10-54.

$\geqq \underline{r'}$	$\geqq \underline{r'}$		
$\geqq \underline{r'}$	$\geqq \underline{r'}$		$\geqq \underline{r'}$
1			
	1	1	

Fig. 10-48

1	1		
1	1		1
1	$\leqq \underline{r'}$	$\leqq \underline{r'}$	$\leqq \underline{r'}$
$\leqq \underline{r'}$	1	1	$\leqq \underline{r'}$

Fig. 10-49

$\geqq \underline{r'}$	$\geqq \underline{r'}$	0	0
$\geqq \underline{r'}$	$\geqq \underline{r'}$	0	$\geqq \underline{r'}$
1	$\leqq \underline{r'}$	$\leqq \underline{r'}$	$\leqq \underline{r'}$
$\leqq \underline{r'}$	1	1	$\leqq \underline{r'}$

Fig. 10-50

$\emptyset$	$\emptyset$	1	1
$\emptyset$	$\emptyset$	1	$\emptyset$
0	$\emptyset$	$\emptyset$	$\emptyset$
$\emptyset$	0	0	$\emptyset$

Fig. 10-51

0	0	0	0
1	1	0	0
1	1	1	0
0	1	1	0

$s' = ad + bc'$

Fig. 10-52

0	0	1	1
0	1	1	0
0	1	1	0
0	0	0	0

$r' = ab + b'cd'$

Fig. 10-53

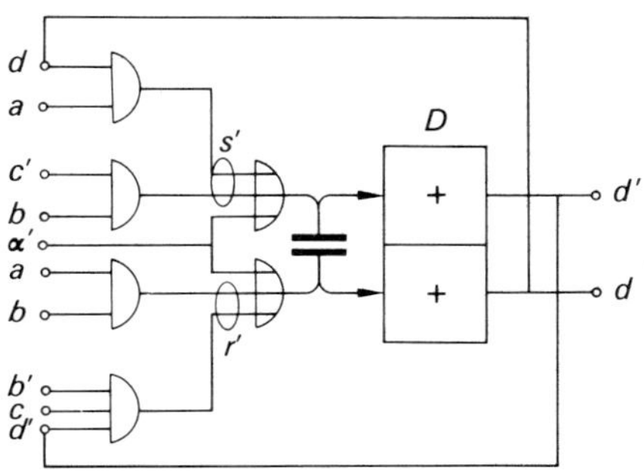

Fig. 10-54

Another solution of the s' and r' maps of the D element of the 16-step-sequence switch of the example is given in Figs. 10-55 and 10-56.

The resulting circuit of the D MS memory element is shown in Fig. 10-57.

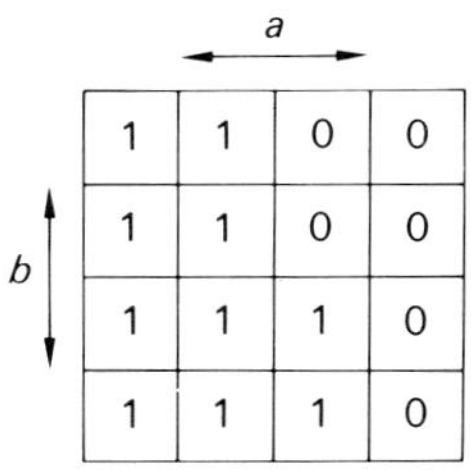

$s' = c' + ad$

Fig. 10-55

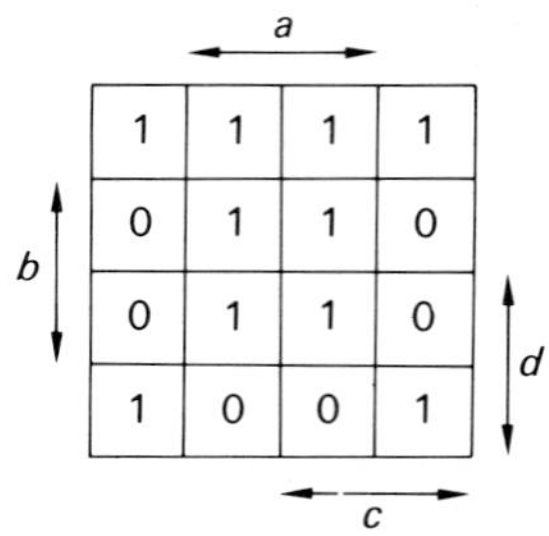

$r' = ab + a'b' + b'd'$

Fig. 10-56

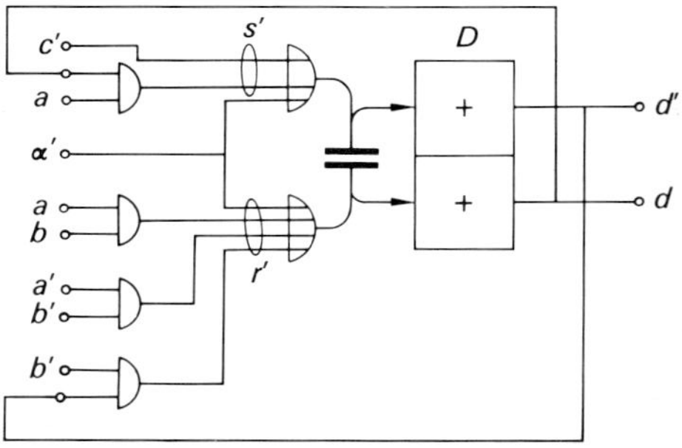

Fig. 10-57

In the solution of Figs. 10-52 and 10-53, only the ad AND gate of the s' signal can be combined with an ad AND gate of the B element, whereas in the solution of Figs. 10-55 and 10-56, gates ad and $a'b'$ can, respectively, be combined with an input gate of the B and C elements. The number of diodes required in the input gates of the four elements A, B, C, and D of the 16-step-sequence switch is now summarized in Table 10-9.

Table 10-9

	A	*B*	*C*	*D*	*Total*
Number of diodes	7	4	4	9	24
	7		13		20

It may be concluded from the above that an important reduction in the number of diodes in the input gates can be obtained when the circuit design is based on MS memory elements instead of predetermined types of flip-flops.

PROBLEMS

1. Show that the differentiating and integrating effects of an RC circuit are complementary phenomena.

2. Show by means of simple circuits that the integrating effect of an RC circuit delays the adaptation of the output signal to the input signal over a certain interval of time, whereas in a differentiating RC circuit with the same dimensioning, an output signal will be obtained only during that interval of time.

3. Write the switching formula for the output signal of a differentiating RC delay network. Draw a diagram to illustrate the operation of this RC circuit.

4. Determine the switching formula for the output signal of a differentiating RC delay network with front and back control. Show the operation of the RC circuit by means of a diagram.

5. Show that a differentiating RC delay network can be used as a bidirectional switching element.

6. Describe the operation of a differentiating RL delay network and draw a circuit.

7. Draw a symbolic circuit of an MS memory element and explain its operation by means of a time chart.

8. Explain the operation of an MS memory element with control signals on front and back gates. Give the formulas for the resulting set and reset signals for the trigger. Give the formula of the state of the trigger in this MS memory element with a set and a reset signal on the front gates and the set signal cross connected to a back gate.

9. Give the formulas of the state of a single-capacitor MS memory element with both output signals cross-connected to the input gates and with an input signal a' connected to both these gates.

10. (*a*) Draw the voltages on points W, X, Y, and Z of the circuit of Fig. 10-58 as a function of time if a square voltage variation from 0 to −6 volts is given on point a. Give this for signal voltages on point b of 0 volts and −6 volts, respectively.

(*b*) Derive a switching formula expressing output signal z in input signal voltages a and b (0 and 1 are, respectively, −6 volts and earth; positive logic).

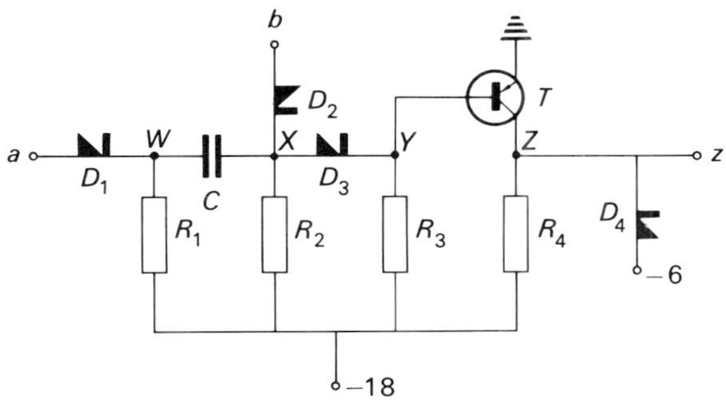

Fig. 10-58

11. In the circuit of Fig. 10–59, a clock pulse α' $^1_0$⊔ controls the timing of all operations. Input signals a and b do not change when $\alpha' = 0$. Determine a switching formula that expresses the state of trigger Z after a clock pulse ($\alpha' = 1$) in the signals a, b, and z during the clock pulse $\alpha' = 0$.

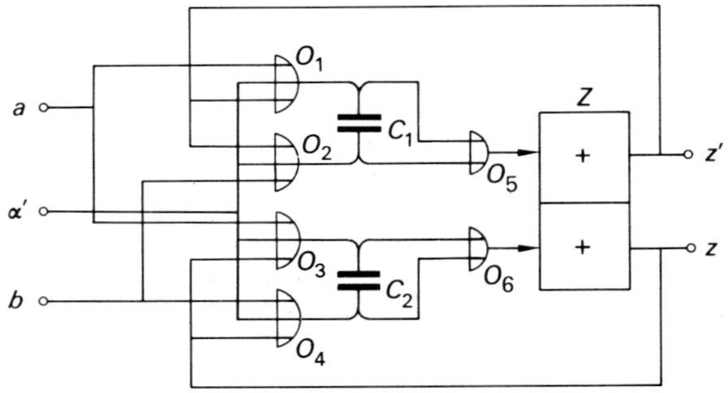

Fig. 10-59

12. In the circuit of Fig. 10-60 a capacitor switching element is shown. Show by means of a truth table for which combinations of input signals a, b, c, and d a positive-going clock pulse α' will generate an output signal z.

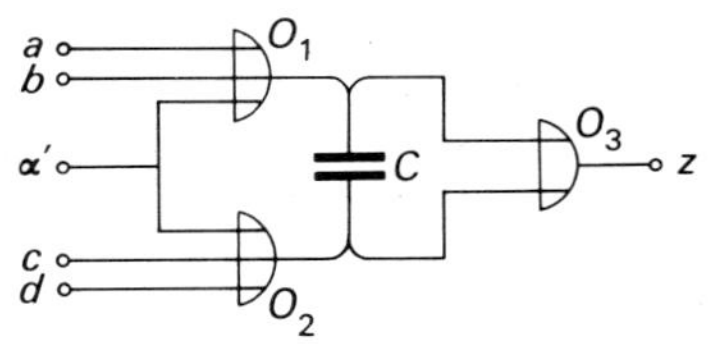

Fig. 10-60

13. The circuit of Fig. 10-61 is fed with a clock pulse α' and a signal a. Signal a does not change when $\alpha' = 0$. The output signal z is in feedback connected to input gate O_1. Devise a circuit having the same operation as that of Fig. 10-61 but without feedback to input gates O_1 and O_2.

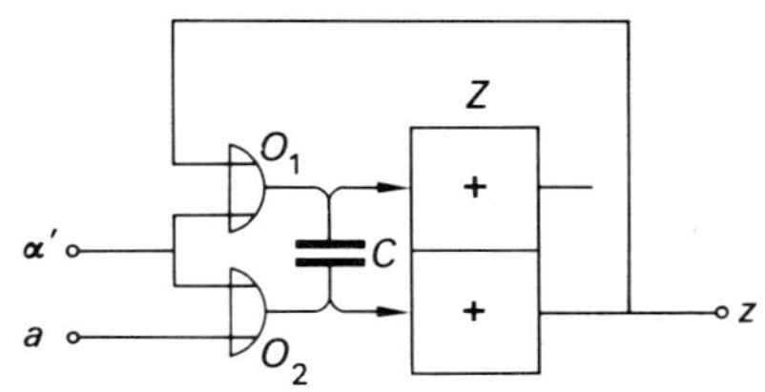

Fig. 10-61

14. The following switching formulas for MS memory elements are given:

$$A = D_{\alpha'}[(cd' + b'c'd)a + (cd' + b'c'd)'a']$$
$$B = D_{\alpha'}[(ad' + a'd + c')b + (ad' + a'd)c']$$
$$C = D_{\alpha'}(a'b' + bd')$$

Determine the input gates of MS memory elements A, B, and C by means of Eqs. (10-61) and (10-62).

SELECTIVE BIBLIOGRAPHY

1. Fleischhammer, W.: Eine Systematik der zusammengesetzte bistabilen Kippstufen (Classification of Composite Flip-flop Circuits), *Elektron. Rechenanlagen,* vol. 10, pp. 34–40, 1968.
2. Lagemann, K.: Ein Vorschlag zur Darstellung asynchron betriebener JK-flip-flops, *Elektron. Rechenanlagen*, vol. 10, pp. 171–176, 1968.
3. Phister, M., Jr.: Memory Element Input Equations, chap. 5 of "Logical Design of Digital Computers," John Wiley & Sons, Inc., New York, 1958.
4. Sparkes, J. J.: Bistable Elements for Sequential Circuits, *Electron. Eng.*, vol. 38, pp. 510–515, 1966.

11
Pulse-frequency-halving Circuits

11-1 RELAY CIRCUITS

Pulse-frequency-halving circuits have numerous switching applications. The terminology for this type of circuit is not well defined: it sometimes is referred to as a "bistable multivibrator"; another suitable name is "binary unit counter," because this basic circuit constitutes a building block for many types of so-called binary counters that count the number of pulses of a pulse series in the binary number system. In this book, we have chosen to use the neologism "pulse halver" or "pulse-halving circuit," because either term describes exactly the purpose of the circuit.

A pulse halver is a circuit that produces one output pulse β for each pair of input pulses α. Each complete cycle of two input pulses consists in fact of two pulses and two pauses between these pulses, i.e., four different states in total. These four states of the pulse halver can be arrived at by means of a minimum of two memory elements (e.g., two relays). The time chart describing the operation of such a circuit is shown in Fig. 11-1. This diagram is fundamental for all pulse-halving circuits irrespective of the components used to implement them. In some applications, the control pulses are relatively long; in others, e.g., in automatic telephony, the duration of pulses and pauses is of the same order of magnitude. In still other applications, the control pulses sent to the pulse-halving circuit instead of forming one train of pulses may consist of two interleaved series of pulses. These differences are of minor importance with respect to the design of pulse-halving circuits.

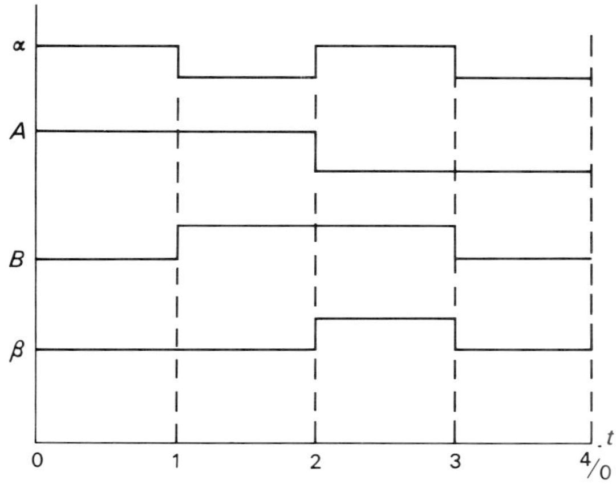

Fig. 11-1

Through use of Eq. (9-2), the state equations of a pulse halver can be determined as follows:

$$A = s_a + ar_a' = \alpha b' + a(\alpha b)' = \alpha b' + \alpha' a + ab' \tag{11-1}$$

$$B = s_b + br_b' = \alpha' a + b(\alpha' a')' = \alpha' a + \alpha b + ab \tag{11-2}$$

These equations can be written as

$$A = (\alpha + a)\, b' + \alpha' a \tag{11-3}$$

$$B = (\alpha + a)\, b + \alpha' a \tag{11-4}$$

Equations (11-3) and (11-4) represent one possible form of a pulse halver. The corresponding relay circuit is shown in Fig. 11-2. Relays with two windings that allow an easy combination of the equal parts of both equations in one circuit have been used here. The resulting relay circuit has two attractive features.

1. The transition periods of the transfer contacts do not influence in any way the reliable operation of the circuit.

2. No demagnetization will occur with a bridged secondary winding. If a winding of a releasing relay is bridged by a resistor or a winding belonging to another relay, release times will increase.

Equations (11-3) and (11-4) can also be written in another form:

$$A = (\alpha + a)(\alpha' + b') \tag{11-5}$$

$$B = (\alpha + a)(\alpha' + b) \tag{11-6}$$

This set of state equations gives a circuit with single-winding relays of the type shown in Fig. 11-3. It has the same features as the preceding circuit. The first factor of both equations can be combined into one contact circuit for both relays. The circuits shown in Figs. 11-2 and 11-3 have the same number of contacts.

Equations (11-1) and (11-2) can be simplified further by means of Theorem 12*b* (Chap. 2):

$$A = \alpha b' + \alpha' a \tag{11-7}$$

$$B = \alpha b + \alpha' a \tag{11-8}$$

The resulting circuit is shown in Fig. 11-4; this circuit differs from Fig. 11-2 in that contact a^2 is missing. This contact bridges the transfer time of the controlling contact α in the circuit of Fig. 11-2. This simplification does not lead to difficulties as long as the release time of relays A and B is large compared with the transition time of α.

The state equations of a pulse halver based on Eq. (9-1) yield the same result:

$$A = r_a'(s_a + a) = (\alpha b)'(\alpha b' + a) = \alpha b' + \alpha' a + ab'$$

The right-hand side of this equation is equal to the right-hand side of Eq. (11-1).

11-2 PULSE HALVERS USING TRIGGERS AS MEMORY ELEMENTS

The design of a pulse halver using triggers as memory elements can be very simple if the trigger is considered merely as a building block. The only things to be determined are the set and reset signals of the triggers. They are given to be as follows:

$$s_a = \alpha b' \qquad r_a = \alpha b \tag{11-9}$$

$$s_b = \alpha' a \qquad r_b = \alpha' a' \tag{11-10}$$

The resulting circuit with relay triggers is shown in Fig. 11-5. The trigger portion of this circuit is surrounded by a dotted line; the set and reset circuits conforming to Eqs. (11-9) and (11-10) are outside the dotted line.

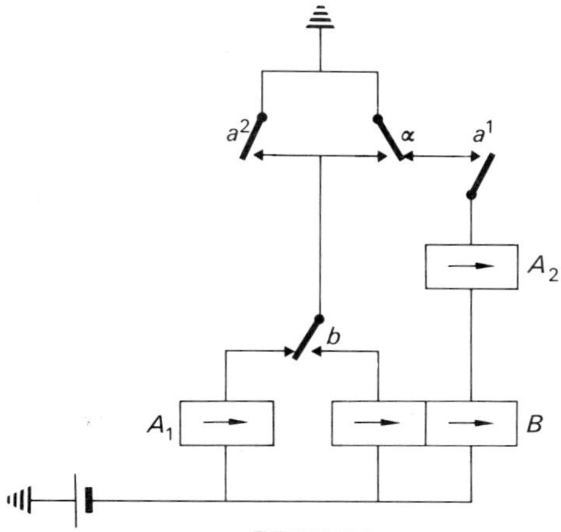

Fig. 11-2

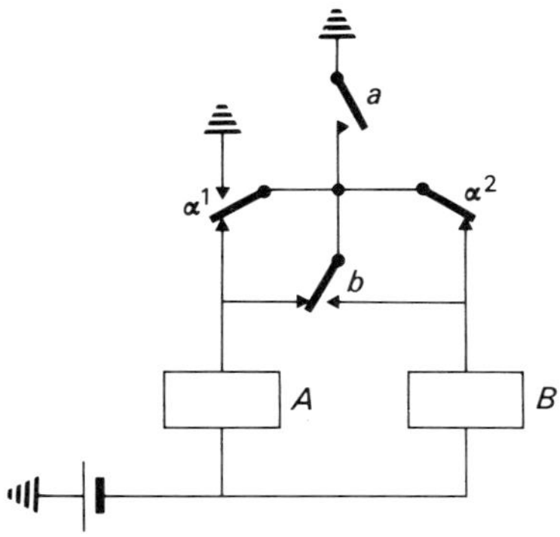

Fig. 11-3

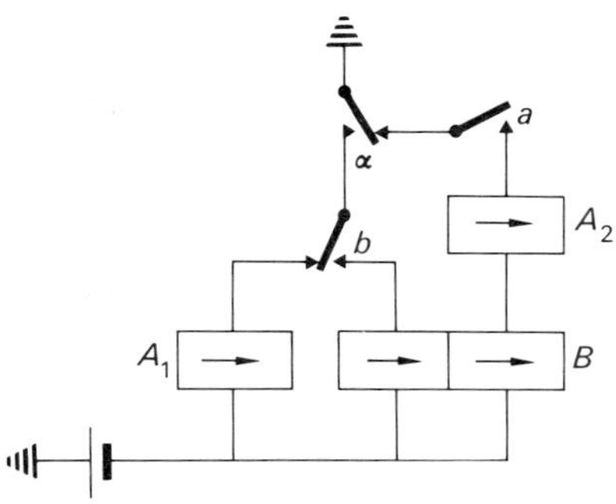

Fig. 11-4

The circuit shown in Fig. 11-5 operates as follows. In the rest condition of the circuit, relays A and B are, say, unoperated; consequently, relays A' and B' are operated. Relay A will be magnetized by the first operation of the controlling contact α. Relay A causes contact a_1 to be opened so that relay A' is then demagnetized. The state of the A trigger is thereby reversed. A circuit is also provided for the operation of the B relay; this relay will

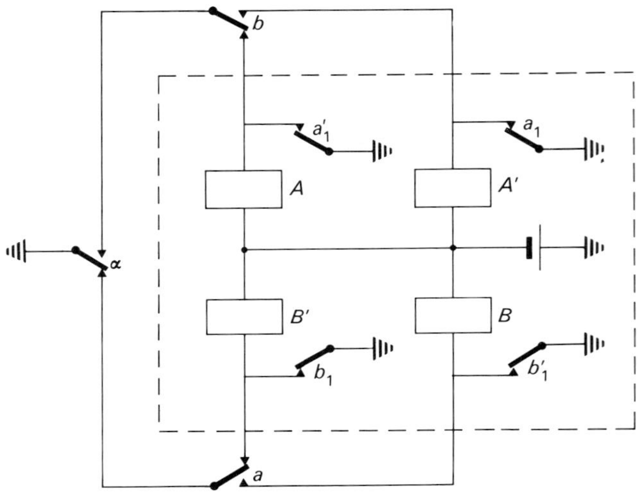

Fig. 11-5

operate on the first release of the α contact. Relay B causes contact b_1 to be opened so that relay B' is demagnetized. The state of the B trigger is now also reversed. Transfer contact b (Fig. 11-5) now prepares the reset circuit for the A trigger via its make side. This reset circuit becomes effective on the second operation of the α contact. Trigger A is reset by the operation of relay A' and the following release of relay A. Transfer contact a is thereby restored to the condition shown in the circuit of Fig. 11-5, so that a circuit is now formed for resetting the B trigger. This reset circuit becomes effective on the second release of the α contact. The pulse halver is now again in the condition it was in initially.

In comparison with the relay circuits shown in the preceding section, this pulse halver may seem to be complicated and uneconomical; but these general design considerations are useful as introduction to the subject of electronic pulse-halving circuits.

The circuit of Fig. 11-5 is shown symbolically in Fig. 11-6. This symbolic circuit is a straightforward reproduction of the set and reset equations (11-9) and (11-10) of the A and B triggers that perform the pulse-halving function in accordance with the time chart of Fig. 11-1.

The serial contact circuits that produce the set and reset signals for both triggers are shown as AND gates in the circuit of Fig. 11-6. Triggers A and B have been indicated by means of their agreed symbols.

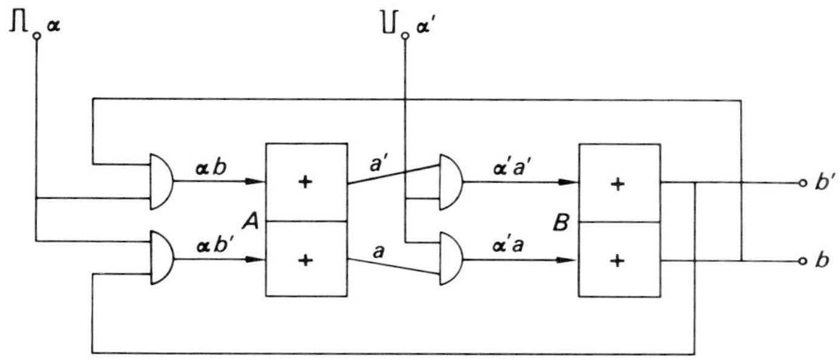

Fig. 11-6

Equations (11-1) and (11-2), which are the state equations of the A and B memory elements derived from the time chart of Fig. 11-1, can also be translated into electronic circuit elements. It is necessary to apply a double negation to both state equations in order to fulfil the feedback and amplification requirements of the electronic circuit.

$$A = \alpha b' + a(\alpha b)' = \alpha b' + (a' + \alpha b)' \tag{11-11}$$

$$B = \alpha' a + b(\alpha' a')' = \alpha' a + (b' + \alpha' a')' \tag{11-12}$$

The straight translation of Eqs. (11-11) and (11-12) into a symbolic (electronic) circuit is shown in Fig. 11-7. Two portions are enclosed by dotted lines in this figure. They correspond to the two parts of the complete pulse-halving circuit that form the A and the B trigger. The gates outside the enclosure provide the necessary set and reset signals for both triggers, so that the symbolic trigger circuit of Fig. 11-7 can actually be considered as corresponding to the symbolic circuit of Fig. 11-6. No fundamentally new circuit will result if one disregards the fact that the A and B memory elements are in fact triggers.

If AND triggers instead of OR triggers are used for the A and B memory elements, the input gates have to be changed from OR gates into AND gates. In addition, the polarity of the controlling pulses α must be reversed.

The mixed diode-transistor trigger circuits shown in Fig. 9-9*a* and *b* are well suited for use as a memory element in such pulse-halving circuits as are shown in Figs. 11-6 and 11-7. The full circuit layout of such a pulse halver using *pnp* transistors is shown in Fig. 11-8.

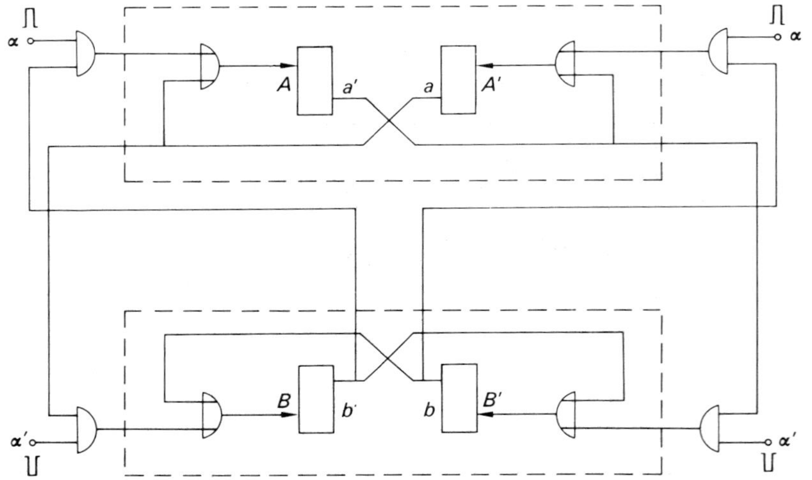

Fig. 11-7

It can be easily ascertained that all input gates are designed to comply with the negated Eqs. (11-9) and (11-10) and that the circuit as a whole conforms to the negated Eqs. (11-11) and (11-12). This type of pulse-halving circuit is well suited for operation with control pulses and pauses of equal duration.

There are no speed limitations in this type of circuit other than the fall and rise times of the currents in the transistors and diodes. All means described in Chap. 6 for increasing operating speeds may be applied to the circuit of Fig. 11-8. They have been excluded in this circuit in order to distinguish clearly between what is necessary from the logical point of view on the one hand and what is desirable for the dimensioning of the circuit on the other.

It is not essential that a pulse halver be controlled by means of α and α' pulses. If an AND trigger is used for the B memory element instead of the OR trigger, as is indicated in Fig. 11-9, the set and reset equations of this trigger can be obtained by inverting Eqs. (11-10):

$$s_b' = \alpha + a' \qquad r_b' = \alpha + a \tag{11-13}$$

The resulting symbolic circuit is shown in Fig. 11-9.

Pulse-halving circuits can also be formed by means of asymmetrical triggers used as memory elements. For example, if triggers that are set by an α pulse and reset by an α' pulse are used, the equations for the set signals

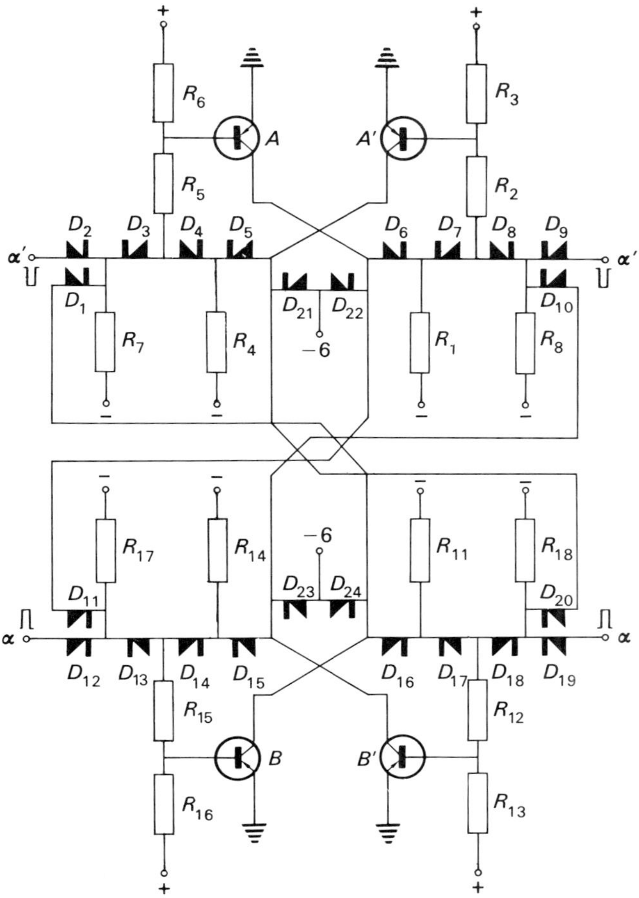

Fig. 11-8

remain the same as those already given in Eqs. (11-9) and (11-10), whereas the equations for the reset signals can be derived by a negation process from the corresponding Eqs. (11-9) and (11-10).

$$s_a = \alpha b' \qquad r'_a = \alpha' + b' \tag{11-14}$$

$$s_b = \alpha' a \qquad r'_b = \alpha + a \tag{11-15}$$

The resulting diagram of a pulse halver is shown in Fig. 11-10 in its general symbolic form.

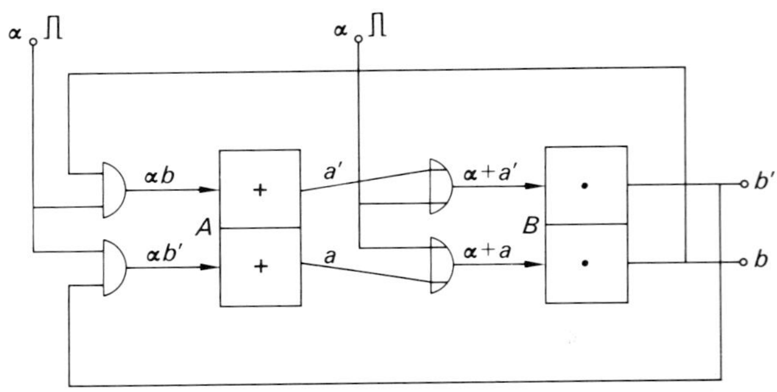

Fig. 11-9

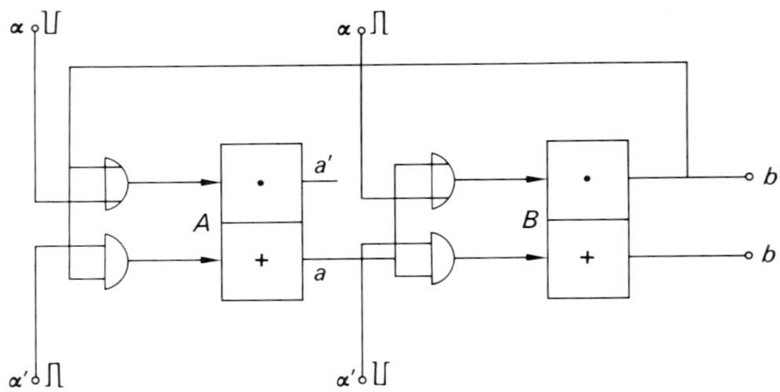

Fig. 11-10

The fact that only one output terminal of each trigger is used in this form of pulse halver does not imply that the total circuit can be simplified as far as the number of transistors is concerned.

State equations (11-1) and (11-2) lead more or less automatically to symmetrical types of pulse-halving circuits. Nonsymmetrical pulse-halving circuits are also possible if state equations of the form given in Eqs. (11-5), (11-6), (11-7), or (11-8) are used. However, some caution should be exercised in the handling of these state equations, which are already somewhat simplified. The minterm ab' of the complete state equation of the A element as given by Eq. (11-1) does away with the difficulties that can be caused by a nonideal α transfer contact with a gap between the periods that $\alpha = 1$ and $\alpha' = 1$ in the circuit condition where $a = 1$ and $b' = 1$ (see also Theorem 12b, Chap. 2). An overlap between these two α periods causes a shift away from the theoretical moments of the set and reset moments of the A element in the circuit condition where $a = 0$ and $b' = 1$ or $a = 1$ and $b' = 0$.

The inverse holds for the state equation of the A element in the form of a logical product of logical sums; thus

$$A = (\alpha + a)(\alpha' + b')(a + b') = (\alpha + a)(\alpha' + b')$$

In relay circuits, a gap in the operation of a transfer contact or the overlap in the operation of a make-before-break contact does not give much trouble because these periods are in general too short to affect the condition of a relay. In electronic circuits, however, if an α' signal is generated with a transistor that in its turn is controlled by an α signal, an overlap between these two signals will arise after the transition from $\alpha = 0$ to $\alpha = 1$; but a gap will follow the transition in the reverse direction. The reverse will happen if a transistor is driven by an α' signal and if an α signal is generated.

These gaps and overlaps in electronic circuitry have a duration of the order of the operating times of the triggers, so that difficulties may be expected if Theorems 12a and 12b of Chap. 2 remain valid in these types of circuits. For example, state equations (11-5) and (11-6) are to be implemented by means of electronic circuits. A double negation of these equations is required:

$$A = \{[(\alpha + a)\cdot(\alpha' + b')]'\}' \tag{11-16}$$

$$B = \{[(\alpha + a)\cdot(\alpha' + b)]'\}' \tag{11-17}$$

These state equations show that part of the logical gates can be used for controlling both memory elements. The corresponding symbolic circuit is shown in Fig. 11-11. A possible implementation of the symbolic circuit of

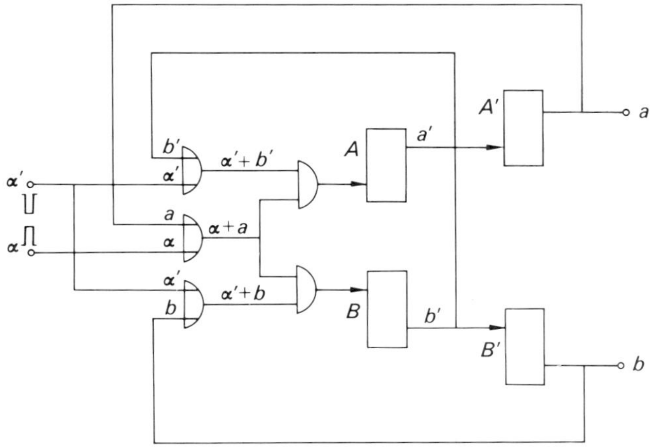

Fig. 11-11

Fig. 11-11 by means of diodes and *pnp* transistors is given in the diagram of Fig. 11-12.

The circuit of Fig. 11-12 operates in accordance with the time chart of Fig. 11-1. It is assumed that in the rest condition of the pulse halver, transistors A (T_2) and B (T_3) are made conducting by a 0 signal applied on

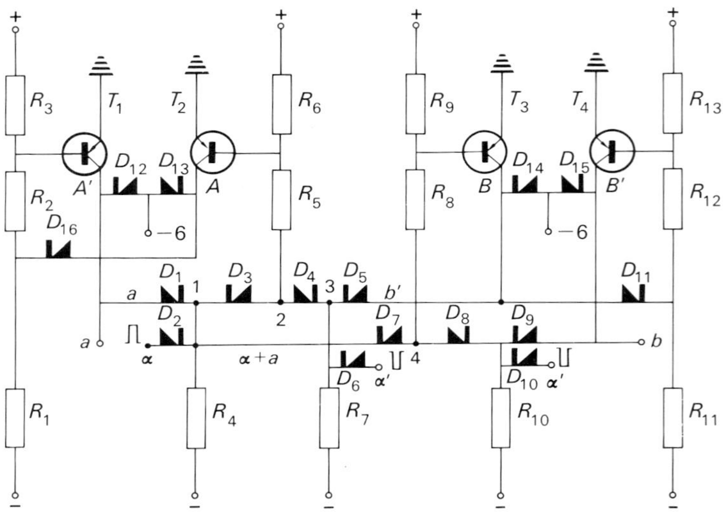

Fig. 11-12

their base. Transistors A and A' as well as transistors B and B' combine to form triggers, so that it can be readily seen that if the state of one of the transistors of a triggered pair is changed, the other will be reversed.

The state of the A trigger is exclusively controlled via the base of the A transistor. An analogous situation exists in the control of the B trigger. Hence, in the assumed rest condition of the circuit, point 1 is 0, point 3 is 1, and thus point 2 is 0. The first control pulse α marks point 1 with a 1, so that point 2 also receives a 1. This blocks transistor A, and the state of the A trigger is reversed.

Signal $\alpha + a$, which is passed on to the B trigger, is now 1, so that at the end of the first control pulse, when α' reverts to 1, the B trigger is reversed. Consequently, after one control pulse the A and B triggers are in their assumed operated state. During the pause after the first control pulse, a signal $b' = 0$ is received on diode D_5, so that point 3 can change from 1 to 0 by the second controlling pulse when $\alpha' = 0$. Point 2 also changes from 1 to 0, so that transistor A again becomes conducting. The A trigger is now reversed to its assumed rest condition.

The a signal changes to 0 during the second controlling pulse, but $\alpha = 1$ so that point 1 is kept in the 1 condition during that pulse. Point 2 retains its 0 state under the influence of point 3. During the following change of α from 1 to 0 and α' from 0 to 1, point 2 will have a potential somewhere between the symbolic 1 and 0 levels for a very short moment the duration of which will depend on the squareness of the controlling pulse. This tends to affect the reliable operation of the circuit.

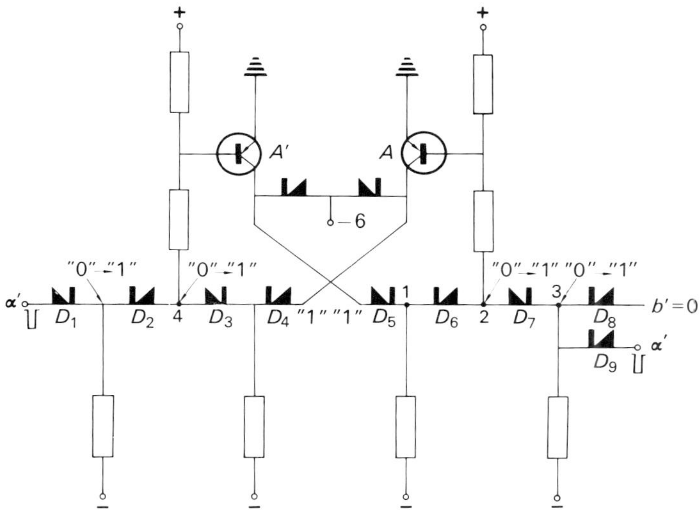

Fig. 11-13

If one tries to modify the circuit of Fig. 11-12 in such a way that it is controlled by an α' pulse only, an intolerable situation arises as a result of the use of ideal-state equations. State equation (11-16) of the A-memory element can be written as

$$A' = [(\alpha' \cdot a')' \cdot (\alpha' + b')]' \qquad (11\text{-}18)$$

The A memory circuit of Fig. 11-12 modified in accordance with state equation (11-18) is shown in Fig. 11-13. The difficult situation which arose in the pulse halver of Fig. 11-12 at the end of the second controlling pulse is illustrated further in Fig. 11-13 by means of 0 and 1 notations at various points of the circuit.

The situation in the A trigger during the second control pulse is such that the A transistor is made conducting by a 0 signal, so that this transistor passes a 1 signal to diode D_4. However, the control signal being $\alpha' = 0$, point 4 of the circuit remains at the 0 signal level. Transistor A' cannot even change its condition, so that point 1 of the circuit retains its 1-signal level. The feedback path of the A trigger is not closed in this situation. During the trailing edge of the control pulse, the signal level of point 2 will start to change from 0 to 1; this change cannot take place normally because the feedback path should have locked diode D_7. Simultaneously, the signal level of point 4 of the circuit will also start to change from 0 to 1. The first transistor of the two to change its condition in this situation will determine, perhaps incorrectly, the final state of the A trigger at the end of the second controlling pulse.

To conclude, it should be reemphasized that the term ab' is redundant in the ideal case only and that it may not be omitted in the implementation. In the circuit of Fig. 11-12, this term is formed in the diode-gate structure that controls the A trigger, but this term is no longer present in the circuit of Fig. 11-13.

11-3 SINGLE-TRIGGER PULSE HALVERS

The pulse halvers described in the preceding sections all consist of two triggers and four input gates. However, the earliest pulse-halving circuits consisted in principle of one double-triode trigger and two capacitors functioning as memory elements in that during the transition of the circuit from one state into the other, they indicate to which state the transition had to be completed. The presence in a pulse-halving circuit of a memory element other than the trigger is essential; otherwise the trigger cannot be brought safely through the midcondition.

The memory function of a capacitor follows from the fact that this element can stay charged for a certain period of time after it has been put in that condition. This memory function of a capacitor is to a certain extent analogous to that of a relay with a holding winding, or even to that of a trigger. The idea of a capacitor memory can be applied successfully to such electronic pulse-halving circuits as were described in the preceding section.

The electronic pulse-halving circuits are as a rule controlled by pulses of short duration followed by a relatively long pause, as shown in the time chart of Fig. 11-14. The operation of the B memory element of a pulse halver can be considered as occurring with a delay corresponding to the duration of the α control pulse after the operation of the A memory element. However, there is no fundamental objection against introducing a much larger delay in the operation of the B memory element. The operation of the B memory element has to be completed before the next α pulse begins. This possibility in the operation of the B memory element is provided for in the time chart of Fig. 11-15. The delay in the operation of the B memory element is not bound within narrow limits, so that this delay could be produced by fairly

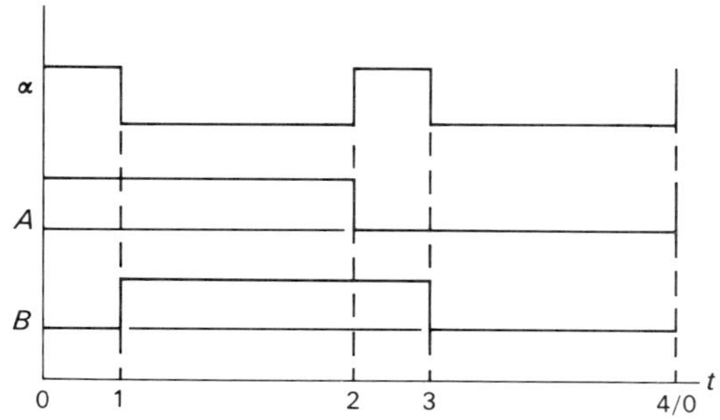

Fig. 11-14

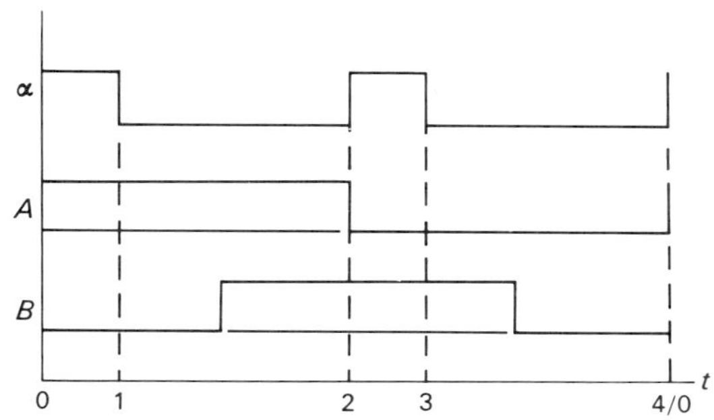

Fig. 11-15

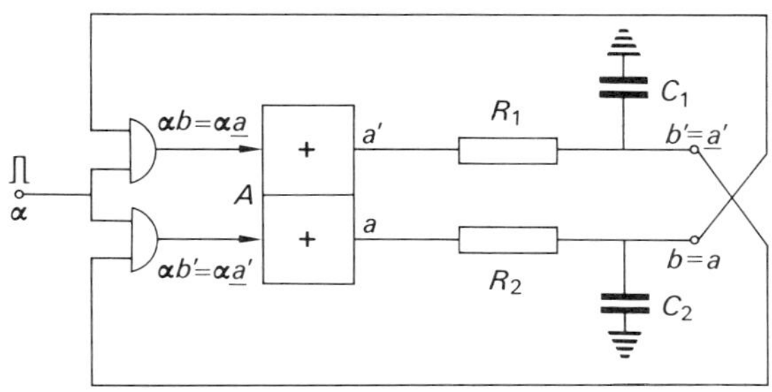

Fig. 11-16

inaccurate, cheap memory elements. However, the more inaccurately these memory elements operate, the more will the speed of the pulse halver be limited.

The B memory element of a pulse halver can be replaced by two integrating RC networks. The A part of the pulse halver of Fig. 11-6 produces an a and an a' signal, which are always followed, respectively, by the b and b' signal with a delay corresponding to the operation of the B part of the circuit. This can be indicated by the following expressions:

$$b = \underline{a} \tag{11-19}$$

and

$$b' = \underline{a}' \tag{11-20}$$

The underlined terms correspond to the variables that are delayed, i.e., as they appear at the output of the network with the delayed memory action. The symbolic diagram of a pulse-halving circuit in which the B memory element has been replaced by two integrating RC networks is shown in Fig. 11-16. This type of delay network can only be applied in those cases where the output is not heavily loaded—for example, in trigger circuits with voltage-controlled high-vacuum tubes. It cannot be used whenever the delay network is loaded by the base of a conducting transistor during the charging of the capacitor. A suitable dimensioning of the RC network cannot be obtained in such a case.

In the pulse halver of Fig. 11-10, which contains asymmetrical triggers, the replacement of the B memory element does not appear to be quite as simple, because the A memory element of this pulse halver produces only an a signal for controlling the B memory element, while the b' signal produced by the B memory element serves only to control the A memory element. However, an a' output will always be available in the A trigger, so that the desired b' signal can be generated as follows:

$$b' = \underline{a}' \tag{11-21}$$

In view of the fact that in the type of circuit shown in Fig. 11-10 only one variable is necessary for interconnecting the A and B memory elements, it is implied that only one RC-delay network is necessary. The resulting symbolic circuit is shown in Fig. 11-17.

The transformation of the symbolic circuits of Figs. 11-16 and 11-17 into actual transistor circuits can be performed by using as a starting point the trigger circuits shown in Figs. 9-9, 9-10, and 9-11.

Mixed diode and transistor pulse halvers hardly lend themselves to the use of RC-integrating delay networks, since one of the transistors of a trigger will always be conducting. The base current required for the conducting transistor puts too heavy a load on the output of the RC-integrating delay

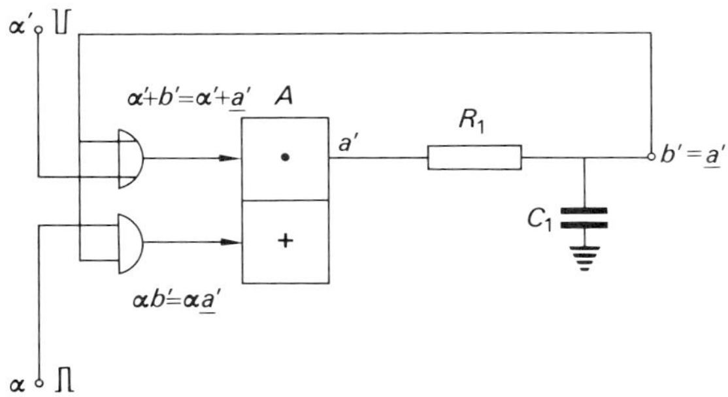

Fig. 11-17

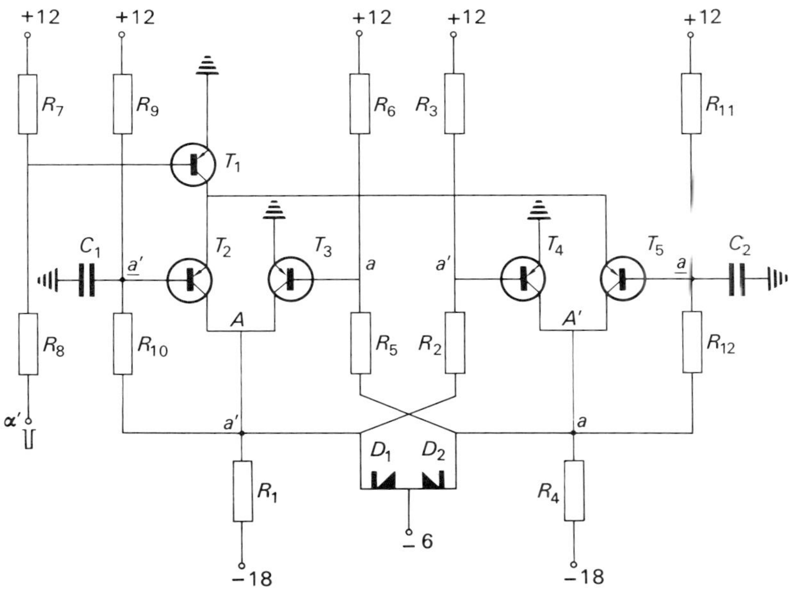

Fig. 11-18

network. This need not be the case in all-transistor triggers or pulse halvers using this type of trigger.

A pulse-halving circuit based on the all-transistor trigger circuit of Fig. 9-11 is given in Fig. 11-18. Resistor R_{10} of one RC network can be connected directly to the pair of collectors of the A part of the trigger, whereas resistor R_{12} of the other RC network can be linked to the pair of collectors of the A' part of the trigger. The capacitors must be connected to the bases of transistors T_2 and T_5 in the way shown in Fig. 11-18.

The state equation describing the complete pulse halver shown in Fig. 11-18 can be derived from the state equation of the A memory element of a two-trigger pulse halver given in Eq. (11-11), Eqs. (11-19) and (11-20) being substituted in Eq. (11-11) after double inversion of that equation:

$$A = \{[(\alpha' + \underline{a}')\cdot a]'\cdot(\alpha' + \underline{a})\}' \tag{11-22}$$

It can be easily ascertained that state equation (11-22) indeed corresponds to the circuit shown in Fig. 11-18.

The pulse halvers shown above have some practical advantages over the well-known Eccles-Jordan pulse-halving circuit in that they are to a rather large extent insensitive to transients in the power supply.

Equation (11-22) is in fact the state equation for the memory element (A trigger) in the pulse halver of Fig. 11-18. When reduced to its basic form, it becomes

$$A = \alpha\underline{a}' + a(\alpha\underline{a})' = s_a + a(r_a)' \tag{11-23}$$

Hence,

$$s_a = \alpha\underline{a}' \qquad \text{and} \qquad r_a = \alpha\underline{a} \tag{11-24}$$

The delay in the generation of the $\underline{a}$ and $\underline{a}'$ signals with respect to the a and a' signals must be longer than the α pulse.

The A trigger of a two-trigger pulse-halving circuit can also be replaced by a pair of memory elements in the form of delay networks. It is possible, of course, to solve this problem by replacing all symbols a by b, and vice versa; but this is not the purpose.

The time chart of a pulse-halving circuit with two memory elements (triggers) as shown in Fig. 11-19 shows that the A element operates under control of the leading edge of the α pulse and that the trailing edge of this pulse controls the operation of the B element.

Hence, the B trigger can be replaced by an MS memory element, which makes the A trigger superfluous. The signals for the front gates of the B MS memory element become as follows:

$$s_b = \alpha' + b \qquad r_b = \alpha' + b' \tag{11-25}$$

The corresponding symbolic circuit is shown in Fig. 11-20.

Advantage may be taken of the possibility of back input gates. The actual set and reset pulses fed to the trigger in the MS memory element become

$$\begin{aligned} s_b &= (\alpha' + b)\,(\underline{\alpha' + b})' \\ &= (\alpha' + b)\cdot\underline{\alpha b'} \\ &= \alpha'\cdot\underline{\alpha}\cdot\underline{b}' + \underline{\alpha}\cdot b\cdot\underline{b}' \end{aligned} \qquad \text{and} \qquad \begin{aligned} r_b &= (\alpha' + b')\,(\underline{\alpha + b'})' \\ &= (\alpha' + b')\cdot\underline{\alpha b} \\ &= \alpha'\cdot\underline{\alpha}\cdot\underline{b} + \underline{\alpha}\cdot b'\,\underline{b} \end{aligned}$$

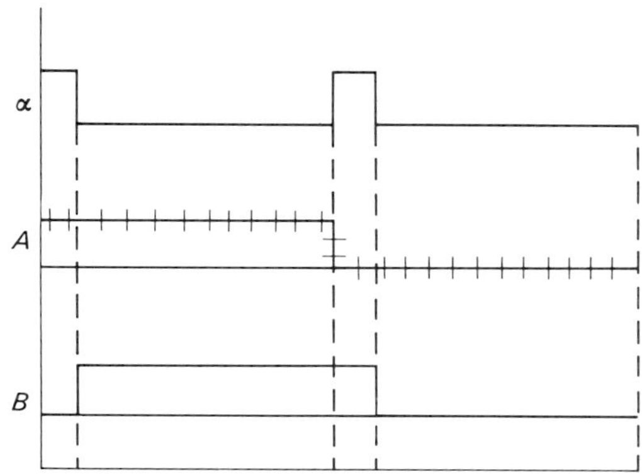

Fig. 11-19

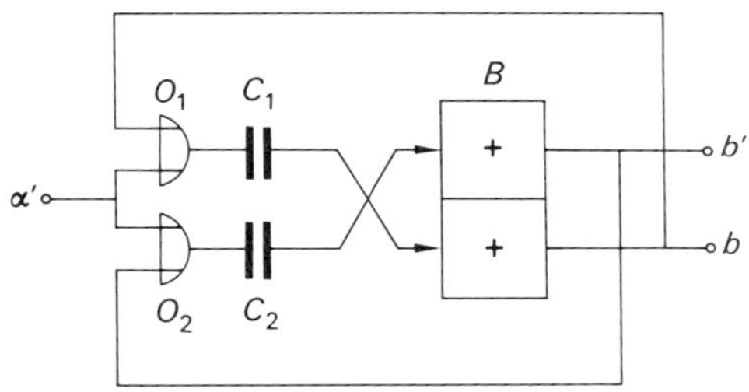

Fig. 11-20

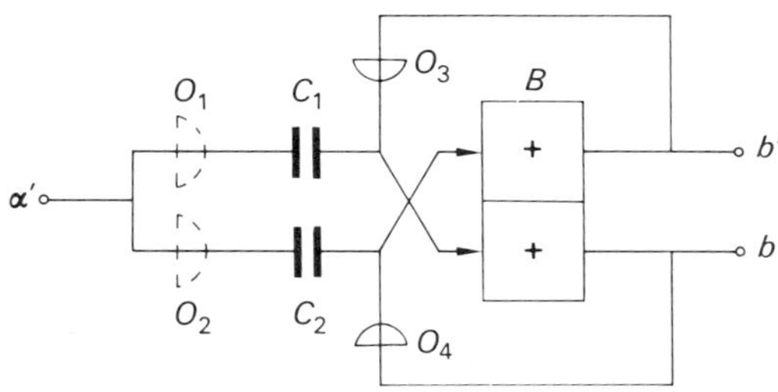

Fig. 11-21

Because the state of the trigger in the B MS memory element will not change during the $\underline{\alpha}$ pulse, $\alpha \cdot b \cdot \underline{b}' = 0$ and $\alpha \cdot b' \cdot \underline{b} = 0$.

$$s_b = \alpha' \cdot \underline{\alpha} \cdot \underline{b}' \qquad \text{and} \qquad r_b = \alpha' \cdot \underline{\alpha} \cdot \underline{b} \tag{11-26}$$

Equations (11-26) can be split up into parts and applied to the front gates and other parts to be fed into the back gates:

$$s_{b'} = \alpha' \qquad s_{b''} = b' \qquad r_{b'} = \alpha' \qquad r_{b''} = b \tag{11-27}$$

The resulting symbolic circuit is shown in Fig. 11-21.

The general state equation of this type of pulse halver becomes

$$B = s_b + b(r_b)' = \alpha' \underline{\alpha b}' + b(\alpha' \underline{\alpha b})' \tag{11-28}$$

Equation (11-28) does not fulfil the feedback requirements of an electronic circuit with respect to the b signal. Application of De Morgan's theorem to the last term of Eq. (11-28) results in a double negation of the b variable, which corresponds to the required amplification and the correct phase of this feedback signal. The resulting state equation is as follows:

$$B = \alpha' \underline{\alpha b}' + (b' + \alpha' \underline{\alpha b})' \tag{11-29}$$

The symbolic circuits of Figs. 11-20 and 11-21 corresponding to those already given in Fig. 10-28*b* represent the most general forms of pulse-halving circuits of this type. For medium speeds, a single-capacitor dynamic memory may be used in the circuit of Fig. 11-20. If the load of the pulse source does not matter, the front input gates O_1 and O_2 in the circuit of Fig. 11-21 may be omitted, which simplifies the circuit.

A complete pulse halver with front control only, operating in accordance with state equation (11-29), is shown in Fig. 11-22.

At first sight, the set and reset functions as stated by Eq. (11-27) lead to a strange situation when using the back gates of the capacitor elements. In Fig. 11-23, this situation is shown as an actual circuit. The output of the RC network is connected via diode D_3 to the base of transistor B. The collector of this transistor produces the b' signal, which must be connected in this case to diode D_1, so that the wanted signal pulse

$$s = \alpha' \underline{\alpha} \cdot \underline{b}'$$

will be generated at point 2 of the circuit of Fig. 11-23. This apparently results in the signal on the base of the transistor being immediately counteracted by the b' signal produced by the collector of the same transistor.

The RC delay network (capacitor C, resistors R_4 and R_5, and diode D_1) feeds a positive pulse to the base of transistor B for $b' = 1$. Transistor B, which is initially conducting, is made nonconducting by that pulse. Just before transistor B became conducting, a signal level of 0 volts was connected to diode D_1, this level being immediately changed to -6 volts once the state of this transistor has been changed. This does not affect the pulse that is passed from the RC delay network to the base of transistor B, because diode D_1 in Fig. 11-23 already performs a blocking function from the beginning of the steep leading edge of that pulse onwards. The change of the b' signal from 1 to 0 cannot influence the state of point 2 of that circuit until the output voltage of the RC network has decreased to 0 volts. As shown in line 2 of Fig. 11-24, the potential of point 2 of the circuit will decrease further to -6 volts.

The set and reset functions of an MS memory element operating as a pulse halver can also be implemented by means of a single-capacitor MS

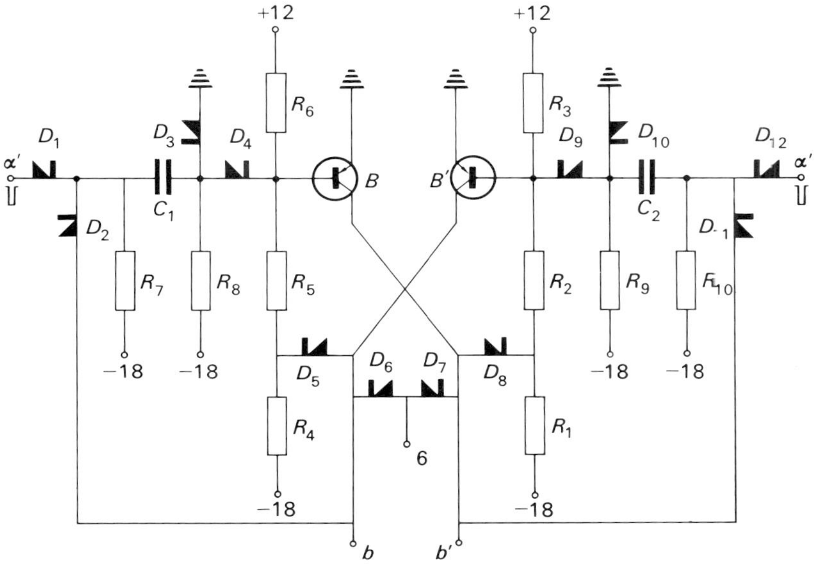

Fig. 11-22

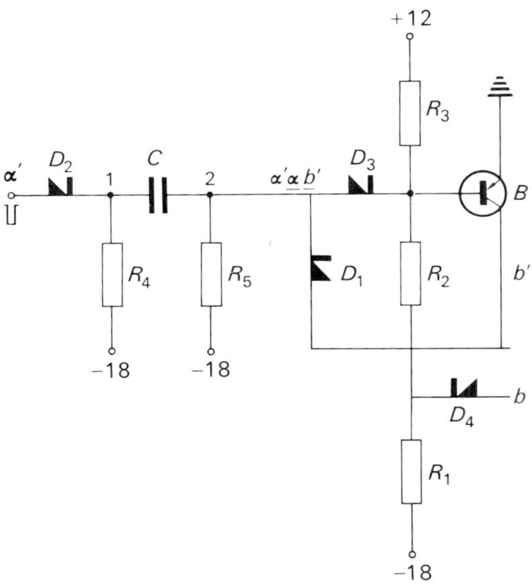

Fig. 11-23

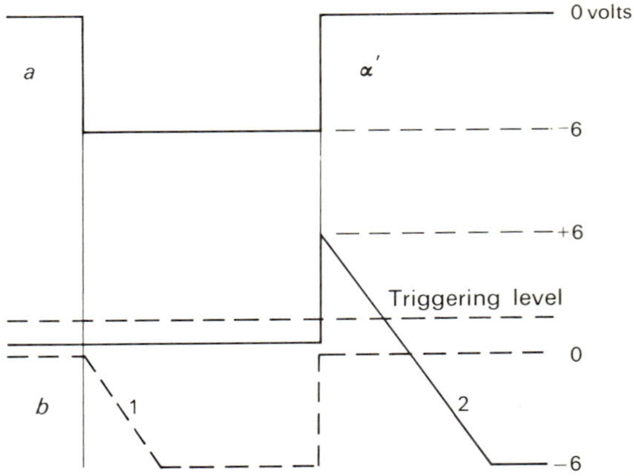

Fig. 11-24

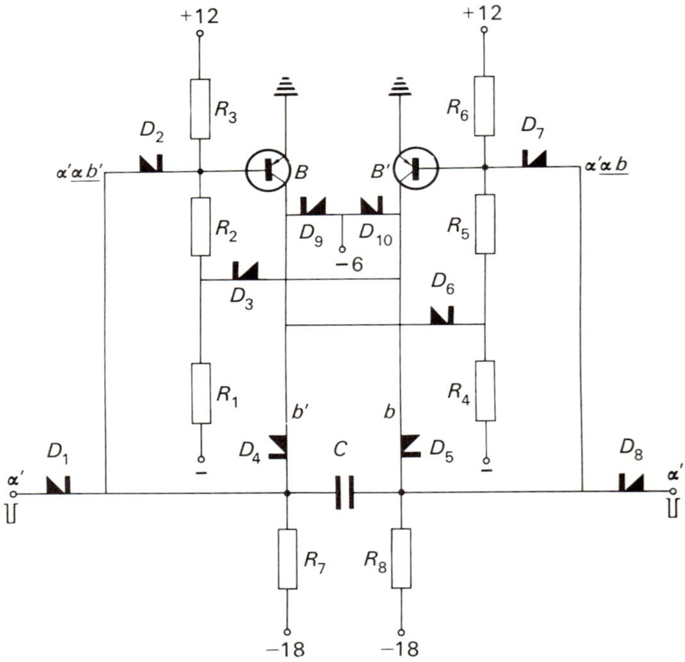

Fig. 11-25

memory element. The complete circuit of such a pulse halver is shown in Fig. 11-25. The various means available for increasing operating speeds (see Chap. 6) can also be applied to the pulse halvers shown in Figs. 11-22 and 11-25.

PROBLEMS

1. Design a pulse-halving circuit with two relays. Draw up the time chart describing the operation of this circuit. Are the memory features of the pulse halver in the circuits of the individual relays or in the circuit combination of both relays?

2. Draw up the switching formulas and the symbolic circuit for a pulse halver with two triggers.

3. Determine the switching formulas and draw up the symbolic circuit corresponding to a pulse halver operating on negative (α') pulses only.

4. Design a two-trigger all-transistor pulse halver operating on interleaved α and β clock pulses given on two different wires.

5. Design a two-trigger, all-transistor pulse halver operating on positive (α) pulses only.

6. Show that one trigger of a two-trigger pulse halver can be replaced by an RC delay circuit:

(*a*) The B trigger by an integrating RC circuit
(*b*) The A trigger by a differentiating RC circuit

Determine for each case the state equation that describes the pulse halver.

7. Prove that the feedback which exists in a single-trigger pulse halver between the collector of one of the transistors and the back input terminal of the differentiating RC delay network does not affect the state change of the trigger.

8. Show that one memory capacitor is adequate in a symmetrical pulse-halving circuit.

9. Show that it is possible to design a pulse halver to operate on α and α' pulses by means of a single trigger and a single capacitor.

10. Design an all-transistor pulse halver with integrating RC delay circuits. Give the state equation corresponding to this circuit.

SELECTIVE BIBLIOGRAPHY

1. Barnes, R. C. M.: Relay Scale-of-two Circuits, *Electron. Eng.*, vol. 26, pp. 432–435 and pp. 493–497, 1954.
2. Corbett, B. D.: A Single Relay Scale-of-two Counting Unit, *J. Sci. Instr.*, vol. 29, p. 270, 1952.
3. Corbett, B. D.: Relay Scale-of-two Circuits, *Electron. Eng.*, vol. 28, p. 130, 1956.
4. Ericcson, L. M.: Netherlands Patent 17263, 1927.
5. Kendall, J. S.: Relay Scale-of-two Circuits, *Electron. Eng.*, vol. 27, p. 39, 1955.
6. Montgomerie, G. A.: Sketch for an Algebra of Relay and Contactor Circuits, *J. Inst. Elec. Engrs.* (*London*), vol. 95, pp. 303–312, 1948.
7. Oberman, R. M. M.: Relay Register, Netherlands Patent 56192, filed July 6, 1940.
8. Oberman, R. M. M.: Switching Algebra (Schakelalgebra), *Het P.T.T.-Bedrijf*, vol. 4, pp. 125–175, 1952 (in Dutch).
9. Oberman, R. M. M.: The Design of a Circuit (Het Ontwerpen van een Shakeling), *De Ingenieur*, vol. 66, pp. E 125–130, 1954 (in Dutch).
10. Oberman, R. M. M.: The Design of Triggers, Counters and Distributors, *Het. P.T.T.-Bedrijf*, vol. 6, pp. 58–74, 1955 and vol. 7, pp. 33–45, 1956.
11. Pressman, R.: How to Design Bistable Multivibrators, *Electronics*, vol. 26, pp. 164–168, 1953.
12. The Staff of the Computation Laboratory: Triggers, Rings and Digit Counters, chap. 8 of "Synthesis of Electronic Computing and Control Circuits," Harvard University Press, Cambridge, Mass., 1951.
13. Ware, W. H.: The Logical Principle of a New Kind of Binary Counter, *Proc. IRE*, vol. 41, pp. 1429–1437, 1953.
14. Wicke, R. G.: Relay Scale-of-two Circuits, *Electron. Eng.*, vol. 27, p. 135, 1955.

12
Transformation of Relay Circuits

12-1 PARALLEL CONTROL AND DIFFERENTIAL RELAYS

The symbolic values 0 for an open circuit and 1 for a closed circuit and the signs + for a parallel connection of contacts and · for a series connection of contacts have been used consistently throughout the preceding chapters. It has already been pointed out that the system used in this textbook is one out of four possible systems. A number of authors use different systems, which complicates somewhat the study of the literature on switching algebra.

All four possible systems lead to equivalent mathematical systems; but from a practical viewpoint, one of these systems has a slight advantage over the others. The switching algebra dealt with in the previous chapters in fact only describes the laws of logical circuits. If the winding of a relay is magnetized in series with a contact circuit, the condition of the relay can be ascertained by looking at the condition of the controlling contact circuit.

However, the condition of a relay controlled by a contact circuit connected to the relay winding, either in parallel or partly in series and partly in parallel, cannot be directly related to the laws as given above.

A differential relay can be represented indirectly by means of the laws given above, but the equations that can be derived from them are not directly usable for drawing up the various control circuits.

Figure 12.1 shows a single-winding relay Z, which is controlled in principle by a make contact a_0 and magnetized by a voltage source E in the closed condition of that contact. The voltage source E is supposed to have

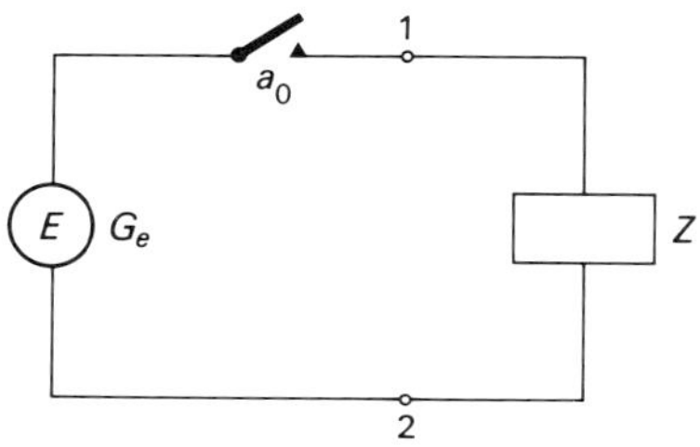

Fig. 12-1

ideal characteristics. Its conductance $G_e = \infty$ may also be represented symbolically by the cardinal number **1**. The condition of relay Z coincides with the conductance of the controlling contact circuit as measured on terminals 1 and 2, voltage source included. The total conductance of the controlling circuit of Fig. 12-1 is then:

$$G_t = \frac{G_e \times G_a}{G_e + G_a} = G_e \cdot G_a \tag{12-1}$$

As these conductances only take the value 0 or 1, the total conductance can in fact be reduced to the right-hand side of Eq. (12-1). The conductance of the ideal-voltage source is $G_e = \mathbf{1}$. Hence the condition of relay Z can be represented symbolically by

$$Z = \mathbf{1} \cdot a \tag{12-2}$$

The boldface **1** in Eq. (11-2) represents the voltage source E in the circuit of Fig. 12-1. As could be expected, this symbolic representation of the conductance of the voltage source in the switching formula does not affect the condition of relay Z.

The inverse of the control circuit of Fig. 12-1, where a relay is controlled by a break contact a_1 linked in parallel to the relay winding, is shown in Fig. 12-2. The term inverse circuit implies that the ideal-voltage source must be replaced by an ideal-current source with a conductance $G_i = \mathbf{0}$. An algebraic description of relay Z in this circuit controlled by relay A (not shown in Fig. 12-2) can be arrived at by a double inversion of Eq. (12-2), its operation remaining thereby unaffected:

$$Z = [(\mathbf{1} \cdot a)']' = (\mathbf{0} + a')' \tag{12-3}$$

The control circuit of relay Z is represented by the bracketed right-hand portion of Eq. (12-3). The boldface **0** represents the current source I to which the break contact a_1 is connected in parallel. The boldface **0** obviously has no significance with respect to the operation of the circuit itself, but it acts as a reminder for the circuit designer, where attention is thereby called to the fact that a parallel control circuit has been obtained.

It should be emphasized here that conductances **0** and **1** of the current and voltage source are part of the same symbolic system used to describe open and closed contact circuits with 0 and 1. They follow the same set of laws.

The series-parallel control of a relay can be expressed by means of Eqs. (12-2) and (12-3). Figures 12-3 to 12-6 show the various control possibilities of a relay Z by means of two contacts a and b. The diagram of Fig. 12-3 can be represented as follows, using contact symbols:

$$Z = \mathbf{1} \cdot a_0 b_0 \tag{12-4}$$

The equation describing the circuit shown in Fig. 12-4 can be derived from Eq. (12-4) by converting only the b variable in accordance with the process shown in Eq. (12-3):

$$Z = \mathbf{1} \cdot a_0 b_0 = \mathbf{1} \cdot a_0 \cdot \mathbf{1} \cdot b_0 = \mathbf{1} \cdot a_0 \cdot (\mathbf{0} + b_1)' \tag{12-5}$$

The sequence of the variables in Eq. (12-5) suggests that make contact a_0 is connected in series to the voltage source and that break contact b_1 is connected in parallel to the first part of the circuit.

It should be stressed here that neither an ideal-voltage source nor an ideal-current source can in fact be used for a series-parallel control as shown in Figs. 12-4 and 12-5. However, this does not lead to difficulties in terms of symbolic representation.

The commutative law may be applied to Eq. (12-5):

$$Z = (\mathbf{0} + b_1)' \cdot \mathbf{1} \cdot a_0 \tag{12-6}$$

Equation (12-6) describes the circuit shown in Fig. 12-5. The b_1 contact is now directly connected in parallel to a current source, and this circuit now acts as a voltage source for the series-connected make contact a_0.

Finally, the circuit shown in Fig. 12-6 can be represented by the following expression:

$$Z = \mathbf{1} \cdot a_0 b_0 = [\mathbf{0} + (a_0 b_0)']' = (\mathbf{0} + a_1 + b_1)' \tag{12-7}$$

This equation clearly describes the construction of this circuit.

If a relay has more than one winding, each winding may be considered as having its own power source. The representation of that source by means of **1** and **0**—nonsignificant as it is for the operation of the circuit—may be used to indicate the group of contacts controlling a specific winding.

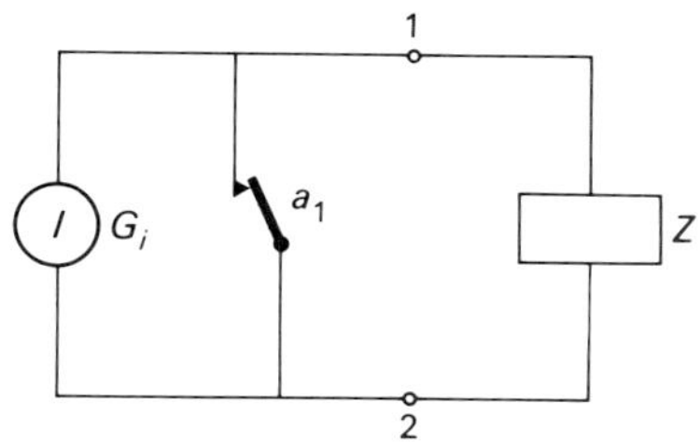

Fig. 12-2

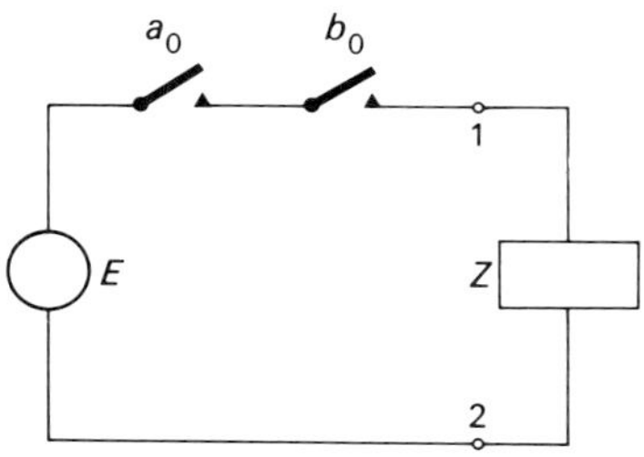

Fig. 12-3

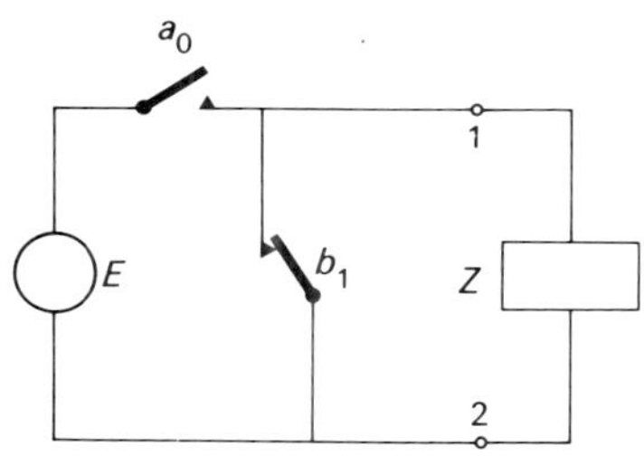

Fig. 12-4

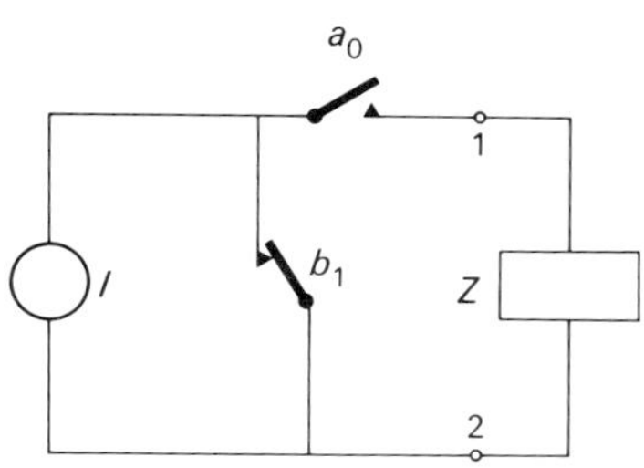

Fig. 12-5

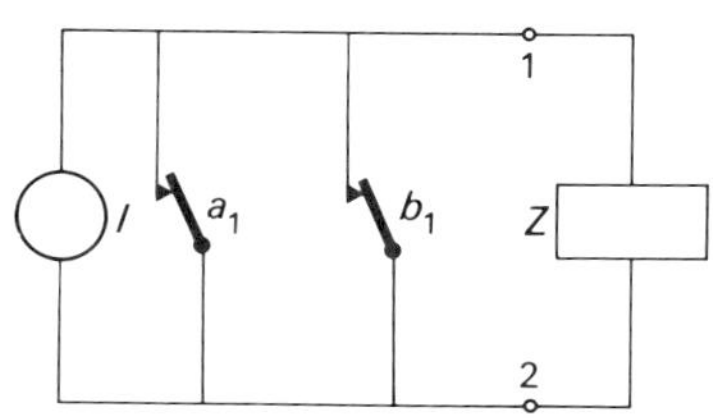

Fig. 12-6

A relay with two windings can be so connected as to have the same direction (Fig. 12-7*b*) or opposite directions of magnetization (Fig. 12-8*b*) in these two windings. The latter type of relay is generally called a differential relay.

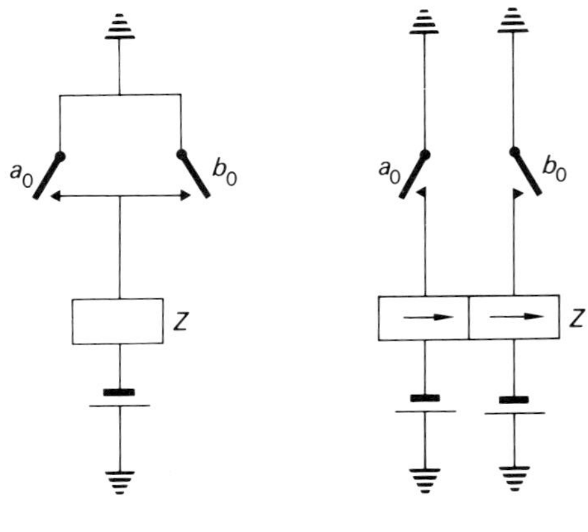

Fig. 12-7

It is assumed that the windings of the differential relay have opposite but equal ampere-turn values. Furthermore, it is assumed that saturation of the magnetic circuit of the relays is already found to occur as a result of the magnetization of only one winding. The circuits of Fig. 12-7, whose operation is identical, may be represented by the following equation:

$$Z = \mathbf{1} \cdot (a_0 + b_0) = \mathbf{1} \cdot a_0 + \mathbf{1} \cdot b_0 \tag{12-8}$$

The + sign in the first part of this equation corresponds to the ordinary logical OR connection; the + sign in the second part represents the addition of two magnetic fields in the yoke of the relay. These two operating signs will conform to the same laws provided that the relay armature is already energized by the magnetization of one winding only.

The use of the $\cdot$ sign for the representation of a parallel circuit of two contacts in the left-hand side of Eq. (12-8) would not make the form of that equation unacceptable. The + sign in the right-hand side, however, may not be replaced by a $\cdot$ sign because this part represents the physical addition of two magnetic fields in the relay core. Therefore, the + sign as a symbol for the parallel connection of two contacts has some advantage.

In order to arrive at an arithmetic system akin to ordinary algebra, it is necessary to use 0 for an open contact circuit and 1 for a closed contact circuit. The reverse would lead to very unattractive deviations from the rules of ordinary algebra.

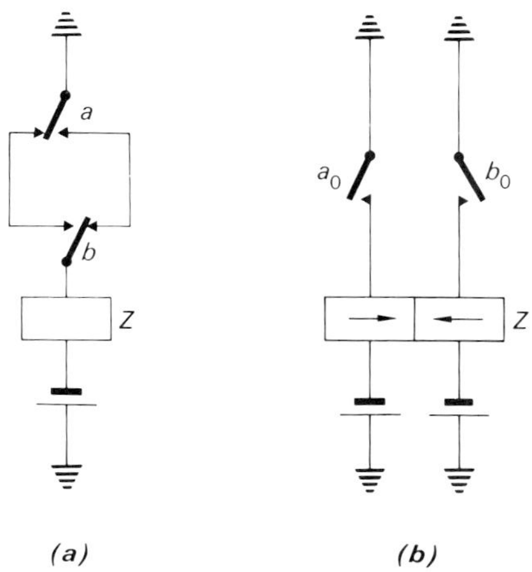

Fig. 12-8

The differential relay circuit shown in Fig. 12-8*b*, in which a subtraction of two magnetic fields is performed in the core of the relay, can be represented by the following formula:

$$Z = \mathbf{1} \cdot a_0 \oplus \mathbf{1} \cdot b_0 \tag{12-9}$$

The indication of the power source shows that the equation relates to a two-winding relay. The $\oplus$ sign indicates the addition mod 2 ($0 \oplus 0 = 0$, $0 \oplus 1 = 1$, $1 \oplus 0 = 1$, $1 \oplus 1 = 0$) of symbolic values of the magnetic fields of the two windings. The distributive law cannot be applied, and thus simplify Eq. (12-9) to $\mathbf{1} \cdot (a_0 \oplus b_0)$, because $a_0 \oplus b_0$ has no meaning in contact circuitry.

Equation (12-9) can be developed by means of law 2-4*a*:

$$\begin{aligned} Z &= \mathbf{1} \cdot a_0 \oplus \mathbf{1} \cdot b_0 = a_0(\mathbf{1} \cdot 1 \oplus \mathbf{1} \cdot b_0) + a_1(\mathbf{1} \cdot 0 \oplus \mathbf{1} \cdot b_0) \\ &= a_0 b_0(\mathbf{1} \cdot 1 \oplus \mathbf{1} \cdot 1) + a_0 b_1(\mathbf{1} \cdot 1 \oplus \mathbf{1} \cdot 0) + a_1 b_0(\mathbf{1} \cdot 0 \oplus \mathbf{1} \cdot 1) \\ &\quad + a_1 b_1(\mathbf{1} \cdot 0 \oplus \mathbf{1} \cdot 0) \\ &= a_0 b_1 \cdot (\mathbf{1} \oplus 0) + a_1 b_0 \cdot (0 \oplus \mathbf{1}) = a_0 b_1(\mathbf{1} \oplus 0) + a_1 b_0(\mathbf{1} \oplus 0) \\ &= \mathbf{1} \cdot (a_0 b_1 + a_1 b_0) \end{aligned} \tag{12-10}$$

Equation (12-10) shows the equivalence of both circuits shown in Fig. 12-8. The right-hand side of the equation corresponds to a single-winding relay controlled by a series circuit consisting of two transfer contacts.

If $b_0 = 1$, Eq. (12-10) simplifies to:

$$\mathbf{1} \cdot a_1 = \mathbf{1} \cdot a_0 \oplus \mathbf{1} \cdot 1 \qquad \text{or} \qquad \mathbf{1} \cdot a_0 = \mathbf{1} \cdot 1 \oplus \mathbf{1} \cdot a_1 \tag{12-11}$$

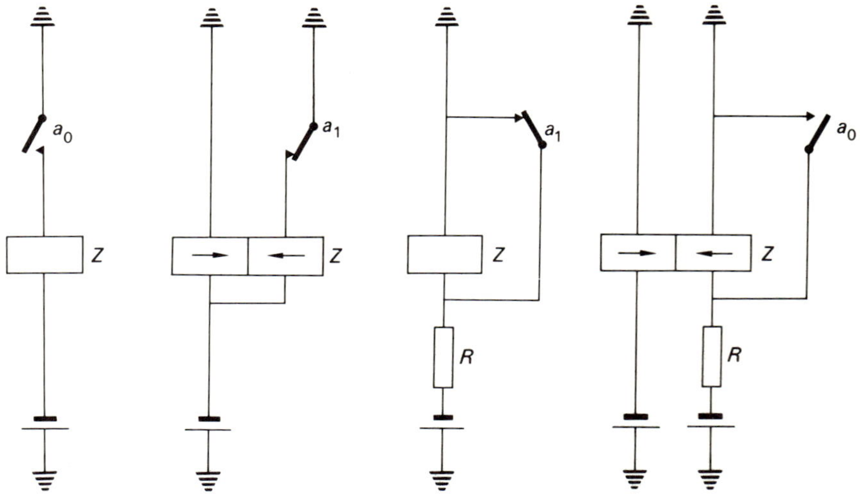

Fig. 12-9 Fig. 12-10 Fig. 12-11 Fig. 12-12

The above considerations lead to four possible modes of relay operation, which are illustrated graphically in Figs. 12-9 to 12-12. These four circuits, which are equivalent with respect to their operation, can be represented by the following equation:

$$Z = \mathbf{1} \cdot a_0 = \mathbf{1} \cdot 1 \oplus \mathbf{1} \cdot a_1 = (\mathbf{0} + a_1)' = \mathbf{1} \cdot 1 \oplus (\mathbf{0} + a_0)' \tag{12-12}$$

It should be pointed out that a resistor R has to be connected in series to the current source in case of parallel control of a relay by contact circuits. This resistor is found in Figs. 12-11 and 12-12.

By means of these transformation equations, relay circuits of one kind can easily be transformed into circuits of another kind, or even into mixed circuits. This is important for persons who are called upon to examine relay-circuit patent applications; this method affords an easy means for checking whether "new" circuits are in fact what they claim to be.

A number of transformations of relay circuits will be shown in the next section.

12-2 EXAMPLES OF CIRCUIT TRANSFORMATIONS

In addition to the transformation of a controlling circuit consisting of two series-connected contacts into the various possible arrangements of series-parallel control of a relay, the transformation of this same type of circuit into the various possible differential relay types will be examined.

$$Z = \mathbf{1}\cdot a_0 b_0 = \mathbf{1}\cdot 1 \oplus \mathbf{1}(a_0 b_0)' = \mathbf{1}\cdot 1 \oplus \mathbf{1}\cdot(a_1 + b_1) \tag{12-13}$$

$$Z = \mathbf{1}\cdot a_0 b_0 = a_0(\mathbf{1}\cdot b_0) = a_0(\mathbf{1}\cdot 1 \oplus \mathbf{1}\cdot b_1) \tag{12-14}$$

$$Z = \mathbf{1}\cdot a_0 b_0 = \qquad = b_0(\mathbf{1}\cdot 1 \oplus \mathbf{1}\cdot a_1) \tag{12-15}$$

The corresponding circuits are shown in Figs. 12-13 to 12-16.

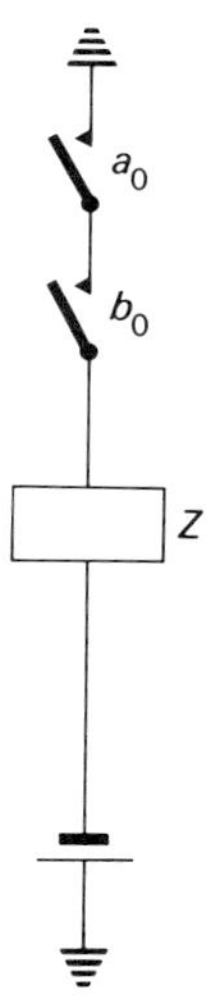

Fig. 12-13

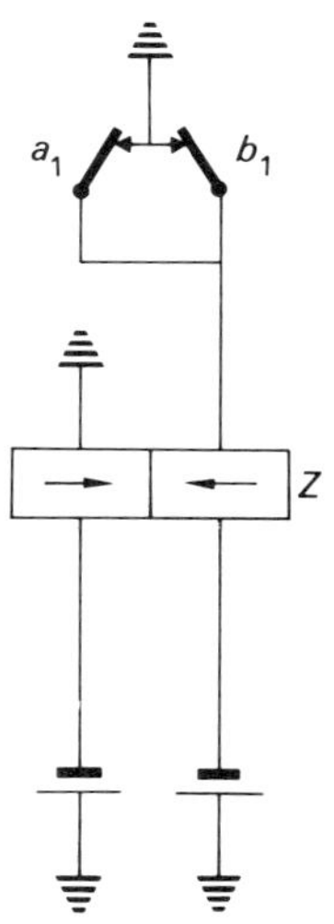

Fig. 12-14

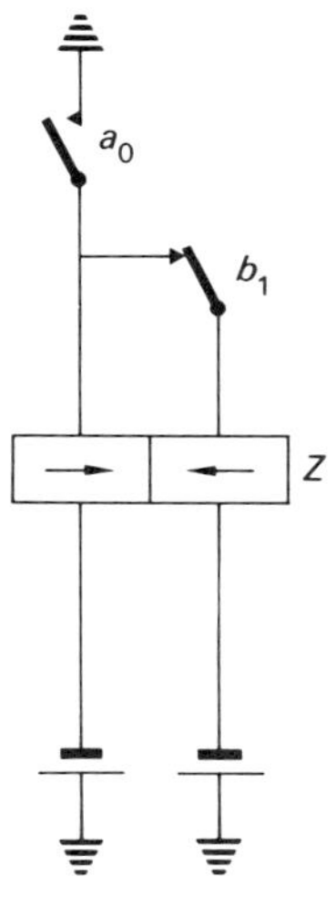

Fig. 12-15

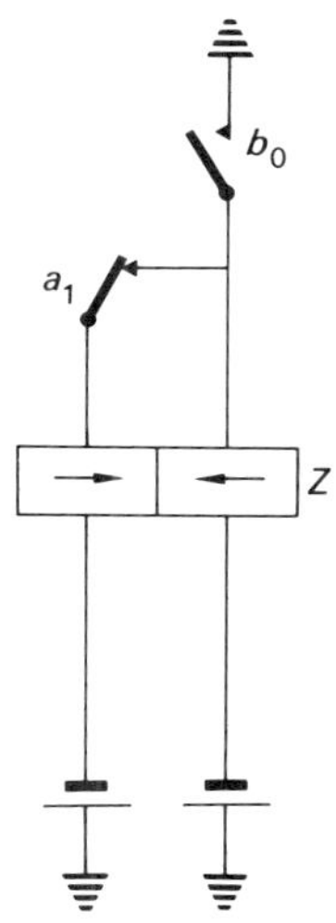

Fig. 12-16

These transformations can be applied to all types of contact circuits. More interesting applications of this transformation process can be found in pulse-halving circuits. A pulse-halving relay circuit operates in principle as indicated in the time chart of Fig. 12-17. The equations obtained for both switching elements are:

$$A = \alpha_0 b_1 + \alpha_1 a_0 \tag{12-16}$$

and

$$B = \alpha_1 a_0 + \alpha_0 b_0 \tag{12-17}$$

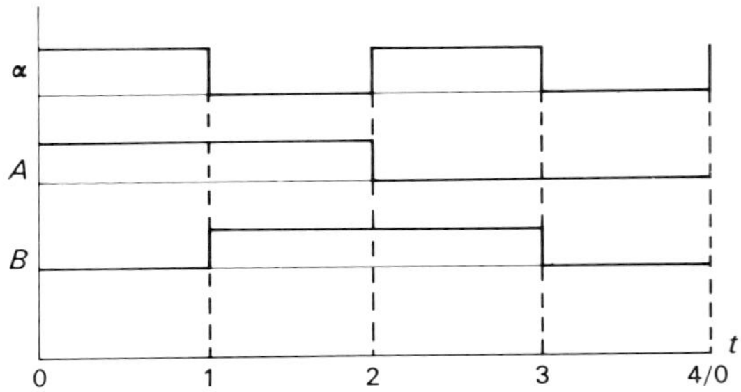

Fig. 12-17

The resulting diagram is reproduced here as Fig. 12-18. The contact circuits are connected in series to the relay windings. Relays A and B have one winding only.

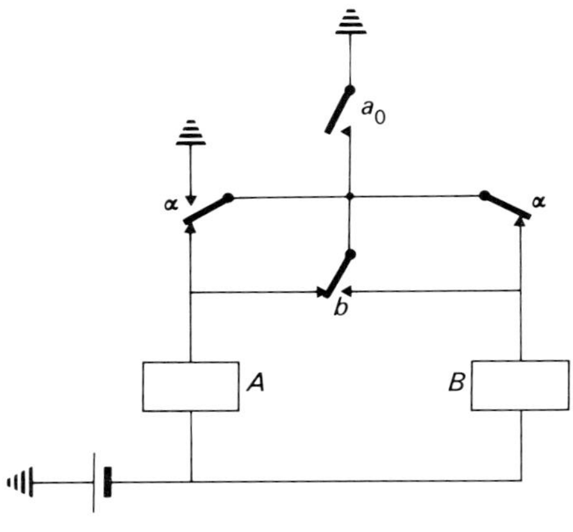

Fig. 12-18

The transformation of both relays into differential relays will now be attempted. An important decrease in the number of contacts required can be obtained in some types of circuits by introducing differential relays. The best transformation process is based on Eq. (12-10), which reads as follows in the reverse direction:

$$\mathbf{1}\cdot(y\cdot z' + y'\cdot z) = \mathbf{1}\cdot y \oplus \mathbf{1}\cdot z \tag{12-18}$$

Equations (12-16) and (12-17) can easily be made to conform by adding redundant terms.

$$\begin{aligned} A &= \mathbf{1}\cdot(\alpha_0 b_1 + \alpha_1 a_0) = \mathbf{1}\cdot(\alpha_0 + a_0)(\alpha_1 + b_1) \\ &= \mathbf{1}\cdot(\alpha_0 + a_0)(\alpha_1 + b_1) + \mathbf{1}\cdot(\alpha_1 a_1)(\alpha_0 b_0) \\ &= \mathbf{1}\cdot yz' + \mathbf{1}\cdot y'z = \mathbf{1}\cdot y \oplus \mathbf{1}\cdot z = \mathbf{1}\cdot(\alpha_0 + a_0) \oplus \mathbf{1}\cdot(\alpha_0 b_0) \end{aligned} \tag{12-19}$$

$$\begin{aligned} B &= \mathbf{1}\cdot(\alpha_1 a_0 + \alpha_0 b_0) = \mathbf{1}\cdot(\alpha_0 + a_0)(\alpha_1 + b_0) \\ &= \mathbf{1}\cdot(\alpha_0 + a_0)(\alpha_1 + b_0) + \mathbf{1}\cdot(\alpha_1 a_1)(\alpha_0 b_1) \\ &= \mathbf{1}\cdot yz' + \mathbf{1}\cdot y'z = \mathbf{1}\cdot y \oplus \mathbf{1}\cdot z = \mathbf{1}\cdot(\alpha_0 + a_0) \oplus \mathbf{1}\cdot(\alpha_0 b_1) \end{aligned} \tag{12-20}$$

The controlling α_0 contact can be freed out of the parallel connection, so that all four α variables in both equations can be combined into one contact:

$$A = \mathbf{1}\cdot(\alpha_0 + a_0) \oplus \mathbf{1}\cdot\alpha_0 b_0 = \mathbf{1}\cdot(a_1 \alpha_0 + a_0) \oplus \mathbf{1}\cdot\alpha_0 b_0 \tag{12-21}$$

$$B = \mathbf{1}\cdot(\alpha_0 + a_0) \oplus \mathbf{1}\cdot\alpha_0 b_1 = \mathbf{1}\cdot(a_1 \alpha_0 + a_0) \oplus \mathbf{1}\cdot\alpha_0 b_1 \tag{12-22}$$

The resulting circuit is shown in Fig. 12-19. The left-hand windings of both relays are controlled by the same contact circuit. With this type of parallel connection, a rather slow release of the relays is obtained. This circuit has been known for many years; but without the transformation shown here, it was hard to prove that fundamentally it was in fact the same circuit as the seemingly entirely different circuit shown in Fig. 12-18.

The basic equations (12-16) and (12-17) for the relay pulse halvers can be transformed into a circuit wherein the relays are controlled partially in parallel. It serves no purpose to draw the identical parts of the equations into the transformation process, since these parts can already be combined into one circuit for both relays. This would not be possible after the transformation into a parallel control network.

$$A = \mathbf{1}\cdot(\alpha_0 + a_0)\cdot(\alpha_1 + b_1) = \mathbf{1}\cdot(\alpha_0 + a_0)\cdot(\mathbf{0} + \alpha_0 b_0)' \tag{12-23}$$

$$B = \mathbf{1}\cdot(\alpha_0 + a_0)\cdot(\alpha_1 + b_0) = \mathbf{1}\cdot(\alpha_0 + a_0)\cdot(\mathbf{0} + \alpha_0 b_1)' \tag{12-24}$$

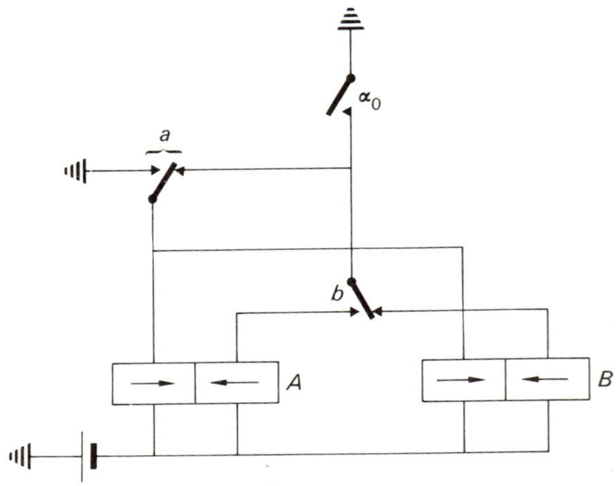

Fig. 12-19

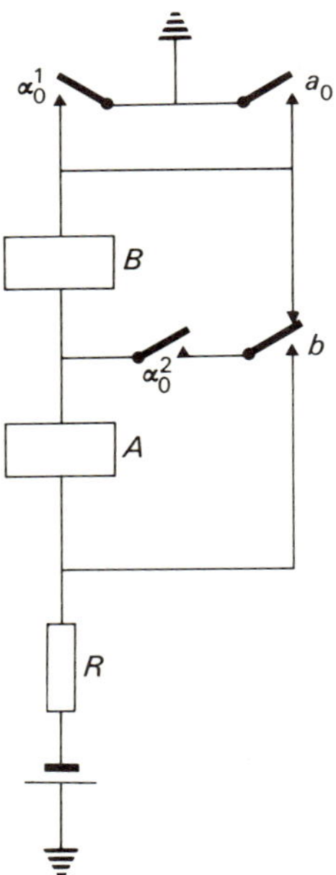

Fig. 12-20

This transformation process changes the circuits consisting of parallel-connected contacts $\alpha_1 + b_1$ and $\alpha_1 + b_0$ into serial-contact circuits which allow α_0 to be combined. The resulting circuit is shown in Fig. 12-20.

Another possible transformation would be to replace relay A by a differential relay and use a parallel control for relay B. Equation (12-21) can be used for relay A, and Eq. (12-24) is suitable for relay B. The latter equation, however, has to be rearranged with a view to the common use of the α contact. The equations for this pulse halver are:

$$A = \mathbf{1}\cdot(\alpha_0 a_1 + a_0) \oplus \mathbf{1}\cdot(\alpha_0 b_0) \tag{12-25}$$

$$B = \mathbf{1}\cdot(\alpha_0 + a_0)\cdot(\mathbf{0} + \alpha_0 b_1)' = \mathbf{1}\cdot(\alpha_0 a_1 + a_0)\cdot(\mathbf{0} + \alpha_0 b_1)' \tag{12-26}$$

Figure 12-21a and b shows the resulting circuits for relays A and B separately. The circuit for the left-hand winding of differential relay A and the serial part of the control circuit of relay B can be combined into one circuit. However, in the combination the parallel control circuit of relay B may not disturb the operation of relay A.

This combination of control circuits is possible provided that the left-hand winding of relay A is so arranged as to act as a series resistor for relay B, and provided that $a_0 + \alpha_0 a_1 + \alpha_0 b_1$ can be used as the control circuit of the left-hand winding of relay A. This does not raise any difficulties, as will be demonstrated below:

$$a_0 + \alpha_0 a_1 + \alpha_0 b_1 = a_0 + \alpha_0 + \alpha_0 b_1 = a_0 + \alpha_0 \tag{12-27}$$

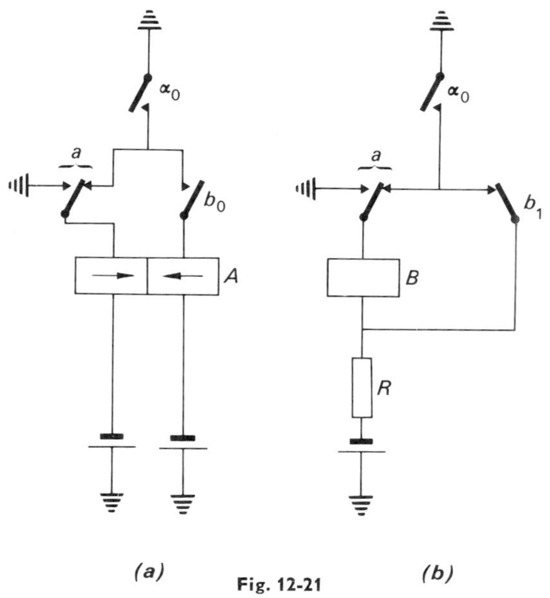

Fig. 12-21

The original form of the control circuit for the left-hand winding reappears after this manipulation [see Eq. (12-23)]. The combined pulse-halving circuit is shown in Fig. 12-22.

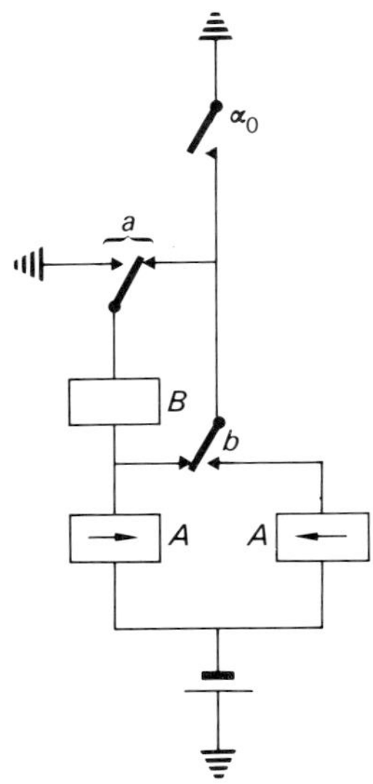

Fig. 12-22

Another interesting example of circuit transformation is afforded by the relay trigger with relays Z_1 and Z_0' shown in Fig. 9-7*a*. This circuit can be transformed into a single differential relay A operating on an α_0 and a β_0 contact for the set and reset functions, respectively.

The condition of relay A is given by

$$A = s_a + a \cdot r_a' = \alpha_0 + a_0 \beta_1 \tag{12-28}$$

This transformation process, based on the addition of nonsignificant terms, is as follows:

$$\alpha_0 + a_0 \beta_1 = \alpha_0 \cdot (a_0 \beta_1)' + a_0 \beta_1 = \alpha_0 (a_1 + \beta_0) + a_0 \beta_1$$

Since $\alpha_0 \beta_0 = 0$, this equation can be simplified to:

$$\begin{aligned} \alpha_0 a_1 + \beta_1 a_0 &= (a_0 + \alpha_0) \cdot (a_1 + \beta_1) + (a_1 \alpha_1) \cdot (a_0 \beta_0) \\ &= \mathbf{1} \cdot (a_0 + \alpha_0) \oplus \mathbf{1} \cdot a_0 \beta_0 \end{aligned} \tag{12-29}$$

The resulting circuit is shown in Fig. 12-23.

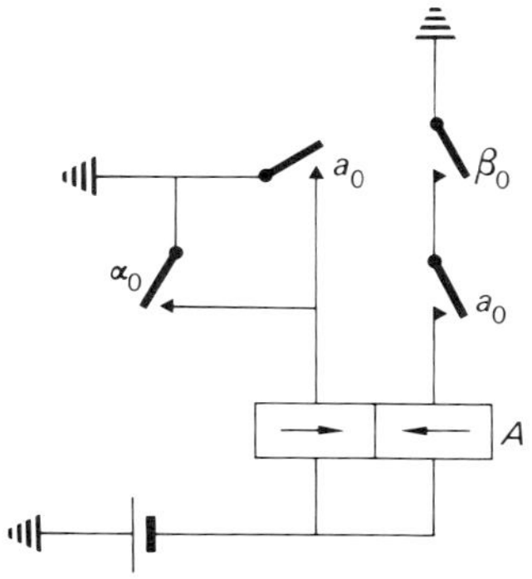

Fig. 12-23

PROBLEMS

1. Transform $Z = a_0b_0 + a_1b_1$ into a relay circuit with two make contacts only.

2. Find two transformations of $Z = a + b$ leading to a parallel and to a partly parallel-controlled relay circuit.

3. Using the time chart of Fig. 12-17, design a relay pulse halver consisting of a series-parallel-controlled A relay and a differential B relay.

4. Prove that the circuit shown in Fig. 12-24 conforms to the basic switching equations of a pulse halver.

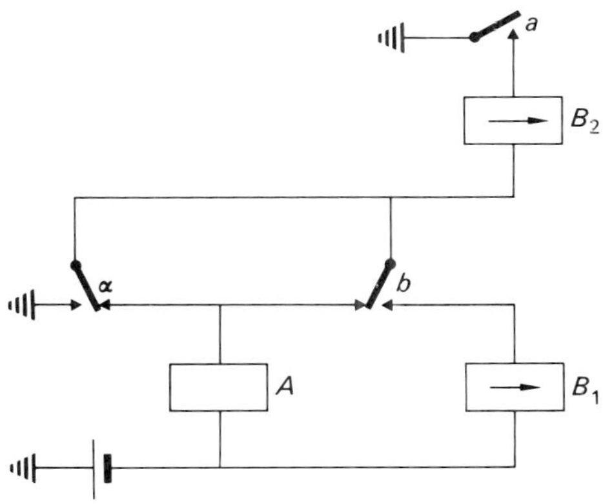

Fig. 12-24

5. Design a single-relay trigger circuit operating and releasing on an α_0 and β_0 contact, respectively, without using the differential effect.

BIBLIOGRAPHY

1. Oberman, R. M. M.: The Operating Signs in Switching Algebra (De Bewerkingstekens in de Schakelalgebra), *Het P.T.T.-Bedrijf*, vol. 6, pp. 1–10, 1954 (in Dutch).

13
Relay Counters

13-1 TYPES OF COUNTING CHAINS

There are numerous methods for counting pulse trains of up to ten pulses. The choice of a counting system depends largely on the use to be made of the result of the counting operation. It is not intended in this textbook to give a complete survey of all possible counting chains for series of up to ten pulses. We will deal with the design of the most important types of counters. The examples are chosen in such a way that circuits having other features can easily be extrapolated.

The following counting methods will be considered:

1. Counting with pairs of relays
   a. With ten pairs, up to ten (decimal code)
   b. With six pairs of relays, up to ten (double use of relays)
   c. With five pairs of relays, up to ten (two-out-of-five code)
   d. With n pairs, up to 2^n (binary code) or $2^n - 1$

2. Counting with single relays used as basic switching elements
   a. With ten plus one relays, up to ten (decimal code)
   b. With six plus one relays, up to ten (double use of relays)
   c. With five relays, up to ten (two-out-of-five code)
   d. With five relays, up to ten (binary code)
   e. With n plus 1 relays, up to 2^n (reflected binary code); see Chap. 15

With these circuits, just as with all other types of circuits, the design process must start with the definition of a set of operating requirements and with the drawing up of a time chart. This is not very difficult for practically all counting chains dealt with in this chapter, because the operating requirements will normally be dictated by the desired method of counting.

The counting chains treated in the following sections are mostly of the type in which one external change in condition (one operation of the controlling contact α, make or break) is followed by one internal change of condition (one relay being operated or released).

13-2 COUNTING WITH PAIRS OF RELAYS

WITH TEN PAIRS OF RELAYS, UP TO TEN

A time chart of a relay counting chain that is to count series of up to ten pulses by means of ten pairs of relays is shown in Fig. 13-1. One pair of relays

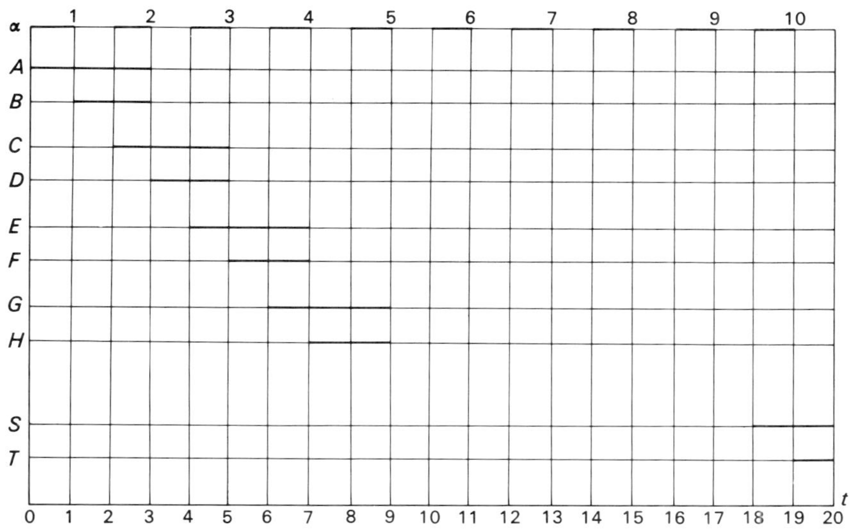

Fig. 13-1

is provided for each pulse to be received, and the nth pair of relays remains operated during the pause after the nth pulse.

Before attempting to design such a counting device, it should be pointed out that when using relay pairs only, a very small number of the possible operating combinations of the relays is used. The simplification of the switching formulas had therefore better be undertaken by some form of logical reasoning rather than by attempting to simplify the switching formulas by means of a very extensive set of formulas corresponding to the don't-care conditions.

While reasoning thus, it should be borne in mind that the circuits to be built must consist of sections of the same design. These types of circuits require contact chains to control them. The relays of a section have often to be controlled via a chain of contacts of all relays preceding and/or following the section concerned.

The diagram shows the grouping of relays in pairs A/B, C/D, E/F, etc., up to S/T. The switching formulas for all relays must be derived from the general formula

$$A = s_a + ar'_a \tag{13-1}$$

The switching formulas for the set function of the first or counting relay of each pair are

$$s_a = b_1 c_1 d_1 e_1 f_1 \cdots \alpha_0 \tag{13-2}$$

$$s_c = a_0 b_0 d_1 e_1 f_1 \cdots \alpha_0 \tag{13-3}$$

$$s_e = a_1 b_1 c_0 d_0 f_1 \cdots \alpha_0 \tag{13-4}$$

Since a sectional circuit consisting of pairs of relays, which act more or less as pulse halvers, is wanted, it is not essential that the set formulas of the relays of a pair contain variables found in both relays. It is sufficient to use the variables belonging to the second auxiliary relay of a pair, since only these relays can indicate which pair of the chain has to be operated; moreover, only these relays indicate whether the operation of a relay pair is completed.

Furthermore, in counting chains where only one relay pair remains magnetized after each pulse, it is not necessary to have the switching formula state the condition of all relay pairs in front and behind the operated pair. One side is sufficient. In this example the variables of the sections with the higher numbers in the counting chain have to be chosen so as to meet the requirements shown in the time chart of Fig. 13-1. Each pair of relays must be released when the next pair is operated.

This requirement cannot be fulfilled when the alternative group of controlling variables is used. With this latter choice, the counting chain would contain as many operated relay pairs as there are pulses received.

Switching equations (13-2) to (13-4) can be simplified to:

$$s_a = b_1 d_1 f_1 \cdots \alpha_0 \tag{13-5}$$

$$s_c = b_0 d_1 f_1 \cdots \alpha_0 \tag{13-6}$$

$$s_e = \quad d_0 f_1 \cdots \alpha_0 \tag{13-7}$$

In order to obtain a common chain of transfer contacts so designed as to preset the operating circuit of the counting relay (A, C, E, etc.) of a pair following the operation of the auxiliary relay of the preceding pair, the first variable in each equation must be a make contact of the latter relay. The second variable in each of these equations isolates the chain of transfer contacts from the pair which was last operated. Equations (13-5) to (13-7) thus represent a chain of transfer contacts which is extended to the controlling contact α and which branches off for all relay pairs to be operated.

The set equations for all auxiliary relays can be made very simple, because it is obvious that when a counting relay is magnetized, its auxiliary relay has to be operated immediately thereafter.

$$s_b = \alpha_1 a_0 \tag{13-8}$$

$$s_d = \alpha_1 c_0 \tag{13-9}$$

$$s_f = \alpha_1 e_0 \tag{13-10}$$

It is possible to control the set circuits of the relays of the counting chain to be designed by means of an α transfer contact. However, a more interesting type of counting chain can be arrived at by using an α_0 make contact only.

The auxiliary relays B, D, F, etc., can be controlled by means of an α_0 make contact when advantage is taken of the transformation methods treated in Chap. 12. In case of parallel control, equations (13-8) to (13-10) are changed as follows:

$$s_b = a_0 \cdot (\mathbf{0} + \alpha_0)' \tag{13-11}$$

$$s_d = c_0 \cdot (\mathbf{0} + \alpha_0)' \tag{13-12}$$

$$s_f = e_0 \cdot (\mathbf{0} + \alpha_0)' \tag{13-13}$$

This transformation of serial circuits into parallel circuits involves the use of an individual α_0 contact for each auxiliary relay. This is certainly not a practical solution. This individual controlling contact can be avoided, however, if the α_0 variable in each of the Eqs. (13-11) to (13-13) is replaced by the set equations of the counting relays of the relay pair concerned, so that all auxiliary relays make a common use of the single α_0 contact of Eqs. (13-5) to (13-7).

The complete set equations for the auxiliary relays of the counting chain then become

$$s_b = a_0 \cdot (\mathbf{0} + b_1 d_1 f_1 \cdots \alpha_0)' \tag{13-14}$$

$$s_d = c_0 \cdot (\mathbf{0} + b_0 d_1 f_1 \cdots \alpha_0)' \tag{13-15}$$

$$s_f = e_0 \cdot (\mathbf{0} + \quad d_0 f_1 \cdots \alpha_0)' \tag{13-16}$$

These equations indicate that the winding of an auxiliary relay is connected in series to that of the corresponding counting relay. This last relay acts as a protecting resistor for its auxiliary relay. It should be pointed out that Eqs. (13-14) to (13-16) contain a break contact on the relay to be operated. This contact is opened when the relay is operated. This is not objectionable because the auxiliary relays are operated by the opening of a contact circuit connected in parallel to the winding.

The relays of each pair can be held and released as part of a common circuit. The time chart requires that a pair of relays be released, e.g., by the operation of the auxiliary relay of the next pair. The reset equations for the relay pairs with a common holding circuit (series-connected windings) then become

$$r'_{a/b} = d_1 \tag{13-17}$$

$$r'_{c/d} = f_1 \tag{13-18}$$

$$r'_{e/f} = h_1 \tag{13-19}$$

It should be noted that the resulting diagram of Fig. 13-2 has a secondary internal operation, viz., the release of a relay pair is not controlled by the α contact but by the operation of the auxiliary relay of the next pair. This does not lead to any particular difficulties in the case considered above because the released relays do not play a part in the operation of the pairs still to be operated. Furthermore, it should be recalled that this type of counting chain had already found widespread recognition long before the first application of symbolic logic to contact circuits.

WITH SIX PAIRS OF RELAYS, UP TO TEN

The circuit shown in Fig. 13-2 uses the decimal code; this means that 10 relay pairs are required for a series of 10 pulses. In this case, only one relay pair remains in the operated state. This process can be called a one-out-of-ten code; it is by no means an economical way of registering pulse trains. Nevertheless, it can be very useful when the only information required is a decimal indication.

The registration of pulse series can be performed in a more economical way by means of a repeated use of a portion of the counting chain, as is shown

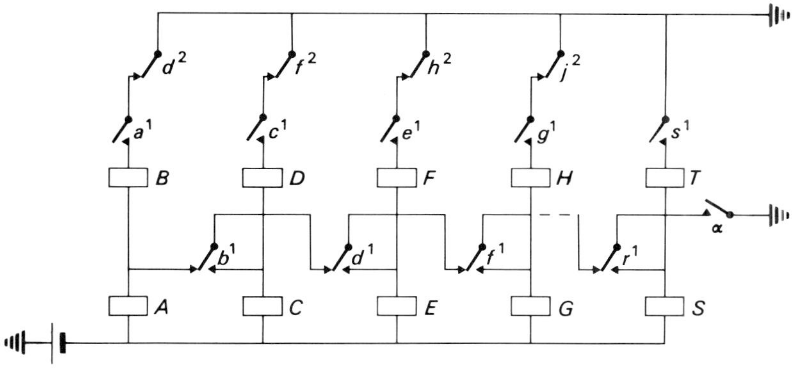

Fig. 13-2

Number of pulses received	1, 2, 3, 4, 5, 6, 7, 8, 9, 10
n° of operated relay pairs	1, 2, 3, 4, 5, 6, $\overset{6}{1'}$ $\overset{6}{2'}$ $\overset{6}{3'}$ $\overset{6}{4}$

Fig. 13-3

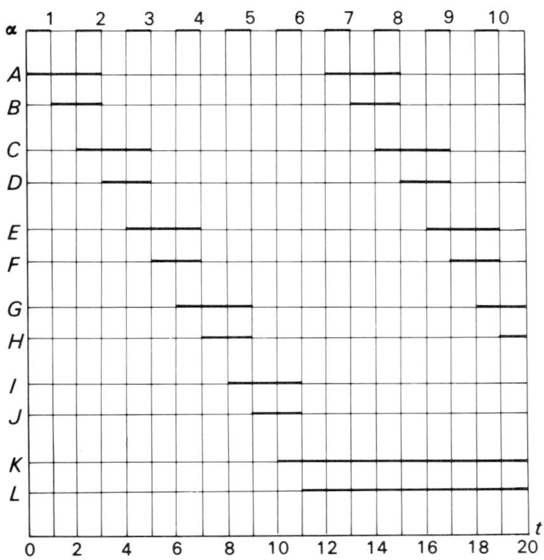

Fig. 13-4

schematically in the diagram of Fig. 13-3. In this repetition process, relay pair 6 remains in the operated state for pulse series of six pulses or more, counting relay 1, etc., being operated again on the seventh, etc., pulse. This mode of operation is illustrated in the time chart of Fig. 13-4. The set equation for relay A is

$$\begin{aligned} s_a &= \alpha_0 b_1 d_1 f_1 h_1 j_1 (l_1 + l_0) \\ &= \alpha_0 b_1 d_1 f_1 h_1 j_1 \end{aligned} \tag{13-20}$$

This equation shows that the repeated operation of the A/B relay pair, and therefore of the other pairs, will be obtained by leaving the l contact out of controlling contact chain.

A circuit conforming to the time chart of Fig. 13-4 can now easily be drawn. The resulting circuit is shown in Fig. 13-5.

It is not necessary to store the number of pulses of a series in the same code as the one used upon receiving this series on the counting chain. A code converter which converts the six-unit code of the counting chain into a four-unit code with digit values 1, 2, 4, 6 is provided in Fig. 13-5. Such a code converter can easily be designed on the basis of the indications given in Chap. 8.

WITH FIVE PAIRS OF RELAYS, UP TO TEN (TWO-OUT-OF-FIVE CODE)

When using a kind of shifting operation, the counting cycle of 10 steps can be split up into two 5-step shifting cycles as shown in Table 13-1. In each

Table 13-1

No.	*V*	*W*	*X*	*Y*	*Z*
1	1	1	0	0	0
2	0	1	1	0	0
3	0	0	1	1	0
4	0	0	0	1	1
5	1	0	0	(0)	(1)
6	(0)	1	0	0	(1)
7	1	0	1	0	0
8	0	1	0	1	0
9	0	0	1	0	1
0	1	0	0	(1)	(0)
1	(1)	1	0	0	(0)

of these cycles, the next combination can be derived from the previous one merely by shifting the 1's and 0's one element to the right. The 20 circuit states required to count 10 pulses can be obtained in a very simple way by means of 5 relay pairs, as is illustrated by the time chart of Fig. 13-6.

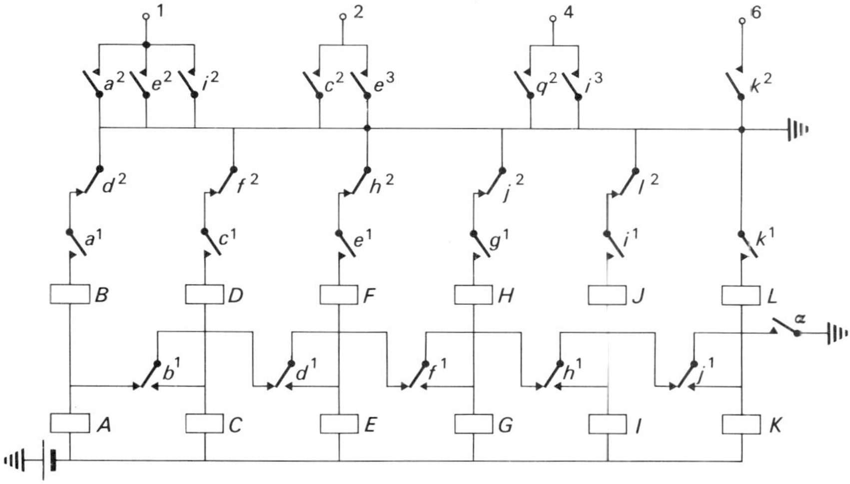

Fig. 13-5

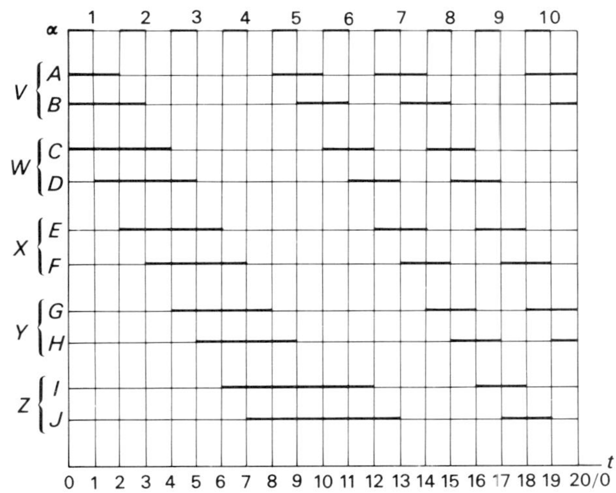

Fig. 13-6

The first code symbol or combination of one group of five can be obtained from the last symbol of the other group by replacing the first element by the inverted fifth element and the fifth element by the inverted fourth element. This is indicated in Table 13-1 by means of two pairs of enclosures.

The time chart of Fig. 13-6 translates Table 13-1 in such a way that the first relay of all relay pairs operates when required upon receiving the counting pulse α, whereas the second relay of each pair operates on the α' part of the counting pulse.

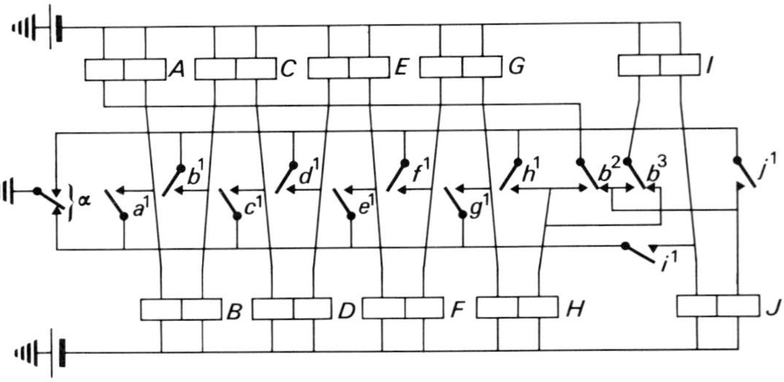

Fig. 13-7

Using the relay equation $Z = s + zr'$ in which the set signal is predominant, the following group of state equations can be derived from the time chart:

$$
\begin{aligned}
A &= \alpha_0(b_1 j_0 + b_0 h_0) + \alpha_1 a_0 & B &= \alpha_1 a_0 + \alpha_0 b_0 \\
C &= \alpha_0 b_0 + \alpha_1 c_0 & D &= \alpha_1 c_0 + \alpha_0 d_0 \\
E &= \alpha_0 d_0 + \alpha_1 e_0 & F &= \alpha_1 e_0 + \alpha_0 f_0 \\
G &= \alpha_0 f_0 + \alpha_1 g_0 & H &= \alpha_1 g_0 + \alpha_0 h_0 \\
I &= \alpha_0(b_1 h_0 + b_0 j_0) + \alpha_1 i_0 & J &= \alpha_1 i_0 + \alpha_0 j_0
\end{aligned}
\tag{13-21}
$$

The resulting circuit is shown in Fig. 13-7. This diagram clearly illustrates the very simple basic organization of the circuit, which will be interrupted by the two transfer contacts b^2 and b^3 for the change in basic code symbols to be shifted around.

The number of relays involved in the counting circuit of Fig. 13-7 amounts to 10. In addition to pulse contact α, there are only 14 contacts. This relay circuit cannot start by itself; it must be preconditioned, i.e., put in one of its possible ten states. A starting strap can be provided.

A check circuit on the two-out-of-five operation of this counter can easily be introduced so that possible faults in the internal operation can be detected.

An even more simple circuit for counting series of 10 pulses will be discussed in Chap. 16.

WITH n PAIRS OF RELAYS, UP TO 2^n OR $2^n - 1$

The counting chains dealt with so far consisted in fact of imperfect pulse-halving circuits, which in principle could operate only once per counting cycle. It is possible to reduce substantially the number of relays required by using pulse halvers as basic switching elements. If these pulse-halving circuits are so coupled that each section controls the next one with half the number of pulses as had been received by that section, a circuit with n sections will receive series of up to 2^n pulses. The time chart of such a circuit is shown in Fig. 13-8.

Let us assume that the pulse halvers to be used are controlled by a make contact α_0; the control circuits of the various sections could accordingly be described by the following expressions:

$$P_1 = \alpha_0 \tag{13-22}$$

$$P_2 = \alpha_0 b_0 \tag{13-23}$$

$$P_4 = \alpha_0 b_0 d_0 \tag{13-24}$$

$$P_8 = \alpha_0 b_0 d_0 f_0 \tag{13-25}$$

These switching formulas could be combined into a common contact chain, as is illustrated in the resulting circuit of Fig. 13-9. In this figure, the pulse halvers are represented symbolically as unit circuits. In principle, any type of pulse halver could be used for that purpose.

All pulse halvers operate under the direct control of pulse contact α_0. With some types of pulse-halving circuits, the direct coupling of the sections

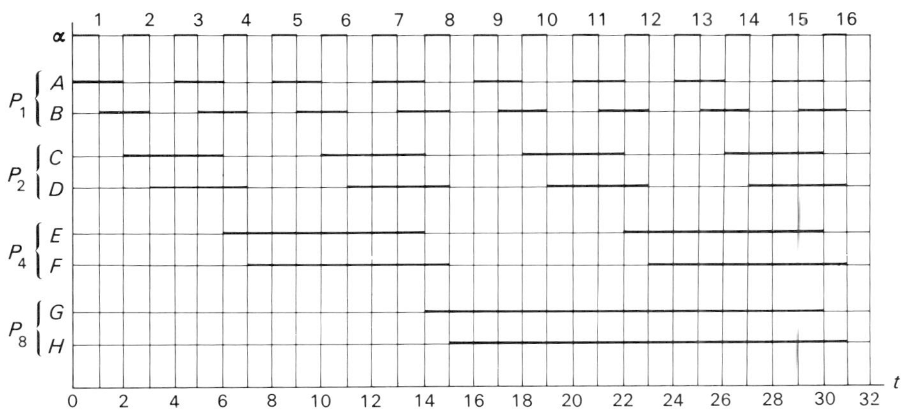

Fig. 13-8

to one controlling pulse contact can decrease the operating speed of the circuit during the releasing process because the relay windings are connected in parallel. If the counting chain steps, say, from position 7 into position 8, the relays of the first three sections must be released simultaneously. The windings of these relays remain connected in parallel during this release, and this results in an increase in release time.

This unwanted influence of the consecutive sections can be eliminated altogether by splitting up the controlling contact chain and by controlling each binary-unit counter by means of a contact of the auxiliary relay contained in the preceding section, as is shown in Fig. 13-10.

The consecutive sections of such a counter are no longer directly controlled by the α contact; but they operate with a certain delay, which increases with the number of sections involved and which is determined by the operating speed of the relays.

A very simple but rather slow-working.counting chain can be obtained by means of the type of pulse halver shown in Fig. 12-19. The diagram of such a counter is shown in Fig. 13-11. The controlling contact chain is provided with a group of k contacts. The counting chain performs the counting operation in accordance with the time chart of Fig. 13-8 when the k contacts are in the position shown. It counts backward, 0, 15, 14, 13, etc., with the k contacts in the reversed position.

The time chart of the circuit in the backward-counting condition is shown in Fig. 13-12. This backward counting can be of importance for the design of automatic-telephone registers that receive pulse series and have to retransmit these pulse trains at a later moment.

When pulse-halving circuits are used as basic switching elements for binary counters, the resulting diagram has a cyclic operation. Position 0 of the chain comes immediately after position $2^n - 1$, that is, there is no lockup in position $2^n - 1$. However, a large number of counting chains do not use full pulse-halving circuits as basic elements; instead, they contain relay pairs that act as a binary counting element only in combination with the higher-order pairs of the circuit. Such a type of counting circuit locks in the "full" condition after receiving $2^n - 1$ pulses because there is no pair number $(n + 1)$ in such chains. The time chart for such a type of counting chain corresponds to the one shown in Fig. 13-8 with the exception of the final operating period. The resulting time chart is shown in Fig. 13-13.

The operating formulas for counting relays A, C, E, and G are

$$\begin{aligned} s_a = &\ \alpha_0 b_1 c_1 d_1 e_1 f_1 g_1 h_1 + \alpha_0 b_1 c_0 d_0 e_1 f_1 g_1 h_1 \\ &+ \alpha_0 b_1 c_1 d_1 e_0 f_0 g_1 h_1 + \alpha_0 b_1 c_0 d_0 e_0 f_0 g_1 h_1 \\ &+ \alpha_0 b_1 c_1 d_1 e_1 f_1 g_0 h_0 + \alpha_0 b_1 c_0 d_0 e_1 f_1 g_0 h_0 \\ &+ \alpha_0 b_1 c_1 d_1 e_0 f_0 g_0 h_0 + \alpha_0 b_1 c_0 d_0 e_0 f_0 g_0 h_0 \\ = &\ \alpha_0 b_1 \end{aligned} \tag{13-26}$$

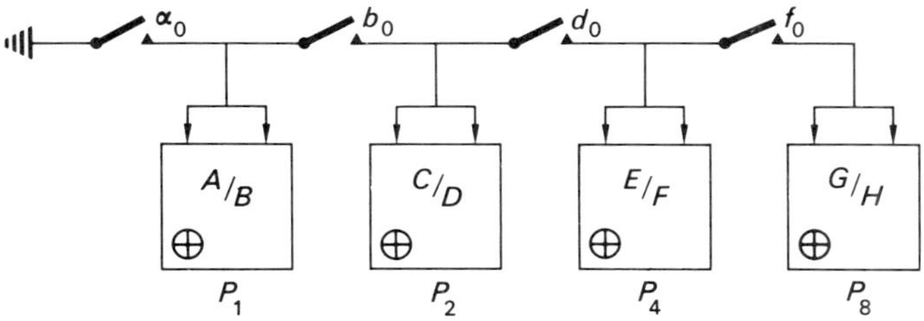

Fig. 13-9

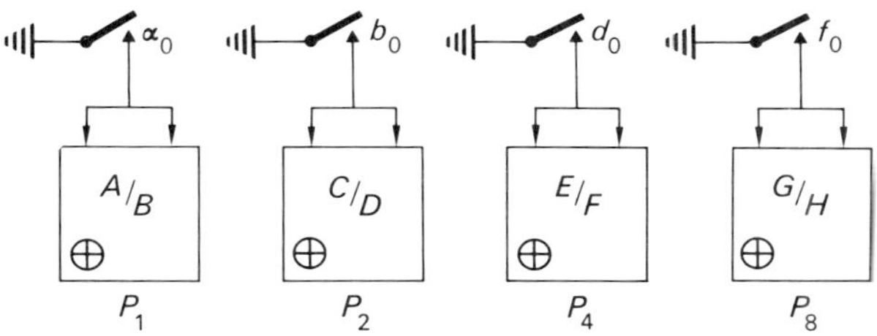

Fig. 13-10

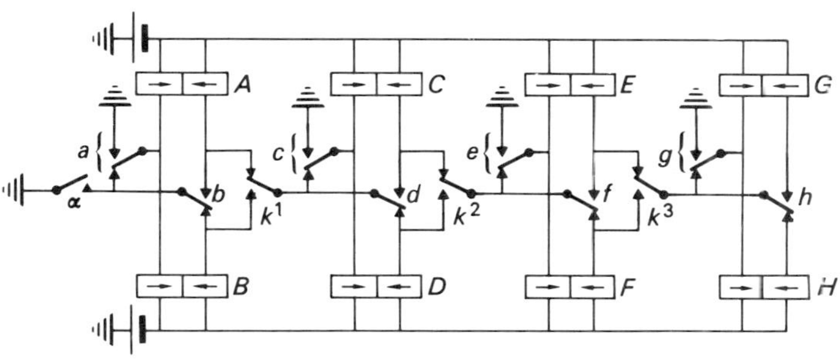

Fig. 13-11

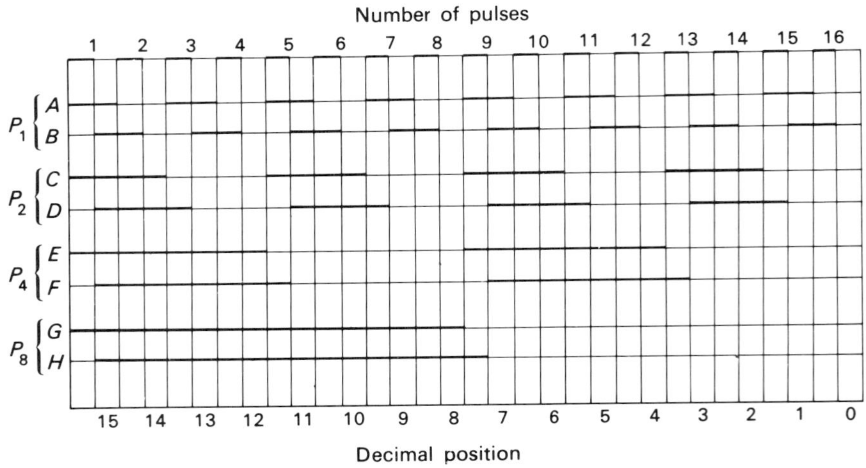

Fig. 13-12

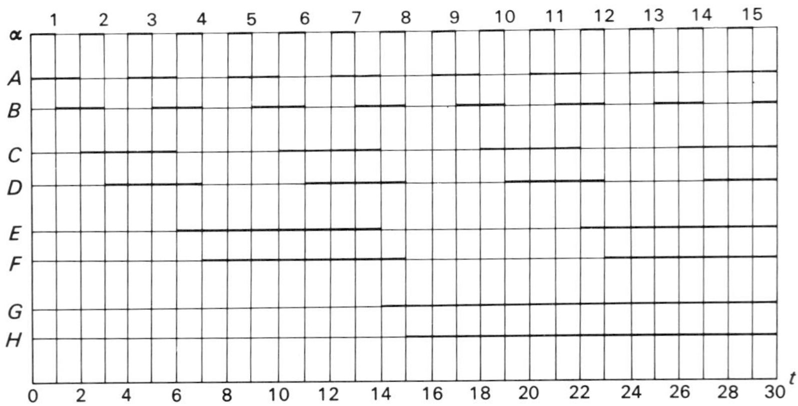

Fig. 13-13

Since relays operate in pairs, there is no need to mention both variables representing the relay pairs in the equation of the operating contact chain. This chain merely indicates the relay pair to be operated, and this function must always be performed via a contact of the auxiliary relay of the particular relay pair involved. The operating equation for relay A can then be simplified as shown in Eq. (13-26). Similarly, the set equations for relays C, E, and G then become

$$s_c = \alpha_0 b_0 d_1 \tag{13-27}$$

$$s_e = \alpha_0 b_0 d_0 f_1 \tag{13-28}$$

$$s_g = \alpha_0 b_0 d_0 f_0 h_1 \tag{13-29}$$

The operating circuits of the auxiliary relays are entirely and exclusively defined by their respective counting relays. The set equations for these relays may be written as follows:

$$s_b = \alpha_1 a_0 = a_0 \cdot (\mathbf{0} + \alpha_0)' = a_0 \cdot (\mathbf{0} + \alpha_0 b_1)' \tag{13-30}$$

$$s_d = \alpha_1 c_0 = c_0 \cdot (\mathbf{0} + \alpha_0)' = c_0 \cdot (\mathbf{0} + \alpha_0 b_0 d_1)' \tag{13-31}$$

$$s_f = \alpha_1 e_0 = e_0 \cdot (\mathbf{0} + \alpha_0)' = e_0 \cdot (\mathbf{0} + \alpha_0 b_0 d_0 f_1)' \tag{13-32}$$

$$s_h = \alpha_1 g_0 = g_0 \cdot (\mathbf{0} + \alpha_0)' = g_0 \cdot (\mathbf{0} + \alpha_0 b_0 d_0 f_0 h_1)' \tag{13-33}$$

The left-hand parts of these set equations involve the use of an α_1 contact; but what one aims at is to design a counting chain operating on an α_0 contact only. By means of a parallel control transformation, the α_1 contact can be converted into an α_0 contact. This transformed circuit, which corresponds to the middle part of Eqs. (13-30) to (13-33), can then be combined with the operating circuits already obtained for counting relays A, C, E, and G.

The release of the relays at the exact moment is a more complex problem. The holding and reset circuit of a relay Z is represented by $z_0 \cdot r_z'$ in the general switching equation of a relay. For relay A, the corresponding part of the switching equation would be

$$\begin{aligned} a_0 \cdot r_a' = a_0(&c_0 d_1 e_1 f_1 g_1 h_1 + c_1 d_0 e_0 f_1 g_1 h_1 \\ &+ c_0 d_1 e_0 f_0 g_1 h_1 + c_1 d_0 e_1 f_0 g_0 h_1 \\ &+ c_0 d_1 e_1 f_1 g_0 h_0 + c_1 d_0 e_0 f_1 g_0 h_0 \\ &+ c_0 d_1 e_0 f_0 g_0 h_0)' \end{aligned} \tag{13-34}$$

All terms containing the factor $c_0 d_1$ can be reduced to $c_0 d_1$, because no other combinations containing this factor occur in the time chart of Fig. 13-13. The same holds for the two terms containing the factor $e_0 f_1$ and the single term with the factor $g_0 h_1$. Hence, Eq. (13-34) reduces to

$$\begin{aligned} a_0 \cdot r_a' &= a_0(c_0 d_1 + e_0 f_1 + g_0 h_1)' \\ &= a_0[(c_1 + d_0) \cdot (e_1 + f_0) \cdot (g_1 + h_0)] \end{aligned} \tag{13-35}$$

The hold and release circuits for relays C and E can be arrived at in the same way:

$$c_0 \cdot r_c' = c_0[(e_1 + f_0) \cdot (g_1 + h_0)] \tag{13-36}$$

$$e_0 \cdot r_e' = e_0(g_1 + h_0) \tag{13-37}$$

These three equations together form a contact chain, because the serial character of the equations allows a combination of equal terms.

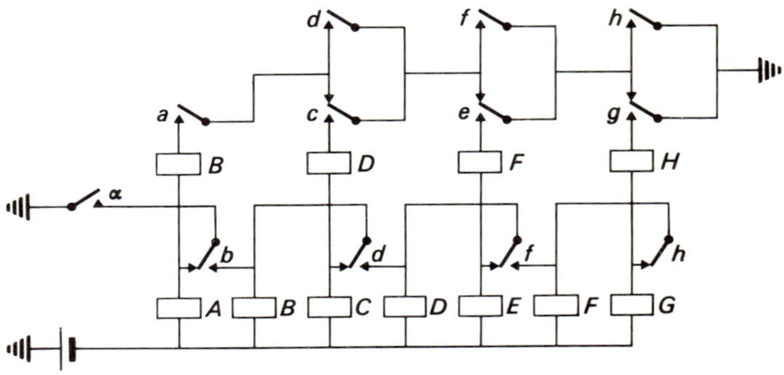

Fig. 13-14

All auxiliary relays of the various relay pairs can be held until the corresponding counting relays in the circuits (13-35) to (13-37) are released. However, the time chart prescribes working periods that are longer by one time element. This can easily be achieved by providing each auxiliary relay with a secondary winding that is directly connected to the pulse contact α_0 via a make contact of that same relay. The complete circuit is shown in Fig. 13-14.

The number of sections that can be connected in succession is limited because the holding windings of relays B, D, and F are briefly connected in parallel to the winding of relay G. If these holding windings do not have enough resistance, relay G will release its armature during the operation of relay H and thereby disturb the operation of the counting chain.

Binary relay counting chains using relay pairs or pulse halvers as switching elements can be reversed into the complementary registration code (complement of $2^n - 1$) following the reception of a series of pulses. This can be useful in the case of registers of automatic-telephone systems, where a received pulse series must subsequently be retransmitted. The retransmission can be performed after inversion of the counting chain and is then completed when all relay pairs are operated. This variant of the counting circuit of Fig. 13-14 is illustrated in Fig. 13-15.

The inversion of the circuit of Fig. 13-15 can be performed by means of one operation of relay X. Contacts x^1 up to x^4 cause the transfer contact chain consisting of contacts b^2, d^2, f^2, and h^2 to be interrupted and connect these transfer contacts to earth potential. This closes a holding circuit for the second winding of each of the operated auxiliary relays and also closes an operating circuit for the counting relays that were not operated at the end of the pulse series. Moreover, contact x^4 breaks the holding circuit of the series-connecting windings of the operated relay pairs. This demagnetizes the

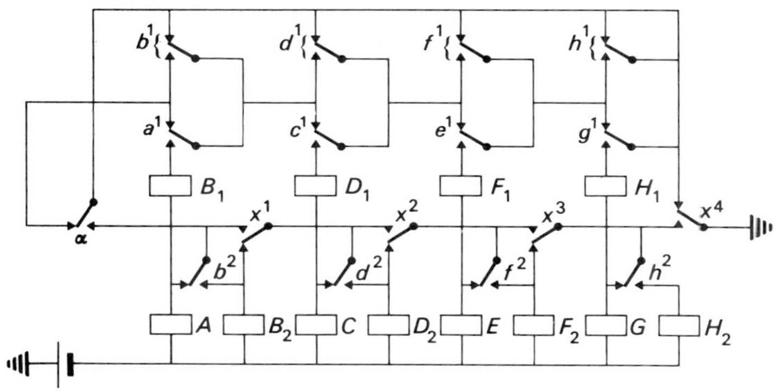

Fig. 13-15

counting relays that were operated at the end of the reception of the pulse series. The auxiliary relays that are operated then remain so until relay X is again released. Then, the auxiliary relays that are still operated release their armature, while the auxiliary relays belonging to the counting relays that were just operated are now operated in series with the windings of these counting relays. The registration code so obtained is the complement of 15 of the received number of pulses.

13-3 COUNTING WITH SINGLE RELAYS USED AS BASIC SWITCHING ELEMENTS

WITH TEN-PLUS-ONE RELAYS, UP TO TEN

It is possible to construct a time chart for a counting chain in which only one relay per pulse is needed. This diagram uses the one-out-of-ten code. The corresponding time chart is shown in Fig. 13-16. This time chart is of the type in which each external change in condition is followed by only one change in the internal condition. The set parts of the various switching equations are as follows:

$s_x =$ already operated in the rest condition of the circuit

$$s_a = \alpha_0 x_0 \tag{13-38}$$

$$s_b = \alpha_0 x_1 a_0 \tag{13-39}$$

$$s_c = \alpha_0 x_1 a_1 b_0 \tag{13-40}$$

These equations are already given in their simplified form. They can be combined into one chain of transfer contacts. The working period of each relay consists of three time elements. The set equations (13-38) to (13-40) correspond to the first of the three time elements. The second time element can be represented by the following set of formulas:

$$X_{\text{hold}} = a_1 b_1 c_1 d_1 e_1 f_1 g_1 h_1 i_1 k_1 \tag{13-41}$$

$$A_{\text{hold}} = a_0 b_1 c_1 d_1 e_1 f_1 g_1 h_1 i_1 k_1 \tag{13-42}$$

$$B_{\text{hold}} = b_0 c_1 d_1 e_1 f_1 g_1 h_1 i_1 k_1 \tag{13-43}$$

$$C_{\text{hold}} = c_0 d_1 e_1 f_1 g_1 h_1 i_1 k_1 \tag{13-44}$$

etc.

These equations also represent a chain of transfer contacts. The third time element of the working period of the relays is the actual release period, at the end of which the relay circuit has to be interrupted under the control of the α_0 contact. The switching equations for the third working period of the various relays can be written as follows:

$$X_{\text{rel}} = \alpha_0 x_0 \tag{13-45}$$

$$A_{\text{rel}} = \alpha_0 x_1 a_0 \tag{13-46}$$

$$B_{\text{rel}} = \alpha_0 x_1 a_1 b_0 \tag{13-47}$$

$$C_{\text{rel}} = \alpha_0 x_1 a_1 b_1 c_0 \tag{13-48}$$

These equations also represent a chain of transfer contacts of the same type as the one described by Eqs. (13-38) to (13-40). There is a shift of one section between these two chains of transfer contacts. The operating requirements of relay A and the release requirements of relay X coincide, as can also be seen in the time chart. These two chains of transfer contacts can easily be combined by connecting the operating winding of a relay in series to the release winding of the preceding relay. Furthermore, to an experienced observer, it will soon be apparent that the operating winding of a relay in this chain can also be used as a holding winding as required by Eqs. (13-41) to (13-44). The complete circuit is shown in Fig. 13-17.

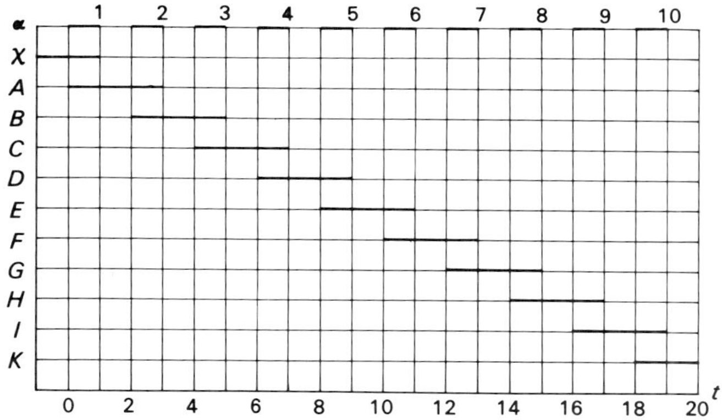

Fig. 13-16

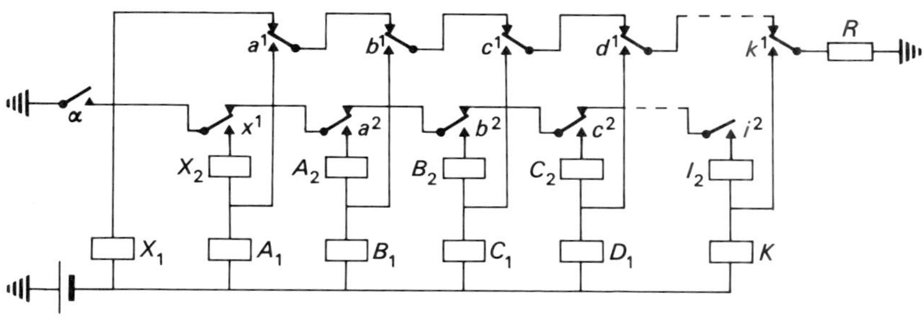

Fig. 13-17

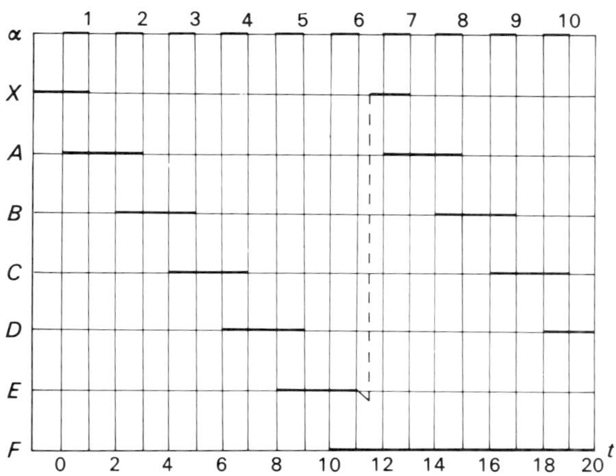

Fig. 13-18

In the design of the circuit shown in Fig. 13-17, the general switching formula for a relay $Z = s_z + z \cdot r_z'$ has not been followed, but the term $z \cdot r_z'$ has been split up into a holding period and a releasing period. This sometimes leads to an improved combination of contact circuits.

Part of the circuit shown in Fig. 13-17 may be recycled as shown in the time chart of Fig. 13-18. All relay actions happen under the direct control of an α_0 contact, with the exception of the operation of the auxiliary relay

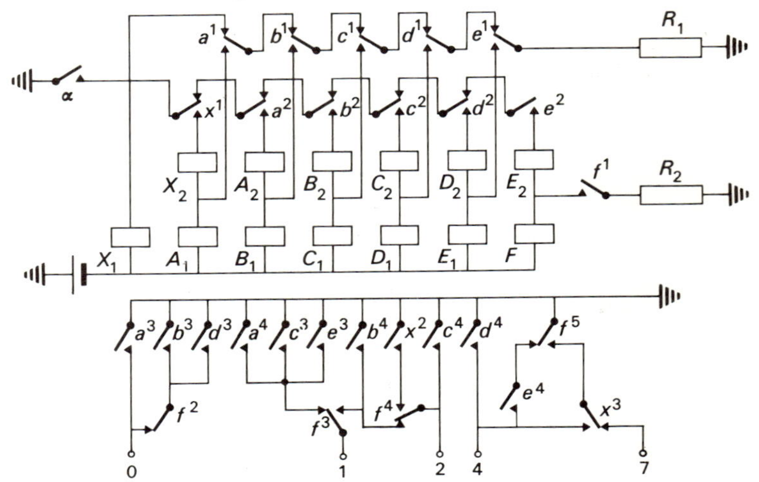

Fig. 13-19

X in the pause following the sixth pulse. This operation is performed by the release of relay E. The release time of relay E plus the operation time of relay X must be shorter than the pause between the sixth and the seventh pulse. Resistors R_1 and R_2 are used to obtain identical dimensioning features for all windings of all relays of the circuit shown in Fig. 13-19.

The pulse series received by means of a relay chain of the type shown in Fig. 13-19 can easily be stored in a more compact code. It is not necessary to resort to the known 1246 code for this purpose. The code could, for instance, be converted into a two-out-of-five code, which can find further use in certain error-detection applications.

A commonly used two-out-of-five code is the so-called 01247 code. The sum of the values of the operated elements gives the number of pulses received, except for the $7 + 4$ combination, which corresponds to 10 and not 11 pulses.

The switching equations for this code conversion can be derived from the time chart of Fig. 13-18. They are:

$$0 = a_0 + (b_0 + d_0) \cdot f_1 \tag{13-49}$$

$$1 = (a_0 + c_0 + e_0) \cdot f_1 + b_0 f_0 \tag{13-50}$$

$$2 = c_0 + b_0 f_1 + f_0 x_0 \tag{13-51}$$

$$4 = d_0 + e_0 f_1 + f_0 x_0 \tag{13-52}$$

$$7 = f_0 x_1 \tag{13-53}$$

This code converter is identical to the circuit shown in Fig. 13-19. Other types of code conversions may be applied as well, e.g., a conversion into a binary registration, this form being the most condensed registration obtainable.

WITH FIVE RELAYS, UP TO TEN IN A TWO-OUT-OF-FIVE CODE

The examples given in the preceding discussion raise the question of the feasibility of counting directly with single relays as basic switching elements in the two-out-of-five code. The 01247 code is not very suitable for this purpose, because in going from position 6 to position 7, the relays with the values 2 and 4 must be released and the relays with the values 0 and 7 must be operated. This involves too many changes in condition to meet the design requirement that one relay of the chain be operated or released following a change in condition of the controlling pulse contact.

As can be seen in the time chart of Fig. 13-20, one could conceivably devise an operating sequence of five relays in a two-out-of-five code whereby one relay at a time would be operated or released. The operation would then

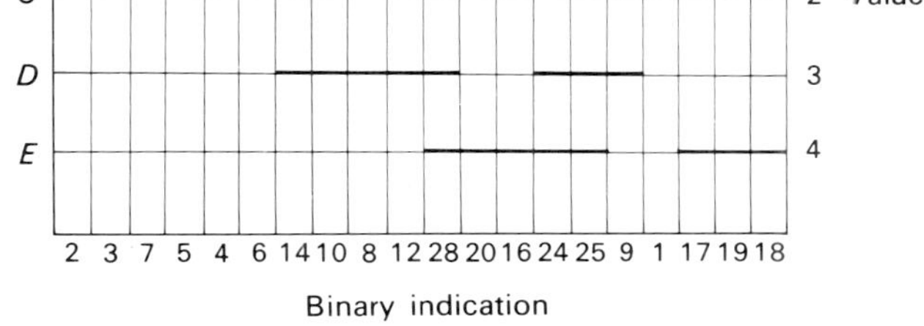

Fig. 13-20

have a cyclic character. The number of pulses received could easily be derived from the combination of the operated relays. For that purpose, the relays must be made to correspond to the following values: $A = 0$ or 5, $B = 1$ or 6, $C = 2$, $D = 3$, and $E = 4$. Relays A and B can assume two different values. The choice of the value to be awarded to A or B will depend on the number of pulses already received. Their values will be, respectively, 0 and 1 for pulse series less than 7 pulses, and 5 and 6 for series of more than 7 pulses.

This time chart has a rather peculiar form, in that all five relays have the same number of operated periods of equal duration; but these periods are shifted a constant number of time elements. This implies that the circuits of all five relays will in principle be identical but will have cyclically shifted contact indications. It can also be readily seen that the contact circuits will not be simple, because the operated periods of a relay are not of equal duration, nor will they occur symmetrically in the time chart.

This example of a counting chain is especially interesting from the point of view of simplification of the state equations by means of don't-care conditions. Let us look first at the set equations. It is sufficient to find the set function for one of the relays, e.g., the A relay, because the functions for the other relays can be obtained by cyclic permutation of the variables. The set function for relay A is derived from the time chart of Fig. 13-20:

$$s_a = \alpha_1 b_0 c_1 d_1 e_1 + \alpha_0 b_1 c_1 d_0 e_0 \tag{13-54}$$

Since the counting chain to be designed has to count series of 10 pulses at most, only 20 of the operating combinations of the 5 relays plus 1 rest condition out of the possible 32 combinations will be used, so that there are 11 unused relay combinations which can be used as don't-care conditions with a view to simplifying the set functions. The combinations used in the time chart of Fig. 13-20 have been provided with binary numbers ($a = 1$, $b = 2$, $c = 4$, $d = 8$, $e = 16$ and $a' = b' = c' = d' = e' = 0$). The unused combinations 11, 13, 15, 21, 22, 23, 26, 27, 29, 30, and 31 represent the don't-care conditions.

Figure 13-21 shows a Karnaugh map in which the minterms of Eq. (13-54) have been indicated—the α variable being left out—by means of 1 (four combinations) and the redundant terms by d (11 terms). It is therefore permissible to simplify the terms of Eq. (13-54) further by combining them with each other.

It follows from the Karnaugh map of Fig. 13-21 that the set function of relay A can be written as follows:

$$s_a = c_1(d_1 b_0 e_1 \alpha_1 + d_0 e_0 \alpha_0) \tag{13-55}$$

The holding and reset part of the state equation for relay A, which can be derived from the time chart of Fig. 13-20, is:

$$a_0 r_a' = a_0(\alpha_0 b_1 c_0 d_1 e_1 + \alpha_1 b_0 c_1 d_1 e_0)' \tag{13-56}$$

The group of unused operating combinations 11, 13, 15, 21, 22, 23, 26, 27, 29, 30, and 31 can also serve to simplify Eq. (13-56). Indeed, one could also use those combinations of the time chart of Fig. 13-20 in which relay A is not operated but for which there is no operating combination of the B, C, D, and E relays that is identical to any one of the operating combinations in which relay A is operated. These combinations correspond to the numbers 10, 12, 14, 20, 24, and 28.

	0	4	12	8	24	28	20	16
2	1	0	0	0	d	d	d	0
3	1	0	d	d	d	d	d	0
1	0	0	d	0	1	d	d	0
0	0	0	0	0	1	0	0	0

Fig. 13-21

	0	4	12	8	24	28	20	16
2	0	0	D	D	d	d	d	1
3	0	0	d	d	d	d	d	1
1	0	1	d	0	0	d	d	0
0	0	1	D	0	D	D	D	0

Fig. 13-22

Figure 13-22 shows a Karnaugh map in which the bracketed minterms of Eq. (13-56) have been indicated by 1 (four terms), the unused operating combinations by means of d (11 terms), and the 6 extra don't-care conditions by D (6 terms). Since the terms of Eq. (13-56) indicated in the map of Fig. 13-22 leave out the α variable, it is not permissible to simplify these terms by combining them with each other.

The Karnaugh map of Fig. 13-22 shows that formula (13-56) can be simplified to

$$a_0 r'_a = a_0(\alpha_0 b_1 c_0 + \alpha_1 b_0 e_0)' = a_0(\alpha_1 + b_0 + c_1)(\alpha_0 + b_1 + e_1) \qquad (13\text{-}57)$$

If one α transfer contact has to be used for controlling the complete counting chain, Eq. (13-57) is not suitable. The α contacts are still connected in parallel to other contacts of the circuit. Equation (13-57) can be transformed as follows:

$$\begin{aligned} a_0 r'_a &= a_0(\alpha_1 b_1 + \alpha_1 e_1 + \alpha_0 b_0 + b_0 e_1 + \alpha_0 c_1 + b_1 c_1 + c_1 e_1) \\ &= a_0[b_0(\alpha_0 + e_1) + b_1(\alpha_1 + c_1)] \\ &= a_0[b_0(\alpha_0 e_0 + e_1) + b_1(\alpha_1 c_0 + c_1)] \end{aligned} \qquad (13\text{-}58)$$

It should be pointed out that the terms $b_1 c_1$, $b_0 e_1$, and $c_1 e_1$ on the first line of Eq. (13-58) are redundant. From another possible combination of terms, it follows that $\alpha_1 e_1$ and $\alpha_0 c_1$ can also be considered redundant. If these last terms are eliminated, the combinations $b_1 c_1$ and $b_0 e_1$ are not superfluous. The terms $b_0 c_1$, $\alpha_1 e_1$, and $\alpha_0 c_1$ are used as redundant terms. The resulting equation (13-58) splits the circuit up into a holding and releasing circuit. The α contacts are part of the release circuit.

The analogous equations for the other relays are obtained by cyclic permutation of the variables in Eqs. (13-55) and (13-58). The complete circuit of the counting chain is shown in Fig. 13-23. The circuit of relay *B* is slightly different from the other relay circuits. This difference causes the relay circuit to start from the condition of rest with the operation of the *B* relay in accordance with the time chart. After this start, the circuit takes up its cyclic operation.

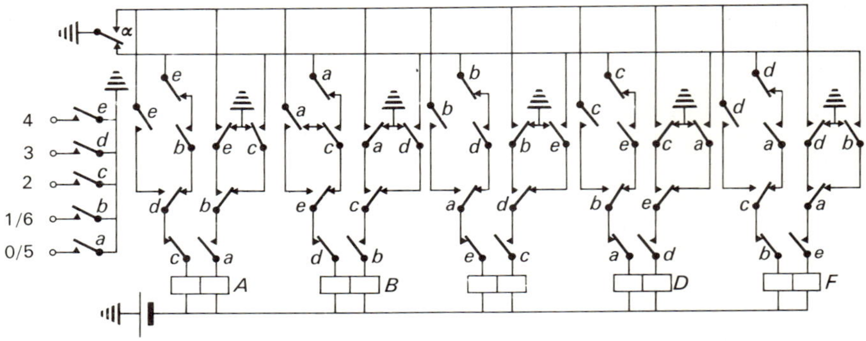

Fig. 13-23

The counting process given in the time chart of Fig. 13-20 is in certain respects highly systematic, but it leads to a rather complex circuit with five heavily loaded relays. It contains 65 contacts over and above the special starting strap in the circuit of relay *B*.

The number of relays involved in the counting circuit of Fig. 13-7 amounts to 10, that is, twice as many relays as were required for the solution of the two-out-of-five counting problem of Fig. 13-23. However, there are only 14 contacts (in addition to pulse contact α) in the circuit of Fig. 13-7, as opposed to 65 in the circuit of Fig. 13-23. This results in a saving of 51 contacts; so it is very probable that both from an economic and a technical point of view, the solution with 10 very small relays is preferable to the solution with 5 large relays.

WITH FIVE RELAYS, UP TO TEN IN THE BINARY CODE

It is not possible to design a counting chain to count directly in the binary 1248 code by means of which each external change in condition would be followed by one internal change in condition only. In the process of stepping from 7 to 8, at least three relays must release their armature, while one relay must be operated. All these changes must be caused by two external changes in condition, and this does not meet the requirement.

It is possible, however, to design a fairly good time chart for a binary five-relay counting chain if some additional internal operations are allowed, as shown in Fig. 13-24.

Relays X, A, and B combine to form a double pulse halver cyclically operating. Relays C and D expand the operation to 11 pulses. It is necessary if this circuit is to function correctly that relay C does not operate at moment t_6 until relay B has released its armature, while relays B and C cannot release

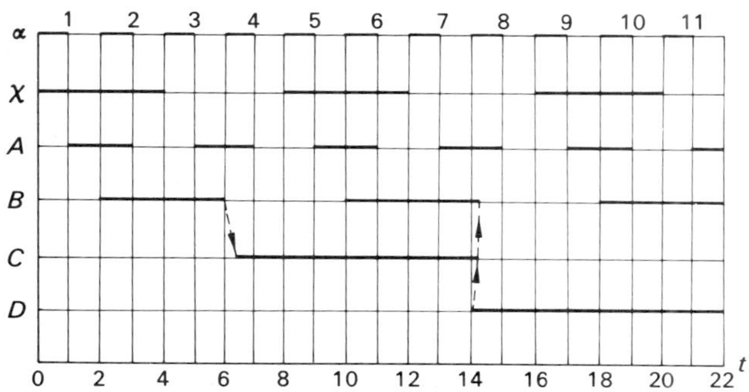

Fig. 13-24

their armature at moment t_{14} before relay D is operated. The switching equations derived from this time chart, without taking into account either the parallel connection of windings of different relays or the transfer times of contacts, are as follows:

$$X = \alpha_0 a_1 b_1 + x_0(\alpha_0 a_1 b_0)' \tag{13-59}$$

This equation may be simplified to:

$$X = \alpha_0 a_1 b_1 + x_0(\alpha_0 a_1)' \tag{13-60}$$

because the release part of the circuit need not be closed during the first element of the working periods of relay X. Equation (13-60) can be manipulated in such a way that the α contacts can be used in common by all relays of the circuit:

$$X = \alpha_0 a_1 b_1 + x_0(a_0 + a_1 \alpha_1) \tag{13-61}$$

The switching equation for relay A is:

$$A = \alpha_1(x_0 b_1 + x_1 b_0) + a_0(\alpha_1 x_0 b_0 + \alpha_1 x_1 b_1)' \tag{13-62}$$

Since the holding and reset circuit of relay A need not be closed during all periods $x_1 b_0$ and $x_0 b_1$, Eq. (13-62) then becomes

$$A = \alpha_1(x_0 b_1 + x_1 b_0) + a_0 \alpha_0 \tag{13-63}$$

The formula for relay B shows some peculiarities caused by relays C and D:

$$B = \alpha_0 x_0 a_0 + b_0(\alpha_0 x_1 a_0)' + b_0 c_0 d_1 \tag{13-64}$$

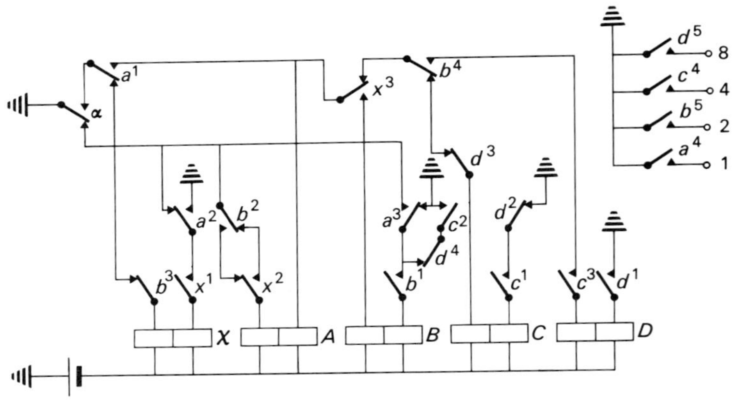

Fig. 13-25

The reset part of this equation can again be simplified in view of the fact that the circuit need not be closed during the first time element of a working period of a relay. The equation for relay B then becomes:

$$B = \alpha_0 x_0 a_0 + b_0(a_0 \alpha_1 + a_1) + b_0 c_0 d_1 \tag{13-65}$$

The term $b_0 c_0 d_1$ shows that the normal operation of relay B of the double-pulse halver is interrupted during the time relay C is operated and relay D is not.

The switching equation for relay C becomes

$$C = \alpha_0 x_1 a_0 b_1 d_1 + c_0 d_1 \tag{13-66}$$

The equation for relay D can be written as follows:

$$D = \alpha_0 x_1 a_0 b_0 c_0 + d_0 \tag{13-67}$$

The diagram corresponding to switching equations (13-61), (13-63), (13-65), (13-66), and (13-67) is shown in Fig. 13-25.

It should be pointed out in this connection that the simplification method used above (whereby the holding circuit of a relay need not be closed during the first element of a working period of a relay) implies that this relay may not release its armature during the transfer time of the pulse contact α.

Another interesting example of a counting chain with relays will be dealt with separately in Chap. 15.

PROBLEMS

1. (*a*) How many bistable memory elements at the least are required for the registration of the decimal digits 0 up to 9?

(*b*) How many bistable memory elements at the least are required for the reception of a pulse series of up to 10 pulses?

2. Explain why it is not possible to count with 5 relays up to 15, with 4 relays operating in the additive 8421 code.

3. Design the time chart and the circuit of a relay counting chain consisting of pairs of relays, n pairs of which are operated after the reception of a series of n pulses.

4. Design the time chart and the diagram of a single relay counting chain which contains no operated relays in the rest condition and which operates on an α transfer contact. Two relays only remain in the operated condition during the pause after each pulse.

5. Design the time chart and the circuit of a double-pulse halver with three relays (pulse division by four). Try to find the solution in which none of the windings of the various relays are connected in parallel during the release of the relays.

6. Redesign the pulse contact chain of the binary counter shown in Fig. 13-9 so that it will recycle after a series of 10 pulses.

7. Design the time chart and the circuit of a counting chain with four pairs of relays (weight of the pairs 7, 4, 2, 1, 0) so that it counts series of up to 10 pulses in the additive 7421 two-out-of-five code (7 & 4 = 10). The 0 output circuit which represents the parity check bit must be added as a separate circuit.

8. Design the time chart and the circuit of a counting chain with six relays. Five of these relays have to operate in the additive 74210 code. Race conditions may not coincide with stable conditions in the circuit.

9. Design the time chart and the circuit of a relay counting chain with five relays, four of which operate in the additive 4321 code. One external change in condition may be followed by only one internal change.

10. Design the time chart and the circuit of a counting chain with four relays that can count series of up to six pulses, so that the circuits of all four relays are similar except for the start of the circuit cycle.

11. Design the time chart and the circuit of a counting chain with six pairs of relays that can count series of up to 15 pulses in the two-out-of-six code.

SELECTIVE BIBLIOGRAPHY

1. Keister, W., A. E. Ritchie, and S. H. Washburn: The Design of Switching Circuits, chap. 11, D. Van Nostrand Company, Inc., New York, 1951.
2. Lubberger, F.: "Die Fernsprechanlagen mit Wählerbetrieb," Fig. 244, R. Oldenbourg Verlag, Munich, 1941.
3. Myers, Oscar: Codes and Translations, *Trans. AIEE*, pt. 1, vol. 68, pp. 592–597, 1949.
4. Oberman, R. M. M.: Relay Circuit with a Number of Counting Relays for the Registration of Pulse Series (Relaisschakeling met een aantal telrelais voor het registreren van het aantal sluitingen gemaakt door een impulscontact), Netherlands Patent 52592, Staatsbedrijf der P.T.T., filed Feb. 14, 1939.

5. Oberman, R. M. M.: Relay Register (Relaisregister), Netherlands Patent 56192, Staatsbedrijf der P.T.T., filed July 6, 1940.

6. Oberman, R. M. M.: Circuit for Telephone Switching Systems with Senders (Shakeling voor telefoonstelsels en dergelijke met registers), Netherlands Patent 58324, Staatsbedrijf der P.T.T., filed July 25, 1940.

7. Oberman, R. M. M.: Relay Circuit with a Number of Counting Relays for the Registration of Pulse Series (Relaischakeling met een aantal telrelais voor het registreren van het aantal sluitingen, gemaakt door een impulscontact), Netherlands Patent 60294, Staatsbedrijf der P.T.T., filed Dec. 9, 1943.

8. Oberman, R. M. M.: Electrical Counting Circuit with a Group of Counting Relays (Electrische telschakeling met een groep telrelais), Netherlands Patent 69573, Staatsbedrijf der P.T.T., filed July 9, 1947.

9. Oberman, R. M. M.: Relay Counting Chains, Codes, and Translations in Dial Pulse Register Circuits, *Proc. AIEE*, vol. 70, pp. 1–5, 1951.

10. Oberman, R. M. M.: Switching Algebra (Schakelalgebra), *Het P.T.T.-Bedrijf*, vol. 4, pp. 125–175, 1952 (in Dutch).

11. Oberman, R. M. M., and A. Snijders: Relay Counting Chain Using Double Coil Relays, U.S. Patent 2,636,932, filed Aug. 4, 1947.

12. Schreiber, L., and W. Hatton: The 7A2 Rotary Automatic Telephone System, *Elec. Commun.*, vol. 12, p. 131, 1933.

13. Snijders, A.: Distributor for a Signaling System (Verdeelschakeling voor Seinstelsels, in het byzonder telegraafstelsels), Netherlands Patent 75737, Staatsbedrijf der P.T.T., filed July 7, 1947.

14. Staehler, R.: An Application of Boolean Algebra to Switching Circuit Design, *Bell System Tech. J.*, vol. 31, pp. 280–305, 1952.

14

The Design of Electronic Counters and Distributors

14-1 COUNTERS AND DISTRIBUTORS

Various types of electronic counting circuits can now be designed by means of the unit circuits, or "building blocks," described in Chaps. 9 and 10. The circuits that can be used for this purpose are:

1. The MS memory element
2. The trigger
3. The inverter

These switching elements can be considered as the equivalents, respectively, of a relay pair, a single relay with make and break contacts, and a single relay with break contacts only. Counting circuits in which the above basic elements are used can be designed in a way analogous to that of the corresponding relay circuits.

The design of the counting circuits to be discussed below will be based on their logical principles, and no effort will be made to limit the number of components to the absolute minimum. For special applications, these networks can sometimes be somewhat simplified, mostly at the cost of the operating margins. Methods for speeding up the operation of the counting chains have not been considered in the various examples. In most cases, the means developed in Chap. 6 will be adequate for that purpose.

The various types of counters that have come into existence have mainly been devised with two aims in mind:

1. To count series of up to 10 pulses or very large series of pulses. Series of up to 10 pulses are, for example, used for dialing purposes in automatic-telegraph and telephone switching systems. Very large series of pulses are mostly counted by means of so-called decimal counters, which contain a number of units designed to count series of 10 pulses and then revert to 0 while sending a pulse to the next decimal section of the counter. The unit circuits for the counting of 10 pulses (decimal section) operate mostly in a coded mode in order to reduce the number of active circuit elements such as tubes or transistors.
2. To distribute incoming code symbols to corresponding memory elements, or to retransmit code symbols stored in a memory in a predetermined order. A distributor circuit establishes a transmission path between an input and one of a number of outputs, or between one of a number of inputs and one output. The function of this type of circuit corresponds to that of the mechanical distributor that can be found in various types of telegraph systems. It is therefore advisable to use the term "distributor" or "ring counter" in order to distinguish between these straightforward or linear types of counting circuits and the coded type of counting circuit mentioned under 1.

A number of circuits which are particularly suitable for use as distributors or ring counters will be discussed in the chapter on shift registers.

A further differentiation in the types of counters is possible, viz.:

1. Counters in which only the first section operates under the direct control of the counting pulse α.
2. Counters in which all sections, if required, are directly controlled by the counting pulse α. These counters are synchronized on the counting pulse, in the same way that all actions in a clock pulse circuit are performed under the direct control of the clock pulse.

The first type of counter is adequate for a number of applications where counting is the only requirement; however, where high-speed counting is used to stop a pulse source after the reception of a predetermined number of pulses, counters of type 2 are of importance. In an asynchronous counter, the effect of a counting pulse received by the first section is gradually transmitted through the various sections, so that it may very well happen that this propagation is not completed by the time the next counting pulse is received. In such a case, the counter cannot cause the driving pulse source to stop correctly.

However, when counting backward with an asynchronous counter from a predetermined state to 0, a driving pulse source can be stopped at the correct moment because on the last pulse only the first pulse halver of the counter steps back from 1 to 0, in which case there is no pulse propagation through the counter.

Another important feature of electronic counters which should be mentioned here is that most types of counting circuits have no means for checking the correct internal operation. However, it is possible to provide a counting circuit with means for such a check. These types of counters are characterized by the fact that they contain more MS memory elements than would be strictly required for the actual counting cycle.

14-2 ASYNCHRONOUS BINARY COUNTERS

A counting circuit consisting of a sequence of n binary unit counters can count series of up to 2^n pulses. In most cases, the 2^nth pulse of a series restores the counter to 0, in which case the counter has a cyclic character. The time chart, or sequence diagram, of such a counter containing three pulse halvers is given in Fig. 14-1. This time chart is drawn for pulse halvers consisting of two triggers. The triggers of each pair (A/B, C/D, E/F) operate on the leading and trailing edge, respectively, of their control pulse.

The time chart shows that each pulse halver produces a control pulse for the next one, at half the repetition rate of the received control pulse. The pulse halvers can be controlled in the following way:

$$\beta = a'b \qquad \gamma = c'd \qquad \text{etc.} \tag{14-1}$$

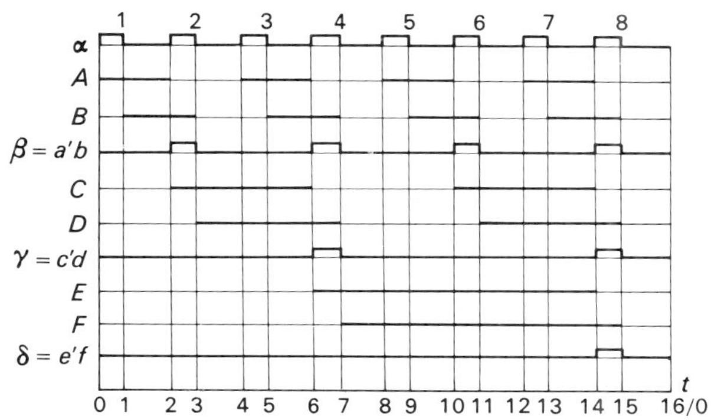

Fig. 14-1

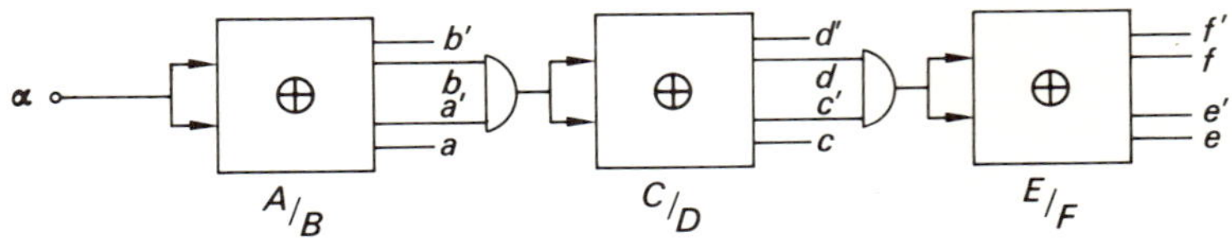

Fig. 14-2

If block symbols are used to represent pulse halvers, a binary counter conforming to the time chart of Fig. 14-1 and to Eqs. (14-1) can be described schematically by Fig. 14-2.

The units, or sections, of a binary counter having this type of interconnection show a gradual propagation of the effect of the counting pulse α through the circuit. This can lead to an unacceptable difference between the external control and the internal state of the counter.

MS memory elements can successfully be used as pulse halvers in the successive sections of a binary counter of the type discussed above. The time chart of Fig. 14-3 shows the operating requirements of such a counter with two-capacitor MS memory elements. It might be added that single-capacitor MS memory elements could also be used for this purpose; but the operation of the complete counter can be demonstrated in a more accessible way if two-capacitor MS memory elements are used.

The time chart of Fig. 14-3 shows the operation of triggers B, D, F, etc., on the rise of the capacitor voltage. The capacitor voltages of CA_1 and CA_2 are directly determined by the counting pulse α' and by the state of trigger B.

Since the operation of trigger B coincides with the trailing edge of the counting pulse α', the state of trigger B can directly be used to control capacitors CC_1 and CC_2 of the second pulse halver. The voltages of capacitors CC_1 and CC_2 are thus directly determined by the state of the B and D

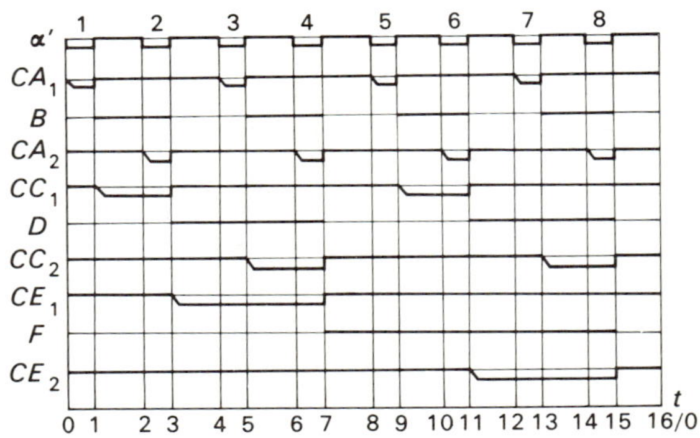

Fig. 14-3

triggers. These voltages are shown on the corresponding lines of the time chart of Fig. 14-3. The voltage of capacitor CC_1 controls the setting of trigger D whereas the voltage of capacitor CC_2 resets trigger D.

All following MS memory elements can be controlled in this same way by the trigger of the preceding MS memory element. The resulting circuit is shown in Fig. 14-4.

The circuit of Fig. 14-4 can be made out of any of the many elements of the same construction used for a number of other applications.

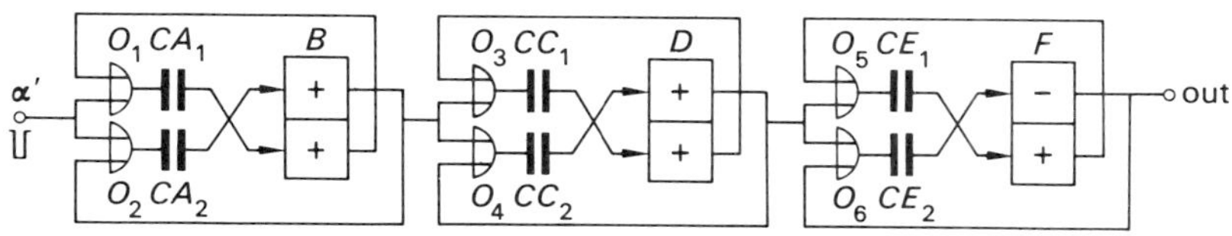

Fig. 14-4

DECIMAL-BINARY COUNTERS

A binary counter with four-unit counters can be made to count 10 pulses, then produce an output pulse to the next higher decimal order, and finally return to 0. This type of counter is a decimal-binary counter

There are many different ways to achieve this. One might provide feedback connections from the output of a pulse halver to the input of one or more preceding sections. A feedback to the input of the first pulse halver in the form of a pulse must be avoided, for it would reduce the maximum operating speed of the counter by a factor of 2 because the feedback pulse would have to be interleaved with the incoming counting-pulse series.

Moreover, the feedback pulse system disturbs the binary coding of the received pulse series in the four pulse halvers of a decimal section. It is also possible to advance the tenth pulse of a series to the second and third order of a decimal section, thus adding in one pulse the six auxiliary pulses which are required to bring the counter from position 10 to 16 or 0.

A more practical way of limiting a four-order binary counter to 10 positions can be found in the time chart of Fig. 14-5.

This time chart is based on the fact that the binary counter has to return from position $9 = 1\ 0\ 0\ 1$ to position $0 = 0\ 0\ 0\ 0$. If the triggers in the binary unit counters are called H, F, D, and B (decreasing weight of the binary digit), then nothing has to be done to control the B trigger, since this trigger will normally return to 0 on the tenth pulse of a series. Triggers D and F must remain in the 0 state during the return of the decimal counter to 0. When using MS memory elements as basic building stones, triggers D and F can be blocked in the 0 state by blocking trigger D by means of trigger H. Finally, on the tenth pulse, trigger H must be reset to 0.

The time chart of Fig. 14-5 is drawn for two-capacitor MS memory elements, but the decimal cutoff described above also holds for single-capacitor circuits.

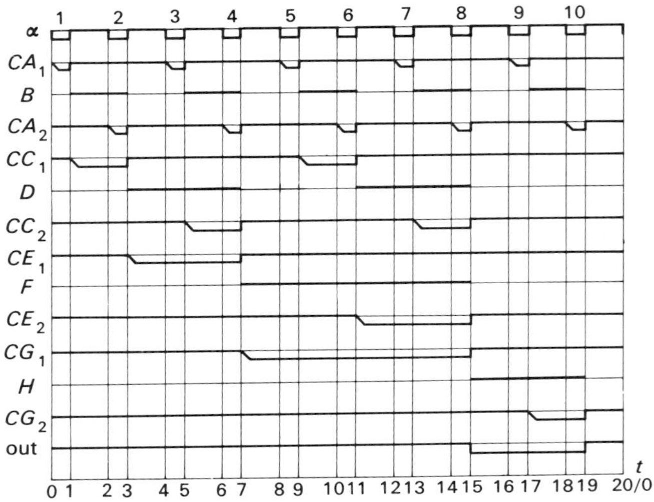

Fig. 14-5

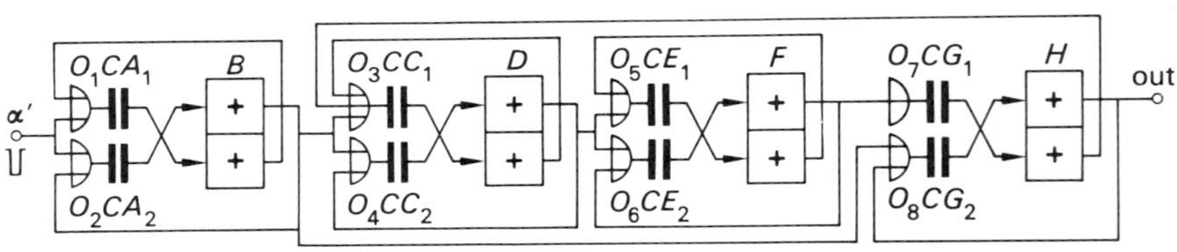

Fig. 14-6

The blocking signal for trigger D is $h = 1$. The reset signal for trigger H at the input gate of the MS memory element is $b' + h' = 0$. The state of trigger H can be used to drive the next decimal section, since this trigger is reset at the trailing edge of each tenth pulse of a series. The resulting circuit is shown in Fig. 14-6.

The circuits of Figs. 14-4 and 14-6 have the same number of diodes. The circuit of Fig. 14-4 can easily be converted into that of Fig. 14-6 by disconnecting two diodes and connecting them to other points in the circuit, and vice versa. A binary counter with many sections can be so designed as to allow switching over to decimal-binary operation. It should be pointed out that the registration of the series of pulses in a decimal section remains purely binary. The decoding of the state of a decimal-binary section can be performed by means of the standard decoding trees.

Decimal-binary counters need not necessarily operate in the ordinary 8421 binary code. Other codes of registration can easily be used, such as the 5421 code.

14-3 SYNCHRONOUS BINARY COUNTERS

The time chart of Fig. 14-3 can also be used when designing a synchronous binary counter, i.e., a counter in which all successive pulse halvers operate under the direct control of the counting pulse α. The following set and reset signals for the B, D, and F triggers of the respective MS memory elements can be derived from Fig. 14-3.

$$\begin{aligned} s_b &= \alpha b' & r_b &= \alpha b \\ s_d &= \alpha b d' & r_d &= \alpha b d \\ s_f &= \alpha b d f' & r_f &= \alpha b d f \end{aligned} \tag{14-2}$$

The equations for the various input gates of the pulse halvers can be obtained in this case from Eqs. (14-2) by negation:

$$\begin{aligned} (s_b)' &= \alpha' + b & (r_b)' &= \alpha' + b' \\ (s_d)' &= \alpha' + b' + d & (r_d)' &= \alpha' + b' + d' \\ (s_f)' &= \alpha' + b' + d' + f & (r_f)' &= \alpha' + b' + d' + f' \end{aligned} \tag{14-3}$$

The resulting circuit (Fig. 14-7) can be implemented by means of MS memory elements whose input gates have increasing numbers of input terminals. No extra diode order for the interconnection of the successive pulse halvers is required to synchronize the operation of all sections. However, the price to be paid is the impossibility of using uniform building blocks.

Equations (14-3) make it possible to design a synchronous binary counter by means of standard MS memory elements at the cost of some additional diode gates with increasing numbers of inputs. The OR gates of

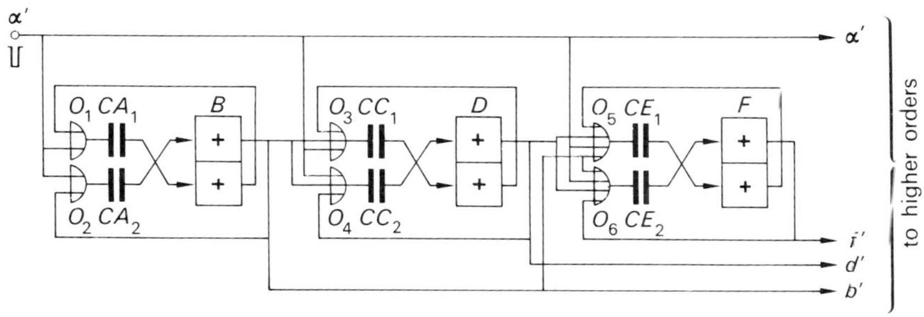

Fig. 14-7

the various pulse halvers can be directly connected to the α' counting pulse, to the internal feedback signals, and to the output of the external diode gate to be added. This could take the form of a single gate for each pulse halver of the counter because the remaining input signal is the same for both the set and the reset input gate. The signal produced by the additional diode gate acts as a blocking signal, blocking the operation of the corresponding pulse halver whenever this is required.

The blocking signals are:

$$\begin{array}{ll} \text{Input gates of } D: & b' \\ \text{Input gates of } F: & b' + d' \\ \text{Input gates of } H: & b' + d' + f' \end{array} \tag{14-4}$$

The resulting circuit with four sections is shown in Fig. 14-8. The pulse halvers are indicated as squares in this figure. These squares may contain any one of the circuits suitable for this purpose. The increasing number of inputs of the blocking gates may be considered a disadvantage in this type of circuit. This disadvantage can be compensated by using a series connection of OR gates as shown in the circuit of Fig. 14-9. There is no objection to connecting in series a limited number of gates of the same type in such a blocking circuit. If so desired, it could be replaced by a logical circuit with transistors connected in series.

The counting circuit of Fig. 14-9 can easily be converted into a synchronous decimal section consisting of four equal groups of components. This decimal counter will operate in accordance with the time chart of Fig. 14-5. The state 1 0 0 1, which is obtained after the ninth counting pulse, must be changed into 0 0 0 0 by the tenth pulse. As may follow from the time chart of Fig. 14-5, the blocking signal for MS memory element D then must be $b' + h$, and the reset signal for MS memory element H coinciding with the tenth pulse of each series must be $b' + h'$. Signal h' can be used as an output signal of this decimal section to the next higher order. The resulting circuit is shown in Fig. 14-10.

14-4 SELF-CHECKING BINARY COUNTERS

In the following paragraphs, the notation of the successive binary unit counters or pulse halvers of a binary counter will be changed from B, D, F, H, etc., to Z, Y, X, W, etc., because MS memory elements rather than two-trigger pulse halvers will be used in the examples to follow.

By adding an additional MS memory element to a binary counter, it is possible to organize the operation of the whole circuit in such a way that an even number of these elements is always in the 1 state. Faulty operation of one of the memory elements then puts the counting circuit in a state with

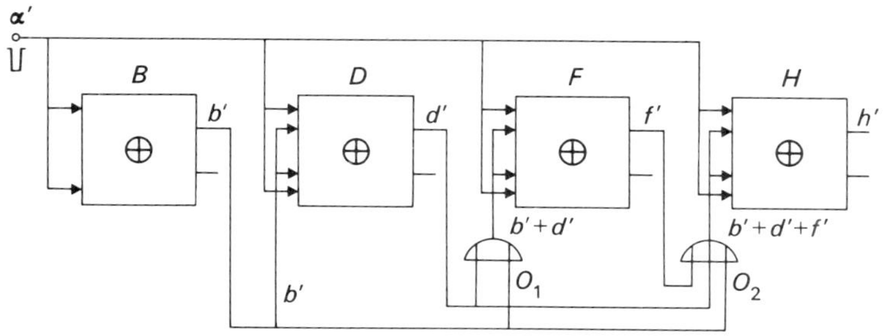

Fig. 14-8

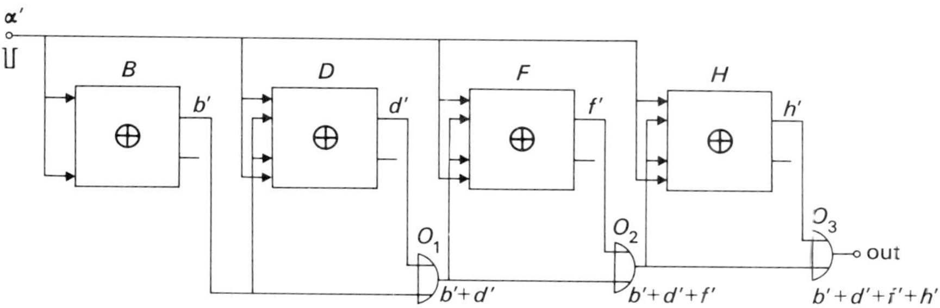

Fig. 14-9

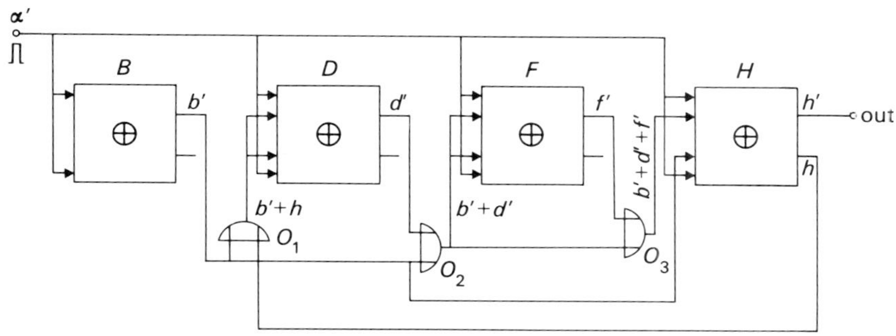

Fig. 14-10

an odd number of 1's. This can easily be detected so that the counter can be stopped after a fault has occurred.

Table 14-1 shows in columns W, X, Y, and Z the operation of an

Table 14-1

No.	W	X	Y	Z	H
0	0	0	0	0	0
1	0	0	0	1	1
2	0	0	1	0	1
3	0	0	1	1	0
4	0	1	0	0	1
5	0	1	0	1	0
6	0	1	1	0	0
7	0	1	1	1	1
8	1	0	0	0	1
9	1	0	0	1	0
10	1	0	1	0	0
11	1	0	1	1	1
12	1	1	0	0	0
13	1	1	0	1	1
14	1	1	1	0	1
15	1	1	1	1	0

ordinary binary counter. Element H is planned to operate in the manner indicated above. The design of the binary-counter part of this self-checking counter can be based on any one of the methods described above.

The operation of auxiliary MS memory element H, which is in principle connected as a pulse halver, must be blocked on the counting pulse given after the states 1, 5, 7, 9, 13, or 15 have been reached. The blocking signal of the pulse halver H thus is determined by:

$$wxyz + wxy'z + wx'y'z + w'xyz + w'xy'z + w'x'y'z = xz + y'z$$
$$= z(x + y') \quad (14\text{-}5)$$

The resulting circuit of the auxiliary pulse halver is given in Fig. 14-11.

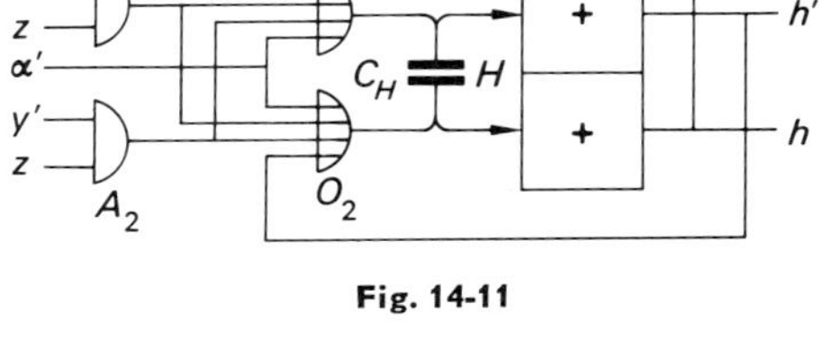

Fig. 14-11

Two diode AND gates are required in addition to an extension of the input OR gates of the MS memory element itself.

The operation of memory element H has been chosen so as to ensure that the number of elements in the 1 state is always even, so that a single fault in the operation makes this number odd.

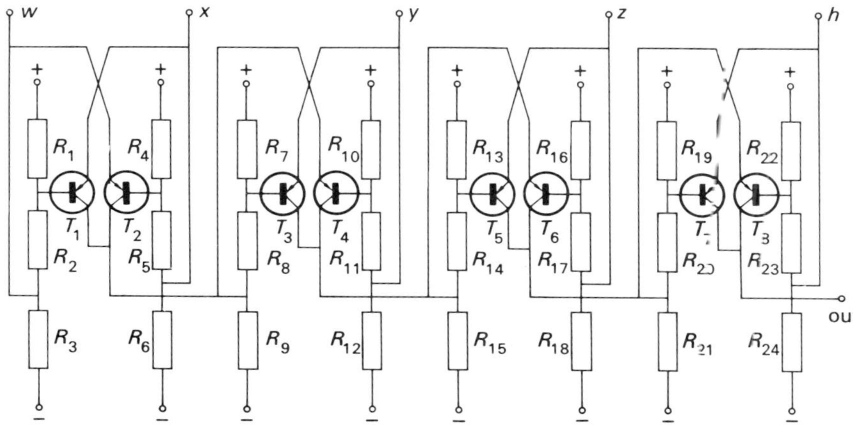

Fig. 14-12

A blocking signal that can at the same time be used as an alarm signal for the pulse source could be defined as follows:

$$S = w \oplus x \oplus y \oplus z \oplus h = 1 \qquad (14\text{-}6)$$

This signal S is 1 for all odd numbers of the variables. Equation (14-6) can be implemented by means of four mod 2 adders each consisting of two transistors, as shown in Fig. 14-12. The interconnection of these four mod 2 adders can be arranged in various ways.

Finally, it should be pointed out that the self-checking principle developed above can also be applied to the decimal variants of the four-section binary counter.

14-5 SYNCHRONOUS SELF-CHECKING MAXIMUM-VARIATION BINARY COUNTER

By means of MS memory elements connected as pulse halvers, an interesting type of counting code can be devised which, instructive though it may be, hardly has any practical value.

The problem can be stated as follows: it is requested to set up a counting code (and counter) which ensures that a maximum number of memory elements change their state on each successive counting pulse. This design requirement is the inverse of the problem of the reflected binary counting code, in which on each counting pulse only one memory element changes its state.

The problem of the maximum-variation counting code can be solved by using the reflected binary code (see Chap. 15) as a starting point. If the first eight code symbols of a four-element reflected binary code are interleaved with the complementary code symbols, the desired maximum-variation sequence of code symbols is obtained, as may follow from Table 14-2 (elements W, X, Y, Z). In this way, the maximum variation between

Table 14-2

No.	W	X	Y	Z	H
0	0	0	0	0	0
1	1	1	1	1	0
2	0	0	0	1	1
3	1	1	1	0	1
4	0	0	1	1	0
5	1	1	0	0	0
6	0	0	1	0	1
7	1	1	0	1	1
8	0	1	1	0	0
9	1	0	0	1	0
10	0	1	1	1	1
11	1	0	0	0	1
12	0	1	0	1	0
13	1	0	1	0	0
14	0	1	0	0	1
15	1	0	1	1	1

successive code symbols is obtained because in an n-element code there will always be n differences between a code symbol and its complement, whereas this symbol will differ in $n - 1$ elements with the next one. The cycle of this sequence of code symbols will have a length of 2^n steps, because it consists in fact of two standard cycles, viz., a 2^{n-1} step reflected binary cycle and an equally long cycle of the complements.

Code table 14-2 can be implemented in a rather simple way by means of pulse halvers operating directly on the counting pulse provided an auxiliary pulse halver H is added. This counter must operate as a binary-unit counter following section W.

The operation of pulse halvers X, Y, and Z can best be defined in terms of a blocking of their regular operation as pulse halvers.

The regular operation of the pulse halver Z must be blocked on the counting pulses which follow after state 1, 5, 9, or 13 has been reached. This blocking can be performed by a signal defined by

$$\begin{aligned} b_z &= wxyh' + wxy'h' + wx'y'h' + wx'yh' \\ &= wh' \end{aligned} \tag{14-7}$$

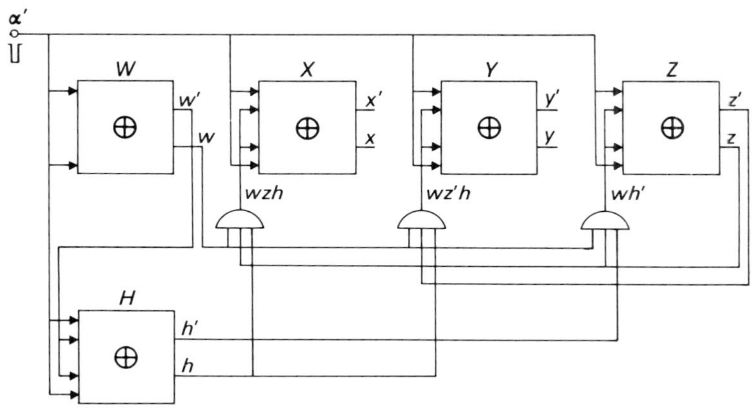

Fig. 14-13

The regular operation of Y must be blocked on counting pulse α', which follows after state 3 or 11 has been achieved. The required blocking signal is then

$$b_y = wxz'h + wx'z'h = wz'h \tag{14-8}$$

In the same way, the blocking signal for X is found to be

$$b_x = wzh \tag{14-9}$$

The resulting circuit is shown in Fig. 14-13.

Table 14-2 illustrates the operation of the maximum-variation counter, where there is always an even number of elements in the 1 state. A faulty operation of one of the elements of the circuit will cause its condition to be such that an odd number of elements is in the 1 state. This can be checked in the same manner as described in the preceding section.

If for some reason the AND blocking gates A_1 to A_3 are undesirable, they may be replaced by OR blocking gates provided that two-capacitor MS memory elements are used in combination with a connection of the blocking signals to the output sides of the capacitors.

The introduction of auxiliary pulse halver H causes the counter with an odd total number of elements to run through a cycle of states with an even number of elements. This is not the case for counters of the maximum-variation type with an even total number of pulse halvers.

An advantage of this type of counter is that throughout its operation the same number of elements $n - 1$ change their state, either from 0 to 1 or the reverse, so that the load on the pulse source is equal during the whole operating cycle of the counter.

14-6 TWO-OUT-OF-FIVE COUNTERS

An interesting group of counters is the class of so-called n-out-of-m counters. A more general survey of this category of counters will be given in Sec. 14-9 and in Chap. 16. However, one variant that falls outside the general scope of Sec. 14-9 and Chap. 16 will be illustrated here.

The two-out-of-five counting scheme given in the time chart of Fig. 13-20 and restated here in the form of Table 14-3 can be implemented in an

Table 14-3

No.	*A* 0/5	*B* 1/6	*C* 2	*D* 3	*E* 4
1	1	1	0	0	0
2	1	0	1	0	0
3	0	1	1	0	0
4	0	1	0	1	0
5	0	0	1	1	0
6	0	0	1	0	1
7	0	0	0	1	1
8	1	0	0	1	0
9	1	0	0	0	1
10	0	1	0	0	1

extremely simple way by means of MS memory elements if it is recalled that when an element is set, the preceding element is reset. It follows from the analysis of Table 14-3 that the set signal of element B is $ab'd'$ (the common parts of code symbols 2 and 9), whereas the reset signal of the A element can be represented by the same equation.

An MS memory element of the type shown in the symbolic diagram of Fig. 10-25 or 10-27 is controlled via OR gates. The same combination of variables as the one given above, but in negated form, must be connected to the input terminals of such an OR gate. The set and reset functions of the input gates of the other MS memory elements will be derived from the ones given above by cyclic permutation. The resulting group of input equations is:

$$\begin{aligned} S_a &= R_e = \alpha' + a + c + e' \\ S_b &= R_a = \alpha' + b + d + a' \\ S_c &= R_b = \alpha' + c + e + b' \\ S_d &= R_c = \alpha' + d + a + c' \\ S_e &= R_d = \alpha' + e + b + d' \end{aligned} \qquad (14\text{-}10)$$

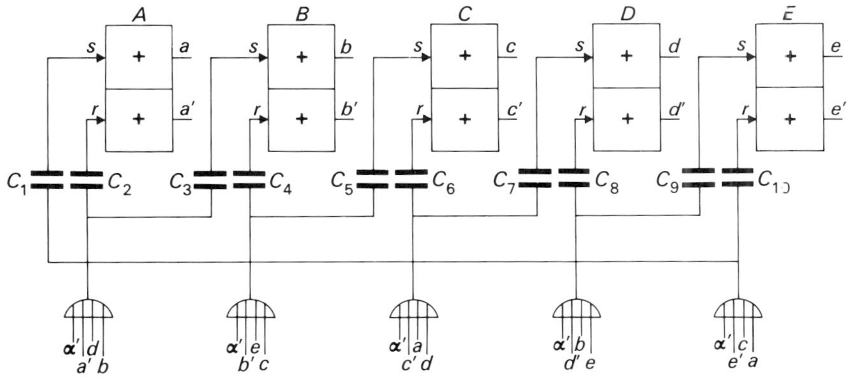

Fig. 14-14

Each of these five OR gates will control two memory capacitors, as shown in the symbolic diagram of Fig. 14-14.

The circuit shown in Fig. 14-14 can start on any one of the 10 two-out-of-five code symbols of Table 14-3 because of the cyclic nature of the counter. The circuit has some error-detecting features. If any of the 10 two-out-of-five code combinations of Table 14-3 is changed into a one-out-of-five code symbol by some faulty operation of the MS memory elements, this combination will further change into 5×0. The counter stops its operation in that state. If the state of that counter is changed into a three-out-of-five combination without a closed group of 1's, the counter will also stop. In case the state of the counter is changed into a closed group of three 1's, the ten-step operating cycle degenerates into a five-step operating cycle. In all other combinations (four-out-of-five and five-out-of-five) the counter will also stop. This useful, although somewhat incomplete, error-detecting feature is not found in the normal types of natural and reflected binary counters.

This extremely simple decimal counter operates with all its sections directly synchronized on the counting pulse α'. The code translation of the various states into a purely decimal code can be performed by means of a decoding tree with 10 two-diode leaves, as shown in Fig. 14-15.

14-7 DISTRIBUTORS WITH TRIGGERS AS BASIC ELEMENTS

Distributors using triggers as basic elements are analogous to single-relay counting chains. Part of a possible sequence diagram of the trigger operation in a 10-step distributor or counter is given in Fig. 14-16. In the pauses between clock pulses α, one trigger only is in the operated state, whereas the trigger which was operated in the first instance is reset in the rest condition during the pause between clock pulses α. The overlap in the operated state of two successive triggers is necessary to avoid unstable conditions.

It is quite feasible to design sequence diagrams in which a number of consecutive triggers are in the operated condition. Locking against false operating combinations is not provided in this type of circuit. Each of the building stones has a self-contained locking safety, but not in combination with other elements of the circuit.

The switching equations for the B trigger in the time chart of Fig. 14-16 are

$$s_b = \alpha j' a \qquad r_b = \alpha' c \tag{14-11}$$

The set and reset equations for the other triggers of this type of distributor can be formed by means of cyclic permutation of the variables in Eq. (14-11). The block diagram of the circuit that corresponds to Eq. (14-11) is given in Fig. 14-17. This circuit can be implemented by means of all types of switching elements that meet the basic requirements of the trigger and gate symbols shown.

14-8 INVERTERS AS BASIC SWITCHING ELEMENTS FOR COUNTERS AND DISTRIBUTORS

So far there has been found no practical relay equivalent to counters and distributors belonging to this class. These circuits can be developed as an extension of the multiposition trigger circuits. The two-state trigger has two locking or feedback paths between the two component inverter elements. A pulse halver can be built by means of two such trigger circuits, one being the memory for the other to determine the next state to be occupied during its change of state occurring as a result of a clock pulse. One of these triggers can be replaced by cheaper switching devices such as RC delay storage elements. The same procedure can be followed in the design of the counters and distributors to be discussed in this section.

The time chart of Fig. 14-18 shows some of the possible states of a one-out-of-ten distributor consisting of two 10-position triggers with 10

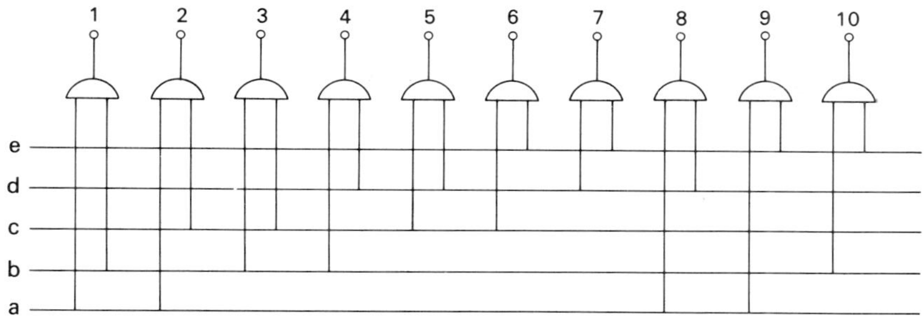

Fig. 14-15

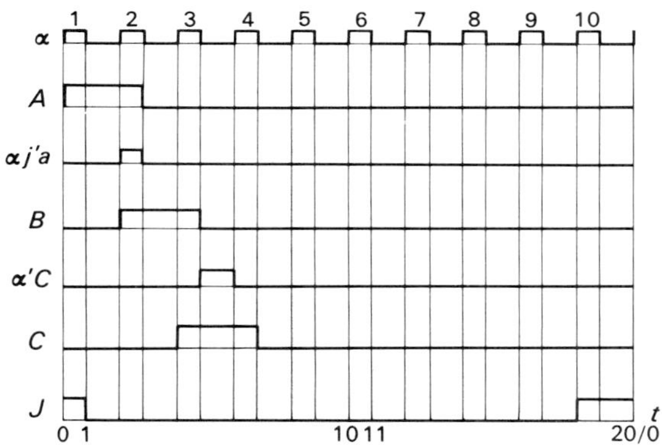

Fig. 14-16

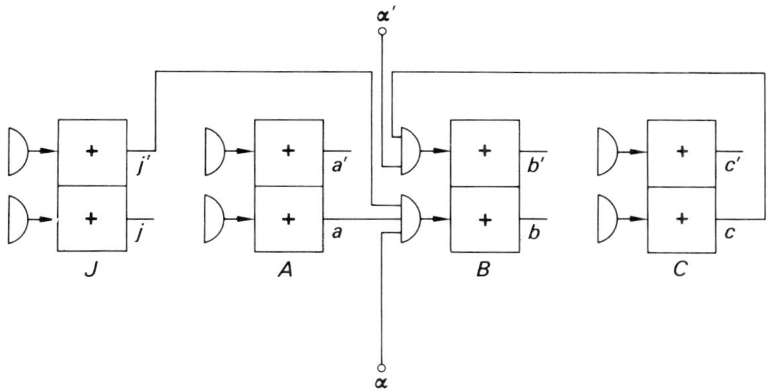

Fig. 14-17

inverters each. The distributor will operate on α' pulses. The first 10-position trigger consists of inverters A', C', E', ..., S'; the second, of inverters B, D, F, ..., T.

The state of the inverters of one multiposition trigger can be considered as the complement of the state of the inverters in the other multiposition trigger. The lines in the diagram of Fig. 14-18 show the symbolic values of the controlling signals at the inverter input. The inverters in the first multiposition trigger have been provided with negated indications. The operated and rest conditions of these inverters have been interchanged. Accordingly, one inverter of the first multiposition trigger is always operated and the other nine are in the rest condition.

The second 10-position trigger has normal operating conditions. In this trigger also, one inverter is always operated while the other nine are in the rest condition.

The multiposition trigger containing inverters A', C', E', ..., S' can be of the same type as the one shown in Fig. 9-18 and defined by state equations (9-23), (9-24), and (9-25). The multiposition trigger with inverters B, D, F, ..., T can thus be described in principle by the same equations, but in their negated form. By way of an example, the state equations for one inverter of each multiposition trigger as defined by the time chart of Fig. 14-18 are

$$C' = (\alpha' + b')\cdot(a + e + g + \cdots + s) \tag{14-12}$$

$$D = \alpha'\cdot c + b'\cdot f'\cdot h' \cdots t' \tag{14-13}$$

The state equations of the other inverters can be derived from these equations by cyclic permutation of the variables. The resulting circuit is shown in Fig. 14-19. It contains in fact nothing else than the circuit of the multiposition trigger shown in Fig. 9-18 for the A', C', E', ..., S' trigger and the same circuit but with reversed polarity of the gates for the B, D, F, ..., T trigger.

The 10-step distributor defined by the time chart of Fig. 14-18 can also be implemented by means of the multiposition trigger defined in the circuit of Fig. 9-20. On the analogy of state equation (9-31), the C' inverter will be described by the following expression

$$C' = (\alpha' + b)\cdot a\cdot e\cdot g \cdots s \tag{14-14}$$

The state equation of the D inverter can be derived by means of an analogous transformation from the state equation of the type found in Eq. (9-31):

$$D = \alpha' c + b' + f' + h' + \cdots + t' \tag{14-15}$$

The state equations of the other inverters of both triggers can be

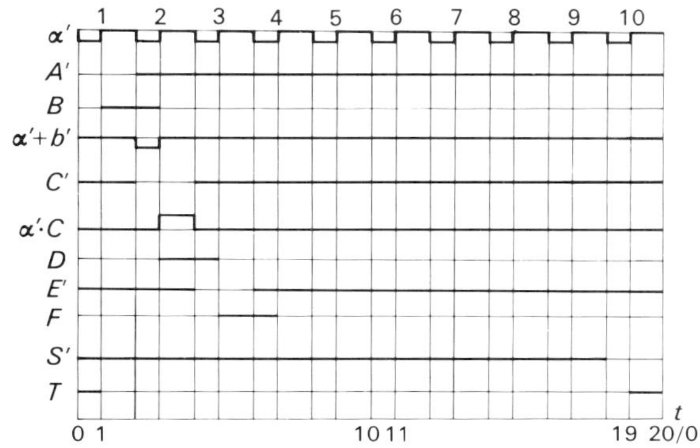

Fig. 14-18

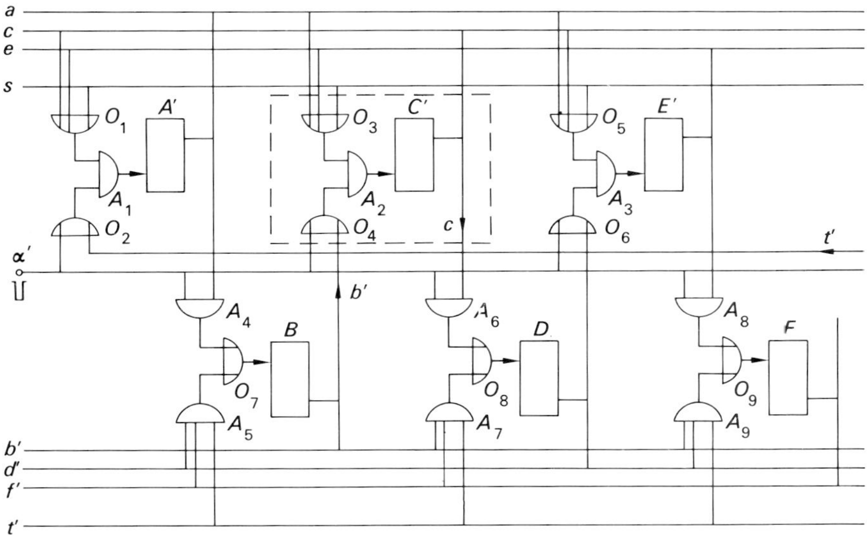

Fig. 14-19

obtained by cyclic permutation of the variables in these two equations. The resulting circuit is shown symbolically in Fig. 14-20.

In Figs. 14-19 and 14-20, the delay between the operation of inverters of the second multiposition triggers B, D, F, ..., T and the corresponding inverters A', C', E', ..., S' of the first trigger is equal to the duration of the

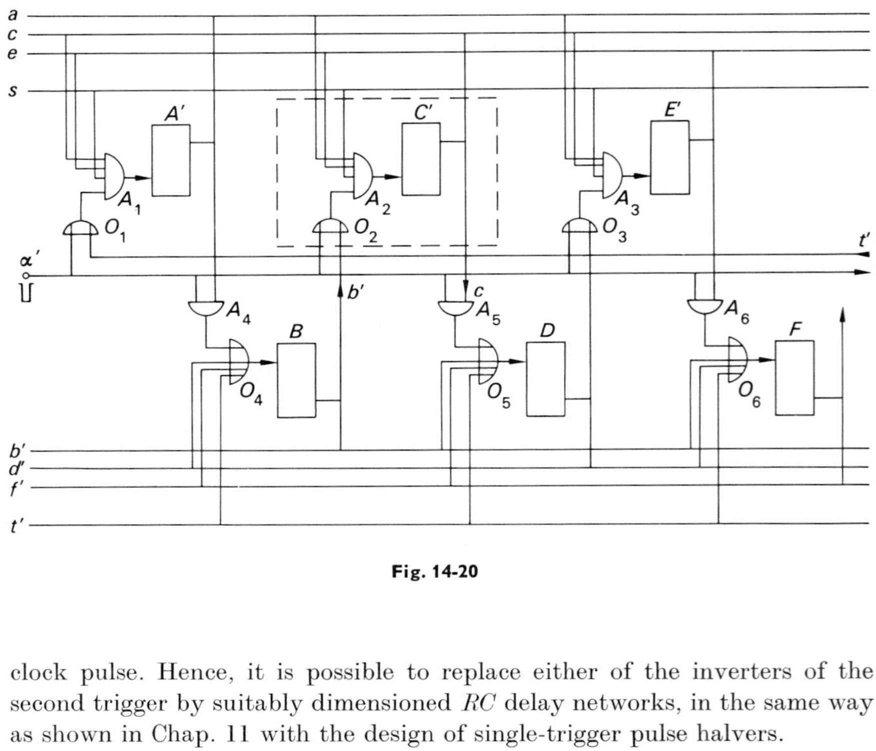

Fig. 14-20

clock pulse. Hence, it is possible to replace either of the inverters of the second trigger by suitably dimensioned RC delay networks, in the same way as shown in Chap. 11 with the design of single-trigger pulse halvers.

The replacement of the 10 inverters of the first trigger by 10 RC delay networks of the "differentiating" type can be performed in the same way for both circuits of Figs. 14-19 and 14-20. Inverter C' receives a b' signal from the B inverter and sends a c signal to the D inverter. The C' inverter is operated on the leading edge of an α' clock pulse, whereas the operation of the D inverter coincides with the trailing edge of this same α' clock pulse. The same type of operation is found in the RC delay network of Fig. 10-9. The C' inverter with its b' input and c output signal is surrounded by a dotted line in Figs. 14-19 and 14-20 to illustrate the possibility of replacement.

The resulting circuit in which all inverters of the first trigger of Fig. 14-19 are replaced by RC delay networks of the type shown in Fig. 10-9 is illustrated symbolically in Fig. 14-21.

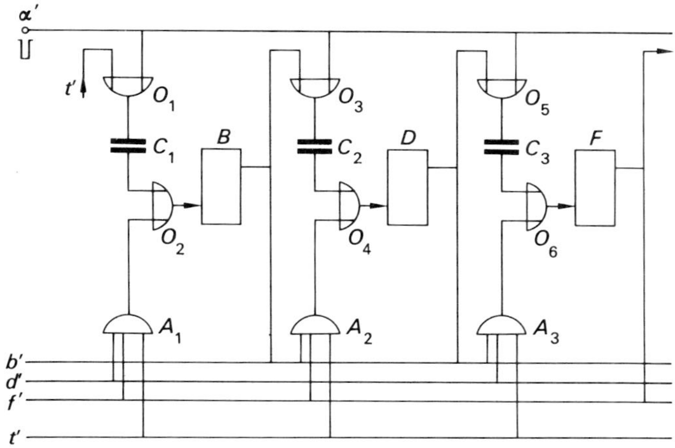

Fig. 14-21

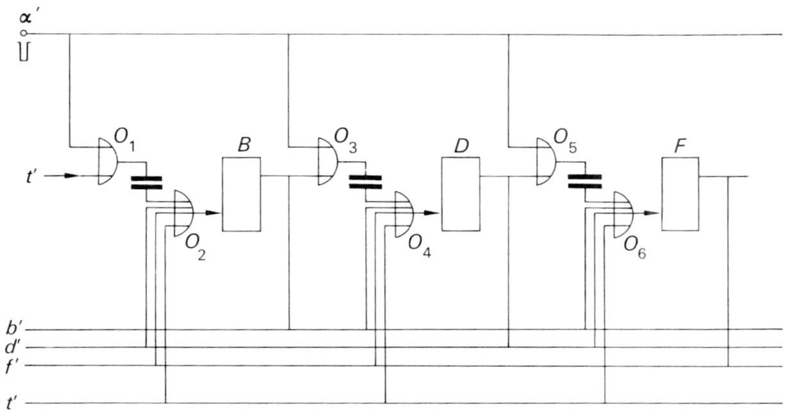

Fig. 14-22

The corresponding circuit in which all inverters of the first trigger of Fig. 14-20 are replaced by RC delay networks of the type presented in Fig. 10-9 is shown symbolically in Fig. 14-22.

The distributor circuit of Fig. 14-22 shows the correct form—from the point of view of switching—of a rather well-known type of n step distributor with n tubes and a marginal interlocking resistor network instead of gate circuits O_2, O_4, O_6, ... That circuit was controlled by a clock pulse applied to a common cathode lead.

It should be pointed out that the distributor circuits of Figs. 14-21 and 14-22 become identical for $n = 2$, that is, for the pulse-halver case. The AND gates A_1, A_2, A_3, etc., then reduce to single-input gates, which can be omitted. The OR gates of the circuit shown in Fig. 14-22 reduce to double-input OR gates, already found in the circuit of Fig. 14-21. The pulse halver may therefore be considered as a special case of the n-step distributors dealt with in this section.

14-9 COUNTERS WITH PART OR ALL OF THE COMBINATIONS OF A CONSTANT NUMBER OF OPERATED INVERTERS

The sequence diagram of Fig. 14-23 shows the operation of a counter consisting of two 4-position triggers which pass through all possible operating

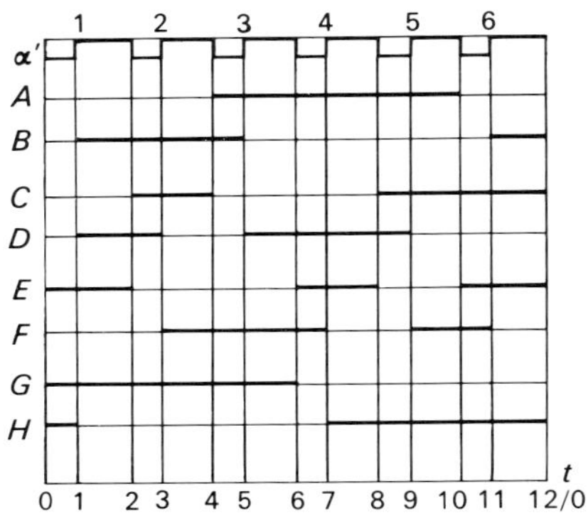

Fig. 14-23

combinations of two-out-of-four. The operating sequences can be chosen arbitrarily. The number of steps in a complete cycle is described by the equation

$$P_m^n = \frac{n!}{(n-m)!\cdot m!} = \frac{4!}{(4-2)!\cdot 2!} = 6 \tag{14-16}$$

The design method used for this example holds for this whole group of counters.

The time chart of Fig. 14-23 shows that the operation of the corresponding inverters of each of the two n-position triggers is essentially similar to that of the one-out-of-n counter or distributor shown in the time chart of Fig. 14-18. This also means that the A, C, E, and G inverter can be replaced by RC delay memory elements in a way analogous to the method used for converting the circuit of Fig. 14-19 into Fig. 14-21 and the circuit of Fig. 14-20 into Fig. 14-22.

The operation of the two-out-of-four counter given as an example can be further illustrated by means of Table 14-4.

Table 14-4

No.	*Present state*				*Next state*							
	B	*D*	*F*	*H*	*A*	*C*	*E*	*G*	*B*	*D*	*F*	*H*
1	1	1	0	0	0	1	0	1	1	0	1	0
2	1	0	1	0	1	0	0	1	0	1	1	0
3	0	1	1	0	1	0	1	0	0	1	0	1
4	0	1	0	1	1	1	0	0	0	0	1	1
5	0	0	1	1	0	1	1	0	1	0	0	1
6	1	0	0	1	0	0	1	1	1	1	0	0

It is required to design this counter circuit by means of one 4-position trigger as shown in Fig. 9-46 and four *RC* delay memory elements. It is not practical to design first a two-out-of-four counter by means of two of these triggers. The design technique used for the distributor circuit of Fig. 14-21 can be directly applied to the present case.

The desired counting circuit can be described by a group of four state equations. However, it is more practical to replace these state equations by two groups of four switching equations describing the four input gates of the four inverters and the four input gates of the *RC* delay memory elements. A switching equation of the first group describes the output signal of the corresponding *RC* delay memory element and the associated locking signals, while the corresponding switching equation of the second group describes the input signals of the *RC* delay memory element which lead to a new state of the corresponding inverter.

If a, c, e, and g are the output signals of the *RC* delay memory elements, it then follows from the present-state column of Table 14-4 that the input gates of inverters B, D, F, and H can be described by the following group of equations:

$$B = a + d'f' + d'h' + f'h' \tag{14-17}$$

$$D = c + b'f' + b'h' + f'h' \tag{14-18}$$

$$F = e + b'd' + b'h' + d'h' \tag{14-19}$$

$$H = g + b'd' + b'f' + d'f' \tag{14-20}$$

In accordance with the next-state columns of Table 14-4, the equations leading from a given present state to the desired next state can be determined as follows. Since the outputs of inverters B, D, F, and H will change only

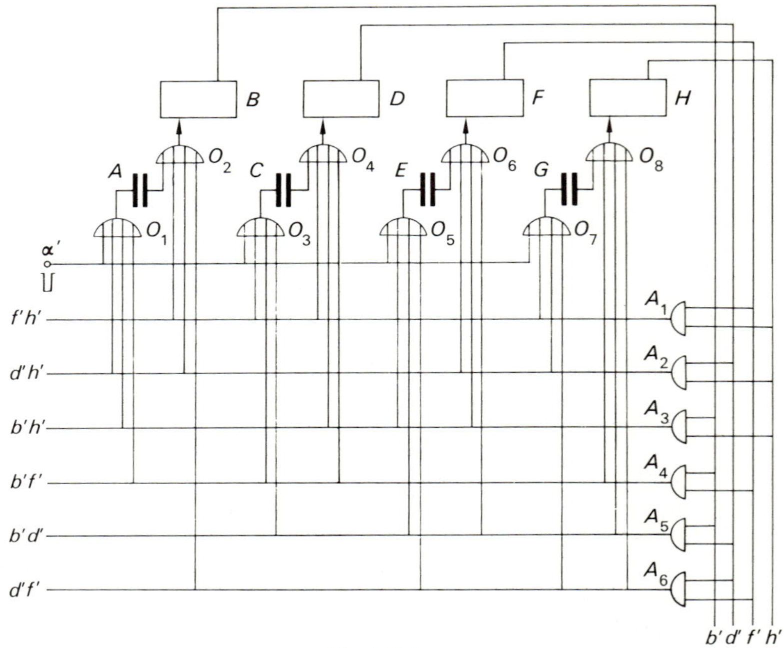

Fig. 14-24

at the end of the clock pulse and not during the clock pulse α', outputs a, c, e, and g of the RC delay memory elements A, C, E, and G can be described by the following equations:

$$a = \alpha' \cdot \underline{\alpha} \cdot \underline{A'} \tag{14-21}$$

$$c = \alpha' \cdot \underline{\alpha} \cdot \underline{C'} \tag{14-22}$$

$$e = \alpha' \cdot \underline{\alpha} \cdot \underline{E'} \tag{14-23}$$

$$g = \alpha' \cdot \underline{\alpha} \cdot \underline{G'} \tag{14-24}$$

on the basis of Eq. (10-16).

Since the B inverter must be operated in states 1, 2, and 6 of the counter, signal $\underline{A}'$ must equal 1 at the beginning of these states. So signal A' must equal 1 during states 6, 1, and 5 (see Table 14-1), which leads to the following equation:

$$A' = bh + bd + fh$$

or

$$\begin{aligned} A &= (b' + h')(b' + d')(f' + h') = (b' + d'h')(f' + h') \\ &= b'f' + b'h' + d'h' \end{aligned} \tag{14-25}$$

This equation describes the construction of the input OR gate of the RC delay memory element A.

Correspondingly, the equations for the input gates of the other RC delay memory elements are

$$C = b'd' + b'f' + f'h' \tag{14-26}$$

$$E = b'd' + b'h' + d'f' \tag{14-27}$$

$$F = d'f' + d'h' + f'h' \tag{14-28}$$

Equations (14-17) to (14-28) lead to a symbolic circuit of a two-out-of-four counter such as the one shown in Fig. 14-24. The indices of AND gates A_1 to A_6 represent their sequence of operation as shown in Table 14-4. Any other sequence of operation can be obtained by rearranging the inputs of the OR gates O_1 to O_8.

The m-out-of-n counter yields the largest possible number of operating combinations of inverters at the cost of a fairly large amount of hardware

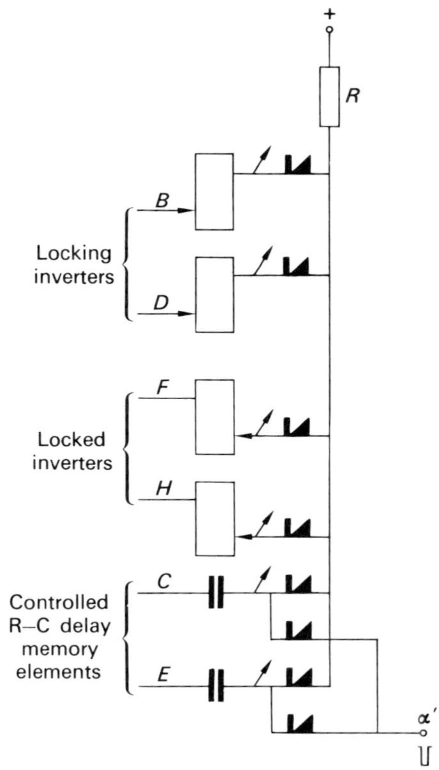

Fig. 14-25

in logical circuitry. The circuit diagram of an m-out-of-n counter can be drawn in a completely general way so that it holds for all possible types of such counters. A section of this generalized circuit is shown in Fig. 14-25. The number of inverters and the nomenclature have been made to match the circuit of Fig. 14-24. The generalized circuit of Fig. 14-25 shows that a number of inverters of an m-out-of-n counter together form a locking signal which locks the complementary combination of inverters during the rest condition of the circuit, and that this combination of locking inverters provides the information for not operating a number of RC delay memory elements on the next clock pulse.

PROBLEMS

1. Explain the similarities and differences between the two main types of counting circuits called counter and distributor. Which basic building blocks can be used for these circuits and what are the analogs in relay circuits.

2. Draw the time chart of a binary counter with three sections and design a symbolic circuit of

(*a*) An asynchronous type of counter
(*b*) A synchronous type of counter

What are the disadvantages of both types of counters?

3. Draw the time chart of an asynchronous binary-coded decimal counter and derive the switching formulas of the inputs of the MS memory elements of a decimal section. Draw a symbolic circuit.

4. Design the symbolic circuit of a synchronous binary-coded decimal counter with blocking gates between the successive sections having a constant number of inputs.

5. Devise the table of code combinations of a self-checking binary counter and derive the switching formulas of the required auxiliary MS memory element. Draw a symbolic circuit of that auxiliary memory element and its control gates. Derive the switching formula for the checking circuit and give an example of this circuit.

6. Devise the counting code of a maximum-variation counter of four counting elements and one auxiliary memory element. Derive the switching formulas using MS memory elements and draw the symbolic circuit.

7. Give the counting code of a two-out-of-five counter in which in the next code combination one element is set and one element is reset with respect to the present code combination. Derive the switching formulas of MS memory elements and draw a symbolic diagram.

8. Draw the time chart of a distributor using triggers as basic switching elements. Derive the set and reset formulas of one of the triggers and draw a symbolic circuit.

9. Draw the time chart of a distributor using inverters only as basic switching elements. Derive the state equation of an inverter of both halves of the distributor circuit and draw a symbolic circuit for

(*a*) Predominant set signals
(*b*) Predominant reset signals

10. Replace in the circuits resulting from Prob. 9 half of the inverters by RC delay elements.

11. Draw the time chart of a two-out-of-three counter with three inverters. Derive the state equations of these inverters and draw a symbolic circuit.

12. Draw the time chart and the symbolic circuit of a binary counting circuit operating in cycles of six pulses and coded in the binary system.

13. Draw the time chart of a binary operating counter using MS memory elements which can count cycles of nine pulses. Draw the symbolic circuit of this counter.

14. Draw the time chart of a binary operating counter using MS memory elements which can count cycles of 12 pulses. Draw the symbolic circuit. Devise a checking circuit giving an output signal 0 when the counter is in one of the combinations 0 to 11 inclusive and an output signal 1 when the counter runs erroneously in one of the states 12 to 15.

15. Draw the time chart of a binary operating counter with MS memory elements operating in cycles of three. Draw the symbolic circuit.

SELECTIVE BIBLIOGRAPHY

1. Brown, R. M.: Some Notes on Logical Binary Counters, *IRE Trans.*, vol. EC 4, pp. 67–69, 1955.
2. Cinema and Television Ltd.: British Patent 552,301, priority Sept. 29, 1941.
3. Cinema and Television Ltd.: British Patent 572,884, priority Feb. 15, 1947.
4. Dijl, B. van: Frequency Measuring and Frequency Dividing by means of Trigger Circuits, *Het P.T.T.-Bedrijf*, vol. 1, pp. 133–141, 1947–1948 (in Dutch).
5. Duuren, H. C. A. van, and A. Snijders: Electronic Counting Circuit, British Patent 794,190, Netherlands General Post Office, priority Sept. 10, 1953.
6. Fischmann, A. F.: Difference Counters, *Electron. Eng.*, vol. 29, pp. 456–550, 1957.
7. Foote, H. L.: Cold Cathode Counting Circuits, *Trans. AIEE*, pt. 1, vol. 74, pp. 161–164, 1955.
8. Golay, M. J. E.: The Logic of Bidirectional Binary Counters, *IRE Trans.*, vol. EC-6, pp. 1–4, 1957.
9. Haidekker, A.: Transistor bestückte Zählschaltungen mit Sichtanzeige, *Elektronik*, vol. 7, pp. 211–213, 1958.
10. Hoffman, G. R., and M. A. Maclean: Quiescant Core-transistor Counters, *Proc. IRE*, vol. 103, pp. 418–421, 1956.
11. Keister, W., A. E. Ritchie, and S. H. Washburn: "The Design of Switching Circuits," chap. 11, D. Van Nostrand Company, Inc., Princeton, N.J., 1951.
12. Murray, D. B.: A Variable Binary Scaler, *IRE Trans.*, vol. EC-4, pp. 70–74, 1955.
13. Oberman, R. M. M.: The Design of Triggers, Counters and Distributors, *Het P.T.T.-Bedrijf*, vol. 6, pp. 58–74, 1955, and vol. 7, pp. 33–45, 1956.
14. Richards, A. K.: Counting, Binary and Decimal, chap. 7 of "Arithmetic Operations in Digital Computers," D. Van Nostrand Company, Inc., Princeton, N.J., 1955.
15. Snijders, A.: Electronic Transmitter, Receiver and Regenerative Repeater for Telegraph Signals, British Patent 765,953, Netherlands General Post Office, priority Nov. 24, 1951.
16. Specht, W.: Frequenzteiler mit Kaltkatodenröhren, *Elektronik*, vol. 6, p. 15, 1957.
17. Staatsbedrijf, P. T. T. der: A Cyclic Electronic Counting Circuit Arrangement, British Patent 799,930, priority Jan. 27, 1954.
18. The Staff of the Computation Laboratory: Triggers, Rings and Digit Counters, chap. 8 of "Synthesis of Electronic Computing and Control Circuits," Harvard University Press, Cambridge, Mass., 1951.
19. Steingass, H.: Voreinstellbarer Zwei-Dekadenzähler mit Kaltkathoden Röhren, *Elektronik*, vol. 7, pp. 176–178, 1958.
20. Trent, R. L.: Transistor Reversible Binary Counter, *Proc. IRE*, vol. 40, pp. 1562–1572, 1952.
21. Ware, W. H.: The Logical Principle of a New Kind of Binary Counter, *Proc. IRE*, vol. 41, pp. 1429–1437, 1953.

15
Reflected Binary Codes

15-1 PROGRESSIVE CODES

The so-called progressive codes, i.e., codes in which one element only changes its value for each successive combination, are of special interest in the general field of switching techniques. These types of codes are essential when it is necessary to read a continuously varying quantity in digital form. When a coding system is used in which several digits change simultaneously during the transitional stage, ambiguities may arise. A progressive code gives the least possible ambiguity, since only one element changes at a time. The evaluation of these code combinations is such that the system then hesitates between two adjacent readings.

These codes are also of importance for switching applications proper, since circuits in which the state of one internal element is changed on each successive external control signal can easily be designed and are extremely reliable in operation. Race conditions cannot occur in such circuits. This applies to both contact and electronic circuits.

There are many possible ways of developing progressive codes. Their number is drastically reduced, however, when all operating combinations of the switching elements (relays or triggers) are to be used cyclically. The numbers of basically different progressive codes for 1, 2, 3, and 4 switching elements are, respectively, 1, 1, 1, and 55. Furthermore, a desired feature is the possibility of finding the decimal number of a progressive code combination by means of addition and/or subtraction of constant bit values, as in the well-known case of the binary 1248 code. In this case, there is only one form of the progressive code—the so-called reflected binary code (Gray code), which can be constructed as shown in Figs. 15-1 to 15-3 using the time chart of a pulse halver as a starting point.

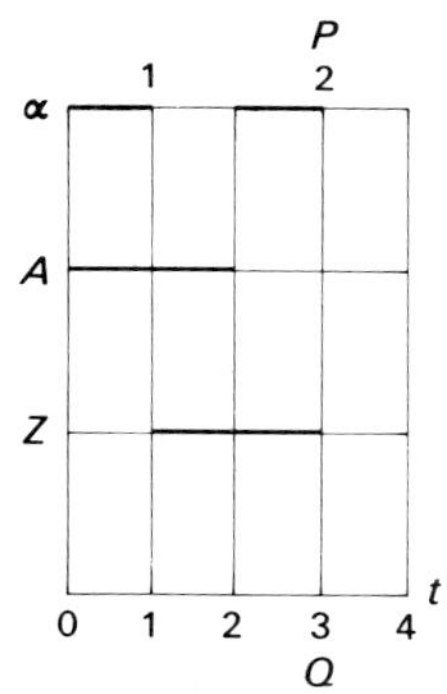

Fig. 15-1

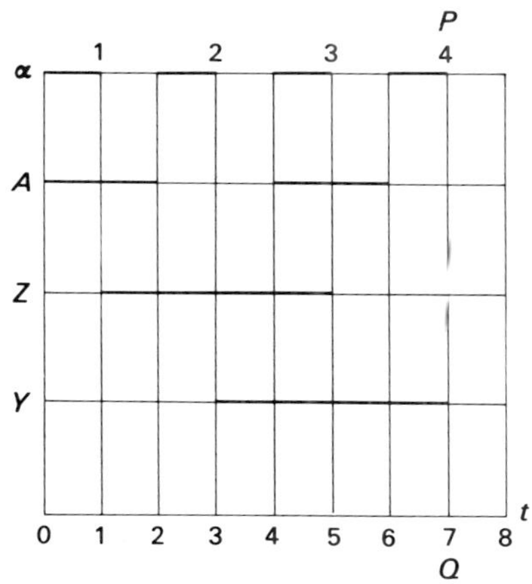

Fig. 15-2

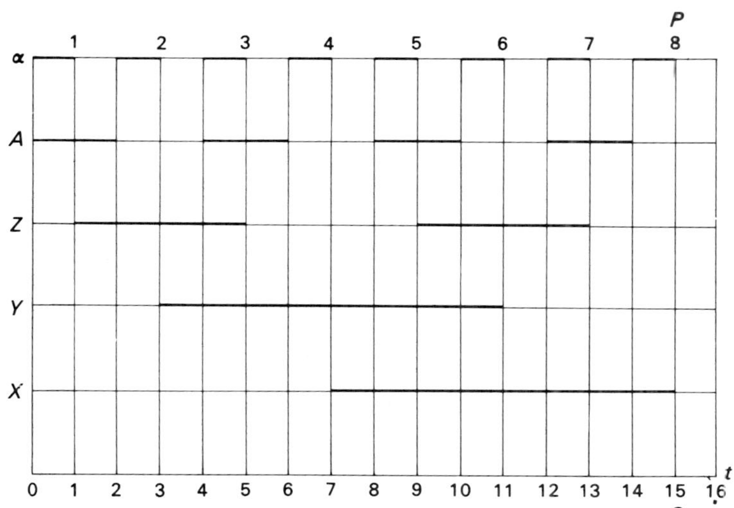

Fig. 15-3

The time chart of a pulse halver with two switching elements is shown in Fig. 15-1. This diagram is in fact both of the natural and reflected binary type in its sequence. The reflected binary diagram for three switching elements can be obtained by reflecting the operating periods of elements A and Z in Fig. 15-1 against a mirror placed on axis PQ and by extending the time chart with a third switching element Y. The total number of time elements in a reflected diagram is always 2^n, where n is the number of switching elements. The resulting diagram is shown in Fig. 15-2. This extension can be continued indefinitely. Figure 15-3 shows a reflected binary diagram for four elements.

Reflected codes—and not necessarily the special reflected binary codes—are frequently used in analog-to-digital-conversion techniques. A discussion of these techniques falls outside the scope of this textbook.

15-2 COUNTING OF PULSE SERIES IN THE REFLECTED BINARY CODE

PULSE HALVERS USED AS BASIC SWITCHING ELEMENTS

Figure 15-4 shows a possible time chart for the series-parallel conversion of a pulse series of eight pulses at most into a reflected binary registration on four pulse-halving circuits (MS memory elements).

The reflected binary operation of the four pulse-halving MS memory elements is obtained simply by shifting the working periods of the successive pulse halvers relative to their operation in the natural binary time chart. The operation of the switching elements in the time charts of Figs. 15-3 and 15-4 is identical.

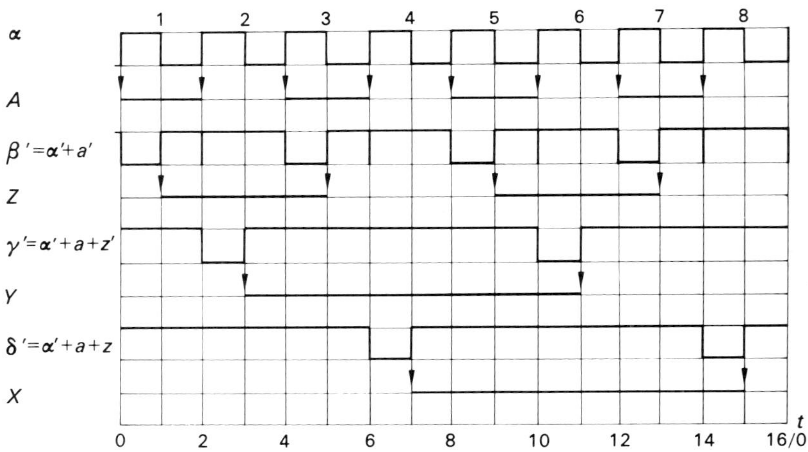

Fig. 15-4

The first pulse-halving element A, which is superfluous with respect to the coding of the number of pulses, operates directly on the α counting pulses.

By means of the first pulse halver A, a pulse

$$\beta' = \alpha' + a' \tag{15-1}$$

can be produced for the control of the second pulse-halving MS memory element Z. Pulse halvers A and Z together with the controlling pulse α' form a pulse

$$\gamma' = \alpha' + a + z' \tag{15-2}$$

for the control of the third pulse-halving element Y, so that this memory element will operate in a reflected mode with respect to element Z. This process may be continued indefinitely. However, for the last section of the counting chain, this system of pulse forming will have to be changed, because if the controlling pulse of the last section would be, for example,

$$\delta' = \alpha' + a + z + y' \tag{15-3}$$

the counting chain would count up to eight in the reflected binary code and then count back to 0 on pulses 9 to 16. The counting chain would then be back in its rest condition after 16 pulses and all combinations of the memory elements (X, Y, Z) would have been used twice.

The working cycle of the reflected binary counter of the example can be arranged in such a way that it has a length of eight counting pulses. The control pulse of the fourth and last section must then be

$$\delta' = \alpha' + a + z \tag{15-4}$$

as is shown in Fig. 15-4.

The switching formulas of both input gates of a pulse-halving MS memory element coincide with the corresponding pulse formulas given above, except that they are extended with the required feedback variable for the pulse-halving operation. The circuit of Fig. 15-5 is then obtained.

The circuit of Fig. 15-5 has some special properties introduced by the superfluous (from the point of view of pulse counting) element A. This

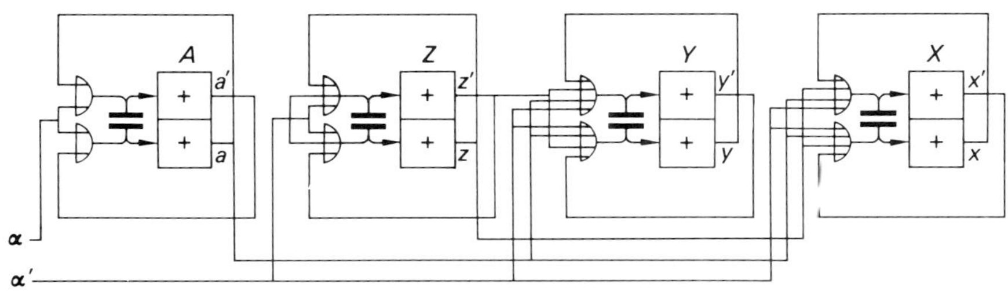

Fig. 15-5

auxiliary element makes it possible for the operation of the whole circuit to be checked. As can easily be verified in the time chart of Fig. 15-4, and in the corresponding Table 15-1, the number of elements in the 1 state during the pause between two pulses is always even. A further special property of this circuit is that a reverse of the state of element A makes the counting circuit count in the reverse direction, as is shown in Table 15-2. During the reverse count, an odd number of elements will always be in the 1 state in the pause between pulses. The operation of the counter in both directions of operation can thus be easily checked.

Table 15-1

No.	*X*	*Y*	*Z*	*A*
0	0	0	0	0
1	0	0	1	1
2	0	1	1	0
3	0	1	0	1
4	1	1	0	0
5	1	1	1	1
6	1	0	1	0
7	1	0	0	1

Table 15-2

No.	*X*	*Y*	*Z*	*A*
0	0	0	0	1
7	1	0	0	0
6	1	0	1	1
5	1	1	1	0
4	1	1	0	0
3	0	1	0	0
2	0	1	1	1
1	0	0	1	0

From a logical point of view, the time chart of Fig. 15-4 seems to be correct; however, from the point of view of implementing this type of circuit in accordance with this time chart, difficulties arise in the form of the generation of sneak pulses at moments 2, 6, 10, and 14 on the β' pulse line. Even with ideally shaped α' and α pulses, these sneak pulses will be generated at said moments because α' will become 0 before a' becomes 1. The generation of these pulses can be prevented by making an α' pulse from an α pulse by means of an inverter. Because the α' pulse is then delayed slightly with respect to the α pulse, MS memory element A, operating under the control of the α pulse, will have been reset to 0 ($\alpha' = 1$) at about the same moment the α' signal changes from 1 to 0. The corresponding circuit is shown in Fig. 15-6.

The difficulty mentioned above in the implementation of the circuit of Fig. 15-5 can be avoided by shifting the moments of operation of the auxiliary element A, as is shown in the time chart of Fig. 15-7. In that time chart, all MS memory elements operate under the control of α' pulses, which means that no sneak pulses can be generated. There is no reflected binary operation in the diagram of element A with respect to elements X, Y, and Z. But the reflected operation of elements X, Y, and Z, which are important for the counting code, remains unchanged. However, the equations of the

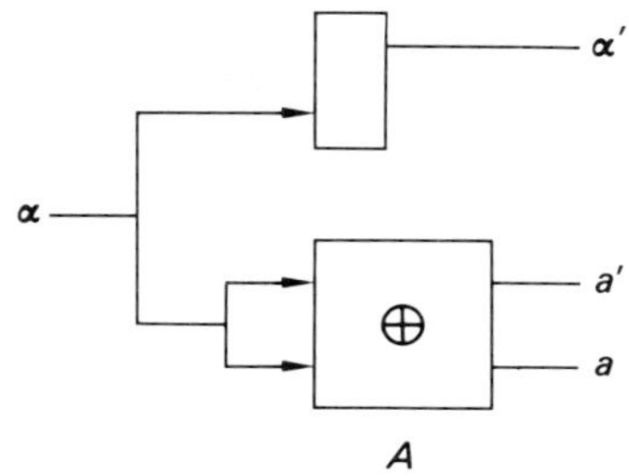

Fig. 15-6

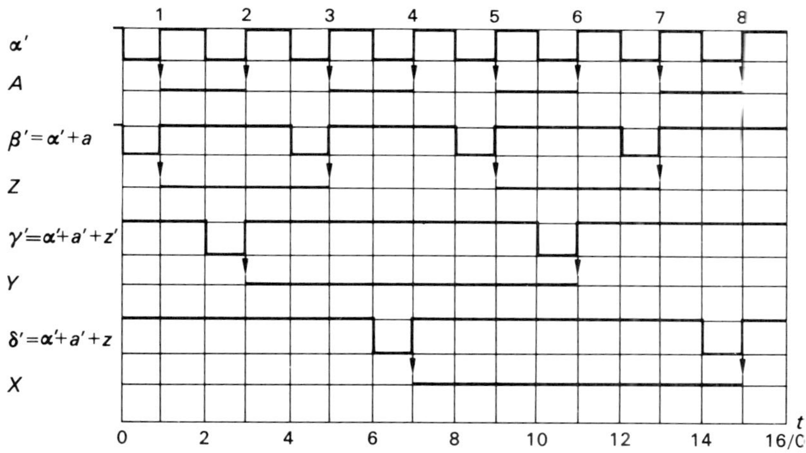

Fig. 15-7

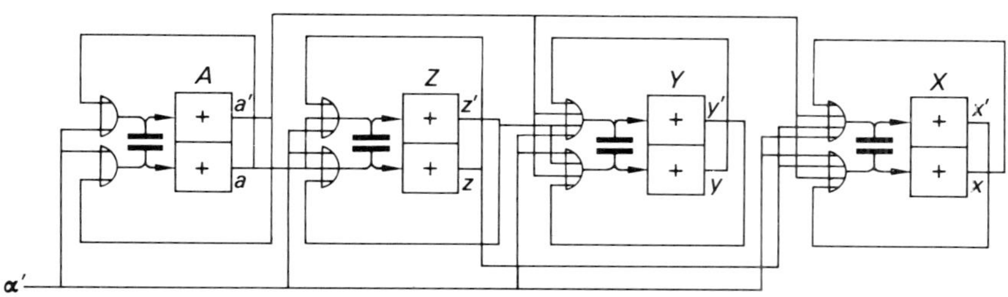

Fig. 15-8

pulse controlling the successive MS memory elements have been changed:

$$\text{Controlling pulse of } A: \quad \alpha' \tag{15-5}$$

$$\text{Controlling pulse of } Z: \quad \beta' = \alpha' + a \tag{15-6}$$

$$\text{Controlling pulse of } Y: \quad \gamma' = \alpha' + a' + z' \tag{15-7}$$

$$\text{Controlling pulse of } X: \quad \delta' = \alpha' + a' + z \tag{15-8}$$

The corresponding circuit is shown in Fig. 15-8.

The direction of operation of this last type of counter can also be reversed by reversing the state of the auxiliary element A. It has the same self-checking character as that shown in Fig. 15-5.

It follows from the reflected binary counting circuits given in Figs. 15-5 and 15-8 that as we go along from one MS memory element to the next, the input gates (except the last) have an ever-increasing number of inputs. This is prohibitive as regards the use of memory elements of standard construction with a fixed and not-too-large number of inputs.

To some extent this problem can be alleviated by the use of two levels of input gates of the same kind, in which case the first level is a single gate blocking the operation of the two gates of the second level when required. These last gates are the standard input gates of the MS memory element connected as pulse halver. However, the single blocking gate will now have an ever-increasing number of inputs. This disadvantage can be avoided by using a series circuit of transistors.

The equations of the controlling pulse of the successive sections of the counter operating as is shown in the time chart of Fig. 15-7 can be split up into a part relating to the direct interconnection of two successive sections as binary counting elements and another part relating to the blocking of this operation in order to yield the desired reflected binary operation.

$$\begin{array}{lll} & \textit{Binary} & \textit{Blocking} \\ \text{Controlling pulse of } Z: & (\alpha' + a) & \\ \text{Controlling pulse of } Y: & (\alpha' + z') + & (a)' \\ \text{Controlling pulse of } X: & (\alpha' + y') + & (az')' \\ \text{Controlling pulse of } W: & & (az'y')' \\ & \text{etc.} & \end{array} \tag{15-9}$$

The resulting series transistor circuit is shown symbolically in Fig. 15-9.

The inverters I_y, I_x, I_w, etc., can be omitted when using two-capacitor MS memory elements. The blocking pulse line then has to be connected to the rear side of the capacitors.

A reflected binary counter has a cycle length of 2^n pulses, n being the number of memory elements significant for the counting code. As is shown in Table 15-3, the cycle length of this type of counter can be changed into

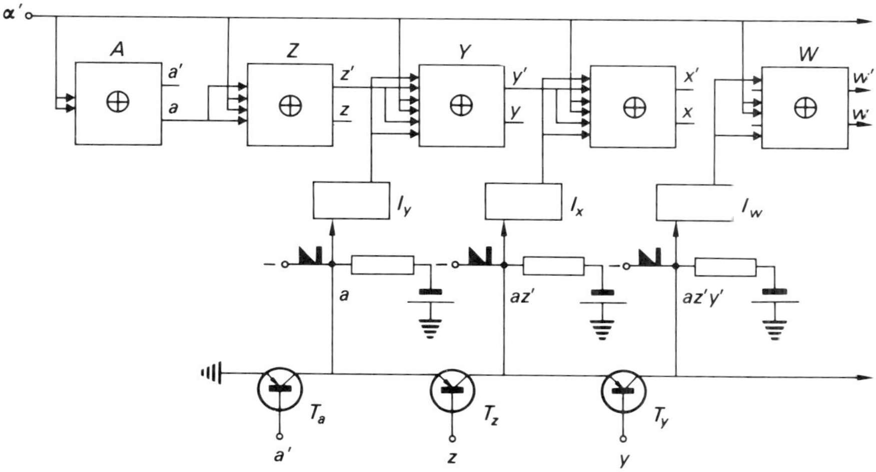

Fig. 15-9

Table 15-3

Number of code combination	*W*	*X*	*Y*	*Z*	*Cycle length*	*Detection combination* *W*	*X*	*Y*	*Z*	*Number of code combination*
15	1	0	0	0	2 ⟵	0	0	0	0	0
14	1	0	0	1	4 ⟵	0	0	0	1	1
13	1	0	1	1	6 ⟵	0	0	1	1	2
12	1	0	1	0	8 ⟵	0	0	1	0	3
11	1	1	1	0	10 ⟵	0	1	1	0	4
10	1	1	1	1	12 ⟵	0	1	1	1	5
9	1	1	0	1	14 ⟵	0	1	0	1	6
8	1	1	0	0	16 ⟵	0	1	0	0	7

any even number of counting steps smaller than 2^n, the counting code remaining progressive. In this table, the complete cycle of a four-element reflected binary code has been divided into two halves. The eight-code combinations in column 4 are the detection combinations from which a jump will be made to the code combination on the same line in column 2 by

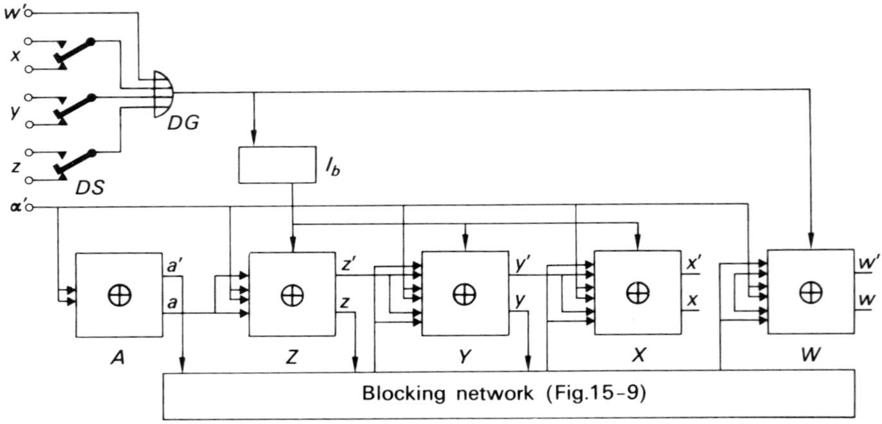

Fig. 15-10

reversing the state of the highest-valued element (that is, W) from 0 to 1. In this way, all even cycle lengths smaller than 2^n can be formed. The sequence of code combinations in all cycles of even length remains progressive, as can easily be verified in Table 15-3.

The code jump can be performed in electronic reflected binary counters by blocking all counting sections when the detection combination is reached and setting the last section (that is, W) from 0 to 1. The implementation of this cycle-length switch is shown in the circuit of Fig. 15-10.

It is not possible to change the cycle length into an odd number of steps without making a nonprogressive step.

The diagram of the counting part of Fig. 15-10 is the same as that shown in Fig. 15-9. A group of switches DS is added (DS_x, DS_y, and DS_z) by means of which the detection combination, from which the jump in the cycle will be made, can be indicated. Reaching this combination, detection gate DG produces a signal for element W so that it is set to 1 on the next counting pulse α'; but this signal, inverted by I_b, is used as a blocking signal for elements X, Y, and Z during the jump.

RELAYS USED AS BASIC SWITCHING ELEMENTS

As was already demonstrated in Chap. 13, the counting of pulse series can be performed not only by means of pulse halvers but also by means of relays or triggers used as basic switching elements. This leads to a decrease in the number of relays involved in the circuit at the expense of some added complexity with respect to the principles of the circuit. The choice of the type of circuit to be used will depend on such a factor as the availability to the user of a highly skilled maintenance staff.

Let us now design a relay counting chain with single relays operating in a reflected binary manner. The following considerations apply only partly to the electronic equivalent using triggers as basic switching elements, because the general type of trigger differs from a relay in that it is a symmetrical apparatus with a set and a reset input terminal.

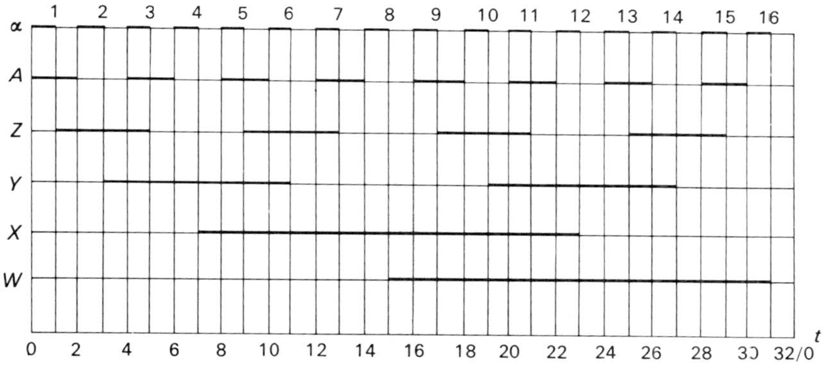

Fig. 15-11

The time chart for the relay circuit to be designed has the same general form as the charts shown in Figs. 15-2 and 15-3. An example of a time chart for five relays is shown in Fig. 15-11.

The switching formulas for the operating circuits of relays A and Z to W can be derived from this time chart in the following way:

$$\begin{aligned} s_a &= \alpha\cdot(z'y'x'w' + zyx'w' + z'yxw' + zy'xw' \\ &\quad + z'y'xw + zyxw + z'yx'w + zy'x'w) \\ &= \alpha\cdot\{z'[y'(x'w' + xw) + y(xw' + x'w)] \\ &\quad + [y(x'w' + xw) + y'(xw' + x'w)]\} \end{aligned} \tag{15-10}$$

The equations corresponding to the other relays can be found in a similar way; the outcome after some rearranging of the various terms involved is as follows:

$$s_z = \alpha' a\cdot[y'(x'w' + xw) + y(xw' + x'w)] \tag{15-11}$$

$$s_y = \alpha' a' z\cdot(x'w' + xw) \tag{15-12}$$

$$s_x = \alpha' a' z' y\cdot w' \tag{15-13}$$

$$s_w = \alpha' a' z' y' x \tag{15-14}$$

These equations only define the respective relay circuits in terms of the first time element of their operating period. The parts of Eqs. (15-10) to (15-14) left of the multiplication sign together form a series chain of transfer contacts of the general type shown in Fig. 4-52.

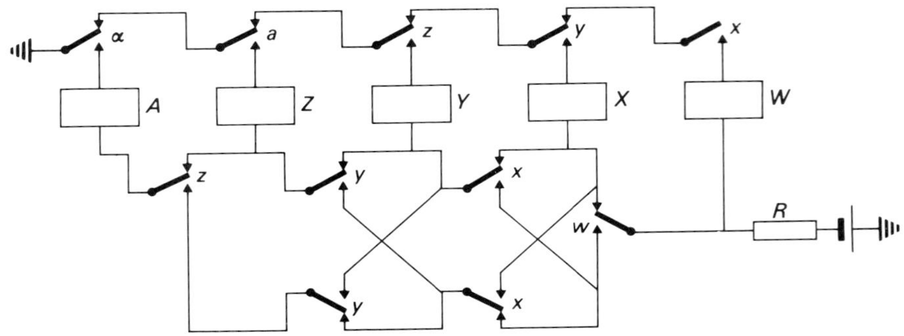

Fig. 15-12

The right-hand portions of these equations can also be combined into a slightly more sophisticated chain circuit of transfer contacts. These two chains must be connected to different sides of the operating windings of the various relays, as is shown in Fig. 15-12. This type of contact circuit is very attractive in that it gives a sectional character to the complete diagram; that is, with a number of identical sections added, no redesigning of the already existing part of the circuit is involved.

These chain circuits cannot be connected to only one side of the operating windings of the relays in order to avoid the generally less desirable feature of having contact circuits on the battery side of the relay windings. These types of circuits require a current-limiting resistor R as a protection against short circuits in the contact chain, because these can easily occur during maintenance operations. This resistor R is shown in Fig. 15-12.

The release portions of the switching equations of the various relays are as follows:

$$ar_a' = a\alpha' \tag{15-15}$$

$$zr_z' = z(a' + \alpha) = z(a' + a\alpha) \tag{15-16}$$

$$yr_y' = y(z' + a + \alpha) = y[z' + z(a + a'\,\alpha)] \tag{15-17}$$

$$xr_x' = x(y' + z + a + \alpha) = x\{y' + y[z + z'(a + a'\,\alpha)]\} \tag{15-18}$$

$$wr_w' = w(y + z + a + \alpha) = w\{y + y'[z + z'(a + a'\,\alpha)]\} \tag{15-19}$$

The full release equation of relay A consists of eight terms, out of which all variables of higher order disappear. The same also holds for relays Z and Y. The release equation of relay X (15-18) can be put in the same general form as the equations for the lower-order relays A, Z, and Y by recalling that this reset circuit need not be closed during time interval $t_{7/8}$. This eliminates the w variable from the release equation. The x variable is eliminated from Eq. (15-19) by applying an analogous reasoning for time interval $t_{15/16}$.

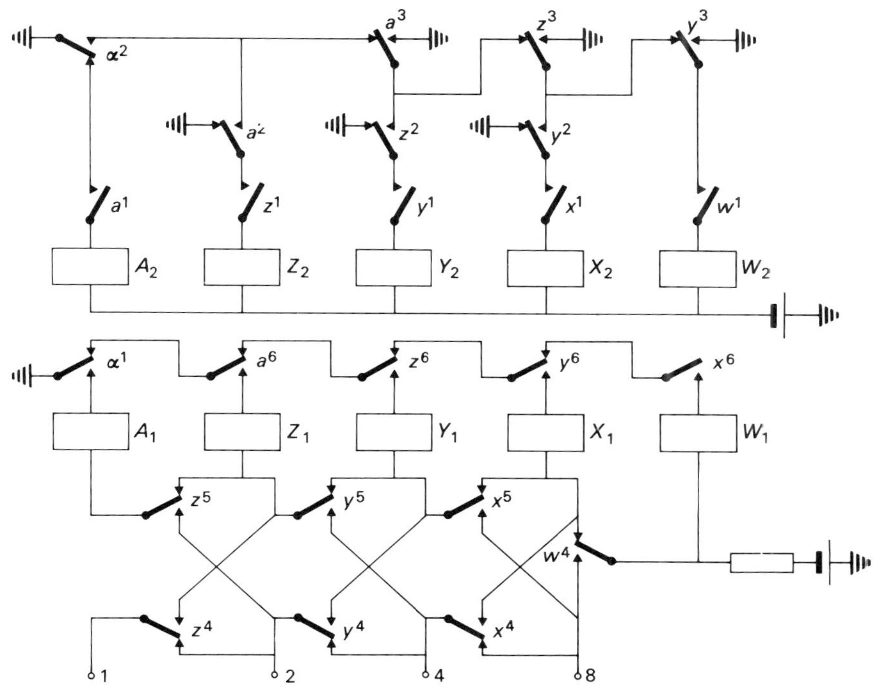

Fig. 15-13

Release equations (15-15) to (15-19) together form a rather complex contact chain, which is shown in the upper half of the complete diagram of Fig. 15-13.

The circuit consisting of transfer contacts z^4, z^5, y^4, y^5, x^4, x^5, and w^4 constitutes a built-in code converter for translation of the reflected binary code into the natural binary code. As will be shown in the next section, it is possible to give a fixed value to the bits or elements of the reflected binary code. The decimal equivalent, however, cannot be found merely by means of the addition process, as is the case with the conventional binary code. In order to make it possible to check easily the number of pulses received in the reflected binary counting chain of Fig. 15-13, a code converter from reflected binary into natural binary can be built in at the cost of an extra z contact. The correctness of the built-in code converter can be proved by means of the time chart. The following equations, which describe the conversion from reflected into conventional binary, can be derived from that time chart:

$$\text{Output 1:} \quad z[y(x'w + xw') + y'(xw + x'w')] + z'[y'(x'w + xw') + y(xw + x'w')] \tag{15-20}$$

$$\text{Output 2:} \quad y(xw + x'w') + y'(x'w + xw') \tag{15-21}$$

$$\text{Output 4:} \quad xw' + x'w \tag{15-22}$$

$$\text{Output 8:} \quad w \tag{15-23}$$

These four switching equations are analogous in nature to the parts of Eqs. (15-10) to (15-14) to the right of the · sign, the difference being that they correspond to the inverse of those equations. Taps on the bottom side of the network of contacts z^4 to w^4 in Fig. 15-13 provide for the desired code conversion. These code-converting outlets can be used to light control lamps or to transfer in conventional binary coding the received pulse series into a register. The counting chain of Fig. 15-11 with code converter is unique in that it is the only systematic way known to the author to count pulse series of 2^{n-1} pulses with a relay chain consisting of n single-relay sections. The chain can be extended indefinitely with relays of the same type.

TRIGGERS USED AS BASIC SWITCHING ELEMENTS

The time chart of Fig. 15-11 can also be used when triggers serve as basic switching elements in the reflected binary counting circuit. Because these elements need set and reset signals of the same character while retaining their state between set and reset signals, another group of control signals must be provided. The set signals for this electronic circuit are the same as those given for the relay circuit, i.e., Eqs. (15-10) to (15-14).

The reset signals for the electronic circuit are as follows:

$$r_a = \alpha \cdot z[y'(x'w' + xw) + y(xw' + x'w)] + z'[y(x'w' + xw) + y'(xw' + x'w)] \tag{15-24}$$

$$r_z = \alpha' a \cdot [y(x'w' + xw) + y'(xw' + x'w)] \tag{15-25}$$

$$r_y = \alpha' a' z \cdot (xw' + x'w) \tag{15-26}$$

$$r_x = \alpha' a' z' y \cdot w \tag{15-27}$$

$$r_w = \alpha' a' z' y' \cdot x' \tag{15-28}$$

These five reset equations are analogous to the five set equations. The parts left of the multiplication sign are identical, while the right-hand portions correspond to the inverse, as can be expected for reasons of symmetry. The parts right of the multiplication sign in Eqs. (15-24) to (15-27) correspond to the expressions, which apply to the code conversion from reflected into natural binary code, given as Eqs. (15-20) to (15-23). The inverse parts of set and reset formulas can accordingly be obtained by means of the same network. It is then merely a problem of practical circuit design to have a trigger controlled by means of these set and reset equations that consist of both an equal and an inverse part.

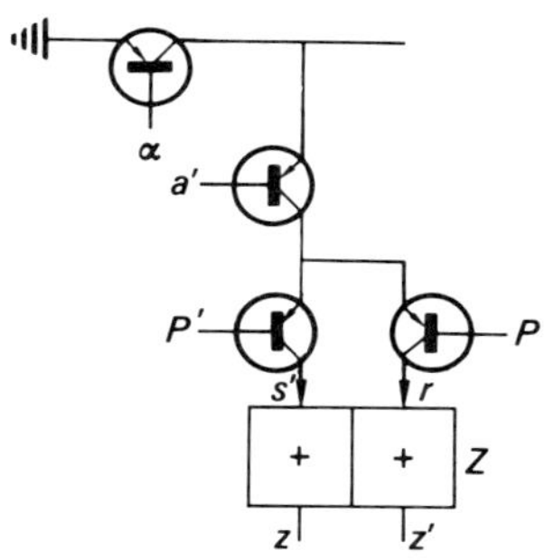

Fig. 15-14

With the trigger circuit shown as a symbol in Fig. 9-5 used as a starting point to keep the final circuit diagram as simple as possible, the Z trigger of the reflected binary counting circuit to be designed can be controlled as shown in Fig. 15-14. The set and reset equations of this Z trigger are

$$s_z = \alpha' a P \tag{15-29}$$

$$r_z = \alpha' a P' \tag{15-30}$$

Furthermore,

$$P = y'(x'w' + xw) + y(xw' + x'w) \tag{15-31}$$

$$P' = y(x'w' + xw) + y'(xw' + x'w) \tag{15-32}$$

Equations (15-31) and (15-32) can be obtained by means of a separate network that generates the analogous parts of all equations. The complete circuit is shown in Fig. 15-15. The operation of this circuit now becomes very simple. The triggers act in a way similar to the relays in Fig. 15-13, whereas the other transistors are similar to the contacts of this same circuit. The control bases of these contact transistors in Fig. 15-15 are shown disconnected from the corresponding trigger outputs, in the same way as in the "detached contact" drawing method in which the contacts of the relays are shown spread all over the drawing but systematically arranged in their circuits. The connection between a control trigger and the corresponding transistor in the controlling network is indicated only by letters.

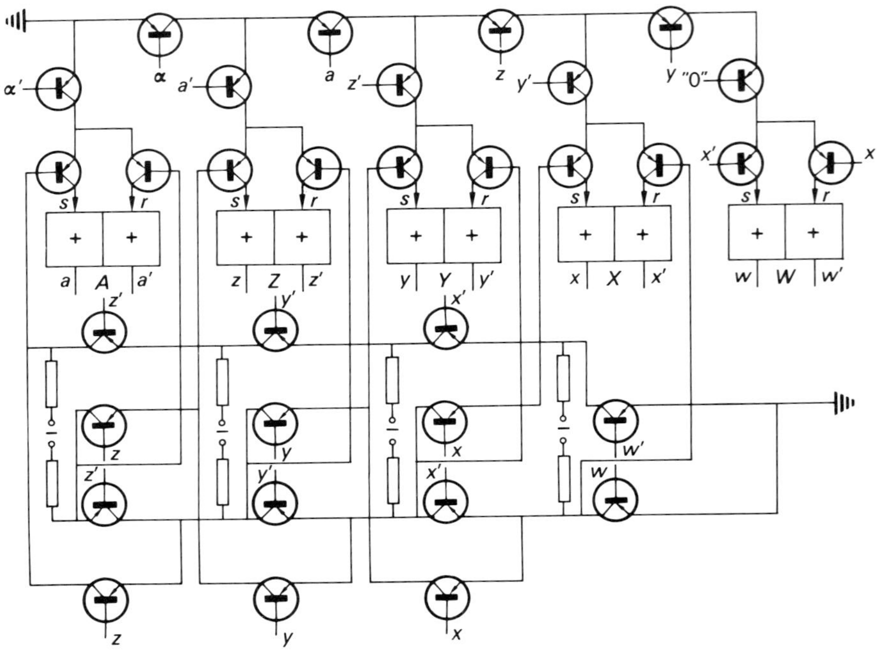

Fig. 15-15

15-3 SERIAL CONVERSION OF REFLECTED INTO NATURAL BINARY CODES, AND VICE VERSA

The problem of parallel conversion of a reflected into a natural binary code does not fall within the scope of this chapter. Indeed a code-converting circuit performing this type of translation can easily be designed by means of the methods given in Chap. 8 and on the basis of Table 15-4.

The serial conversion of a reflected into a natural binary code and back is governed by simplified forms of the general conversion equations for the translation of reflected into natural number systems and back. This simplification results from the fact that all digits in the systems considered here are 0 or 1 and that the addition of these digits is performed on the base mod 2.

Table 15-4

Decimal	*Natural binary* 8	4	2	1	*Reflected binary* 15	7	3	1
0	0	0	0	0	0	0	0	0
1	0	0	0	1	0	0	0	1
2	0	0	1	0	0	0	1	−1
3	0	0	1	1	0	0	1	0
4	0	1	0	0	0	1	−1	0
5	0	1	0	1	0	1	−1	1
6	0	1	1	0	0	1	0	−1
7	0	1	1	1	0	1	0	0
8	1	0	0	0	1	−1	0	0
9	1	0	0	1	1	−1	0	1
10	1	0	1	0	1	−1	1	−1
11	1	0	1	1	1	−1	1	0
12	1	1	0	0	1	0	−1	0
13	1	1	0	1	1	0	−1	1
14	1	1	1	0	1	0	0	−1
15	1	1	1	1	1	0	0	0

Let the digits in the reflected system be as follows, reading from left to right (highest value to lowest value):

$$a_n, a_{n-1}, \ldots, a_2, a_1$$

Let the digits in the natural or conventional system be

$$A_n, A_{n-1}, \ldots, A_2, A_1$$

Let the largest number in the system be L, which number equals base B minus 1. The binary system has a base $B = 2$ and the largest number $L = 1$. It can be shown that

$$A_k = \sum_{i=k}^{n} a_i \ (\mathrm{mod}\ 2) \tag{15-33}$$

In other words, the natural binary digit A_k of a binary number can be found by adding the reflected binary digits in positions k to n on the base mod 2. This result means that a code conversion from a reflected binary code into a natural binary code on a serial basis can only be performed by transmitting the numbers with the most significant digit first. This is contrary to the direction of transmission normally used in serial computers. This code conversion is illustrated by means of Table 15-4, which shows corresponding combinations of a four-bit natural and reflected binary code and their decimal equivalents.

The decimal equivalent of a natural binary code combination can be found by adding the values 1, 2, 4, 8, ... of the 1 digits. That of a reflected binary code combination is more difficult to find. It follows from Eq. (15-33) that the decimal equivalent can be obtained by adding the values 1, 3, 7, 15, ... of the 1 digits, the even-placed 1 digits being taken as negative. The − signs are indicated in Table 15-4.

In view of the fact that the serial conversion of a digit of a reflected binary code into the natural binary code is an addition on base mod 2 of the preceding 1 digits of the reflected binary code combination, it appears that this conversion can be performed by means of a simple pulse-halving circuit, which receives a control pulse for each 1 of the reflected number. The logical diagram of such a serial converter is shown in Fig. 15-16.

The reflected code is given in serial form to one input of an AND gate. The other input of this AND gate is connected to a gating or clock pulse. As is shown in the time chart of Fig. 15-17 on line b, the clock pulse coincides with the theoretical middle of the elements of the reflected code signal. The pulse halver must start from the 0 condition in which the output generates a 0 signal. At the gating moment of the first 1 digit, the pulse halver changes its condition so that its output produces a 1 signal until the second 1 of the reflected binary signal reverses the condition of the pulse halver, this being in fact an addition mod 2 of the first and the second 1 digit of the reflected binary code combination.

This is illustrated in Fig. 15-17 for the translation of the reflected binary code combination 9 shown on line a, into its binary equivalent on line c. There is a shift in the timing of the output elements with respect to the input elements, the duration of this shift being half an element.

If the reflected binary code signal is given in the form of separate pulses that can directly control a pulse halver, there is no time shift between input and output signals. A gating or clock pulse is then no longer necessary.

The translation of the natural into the reflected binary code can be represented by the following expression:

$$a_k = A_k \oplus A_{k+1} \tag{15-34}$$

that is, the reflected binary digit a_k (digit in position k) can be found by adding on base mod 2 the natural binary digits in positions k and $k + 1$. This equation is more difficult than is the reverse case to implement by means of logical circuits. It implies a two-bit memory—one for the digit to be translated and one for the preceding digit. Moreover, an adding circuit is required.

The logical diagram of a serial natural-to-reflected converter is shown symbolically in Fig. 15-18. The operation will be illustrated by means of the example shown in Fig. 15-19. The serial signal in the natural binary code starts with a fictive $(n + 1)$st digit, which is a start element and which is

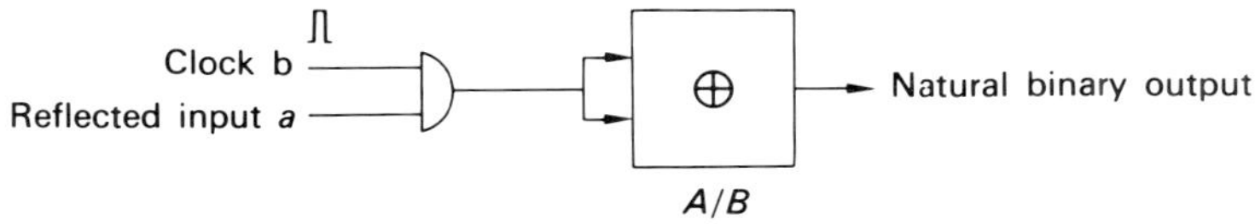

Fig. 15-16

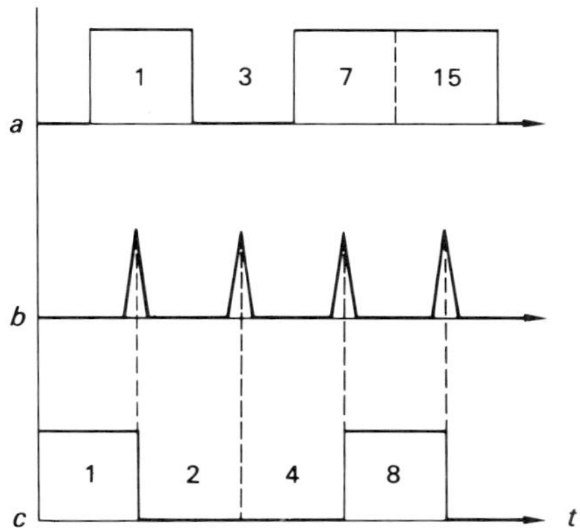

Fig. 15-17

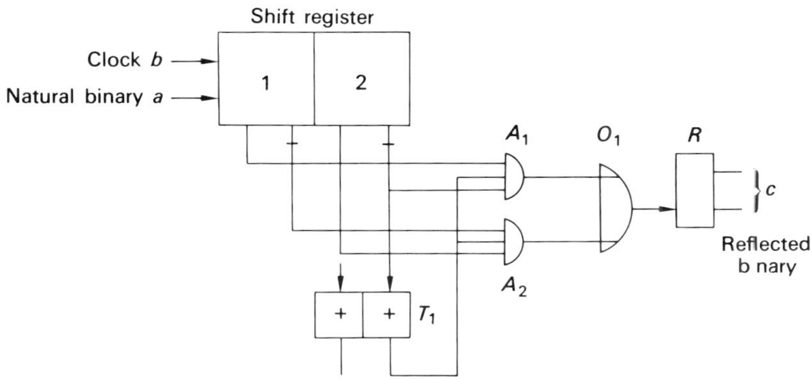

Fig. 15-18

always a 0. This start signal puts an end to the rest condition on the input of the code converter, this condition corresponding to a 1 signal. The memory of the code converter consists of a two-bit shift register, which is in its rest condition with both sections in the 1 condition.

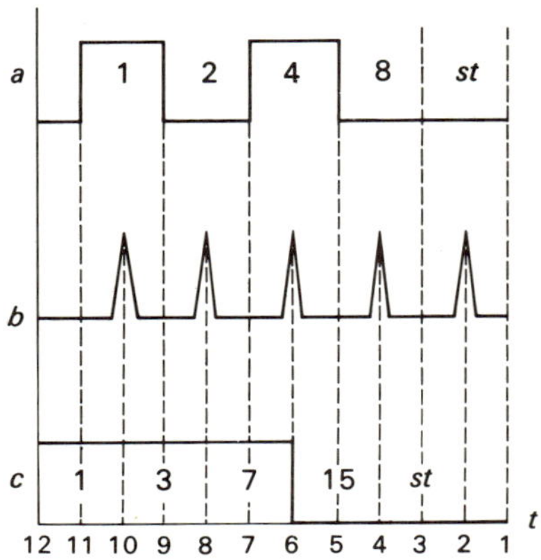

Fig. 15-19

On the first clock pulse, the start element of the natural binary signal to be converted is shifted into the first position of the shift register. The 1 rest signal of the first position is then shifted into the second position. On the second clock pulse, the first and most significant digit is shifted into the first position and the start element into the second position of the register. The second position then changes its condition from 1 to 0, and an inverted output gives a 1 signal to trigger T_1. This trigger changes its condition and unlocks AND gates A_1 and A_2. Both gates are controlled by two inputs connected to outputs of different polarity of the two sections of the shift register. This means that gate A_1 generates an output voltage 1 if section 1 of the shift register is in the 0 condition and section 2 is in the 1 condition. Gate A_2 produces an output voltage 1 in the reverse condition of both sections of the shift register. Gates A_1 and A_2 produce a 0 output voltage for all other conditions of the inputs. The OR gate O_1 transfers the 1 output signal of gate A_1 or of gate A_2 to a repeater R, which restores the power level of the final signal.

At the cost of a slight variation in circuit design, this type of code converter can also operate on code signals consisting of pulse elements.

Since the translation of the natural binary code into the reflected binary code is a question of addition mod 2 of two adjacent digits, the direction of the serial transmission can be reversed and the least significant element can be transmitted first. The first reflected element is then determined by the first and the second element of the natural binary code and the last reflected element by the last natural binary element and a fictitious 0-valued stop element, as can easily be checked by means of Table 15-4.

The serial code conversion from reflected binary to natural binary code is rather easy in a serial representation with the most significant bit first, but it seems impossible to perform this serial conversion if the least significant digit is put in front. When it is required to read directly into a computer a reflected binary code combination resulting from an analog-to-digital conversion process, the procedure with the most significant digit in front is rather unsuitable.

However, the serial conversion of a reflected binary code into a natural binary code with the least significant digit first can be implemented as Gray et al. have shown. This conversion depends on two considerations with regard to the interdependence of the corresponding combinations of both codes.

1. The first or least significant digit of the resulting natural binary code is as follows, in accordance with Eq. (15-33):

$$A_1 = \sum_1^n a_i \ (\text{mod } 2) \tag{15-35}$$

This digit is obtained by reading the reflected binary code to be converted into a pulse halver that has been set initially in its rest condition. Hence, this digit can only be known once the complete reflected binary code combination to be converted has been received.

2. Since $a_k = A_k \oplus A_{k+1}$ [Eq. (15-34)], it is also true that

$$A_{k+1} = A_k \oplus a_k \tag{15-36}$$

because $A_k \oplus a_k = A_k \oplus A_k \oplus A_{k+1} = A_{k+1}$.

Equation (15-36) states that the $(k + 1)$st digit of the desired natural binary code can be formed by the addition mod 2 of digit k of the resulting natural binary code combination as already determined and digit k of the reflected binary code combination to be converted. This can be performed in a serial operation with the least significant digit in front if the first digit of the resulting binary code combination is known in advance. Consideration 1 above shows that this digit depends on all digits of the reflected binary code combination, but since this digit can only be 0 or 1, there are only two possibilities.

The successive digits of the desired natural binary code symbol with an unknown least significant digit z are as follows, in accordance with Eq. (15-36):

$$\begin{aligned} A_1 &= z \\ A_2 &= a_1 \oplus z \\ A_3 &= a_2 \oplus a_1 \oplus z \\ &\cdots\cdots\cdots\cdots \\ A_n &= a_{n-1} \oplus a_{n-2} \oplus \cdots \oplus a_2 \oplus a_1 \oplus z \end{aligned} \qquad (15\text{-}37)$$

And since there is no digit A_{n+1},

$$A_n = a_n \qquad (15\text{-}38)$$

so that

$$a_n = a_{n-1} \oplus a_{n-2} \oplus \cdots \oplus a_2 \oplus a_1 \oplus z \qquad (15\text{-}39)$$

or

$$a_n \oplus a_{n-1} \oplus a_{n-2} \oplus \cdots \oplus a_2 \oplus a_1 \oplus z = 0 \qquad (15\text{-}40)$$

The two possible natural binary code symbols can be generated simultaneously by means of a pulse halver that performs the serial mod 2 addition required by Eq. (15-36). Following the generation of both code symbols, the correct solution will be the one for which Eq. (15-40) holds. Since z is the rest state of the pulse halver performing the addition mod 2 in accordance with Eq. (15-36), the correct code symbol will be the one that is generated by the output terminal of the pulse halver, which produces a 0 signal at the end of the conversion process. This is independent of the initial rest condition of this pulse halver. The symbolic diagram of a code converter performing this operation is shown in Fig. 15-20. The time chart for the conversion of the mth symbol of a series of reflected binary code symbols into the equivalent natural binary code symbol is given in Fig. 15-21.

The reflected binary code symbol to be converted is read into pulse halver P via input gates O_3 and O_4, which sense the digits of the reflected binary signal by means of clock pulse α' connected to input gates O_1 and O_2 of P. The arbitrary rest condition of pulse halver P represents the least significant digit of the mth natural binary-code symbol to be generated. The two-phase result of the code conversion performed by pulse halver P is read into a so-called shift register (see Chap. 16) consisting of as many sections as there are digits in the code combinations to be converted. The information content of this shift register is shifted one place to the right under the control of each clock pulse α'. Since the shift register also has two output terminals with opposite phases, both solutions of the code conversion will appear at the output of this shift register S. After the sensing of the last digit of a reflected binary code symbol, trigger T can be set or reset depending on the

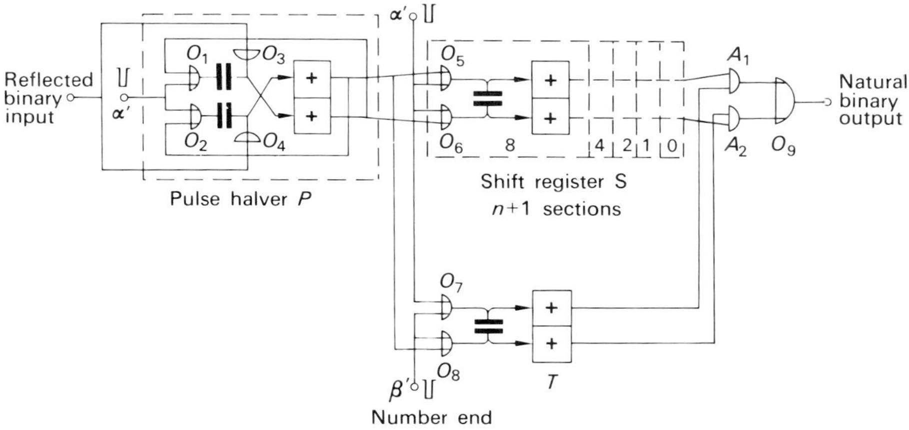

Fig. 15-20

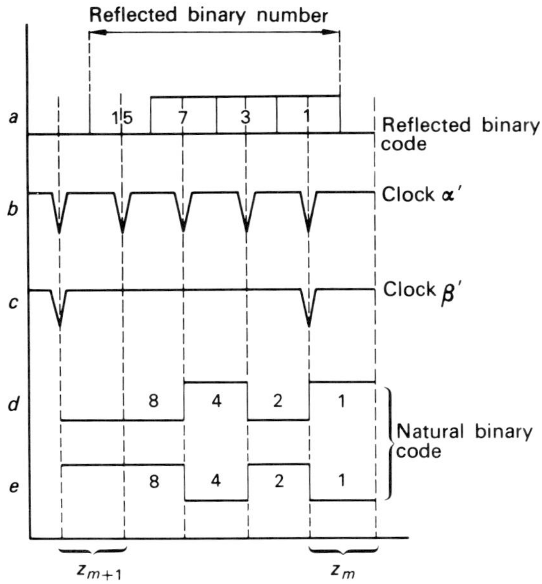

Fig. 15-21

0 output terminal of pulse halver P, as a result of which the correct gate of the two gates A_1 and A_2 will be opened.

The code symbol to be converted is shown on line a and the sensing clock pulse α' on line b in the diagram of Fig. 15-21. This clock pulse is given at one-digit intervals in the middle of the time slot in order to ensure the best operating margin. Trigger T, which will determine the required solution of the code-converting process, is positioned once per code combination, as indicated on line c of the time chart. The two possible results of the code-converting process are shown on lines d and e of the time chart. The first digit of the mth natural binary signal to be generated is indicated by z_m. The line—d or e—which will carry the 1 or 0 signal will be determined by the preceding code combination.

PROBLEMS

1. Define a reflected binary code and show the basic difference between this code and the conventional binary code.

2. Draw a time chart of a reflected binary counter built up of pulse halvers. Derive the control equations for the successive pulse halvers of such a counter and draw a symbolic circuit for

(*a*) An asynchronous operating circuit
(*b*) A synchronous operating circuit
(*c*) A synchronous operating self-checking circuit

3. Draw the time chart of a reflected binary counter with five relays and derive the switching formulas. Show that this counter can also produce a conventional binary output.

4. Give equations for the conversion of the digits of

(*a*) Reflected binary numbers into conventional binary numbers
(*b*) Conventional binary numbers into reflected binary numbers

Determine weights for the digits of a reflected binary number.

5. Draw the time chart of a serial conversion of a conventional binary number into a reflected binary number and give a symbolic circuit for such a conversion with a description of its operation.

6. Draw the time chart and symbolic circuit of a serial conversion of a reflected binary number into a conventional binary number. Show that this type of serial code conversion can also be performed with the lowest-valued digit in front.

7. Explain the construction of the time chart of a reflected binary counter with five relays. What are the values of the elements of the reflected binary code and in what manner can each code symbol be characterized by a decimal number?

SELECTIVE BIBLIOGRAPHY

1. Fischmann, A. F.: A Gray Code Counter, *IRE Trans.*, vol. EC-6, p. 120, 1957.
2. Flores, Ivan: Reflected Number Systems, *IRE Trans.*, vol. EC-5, pp. 79–82, 1956.
3. Fuss, F. A.: The Use of a Reflected Code in Digital Control Systems, *IRE Trans.*, vol. EC-3, pp. 1–7, 1954.
4. Gilbert, E. N.: Gray Codes and Paths on the N-cube, *Bell System Tech. J.*, vol. 37, no. 3, pp. 815–816, 1958.
5. Gray, H. J., Jr., P. V. Levonian, and M. Rubinoff: An Analogue to Digital Converter for Serial Computing Machines, *Proc. IRE*, vol. 41, pp. 1462–1465, 1953.

6. Oberman, R. M. M.: A Relay Counting Chain in the Reflected Binary Code, *Electrotechniek*, vol. 32, no. 15, pp. 295–297, 1954 (in Dutch).
7. Oberman, R. M. M.: Reflected Binary Codes, Symposium Digitale Technieken bij Meten en Regelen, Royal Institution of Engineers, 1958 (in Dutch).
8. Tootill, G. C.: The Use of Cyclic Permuted Codes in Relay Counting Circuits, *Proc. Inst. Elec. Engrs.* (*London*), vol. 103, pp. 432–436, 1956.
9. Yeager, R. E.: The Gray to Binary Translator and Shift Register, The Transistor, pp. 611–626, Bell Telephone Laboratories, 1951.

16
Shift Registers

16-1 SHIFT REGISTERS WITH RELAYS AS MEMORY ELEMENTS

A shift register is a device by means of which digital information can be stored temporarily while—during the processing operation—the information is transferred from one memory element to the next under control of a clock pulse. The time chart of a three-section shift register receiving, for example, the binary number 110011 is shown in Fig. 16-1. Clock pulse α controls all operations.

A shift register consists therefore of a number of memory elements that pass the information to be shifted from one element to the next. When doing so, it is not possible to pass the information directly from one shift section to the next. In such a case, difficulties would arise because while a shift section was receiving a new digit from the preceding section it would simultaneously transport the stored digit to the next section. It is impossible to perform these two operations on a memory element without mutual interference. These two operations can be separated by introducing an auxiliary memory element before each main memory element. The information to be shifted is read into the auxiliary memory elements during the clock pulse and shifted into the main memory elements during the pause after the clock pulse, so that the shift register can function as a store when the series of clock or counting pulses is stopped.

In Fig. 16-1, switching elements A, C, E, etc., are the auxiliary memory elements and switching elements B, D, F, etc., are the main memory elements. An auxiliary element and the following main memory element, for example

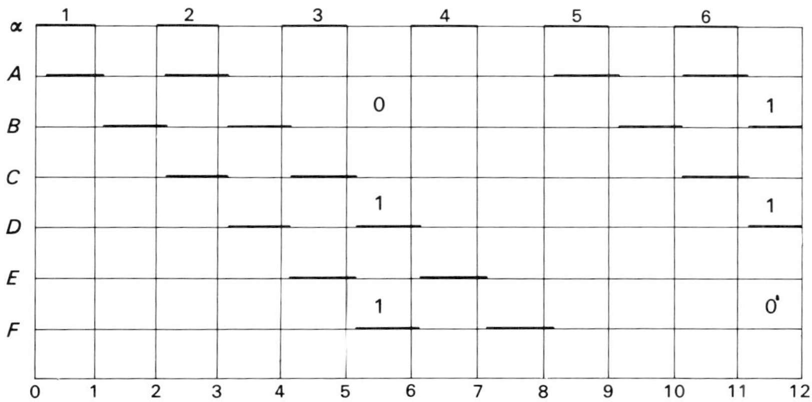

Fig. 16-1

A/B, C/D, E/F, together form a section of the shift register. Since the diagram of Fig. 16-1 will be implemented by means of relay circuits, the control pulse α and the pause between the α pulses will be of equal duration.

Auxiliary memory element A must be positioned under the control of the α clock contact at moment 0 in accordance with the first digit 1 of the example (110011) of Fig. 16-1. This auxiliary memory element then conditions the circuit for the main memory element B. This memory element B operates at moment 1, when the controlling α clock contact has been restored to its rest position. The time chart of Fig. 16-1 requires that auxiliary memory element A be restored to its rest condition after the main memory element B has been operated. This can be achieved by means of a break contact of switching element B in the holding circuit of element A. The first section of the shift register is now ready to receive the second digit of the binary number to be stored, while the second section of the shift register is ready to receive the information stored in section 1.

The second operation of section 1 of the shift register and the first operation of section 2 (in fact, this is also the second operation for section 2, because during the first cycle a 0 digit was shifted from section 1 to section 2!) are performed in the same way as shown above. At moment 2, the second

1 digit stored in the main memory element B positions auxiliary memory element C, which in turn resets the main memory element B to 0. At moment 3, auxiliary memory element A positions the main memory element B, and auxiliary memory element C positions the main memory element D.

At moment 4, the third digit (0) leaves auxiliary memory element A unoperated, while auxiliary memory elements C and E are operated by the preceding main memory elements, which are in the 1 position. The main memory element B remains unoperated at moment 5, while the main memory elements D and F are put in their operated condition (1) under control of auxiliary memory elements C and E.

It should be pointed out here that both the shifting of the information from an auxiliary memory element to a main memory element and the shifting from a main memory element to the following auxiliary memory element are in fact analogous operations that can be performed by means of the same type of circuit.

The time chart of Fig. 16-1 shows that at moment 6, three more digits push the registered information 110 out of the shift register so that at moment 11, the last three digits 011 of the code combination involved remain registered in the three sections A/B, C/D, and E/F of the shift register.

The time chart of Fig. 16-1 leads to the following switching equations for auxiliary memory element A and for the main memory element B:

$$A = s_a + ar_a' = \alpha i + ab' \tag{16-1}$$

$$B = s_b + br_b' = \alpha' a + bc' \tag{16-2}$$

The switching equations for the other memory elements can be derived from the above by cyclic permutation of the variables.

If the α_0 contact must be common to all auxiliary memory elements and the α_1 contact to all main memory elements, the relays must be provided with an operating and a holding winding in order to avoid unwanted sneak paths. The circuit resulting from switching equations (16-1) and (16-2) is shown in Fig. 16-2.

The winding of relay I receives the information to be shifted into the register. The serial transmission of information to the winding of relay I must be synchronized by means of clock α.

A second type of relay shift register can be designed on the basis of the time chart of Fig. 16-3. This diagram also illustrates the operation of a shift register consisting of three sections $A/B, C/D$, and E/F into which the number 110011 is read in. The various memory elements remain operated as long as a sequence of 1 digits is read in and are reset in the 0 condition only if a 0 is to be read into the memory element involved. This type of operation is comparable to the "superimposing" of information as is normally done on magnetic tape.

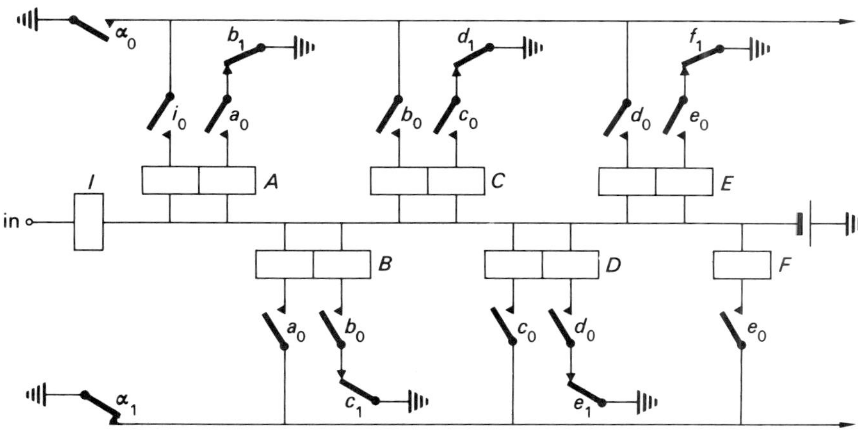

Fig. 16-2

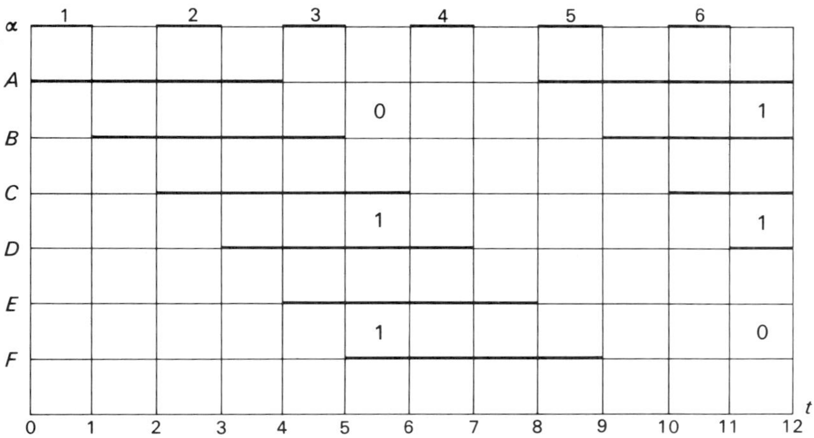

Fig. 16-3

It is possible to design a relay shift register to operate in this way by bringing all relay circuits under the control of clock contact α.

The switching equations of the various memory elements follow from the time chart, that is,

$$A = s_a + ar'_a = \alpha i + a\alpha' \tag{16-3}$$

$$B = s_b + br'_b = \alpha' a + b\alpha \tag{16-4}$$

$$C = s_c + cr'_c = \alpha b + c\alpha' \tag{16-5}$$

$$D = s_d + dr'_d = \alpha' c + d\alpha \tag{16-6}$$

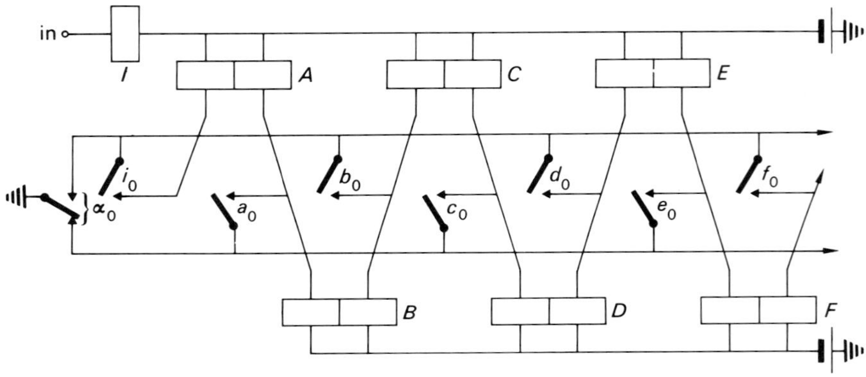

Fig. 16-4

It should be noted that these equations bear a great resemblance to Eqs. (11-7) and (11-8), which describe a pulse halver.

Switching equations (16-3) to (16-6) describe two sections of the shift register. It is not necessary to take a larger number of such sections into consideration because the operation of all other sections can be described by analogous equations.

The diagram derived from Eqs. (16-3) to (16-6) is shown in Fig. 16-4.

Make-before-break clock contact α can operate relay A (contact i closed); this latter relay closes its contact a so that it finds a holding circuit via the break side of clock contact α during the pause between two operations of this contact. This holding circuit for relay A is at the same time the operating circuit for relay B. Contact b closes the holding circuit for relay B during the working period of clock contact α; the circuit is at the same time the operating circuit for relay C, etc.

If a 0 digit is to be shifted in the register after a 1, contact i is open, while the holding circuit of relay A—still in the operating condition at the beginning of this time period—is opened by the break side of clock contact α. Relay A releases its armature and is reset in the 0 condition as required. The shifting of a 0 digit into the following memory elements is performed in an analogous way.

Coming back to the above remark concerning the resemblance between switching equations for pulse-halving circuits and shift registers, it can be readily seen that Eqs. (16-3) and (16-4) for memory elements A and B of the shift register assume the same form as the equations describing switching elements A and B of a pulse halver for

$$i_0 = b_1 \tag{16-7}$$

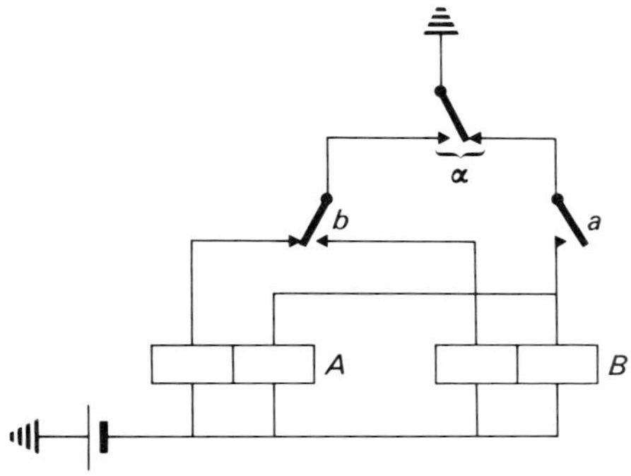

Fig. 16-5

If read-in contact i_0 is replaced by a break contact b_1, the circuit of the first two memory elements of the shift register in fact becomes a pulse-halving circuit.

$$A = \alpha b' + \alpha' a \tag{16-8}$$

$$B = \alpha' a + \alpha b \tag{16-9}$$

The resulting diagram illustrated in Fig. 16-5 is the same as the circuit of Fig. 11-4. The feedback of the output of a shift register memory element into the first memory element can also be extended to several of the sections. The shift register then changes into a distributor which can move with a variable information content. This property is inherent to all shift registers that operate in accordance with the diagram of Fig. 16-3.

Relay triggers can also be used as memory elements in shift registers. The use of triggers for this purpose will be described for shift registers using electronic components.

16-2 SHIFT REGISTERS WITH TRIGGERS AS MEMORY ELEMENTS

A shift register using triggers as memory elements can be based either on the time chart of Fig. 16-1 or on the chart given in Fig. 16-3. The set and reset equations for a shift register with triggers as memory elements as derived from the time chart of Fig. 16-1 are as follows:

$$s_a = \alpha \cdot i \qquad r_a = \alpha' b \tag{16-10}$$

$$s_b = \alpha' a \qquad r_b = \alpha \cdot c \tag{16-11}$$

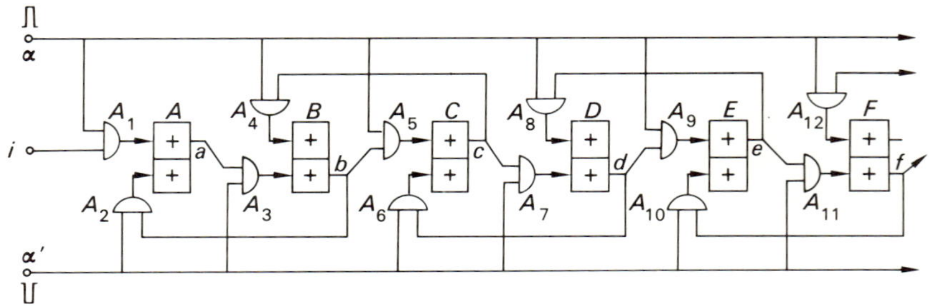

Fig. 16-6

The switching equations describing the memory elements of the following sections of the shift register can be found by cyclic permutation of the variables.

It should be pointed out in this connection that one might expect $r_a = b$ and $r_b = c$ to be sufficient for the reset signals. This is true for a relay circuit in which reset signals are interrupted by a make contact of the relay to be released. However, in the case of electronic triggers this type of reset signal is a continuous signal that leads to a condition in which both set and reset signals are applied simultaneously to the trigger. This type of condition is normally unacceptable in the design of a trigger.

Equations (16-10) and (16-11) lead to the diagram shown in Fig. 16-6. The triggers can consist of tubes, transistors, or other types of equivalent components.

The time chart of Fig. 16-3, which describes a shift register in which the next digit to be stored is superimposed on the information content of a memory element, leads to the following group of set and reset signals for the various memory elements:

$$s_a = \alpha i \qquad r_a = \alpha i' \tag{16-12}$$

$$s_b = \alpha' a \qquad r_b = \alpha' a' \tag{16-13}$$

$$s_c = \alpha b \qquad r_c = \alpha b' \tag{16-14}$$

$$s_d = \alpha' c \qquad r_d = \alpha' c' \tag{16-15}$$

$$s_e = \alpha d \qquad r_e = \alpha d' \tag{16-16}$$

On clock pulse α, the incoming i signal sets the A trigger in condition 1; this trigger remains in the condition stated until an i' signal resets the A trigger in condition 0 on one of the following clock pulses α. A memory element in this type of shift register remains in any given condition until an inverse digit is fed to it to be stored there. The same holds for all memory elements of the shift register. Equations (16-12) to (16-16) lead to Fig. 16-7; this diagram needs no further explanation.

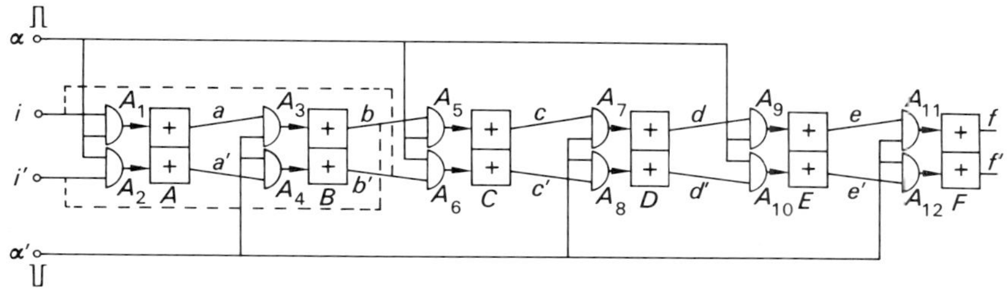

Fig. 16-7

If feedback of the output signals of the B memory element to the inputs of the A memory element is provided as shown in dotted lines in Fig. 16-7 a pulse-halving circuit is obtained in the same way as with the relay shift register of Fig. 16-4. The pulse halver thus obtained corresponds to the one shown in Fig. 11-6.

16-3 SHIFT REGISTERS USING MS MEMORY ELEMENTS

It has been shown in Chap. 11 that a pulse halver consisting of two triggers can be replaced by a circuit containing a single MS memory element. These MS memory elements can also be used in shift registers. Both types of MS memory elements (single-capacitor as well as double-capacitor) may be used for this purpose. It follows from the similarity between pulse halvers and shift-register sections that all pulse-halving circuits may be used as shift-register sections. The input gates must then be provided with an extra information input terminal.

The time chart of the operation of a shift register with MS memory elements is shown in Fig. 16-8. The clock pulses are indicated as short-duration pulses, as is normal in electronic circuitry. The number 110011 is shifted through the register. Only the state of the triggers in the various sections of the shift register is indicated in the time chart of Fig. 16-8.

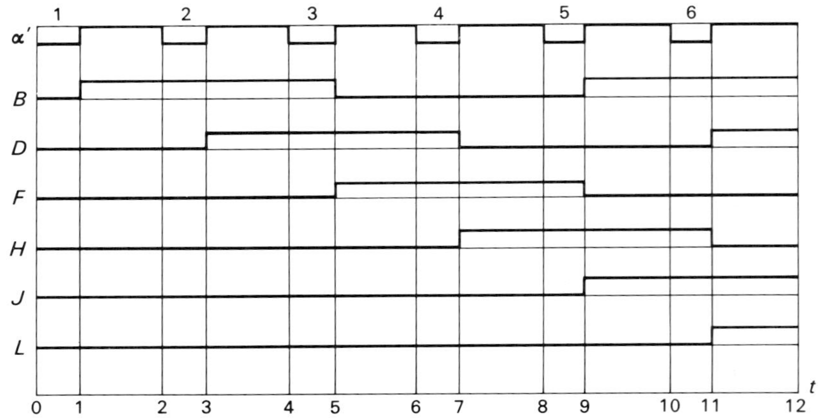

Fig. 16-8

Six sections are shown in the time chart of Fig. 16-8, so that after six clock pulses the complete number 110011 will have been shifted in the register. This is illustrated in the diagram between moments t_{11} and t_{12}. If the information signal at the input terminals of the shift register is called i, the set and reset functions of the trigger in the first MS memory element can be derived from the following truth table describing the desired operation:

i	b	B
0	0	0
0	1	0
1	0	1
1	1	1

The state equation for the first MS memory element (B) of the shift register corresponding to this truth table is

$$B = D_{\alpha'}(ib' + ib) = D_{\alpha'}(i) \tag{16-17}$$

The state equations for the MS memory elements of the following sections of this shift register can be obtained by replacing i by d, f, h, j. State equation $B = D_{\alpha'}(i)$ can be implemented by means of the circuits of Fig. 10-29. In the symbolic circuit of Fig. 16-9, the single-input solution given by the second column of Fig. 10-29a has been used.

In the following chapters the MS memory element will be used exclusively for shifting purposes.

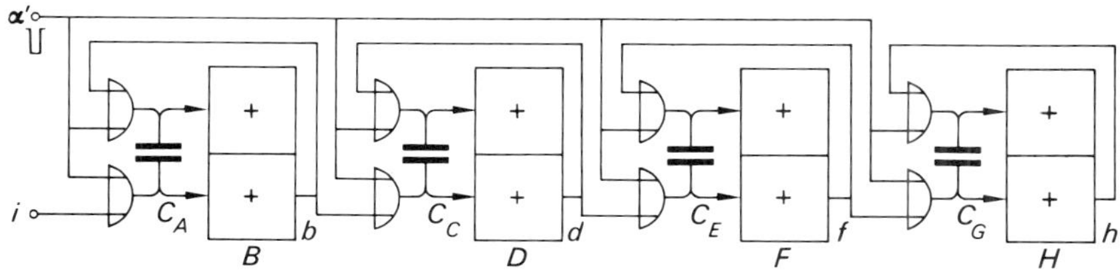

Fig. 16-9

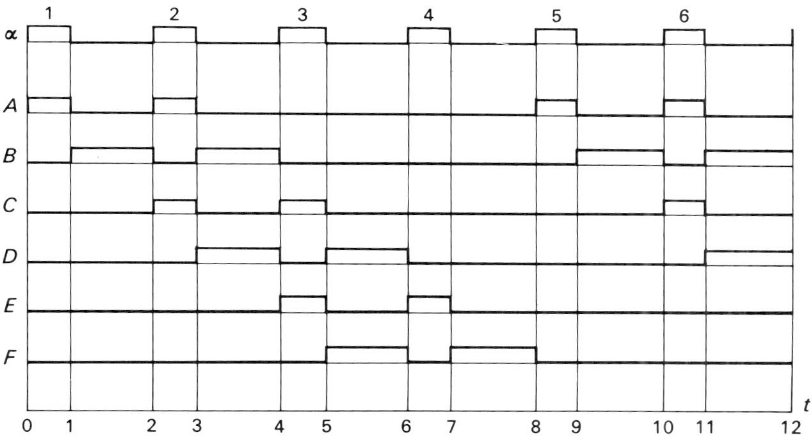

Fig. 16-10

16-4 SHIFT REGISTERS USING CAPACITORS AS COUPLING ELEMENTS BETWEEN TRIGGERS

The time chart of a shift register still allows for another variant, as is shown in Fig. 16-10. This time chart again shows the number 110011 being read into and shifted through the register, this code combination being the same as the one used in the previous examples.

The time chart given in Fig. 16-10 is in fact the time chart of a shift register with magnetic cores, which will be treated in Sec. 16-5. This type of time chart can also be implemented by means of triggers as memory elements and capacitors as coupling elements. Figure 16-10 is more or less analogous to the chart shown in Fig. 16-1. In Fig. 16-1, the B memory element coming into the 1 condition resets the A memory element into the 0 condition; whereas in the time chart of Fig. 16-10, the next memory element, e.g., the B element, is set in the 1 condition by a pulse generated by the reset of the preceding memory element A. During the cancelation, or "destruction," of the information in the A element, a 1 pulse, which is used to set the next memory element, is generated. There is no transmission of information between successive memory elements for 0 digits. This destructive method of information transport can also be performed with trigger circuits interconnected by means of capacitors. It is a possible, but not so commendable, switching method.

The time chart of Fig. 16-10 shows that the a' output terminal of the A memory element must be used to position the B memory element. If interconnected via a capacitor, a pulse $a' \cdot \underline{a}$ is transmitted from the A memory

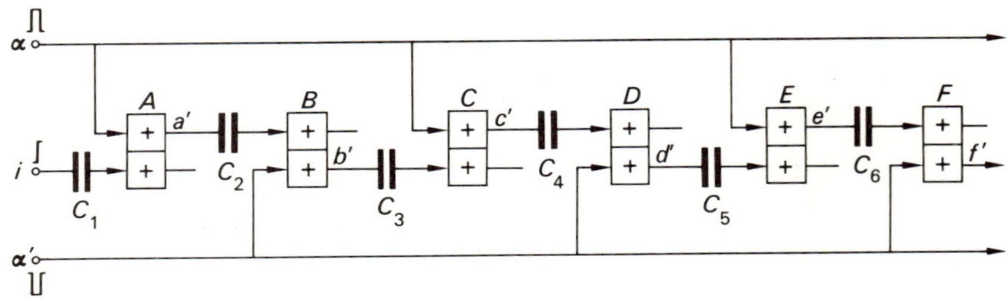

Fig. 16-11

to the B memory element. The B memory element is positioned by the leading edge of this pulse, but its trailing edge cannot influence the condition of this memory element. The circuits for interconnecting the other memory elements of this type of shift register can be found in the same way. The resulting circuit is shown in Fig. 16-11.

The shift register of Fig. 16-11 operates as follows. The transition of the incoming i signal from 0 to 1 brings the A memory element—a trigger—into the 1 condition. The output terminal a' then changes from 1 to 0, thereby generating a negative pulse that has no influence on the set terminal of the B trigger. The α clock pulse, which is connected to the reset terminals of the A, C, E, ... memory elements, resets all of these elements with a 1 condition. The triggers which are already in the 0 condition are not affected by this pulse. If the A trigger changes its condition from 1 to 0, the a' output terminal changes from 0 to 1, so that condenser C_2 feeds a short pulse $a' \cdot \underline{a}$ to the set input terminal of the B trigger. The information transport between the B and C memory elements is performed in the same way on the α' clock pulse.

The circuit of the type shown in Fig. 16-11 raises some difficulties in that there are no gates controlled by clock pulses between the memory elements. A setting memory produces a transition of the outgoing signal, which results in a negative pulse in this circuit being fed to the next trigger. This pulse has no influence on the condition of that trigger; it cannot disturb the correct operation of the shift register.

16-5 SHIFT REGISTERS USING MAGNETIC CORES AS MEMORY ELEMENTS

The time chart of Fig. 16-10 is in principle also applicable to shift registers using magnetic cores as memory element. A magnetic core gives only an output signal when its magnetic condition is reversed. The same principle was used for the elements of the shift register described in the preceding section.

A special type of problem is found to occur in this type of magnetic-core shift register. A magnetic core used as part of a shift register will be connected not only to a core in a following section but also to a core in a preceding section. If the condition of a magnetic core is reversed to produce a setting pulse for a core in the next section, another unwanted pulse will also be generated in the circuit connecting this section with the core in the preceding section. Moreover, the core in the next section transmits while setting an unwanted pulse to the core in the following section.

The connecting circuits between cores in the successive sections of the shift register present a problem that is more difficult to solve than the functional switching problem, the latter being essentially the same as for the shift register shown in Fig. 16-11.

The time chart of the magnetic-core shift register shown in Fig. 16-13 is given in Fig. 16-12. The shift register operates under the control of two interleaved series of α clock pulses, which have the same effect as a single series of α clock pulses given as reversals. The first 1 digit is stored in core A at moment 0. The α_1 clock pulse at moment 1 resets core A in the 0 condition. This produces a pulse in winding n_2 which is transmitted to winding n_1

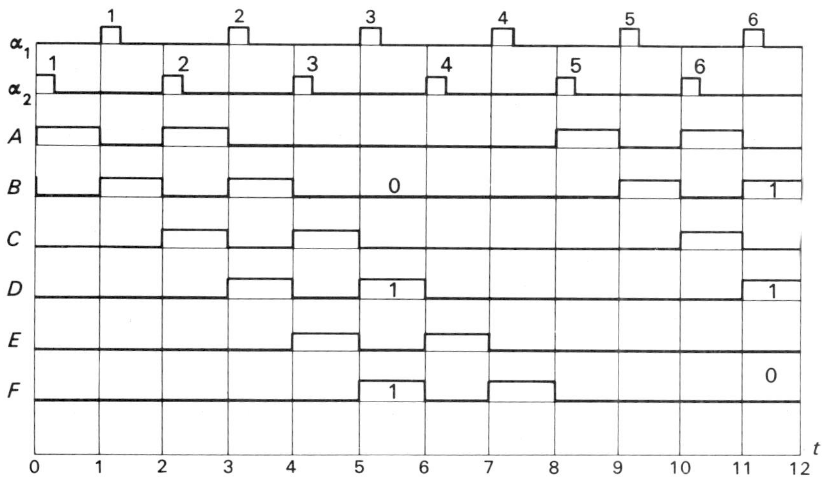

Fig. 16-12

of core B. At moment 2, a new 1 digit can be registered in core A, and the α_2 clock pulse resets core B in the 0 condition, so that a pulse is produced in winding n_2 which is transmitted to winding n_1 of core C, etc. A 0 digit leaves the A core in the 0 condition, so that there is no information transfer on the next α_1 pulse to core B. It should be noted that the transfer of 1 digits from one section to another in this type of shift register is again based on the destructive principle.

The direction of the normal information transfer current i_1 is shown as a dotted line and arrow in the loop between cores B and C. Current i_1 flows in the forward direction through diode D_3 but cannot flow through diode D_4 because it is blocking for that direction of the current. Current i_1 is mainly limited by resistor R_2. The unwanted back current i_2 flows in the direction of the dotted arrow in the loop between cores B and A. Diode D_2 is conducting for that direction of the current, so that a short circuit is made to protect winding n_2 against current i_2. Finally, the transfer of a 1 digit from core B to core C produces an unwanted current i_3 in the loop between cores C and D, which is also shown as a dotted arrow in Fig. 16-13. Diode D_5 is nonconducting for that direction of the current. Core D is therefore unaffected. The winding ratio $n_2 : n_1$ is normally taken as 2:1.

The magnetic cores of the auxiliary memory elements of each section of a shift register may be replaced by capacitor memories. This type of shift register, which can be controlled by a single α clock pulse, operates in accordance with the time chart of Fig. 16-14. The combination 110011 is read into and shifted through this register. A digit to be shifted through the circuit of this shift register (Fig. 16-15) will be received during the pause between two α clock pulses (e.g., between moments 1 and 2 in Fig. 16-14). The α clock pulse starting at moment 2 resets the A core into the 0 condition. This induces a pulse in winding n_2 of core A. This pulse is temporarily stored in capacitor C_1. Inductance L_1 gives a periodic character to the delayed discharge of capacitor C_1 (the negative parts are cut off by diode D_2), thereby bringing core C, via its winding n_1, into the 1 condition between moments 3 and 4. Core A can be set again in the 1 condition during the working cycle of clock pulse 2.

There is no information transfer on 0 digits in this shift register; a "noise" current is generated instead.

There is no ready answer to the question of whether the shift register of Fig. 16-15 is more economical and more reliable than the register shown in Fig. 16-13. From the point of view of switching, both types have against them their low efficiency as regards the energy transfer between the windings of successive cores caused by the relatively high resistor R ($R = 30$ ohms). This requires a rather high energy level for clock pulse α.

This difficulty has been overcome by using shift registers with a split-winding information transfer loop, as was already discussed in principle in

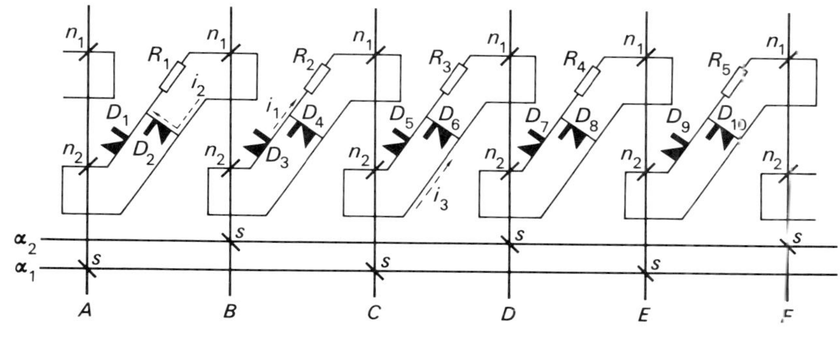

Fig. 16-13

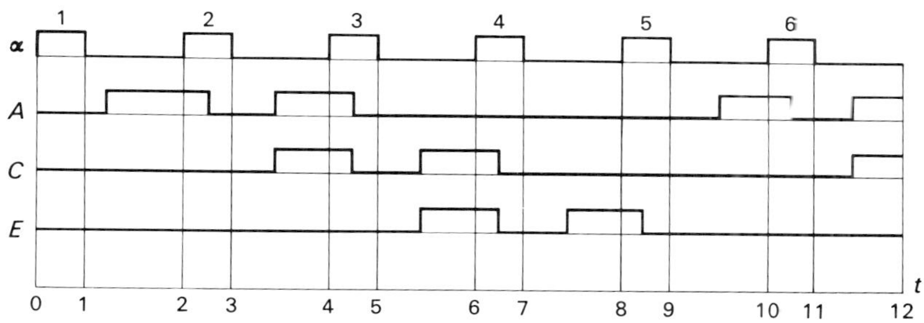

Fig. 16-14

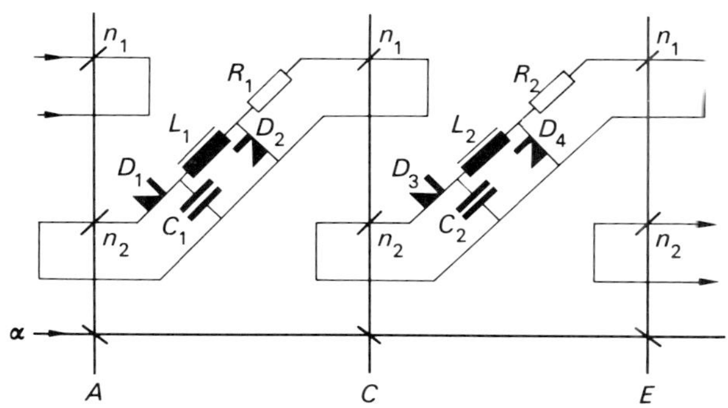

Fig. 16-15

Chap. 7. The operation of these shift registers is illustrated by the time chart of Fig. 16-12. The circuit representing this type of shift is shown in Fig. 16-16. Each core is connected to the next one by means of a split-winding information transfer loop that consists of two parallel paths. The first path comprises a winding on the core to be reset by clock pulse α and a diode and a winding on the core to be set; the second path consists of a diode and a winding on the core to be set. The information-receiving windings on the core to be set have opposite directions of magnetization, but they have an equal number of turns.

Clock pulse α_1 is transmitted through one group of split-winding transfer loops connected in series; clock pulse α_2 passes through a second group of such transfer loops, which is shifted by one core with respect to the first group.

If a clock pulse is fed to a transfer loop with the transmitting and the receiving cores in the 0 condition, the impedance of the winding on the transmitting core is nearly 0, so that the clock current is almost equally divided over both parallel circuits of the transfer loop. The receiving core is then nearly equally magnetized over both opposing windings, so that the net magnetizing force of the receiving core is nearly 0. The effect of small imbalances in the diodes can be reduced by connecting a small resistor in series to each of these diodes. Moreover, the *BH* characteristic of a core is reversible for small magnetizing forces.

If a clock pulse is applied to a transfer loop with the transmitting core in the 1 condition and the receiving core in the 0 condition, the impedance of the transfer circuit containing the winding of the transmitting core is large relative to that of the other transfer circuit. The receiving core will be magnetized by $(i_1 - i_2) \cdot n_2$ amp-turns. This must be sufficient to set the receiving core in the 1 condition. Current i_2 must be sufficient to reset the transmitting core in the 0 condition, because it is only during the reset process of the transmitting core that there is a difference between the currents in both parallel paths of the transfer loop. The receiving core must have been set during the reset time of the transmitting core.

There is neither an unwanted backward current nor an undesired forward current in this type of information transfer loop when used in a shift register with magnetic cores. The two diodes D_1 and D_2 in both parallel paths of the transfer loop prevent the transfer of information between successive cores by switching individually the cores back and forth between the 0 and the 1 condition. This means that the information can be read into this type of shift register in parallel form or extracted in that same form without there being simultaneously a shift—highly undesired at that moment —of the information in the shift register.

The efficiency of the energy transfer in the split-winding loop is rather high, so that one core can switch a number of other cores.

The shift register of Fig. 16-16 has a number of attractive features, but it has the following two disadvantages: there is no information transfer for 0 digits and the clock-pulse current varies with the information to be transferred from one section of the shift register to the next. The variation of the clock-pulse current causes the load of the power supply to vary, and this may be undesirable in some cases. The lack of information transfer during the

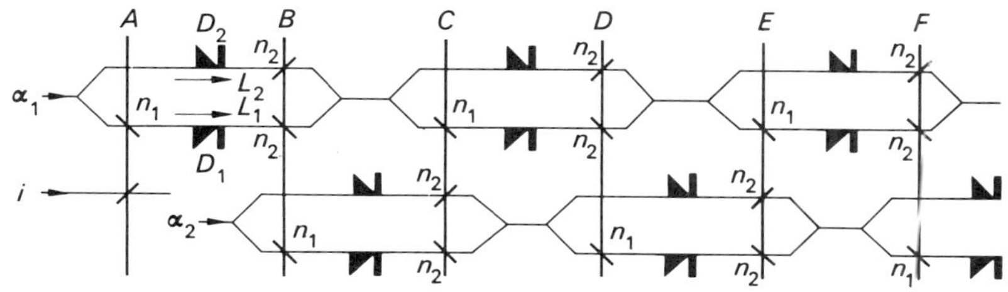

Fig. 16-16

shifting of 0 digits may leave a fault undetected for some time. Both disadvantages can be remedied by the doubling of the number of cores in the shift register of Fig. 16-16, as shown in the shift register of Fig. 16-17. The cores are grouped in pairs which together form one memory element. One of the cores of each pair is always in the 1 condition, the other being then in the 0 condition.

Once a pair of cores has transferred its information to the next pair, the transmitting pair remains in a neutral condition, both cores being then in the 0 condition. A clock pulse α_1 can be applied in such situation where, for example, the A pair of cores is in the 0 condition (for example, $A_1 = 0$ and $A_2 = 1$) and the B pair in the neutral condition ($B_1 = 0$ and $B_2 = 0$). This clock pulse resets core A_2 into the 0 condition, while core A_1 remains in its 0 condition. This causes a difference in impedance between the two paths of the transfer loop, so that transfer current i_1 is large and i_2 is small. Current i_1 will leave core B_1 in the 0 condition and set core B_2 in the 1 condition, which together represent the 1 condition of the memory element B of the shift register.

If clock pulse α_1 were applied in a situation where the A pair of cores is in the 1 condition ($A_1 = 1$ and $A_2 = 0$), it would reset core A_1 so that current i_2 would then be large. Current i_2 would then set core B_1 in the 1 condition and leave core B_2 in the 0 condition. The B pair would then represent the 1 condition.

Because transfer currents i_1 and i_2 are no longer equal for the transfer of a 0, diodes D_1 and D_2 need no longer be equal either.

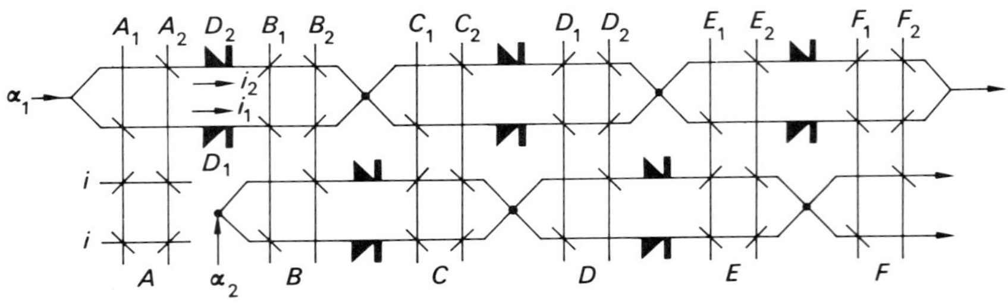

Fig. 16-17

Transistors are very useful to separate the successive memory elements of a core shift register. The double-diode transfer loop of Fig. 16-13 can be replaced by the single-transistor transfer loop shown in Fig. 16-18. The α_1 clock pulse resets, for example, core A into the 0 condition, so that a pulse is induced in winding n_2 of that core. This negative pulse opens transistor T_1 for a short moment, so that the collector current sets core B via winding n_1. The α_2 clock pulse transfers this information to core C. The operation of this type of shift register has already been illustrated by means of the time chart of Fig. 16-12. The setting of core B has no influence on core C, because the polarity of the voltage induced in winding n_2 of core B is such that transistor T_2 is blocked. A backward current flow from a transmitting core to the preceding core is impossible via the interconnected transistor.

A variant of the shift register of Fig. 16-18 is shown in Fig. 16-19. The operational reliability of the auxiliary-capacitor memory element depends on the correct dimensioning of this component and of the self-inductance of L_1 (Fig. 16-15). The transfer loop shown in Fig. 16-19, by using a transistor, eliminates this dependence on the dimensioning of the components.

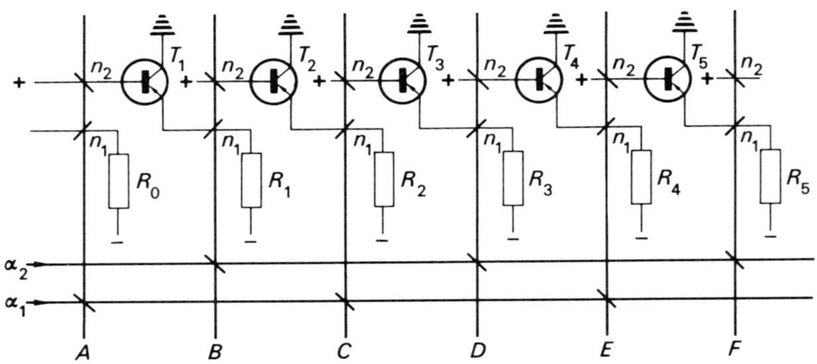

Fig. 16-18

Clock pulse α, when resetting a core in the 0 condition, induces a pulse in winding n_2 of a core, so that the corresponding capacitor is charged. The diode in the circuit prevents the charging of the capacitor whenever a core is set via winding n_1.

During the α clock pulse, the capacitor voltage ($C_1, C_2, \ldots$) cannot make the base of the transistor ($T_1, T_2, \ldots$) negative relative to the base because the α clock pulse renders the capacitor circuits sufficiently positive via diodes D_2, D_4, ... After the end of the α clock pulse, the bases of the various

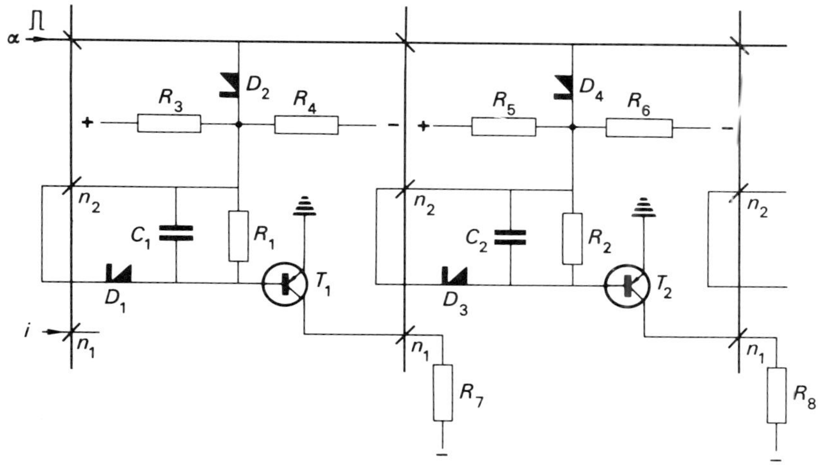

Fig. 16-19

transistors connected to the auxiliary memory elements are controlled by the capacitor voltages. Only those capacitors which are charged during the reset process of a core from the 1 condition into the 0 condition will open the corresponding transistor so that, via the winding n_1, the core of the next section of the shift register will be set.

The advantage of this type of shift register compared with the register shown in Fig. 16-15 is that its operation does not entirely depend on a correct dimensioning of the components. More types of shift registers can be designed, but the examples given above will be sufficient for the proper understanding of other types.

16-6 COUNTING SHIFT REGISTERS WITH FEEDBACK FROM THE LAST SECTION ONLY

A shift register is primarily a device through which (binary) information or (binary) code symbols are shifted for some information-processing purpose. The information or code symbols originate then from an outside source.

A shift register can also be used as a counter with constant or variable information contents by using a feedback from the last section to the first section with no external variable information input. The counting pulses then act as clock pulses.

COUNTING SHIFT REGISTER WITH STRAIGHT FEEDBACK

The general form of this type of counter using MS memory elements is given in Fig. 16-20. The theory of the following chapters applies equally well to all other types of shift registers. Counting shift registers of this class must be preconditioned by means of a code combination not consisting of $n \times 0$ or $n \times 1$. Any other combination will be shifted around in n counting steps, n being the number of shift sections. However, when the number of sections is not a prime number, some code combinations will lead to a degeneration of the number of different counting steps

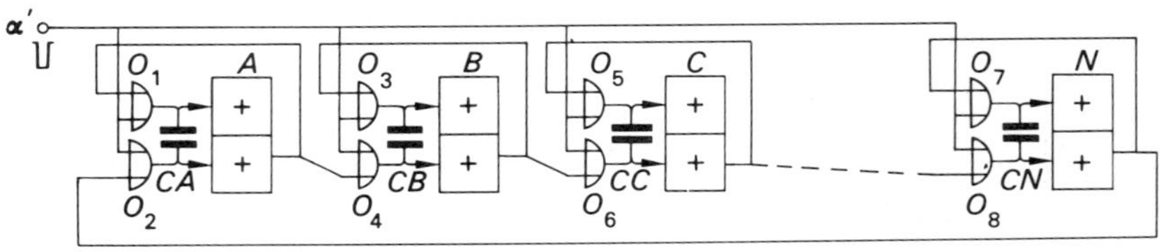

Fig. 16-20

For example, a six-section shift register preconditioned by means of code symbol 100100 counts in a three-step cycle which is a degeneration of the possible six-step cycle. The three-step cycle is shown in Table 16-1.

Table 16-1

No.	*U*	*V*	*W*	*X*	*Y*	*Z*
1	1	0	0	1	0	0
2	0	1	0	0	1	0
3	0	0	1	0	0	1

Degeneration of the length of the counting cycle will not occur for all n-out-of-m code combinations in which is $n \neq 0$ or $n \neq m$ and in which the 1 digits form a closed group.

The disadvantage of using this type of shift register as a counter is that there is no lockout circuit on the various sections of the shift register to ensure that no code symbol other than the correct one is allowed to circulate.

When a decimal decoding tree is required in which the kth terminal of this tree is marked after counting k pulses, the best initial code symbol is 10000 .... In relay shift registers, this kth contact circuit can be formed by means of a single contact. In electronic circuits, a single diode connected to an output terminal of the kth section of the shift register is required.

COUNTING SHIFT REGISTER WITH INVERTED FEEDBACK

When the output terminals of an n-section shift register are cross-connected to the input terminals, the counting cycle of the circuit will be extended (degenerations excepted) at most to $2n$ steps, n being the number of sections of the shift register.

A well-known example of this type of counting circuit is the pulse halver. This circuit is in fact a one-section shift register with the output terminals cross-connected to the information input terminals, so that a two-step counting cycle is obtained.

One of the simplest types of decimal counters is in fact a five-section counting shift register of this type. The code table of this circuit starting from the code combination 00000 is given below in Table 16-2. Implemented

Table 16-2

No.	*V*	*W*	*X*	*Y*	*Z*
0	0	0	0	0	0
1	1	0	0	0	0
2	1	1	0	0	0
3	1	1	1	0	0
4	1	1	1	1	0
5	1	1	1	1	1
6	0	1	1	1	1
7	0	0	1	1	1
8	0	0	0	1	1
9	0	0	0	0	1

with relay pairs, this counter is one of the simplest of its kind. Contact i_0 of the shift register of Fig. 16-4 must be replaced by a break contact j of the second relay of the fifth pair in order to obtain a counter operating in accordance with Table 16-2.

The contact tree for the translation of the code symbols of this table into the decimal code is also very simple. The code-translating circuit is shown in Fig. 16-21.

A decimal relay counter operating in accordance with Table 16-2, including the decimal contact tree, can be constructed by means of ten relays of standard design with two windings and two transfer contacts each.

It should be pointed out that other code combinations than those given in Table 16-2 can be used as a starting combination in a decimal counter of this type; however, the group of code symbols given in that table leads to the most systematic contact tree.

Furthermore, the counting shift registers of the type shown above go out-of-step through faulty operation of one of the relays, the group of code symbols originally inserted then being replaced by an unwanted group. This results in a complete malfunction of the register.

If implemented by electronic means, the code-translating circuit required for the decimal translation of Table 16-2 can best be made by means of ten diode gates with two diodes each, as shown in Fig. 16-22.

Counters of this type with an odd number of M steps can be derived from the circuit with an even number of N steps, where

$$M = N - 1 \qquad (16\text{-}18)$$

A counter operating basically in accordance with Table 16-2 in an even-step cycle can be changed into an odd-step-cycle counter by omitting one of the code symbols from the cycle. Any one of the code combinations of the even-step cycle can be skipped. The design of an odd-step circuit will be demonstrated for the case where code symbol 9 of Table 16-2 is skipped, so that a nine-step cyclic counting circuit will be obtained. The counter must then step from code symbol 00011 to 00000 instead of 00001. The skipping of the last code symbol only affects the gate circuits of the MS memory elements of the last section.

It follows from an analysis of Table 16-2 that the set-input gate of section Z must be blocked after code symbol 8 has been reached, whereas a reset signal must then be provided to the reset input gate of that section. This situation is characterized by $x = 0$ and $y = 1$.

The desired operation will be obtained by adding a blocking signal x' to the set input gate O_3 (Fig. 16-23) so that the set signal for Z is inhibited on the tenth pulse of a series, whereas an AND gate A_1 must be introduced in the reset path of Z in order to allow the generation of a reset signal $x \cdot y$ for Z to be reset at the correct moment.

The resulting circuit (for the last sections only) is shown in Fig. 16-23.

The elimination of any of the other code symbols leads to the same type of circuit, which is, however, incorporated in a different part of the shift register.

16-7 COUNTING SHIFT REGISTERS WITH CONSTANT-RATIO CODE SYMBOLS

The example of the two-out-of-five counter shown in Fig. 14-14 was given independently of the general solution of all two-out-of-n counters by means of shift registers. When the use of shift registers for the purpose of counting

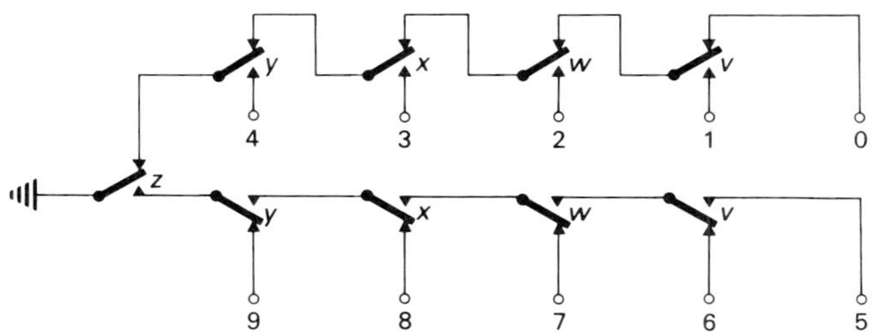

Fig. 16-21

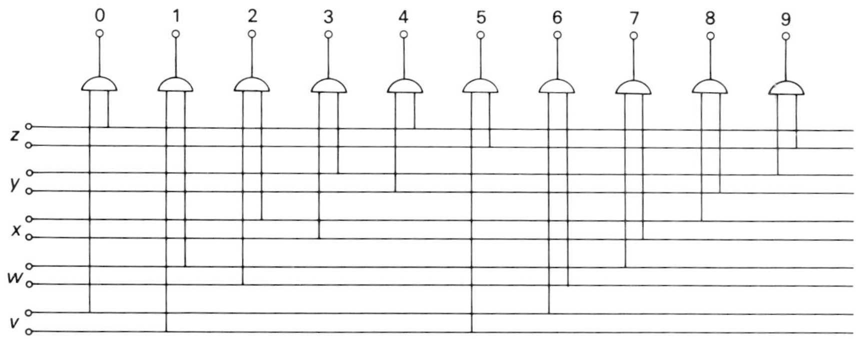

Fig. 16-22

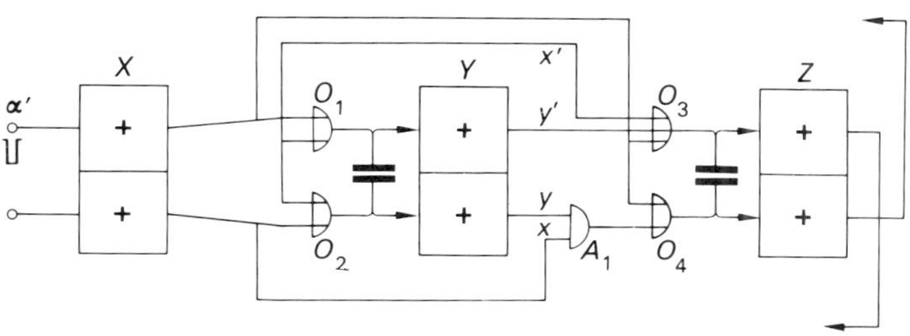

Fig. 16-23

in a two-out-of-n code is considered, a rather obvious way is to arrange the code combinations in groups in such a way that the code symbols in these groups are derived from each other by a simple shifting operation, whereas a special operation must be performed to change the basic code symbol from one group into the basic combination for the next group.

This general method of circuit design, i.e., the use of a simple basic organization that fulfils a large part of the given requirements and conforms to the specified exceptions to the basic organization, leads generally to the simplest circuit, which can most readily be understood. The counters given in this section are a perfect illustration of this concept of circuit design.

The number of code combinations in a two-out-of-n code is

$$N = \frac{n(n-1)}{2} \tag{16-19}$$

This number of code combinations can always be subdivided into

$$\frac{n(n-1)}{2} : n = \frac{n-1}{2}$$

groups of n symbols, which are derived from each other by means of a shifting operation. For all cases where n is even, one of the groups of code symbols degenerates to half-length because of symmetry.

It has been shown that all two-out-of-n codes can be arranged as shown above in basically the same way, so that a general block diagram can be drawn for all two-out-of-n shift registers operating in accordance with that mode.

However, this most obvious way does not lead to the simplest circuit of which the operation can also be easily explained.

The combinations of any two-out-of-n code can be arranged in such a sequence that the successive symbols follow from each other by two different modes of shifting operations. The first mode of operation is the ordinary shifting operation indicated symbolically in Fig. 16-24, and the second mode will occur under the control of the first k sections of the shift register as indicated schematically in Fig. 16-25. For $n = 5$ or 6, $k = 2$. The exception signal is thus in both cases $a'b'$ (0 0). In general, the exception signal that changes the operation from mode 1 into mode 2 is $k \times 0$, in which $k = [(n-1)/2]$,

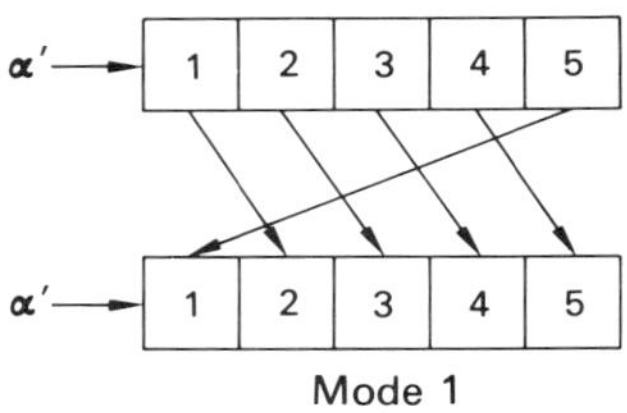

Fig. 16-24

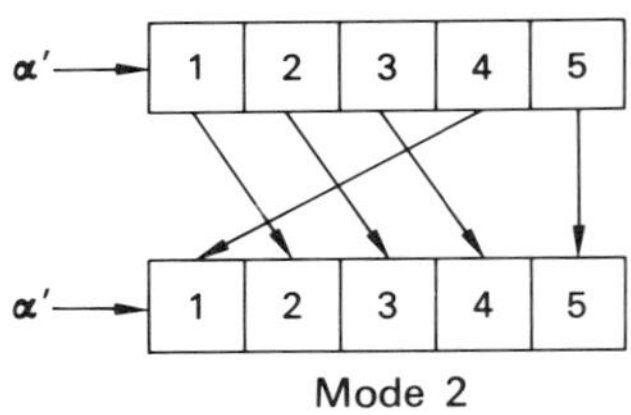

Fig. 16-25

where the brackets indicate that the integral part of the enclosed quantity must be taken into account. Table 16-3 shows the sequence of symbols for the two-out-of-five code, and Table 16-4 shows the sequence for the two-out-of-six code, arrived at by means of the above method. When the circuit operates in mode 2, the last section is blocked and the information contents of the last section but one is shifted into the first section. The operation of this basic type of two-out-of-n counter can be described by a general set of state equations. A capital indicates the state of a section of the circuit; and the index 0 refers to the present state and the index 1 to the next state of the circuit. This set of basic equations is as follows:

Table 16-3

No.	*A*	*B*	*C*	*D*	*E*
1	1	1	0	0	0
2	0	1	1	0	0
3	0	0	1	1	0
4	1	0	0	1	0
5	0	1	0	0	1
6	1	0	1	0	0
7	0	1	0	1	0
8	0	0	1	0	1
9	0	0	0	1	1
10	1	0	0	0	1

Table 16-4

No.	*A*	*B*	*C*	*D*	*E*	*F*
1	1	1	0	0	0	0
2	0	1	1	0	0	0
3	0	0	1	1	0	0
4	0	0	0	1	1	0
5	1	0	0	0	1	0
6	0	1	0	0	0	1
7	1	0	1	0	0	0
8	0	1	0	1	0	0
9	0	0	1	0	1	0
10	1	0	0	1	0	0
11	0	1	0	0	1	0
12	0	0	1	0	0	1
13	0	0	0	1	0	1
14	0	0	0	0	1	1
15	1	0	0	0	0	1

$$
\begin{aligned}
A_1 &= (A_0 + B_0 + C_0 + \cdots) N_0 + A_0' B_0' C_0' \cdots M_0 \\
B_1 &= A_0 \\
C_1 &= B_0 \\
&\cdots\cdots\cdots\cdots\cdots\cdots \\
N_1 &= (A_0 + B_0 + C_0 + \cdots) M_0 + A_0' B_0' C_0' \cdots N_0
\end{aligned}
\qquad (16\text{-}20)
$$

Equations (16-20) can be implemented by means of a shift register consisting of MS memory elements. The symbolic diagram is shown in Fig. 16-26. The circuit design is extremely simplified if in the mode 2 operation, the first and last sections of the circuit are completely blocked against the mode 1 operation, i.e., the normal shift operation. A special circuit must be provided for setting the first section in the mode 2 operation under control of the last section but one. This circuit part is indicated in Fig. 16-26 by means of a special input to section A.

It should be pointed out that the basic type of counting shift register as given above can be converted into different numbers of steps by modifying the interstage logic and/or the number of 1's in the code symbol.

The proof that this general design method of a two-out-of-n counter always yields all these combinations is slightly different according to whether n is odd or even.

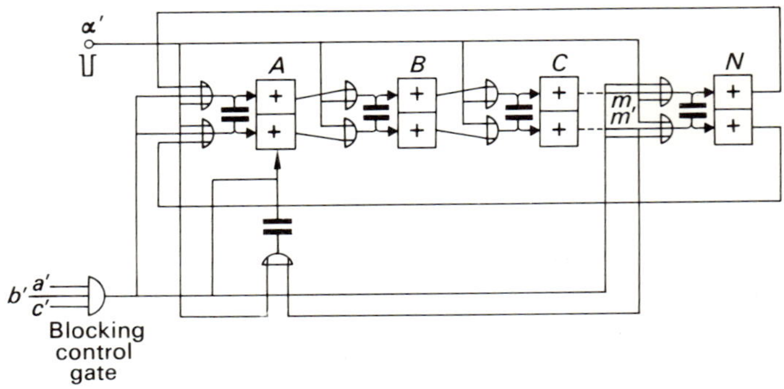

Fig. 16-26

n **is odd** The shift register operates in two modes. The first is the ordinary shift operation. In the second, after the detection of $[(n-1)/2] \times 0$ starting from the left, section n is locked and bypathed. Starting from code symbol 10...01, $n-1$ steps must always occur before the mode 2 operation has any visible effect (code symbol 2). Then a 1 arrives for the first time in section $n-1$ and is shifted through to section 1, as is indicated by code symbol 3.

This is repeated $(n-1)/2$ times. The last symbols of these groups of $n-1$ symbols are, respectively, 00 . . . 00110, 00 . . . 01010, 00 . . . 010010, etc. The operation via the bypath stops on code symbol 5 in Table 16-5 when the number of successive 0's (starting from the left) has decreased to $[(n-1)/2]-1$.

The distance between the two 1's is then also $[(n-1)/2]-1$. On the next step, code symbol 5 is shifted into code symbol 6. At that moment, a 1 is found in section n; the number of successive 0's (starting from the left) is $(n-1)/2$, so that the right-hand 1 is trapped in the last section. On each following counting pulse, the distance between the two 1's is decreased by 1, so that in the group of symbols starting with 6 and ending with 7, $(n-1)/2$ symbols will be generated. Table 16-5 will accordingly contain

$$N = \frac{n-1}{2} \cdot (n+1) + \frac{n-1}{2} = \frac{n(n-1)}{2} \tag{16-21}$$

symbols.

Table 16-5

1	100.......001	
	110.......000	
	011.......000	$n-1$
	....	
2	000......0110	
3	100......0010	
	010......0001	
	101......0000	$n-1$
	....	
4	000......1010	
	100......0100	
	....	$n-1$
	000.....10010	
	....	
	....	$n-1$
5	00..010..0010	
6	00..001..0001	
	00..0001..001	
	00..00001..01	$\frac{n-1}{2}$
	....	
7	00..0000..011	

n **is even** The proof for counting shift registers with an even number of sections is similar. The detection for the mode 2 operation must occur on $[(n-2)/2] \times 0$ counting from the left. There are now $(n-2)/2$ groups of $n-1$ symbols to begin with. After that, a group of $(n-2)/2$ symbols with $(n-2)/2$ 0's between the 1's will be generated, this group ending with the combination $000 \ldots 1000 \ldots 1$. A 1 is now trapped in the last section, so that a final group of $\frac{1}{2}n$ symbols will be generated in which the distance between the two 1's is decreased to 0. The total number of combinations will then be

$$N = \frac{n-2}{2} \cdot (n-1) + \frac{n-2}{2} + \tfrac{1}{2}n = \frac{n(n-1)}{2} \tag{16-22}$$

A further improvement in the combinational use of the various sections of the counting shift register will be realized by use of three-out-of-n code symbols. The decoding trees of these counters and their internal logic are of a more complex nature, so that their practical value becomes questionable. Kautz has shown that the design of three-out-of-n counting shift registers can be generalized by means of three modes of operation, viz., the ordinary shifting operation and two special transition modes. However, it has been found that for $n = 7$ and $n = 11$ it is possible to use only two modes of operation. Table 16-6 describes the operation of a three-out-of-seven counting shift register.

Table 16-6

No.	*A*	*B*	*C*	*D*	*E*	*F*	*G*
1	1	1	1	0	0	0	0
2	0	1	1	1	0	0	0
3	0	0	1	1	1	0	0
4	1	0	0	1	1	0	0
5	0	1	0	0	1	1	0
6	0	0	1	0	0	1	1
7	0	0	0	1	0	1	1
8	1	0	0	0	1	0	1
9	1	1	0	0	0	1	0
10	0	1	1	0	0	0	1
11	1	0	1	1	0	0	0
12	0	1	0	1	1	0	0
13	0	0	1	0	1	1	0
14	1	0	0	1	0	1	0
15	0	1	0	0	1	0	1
16	1	0	1	0	0	1	0
17	0	1	0	1	0	0	1
18	1	0	1	0	1	0	0
19	0	1	0	1	0	1	0
20	0	0	1	0	1	0	1
21	1	0	0	1	0	0	1
22	1	1	0	0	1	0	0
23	0	1	1	0	0	1	0
24	0	0	1	1	0	0	1
25	0	0	0	1	1	0	1
26	1	0	0	0	1	1	0
27	0	1	0	0	0	1	1
28	1	0	1	0	0	0	1
29	1	1	0	1	0	0	0
30	0	1	1	0	1	0	0
31	0	0	1	1	0	1	0
32	0	0	0	1	1	1	0
33	0	0	0	0	1	1	1
34	1	0	0	0	0	1	1
35	1	1	0	0	0	0	1

Mode 2 operation occurs under the control of signal $a'b'c$. Sections F and G are then blocked and E is shifted to A. This is basically the same type of operation as in the general type of two-out-of-n counting shift registers. The resulting circuit has the same character as the network of Fig. 16-26.

16-8 COUNTING SHIFT REGISTERS WITH MOD 2 FEEDBACK FROM A NUMBER OF SECTIONS

More complex feedback networks than a simple cross-connection from the last section to the first will cause a k-section shift register to generate series of 0's and 1's longer than $2k$. The longest sequence is of course a series of 2^k digits. This is quite feasible, but of special interest are those sequences of length $2^k - 1$ which can be generated with mod 2 feedback networks (combinations of exclusive OR gates) only. These series are called maximum-length recursive sequences or m sequences.

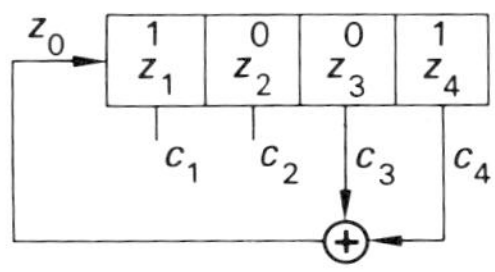

Fig. 16-27

The digit that will be generated in a certain feedback network in order to be fed back into the shift register on the next clock pulse can be represented by the following equation:

$$z_0 = c_1 z_1 \oplus c_2 z_2 \oplus c_3 z_3 \oplus \cdots \oplus c_k z_k \tag{16-23}$$

Variables z_1 to z_k represent the information contents of the shift register at a given moment. Coefficients c_1 to c_k indicate whether or not a specific section of the shift register is connected to the mod 2 adders of the feedback network. These coefficients are either 0 or 1.

An example is given in Fig. 16-27. The four-section shift register contains at the moment shown code symbol 1 0 0 1, whereas sections 3 and 4 are connected to a mod 2 adder. The digit to be generated in that case is determined by $z_1 = 1$, $z_2 = 0$, $z_3 = 0$, $z_4 = 1$, $c_1 = 0$, $c_2 = 0$, $c_3 = 1$, and $c_4 = 1$, so that

$$z_0 = 0 \cdot 1 \oplus 0 \cdot 0 \oplus 1 \cdot 0 \oplus 1 \cdot 1 = 1 \tag{16-24}$$

The next code combination in the shift register will then be 1 1 0 0. Starting from any code combination of k digits except $k \times 0$, a shift register generating a maximum-length sequence will run through all possible code symbols of k digits except $k \times 0$. When the shift register is in position $k \times 0$, the feedback network will always generate 0, so that the information contents of the shift register can never change.

The $2^k - 1$ code symbols of a maximum-length-sequence shift register follow each other in a sequence which is not of the ordinary binary type. Combined into a table, the k streams of 0's and 1's in the k columns show the *m* sequence as illustrated in Table 16-7. This table gives the sequence of the 15 code symbols generated by the four-section shift register shown in Fig. 16-27. The numbering of the code symbols in Table 16-7 is arbitrary, since constant-weight coding is not possible.

Table 16-7

No.	z_1	z_2	z_3	z_4
1	1	0	0	1
2	1	1	0	0
3	0	1	1	0
4	1	0	1	1
5	0	1	0	1
6	1	0	1	0
7	1	1	0	1
8	1	1	1	0
9	1	1	1	1
10	0	1	1	1
11	0	0	1	1
12	0	0	0	1
13	1	0	0	0
14	0	1	0	0
15	0	0	1	0

The fact that only symbol $k \times 0$ is missing in the series of $2^k - 1$ symbols leads to some important properties of the maximum-length recursive sequences.

The number of 1's in Table 16-7 or any other table for other length of *m* sequences is $(2^k \cdot k)/2 = k \cdot 2^{k-1}$. Hence, the number of 1's in a maximum-length recursive sequence will be

$$N = 2^{k-1} \tag{16-25}$$

in which *k* is the number of digits in the generated symbol or the number of sections of the generating shift register.

Furthermore, the sum mod 2 of all digits of an *m* sequence will be 0:

$$z_1 \oplus z_2 \oplus z_3 \oplus \cdots \oplus z_{2^k-1} = 0 \tag{16-26}$$

Two important questions must be solved in this section:

1. How can *m* sequences be constructed by means of a *k*-section shift register with its corresponding suitable feedback network, and what are these sequences?
2. What are the required feedback connections of the shift register?

The construction of m sequences can be based in all cases on the same type of numbering process, which will be given here without proof. The digits of the desired m sequence are written on a row, and below, in a second row, these digits are written in a sequence with the indices increasing by two. This becomes for a $2^k - 1 = 15$-digit recurrent series:

$$\begin{array}{ccccccccccccccc} z_1 & z_2 & z_3 & z_4 & z_5 & z_6 & z_7 & z_8 & z_9 & z_{10} & z_{11} & z_{12} & z_{13} & z_{14} & z_{15} \\ z_1 & z_3 & z_5 & z_7 & z_9 & z_{11} & z_{13} & z_{15} & z_2 & z_4 & z_6 & z_8 & z_{10} & z_{12} & z_{14} \end{array} \qquad (16\text{-}27)$$

The value (0 or 1) of each pair of variables in the columns above must be the same if a recurrent series is to be obtained. These variables z_1 to z_{15} can be arranged, as shown in Table 16-8, in groups with the same value.

Table 16-8

1	2	3	4	5	6	7	8	9	10	11	12	13	14	15
z_1														
	z_2	z_3		z_5				z_9						
			z_4			z_7			z_{10}			z_{13}		
					z_6					z_{11}				
							z_8				z_{12}		z_{14}	z_{15}

Since the desired sequence of length $2^k - 1 = 15$ must contain $2^{k-1} = 8$ 1 digits, there are three ways to obtain this. Table 16-8 shows three groups of four variables, which may be chosen as 1 or 0. There are three possibilities to do so. The choice, however, is limited by the fact that a group of k successive 1 digits must be present in the sequence. In our example, a group of four successive 1's must be formed by a suitable choice of the values of the variables of the three groups of four. The two possible choices are indicated by enclosures in Table 16-8. The two resulting m sequences are

$$0\ 1\ 1\ 1\ 1\ 0\ 1\ 0\ 1\ 1\ 0\ 0\ 1\ 0\ 0 \qquad (16\text{-}28)$$

and

$$0\ 0\ 0\ 1\ 0\ 0\ 1\ 1\ 0\ 1\ 0\ 1\ 1\ 1\ 1 \qquad (16\text{-}29)$$

In the sequence of Eq. (16-28), the digits are in the reverse order of the sequence of Eq. (16-29). These sequences can be considered as being fundamentally the same.

The desired feedback connections between the shift register and the mod 2 adder can be determined when $2k$ successive digits of the maximum-length sequence are given. It follows from Eq. (16-23) that

$$z_0 \oplus c_1 z_1 \oplus c_2 z_2 \oplus c_3 z_3 \oplus \cdots \oplus c_k z_k = 0 \tag{16-30}$$

This equation is true for any part of an m sequence and the next digit to be generated, or the digit generated immediately before. Hence, in addition to Eq. (16-30), the following set of equations will be obtained:

$$z_1 \oplus c_1 z_2 \oplus c_2 z_3 \oplus \cdots \oplus c_k z_{k+1} = 0 \tag{16-31}$$

$$z_2 \oplus c_1 z_3 \oplus c_2 z_4 \oplus \cdots \oplus c_k z_{k+2} = 0 \tag{16-32}$$

. . . .

$$z_k \oplus c_1 z_{k+1} \oplus c_2 z_{k+2} \oplus \cdots \oplus c_k z_{2k} = 0 \tag{16-33}$$

The c coefficients determining the required feedback connections between the shift register and the mod 2 adder can be determined from the set of Eqs. (16-31) to (16-33) when $2k$ successive digits of the m sequence are given. The solution of this set of k equations will be simplest if the given part of this sequence contains a maximum number of 0's.

For example, this set of equations becomes, for the first eight digits of the sequence given in Eq. (16-28),

$$\begin{aligned} c_1 \oplus c_2 \oplus c_3 \oplus c_4 &= 0 \\ 1 \oplus c_1 \oplus c_2 \oplus c_3 &= 0 \\ 1 \oplus c_1 \oplus c_2 \oplus c_4 &= 0 \\ 1 \oplus c_1 \oplus c_3 \qquad &= 0 \end{aligned} \tag{16-34}$$

and for the last eight digits of that sequence, the set of equations then becomes

$$\begin{aligned} c_1 \oplus c_2 \qquad &= 0 \\ 1 \oplus c_1 \oplus c_4 &= 0 \\ 1 \oplus c_3 \qquad &= 0 \\ c_2 \qquad &= 0 \end{aligned} \tag{16-35}$$

The resulting feedback coefficients are $c_1 = c_2 = 0$ and $c_3 = c_4 = 1$.

The reverse sequence of Eq. (16-29) requires another feedback pattern. The corresponding set of equations for the c coefficients is then as follows:

$$\begin{aligned} c_3 \qquad &= 0 \\ c_2 \qquad &= 0 \\ c_1 \oplus c_4 \qquad &= 0 \\ 1 \oplus c_3 \oplus c_4 &= 0 \end{aligned} \tag{16-36}$$

The resulting c coefficients are

$$c_2 = c_3 = 0 \qquad \text{and} \qquad c_1 = c_4 = 1$$

It should be pointed out that the coefficient c_k will always be 1, because otherwise the sequence will degenerate into a sequence which can be generated by a shift register with one section less. With a number of sections larger than four, a number of basically different m sequences can be generated. Some are listed below as Table 16-9. More complete lists can be found in the literature.

Table 16-9

k	$2^k - 1$	Tap numbers on shift register: *Sequence*	*Reverse sequence*
2	3	1, 2	
3	7	2, 3	1, 3
4	15	3, 4	1, 4
5	31	3, 5 2, 3, 4, 5 1, 3, 4, 5	2, 5 1, 2, 3, 5 1, 2, 4, 5
6	63	5, 6 2, 3, 5, 6 1, 4, 5, 6	1, 6 1, 3, 4, 6 1, 2, 5, 6
7	127	6, 7 4, 7 4, 5, 6, 7 3, 4, 5, 7 2, 5, 6, 7 2, 4, 6, 7 1, 4, 6, 7 2, 3, 4, 5, 6, 7 1, 2, 4, 5, 6, 7	1, 7 3, 7 1, 2, 3, 7 2, 3, 4, 7 1, 2, 5, 7 1, 3, 5, 7 1, 3, 6, 7 1, 2, 3, 4, 5, 7 1, 2, 3, 5, 6, 7
8	255	4, 5, 6, 8 3, 5, 7, 8 3, 5, 6, 8 2, 5, 6, 8 2, 3, 7, 8 1, 6, 7, 8 2, 4, 5, 6, 7, 8 1, 2, 5, 6, 7, 8	2, 3, 4, 8 1, 3, 5, 8 2, 3, 5, 8 2, 3, 6, 8 1, 5, 6, 8 1, 2, 7, 8 1, 2, 3, 4, 6, 8 1, 2, 3, 6, 7, 8

16-9 POLYNOMIAL REPRESENTATION OF SEQUENCES GENERATED BY A MOD 2 FEEDBACK SHIFT REGISTER

POLYNOMIAL REPRESENTATION OF SEQUENCES

It has been shown in the preceding section that sequences of 0's and 1's can be generated by mod 2 feedback shift registers. But not every shift register with a mod 2 feedback network generates an m sequence. As an example, Tables 16-10 and 16-11 show the sequences of state combinations through

Table 16-10

n	S	Z_1	Z_2	Z_3	Z_4
1	0	0	0	0	0
1	1	0	0	0	1
2	0	1	0	0	0
3	0	0	1	0	0
4	1	0	0	1	0
5	1	1	0	0	1
6	0	1	1	0	0
7	1	0	1	1	0
8	0	1	0	1	1
9	1	0	1	0	1
10	1	1	0	1	0
11	1	1	1	0	1
12	1	1	1	1	0
13	0	1	1	1	1
14	0	0	1	1	1
15	0	0	0	1	1

Table 16-11

n	S	Z_1	Z_2	Z_3	Z_4
1	0	0	0	0	0
1	1	0	0	0	1
2	0	1	0	0	0
3	1	0	1	0	0
4	0	1	0	1	0
5	0	0	1	0	1
6	0	0	0	1	0
1	1	0	0	1	1
2	1	1	0	0	1
3	1	1	1	0	0
4	1	1	1	1	0
5	0	1	1	1	1
6	0	0	1	1	1
1	1	0	1	1	0
2	1	1	0	1	1
3	0	1	1	0	1

which the networks of Figs. 16-28 and 16-29 run. These tables indicate that, depending on the initial state, the network of Fig. 16-28 generates one sequence of length 1 and one m sequence of length 15, whereas the network of Fig. 16-29 generates one sequence of length 1, one of length 3, and two of length 6.

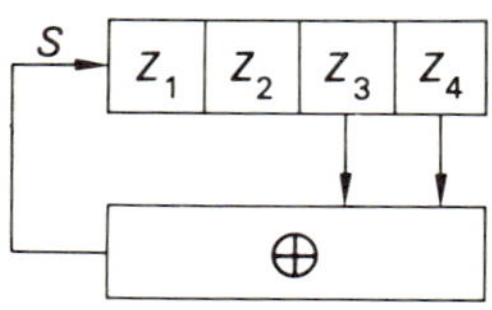

Fig. 16-28

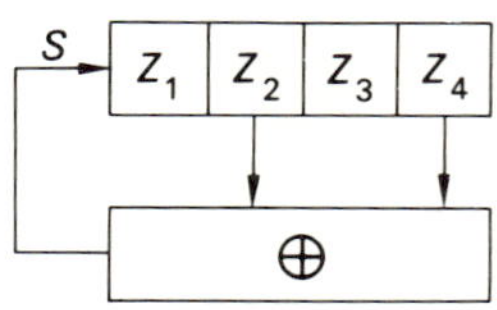

Fig. 16-29

Let the initial state of the shift register of Fig. 16-29 be 0 0 0 1. Then it follows from Table 16-11 that sequence S, which is fed into the input, is

$$1\,0\,1\,0\,0\,0\,1\,0\,1\,0\cdots \tag{16-37}$$

The order of the bits in sequence (16-37) is such that the left-hand bit is shifted into the register on the first clock pulse. This is illustrated in Fig. 16-30, which shows the input signal as a function of time.

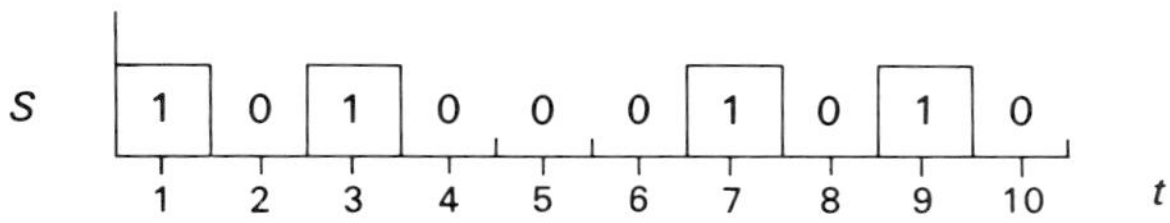

Fig. 16-30

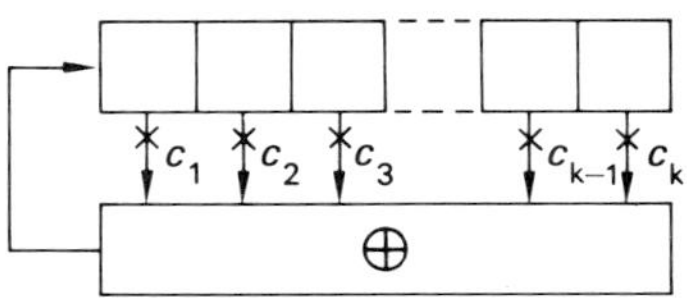

Fig. 16-31

The general form of a mod 2 feedback shift register is given in Fig. 16-31. Each section is connected via a switch c_i to the mod 2 adding feedback network, the output of which is connected to the input of the shift register. If the number of sections is indicated by k, c_k will always be 1, as has been pointed out in the preceding section.

The topic of this and the next section is to determine the number and lengths of the sequences which will be generated if the combination $c_1, c_2, \ldots, c_{k-1}$ is given.

A sequence S of infinite length can be given in the following form:

$$S = y_0, y_1, y_2, \ldots \tag{16-38}$$

This sequence can also be represented by a polynomial of the following form:

$$S(x) = y_0 \oplus y_1 x \oplus y_2 x^2 \oplus \cdots \tag{16-39}$$

$$= \sum_{i=0}^{\infty} y_i x^i \tag{16-40}$$

In polynomial (16-39) use has been made of the $\oplus$ sign to indicate that the coefficients must be taken mod 2. It should be noted that in mod 2 arithmetic $a = -a$ ($0 = -0$ and $1 = -1$). Hence $a - b = a \oplus b$ and $a \oplus a = 0$.

Polynomials the coefficients of which are subject to the laws of mod 2 arithmetic have the same properties as those subject to the laws of ordinary algebra. For example,

$$\begin{aligned}(1 \oplus x)(1 \oplus x^2 \oplus x^3) &= 1 \oplus x^2 \oplus x^3 \oplus x \oplus x^3 \oplus x^4 \\ &= 1 \oplus x \oplus x^2 \oplus (1 \oplus 1)\,x^3 \oplus x^4 \\ &= 1 \oplus x \oplus x^2 \oplus x^4 \qquad (16\text{-}41)\end{aligned}$$

Furthermore,

$$\begin{array}{r|rrrr|l} x^2 \oplus x \oplus 1 & x^4 \oplus & & x^2 \oplus x \oplus 1 & & x^2 \oplus x \oplus 1 \\ & x^4 \oplus x^3 \oplus x^2 & & & & \\ \hline & x^3 \oplus & & x & & \\ & x^3 \oplus x^2 \oplus x & & & & \\ \hline & & x^2 \oplus & & 1 & \\ & & x^2 \oplus x \oplus 1 & & & \\ \hline & & & x & & \end{array}$$

Hence, in the division of $x^4 \oplus x^2 \oplus x \oplus 1$ by $x^2 \oplus x \oplus 1$, the quotient q equals $x^2 \oplus x \oplus 1$ and the remainder $r = x$.

From sequence S of Eq. (16-38), a sequence S_a can be derived starting with element y_a of S $(a > 0)$.

$$S_a = y_a, y_{a+1}, y_{a+2}, \ldots \qquad (16\text{-}42)$$

$$S_a(x) = y_a \oplus y_{a+1}\,x \oplus y_{a+2}\,x^2 \oplus \cdots \qquad (16\text{-}43)$$

$$\begin{aligned} &= \frac{1}{x^a}(y_0 \oplus y_1 x \oplus \cdots \oplus y_{a-1}x^{a-1}) \oplus \frac{1}{x^a}(y_a x^a \oplus y_{a+1}x^{a+1} \oplus \cdots) \\ &\quad \oplus \frac{1}{x^a}(y_0 \oplus y_1 x \oplus \cdots \oplus y_{a-1}x^{a-1}) \\ &= \frac{1}{x^a}\left[S(x) \oplus \sum_{i=0}^{a-1} y_i x^i\right] \qquad (16\text{-}44)\end{aligned}$$

It is also possible to derive from sequence S a sequence S_{-a} starting with element y_{-a} by introducing a number of new elements.

$$S_{-a} = y_{-a}, y_{-a+1}, \ldots, y_{-1}, y_0, y_1, \ldots \qquad (16\text{-}45)$$

Then the polynomial representation of S_{-a} is as follows:

$$S_{-a}(x) = y_{-a} \oplus y_{-a+1}\,x \oplus \cdots \oplus y_{-1}\,x^{a-1} \oplus y_0\,x^a \oplus y_1\,x^{a+1} \oplus \cdots \qquad (16\text{-}46)$$

$$= \sum_{i=0}^{a-1} y_{-a+i} x^i \oplus x^a \sum_{i=0}^{\infty} y_i x \tag{16-47}$$

$$= x^a S(x) \oplus \sum_{i=0}^{a-1} y_{-a+i} x^i \tag{16-48}$$

It should be noted that S_a is uniquely determined by S but sequence S_{-a} is not.

A sequence S is called periodic over L if the following relation holds for every element y_i of S:

$$y_i = y_{i+L} \tag{16-49}$$

Sequence S can then be written in the following form.

$$S = y_0, y_1, \ldots, y_{L-1}, y_0, y_1, \ldots, y_{L-1}, y_0, y_1, \ldots \tag{16-50}$$

Hence

$$S(x) = \sum_{i=0}^{L-1} y_i x^i \oplus \sum_{i=0}^{L-1} y_i x^{L+i} \oplus \sum_{i=0}^{L-1} y_i x^{2L+i} \cdots$$

$$= (1 \oplus x^L \oplus x^{2L} \oplus \cdots) \sum_{i=0}^{L-1} y_i x^i \tag{16-51}$$

In this equation, the second factor represents one period of sequence S and the first factor indicates that this period is repeated. The first factor can also be written in the following form:

$$\sum_{j=0}^{\infty} x^{jL} = \frac{1 \oplus x^L}{1 \oplus x^L} \sum_{j=0}^{\infty} x^{jL}$$

$$= \frac{1}{1 \oplus x^L} \left(\sum_{j=0}^{\infty} x^{jL} \oplus \sum_{j=0}^{\infty} x^{(j+1)L} \right)$$

$$= \frac{1}{1 \oplus x^L} \left(\sum_{j=0}^{\infty} x^{jL} \oplus \sum_{j=1}^{\infty} x^{jL} \right)$$

$$= \frac{1}{1 \oplus x^L} \tag{16-52}$$

Hence

$$S(x) = \frac{1}{1 \oplus x^L} \sum_{i=0}^{L-1} y_i x^i = \frac{f(x)}{1 \oplus x^L} \tag{16-53}$$

or

$$\sum_{i=0}^{L-1} y_i x^i = (1 \oplus x^L) \cdot S(x) \tag{16-54}$$

For example, the sequence $S = 0\ 1\ 0\ 0\ 1\ 0\ 0\ 1\ 0$ is periodic over 3 and can be represented by the following polynomial:

$$S(x) = x \oplus x^4 \oplus x^7 \oplus x^{10} \oplus \cdots$$
$$= (1 \oplus x^3 \oplus x^6 \oplus x^9 \oplus \cdots)\, x \tag{16-55}$$

Multiplication by $1 \oplus x^L = 1 \oplus x^3$ gives

$$(1 \oplus x^3)\cdot S(x) = (1 \oplus x^3)\,(1 \oplus x^3 \oplus x^6 \oplus x^9 \oplus \cdots)\, x = x \tag{16-56}$$

Hence

$$S(x) = \frac{x}{1 \oplus x^3} \tag{16-57}$$

It is obvious that if a sequence S is periodic over L, the relation

$$S_L(x) = S(x) \tag{16-58}$$

is satisfied. This can easily be proved by substitution of $a = L$ and Eq. (16-54) in Eq. (16-44).

Equation (16-54) indicates that every sequence periodic over L can be represented by an expression of this form. In Eq. (16-53), $f(x)$ is at most of degree $L-1$. Mutually every polynomial $S(x)$ which can be written in the form of Eq. (16-53) with $f(x)$ of a degree smaller than L represents a sequence which is periodic over L.

$$S(x) = \sum_{j=0}^{\infty} x^{jL}\cdot f(x) = (1 \oplus x^L \oplus x^{2L} \oplus \cdots)\, f(x)$$
$$= f(x) \oplus x^L f(x) \oplus x^{2L} f(x) \oplus \cdots \tag{16-59}$$

The degree of $f(x)$ being smaller than L, this formula indicates that the y coefficient of x^i equals that of the coefficient of x^{L+i} for all nonnegative values of i. Hence S is periodic over L.

If the degree of $f(x)$ in Eq. (16-53) exceeds $L-1$, polynomial $S(x)$ can be written as follows:

$$S(x) = F(x) \oplus \frac{r(x)}{1 \oplus x^L} \tag{16-60}$$

In this formula, the degree of remainder $r(x)$ is less than L. Hence $[r(x)]/(1 \oplus x^L)$ represents a sequence which is periodic over L, whereas $F(x)$ is a polynomial of finite degree d representing a sequence of which all elements y_i with $i > d$ are 0.

Equation (16-60) illustrates that sequence S can be derived by adding the corresponding elements of these two sequences and is nonperiodic although it contains a periodic part starting from element y_{d+1}. For example,

$$S(x) = \frac{1 \oplus x^3 \oplus x^4}{1 \oplus x^3} = 1 \oplus x \oplus \frac{x}{1 \oplus x^3} \tag{16-61}$$

The two sequences involved are as follows:

$$\frac{x}{1 \oplus x^3} = x \oplus x^4 \oplus x^7 \oplus x^{10} \oplus \cdots = 0\,1\,0\,0\,1\,0\,0\,1\,0 \cdots$$
$$1 \oplus x = 1\,1\,0\,0\,0\,0\,0\,0\,0 \cdots$$
$$S = 1\,0\,0\,0\,1\,0\,0\,1\,0 \cdots$$

This sequence is nonperiodic; although starting from bit y_2, it is periodic over 3.

The least positive integer l over which a sequence S is periodic is called the cycle length of S. One cycle length of S will be represented by

$$s(x) = \sum_{i=0}^{l-1} y_i x^i \tag{16-62}$$

Hence, polynomial $S(x)$ can be written as follows:

$$S(x) = \frac{s(x)}{1 \oplus x^l} \tag{16-63}$$

INITIAL-STATE POLYNOMIALS

Now consider the shift register of Fig. 16-32 connected to a mod 2 adding feedback network. The initial state of the successive sections can be indicated as follows:

$$y_{-1}, y_{-2}, y_{-3}, \ldots, y_{-k} \tag{16-64}$$

The following time sequence S will be fed into this shift register:

$$S = y_0, y_1, y_2, \ldots, y_i, \ldots \tag{16-65}$$

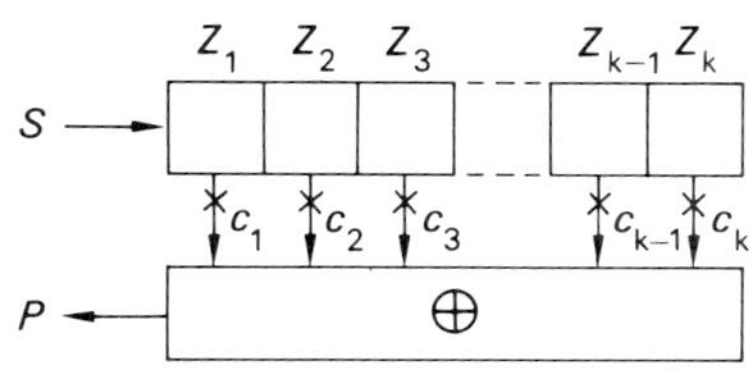

Fig. 16-32

The sequences passing through the successive sections Z_1 to Z_k of this primed shift register are then as follows:

$$\begin{array}{l} Z_1 :: S_{-1} = y_{-1}, y_0, \ldots, \quad y_{-1+i}, \ldots \\ \cdots \\ Z_j :: S_{-j} = y_{-j}, y_{-j+1}, \ldots, \quad y_{-j+1}, \ldots \\ \cdots \\ Z_k :: S_{-k} = y_{-k}, y_{-k+1}, \ldots, y_{-k+1}, \ldots \end{array} \tag{16-66}$$

The output sequence P of the mod 2 adding feedback network can be derived from Eq. (16-66) in the following way:

$$P = \sum_{j=1}^{k} c_j y_{-j}, \sum_{j=1}^{k} c_j y_{-j+1}, \ldots, \sum_{j=1}^{k} c_j y_{-j+i}, \ldots \tag{16-67}$$

The polynomial representation $P(x)$ of sequence P will be as follows:

$$P(x) = \sum_{j=1}^{k} c_j \cdot S_{-j}(x) \tag{16-68}$$

$$P(x) = \sum_{j=1}^{k} c_j \left[x^j S(x) \oplus \sum_{i=0}^{j-1} y_{-j+i} x^i \right] = S(x) \sum_{j=1}^{k} c_j x^j \oplus \sum_{j=1}^{k} c_j \sum_{i=0}^{j-1} y_{-j+i} x^i \tag{16-69}$$

In this equation, the second term is a polynomial in x, the coefficients of which depend on the initial state $y_{-1}, y_{-2}, y_{-3}, \ldots, y_{-k}$ It is called the initial-state polynomial and will be denoted by $m(x)$.

$$m(x) = \sum_{j=1}^{k} c_j \sum_{i=0}^{j-1} y_{-j+i} x^i \tag{16-70}$$

When the output of the mod 2 adding network is connected to the input of the shift register, sequence S will be identical to sequence P. Then it follows from Eqs. (16-65) and (16-67) that

$$y_i = \sum_{j=1}^{k} c_j y_{-j+i} \tag{16-71}$$

This equation holds for all nonnegative values of i, which can also be seen by inspection of Fig. 16-32. Using the identity $S(x) = P(x)$, it follows furthermore from Eq. (16-69) that

$$S(x) = S(x) \cdot \sum_{j=1}^{k} c_j x^j \oplus m(x) \tag{16-72}$$

so that

$$S(x) \cdot \left(1 \oplus \sum_{j=1}^{k} c_j x^j \right) = m(x) \tag{16-73}$$

In Eq. (16-73)

$$1 \oplus \sum_{j=1} c_j x^j = \phi(x) \tag{16-74}$$

represents the input connection of the shift register and the feedback conditions to the mod 2 adding network. It is called the characteristic polynomial of the mod 2 feedback shift register. Because c_k is always 1, the degree of $\phi(x)$ is k. Equation (16-73) can now be written as follows:

$$S(x) \cdot \phi(x) = m(x) \tag{16-75}$$

This equation is called the "basic equation of the mod 2 feedback shift register."

In order to find for a given initial state $y_{-1}, y_{-2}, y_{-3}, \ldots, y_{-k}$ the initial-state polynomial $m(x)$, it is convenient to write Eq. (16-70) as follows:

$$\begin{aligned} m(x) &= \sum_{j=1}^{k} \sum_{i=0}^{j-1} c_j y_{-j+i} x^i = \sum_{i=0}^{k-1} x^i \sum_{j=i+1}^{k} c_j y_{-j+i} \\ &= \sum_{i=0}^{k-1} m_i x^i \end{aligned} \tag{16-76}$$

This result can easily be checked by using the original polynomials and regrouping their terms into other polynomials having the above indicated ranges.

In Eq. (16-76), coefficient m_i is

$$m_i = \sum_{j=i+1}^{k} c_j y_{-j+i} \tag{16-77}$$

Example Consider, for instance, the circuit of Fig. 16-29 in which the feedback connections are chosen as follows:

$$c_1 = 0 \qquad c_2 = 1 \qquad c_3 = 0 \qquad c_4 = c_k = 1$$

Hence, the characteristic polynomial $\phi(x)$ of this mod 2 feedback shift register is as follows:

$$\phi(x) = 1 \oplus x^2 \oplus x^4$$

The initial state of the shift register is chosen as follows:

$$y_{-1} = 0 \qquad y_{-2} = 0 \qquad y_{-3} = 0 \qquad y_{-4} = 1$$

The coefficients m_i resulting from Eq. (16-77) are

$$\begin{aligned} m_0 &= c_1 y_{-1} \oplus c_2 y_{-2} \oplus c_3 y_{-3} \oplus c_4 y_{-4} = 1 \\ m_1 &= c_2 y_{-1} \oplus c_3 y_{-2} \oplus c_4 y_{-3} \qquad = 0 \\ m_2 &= c_3 y_{-1} \oplus c_4 y_{-2} \qquad\qquad\quad = 0 \\ m_3 &= c_4 y_{-1} \qquad\qquad\qquad\qquad\quad\; = 0 \end{aligned}$$

These equations show that the influence of the initial state $y_{-1}, y_{-2}, y_{-3}, \ldots, y_{-k}$ and consequently of the initial-state polynomial on polynomial $P(x)$ decreases to 0 in k successive steps.

$$m(x) = \sum_{i=0}^{k-1} m_i x^i = m_0 x^0 \oplus m_1 x^1 \oplus m_2 x^2 \oplus m_3 x^3 = 1$$

Substitution of the characteristic polynomial $\phi(x) = 1 \oplus x^2 \oplus x^4$ and the initial-state polynomial $m(x) = 1$ in the basic equation of the mod 2 feedback shift register leads to

$$S(x)\,(1 \oplus x^2 \oplus x^4) = 1$$

Hence

$$S(x) = \frac{1}{1 \oplus x^2 \oplus x^4} = 1 \oplus x^2 \oplus x^6 \oplus x^8 \oplus x^{12} \oplus \cdots = \frac{1 \oplus x^2}{1 \oplus x^6}$$

This result represents the sequence

$$S = 1\,0\,1\,0\,0\,0\,1\,0\,1\,0\,0\,0\,1\cdots$$

This sequence is in accordance with the first sequence of length 6 shown in Table 16-11.

The initial-state polynomial corresponding to the initial state 0 1 1 0 will be as follows:

$$m(x) = 1 \oplus x \oplus x^2$$

Hence, in this case,

$$S(x) = \frac{1 \oplus x \oplus x^2}{1 \oplus x^2 \oplus x^4} = \frac{1}{1 \oplus x \oplus x^2} = \frac{1 \oplus x}{1 \oplus x^3}$$

From this equation it follows that the corresponding sequence S has a cycle length of three of the form 1 1 0, which is in accordance with Table 16-11.

The initial-state polynomial $m(x)$ is uniquely determined by the initial state and mutually the initial state is uniquely determined by $m(x)$. This can be seen as follows. Let the coefficients $m_0, m_1, \ldots, m_{k-1}$ be given. Then for $i = k - 1$, Eq. (16-77) gives

$$m_{k-1} = c_k y_{-1} \tag{16-78}$$

and for $i = 0, 1, \ldots, k-2$

$$m_i = \sum_{j=i+1}^{k-1} c_j y_{-j+i} \oplus c_k y_{-k+i} \tag{16-79}$$

And since $c_k = 1$, this pair of equations leads to the following pair:

$$y_{-1} = m_k - 1 \tag{16-80}$$

and

$$y_{-k+i} = m_i \oplus \sum_{j=i+1}^{k-1} c_j y_{-j+i} \tag{16-81}$$

Hence, for a series of values of i, the following group of equations will be obtained:

$$
\begin{aligned}
i &= k-1: & y_{-1} &= m_{k-1} \\
i &= k-2: & y_{-2} &= m_{k-2} \oplus c_{k-1} y_{-1} \\
i &= k-3: & y_{-3} &= m_{k-3} \oplus c_{k-2} y_{-1} \oplus c_{k-1} y_{-2} \\
&\cdots & &\cdots \\
i &= 0 \quad : & y_{-k} &= m_0 \oplus c_1 y_{-1} \oplus c_2 y_{-2} \oplus \cdots c_{k-1} y_{-k+1}
\end{aligned}
\tag{16-82}
$$

By means of these equations the values of $y_{-1}, y_{-2}, \ldots, y_{-k}$ can be calculated successively.

Example For instance, consider again the circuit of Fig. 16-29 with

$$c_1 = 0 \qquad c_2 = 1 \qquad c_3 = 0 \qquad \text{and} \qquad c_4 = 1$$

and let $m(x)$ be

$$m(x) = 1 \oplus x \oplus x^2$$

so that

$$m_0 = 1 \qquad m_1 = 1 \qquad m_2 = 1 \qquad \text{and} \qquad m_3 = 0$$

Substitution of these values in Eqs. (16-82) leads to the following result:

$$
\begin{aligned}
y_{-1} &= m_3 && = 0 \\
y_{-2} &= m_2 \oplus c_3 y_{-1} && = 1 \\
y_{-3} &= m_1 \oplus c_2 y_{-1} + c_3 y_{-2} && = 1 \\
y_{-4} &= m_0 \oplus c_1 y_{-1} + c_2 y_{-2} \oplus c_3 y_{-3} && = 0
\end{aligned}
$$

In this example, the shift register of Fig. 16-29 was primed with the initial combination 0 1 1 0.

The above considerations indicate that for every combination m_0, m_1, ..., m_{k-1} and hence for every polynomial of degree less than k, just one corresponding initial state can be found. This is in accordance with the fact that there exist 2^k different initial-state polynomials of degree less than k and also 2^k corresponding different initial states for a shift register with k sections.

It follows from Eqs. (16-82) that initial-state polynomial $m(x) = 0$ corresponds to initial state 0, 0, ..., 0, 0, and since $c_k = 1$ the $m(x) = 1$ corresponds to initial state 0, 0, ..., 0, 1. It is noted that these two initial-state polynomials are independent of the feedback conditions.

Now consider again Fig. 16-31 with the initial state

$$y_{-1}, y_{-2}, \ldots, y_{-k}$$

If sequence $S = y_0, y_1, \ldots$ is generated, then the contents of the shift register after a series of a shift pulses will be as follows:

$$S_a = y_{-1+a}, y_{-2+a}, \ldots, y_{-k+a} \tag{16-83}$$

It is obvious that if this contents is used as initial state, the sequence S_a given by Eq. (16-44) will be generated. Because polynomial $S_a(x)$ can be expressed in $S(x)$ by means of Eq. (16-44), the initial-state polynomial $m_a(x)$ corresponding to the initial state (16-83) can be expressed in $m(x)$. This can be shown as follows, starting from Eq. (16-44):

$$S_a(x) = \frac{1}{x^a}\left[S(x) \oplus \sum_{i=0}^{a-1} y_i x^i\right]$$

$$S_a(x) \cdot \phi(x) = \frac{1}{x^a}\left[S(x) \cdot \phi(x) \oplus \phi(x) \cdot \sum_{i=0}^{a-1} y_i x^i\right]$$

$$= \frac{1}{x^a}\left[m(x) \oplus \phi(x) \cdot \sum_{i=0}^{a-1} y_i x^i\right] = m_a(x) \tag{16-84}$$

PROPERTIES OF SEQUENCES

Now let U and V be sequences which are generated by the same shift register starting from the initial states

$$y_{-1}^{(u)}, y_{-2}^{(u)}, \ldots, y_{-k}^{(u)} \tag{16-85}$$

and

$$y_{-1}^{(v)}, y_{-2}^{(v)}, \ldots, y_{-k}^{(v)} \tag{16-86}$$

The corresponding initial-state polynomials can then be represented by $m^{(u)}(x)$ and $m^{(v)}(x)$. Then sequence W, which is generated starting from the initial state

$$y_{-1}^{(u)} \oplus y_{-1}^{(v)}, y_{-2}^{(u)} \oplus y_{-2}^{(v)}, \ldots, y_{-k}^{(u)} \oplus y_{-k}^{(v)} \tag{16-87}$$

can be found by adding the corresponding elements of sequences U and V, which can be proved as follows, starting with Eq. (16-76):

$$m^{(w)}(x) = \sum_{i=0}^{k-1} x^i \sum_{j=i+1}^{k} c_j y_{-j+i}^{(w)} = \sum_{i=0}^{k-1} x^i \sum_{j=i+1}^{k} c_j(y_{-j+i}^{(u)} \oplus y_{-j+i}^{(v)})$$

$$= \sum_{i=0}^{k-1} x^i \left(\sum_{j=i+1}^{k} c_j y_{-j+i}^{(u)} \oplus \sum_{j=i+1}^{k} c_j y_{-j+i}^{(v)}\right)$$

$$= \sum_{i=0}^{k-1} (m_i^{(u)} \oplus m_i^{(v)})\, x^i = m^{(u)}(x) \oplus m^{(v)}(x) \tag{16-88}$$

It follows from the basic equation of the shift register that

$$W(x) = \frac{m^{(w)}(x)}{\phi(x)} = \frac{m^{(u)}(x)}{\phi(x)} \oplus \frac{m^{(v)}(x)}{\phi(x)} = U(x) \oplus V(x) \tag{16-89}$$

Equations (16-88) and (16-89) indicate that the sum mod 2 of two sequences which can be generated by a shift register with characteristic polynomial $\phi(x)$

can also be generated by this shift register, namely, starting from the sum of the corresponding initial states.

Because there exist only 2^k combinations of k bits, the shift register of Fig. 16-31 must contain a combination that it already contained after at most 2^k shift pulses.

It follows from the above that for each sequence S, the following relation must hold for all $i = -k, -k+1, \ldots, -1, 0, 1, 2, \ldots$

$$y_{i+n} = y_i$$

because the combination of the initial state and the following states of the shift register will be repeated after n shift pulses. Relation $y_{i+n} = y_i$ does not indicate that all sequences generated by a k-section shift register with a certain characteristic polynomial $\phi(x)$ need to have the same cycle length n.

There is an algebraic theorem stating that for every polynomial $f(x)$ there exists at least one positive integer N such that $f(x)$ divides $1 \oplus x^N$. The least one of these integers, which is indicated by n, is called the period of $f(x)$.

Example

$$f(x) = 1 \oplus x^2 \oplus x^4$$

$$\begin{array}{l} 1 \oplus x^2 \oplus x^4 \,|\, 1 \oplus x^n \,|\, 1 \oplus x^2 \\ \qquad\qquad\quad \underline{1 \oplus x^2 \oplus x^4} \\ \qquad\qquad\qquad\quad x^2 \oplus x^4 \oplus x^n \\ \qquad\qquad\qquad\quad \underline{x^2 \oplus x^4 \oplus x^6} \\ \qquad\qquad\qquad\qquad\qquad x^6 \oplus x^n \end{array}$$

If $n = 6$, the remainder of this division equals 0; hence, 6 is the period of $f(x) = 1 \oplus x^2 \oplus x^4$.

Sequence $S(x)$ starting from the initial state $y_{-1}, y_{-2}, \ldots, y_{-k+1}, y_{-k} = 0, 0, \ldots, 0, 1$ corresponding with initial-state polynomial $m(x) = 1$ is called the Q sequence. It follows from the basic equation of a mod 2 feedback shift register that

$$Q(x) = \frac{m(x)}{\phi(x)} = \frac{1}{\phi(x)} \tag{16-90}$$

The basic equation of a shift register can now be written as follows:

$$S(x) = \frac{m(x)}{\phi(x)} = \left(\sum_{i=0}^{k-1} m_i x^i\right) \cdot Q(x) = \sum_{i=0}^{k-1} m_i [x^i Q(x)] \tag{16-91}$$

Using Eq. (16-48) sequence Q_{-a} can be indicated as follows:

$$Q_{-a}(x) = x^a Q(x) \oplus \sum_{i=0}^{a-1} y_{-a+i} x^i$$

or

$$x^aQ(x) = Q_{-a}(x) \oplus \sum_{i=0}^{a-1} y_{-a+i}x^i$$

$$= Q_{-a}(x) \oplus \sum_{i=-a}^{-1} y_i x^{i+a}$$

$$= Q_{-a}(x) \oplus \sum_{i=1}^{a} y_{-i} x^{-i+a} \tag{16-92}$$

Since the Q sequence starts from the initial state

$$y_{-1}, y_{-2}, \ldots, y_{-k+1}, y_{-k} = 0, 0, \ldots, 0, 1$$

$$\sum_{i=1}^{a} y_{-i} x^{-i+a} = 0 \qquad \text{for } a = 1, 2, \ldots, k-1 \tag{16-93}$$

Hence, in case $m(x) = 1$ and $1 \leqslant a < k$,

$$x^a Q(x) = Q_{-a}(x) \tag{16-94}$$

Equation (16-94) can be substituted in Eq. (16-91):

$$S(x) = \sum_{i=0}^{k-1} m_i \cdot [x^i Q(x)] = \sum_{i=0}^{k-1} m_i Q_{-i}(x) \tag{16-95}$$

in which $Q_0(x) = Q(x)$.

Sequence S will now be represented by the following equation:

$$S = \sum_{i=0}^{k-1} m_i Q_{-i} \tag{16-96}$$

with $Q_0 = Q$, the $\sum$ sign indicating that sequence S will be obtained by addition of the corresponding elements of all sequences Q_{-i} involved. This last relation also shows that the 2^k different sequences S, which can be generated starting from the 2^k possible initial states, are the linear combinations of the k sequences $Q, Q_{-1}, \ldots, Q_{-(k-1)}$.

Example Now consider the shift register of Fig. 16-29 with characteristic polynomial $\phi(x) = 1 \oplus x^2 \oplus x^4$ and initial state 0 1 1 0.

It follows from Eqs. (16-63) and (16-90) that the first cycle $q(x)$ of $Q(x)$ can be obtained from the following equation:

$$q(x) = \frac{1 \oplus x^n}{\phi(x)} = \frac{1 \oplus x^n}{1 \oplus x^2 \oplus x^4} = 1 \oplus x^2 \tag{16-97}$$

with $n = 6$. The corresponding cycle of the Q sequence is 1 0 1 0 0 0. The initial-state polynomial corresponding to initial state 0 1 1 0 is $1 \oplus x \oplus x^2$, as has already been calculated. Thus $m_0 = 1$, $m_1 = 1$, $m_2 = 1$, and $m_3 = 0$. Hence, sequence S, which is generated starting from the given initial state, will be as follows:

$$S = \sum_{i=0}^{k-1} m_i Q_{-i} = Q \oplus Q_{-1} \oplus Q_{-2} \tag{16-98}$$

When determining sequence S, it is sufficient to determine the first six elements of S only because 6 is the period of $\phi(x)$.

Hence,

$$\begin{array}{ll} Q & = 1\ 0\ 1\ 0\ 0\ 0 \ldots \\ Q_{-1} & = 0\ 1\ 0\ 1\ 0\ 0 \ldots \\ Q_{-2} & = 0\ 0\ 1\ 0\ 1\ 0 \ldots \\ \hline S & = 1\ 1\ 0\ 1\ 1\ 0 \ldots \end{array}$$

This sequence with period 3 is in accordance with Table 16-11.

16-10 THE NUMBER OF SEQUENCES GENERATED BY A MOD 2 FEEDBACK SHIFT REGISTER AND THEIR CYCLE LENGTHS

In the preceding section, it is shown that the 2^k possible initial states of a mod 2 feedback shift register with k sections lead to 2^k different sequences. Now let $m_0(x)$ be an initial-state polynomial and let l be the cycle length of the corresponding sequence S_0. From this sequence S_0, $l-1$ sequences S_1, S_2, ..., S_{l-1} corresponding to initial-state polynomials $m_1(x)$, $m_2(x)$, ... , $m_{l-1}(x)$, respectively, can be derived. The contents of one cycle of each of the sequences S_0, S_1, S_2, ..., S_{l-1} is a cyclic permutation of the contents of one cycle of the others. Hence, these sequences can be considered as being equivalent.

The 2^k different sequences can be divided into a number of classes, containing equivalent sequences. Each class can be represented by one of its members; so each sequence can be considered as representing the class of which it is a member. Hence, the number of sequences generated by a mod 2 feedback shift register means the number of nonequivalent sequences or, in other words, the number of different classes. This is in accordance with the fact that the shift register of Fig. 16-29, for example, generates one sequence of cycle length 1, one sequence of cycle length 3, and two sequences of cycle length 6 (see Table 16-11).

Suppose that a shift register of k sections generates B (nonequivalent) sequences; b_1 has cycle length l_1, b_2 cycle length l_2, ..., b_t cycle length l_t.

Then

$$b_1 + b_2 + \cdots + b_t = B \tag{16-99}$$

Each sequence with cycle length l_i represents l_i different but equivalent sequences. Hence, the following relation holds:

$$b_1 l_1 + b_2 l_2 + \cdots + b_t l_t = 2^k \tag{16-100}$$

This equation shows the relation between the total number of sequences 2^k and the numbers of nonequivalent sequences with their cycle lengths.

In Fig. 16-29, an example was given of a mod 2 feedback shift register generating sequences of various cycle lengths, this depending on the initial state. The operation of this shift register was shown in Table 16-11. In this example, one sequence of length 1, one sequence of length 3, and two sequences of length 6 can be generated, which is in accordance with Eq. (16-100).

If sequence S has a cycle length l, it follows from Eqs. (16-63) and (16-75) that

$$S(x) = \frac{s(x)}{1 \oplus x^l} = \frac{m(x)}{\phi(x)}$$

or

$$s(x) = \frac{m(x) \cdot (1 \oplus x^l)}{\phi(x)} \tag{16-101}$$

When starting sequence S from the initial state $y_{-1}, y_{-2}, \ldots, y_{-(k-1)}, y_{-k} = 0, 0, \ldots, 0, 1$ so that $m(x) = 1$, Eq. (16-101) reduces to

$$q(x) = \frac{1 \oplus x^l}{\phi(x)} \tag{16-102}$$

Characteristic polynomial $\phi(x)$ has a period n so that

$$q(x) = \frac{1 \oplus x^n}{\phi(x)} \tag{16-103}$$

in which n is the smallest positive integer, so that $\phi(x)$ divides $1 \oplus x^n$. It follows from this definition that

$$l \geqslant n \tag{16-104}$$

On the other hand, it follows from the above that for $m(x) = 1$,

$$S(x) = \frac{1}{\phi(x)} = \frac{q(x)}{1 \oplus x^l} \qquad \text{and} \qquad S(x) = \frac{q(x)\, m(x)}{1 \oplus x^n} = \frac{q(x)}{1 \oplus x^n}$$

It should be noted that the degree of $m(x)$ is at most $k - 1$, the degree of $q(x)$ is at most $n - k$, and thus the degree of $q(x)m(x)$ is at most $n - 1$, so that $S(x)$ is periodic over n. It follows from the definition of the cycle length l of a sequence S that l is the least positive integer over which sequence S is periodic, so that from this point of view

$$l \leqslant n \tag{16-105}$$

Because both Eqs. (16-104) and (16-105) must be true

$$l = n \qquad \text{in case } m(x) = 1 \tag{16-106}$$

and

$$s(x) = q(x) \tag{16-107}$$

Table 16-11 shows that for certain polynomials the sequence length l can be smaller than the period n of the characteristic polynomial $\phi(x)$.

In this case,

$$S(x) = \frac{m(x)}{\phi(x)} = \frac{m(x)\cdot q(x)}{\phi(x)\cdot q(x)} = \frac{m(x)\cdot q(x)}{1 \oplus x^n} = \frac{s(x)}{1 \oplus x^l} \tag{16-108}$$

If $m(x)\cdot q(x)$ and $1 \oplus x^n$ contain common factors $f(x)$ such that $(1 \oplus x^n)/[f(x)] = 1 \oplus x^l$, sequence length l can be smaller than period n of $\phi(x)$ and l must divide n.

Example $\phi(x) = 1 \oplus x^2 \oplus x^4$ and $m(x) = 1 \oplus x \oplus x^2$ (Fig. 16-29 and Table 16-11).

$$S(x) = \frac{1 \oplus x \oplus x^2}{1 \oplus x^2 \oplus x^4} = \frac{(1 \oplus x^2)(1 \oplus x \oplus x^2)}{1 \oplus x^6} = \frac{(1 \oplus x)(1 \oplus x^3)}{(1 \oplus x^3)(1 \oplus x^3)} = \frac{1 \oplus x}{1 \oplus x^3}$$

This equation shows that sequence S with cycle length 3 will be generated by a mod 2 feedback shift register with a characteristic polynomial $\phi(x) = 1 \oplus x^2 \oplus x^4$ with period $n = 6$ when starting from the initial state 0 1 1 0 with $m(x) = 1 \oplus x \oplus x^2$. This result is in accordance with Table 16-11.

This means that for the determination of the possible sequence lengths, use must be made of the factorization properties of polynomials, the coefficients of which are subject to the laws of mod 2 arithmetic.

It is already mentioned that the operations on these polynomials have the same properties as those on polynomials with real coefficients. This implies that the unique factorization theorem also holds. This theorem states that every polynomial can be written as a uniquely determined product of irreducible polynomials. To find such a product, use can be made of Table 16-12, in which all irreducible polynomials of degree 1 to 8 are listed.

The laws of mod 2 arithmetic lead to a number of special properties of polynomials the coefficients of which are subject to these laws. Some of these properties are stated in the following theorems.

Theorem 1

$$[f(x)]^{2^t} = f(x^{2^t}) \tag{16-109}$$

It is obvious that this theorem holds for the case that $t = 1$ and $f(x)$ consists of only two terms. For example,

$$(1 \oplus x)^2 = 1 \oplus x \oplus x \oplus x^2 = 1 \oplus x^2$$

For the case that $t = 1$ and $f(x)$ consists of more than two terms, the theorem can be proved by finite induction on the number of terms. Then, for the general case, the theorem follows from finite induction on t.

Table 16-12 Table of irreducible polynomials

Degree	*Irreducible polynomial*	*Period*
1	x	—
	$1 \oplus x$	1
2	$1 \oplus x \oplus x^2$	3
3	$1 \oplus x \oplus x^3$	7
	$1 \oplus x^2 \oplus x^3$	7
4	$1 \oplus x \oplus x^4$	15
	$1 \oplus x \oplus x^2 \oplus x^3 \oplus x^4$	5
	$1 \oplus x^3 \oplus x^4$	15
5	$1 \oplus x \oplus x^2 \oplus x^3 \oplus x^5$	31
	$1 \oplus x \oplus x^2 \oplus x^4 \oplus x^5$	31
	$1 \oplus x \oplus x^3 \oplus x^4 \oplus x^5$	31
	$1 \oplus x^2 \oplus x^5$	31
	$1 \oplus x^2 \oplus x^3 \oplus x^4 \oplus x^5$	31
	$1 \oplus x^3 \oplus x^5$	31
6	$1 \oplus x \oplus x^6$	63
	$1 \oplus x \oplus x^2 \oplus x^4 \oplus x^6$	21
	$1 \oplus x \oplus x^2 \oplus x^5 \oplus x^6$	63
	$1 \oplus x \oplus x^3 \oplus x^4 \oplus x^6$	63
	$1 \oplus x \oplus x^4 \oplus x^5 \oplus x^6$	63
	$1 \oplus x^2 \oplus x^3 \oplus x^5 \oplus x^6$	63
	$1 \oplus x^2 \oplus x^4 \oplus x^5 \oplus x^6$	21
	$1 \oplus x^3 \oplus x^6$	9
	$1 \oplus x^5 \oplus x^6$	63
7	$1 \oplus x \oplus x^7$	127
	$1 \oplus x \oplus x^2 \oplus x^3 \oplus x^7$	127
	$1 \oplus x \oplus x^2 \oplus x^3 \oplus x^4 \oplus x^5 \oplus x^7$	127
	$1 \oplus x \oplus x^2 \oplus x^3 \oplus x^5 \oplus x^6 \oplus x^7$	127
	$1 \oplus x \oplus x^2 \oplus x^4 \oplus x^5 \oplus x^6 \oplus x^7$	127
	$1 \oplus x \oplus x^2 \oplus x^5 \oplus x^7$	127
	$1 \oplus x \oplus x^3 \oplus x^5 \oplus x^7$	127
	$1 \oplus x \oplus x^3 \oplus x^6 \oplus x^7$	127
	$1 \oplus x \oplus x^4 \oplus x^6 \oplus x^7$	127
	$1 \oplus x^2 \oplus x^3 \oplus x^4 \oplus x^7$	127
	$1 \oplus x^2 \oplus x^3 \oplus x^4 \oplus x^5 \oplus x^6 \oplus x^7$	127
	$1 \oplus x^2 \oplus x^4 \oplus x^6 \oplus x^7$	127
	$1 \oplus x^2 \oplus x^5 \oplus x^6 \oplus x^7$	127
	$1 \oplus x^3 \oplus x^7$	127
	$1 \oplus x^3 \oplus x^4 \oplus x^5 \oplus x^7$	127
	$1 \oplus x^4 \oplus x^7$	127
	$1 \oplus x^4 \oplus x^5 \oplus x^6 \oplus x^7$	127
	$1 \oplus x^6 \oplus x^7$	127

Table 16-12 Table of irreducible polynomials *(continued)*

8	$1 \oplus x \oplus x^2 \oplus x^3 \oplus x^4 \oplus x^5 \oplus x^8$	85
	$1 \oplus x \oplus x^2 \oplus x^3 \oplus x^4 \oplus x^6 \oplus x^8$	255
	$1 \oplus x \oplus x^2 \oplus x^3 \oplus x^4 \oplus x^7 \oplus x^8$	51
	$1 \oplus x \oplus x^2 \oplus x^3 \oplus x^6 \oplus x^7 \oplus x^8$	255
	$1 \oplus x \oplus x^2 \oplus x^4 \oplus x^5 \oplus x^6 \oplus x^8$	85
	$1 \oplus x \oplus x^2 \oplus x^4 \oplus x^6 \oplus x^7 \oplus x^8$	17
	$1 \oplus x \oplus x^2 \oplus x^5 \oplus x^6 \oplus x^7 \oplus x^8$	255
	$1 \oplus x \oplus x^2 \oplus x^7 \oplus x^8$	255
	$1 \oplus x \oplus x^3 \oplus x^4 \oplus x^8$	51
	$1 \oplus x \oplus x^3 \oplus x^4 \oplus x^5 \oplus x^6 \oplus x^8$	85
	$1 \oplus x \oplus x^3 \oplus x^5 \oplus x^8$	255
	$1 \oplus x \oplus x^3 \oplus x^7 \oplus x^8$	85
	$1 \oplus x \oplus x^4 \oplus x^5 \oplus x^6 \oplus x^7 \oplus x^8$	51
	$1 \oplus x \oplus x^5 \oplus x^6 \oplus x^8$	255
	$1 \oplus x \oplus x^5 \oplus x^7 \oplus x^8$	85
	$1 \oplus x \oplus x^6 \oplus x^7 \oplus x^8$	255
	$1 \oplus x^2 \oplus x^3 \oplus x^4 \oplus x^8$	255
	$1 \oplus x^2 \oplus x^3 \oplus x^4 \oplus x^5 \oplus x^7 \oplus x^8$	85
	$1 \oplus x^2 \oplus x^3 \oplus x^4 \oplus x^6 \oplus x^7 \oplus x^8$	85
	$1 \oplus x^2 \oplus x^3 \oplus x^5 \oplus x^8$	255
	$1 \oplus x^2 \oplus x^3 \oplus x^6 \oplus x^8$	255
	$1 \oplus x^2 \oplus x^3 \oplus x^7 \oplus x^8$	255
	$1 \oplus x^2 \oplus x^4 \oplus x^5 \oplus x^6 \oplus x^7 \oplus x^8$	255
	$1 \oplus x^2 \oplus x^5 \oplus x^6 \oplus x^8$	255
	$1 \oplus x^3 \oplus x^4 \oplus x^5 \oplus x^8$	17
	$1 \oplus x^3 \oplus x^4 \oplus x^5 \oplus x^6 \oplus x^7 \oplus x^8$	85
	$1 \oplus x^3 \oplus x^5 \oplus x^6 \oplus x^8$	255
	$1 \oplus x^3 \oplus x^5 \oplus x^7 \oplus x^8$	255
	$1 \oplus x^4 \oplus x^5 \oplus x^6 \oplus x^8$	255
	$1 \oplus x^4 \oplus x^5 \oplus x^7 \oplus x^8$	51

Theorem 2 *Polynomial $x \oplus x^{2^k}$ can be written as the product of all irreducible polynomials the degrees of which divide k—all once. For instance,*

$$x \oplus x^{2^4} = x(1 \oplus x)(1 \oplus x \oplus x^2)(1 \oplus x \oplus x^4)(1 \oplus x^3 \oplus x^4) \times (1 \oplus x \oplus x^2 \oplus x^3 \oplus x^4)$$

The proof of this theorem is given by W. W. Peterson.

Theorem 3 *If a polynomial $f(x)$ divides $1 \oplus x^n$, then $f(x)$ divides $1 \oplus x^{qn}$ for $q = 0, 1, 2, \ldots$.*

Proof $1 \oplus x^0 = 0$. For $q = 0$, this theorem is trivial. If $q = 1, 2, \ldots,$

$$1 \oplus x^{qn} = (1 \oplus x^n)(1 \oplus \cdots \oplus x^{(q-2)n} \oplus x^{(q-1)n}) \qquad \text{(q.e.d.)}$$

Theorem 4 *If $f(x)$ divides $1 \oplus x^t$, then $f(x)$ has a period n which divides t.*

Proof Suppose that there exists no positive integer such that

1. $0 < s < t$

2. $f(x)$ divides $1 \oplus x^s$

Then t is the smallest integer for which $f(x)$ divides $1 \oplus x^t$, so that following the definition $t = n$ is the period of $f(x)$. In this case, the theorem is trivial.

Now suppose that there exist positive integers s_i such that $0 < s_i < t$ and $f(x)$ divides $1 \oplus x^{s_i}$. Let n be the smallest of these integers. Then n is the period of $f(x)$ and integer t can be written as

$$t = qn + r \qquad \text{with } 0 \leqslant r < n \tag{16-110}$$

Hence $f(x)$ divides $1 \oplus x^{qn+r}$. Furthermore, it follows from Theorem 3 that $f(x)$ divides also $1 \oplus x^{qn}$. Then $f(x)$ divides the sum

$$1 \oplus x^{qn+r} \oplus 1 \oplus x^{qn} = x^{qn}(1 \oplus x^r) \tag{16-111}$$

Because $f(x)$ divides $1 \oplus x^t$, it contains no factor x. Hence $f(x)$ divides $1 \oplus x^r$. The period n being the smallest positive integer such that $f(x)$ divides $1 \oplus x^n$, the condition $0 \leqslant r < n$ leads to $r = 0$; therefore, $t = qn$, so that period n divides t. (q.e.d.)

It is obvious that a polynomial $\phi(x)$ which contains a factor x has no period. This follows immediately from the fact that x does not divide $1 \oplus x^n$ for positive values of n. If $\phi(x)$ contains no factor x, it has a period n which can be determined in the following way.

Polynomial $\phi(x)$ can always be written as a uniquely determined product of irreducible factors:

$$\phi(x) = \phi_1^{u_1}(x) \cdot \phi_2^{u_2}(x) \cdots \phi_i^{u_i}(x) \cdots \phi_v^{u_v}(x) \tag{16-112}$$

in which $u_1, \ldots, u_v$ are positive integers and $\phi_1(x), \ldots, \phi_v(x)$ are different irreducible polynomials of degrees $k_1, \ldots, k_v$, respectively, which do not equal x.

It follows from Theorem 2 that $\phi_i(x)$ divides $1 \oplus x^{2k_i-1}$, and it follows from Theorem 4 that $\phi_i(x)$ has a period n_i which divides $2^{k_i} - 1$. Hence, in order to find period n_i, it is sufficient to divide $1 \oplus x^n$ by $\phi_i(x)$ only for those values of n which are divisors of $2^{k_i} - 1$.

An irreducible polynomial $\phi_i(x)$ is called *primitive* if $n_i = 2^{k_i} - 1$ and *nonprimitive* if $n_i < 2^{k_i} - 1$. It is obvious that if $2^{k_i} - 1$ is prime, all irreducible polynomials of degree k_i are primitive. This is, for example, the case with $k_i = 3, 5, 7, \ldots$ (see Table 16-12).

Example $\phi(x) = 1 \oplus x^3 \oplus x^4$. In this case, $2^{k_i} - 1 = 15$. Because $1 \oplus x^3 \oplus x^4$ does not divide $1 \oplus x^3$ and $1 \oplus x^5$, this polynomial is primitive with $n_i = 15$.

$\phi(x) = 1 \oplus x \oplus x^2 \oplus x^3 \oplus x^4$. This polynomial divides $1 \oplus x^5$, hence $n_i = 5$ and it is not primitive (see Table 16-12).

Characteristic polynomial $\phi(x)$ is reducible Characteristic polynomial $\phi(x)$ is reducible but contains no powers of irreducible polynomials. Equation (16-75) can now be written as follows:

$$S(x) = \frac{m(x)}{\phi_1(x) \cdot \phi_2(x) \cdots \phi_j(x)} = \frac{m(x)\, q(x)}{1 \oplus x^n} \tag{16-113}$$

1. $m(x) = 1$

Let $n_1, n_2, \ldots, n_j$ be the periods of irreducible polynomials $\phi_1(x), \phi_2(x), \ldots, \phi_j(x)$, so that $\phi_1(x)$ divides $1 \oplus x^{n_1}$, $\phi_2(x)$ divides $1 \oplus x^{n_2}, \ldots,$ and $\phi_j(x)$ divides $1 \oplus x^{n_j}$; then it follows from Theorem 3 that

$$\phi_1(x) \cdot \phi_2(x) \cdots \phi_j(x) \text{ divides } 1 \oplus x^{\text{lcm}(n_1, n_2, \ldots, n_j)}$$

Hence, period $n = \text{lcm}(n_1, n_2, \ldots, n_j)$, being also the sequence length l.

2. $m(x) \neq 1$

In this case, $m(x)$, being of lower degree than $\phi(x) = \phi_1(x) \cdot \phi_2(x) \ldots \phi_j(x)$, can be chosen in different ways so that Eq. (16-113) can be reduced to any one of the following possible forms:

$$\frac{m_1(x)}{\phi_1(x)} \qquad \frac{m_2(x)}{\phi_2(x)} \qquad \frac{m_{12}(x)}{\phi_1(x)\,\phi_2(x)} \qquad \text{etc.}$$

leading to sequence lengths $l = \text{lcm}(n_1, n_2, \ldots)$. The sequence length l is the lcm of periods n_i of characteristic polynomials $\phi_i(x)$, which are not divided out by the choice of the initial-state polynomial.

Characteristic polynomial $\phi(x)$ is a power of irreducible polynomial $\phi_i(x)$ Let N_i be the period of polynomial $\phi_i^{u_i}(x)$ for $u_i > 1$. Then $\phi_i^{u_i}(x)$ and thus $\phi_i(x)$ divide $1 \oplus x^{N_i}$ and by Theorem 4 the period n_i of $\phi_i(x)$ divides N_i, so that

$$N_i = q_i \cdot n_i \tag{16-114}$$

It follows from Theorem 2 that $1 \oplus x^{2^{k_i}-1}$ contains only one factor $\phi_i(x)$. Because n_i divides $2^{k_i} - 1$, by Theorem 3, $1 \oplus x^{n_i}$ also divides $1 \oplus x^{2^{k_i}-1}$ and thus $1 \oplus x^{n_i}$ contains also only one factor $\phi_i(x)$. Hence, u_i is the least positive integer such that $\phi_i^{u_i}(x)$ divides $(1 \oplus x^{n_i})^{u_i}$.

Let t be the least positive integer such that $2^{t_i} \geqslant u_i$. Then $(1 \oplus x^{n_i})^{u_i}$ and thus $\phi_i^{u_i}(x)$ divide $(1 \oplus x^{n_i})^{2t_i} = 1 \oplus x^{n_i \cdot 2^{t_i}}$ by Theorem 1, and by Theorem

4 period $N_i = q_i n_i$ is a divisor of $n_i \cdot 2^{t_i}$. Hence, q_i is a divisor of 2^{t_i} and thus can be written as

$$q_i = 2^{w_i} \qquad \text{with } w_i \leqslant t_i \tag{16-115}$$

Substitution of Eq. (16-115) into Eq. (16-114) gives

$$N_i = n_i \cdot 2^{w_i} \tag{16-116}$$

This means that $\phi_i^{u_i}(x)$ divides $1 \oplus x^{n_i \cdot 2^{w_i}} = (1 \oplus x^{n_i})^{2^{w_i}}$

Because u_i was the least positive integer such that $\phi_i^{u_i}(x)$ divides $(1 \oplus x^{n_i})^{u_i}$, $2^{w_i} \geqslant u_i$ and thus $w_i \geqslant t_i$ because t_i is the least integer such that $2^{t_i} \geqslant u_i$. It follows from this result and Eq. (16-115) that $w_i = t_i$. Hence,

$$N_i = n_i \cdot 2^{t_i} \tag{16-117}$$

In Table 16-13, N_i is given for different values of u_i.

Table 16-13

u_i	t_i	N_i
2	1	$2n_i$
3–4	2	$4n_i$
5–8	3	$8n_i$
9–16	4	$16n_i$

Characteristic polynomial $\phi(x)$ is a product of powers of irreducible polynomials This is given by Eq. (16-112). A reasoning analogous to that given for a reducible polynomial not containing powers of irreducible polynomials but containing a product of powers of irreducible polynomials $\phi_i(x)$ leads to the result that its period n is the lcm $(N_1, N_2, \ldots, N_j)$, $N_1, N_2, \ldots, N_j$ being the periods of polynomials $\phi_1^{u_1}(x), \phi_2^{u_2}(x), \ldots, \phi_j^{u_j}(x)$.

The sequence length l of the sequence $S(x)$ generated by a shift register with characteristic polynomial $\phi(x)$ can degenerate in the same way as already shown for a reducible polynomial $\phi(x)$.

Example

$$\phi(x) = (1 \oplus x)^3 (1 \oplus x \oplus x^2)^2 (1 \oplus x \oplus x^3)$$

It is obvious that $1 \oplus x$ has a period 1. Hence $(1 \oplus x)^3$ has a period $2^2 \cdot 1 = 4$. Because $2^2 - 1 = 3$ and $2^3 - 1 = 7$ are prime, polynomials $1 \oplus x \oplus x^2$ and $1 \oplus x \oplus x^3$ have periods 3 and 7, respectively. Thus, $(1 \oplus x \oplus x^2)^2$ has a period 6. Period n of $\phi(x)$ is the lcm of 4, 6, and 7; thus $n = 84$.

THE NUMBER OF SEQUENCES AND THEIR CYCLE LENGTH

$\phi(x)$ is irreducible As has been stated in Eq. (16-108),

$$S(x) = \frac{m(x)}{\phi(x)} = \frac{s(x)}{1 \oplus x^l}$$

Hence

$$\phi(x) \cdot s(x) = m(x) \cdot (1 \oplus x^l) \qquad (16\text{-}118)$$

For the case $m(x) = 0$, a sequence $s(x) = 0$ with length 1 will be obtained. When $m(x) \neq 0$, polynomials $\phi(x)$ and $m(x)$ have no common factors because $m(x)$ is of lower degree than $\phi(x)$, and $\phi(x)$ is supposed to be irreducible. Then $\phi(x)$ must divide $1 \oplus x^l$ and hence cycle length l equals period n of $\phi(x)$. Hence, all sequences except the 0 sequence have cycle lengths n. Then Eq. (16-100) leads to

$$1 \cdot 1 + b \cdot l = 2^k$$

in which b is the number of nonequivalent sequences of length l. Thus

$$b \cdot l = 2^k - 1$$

and

$$b = \frac{2^k - 1}{l} \qquad (16\text{-}119)$$

which is in accordance with the fact that period n of an irreducible polynomial of degree k divides $2^k - 1$.

If $\phi(x)$ is primitive ($l = n = 2^k - 1$), then $b = 1$. Only one sequence except the 0 sequence will be generated in this case. This sequence of length $2^k - 1$ is called a maximum-length recurrent sequence (m sequence).

Example

$$\phi(x) = 1 \oplus x \oplus x^4$$

This polynomial is primitive. It thus generates the 0 sequence of length 1 and one m sequence of length 15. This m sequence is in fact the class of all 15 possible non-0 sequences and can be represented by the q sequence:

$$q(x) = \frac{1 \oplus x^n}{\phi(x)} = \frac{1 \oplus x^{15}}{1 \oplus x \oplus x^4}$$

$$= 1 \oplus x \oplus x^2 \oplus x^3 \oplus x^5 \oplus x^7 \oplus x^8 \oplus x^{11}$$

$$q = 1\,1\,1\,1\,0\,1\,0\,1\,1\,0\,0\,1\,0\,0\,0$$

Example

$$\phi(x) = 1 \oplus x \oplus x^2 \oplus x^3 \oplus x^4$$

This polynomial is nonprimitive. It has a period $n = 5$ (see Table 16-12). Eq. (16-100) leads to

$$1 \cdot 1 + b \cdot 5 = 2^4$$

so that $b = 3$. Hence, polynomial $\phi(x) = 1 \oplus x \oplus x^2 \oplus x^3 \oplus x^4$ generates a 0 sequence of length 1 and 3 and other sequences of length 5. Representatives of these last sequences can be found by appropriate choice of $m(x)$ in

$$s(x) = m(x) \cdot q(x) \tag{16-120}$$

which can easily be derived from the basic state equation of a shift register:

$$s(x) = m(x) \cdot \frac{1 \oplus x^5}{1 \oplus x \oplus x^2 \oplus x^3 \oplus x^4} = m(x) \cdot (1 \oplus x)$$

For instance $m(x) = 1$, $m(x) = 1 \oplus x$, and $m(x) = 1 \oplus x^2$ lead to the following three basically different sequences:

$m(x) = 1$ :: 1 1 0 0 0
$m(x) = 1 \oplus x$ :: 1 0 1 0 0
$m(x) = 1 \oplus x^2$ :: 1 1 1 1 0

$\phi(x)$ is a power of an irreducible polynomial Let $\phi(x) = \phi_i{}^u(x)$, in which $\phi_i(x)$ is an irreducible polynomial of degree d with period n and u is an integer such that $u > 1$. The degree k of polynomial $\phi(x)$ is then

$$k = u \cdot d$$

Furthermore,

$$S(x) = \frac{s(x)}{1 \oplus x^l} = \frac{m(x)}{\phi(x)} = \frac{m(x)}{\phi_i{}^u(x)} \tag{16-121}$$

Now consider polynomials $m(x)$ of the following form:

$$m(x) = m_1(x)\, \phi_i^{u-1}(x) \oplus m_2(x)\, \phi_i^{u-2}(x) \oplus \cdots \oplus m_{u-1}(x)\, \phi_i\,(x) \oplus m_u(x) \tag{16-122}$$

in which $m_1(x), m_2(x), \ldots, m_u(x)$ are polynomials of degree less than d, so that there are 2^d different possible polynomials for each $m_j(x)$. Two different choices of polynomials $m_1(x), m_2(x), \ldots, m_u(x)$ lead to two different polynomials $m(x)$, which can easily be proved. Then the number of polynomials $m(x)$ is $(2^d)^u = 2^{du} = 2^k$. Furthermore, it follows from Eq. (16-121) that the degree of $m(x)$ is at most $u \cdot d - 1 = k - 1$. Hence the 2^k different polynomials $m(x)$, the degree of which is at most $k - 1$, are just the 2^k possible initial-state polynomials.

Substitution of Eq. (16-122) into Eq. (16-121) leads to the following result:

$$
\begin{aligned}
S(x) &= \frac{m_1(x)\,\phi_i^{u-1}(x) \oplus m_2(x)\,\phi_i^{u-1}(x) \oplus \cdots \oplus m_u(x)}{\phi_i^u(x)} \\
&= \frac{m_1(x)}{\phi_i(x)} \oplus \frac{m_2(x)}{\phi_i^{\,2}(x)} \oplus \cdots \oplus \frac{m_u(x)}{\phi_i^{\,u}(x)}
\end{aligned}
\qquad (16\text{-}123)
$$

Because $m_1(x), m_2(x), \ldots, m_u(x)$ are of lower degree than $\phi_i(x)$ and $\phi_i(x)$ is irreducible, $m_j(x)$ and $\phi_i(x)$ have no common factors unless $m_j(x) = 0$. Hence, for all $j = 1, 2, \ldots, u$, equation

$$S^{(j)}(x) = \frac{m_j(x)}{\phi_i^{\,j}(x)}$$

represents a periodic sequence the cycle length of which equals the period of $\phi_i^{\,j}(x)$ if $m_j(x) \neq 0$ and the 0 sequence if $m_j(x) = 0$. Hence

$$S(x) = S^{(1)}(x) \oplus S^{(2)}(x) \oplus \cdots \oplus S^{(u)}(x) \qquad (16\text{-}124)$$

Now let h be the greatest value of j for which $m_j(x) \neq 0$. Then Eq. (16-123) reduces to

$$
\begin{aligned}
S(x) &= \frac{m_1(x)\,\phi_i^{u-1}(x) \oplus m_2(x)\,\phi_i^{u-2}(x) \oplus \cdots \oplus m_h(x)\,\phi_i^{u-h}(x)}{\phi_i^{\,u}(x)} \\
&= \frac{m_1(x)\,\phi_i^{h-1}(x) \oplus m_2(x)\,\phi_i^{h-2}(x) \oplus \cdots \oplus m_h(x)}{\phi_i^{\,h}(x)} \\
&= \frac{m_h^*(x)}{\phi_i^{\,h}(x)}
\end{aligned}
\qquad (16\text{-}125)
$$

In Eq. (16-125), $m_h^*(x)$ stands for the numerator in the preceding part of the equation. Because $\phi_i(x)$ does not divide $m_h^*(x)$ and because $\phi_i(x)$ is irreducible, numerator and denominator in Eq. (16-125) have no common factors. Hence, the cycle length of S equals the period of $\phi_i^{\,h}(x)$.

The number P_1 of polynomials $m(x)$ for which $h = 1$, that is, the number of polynomials $m(x)$ for which $m_1(x) \neq 0$ and $m_j(x) = 0$ for all $j > 1$, is

$$P_1 = 2^d - 1$$

The number P_2 of polynomials $m(x)$ for which $h = 2$, that is, for which $m_1(x)$ is arbitrary, $m_2(x) \neq 0$, and $m_j(x) = 0$ for all $j > 2$, is

$$P_2 = 2^d\,P_1 = 2^d(2^d - 1)$$

Now it is obvious that, in general,

$$P_h = 2^{(h-1)\,d}(2^d - 1) \qquad (16\text{-}126)$$

It has already been mentioned that for a given nonzero initial-state polynomial $m(x)$, the cycle length of the corresponding sequence equals the period of $\phi_i^h(x)$, h being the greatest value of j for which $m_j(x) \neq 0$. Then if n is the period of $\phi_i(x)$, it follows from the above (see Table 16-13) that

$$l_h = 2^\alpha \cdot n \tag{16-127}$$

with

$$2^{\alpha-1} < h \leqslant 2^\alpha \tag{16-128}$$

Because $1 \leqslant h \leqslant u$, relation (16-128) can increase as a maximum to

$$2^{\alpha-1} < u \leqslant 2^\alpha = 2^t(\alpha_{\max} = t) \tag{16-129}$$

Three cases arise, depending on the value of α:

1. If $\alpha = 0$, the number M_0 of initial states leading to sequences with cycle length $l_h = 2^0 \cdot n = n$ equals the number of initial-state polynomials for which $h = 1$. Thus,

   $$M_0 = P_1 = 2^d - 1 \tag{16-130}$$

   Hence the number of nonequivalent sequences with cycle length n is

   $$B_0 = \frac{M_0}{n} = \frac{2^d - 1}{n} \tag{16-131}$$

2. If $0 < \alpha < t$, the number M_α of initial states leading to sequences with cycle length $2^\alpha \cdot n$ equals the number of initial-state polynomials for which h satisfies $2^{\alpha-1} < h \leqslant 2^\alpha$:

   $$M_\alpha = P_{2^{\alpha-1}+1} + P_{2^{\alpha-1}+2} + \cdots + P_{2^\alpha} \tag{16-132}$$

   $$= \sum_{h=2^{\alpha-1}+1}^{2^\alpha} P_h \tag{16-133}$$

   Substitution of Eq. (16-126) gives

   $$M_\alpha = \sum_{h=2^{\alpha-1}+1}^{2^\alpha} 2^{(h-1)d}(2^d - 1) = \sum_{h=2^{\alpha-1}+1}^{2^\alpha} 2^{hd} - \sum_{h=2^{\alpha-1}+1}^{2^\alpha} 2^{(h-1)d}$$

   $$= \sum_{h=2^{\alpha-1}+1}^{2^\alpha} 2^{hd} - \sum_{h=2^{\alpha-1}}^{2^\alpha-1} 2^{hd} = 2^{2^\alpha d} - 2^{2^{\alpha-1} d} \tag{16-134}$$

   Hence, the number B_α of nonequivalent sequences with cycle length $2^\alpha \cdot n$ is

   $$B_\alpha = \frac{M_\alpha}{2^\alpha \cdot n} = \frac{2^{2^\alpha d} - 2^{2^{\alpha-1} d}}{2^\alpha \cdot n} \tag{16-135}$$

3. If $\alpha = t$, integer h must satisfy the following relation:

$$2^{t-1} < h \leqslant u \leqslant 2^t \tag{16-136}$$

The number M_t of initial states leading to sequences with cycle length $2^t \cdot n$ follows from Eq. (16-134) by substitution of $\alpha = t$ and $2^t = u$:

$$M_t = 2^{ud} - 2^{2^{t-1}d} \tag{16-137}$$

Hence, the number B_t of nonequivalent sequences with cycle length $2^t \cdot n$ is as follows:

$$B = \frac{M_t}{2^t \cdot n} = \frac{2^{ud} - 2^{2^{t-1}d}}{2^t \cdot n} \tag{16-138}$$

These results are in accordance with Eq. (16-100), which can easily be verified by substitution of Eqs. (16-131), (16-135), and (16-138).

$$B_T = b_1 l_1 + b_2 l_2 + \cdots + b_t l_t = 1.1 + B_0 n + B_1 \cdot 2n + \cdots + B_{t-1} \cdot 2^{t-1} n + B_t \cdot 2^t n$$

$$= 1 + 2^d - 1 + \sum_{\alpha=1}^{t-1} (2^{2^\alpha d} - 2^{2^{\alpha-1} d}) + 2^{ud} - 2^{2^{t-1} d} = 2^{ud} = 2^k$$

Example

$$\phi(x) = (1 \oplus x \oplus x^2)^{10}$$

It follows from Table 16-12 that $1 \oplus x \oplus x^2$ has a period $n = 3$. Substitution of $u = 10$ in Eq. (16-136) gives

$$2^{t-1} < 10 \leqslant 2^t$$

Hence $t = 4$. Then it follows from Eq. (16-129) that

$$0 \leqslant \alpha \leqslant 4$$

Now the number of sequences with cycle lengths $2^\alpha \cdot n$ can be determined using Eqs. (16-131), (16-135), and (16-138). It follows from Eq. (16-131) that the number B_0 of sequences with cycle length 3 is

$$B_0 = \frac{2^2 - 1}{3} = 1$$

It follows from Eq. (16-135) that the numbers B_1, B_2, and B_3 of sequences with cycle length $2^\alpha \cdot n = 6, 12, 24$, respectively, are

$$B_1 = \frac{2^4 - 2^2}{6} = 2$$

$$B_2 = \frac{2^8 - 2^4}{12} = 20$$

$$B_3 = \frac{2^{16} - 2^8}{24} = 2720$$

Finally, Eq. (16-138) gives the number B_4 of sequences with cycle length 48:

$$B_4 = \frac{2^{20} - 2^{16}}{48} = 20{,}480$$

$\phi(x)$ is a reducible polynomial In its most general form, a reducible polynomial can be represented as follows:

$$\phi(x) = \phi_1^{u_1}(x) \cdot \phi_2^{u_2}(x) \cdots \phi_v^{u_v}(x) \tag{16-139}$$

in which $\phi_1(x), \phi_2(x), \ldots, \phi_v(x)$ are different irreducible polynomials of degrees $d_1, d_2, \ldots, d_v$ with periods $n_1, n_2, \ldots, n_v$, respectively, and in which $u_1, u_2, \ldots, u_v$ are positive integers. The degree of $\phi(x)$ will be

$$k = u_1 d_1 + u_2 d_2 + \cdots + u_v d_v \tag{16-140}$$

Furthermore

$$S(x) = \frac{s(x)}{1 \oplus x^l} = \frac{m(x)}{\phi(x)} = \frac{m(x)}{\phi_1^{u_1}(x) \cdot \phi_2^{u_2}(x) \cdots \phi_v^{u_v}(x)} \tag{16-141}$$

The following way of reasoning is more or less analogous to that used in the preceding part of this section. Now consider polynomials $m(x)$ of the following form:

$$\begin{aligned} m(x) = {} & \quad m_1(x) \cdot \phi_2^{u_2}(x) \cdots \phi_v^{u_v}(x) \\ & \oplus \phi_1^{u_1}(x) \cdot m_2(x) \cdots \phi_v^{u_v}(x) \\ & \quad \cdot \quad \cdot \quad \cdot \quad \cdot \quad \cdot \quad \cdot \quad \cdot \quad \cdot \quad \cdot \quad \cdot \\ & \oplus \phi_1^{u_1}(x) \cdot \phi_2^{u_2}(x) \cdots m_v(x) \\ = {} & \sum_{i=1}^{v} m_i(x) \cdot \prod_{j=1;\, j \neq i}^{v} \phi_j^{u_j}(x) \end{aligned} \tag{16-142}$$

In this equation $m_i(x)$ is a polynomial of degree less than $u_i d_i$ for all $i = 1, 2, \ldots, v$. Then for each polynomial $m_i(x)$, there are $2^{u_i d_i}$ possible choices leading to different polynomials $m(x)$, the proof of which can easily be found starting from the equality of two different choices. The total number of possible different polynomials $m(x)$ following from different choices of $m_i(x)$ is

$$2^{u_1 d_1} \cdot 2^{u_2 d_2} \cdots 2^{u_v d_v} = 2^k \tag{16-143}$$

Furthermore, the degree of polynomial $m(x)$ in Eq. (16-141) is at most

$$u_1 d_1 + u_2 d_2 + \cdots + u_v d_v - 1 = k - 1 \tag{16-144}$$

so that there are at most 2^k different polynomials $m(x)$.

Hence, the 2^k different polynomials $m(x)$ following from Eq. (16-142) are just the 2^k different polynomials $m(x)$ required to satisfy all possibilities in the simplification of Eq. (16-141). Substitution of Eq. (16-142) into Eq. (16-141) leads to the following result:

$$S(x) = \frac{\sum_{i=1}^{v} m_i(x) \cdot \prod_{j=1;\, j \neq i}^{v} \phi_j^{u_j}(x)}{\prod_{j=1}^{v} \phi_j^{u_j}(x)} = \frac{m_1(x)}{\phi_1^{u_1}(x)} \oplus \frac{m_2(x)}{\phi_2^{u_2}(x)} \oplus \cdots \oplus \frac{m_v(x)}{\phi_v^{u_v}(x)} \tag{16-145}$$

Equation (16-145) shows that any arbitrary reducible characteristic polynomial $\phi(x) = \prod_{j=1}^{v} \phi_j^{u_j}(x)$ will generate a sequence S which can be written in the form of a mod 2 sum of sequences generated by characteristic polynomials $\phi_i^{u_i}(x)$.

Now the terms of Eq. (16-145) can be substituted by Eq. (16-123):

$$\begin{aligned} S(x) = {} & \frac{m_{11}(x)}{\phi_1(x)} \oplus \frac{m_{12}(x)}{\phi_1^2(x)} \oplus \cdots \oplus \frac{m_{1u_1}(x)}{\phi_1^{u_1}(x)} \\ & \oplus \frac{m_{21}(x)}{\phi_2(x)} \oplus \frac{m_{22}(x)}{\phi_2^2(x)} \oplus \cdots \oplus \frac{m_{2u_2}(x)}{\phi_2^{u_2}(x)} \\ & \cdots\cdots\cdots\cdots \\ & \oplus \frac{m_{v1}(x)}{\phi_v(x)} \oplus \frac{m_{v2}(x)}{\phi_v^2(x)} \oplus \cdots \oplus \frac{m_{vu_v}(x)}{\phi_v^{u_v}(x)} \end{aligned} \tag{16-146}$$

Equation (16-146) shows finally that sequence S generated by reducible polynomial $\phi(x)$ can be obtained from a mod 2 sum of sequences determined by all the irreducible polynomials $\phi_i(x)$ and their powers which are factors of $\phi(x)$.

From this equation, one can try to derive equations for the number of sequences and their cycle lengths in the way that has been demonstrated starting with Eq. (16-123). However, this is a very complicated method because many different cases have to be distinguished. A better way to solve this problem is to construct an algorism for this purpose.

Let

$$S = U \oplus V \qquad \text{with } s_i = u_i \oplus v_i \text{ for all } i = 1, 2, \ldots$$

and

$$S(x) = \frac{m_u(x)}{\phi_u(x)} \oplus \frac{m_v(x)}{\phi_v(x)} \tag{16-147}$$

in which $m_u(x)$ and $m_v(x)$ are of lower degree than $\phi_u(x)$ and $\phi_v(x)$, respectively, and $\phi_u(x)$ and $\phi_v(x)$ are mutually prime. U and V are periodic with cycle lengths l_u and l_v, respectively.

The cycle length l_s of sequence S is the smallest integer so that

$$S_{l_s}(x) = S(x) \tag{16-148}$$

Because $S(x) = U(x) \oplus V(x)$, then also $S_{l_s}(x) = U_{l_s}(x) \oplus V_{l_s}(x)$. It is obvious that $U_{l_s}(x) = U_{q_u l_u}(x)$ and $V_{l_s}(x) = V_{q_v l_v}(x)$ for all $q = 1, 2, \ldots$, which implies that

$$l_s = \text{lcm}\,(l_u, l_v) \tag{16-149}$$

and

$$S_{\text{lcm}\,(l_u,\,l_v)}(x) = U_{\text{lcm}\,(l_u,\,l_v)}(x) \oplus V_{\text{lcm}\,(l_u,\,l_v)}(x) \tag{16-150}$$

Now let β be the class of all sequences equivalent with sequence

$$U_i(x) = \frac{m_i(x)}{\phi_u(x)}$$

Then β contains just l_u phases of U. Let γ be the class of all sequences equivalent with sequence

$$V_j(x) = \frac{m_j(x)}{\phi_v(x)}$$

Then γ contains just l_v phases of V. All these phases can be obtained by a suitable choice of $m_i(x)$ and $m_j(x)$, respectively.

Consider now the following set X of sequences:

$$\begin{aligned} S_{i,j}(x) &= U_i(x) \oplus V_j(x) = \frac{m_i(x)}{\phi_u(x)} \oplus \frac{m_j(x)}{\phi_v(x)} \\ &= \frac{m_i(x)\,\phi_v(x) \oplus m_j(x)\,\phi_u(x)}{\phi_u(x)\,\phi_v(x)} = \frac{m_{ij}(x)}{\phi_u(x)\,\phi_v(x)} \end{aligned} \tag{16-151}$$

with $m_{ij}(x) = m_i(x)\,\phi_v(x) \oplus m_j(x)\,\phi_u(x)$, $m_i(x)$ and $m_j(x)$ being of degree less than d_u and d_v, respectively. Then the degree of polynomial $m_{ij}(x)$ is less than $d_u + d_v$. Furthermore, it can be proved that two different choices of $m_i(x)$ and $m_j(x)$ lead to different polynomials $m_{ij}(x)$ in the same way that has been mentioned before. Hence, the $2^{d_u} \cdot 2^{d_v} = 2^{d_u+d_v}$ different choices of $m_i(x)$ and $m_j(x)$ merely lead to the $2^{d_u+d_v}$ possible polynomials $m_{ij}(x)$ of degree less than $d_u + d_v$. Then by Eq. (16-151), set X consists only of all sequences that can be generated by the mod 2 feedback shift register with characteristic polynomial $\phi_u(x) \cdot \phi_v(x)$. Then the number of sequences $S_{i,j}$ belonging to X is

$$N_x = l_u \cdot l_v \tag{16-152}$$

The sequences of X can be divided into classes of equivalent sequences. It is obvious that all l_s phases of $S_{i,j}$ belong to X. Furthermore, Eq. (16-149) implies that all sequences in X have the same cycle length $l_s = l_{\text{lcm}\,(l_u,\,l_v)}$.

Hence X contains only lcm (l_u, l_v) equivalent sequences. Then the number B_x of nonequivalent sequences is

$$B_x = \frac{N_x}{\mathrm{lcm}(l_u, l_v)} = \frac{l_u \cdot l_v}{\mathrm{lcm}(l_u, l_v)} = \gcd(l_u, l_v) \tag{16-153}$$

Now consider sets X and Y of all sequences $U^{(i)}$ and $W^{(j)}$ that can be generated by mod 2 feedback shift registers with characteristic polynomials $\phi_u(x)$ and $\phi_w(x)$ of degree d_u and d_w, respectively—these polynomials being mutually prime. Let Z be the set of all sequences that can be written as follows:

$$\begin{aligned} S^{(i,j)}(x) &= U^{(i)}(x) \oplus W^{(j)}(x) = \frac{m^{(i)}(x)}{\phi_u(x)} \oplus \frac{m^{(j)}(x)}{\phi_w(x)} \\ &= \frac{m^{(ij)}(x)}{\phi_u(x)\,\phi_w(x)} \end{aligned} \tag{16-154}$$

with $m^{(ij)}(x) = m^{(i)}(x)\,\phi_w(x) \oplus m^{(j)}(x)\,\phi_u(x)$ and $m^{(i)}(x)$, respectively, $m^{(j)}(x)$ being of lower degree than d_u and d_v. From a reasoning completely analogous to that given above, it follows that by Eq. (16-154) set Z consists only of all sequences that can be generated by the mod 2 feedback shift register with characteristic polynomial $\phi_u(x)\,\phi_w(x)$.

Sets X and Y can be divided into classes of equivalent sequences, each class containing only l_s phases of one sequence with cycle length l_s. Now let β and δ be two such classes of l_u and l_w equivalent sequences of X and Y. Combination of these two classes in the way described above leads to gcd (l_u, l_w) different classes of Z, each containing lcm (l_u, l_w) equivalent sequences. It is obvious that a complete subdivision of Z into classes of equivalent sequences can be derived by combining each class of X with each class of Y.

Example Let $\phi_u(x)$ be $(1 \oplus x \oplus x^2)^3$ and let $\phi_v(x)$ be $(1 \oplus x \oplus x^3)^2$. Because $1 \oplus x \oplus x^2$ and $1 \oplus x \oplus x^3$ are irreducible polynomials, their number of nonequivalent sequences (sets X and Y) and their cycle lengths can be determined in the way given on pages 466–469. The results of this are given in Table 16-14.

Table 16-14

X		Y	
Number	*Cycle length*	*Number*	*Cycle length*
1	1	1	1
1	3	1	7
2	6	4	14
4	12		

Because $\phi_u(x)$ and $\phi_v(x)$ are powers of irreducible polynomials, they are mutually prime. Hence combination of the class of X containing the sequence with cycle length 1 (0 sequence) with each class of Y leads to the following classes of Z.

Number	*Cycle length*
1	1
1	7
4	14

Furthermore, the combination of class X containing the three equivalent sequences with cycle length 3 with each class of Y leads to the following classes of Z:

Number	*Cycle length*
1	3
1	21
4	42

In the same way, combination of both classes of X containing six equivalent sequences with cycle length 6 with each class of Y leads to the following classes of Z:

Number	*Cycle length*
2	6
2	42
16	42

the number 16 being the product $2 \cdot 4 \cdot \gcd(6, 14)$.

Finally, combination of the four classes of X containing twelve equivalent sequences with cycle length 12 with each class of Y leads to the following classes of Z:

Number	*Cycle length*
4	12
4	84
32	84

These results are summarized in Table 16-15, which gives the subdivision of set Z into classes of equivalent sequences:

Table 16-15

Z	
Number	*Cycle length*
1	1
1	3
2	6
1	7
4	12
4	14
1	21
22	42
36	84

Because each class contains only equivalent sequences, the number of nonequivalent sequences in Z equals the number of classes. Hence Table 16-15 gives the number of nonequivalent sequences with their cycle lengths generated by mod 2 feedback shift register with characteristic polynomial

$$\phi(x) = (1 \oplus x \oplus x^2)^3 \cdot (1 \oplus x \oplus x^3)^2$$
$$= 1 \oplus x \oplus x^2 \oplus x^8 \oplus x^9 \oplus x^{11} \oplus x^{12}$$

Now consider again Eq. (16-146). Polynomials $\phi_1(x), \phi_2(x), \ldots, \phi_v(x)$ being irreducible and different, the set of sequences represented by $[m_i(x)]/[\phi_i^{u_i}(x)]$ can be divided into classes of equivalent sequences by applying the process described on pages 466–469.

Because $\phi_1^{u_1}(x)$ and $\phi_2^{u_2}(x)$ are mutually prime, set X of all sequences represented by the sum

$$\frac{m_1(x)}{\phi_1^{u_1}(x)} \oplus \frac{m_2(x)}{\phi_2^{u_2}(x)} = \frac{m_{12}(x)}{\phi_1^{u_1}(x) \cdot \phi_2^{u_2}(x)} \tag{16-155}$$

with $m_{12}(x) = m_1(x) \cdot \phi_2^{u_2}(x) \oplus m_2(x) \cdot \phi_1^{u_1}(x)$ contains only all sequences generated by the mod 2 feedback shift register with characteristic polynomial $\phi_1^{u_1}(x) \cdot \phi_2^{u_2}(x)$. Furthermore, application of the above algorism leads to a subdivision of X into classes of equivalent sequences.

Since $\phi_1^{u_1}(x)\cdot\phi_2^{u_2}(x)$ and $\phi_3^{u_3}(x)$ are mutually prime, set Y of all sequences represented by the sum

$$\frac{m_1(x)}{\phi_1^{u_1}(x)} \oplus \frac{m_2(x)}{\phi_2^{u_2}(x)} \oplus \frac{m_3(x)}{\phi_3^{u_3}(x)} = \frac{m_{12}(x)}{\phi_1^{u_1}(x)\cdot\phi_2^{u_2}(x)} \oplus \frac{m_3(x)}{\phi_3^{u_3}(x)}$$

$$\frac{m_{123}}{\phi_1^{u_1}(x)\cdot\phi_2^{u_2}(x)\cdot\phi_3^{u_3}(x)} \tag{16-156}$$

with

$$m_{123}(x) = m_{12}(x)\cdot\phi_3^{u_3}(x) \oplus m_3(x)\cdot\phi_1^{u_1}(x)\cdot\phi_2^{u_2}(x)$$

contains only all sequences generated by the mod 2 feedback shift register with characteristic polynomial $\phi_1^{u_1}(x)\cdot\phi_2^{u_2}(x)\cdot\phi_3^{u_3}(x)$. This set can also be subdivided into classes of equivalent sequences by applying the above algorism.

By repeating this process $v - 1$ times, the subdivision into classes of set Z of all sequences generated by the mod 2 feedback shift register with characteristic polynomial $\phi(x)$ of Eq. (16-146) can be determined.

This means that the problem of determining the number of non-equivalent sequences and their cycle lengths has been solved.

Example The following characteristic polynomial is given:

$$\phi(x) = 1 \oplus x^5 \oplus x^6 \oplus x^{11} = (1 \oplus x)^3\cdot(1 \oplus x \oplus x^2)^2\cdot(1 \oplus x \oplus x^2 \oplus x^3 \oplus x^4)$$

It follows from the discussion on pages 466–469 that the first factor leads to two sequences of cycle length 1, one sequence of cycle length 2, and one sequence of cycle length 4. This can be written symbolically as follows:

2(1), 1(2), 1(4)

In the same way the second factor leads to

1(1), 1(3), 2(6)

Furthermore, it follows from Table 16-12 that the third factor leads to

1(1), 3(5)

By the above algorism, combination of the sequences produced by the first two factors gives

2(1), 1(2), 2(3), 1(4), 9(6), 5(12)

Combination of this result with the sequences produced by the third factor leads to

2(1), 1(2), 2(3), 1(4), 6(5), 9(6), 3(10)
5(12), 6(15), 3(20), 27(30), 16(60)

Example Design a mod 2 feedback shift register that generates a large number of sequences with cycle length 21.

It follows from Table 16-12 that there are two nonprimitive irreducible polynomials of degree 6 with period 21:

$$\phi_1(x) = 1 \oplus x \oplus x^2 \oplus x^4 \oplus x^6 \qquad \text{and} \qquad \phi_2(x) = 1 \oplus x^2 \oplus x^4 \oplus x^5 \oplus x^6$$

Hence, both polynomials yield the following sequences:

1(1), 3(21)

Then $\phi(x) = \phi_1(x) \cdot \phi_2(x)$ leads to the following number of sequences:

$1(1), (3 + 3 + 9 \cdot 21)(21)$

To get twice the number of sequences, characteristic polynomial $\phi_1(x) \cdot \phi_2(x)$ can by multiplied by $1 \oplus x$. The resulting polynomial is as follows:

$$\phi(x) = 1 \oplus x^2 \oplus x^3 \oplus x^5 \oplus x^6 \oplus x^7 \oplus x^8 \oplus x^{10} \oplus x^{11} \oplus x^{13}$$

which yields the following number of sequences:

2(1), 390(21)

which is in accordance with the fact that

$$2 \cdot 1 + 390 \cdot 21 = 2^{13}$$

16-11 THE CONSTRUCTION OF SHIFT REGISTERS WITH MOD 2 FEEDBACK

The general form of a mod 2 feedback shift register with k sections as used so far in this chapter is shown in Fig. 16-33. If an input sequence $S(x)$ is fed

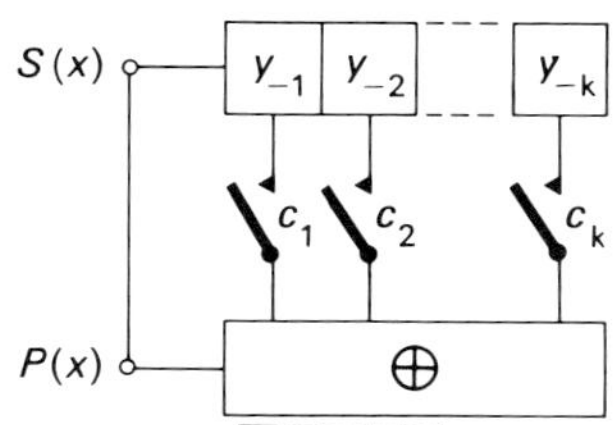

Fig. 16-33

into the shift register, output sequence $P(x)$ of the mod 2 feedback network is given by Eq. (16-69).

$$P(x) = f(x)S(x) \oplus m(x) \tag{16-157}$$

When $S(x) = P(x)$, Eq. (16-157) can be transformed in the following way:

$$S(x)[1 \oplus f(x)] = m(x)$$

or

$$S(x) = \frac{m(x)}{\phi(x)} \qquad \text{with } \phi(x) = 1 \oplus f(x) \tag{16-158}$$

In a number of cases, the mod 2 feedback networks become rather complicated, because it is difficult to construct mod 2 adders with more than 2 inputs. For example, for the construction of the feedback network corresponding to primitive polynomial $\phi(x) = 1 \oplus x^2 \oplus x^3 \oplus x^4 \oplus x^5$, three two-input mod 2 adders are required. In many cases, it is possible to split shift registers into coupled shift registers with, in total, the same number of shift sections but with a reduced number of two-input mod 2 adders in the feedback networks. The general form of the split-up into two smaller shift registers is shown in Fig. 16-34.

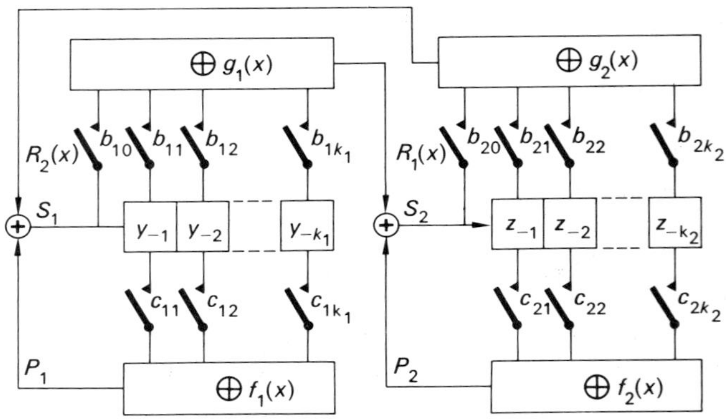

Fig. 16-34

In the split-up of Fig. 16-34, two shift registers and two pairs of feedback networks are shown $f_1(x)$, $g_1(x)$ and $f_2(x)$, $g_2(x)$, respectively, generating feedback signals for the inputs of the originating shift registers and the other shift registers. It should be noted that always one of the pair of switches b_{1k_1}, c_{1k_1} and one of the pair b_{2k_2}, c_{2k_2} must be closed, otherwise a number of sections of the shift register concerned will have a number of redundant sections.

In the diagram of Fig. 16-34,

$$f_1(x) = \sum_{j=1}^{k_1} c_{1j} x^j \qquad f_2(x) = \sum_{j=1}^{k_2} c_{2j} x$$

$$g_1(x) = \sum_{j=0}^{k_1} b_{1j} x^j \qquad g_2(x) = \sum_{j=0}^{k_2} b_{2j} x^j \tag{16-159}$$

Using Eq. (16-157), the output of the four mod 2 adding feeding networks will be as follows:

$$
\begin{aligned}
P_1(x) &= f_1(x)\,S_1(x) \oplus m_{1f}(x) \\
P_2(x) &= f_2(x)\,S_2(x) \oplus m_{2f}(x) \\
R_1(x) &= g_1(x)\,S_1(x) \oplus m_{1g}(x) \\
R_2(x) &= g_2(x)\,S_2(x) \oplus m_{2g}(x)
\end{aligned}
\qquad (16\text{-}160)
$$

The input of both shift registers can now be described by the following pair of equations:

$$
\begin{aligned}
S_1(x) &= P_1(x) + R_2(x) \\
S_2(x) &= P_2(x) + R_1(x)
\end{aligned}
\qquad (16\text{-}161)
$$

Substitution of Eqs. (16-160) into Eqs. (16-161) leads to

$$
\begin{aligned}
S_1(x) &= f_1(x)\,S_1(x) \oplus m_{1f}(x) \oplus g_2(x)\,S_2(x) \oplus m_{2g}(x) \\
S_2(x) &= f_2(x)\,S_2(x) \oplus m_{2f}(x) \oplus g_1(x)\,S_1(x) \oplus m_{1g}(x)
\end{aligned}
\qquad (16\text{-}162)
$$

Now let $1 \oplus f_1(x) = \phi_1(x)$ and $1 \oplus f_2(x) = \phi_2(x)$ and substitute this in Eqs. (16-162), then

$$
\begin{aligned}
\phi_1(x)\,S_1(x) \oplus g_2(x)\,S_2(x) &= m_{1f}(x) \oplus m_{2g}(x) \\
\phi_2(x)\,S_2(x) \oplus g_1(x)\,S_1(x) &= m_{2f}(x) \oplus m_{1g}(x)
\end{aligned}
\qquad (16\text{-}163)
$$

Finally, input sequence $S_1(x)$, which follows easily from Eqs. (16-163), is as follows:

$$
S_1(x) = \frac{\phi_2(x)\,m_{1f}(x) \oplus \phi_2(x)\,m_{2g}(x) \oplus g_2(x)\,m_{2f}(x) \oplus g_2(x)\,m_{1g}(x)}{\phi_1(x)\,\phi_2(x) \oplus g_1(x)\,g_2(x)} \qquad (16\text{-}164)
$$

The shift-register constellations of Figs. 16-33 and 16-34 now must give identical results. For these identical results it is required that

1. $k = k_1 + k_2$. The total number of shift sections involved must be equal in both cases.
2. $S = S_1$. Any sequence S generated in the shift register of Fig. 16-33 must also follow from that of Fig. 16-34.

It follows from the above that Eqs. (16-158) and (16-164) must yield the same result. In that case,

$$
\phi(x) = \phi_1(x)\,\phi_2(x) \oplus g_1(x)\,g_2(x) \qquad (16\text{-}165)
$$

Moreover $\phi_2(x)$ and $g_2(x)$ must be mutually prime, otherwise a degeneration of sequence lengths will happen. The proof of this will not be given here.

Example

$$\phi(x) = 1 \oplus x^2 \oplus x^3 \oplus x^4 \oplus x^5$$

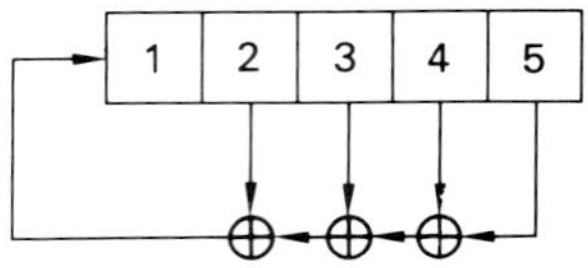

Fig. 16-35

The single shift-register solution with three two-input mod 2 adders is given in Fig. 16-35. Using Eq. (16-65), characteristic polynomial $\phi(x)$ can be written as follows:

$$\phi(x) = 1 \oplus x^2 \oplus x^3 \oplus x^4 \oplus x^5 = \begin{cases} (1 \oplus x^3)(1 \oplus x^2) \oplus x^2 \cdot x^2 \\ (1 \oplus x^3)(1 \oplus x^2) \oplus x^3 \cdot x \end{cases}$$

so that

$$\phi_1(x) = 1 \oplus x^3 \qquad g_1(x) = x^2 \qquad \qquad g_1(x) = x^3$$

or

$$\phi_2(x) = 1 \oplus x^2 \qquad g_2(x) = x^2 \qquad \qquad g_2(x) = x$$

The resulting shift-register circuits are shown in Figs. 16-36 and 16-37. In both variants, the number of two-input mod 2 adders is reduced from three to two.

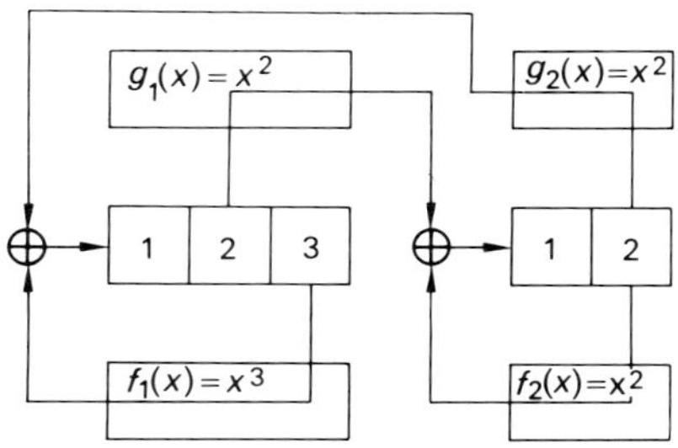

Fig. 16-36

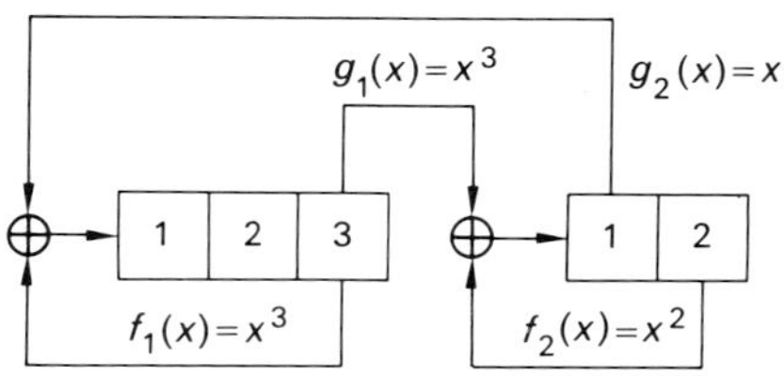

Fig. 16-37

For an advantageous split-up of a mod 2 feedback register in two parts, a factorization of the characteristic polynomial $\phi(x)$ is required without one of the terms x^i.

Another possible special choice of the configuration of Fig. 16-34 is $f_1(x) = 0$, $f_2(x) = 0$, so that $\phi_1(x) = 1$ and $\phi_2(x) = 1$ and $\phi(x) = 1 \oplus g_1(x)g_2(x)$.

$$\phi(x) = 1 \oplus x^2 \oplus x^3 \oplus x^4 \oplus x^5 = \begin{cases} 1 \oplus (1 \oplus x)(x^2 \oplus x^4) \\ 1 \oplus (x \oplus x^2)(x \oplus x^3) \\ 1 \oplus (x^2 \oplus x^3)(1 \oplus x^2) \end{cases}$$

so that

$$\begin{aligned} g_1(x) &= 1 \oplus x \\ g_2(x) &= x^2 \oplus x^4 \end{aligned} \quad \text{or} \quad \begin{aligned} g_1(x) &= x \oplus x^2 \\ g_2(x) &= x \oplus x^3 \end{aligned} \quad \text{or} \quad \begin{aligned} g_1(x) &= x^2 \oplus x^3 \\ g_2(x) &= 1 \oplus x^2 \end{aligned}$$

These three solutions are shown in Fig. 16-38a, b, and c. Solutions of this type were first indicated by Scholefield.

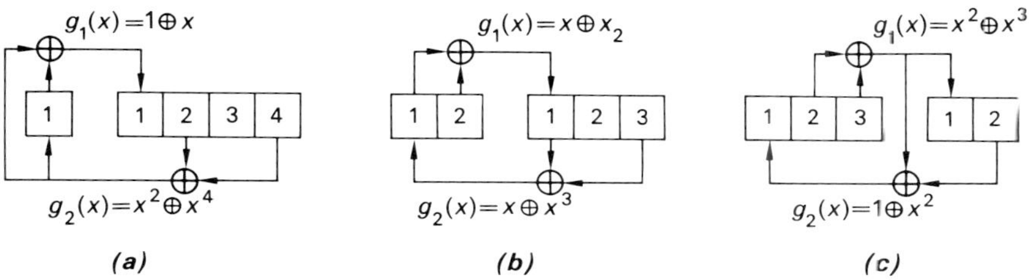

Fig. 16-38

The split-up process in this last case is very simple. It depends entirely on the possibilities of factorization of $f(x)$, that is, the given characteristic polynomial $\phi(x)$ without the 1.

Some further examples will be given to elucidate the use of Eq. (16-165) in the split-up of shift registers into two parts.

Example

$$\phi(x) = 1 \oplus x \oplus x^2 \oplus x^3 \oplus x^4 \oplus x^5 \oplus x^7$$

$$f_1(x) = 0 \qquad f_2(x) = 0 :: \phi_1(x) = 1 \qquad \phi_2(x) = 1$$

$$\phi(x) = \begin{cases} 1 \oplus g_1(x) \cdot g_2(x) \\ 1 \oplus (1 \oplus x^2)(x \oplus x^2 \oplus x^5) \\ 1 \oplus (x \oplus x^3)(1 \oplus x \oplus x^4) \end{cases}$$

leading to

$$\begin{aligned} g_1(x) &= 1 \oplus x^2 \\ g_2(x) &= x \oplus x^2 \oplus x^5 \end{aligned} \quad \text{or} \quad \begin{aligned} g_1(x) &= x \oplus x^3 \\ g_2(x) &= 1 \oplus x \oplus x^4 \end{aligned}$$

The resulting shift registers are shown in Fig. 16-39a and b. In both cases, the result of the split-up of the shift register is a reduction of the number of two-input mod 2 adders from five to three.

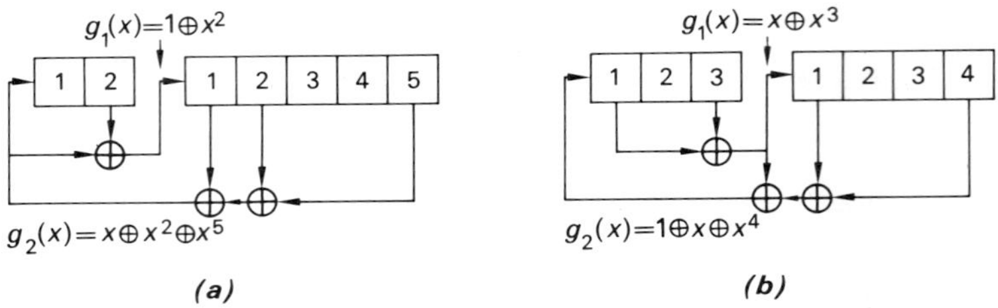

Fig. 16-39

Example

$\phi(x) = 1 \oplus x \oplus x^6 \oplus x^7 \oplus x^8$

This example cannot be solved by the simplified equation used in the immediately preceding example. The split-up of the above polynomial can be performed starting from Eq. (16-165).

$$\phi(x) = \phi_1(x)\phi_2(x) \oplus g_1(x)\,g_2(x)$$

$$\phi(x) = \begin{cases} (1 \oplus x^6)(1 \oplus x) \oplus x^8 \\ (1 \oplus x^7)(1 \oplus x) \oplus x^6 \end{cases}$$

The first possible split-up of $\phi(x)$ leads to:

$f_1(x) = x^6 \qquad f_2(x) = x \qquad\qquad f_1(x) = x^6 \qquad f_2(x) = x$

or

$g_1(x) = x^6 \qquad g_2(x) = x^2 \qquad\qquad g_1(x) = x^7 \qquad g_2(x) = x$

The resulting shift registers are shown in Fig. 16-40a and b.

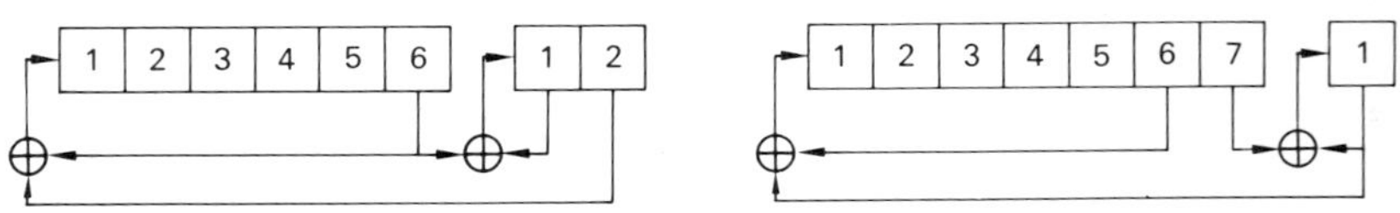

(a) (b)

Fig. 16-40

The second possible split-up of $\phi(x)$ leads to:

$f_1(x) = x^7 \qquad f_2(x) = x \qquad\qquad f_1(x) = x^7 \qquad f_2(x) = x$

or

$g_1(x) = x^5 \qquad g_2(x) = x \qquad\qquad g_1(x) = x^6 \qquad g_2(x) = 1$

The resulting shift registers are shown in Fig. 16-41a and b. In this shift-register split-up the number of two-input mod 2 adders is reduced from three to two.

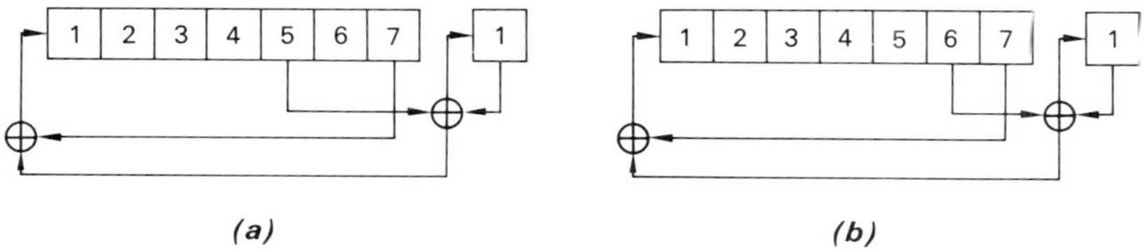

Fig. 16-41

Example

$$\phi(x) = 1 \oplus x^2 \oplus x^3 \oplus x^4 \oplus x^5 \oplus x^6 \oplus x^8 \oplus x^9 \oplus x^{10}$$
$$= (1 \oplus x^2 \oplus x^3 \oplus x^4)(1 \oplus x^6) + x^5$$

It follows from this split-up of characteristic polynomial $\phi(x)$ that:

$f_1(x) = x^2 \oplus x^3 \oplus x^4 \qquad f_2(x) = x^6$

$g_1(x) = x^4 \qquad g_2(x) = x$

or

$g_1(x) = x^3 \qquad g_2(x) = x^2$

or

$g_1(x) = x^2 \qquad g_2(x) = x^3$

or

$g_1(x) = x \qquad g_2(x) = x^4$

or

$g_1(x) = 1 \qquad g_2(x) = x^5$

In all cases, a reduction in the number of two-input mod 2 adders from seven to four is obtained. The first solution given here is shown in Fig. 16-42.

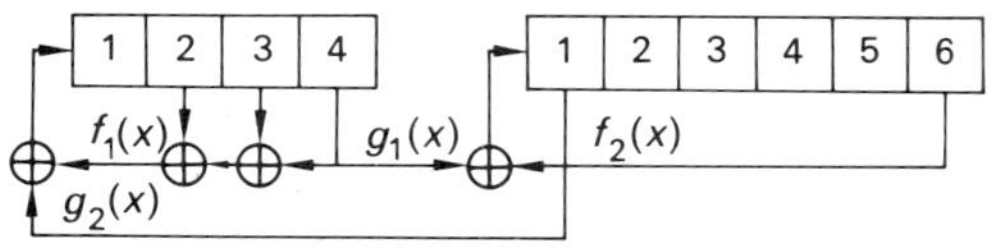

Fig. 16-42

The split-up of a shift register into two parts as shown in Fig. 16-35 can be extended into a generalized series connection of shift-register portions together acting as a single mod 2 feedback shift register. The diagram of such a group of series-connected shift-register portions is shown in Fig. 16-43.

It is not difficult to extend Eq. (16-165) to the generalized series connection of shift-register portions as shown in Fig. 16-43.

$$\phi(x) = \phi_1(x)\,\phi_2(x)\cdots\phi_n(x) + g_1(x)\,g_2(x)\cdots g_n(x) \tag{16-166}$$

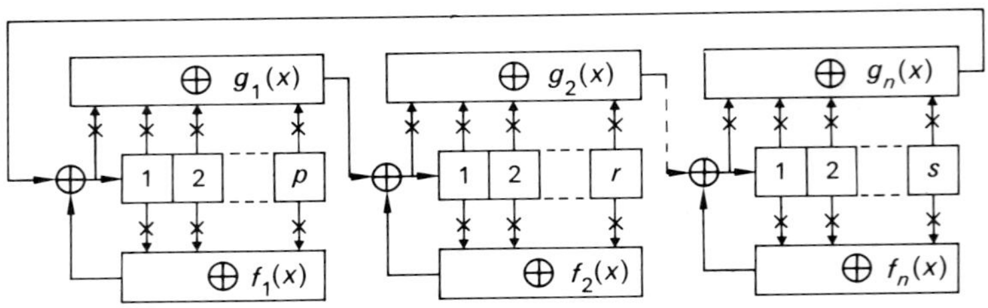

Fig. 16-43

The biggest reduction in the number of two-input mod 2 adders will be obtained in those cases where characteristic polynomial $\phi(x)$ can be written as follows:

For example,

$$\phi(x) = (1 \oplus x^p)\,1 \oplus x^q)\,(1 \oplus x^r) \oplus x^{p+q+r+s}$$

$$\phi(x) = 1 \oplus x^3 \oplus x^5 \oplus x^7 \oplus x^8 \oplus x^{10} \oplus x^{12} \oplus x^{15} \oplus x^{16}$$

$$= (1 \oplus x^3)\,(1 \oplus x^5)\,(1 \oplus x^7) \oplus x^{16}$$

In this example,

$$f_1(x) = x^3 \qquad f_2(x) = x^5 \qquad f_3(x) = x^7$$

and

$$g_1(x) = x^3 \qquad g_2(x) = x^5 \qquad g_3(x) = x^8$$

The resulting series connection of shift-register portions is shown in Fig. 16-44. In examples of this type, the number of two-input mod 2 adders equals the number n of factors $\phi_i(x)$. In the unsplit shift register, a number of $2^n - 2$ two-input mod 2 adders is required.

For all shift sections, MS memory elements can be used as has already been shown in the preceding paragraphs. For the mod 2 adders, transistor circuits as shown in Fig. 14-12 can be very useful. It is also possible to use capacitor elements for this purpose, with both outputs connected to the set or reset input terminal of a trigger. However, the inputs of an MS memory

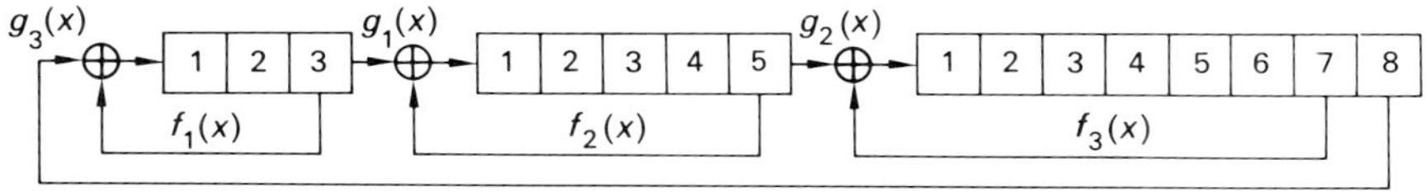

Fig. 16-44

element can be connected in such a way that it functions in accordance with truth table 16-16 and Eq. (16-167).

$$Z = D_{\alpha'}(y'z \oplus yz') = D_{\alpha'}(y \oplus z) \tag{16-167}$$

The input connections of an MS memory element operating in accordance with Eq. (16-167) is shown in Fig. 16-45. This is in fact a JK flip-flop.

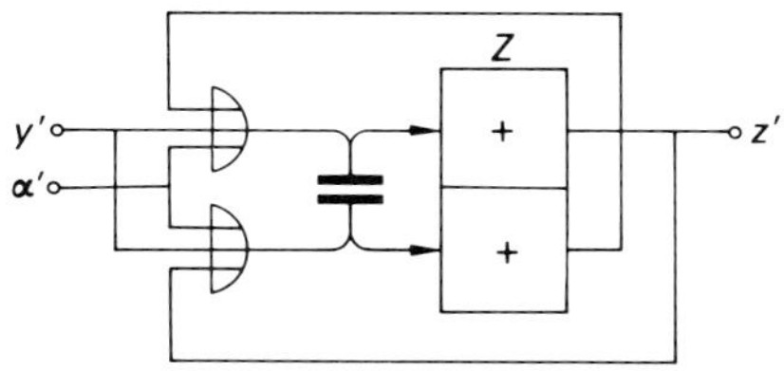

Fig. 16-45

Table 16-16

y	Z
0	z
1	z'

In terms of characteristic polynomials of shift registers, this MS memory element operates in accordance with

$$\phi(x) = 1 \oplus x \tag{16-168}$$

The most simple constructional form of split shift registers will be obtained in those cases where the characteristic polynomial can be written in the following form:

$$\phi(x) = (1 \oplus x)^n \oplus x^k \tag{16-169}$$

The general form of this split shift register is shown in Fig. 16-46.

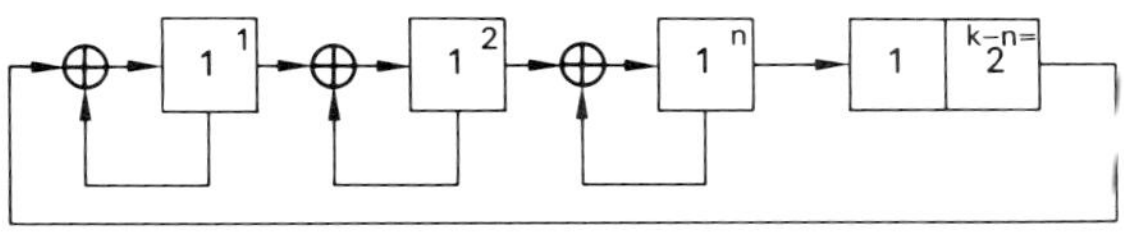

Fig. 16-46

Example

$$\begin{aligned}\phi(x) &= 1 \oplus x \oplus x^2 \oplus x^3 \oplus x^5\\ &= (1 \oplus x)^3 \oplus x^5\\ &= \phi_1(x)\,\phi_2(x)\,\phi_3(x)\,\phi_4(x) + g_1(x)\,g_2(x)\,g_3(x)\,g_4(x)\end{aligned}$$

$$\phi_1(x) = \phi_2(x) = \phi_3(x) = 1 \oplus x \qquad \phi_4(x) = 1$$

$$g_1(x) = g_2(x) = g_3(x) = x \qquad g_4(x) = x^2$$

The resulting shift register is shown in Fig. 16-46.

In the development of the sections of this chapter on mod 2 feedback shift registers, the text started with pure shift registers and external mod 2 adding networks. After that, it was shown that important savings in two-input mod 2 adders could be obtained by splitting the shift register into two shift registers which together have the same number of shift sections. The two-input mod 2 adders are then used as a coupling network between the portions of the complete network. This splitting-up procedure can be continued, leading to the general type of network shown in Fig. 16-47, which in its general form has lost the character of a shift register.

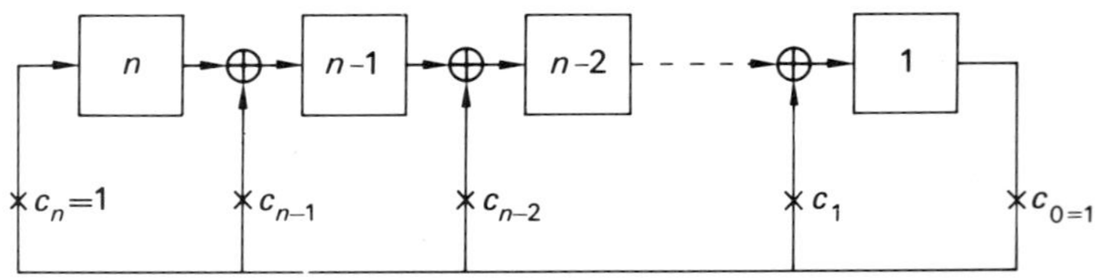

Fig. 16-47

With a suitable combination of closed switches C_i characterized by

$$\phi(x) = \sum_{i=0}^{n} c_i x^i \tag{16-170}$$

the circuit of Fig. 16-47 will produce the same sequence as the circuit shown in Fig. 16-31.

As an example, the four-section shift register of Fig. 16-28 generating a 15-bit maximum-length sequence will be transformed into a circuit of the type shown in Fig. 16-47. This transformation can easily be performed because it leads to a circuit consisting of two shift registers. The procedure used is that given above for the splitting of a shift register into two portions.

The shift register shown in Fig. 16-28 has the following characteristic polynomial:

$$\phi(x) = 1 \oplus x^3 \oplus x^4$$

$$f_1(x) = 0 \qquad f_2(x) = 0 \qquad \phi(x) = 1 \oplus g_1(x) \cdot g_2(x) \tag{16-171}$$

$$= 1 \oplus x^3(1 \oplus x)$$

so that $g_1(x) = x^3$ and $g_2(x) = 1 \oplus x$. The resulting circuit is shown in Fig. 16-48.

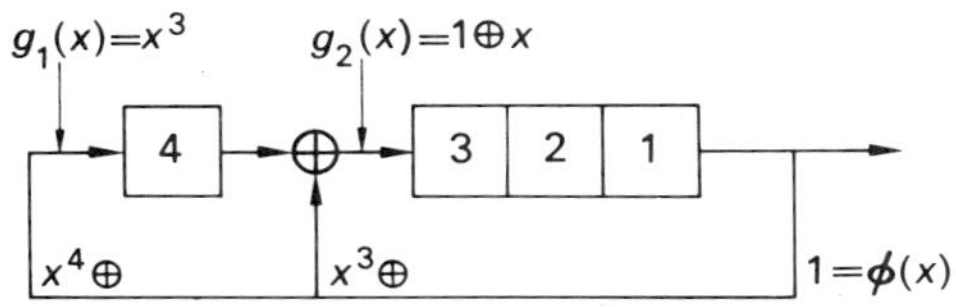

Fig. 16-48

It can easily be checked that the sequence generated by the mod 2 adding feedback network and fed into section 1 of the circuit of Fig. 16-28 is the same as the sequence produced by section 1 of the circuit of Fig. 16-48 and fed into section 4 of that circuit, when starting from the 0 0 0 1 state $[y(x) = m(x)]$ of the circuit.

The information content of the circuit of Fig. 16-47 will be investigated further. It should be noted that sequence $S(x)$ produced by the mod 2 adding feedback of Fig. 16-32 and fed into its shift register could be described by the simple equation

$$S(x) = \frac{m(x)}{\phi(x)}$$

with $m(x)$ representing the initial-state polynomial. This polynomial can be derived from the initial state of the shift register but it is not directly visible as a content of a shift register.

The equation for the sequence $S(x)$ generated by the circuit of Fig. 16-49 can be obtained as follows.

In the circuit of Fig. 16-49, Z is a polynomial representing the sequence of information elements passing the points indicated. The various sections start with an information content y_i.

It follows from the circuit shown in Fig. 16-49 that

$$Z_0(x) = y_0 \oplus x[Z_1(x) \oplus c_1 Z_0(x)]$$

$$= y_0 \oplus xZ_1(x) \oplus c_1 xZ_0(x)$$

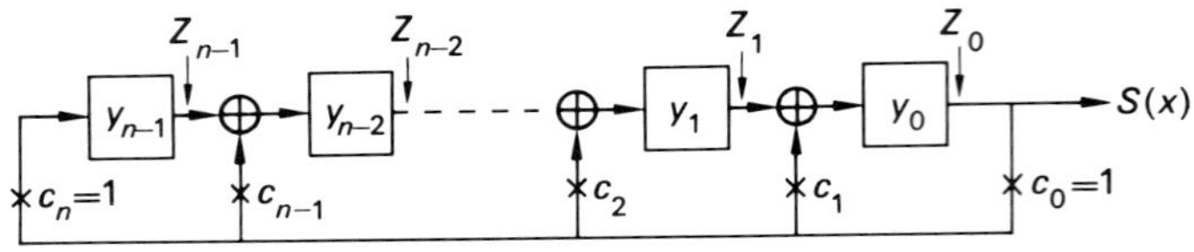

Fig. 16-49

Hence,

$$xZ_1(x) = y_0 \oplus (1 \oplus c_1 x)\, Z_0(x) \tag{16-172}$$

Furthermore,

$$Z_1(x) = y_1 \oplus x[Z_2(x) \oplus c_2\, Z_0(x)]$$

Hence,

$$x^2\, Z_2(x) = y_1 x \oplus xZ_1(x) \oplus c_2 x^2\, Z_0(x) \tag{16-173}$$

Substitution of Eq. (16-172) into Eq. (16-173) leads to

$$\begin{aligned} x^2\, Z_2(x) &= y_1 x \oplus y_0 \oplus (1 + c_1 x)\, Z_0(x) + c_2 x^2\, Z_0(x) \\ &= (y_0 \oplus y_1 x) \oplus (1 \oplus c_1 x \oplus c_2 x^2)\, Z_0(x) \end{aligned}$$

Finally, the following equation is obtained:

$$x^{n-1}\, Z_{n-1}(x) = \sum_{i=0}^{n-2} y_i x^i \oplus \left(1 \oplus \sum_{i=1}^{n-1} c_i x^i\right) Z_0(x) \tag{16-174}$$

It follows from the circuit of Fig. 16-49 that

$$\begin{aligned} Z_{n-1}(x) &= y_{n-1} \oplus c_n xZ_0(x) \\ &= y_{n-1} \oplus c_n xS(x) \end{aligned} \tag{16-175}$$

because in this case $Z_0(x) = S(x)$.

Substitution of Eq. (16-175) into Eq. (16-174) leads to the following result:

$$y_{n-1} x^{n-1} \oplus c_n x^n S(x) = \sum_{i=0}^{n-2} y_i x^i \oplus \left(1 \oplus \sum_{i=1}^{n-1} c_i x^i\right) S(x)$$

Hence,

$$\sum_{i=0}^{n-1} y_i x^i \oplus \left(1 \oplus \sum_{i=1}^{n} c_i x^i\right) S(x) = 0 \tag{16-176}$$

In Eq. (16-176), $\sum_{i=0}^{n-1} y_i x^i$ represents the actually visible information content of the sections of the circuit of Fig. 16-49. This sum will further be indicated by $Y(x)$ and $1 \oplus \sum_{i=1}^{n} c_i x^i = \phi(x)$, so that the operation cf the circuit of Fig. 16-49 can be represented by the following equation:

$$S(x) = \frac{Y(x)}{\phi(x)} \tag{16-177}$$

Equation (16-177) is analogous to the basic shift-register equation (16-75), but it contains the directly visible information content of the memory elements of the circuit instead of the invisible initial-state polynomial.

PROBLEMS

1. Explain the different principles of the transport of information in a shift register. Which of both principles leads to the most general type of circuit and why? Relay and contact circuits can be used in the explanation.

2. Explain the difficulties in the information transport in magnetic-core shift registers which result from the fact that the magnetic core is a bidirectional switching element. What measures can be taken to avoid these difficulties?

3. Draw the diagram of a couple of sections of a magnetic-core shift register with a split-winding information transfer loop. What measures have to be taken to safeguard the set of a receiving core on the temporary difference in the currents of both loop paths during the transport of a 1.

4. Draw the circuit of a couple of sections of a magnetic-core shift register with a separation of the sections by means of transistors. Mention some advantages of this type of circuit.

5. Devise the sequence in which a nine-section shift register can run in three modes of operation through all three-out-of-nine combinations. Draw a symbolic circuit using MS memory elements.

6. Prove that a shift register with any even number of sections can run through all two-out-of-n combinations in two modes of operation. The operation must be cyclic.

7. Determine the feedback connections between a five-section shift register and its mod 2 adding feedback network generating a maximum length recurrent sequence containing the following group of 10 digits:

0 0 0 0 1 0 0 1 0 1

8. A five-section shift register is given with $c_1 = 0$, $c_2 = 0$, $c_3 = 0$, $c_4 = 1$, and $c_5 = 1$. Let the initial state of this shift register be

$$y_{-1} = 0 \qquad y_{-2} = 0 \qquad y_{-3} = 0 \qquad y_{-4} = 1 \qquad y_{-5} = 1$$

Determine initial-state polynomial $m(x)$, the resulting sequence $S(x)$, and the period of this sequence.

9. A five-section shift register is given with $c_1 = 0$, $c_2 = 1$, $c_3 = 0$, $c_4 = 0$, and $c_5 = 1$. Initial-state polynomial $m(x) = 1 \oplus x \oplus x^2$. Determine the initial state of this shift register, sequence $S(x)$, and its period.

10. A six-section shift register is given with characteristic polynomial $\phi(x) = 1 \oplus x^3 \oplus x^6$ and an initial state 0 0 1 1 0 0. Determine sequence $S(x)$ as a mod 2 sum of Q sequences.

11. The following nonprimitive polynomial is given:

$$\phi(x) = 1 \oplus x \oplus x^2 \oplus x^4 \oplus x^6$$

Determine period n of this polynomial, the number, and the length of the sequences generated by a mod 2 feedback shift register with this characteristic polynomial. Determine representatives of the nonzero sequences.

12. Let the following characteristic polynomial be given:

$$\phi(x) = (1 \oplus x \oplus x^3)^6$$

Determine the length and number of the sequences generated by a mod 2 feedback shift register with this characteristic polynomial.

13. The following characteristic polynomial is given:

$$\phi(x) = (1 \oplus x)^4 (1 \oplus x \oplus x^3)^2 (1 \oplus x \oplus x^2 \oplus x^3 \oplus x^4)$$

Determine the number and length of the various sequences generated by a mod 2 feedback shift register with this characteristic polynomial.

14. Design a shift register which generates a large number of sequences with cycle length 31. Sequences with cycle length 31 are generated by the primitive polynomials $\phi(x) = 1 \oplus x \oplus x^2 \oplus x^3 \oplus x^5$, $\phi(x) = 1 \oplus x \oplus x^2 \oplus x^4 \oplus x^5$, $\phi(x) = 1 \oplus x^2 \oplus x^5$, and the corresponding mirror polynomials.

15. A shift register is given with the following characteristic feedback polynomial:

$$\phi(x) = 1 \oplus x \oplus x^2 \oplus x^5 \oplus x^6$$

What possible split-ups in two coupled shift registers can be made with two two-input mod 2 adders and in total the same number of shift sections?

16. A shift register is given with the following characteristic polynomial:

$$\phi(x) = 1 \oplus x \oplus x^5 \oplus x^7 \oplus x^8$$

What are the possible split-ups in two coupled shift registers with, in total, two two-input mod 2 adders as feedback networks?

17. A shift register is given with the following characteristic polynomial:

$$\phi(x) = 1 \oplus x \oplus x^2 \oplus x^3 \oplus x^4 \oplus x^5 \oplus x^6 \oplus x^7 \oplus x^8$$

Determine the split up of this shift register in three coupled shift registers with, in total, the same number of sections and three two-input mod 2 adders.

18. Show that the characteristic polynomial of Prob. 17 can also be implemented with eight MS memory elements without using separate two-input mod 2 adding networks.

SELECTIVE BIBLIOGRAPHY

1. Albert, A. A.: "Fundamental Concepts of Higher Algebra," The University of Chicago Press, Chicago, Ill., 1956.
2. Anderson, J. R.: A New Type of Ferro-electric Shift Register, *IRE Trans.*, vol. EC-5, pp. 184–191, 1956.
3. Birkhoff, G., and S. MacLane: "A Survey of Modern Algebra," The Macmillan Company, New York, 1961.
4. Charney, H., and C. Mengam: Generation of Linear Binary Sequences, *RCA Rev.*, vol. 22, pp. 420–430, 1961.

5. Davies, N. G.: Some Properties of Linear Recursive Sequences, Defense Research Board, Department of National Defense, Ottawa, Canada, 1959.
6. Elspas, B.: The Theory of Autonomous Linear Sequential Networks, *IRE Trans.*, vol. CT-6, pp. 45–60, 1959.
7. Harris, J. R.: A Transistor Shift Register and Serial Adder, *Proc. IRE*, vol. 40, pp. 1597–1602, 1952.
8. Huffman, D. A.: A Linear Circuit Viewpoint on Error-correcting Codes, *IRE Trans.*, vol. IT-2, pp. 20–28, 1956.
9. Kautz, W. H.: Constant weight Counters and Decoding Trees, *IRE Trans.*, vol. EC-9, pp. 231–244, 1960.
10. Kihn, H., and W. E. Barnette: The Megacoder—A High-speed, Large Capacity Microminiature Decoder for Selective Communication, *RCA Rev.*, vol. 20, pp. 153–179, 1959.
11. Kodis, R. D., S. Ruhman, and W. D. Woo: Magnetic Shift Register Using One Core per Bit, *Conv. Record IRE*, pt. 7, pp. 38–42, 1953.
12. Marsh, R. W.: Table of Irreducible Polynomials over GF(2) Through Degree 19, Office of Technical Services, U.S. Department of Commerce, P.B. 161693, Washington, D.C., 1957.
13. Newhouse, V. L., and N. S. Prywes: High Speed Shift Registers Using One Core per Bit, *IRE Trans.*, vol. EC-5, pp. 114–120, 1956.
14. Peterson, W. W.: "Error-correcting Codes," John Wiley & Sons, Inc., New York, 1961.
15. Prywes, N. S.: Diodeless Magnetic Shift Register Utilizing Transfluxors, *IRE Trans.*, vol. EC-7, pp. 316–324, 1958.
16. Sands, E. A.: An Analysis of Magnetic Shift Register Operation, *Proc. IRE*, vol. 41, pp. 993–999, 1953.
17. Scholefield, P. H. R.: Shift Registers Generating Maximum-length Sequences, *Electron. Technol.*, vol. 37, pp. 389–393, 1960.
18. Zierler, N.: Several Binary Sequence Generators, *M.I.T. Lincoln Lab. Tech. Rept.* 95, 1955.

17

The Addition and Subtraction of Binary Numbers

17-1 STATEMENT OF THE PROBLEM

The rules for the addition and subtraction of binary numbers (radix 2) do not differ basically from those for adding and subtracting decimal numbers (radix 10). The basic operation for adding n binary numbers is an operation on the digits of these numbers of the same weight (or in the same column p) by means of which the total number of 1's in each column is subdivided in the possible groups of ..., 8, 4, 2, 1 "ones." The possible groups of ..., 8, 4, and 2 ones represent the carry digits which have to be added to the digits of columns ..., 3, 2, and 1 orders higher, having ..., 8, 4, and 2 times the weight of the digits in the originating column p.

For example, if in a digit column of an addition of n binary numbers eleven digits 1 are counted, a group of eight 1's and a group of two 1's can be formed, representing carries to columns $p + 3$ and $p + 1$. In this example, a 1 representing the sum $s_p = 1$ of the digits in column p remains.

The general concept of the basic binary adder means that in some way or other the following symmetrical functions have to be realized:

$$
\begin{aligned}
S &= S_{1,3,5,7,9,\ldots} \\
C_1 &= S_{2,3,6,7,\ldots} \\
C_2 &= S_{4,5,6,7,\ldots} \\
C_3 &= S_{8,9,\ldots} \\
&\text{etc.}
\end{aligned}
\tag{17-1}
$$

A decoder with m sections can be used for adding at most $2^m - 1$ digits. However, a section of a binary adder not only contains digits of the same weight to be added; it also obtains a number of carry digits of that weight coming from lower-order sections. This means that the effective number of digits n which can be added in one section is at most

$$n = (2^m - 1) - (m - 1) = 2^m - m \tag{17-2}$$

A decoder with m sections produces a sum S and $m - 1$ carries. The number of carry digits $m - 1$ originating from a decoder is equal to the number of carry digits $m - 1$ received from lower-order decoders, except for the lowest-order decoders.

Table 17-1 shows, for some values of m (number of sections), the maximum number of digits n that can actually be added in a section of such an adder.

Table 17-1

Number of decoder sections m	$2^m - 1$	*Number of digits* n	*Number of carries* $m - 1$
2	3	2	1
3	7	5	2
4	15	12	3
5	31	27	4

Table 17-1 shows that the addition of 2, 5, 12, or 27 binary numbers is advantageous with respect to the economical use of the necessary switching means.

The binary-addition process may be illustrated further by the following example in which the binary equivalents of the decimal numbers 11, 12, 13, 14, and 15 will be added.

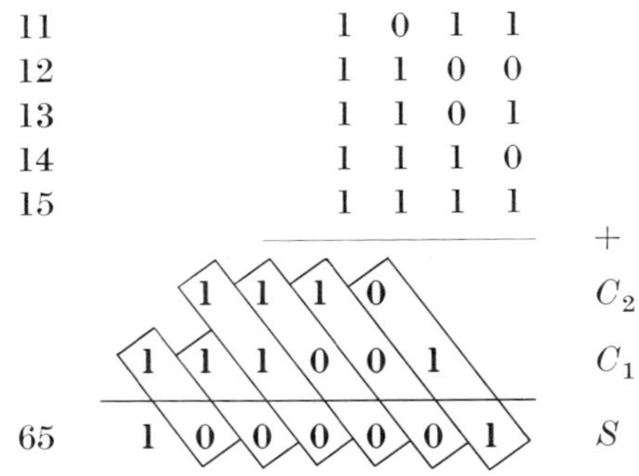

Table 17-1 shows that two carries will be generated in the addition of five binary numbers. In our example, the carries are placed in the digit column of the corresponding weight. The sum S and the carries C_1 and C_2 which will be generated by the same decoder are shown within the same rectangle. A decoder operates on the digits of a certain column as well as on the carries C_1 and C_2 of the same weight received from lower-order columns.

17-2 THE ADDITION OF TWO BINARY NUMBERS BY MEANS OF RELAY CIRCUITS

Table 17-1 shows that in the addition of two binary numbers, only one carry can be generated per column. The decoding device performing the addition of the digits of one column will therefore have three inputs, viz., the addend a, the augend b, and the incoming carry c. The addition can be performed serially or in parallel. The main difference between these two methods is that a parallel adder comprises n decoders (n being the number of digits) and some auxiliary equipment, whereas a serial adder consists mainly of only one such decoder. The operating time of a parallel adder is shorter than that of a serial adder using the same switching elements. The adding circuit of a serial adder is essentially the same as that of one of the sections of a parallel adder; hence, in the two-binary-number case, we will mainly concern ourselves with parallel adders.

The operation of a section of such an adder is defined by truth table 17-2. The various operating combinations are arranged in accordance with the symmetrical functions to which they belong.

The simplest and most straightforward method for designing a section of a parallel binary adder would be to use a three-digit-binary–eight-digit-decimal code converter as shown in Chap. 8. The diagram of Fig. 8-2 is repeated here as Fig. 17-1. The decimal 0 output is omitted in Fig. 17-1

Table 17-2

a	b	c	S	C
0	0	0	0	0
1	0	0	1	0
0	1	0	1	0
0	0	1	1	0
1	1	0	0	1
1	0	1	0	1
0	1	1	0	1
1	1	1	1	1

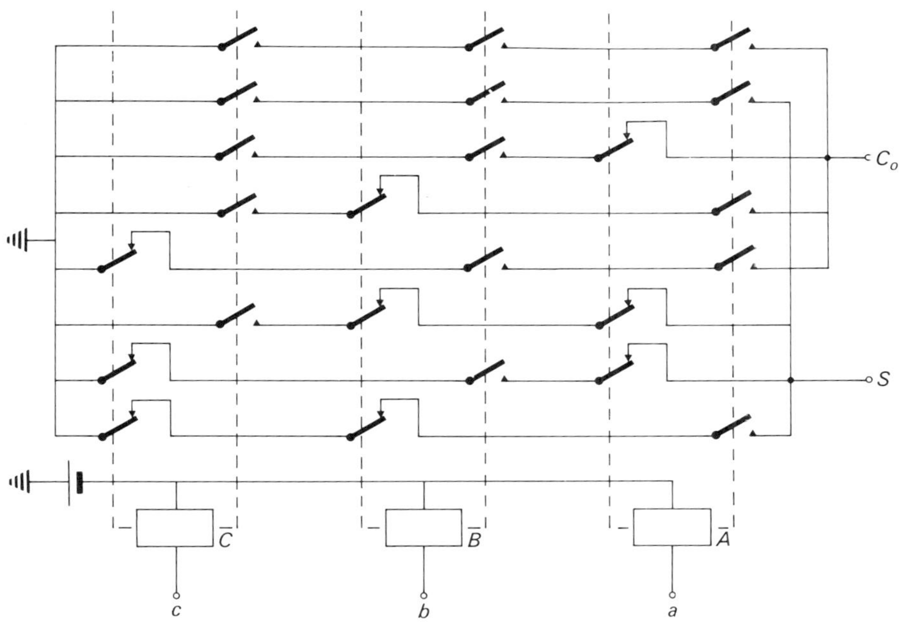

Fig. 17–1

because no S and C_o have to be generated on the corresponding combination of input variables. The decimal seven output has been doubled because a sum digit as well as a carry digit have to be produced on the corresponding combination of input variables. The sum output S can be used to magnetize a relay; the carry output terminal C_o can be connected directly to the carry input c of the next higher-order section of the adder.

The transfer of a carry from one section of the adder to the higher-order sections raises one of the most important problems in the design of parallel adders.

If the digits of addend and augend are applied simultaneously to all sections of a parallel adder, the carry digits are obtained in a sequential operation which is completed within a well-determined maximum time but which varies with the numbers to be added. A block diagram of a four-bit adder using relays is shown in Fig. 17-2. The sum and carry originating from each section are registered in a relay. The contacts of the C relays are controlled by the c digit incoming in the corresponding section. The addition of the binary equivalents of the decimal numbers 1 and 15 is given in Fig. 17-3 as an example of what can happen at most in this type of adder.

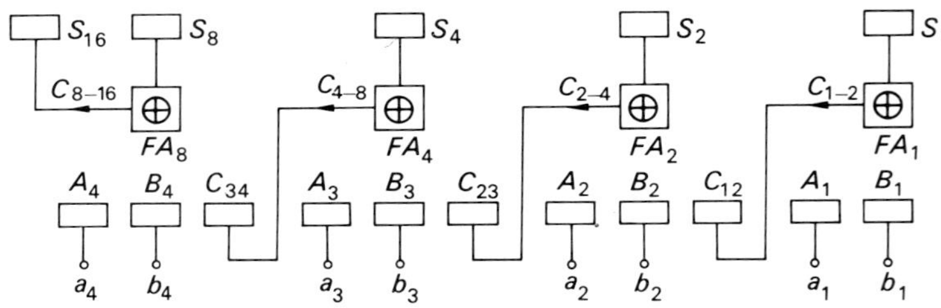

Fig. 17–2

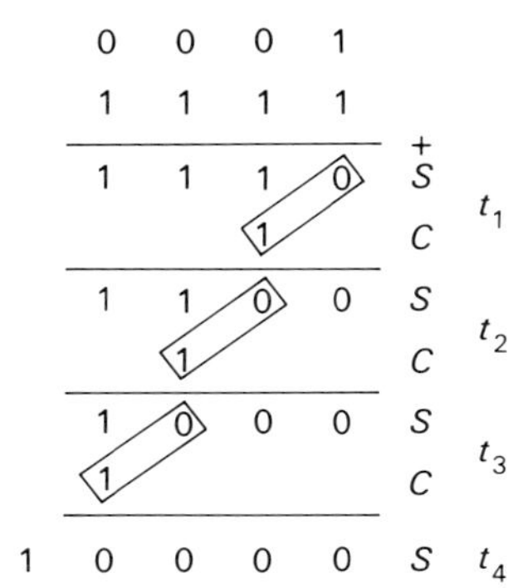

Fig. 17–3

The addition consists of four phases. The digits of addend and augend are applied simultaneously as a combination of open and closed contact circuits to relays $A_1, A_2, A_3, A_4, \ldots, B_1, B_2, B_3, B_4, \ldots$, of the circuit shown in Fig. 17-2. If it is assumed that all relays have the same operating time, line t_1 in Fig. 17-3 shows the states of sum and carry, S and C, after the first time interval. The following pairs of lines show these states after 2, 3, and 4 of these standard time intervals. The carry is gradually transferred through all adding sections; the final sum in this case is only arrived at after the fourth time interval.

It should be pointed out that the sequential operation as described for the relay adder of Fig. 17-2 is not limited to these switching elements only.

Relay adding circuits can be designed without sequential carry transfer between sections. The switching equations, which can be derived from truth table 17-2, are as follows:

$$S = S_{1,3} = ab'c' + a'bc' + a'b'c + abc$$

$$= c(ab + a'b') + c'(ab' + a'b) \qquad (17\text{-}3)$$

$$C = S_{2,3} = abc' + ab'c + a'bc + abc$$

$$c(ab' + a'b) + ab \qquad (17\text{-}4a)$$

$$C' = S_{0,1} = c'(ab' + a'b) + a'b' \qquad (17\text{-}4b)$$

Equations (17-4a) and (17-4b) can be simplified further in other ways, as can easily be seen in a Karnaugh map.

It follows from Eq. (17-3) that it is possible to generate directly the sum S of two binary digits without the aid of an auxiliary relay if the incoming carry c and the incoming no-carry c' are already available in the form of contact circuits. A section of a binary adder must consequently generate an outgoing carry C and an outgoing no-carry C' as given by Eqs. (17-4a) and (17-4b).

It is possible to convert these equations into a matrix with five terminal nodes (c, c', C, C', and S); however, with some experience in the conversion of switching equations into contact circuits, the resulting section diagram (Fig. 17-4) of a two-number binary adder can be obtained in much shorter time.

The first section of a parallel adder based on the diagram of Fig. 17-4 can be somewhat simplified because for that section $c = 0$ and $c' = 1$.

As was shown in Chap. 12, symmetrical contact circuits may—under certain circumstances—be simplified by means of differential relays.

$$S_{1,3} = S = c(ab + a'b') + c'(ab' + a'b) = c \oplus (ab' + a'b) \qquad (17\text{-}5)$$

Equation (17-5) shows that the sum S can also be obtained by introducing a differential relay if only incoming carry c is available. When no outgoing

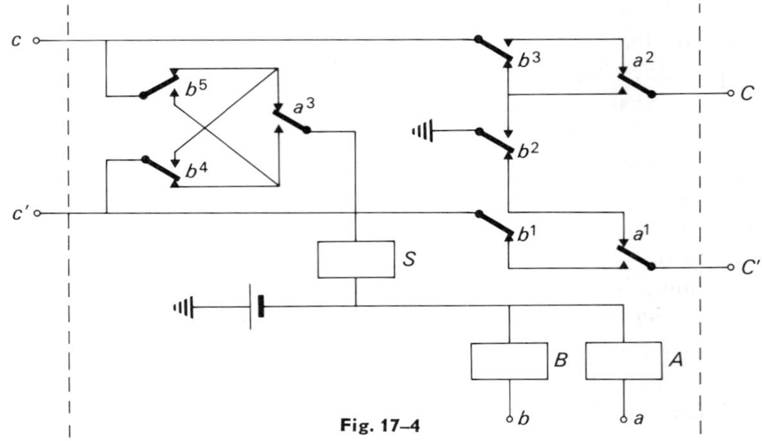

Fig. 17-4

no-carry has to be generated, the number of contacts in the carry network is accordingly halved.

The controlling contact network of the sum relay S is simplified at the expense of a differential two-winding relay S. The simplified circuit of one section of a binary adder is shown in Fig. 17-5. This circuit also has a constant operating time, and there is no delay in the carry transfer.

The two-binary adding circuits are so-called full-adder sections, i.e., sections in which the binary addition is performed with an addend, an augend, and an incoming carry digit. It is also possible to add the augend and addend digits in one operation, thereby obtaining a provisional sum H or an outgoing carry C_1, and to add in a second circuit the incoming carry c and the provisional sum digit h produced in the first circuit, thus arriving at the final sum S or an outgoing carry C_2. Both circuits are called half-adders. The truth tables for both operations are given as Tables 17-3 and 17-4.

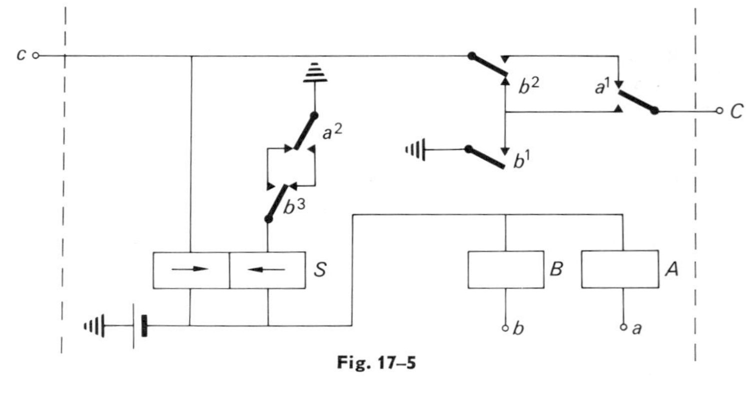

Fig. 17-5

Table 17-3

a	b	H	C_1
0	0	0	0
1	0	1	0
0	1	1	0
1	1	0	1

Table 17-4

h	c	S	C_2
0	0	0	0
1	0	1	0
0	1	1	0
1	1	0	1

Whereas addend a and augend b, on the one hand, and the incoming carry signal c and the provisional sum signal h, on the other, cannot simultaneously be 1, this is also true for partial carries C_1 and C_2. This means that in principle both half-adders can have the same circuit and that partial carries C_1 and C_2 can generate the resulting outgoing carry C by means of an OR gate.

With relay circuits, the half-adder concept leads to an interesting circuit. These circuits are described by the following equations:

$$H = ab' + a'b = a \oplus b \tag{17-6}$$

$$C_1 = ab \tag{17-7}$$

$$S = hc' + h'c = h \oplus c \tag{17-8}$$

$$C_2 = hc \tag{17-9}$$

$$\begin{aligned} C &= C_1 + C_2 = ab + hc = b(ab + a'b') + hc \\ &= bh' + hc \end{aligned} \tag{17-10}$$

The resulting circuit is shown in Fig. 17-6.

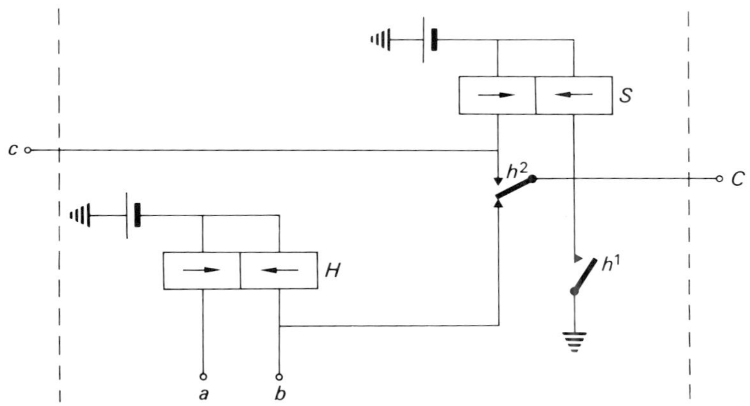

Fig. 17–6

This adding circuit has a constant operating time, and there is no cumulative delay in the transfer of the carry signals.

17-3 THE ADDITION OF BINARY NUMBERS BY MEANS OF ELECTRONIC CIRCUITS

FULL ADDERS

A full adder operates as defined by truth table 17-2 and is described by the following equations:

$$S = ab'c' + a'bc' + a'b'c + abc = (ab' + a'b)c' + (ab + a'b')c \quad (17\text{-}11)$$

$$C_o = abc' + ab'c + a'bc + abc = ab + bc + ac \quad (17\text{-}12)$$

$$C_o' = a'b'c + a'bc' + ab'c' + a'b'c' = a'b' + b'c' + a'c' \quad (17\text{-}13)$$

Equations (17-11) to (17-13) can be implemented by means of diode gates only if input variables a, b, and c and their inverse a', b', and c' are available. In the resulting circuit, the actual input terminals to which the

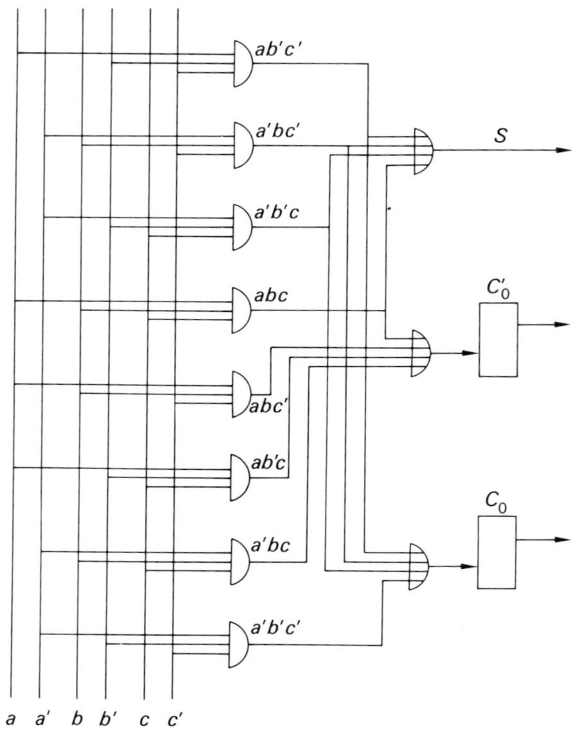

Fig. 17–7

signals corresponding to addend, augend, and incoming carry are connected are immaterial. The circuit corresponding to equations (17-11) to (17-13) is shown in Fig. 17-7, with some rather obvious simplifications having been applied to the equations. It is perhaps not entirely correct to consider this diode-gate circuit as the complete adder, because it is necessary to amplify the resulting outgoing carries C_o and C_o' before they can be used as incoming carry signals c and c' for the addition of the next higher-order binary digits. If a single triode or transistor (inverter) is used for this purpose, the C_0' carry amplifier will generate the c input carry signal of the next order of the adder, and the C_o carry amplifier will generate the c' signal for that section.

If the inverters C and C' are connected in series as a single-input trigger, one of the diode groups, e.g., the group generating the C' carry, may be omitted. This is shown in the diagram of Fig. 17-8. The number of diodes is thereby decreased from 36 (Fig. 17-7) to 29 (Fig. 17-8).

The number of diodes in this type of full adder can be decreased further by taking advantage of the fact that three terms of Eqs. (17-11) and (17-13) are identical, so that a common use of the circuits generating these three terms might prove advantageous. It is of course immaterial whether C or C' is derived from the logical circuit, because the other one will automatically be generated by the single-input trigger.

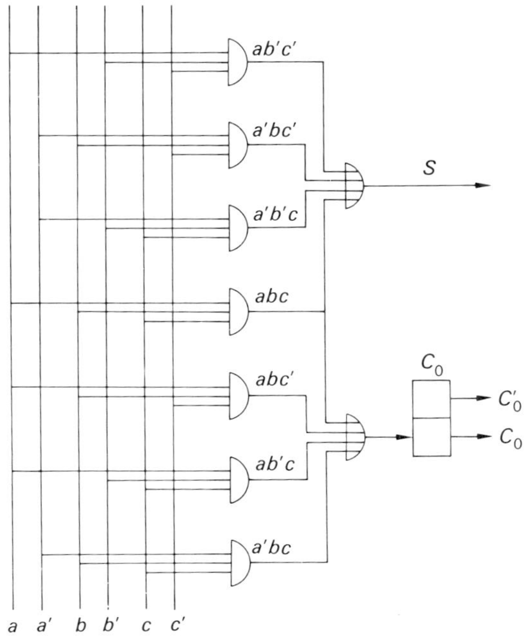

Fig. 17–8

When *pnp* transistors are used to generate the controlling a, b, and c signals, the input gates of the adder circuit are preferably chosen as OR gates. The required maxterm form of the switching equations (17-11) and (17-13) can be written for this purpose as

$$S = (a + b' + c')(a' + b + c')(a' + b' + c)(a + b + c) \tag{17-14}$$

$$C_o' = (a + b' + c')(a' + b + c')(a' + b' + c)(a' + b' + c') \tag{17-15}$$

The resulting circuit is shown in Fig. 17-9. The number of diodes in the circuit of Fig. 17-9 has been decreased to 23. The number of transistors involved has remained the same throughout.

The number of diodes of the full-adder circuit can be reduced even further through a careful factoring of switching equations (17-11) and (17-13):

$$\begin{aligned} S &= (ab' + a'b)c' + (ab + a'b')c = (ab' + a'b + c)(ab + a'b' + c') \\ &= [(a + b)(a' + b') + c][(a + b')(a' + b) + c'] \end{aligned} \tag{17-16}$$

$$\begin{aligned} C_o' &= a'b'(c + c') + (a'b + ab')c' = (a'b' + ab)b' + (a'b + ab')c' \\ &= (a'b + ab' + b')(ab + a'b' + c') = (a' + b')[(a + b')(a' + b) + c'] \end{aligned} \tag{17-17}$$

The resulting circuit is shown in Fig. 17-10. In this circuit all inputs to the adder circuit end on OR gates as is desirable when using *pnp* transistors. The number of diodes has now been decreased to 20. Signals a and b run successively through one OR gate and two AND gates.

A variant of this type of adding circuit can generally be arrived at by replacing all AND gates by OR gates, and vice versa. This does not lead to a further improvement in the number of diodes involved.

The circuit of a full adder can be so arranged that not both phases of the carry are required. For that purpose, the sum S may be written as follows:

$$S = \{[(a' + b)(a + b') + c']' + (a' + b)(a + b')\cdot c'\}' \tag{17-18}$$

The equation for the inverse of the outgoing carry can be written as follows:

$$C_o' = \{[(a' + b)(a + b') + c']' + ab\}' \tag{17-19}$$

The resulting circuit containing as a maximum two gate levels in series after which an inverter is inserted is shown in Fig. 17-11. The output signals of this asynchronous full-adder circuit are well suited for a direct further use because they are produced by inverters.

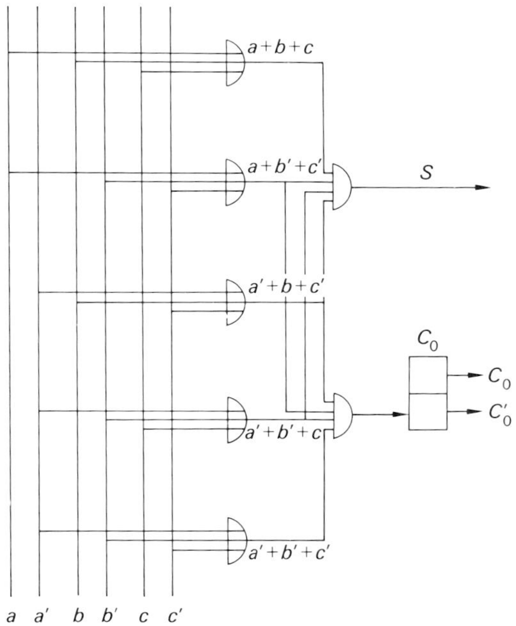

Fig. 17–9

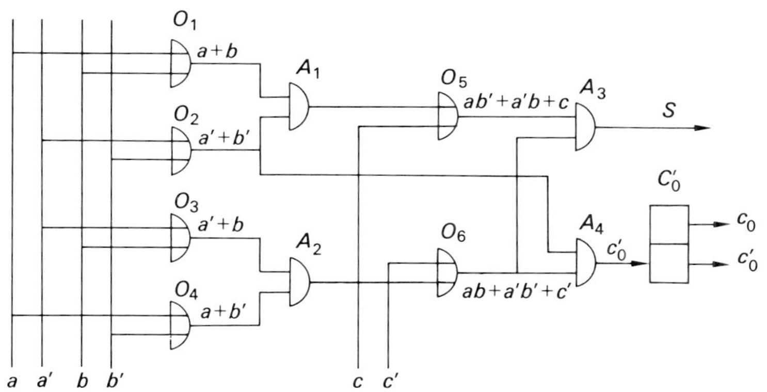

Fig. 17–10

It should be borne in mind that OR gates O_3 to O_5 are required only when the controlling a, b, and c' signals stem from inverters without individual collector resistors. The full-adder circuit of Fig. 17-11 contains only 20 (17) diodes and 3 inverters. This circuit is fully adapted to other types of standard circuits used throughout this book.

It should be further noted that the switching equations of a full-adder circuit can be arranged so that no inverted input variables are required. The following pair of equations can be derived:

$$C_o = ab + ac + bc = (a + b)(a + c)(b + c)$$

$$S = (a + b + c) \cdot C_o + abc \tag{17-20}$$

HALF-ADDERS

An adding circuit can also be designed by means of two half-adder circuits. The first half of the circuit performs the addition of addend and augend, which results in a provisional sum or an outgoing carry; the second half of the circuit performs the addition of the incoming carry and the provisional sum, which results in the final sum S and/or an outgoing carry C_o. Equations (17-6) to (17-10) describe these two circuits. The corresponding circuit using diode gates as switching elements is shown in Fig. 17-12. This circuit also requires negated input variables. It is not very practical, however, to provide a C_o' output in the logical network because the outgoing carry signals have to be amplified before they can be used in the circuit of the next higher-order addition. This is why a single-input trigger C_o is shown in Fig. 17-12.

The circuit of Fig. 17-12 is not very practical, because it requires four orders of diode gates in series in order to arrive at sum S. Amplification between both half-adders is necessary, as is shown in the circuit of Fig. 17-13.

The resulting circuit comprises 16 diodes and 2 single-input triggers. This amount of hardware is less favorable than the requirements for the best full-adder circuit (Fig. 17-11).

If advantage is taken of the special properties of transistors, there are some half-adders design possibilities that are rather unique from the logical point of view but raise some difficulties in the dimensioning of the circuits.

The switching equations for the half-adders can be written as follows:

$$H = ab' + a'b = (a + b)(a' + b') = (a + b)(ab)' \tag{17-21}$$

$$S = hc' + h'c = \quad = (h + c)(hc)' \tag{17-22}$$

$$C_o = ab + hc \tag{17-23}$$

With transistors being used as double-input switching devices—the y signal being applied to the emitter and the z signal to the base—the resulting signal in the case of a *pnp* transistor will be $S = yz'$. The switching circuit

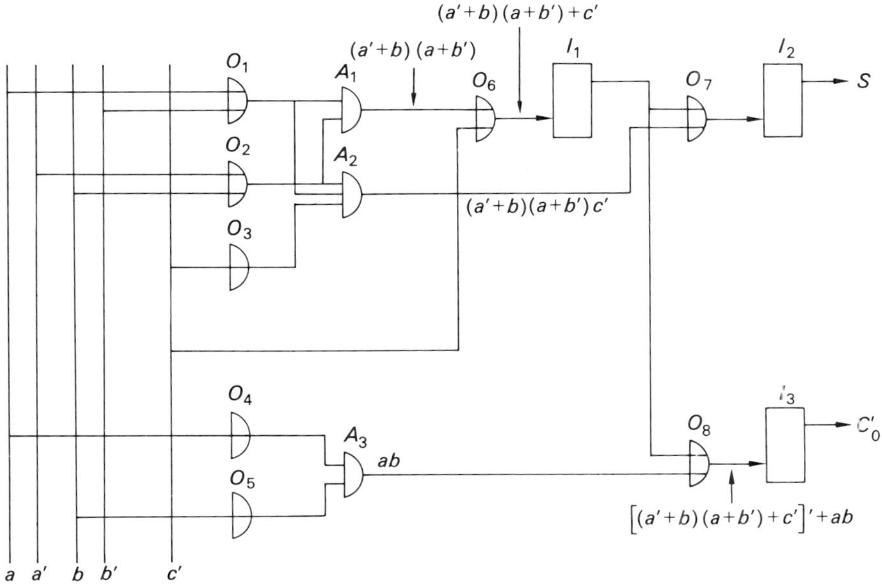

Fig. 17–11

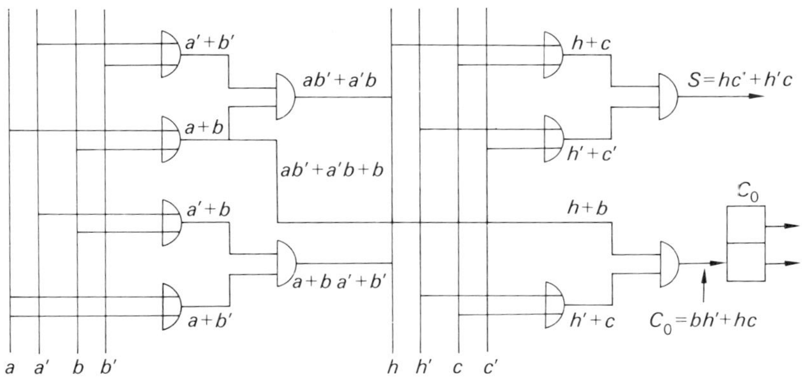

Fig. 17–12

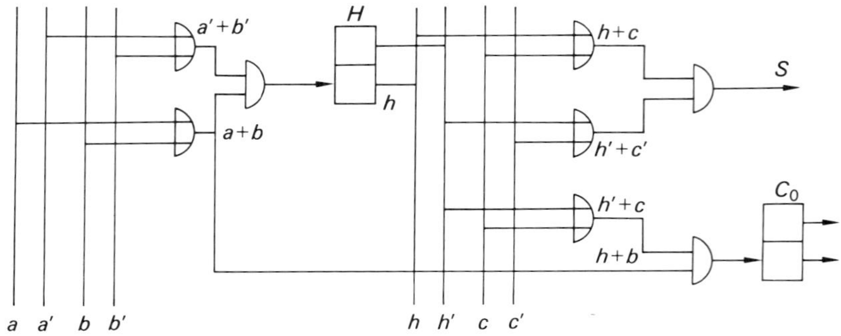

Fig. 17–13

derived from Eqs. (17-21), (17-22), and (17-23), in which transistors are used as double-input switching elements, is shown in Fig. 17-14. The level-shifting voltage dividers are omitted from this diagram.

The ten diodes and two transistors used in the adder circuit of Fig. 17-14 compare favorably with the seventeen diodes and three transistors used in the circuit of Fig. 17-11.

At the expense of some additional switching components, the circuit of Fig. 17-14 can be made out of three identical basic switching circuits. For this purpose, Eq. (17-23) must be transformed as follows:

$$C_o = ab + hc = b(ab + a'b') + h(hc + h'c')$$

$$= bh' + s'h = (b + h)(h' + s') = (b + h)(hs)' \qquad (17\text{-}24)$$

Equation (17-24) has the same general form as Eqs. (17-21) and (17-22), so that the outgoing carry C_o can also be obtained by means of the same type of mixed diode-transistor circuit. The resulting circuit is shown in Fig. 17-15.

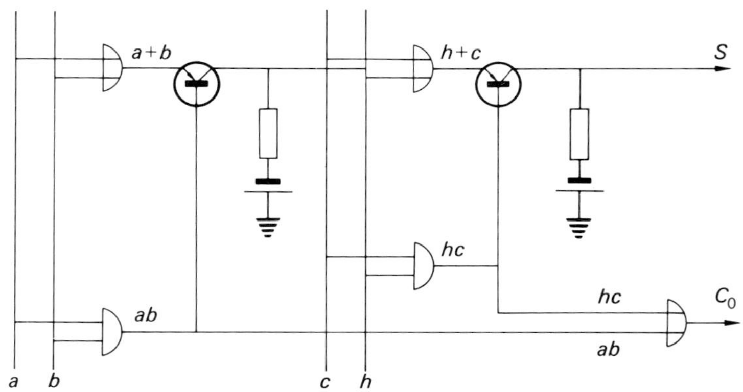

Fig. 17–14

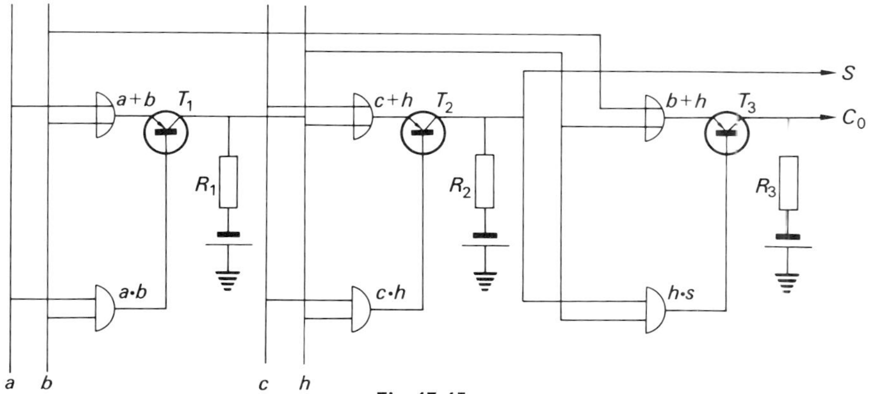

Fig. 17–15

The number of switching elements of an adder can be decreased even further by using transistors as double-input switching elements throughout. This switching element makes it possible to implement the terms of the switching equations of the half-adder in their original form:

$$H = ab' + a'b \qquad \text{and} \qquad S = hc' + h'c \tag{17-25}$$

The corresponding switching equations for the carry can be derived from Eq. (17-24):

$$C_o = bh' + s'h \tag{17-26}$$

The resulting circuit using transistors only is shown in Fig. 17-16. It consists of two half-adder circuits, with three transistors each. It is advisable to clip the h and s signals by means of a diode as in the diagram of Fig. 6-77. It should be pointed out that the usual signal-level-shifting voltage dividers are not shown in the circuit of Fig. 17-16.

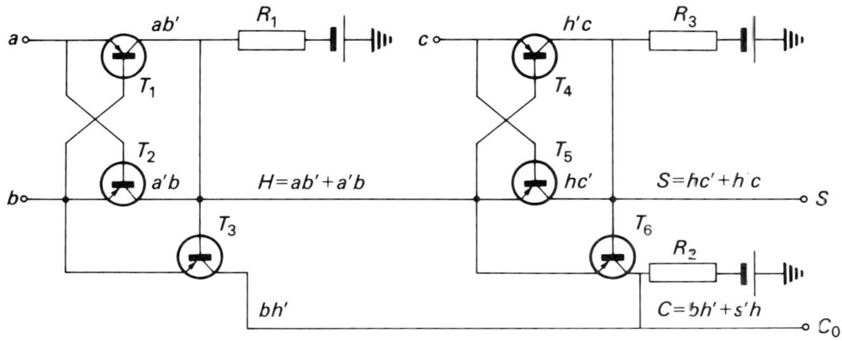

Fig. 17–16

17-4 SYNCHRONOUS ADDITION OF BINARY NUMBERS BY MEANS OF ELECTRONIC CIRCUITS

A great variety of synchronous (clock pulse) adding circuits has been developed. Two variants based on the use of MS memory elements will be discussed here.

THE PULSE HALVER

Any pulse-halving circuit is basically also an adding circuit performing an addition mod 2. For this purpose, however, an information input should be added to the counting-pulse input gates. The clock pulse can be applied to

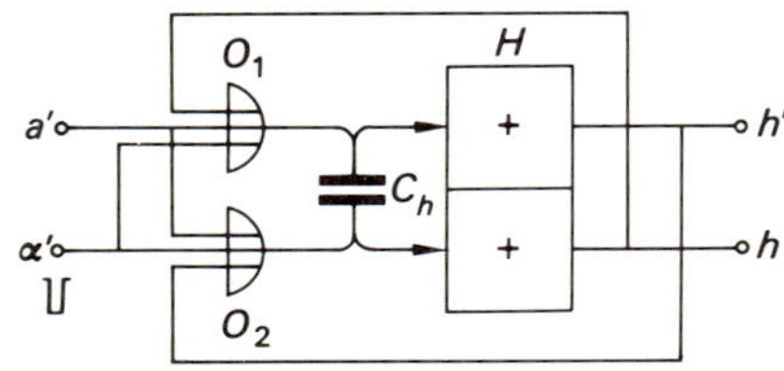

Fig. 17–17

the former counting-pulse input. A circuit layout for such an adder is given in Fig. 17-17. See also the T flip-flop of Fig. 10-34a.

This circuit performs the addition mod 2 of a series of binary digits that are fed into the adder via the information input terminal a' of input gates O_1 and O_2. If the state of trigger H is 0, then $h = 0$ and $h' = 1$. If on the next clock pulse α' a 1 should be added to the information contents of H, then $a' = 0$ is given to O_1 and O_2. During the next clock pulse α', all input signals of input gate O_1 only are 0, so that on the trailing edge of this clock pulse trigger H will be set in the 1 state.

If the next digit $a = 1$, then on the second clock pulse α' all inputs of OR gate O_2 are 0, so that on the trailing edge of this pulse a reset to 0 of trigger H follows and the law of addition mod 2 $1 \oplus 1 = 0$ is fulfilled.

It is rather easy to check the validity of the other possible additions of two binary digits, so this proof will be omitted in this text.

The half-adder of Fig. 17-17 acts as an accumulator in which series of binary digits can be added.

THE SLAVE MEMORY ELEMENT AS MOD 2 ADDER

The slave part of an MS memory element can be connected as a mod 2 adder. This is shown in the circuit of Fig. 17-18. The front and rear outputs of a standard slave memory element (with $R_1 = R_2$) are connected to the set terminal of a preset-to-zero master memory element H. In the case in which $a = b$, capacitor C_h cannot be loaded to a potential difference between the two terminals equal to the clock-pulse voltage. In such a case, no set pulse can be generated on the trailing edge of a clock pulse. Only if $a \neq b$ can capacitor C_h be charged to a potential difference equal to the clock-pulse voltage, so that a set pulse for trigger H will be generated on the trailing edge of the clock pulse. This implies that

$$H = D_{\alpha'}(ab' + a'b) \qquad (17\text{-}27)$$

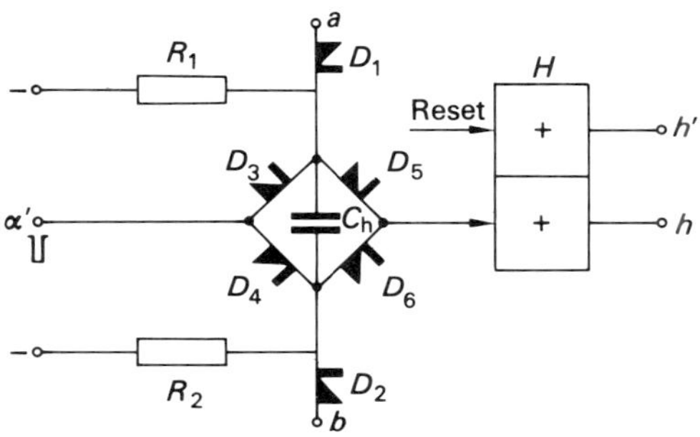

Fig. 17–18

It is noted that the same function will be generated by the circuit of Fig. 17-18 on input signals a' and b', whereas function $H' = D_{\alpha'}(ab + a'b')$ will be generated on input signals a' and b or a and b'.

It is not necessary to reset trigger H for the next addition to be performed when another slave memory is connected to its reset terminal. The input signals of this slave memory element should then be connected to input signals a and b' or a' and b, so that the inverse function of the other slave memory is generated.

SYNCHRONOUS PARALLEL ADDERS

A full-adder section can be formed combining the circuits of Figs. 17-17 and 17-18. The combined circuit is shown in Fig. 17-19. It operates in a two-clock-pulse mode. The slave memory elements connected as mod 2 adders are indicated in this figure by means of symbols MA_1 and MA_2.

In a first operation on clock pulse α', the sum mod 2 of digits a and b is formed and temporarily stored in trigger H/S. In the circuit of Fig. 17-19, two mod 2 adders (MA_1 and MA_2) have been used so that in this case no special reset of trigger H/S is required. After clock pulse α', an outgoing carry

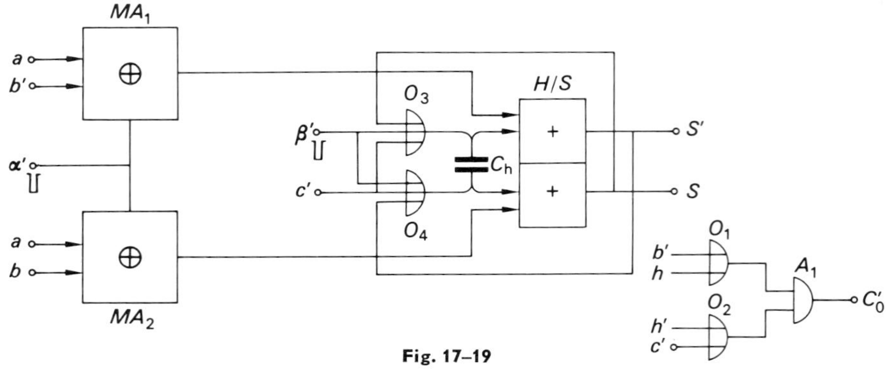

Fig. 17–19

C_o' can be generated by the section or propagated through this section by gates O_1, O_2, and A_1. In Fig. 17-19, the carry circuit has only been indicated in its basic form. A practical form of this carry transfer path will be discussed in the following section.

In the pause between the α' and β' clock pulse, the carry propagation through the transfer path will have come to an end so that one of the terminals of capacitor C_h can—or cannot—be set to 0 during clock pulse β'. Master memory element H/S acts as a pulse halver in association with the slave memory element with capacitor C_h. If $c' = 0$, the state of trigger H/S will be reversed. This state then gives the wanted sum S. The next addition can be made without having to reset H/S.

The accumulator principle of Fig. 17-17 can be applied to a full-adder section, as is indicated in Fig. 17-20. When using master and slave memory elements, two slaves and one master are required for this purpose. On the first clock pulse α', the first digit a' is fed into the circuit. If this digit is a

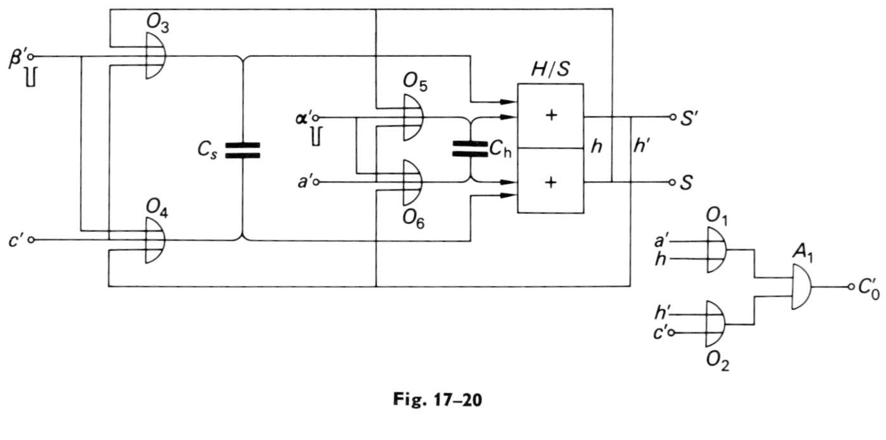

Fig. 17–20

1, it reverses the state of trigger H/S because the latter is connected as a pulse halver. If it is a 0, it leaves the state of H/S unaffected. After that, a second clock pulse β' is given so that the incoming carry digit c is added to the digit already stored in trigger H/S. This yields the final sum S. For $c = 1$, the state of trigger H/S is reversed; for $c = 0$, the state of H/S remains unaffected.

The basic carry-generating and -propagating circuit of the adder section is also shown in Fig. 17-20 in the form of gates O_1, O_2, and A_1.

The accumulating effect of this circuit makes it possible to add series of binary numbers. For a new addition, all triggers H/S have to be reset to 0. This is not indicated in the circuit of Fig. 17-20.

It should be pointed out that it is possible to design single-clock-pulse accumulators in which the sum of two digits is determined under the control of the clock pulse, while the carry transfer is asynchronous.

Many variants of synchronous parallel adders have been designed. The examples given above are limited to those combinations that can easily be implemented by means of the standard building block, the MS memory element used throughout this book.

SYNCHRONOUS SERIAL ADDER

A synchronous serial adder operating in a two-clock-pulse mode can easily be designed on the basis of the principles stated above for synchronous adders. The basic diagrams of such a serial adder for the addition of two binary numbers A and B consisting of digits a_0 to a_i and b_0 to b_i is given in Fig. 17-21. Use is made of various forms of MS memory elements. Resistors R_1 and R_2, diodes D_1 to D_8, and capacitor C_1 form a mod 2 adding circuit similar to the one shown in Fig. 17-18. Resistors R_3 and R_4, diodes D_9 to D_{16}, and capacitor C_2 form another mod 2 adding circuit of the same type. Output s of trigger S is connected to diodes D_3 and D_6 of the former circuit, and output s' to diode D_{11} and D_{14} of the latter mod 2 adding circuit. This is the same type of feedback as has already frequently been used to form a pulse halver from an MS memory element. Its function is the same. Both mod 2 adding circuits have the same inputs, except for the feedback connection from trigger S.

These feedback connections, for $a \oplus b = 1$, cause the state of trigger S to be reversed so that if the primed state of trigger S is called c (carry from the addition of the two previous digits), the state of trigger S can be represented by the following:

$$S = a \oplus b \oplus c \tag{17-28}$$

This is exactly what is required when adding two digits of the binary numbers A and B, taking into account such carries as may originate from the addition of the pair of next-lower-order digits.

Slave memory element SM_2 is connected to output terminal s' of trigger S, so that a change in state of this trigger from 1 to 0 after the end of an α' clock pulse will bring trigger C from the primed 0 state into the 1 state. In this case, a carry must be generated.

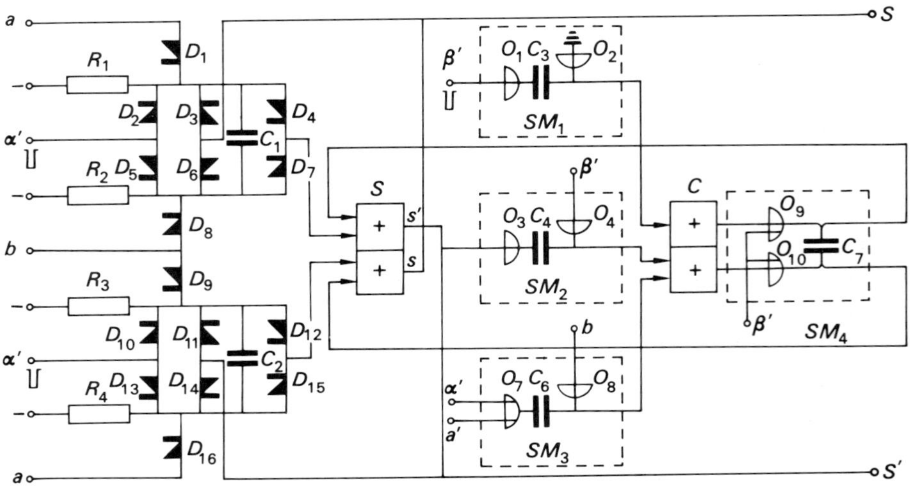

Fig. 17–21

This setting of trigger C can only happen for $a \neq b$ and $c = 1$, so that a provision must be made for the case $a = b = 1$ ($c = 0$ or 1). Slave memory element SM_3 with front control by an α' clock pulse and an a' signal and rear control via the OR gate 0_8 by signal b generates a set pulse for trigger C in this last case.

The output terminals of trigger C are coupled via slave memory element SM_4 to the input terminals of trigger S, so that on clock pulse β' the state of trigger C is shifted into trigger S, and the state of trigger S can be shifted into a sum shift register (not represented in the diagram).

During this shifting operation, a 0 is shifted into trigger C via slave memory element SM_1, while the coupling between the output terminal s' of trigger S and the set input terminal of trigger C is blocked by the β' clock pulse applied to the rear of capacitor C_4; so that during this shifting operation, a state change of trigger S cannot affect the state of trigger C.

17-5 HIGH-SPEED PARALLEL ADDERS

In the design of such electronic adder circuits as were described in the preceding section, more emphasis was put on the use of a minimum number of components and on sectional types of circuits than on high-speed operation, i.e., high-speed carry transfer. If no special means are provided in the circuit design, the carry transfer will severely limit the operating speed of a parallel adder, especially for adders designed to handle a large number of digits.

The following methods are available for obtaining high-speed parallel adders:

1. High-speed carry transfer
2. Conditional-sum logic (Sec. 17-9)
3. Group-carry logic (Sec. 17-10)
4. Special numerical systems (Sec. 17-11)

In the first method, the carry path is a serial path through all adder sections containing individually a minimum number of components. This type of carry path is analogous to that of the relay adders described in Sec. 17-2. A very-high-speed transfer is possible for a limited number of sections at practically no extra cost if primed transistors are used in this carry path.

The second method takes advantage of the fact that in the addition of two binary numbers, only one incoming carry (from the next lower order) can exist at a time; so that for each of the adder sections, the sum and the outgoing carry can be generated for each of the two possibilities, with or without incoming carry ($c_i = 1$ or $c_i = 0$). These sums are conditional sums. The correct sums can be chosen in a binary process (to be described in a separate section) that relates first pairs of columns, then groups of two pairs of columns, then groups of four pairs of columns, and so on. The addition of two binary numbers of, say, 2^n digits can be decreased from 2^n times the carry transfer time of a section to n times that interval. This method involves the use of a fairly large amount of hardware. The organizational principle of this method of conditional-sum logic is interesting in itself; however, there is no need for this rather sophisticated type of carry transfer when the first method can be applied.

In the third method, a group carry is instantaneously generated in a logical bypath on a number of adjacent adder sections. The required logical network is formed by a tree of gates. Consequently, the number of sections on which this principle can be applied is limited. Super-group carry networks can be used to meet this difficulty. Here again the principle as such is interesting, but it is of no use in those circuit techniques where the first method can be applied. The principle will be described in a separate section.

In the fourth method, the numbers to be added are coded in a special number system. In the system of Valach, there are no carries at all between the various columns in the addition; in the system of Avizienis, only a carry can be generated to the next-higher-order column. The principles of both systems will be dealt with in a separate section.

This section will deal further with high-speed carry transfer.

With respect of their operating speed, the relay adders described in Sec. 17-7 are not limited by the carry-propagation time. The addend and augend applied to each section of these adders position the carry transfer path in which the carry is propagated with the speed of an electric pulse in a wire, because contacts and wires have the same properties in this respect.

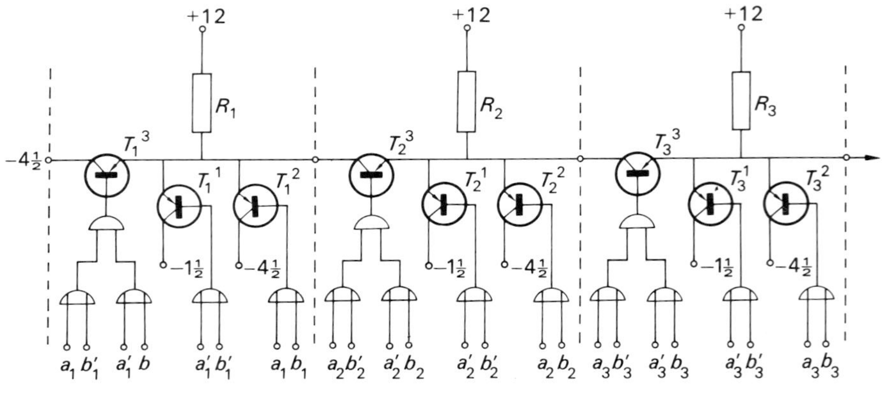

Fig. 17–22

A wire with a number of transistors in series can have similar properties with respect to pulse propagation if the emitter-base junctions of these series of transistors are made conducting before the pulse is applied to the first transistor of the chain.

Figure 17-22 shows a serial-transistor circuit with primed emitter-base junctions through which a pulse can be propagated with negligible delay.

The first three of the many sections of a carry-propagation circuit are shown in Fig. 17-22. They are part of a switching system using the standard power supply of most examples in this textbook of +12, 0, –6, and –18 volts and the signal voltage levels of $0 = -6$ volts and $1 = 0$ volts. In this carry-propagation circuit, special signal voltages have been introduced in order to meet the high-speed requirements. These carry-signal voltage levels are $0 \equiv -4\frac{1}{2}$ volts and $1 \equiv -1\frac{1}{2}$ volts.

As Table 17-5 clearly shows, three different conditions of the carry line have to be met. For $a_n = b_n = 0$, no carry has to be generated by or propagated through the nth section; whereas for $a_n = b_n = 1$, a carry will always originate from that section. In both cases where $a_n \neq b_n$, a carry originating from a lower-order section can be propagated through the section to a higher-order section. A section of the carry circuit contains basically a resistor R_i and three transistors T_i^1, T_i^2, and T_i^3. The last one is connected in series in the carry line; the first two can connect special signal voltages of $-1\frac{1}{2}$ and $-4\frac{1}{2}$ volts, respectively, to the carry line. All three transistors of a section of the carry-line circuit are basically connected as emitter followers with resistor R_i in the emitter circuit. The controlling voltages of the bases of the transistors of all sections can be the standard 0- and −6-volt signals.

Table 17-5

a_n	b_n	C_{n+1}
0	0	0
1	0	C_n
0	1	C_n
1	1	1

If in section 2 of Fig. 17-22, signal $a_2 = b_2 = 0$, the collector of transistor $T_2{}^2$ transfers a voltage of $-4\frac{1}{2}$ volts to the carry line as a no-carry signal to higher-order sections of the carry circuit, this in accordance with the first line of Table 17-5. In this case, the bases of transistors $T_2{}^1$ and $T_2{}^3$ receive a 1 signal (0 volts), which is positive with respect to the carry-line voltage in that section of $-4\frac{1}{2}$ volts, so that these transistors are blocking.

In accordance with the last line of Table 17-5, transistor $T_2{}^1$ will transfer a voltage of $-1\frac{1}{2}$ volts to the carry line when signal $a_2 = b_2 = 1$. Transistors $T_2{}^2$ and $T_2{}^3$ are then blocking because the bases of these transistors receive a 1 signal (0 volts) which is also positive with respect to the applied carry-line voltage of $-1\frac{1}{2}$ volts.

When $a_2 \neq b_2$ the incoming carry signal whether 0 or 1 has to be transferred to the next-higher-order section. In this case, the signal on the base of transistor $T_2{}^3$ is 0, so that this transistor can transfer its collector voltage—being either $-1\frac{1}{2}$ volts or $-4\frac{1}{2}$ volts—to the emitter. Transistors $T_2{}^1$ and $T_2{}^2$ are then blocking because the bases of these transistors are then kept at the 1 signal level.

It should be noted that the signal propagation in the carry line is from left to right, whereas the current in this line flows from right to left. All paths via which controlling voltages are applied to the carry line are low-resistance emitter-collector paths. All series-connected transistors T^3 of the carry line can be simultaneously made conducting in their emitter-base junction, which allows a high-speed propagation of the signal ($-1\frac{1}{2}$ volts or $-4\frac{1}{2}$ volts) applied to the first collector of this chain of transistors. The whole carry line containing a number of series-connected emitter-collector paths has a low resistance, which allows a high-speed propagation of a carry signal less influenced by the capacitance of the components in the various sections.

A more detailed circuit of the nth section of an asynchronous high-speed parallel adder using the carry-line principle of Fig. 17-22 is shown in Fig. 17-23. The control of the generated carry $c_{n+1} = 0$ and $c_{n+1} = 1$ by means of transistors $T_n^{\ 1}$ and $T_n^{\ 2}$ is basically the same as shown in Fig. 17-22. The outputs of input gates $O_n^{\ 1}$ and $O_n^{\ 2}$ are directly connected to the bases of transistors $T_n^{\ 1}$ and $T_n^{\ 2}$ without voltage dividers in order to speed up the operation of these transistors.

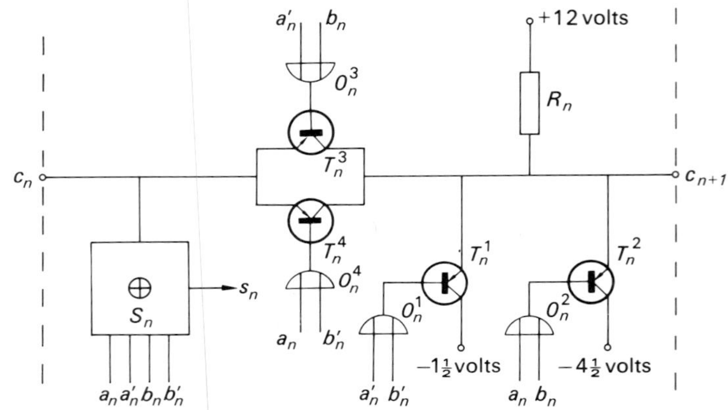

Fig. 17–23

For the control of the propagated carry when $c_{n+1} = c_n$, two transistors $T_n^{\ 3}$ and $T_n^{\ 4}$ have been provided in the carry line, each with its own input gate directly connected to its base. Thus all transistors of a carry-line section are under the same switching conditions. This decreases the priming time of the series-connected transistors in the carry line.

The sum signal s_n is formed in the block S_n of Fig. 17-23 containing, for example, a circuit analogous to the sum-forming part of Fig. 17-11. This adding circuit has to be connected to the carry line via an emitter follower having a high-resistance input.

For transistors with $I_{c_{\max}} = 10$ ma, a maximum of 15 transistors can be connected in series in the carry line; so that after, say, 12 transistors, a current amplifier must be inserted. The circuit of a combined *pnp-npn* emitter-follower amplifier is given in Fig. 17-24. In this circuit, diode D_1 in the input circuit of the transistors shifts both input signal levels so that the output signal levels will resume the original values.

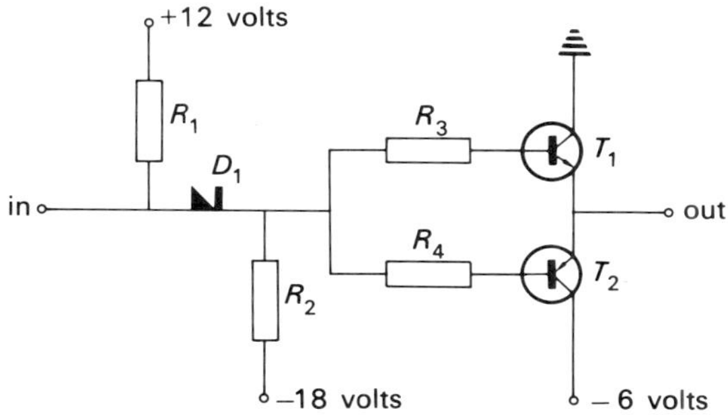

Fig. 17–24

17-6 THE ADDITION OF BINARY NUMBERS BY MEANS OF KIRCHHOFF NETWORKS

The symmetrical functions $S_{1,3,5,7,9,\ldots}$, $S_{2,3,6,7,\ldots}$, $S_{4,5,6,7,\ldots}$, etc., required for the addition of a number of binary digits, can be produced by means of marginal, so-called Kirchhoff networks. A Kirchhoff network for a section of a two-digit binary adder with an addend a, an augend b, and an incoming carry c is shown in Fig. 17-25. It consists of three resistors R_1, R_2, and R_3

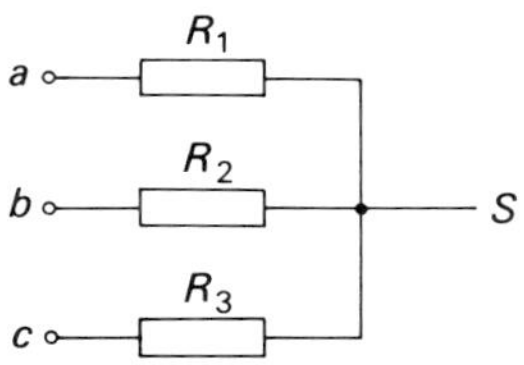

Fig. 17–25

Table 17-6

a	b	c	S	C_0	*No.*	*Volts*
0	0	0	0	0	0	−10
1	0	0	1	0	1	$-3\frac{1}{3}$
0	1	0	1	0	1	$-3\frac{1}{3}$
0	0	1	1	0	1	$-3\frac{1}{3}$
1	1	0	0	1	2	$+3\frac{1}{3}$
1	0	1	0	1	2	$+3\frac{1}{3}$
0	1	1	0	1	2	$+3\frac{1}{3}$
1	1	1	1	1	3	+10

with equal resistance. For the sake of simplicity, the signal voltages of variables a, b, and c are assumed to be −10 volts for 0 and +10 volts for 1. The voltage of the output terminal S of the network of Fig. 17-25 as a function of input variables a, b, and c is given in Table 17-6 for an output current of 0.

Switching levels 1, 2, and 3 in Table 17-6 correspond to the symmetrical functions S_1, S_2, and S_3. These functions characterize the signal level of the output terminal of the Kirchhoff network of Fig. 17-25. The voltage interval between the output voltage with all input terminals in the 0 condition or in the 1 condition is subdivided into n equal intervals, n being the number of input terminals. Thus the more digits there are to be added in one order, the smaller the voltage intervals will be with given-voltage signal levels. The output voltages of a Kirchhoff network thus form an arithmetic series, which has to be decoded into its binary components.

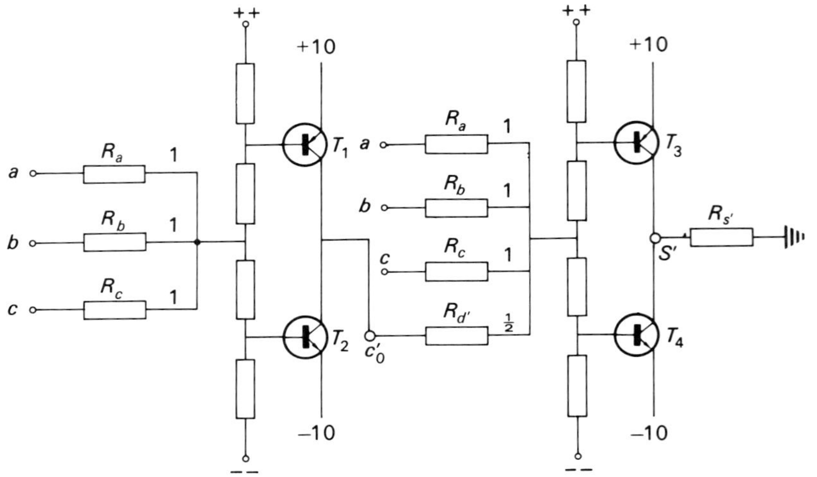

Fig. 17–26

Table 17-6 shows the truth values for input variables a, b, and c arranged according to the switching level, i.e., the number of variables with the value 1. The outgoing carry C_o can be obtained easily by discrimination between $-3\frac{1}{3}$ and $+3\frac{1}{3}$ volts. The resulting sum S has to follow from a detection between switching levels 0 and 1 and a second detection between switching levels 2 and 3, as can be seen from Table 17-6.

A circuit that performs this decoding process is shown in Fig. 17-26. The Kirchhoff network of Fig. 17-25 is connected in Fig. 17-26 to a transistor circuit containing transistors T_1 and T_2. The emitter of the *pnp* transistor T_1 is connected to +10 volts and the emitter of the *npn* transistor T_2 to -10 volts; the collectors of both transistors are connected to each other and to resistor $R_{d'}$. The output terminal of the Kirchhoff network is linked via two level-shifting voltage dividers to the base of transistors T_1 and T_2 in such a way that these transistors cannot be simultaneously conducting. The output signals of this transistor circuit have the wanted standard signal levels and can be used directly as input signals into another Kirchhoff network because they have the same internal resistance for both 0 and 1 signals. These output signals represent the switching function of the inverse of the required carry:

$$C_o' = S_{0,1} \tag{17-29}$$

A second Kirchhoff network is provided in the circuit of Fig. 17-26 to which input variables a, b, c are connected as well as signal c_o' generated in the first section of the decoder. Resistors R_a, R_b, and R_c in the second section of the decoder have the same resistance, whereas resistor $R_{d'}$—to which signal c_o' is connected—has half the value of the other resistors.

This means that the second Kirchhoff network behaves as a network with five equivalent input terminals. If the signal levels of all input variables are again considered to be +10 or −10 volts, the output voltages of the second Kirchhoff network will be +10, +6, +2, −2, −6, and −10 volts with zero current. The specific conditions of the second Kirchhoff network shown in Fig. 17-26 are summarized in Table 17-7.

Table 17-7

Level a, b, c	c_0'	*Total level* a, b, c, c_0'	*Discrimination voltage*	S
0	1	2	−2	0
1	1	3	+2	1
2	0	2	−2	0
3	0	3	+2	1

The input switching level of the a, b, and c signals can be 0, 1, 2, or 3. For levels 0 and 1, $C_o = 0$ and therefore $c_o' = 0$; for levels 2 and 3, $C_o = 1$ and $c_o' = 0$. The influence of the signal voltage c_o' is made twice as large as that of signals a, b, and c, so that the total switching level of the output terminal of the second Kirchhoff network can assume only the values 2 and 3, as is indicated in column 3 of Table 17-7. The corresponding discriminating voltages are −2 and +2 volts, so that this discrimination performed by means of transistors T_3 and T_4 in Fig. 17-26 can be achieved in the same way as with transistors T_1 and T_2 in the first Kirchhoff network. This leads to the inverse of the sum function:

$$S' = S_{0,2} \tag{17-30}$$

It should be noted that the c_o' signal as generated in the circuit of Fig. 17-26 cannot be used directly to control an input terminal c of a decoder of a higher order. For this purpose, a circuit analogous to that of transistors T_1 and T_2 must be connected to the collectors of transistors T_1 and T_2.

The number of digits that can be added in one operation can be larger than two in this type of adder. This number depends only on the possibility of discriminating sharply between the various voltage levels. Sections of a basically similar type can be connected in cascade for the addition of a large number of digits, as is shown in Table 17-1. The possibility of discriminating between the various levels is the limiting factor.

The circuit of Fig. 17-26 does not have a very wide operating margin if it is designed in an analogous fashion by means of high-vacuum tubes; however, a satisfactory arrangement using high-vacuum tubes can be achieved by means of single-input triggers that allow a very sharp voltage discrimination. One controlling Kirchhoff network is then sufficient for the complete decoder. The shift in the voltage discrimination of the second and further sections of the decoder can then be obtained by means of an internal control between the various sections. A section of such a type of adder for, say, five digits can be represented symbolically by the block diagram of Fig. 17-27.

A block diagram of an adder for five four-digit binary numbers in which Kirchhoff networks are used is shown in Fig. 17-28.

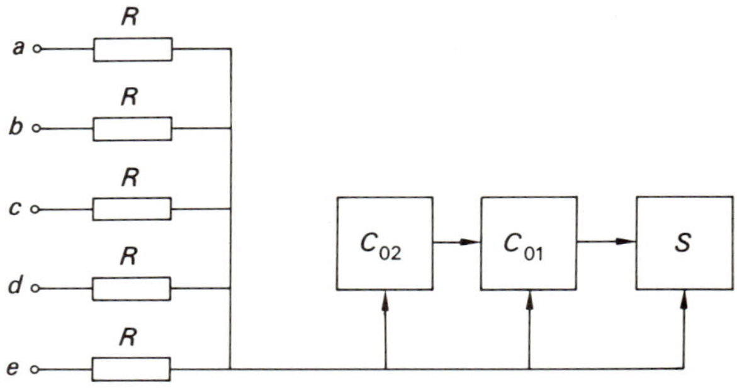

Fig. 17–27

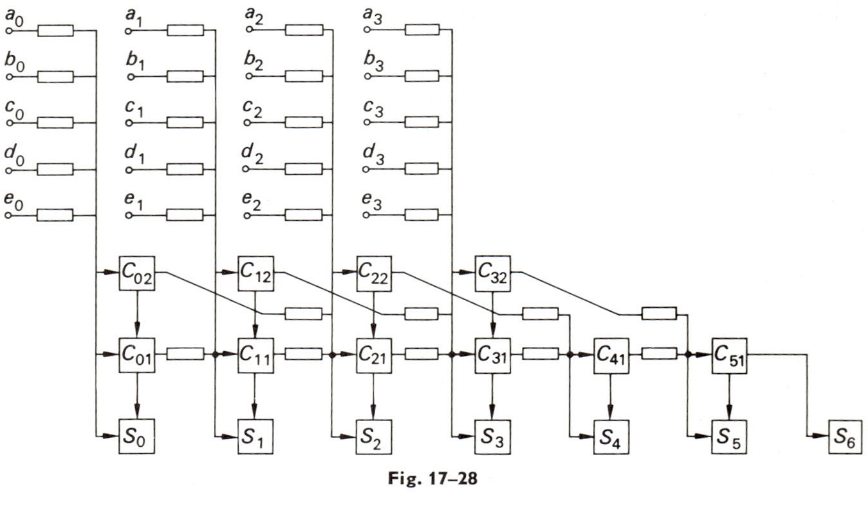

Fig. 17–28

The numbers to be added are a, b, c, d, and e. The signal voltages corresponding to the digits of the same weight are connected to the same Kirchhoff network. Figure 17-28 shows four of these networks. The junction points of these networks are connected to four decoders that consist of three sections in order to generate a second-order carry C_{n2}, a first-order carry C_{n1}, and a sum S_n.

The output of each carry generator C_{n2} is connected to a resistor that forms part of the Kirchhoff network for the digits of the order $n + 2$. The outputs of the carry generator C_{n1} are connected to a resistor that is part of the next-higher-order Kirchhoff network. All these resistors operate as an integral part of the networks to which they are connected. Two additional two-section decoders must be provided (C_{41}/S_4, C_{51}/S_5) in order to generate the sum digits S_4, S_5, and S_6.

17-7 THE SUBTRACTION OF TWO BINARY NUMBERS

The subtraction of two binary numbers is governed by a set of rules similar to the rules for the subtraction of two decimal numbers. The only difference between these two methods is the radix of the number system.

At first sight, a basic difference seems to exist between addition and subtraction. In addition, a carry digit can be generated which is transferred to a higher-order column in the operation. In manual addition of two numbers, the carry is transmitted from one column to the next, i.e., in the direction of operation; whereas in manual subtraction, the borrow seems to be an operation for which an information transfer is required against the "normal" direction of operation. This may be illustrated by the example of the subtraction $9 - 6 = (1\ 0\ 0\ 1) - (0\ 1\ 1\ 0)$, as shown in Fig. 17-29 in which a digit with weight $4 = 2 + 2$ has to be borrowed from the third column to enable the subtraction to be performed in the second column. This borrow is only possible if a digit with weight $8 = 4 + 4$ can be borrowed from the fourth column, as a result of which a digit with weight 4 can be transferred to the second column. This procedure is illustrated in Fig. 17-29. The subtraction in the various columns can now proceed in the usual way.

The binary subtraction, however, can be performed in practically the same way as the binary addition by means of the introduction of a borrow signal, which is an indication for the operation in the next-higher-order column that a digit has been borrowed by the adjacent lower-order column.

Borrow			(1 1)	(1 1)	
9	=	1	0	0	1
6	=	0	1	1	0
3	=	0	0	1	1

Fig. 17–29

This subtraction of a minuend a and a subtrahend b, resulting in a difference digit S_d and a borrow digit C_b, is defined by truth table 17-8.

Table 17-8

a	b	S_d	C_b
0	0	0	0
1	0	1	0
0	1	1	1
1	1	0	0

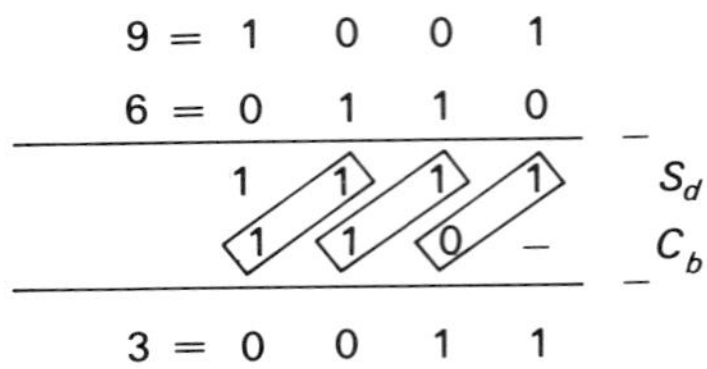

Fig. 17–30

The example of the subtraction of the numbers 9 and 6 is repeated in Fig. 17-30. In this figure, the subtracting operation of the two numbers is presented in accordance with truth table 17-8. The third line of Fig. 17-30 shows the difference S_d of all pairs of digits of minuend and subtrahend; the fourth line shows the various borrow digits and their correspondence with the difference digits S_d. The resulting difference will be obtained by a repeated subtracting operation on the newly formed pairs of digits S_d and C_b.

However, the same problems of borrow transport through a subtracter arise as were described in Sec. 17-1 for the carry transport through an adder. This may be illustrated by means of the subtraction of $14 - 7 = (1\ 1\ 1\ 0) - (0\ 1\ 1\ 1)$, which is shown in Fig. 17-31. This example shows a borrow digit being transferred through all columns of the subtraction.

Subtracters can be designed as full subtracters or half-subtracters. There need not be much difference between adding circuits and subtracters, as can be seen from the design of a subtracter circuit which is equivalent to

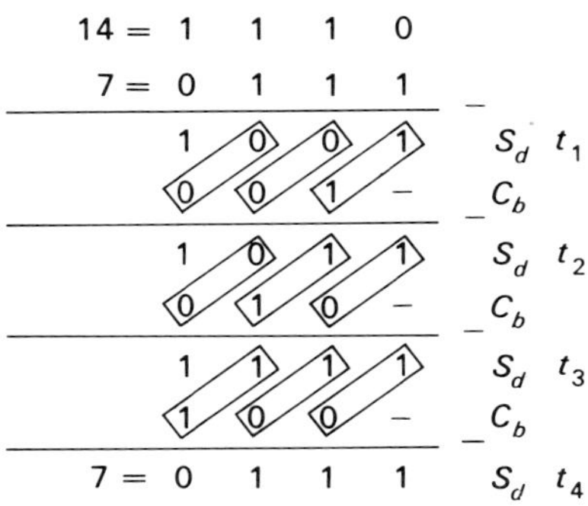

Fig. 17–31

the adding circuit of Fig. 17-6. The two truth tables defining both half operations are given as Tables 17-9 and 17-10.

Table 17-9

a	b	H_d	C_{bo1}
0	0	0	0
1	0	1	0
0	1	1	1
1	1	0	0

Table 17-10

h_d	c_b	S_d	C_{bo2}
0	0	0	0
1	0	1	0
0	1	1	1
1	1	0	0

Truth table 17-9 describes the generation of a provisional difference H_d and a partial borrow C_{bo1}; Table 17-10 describes the forming of the final difference, using as a starting point the provisional difference obtained in the first half-operation and an incoming borrow signal and the production of a second partial-borrow signal C_{bo2}. The switching equations are as follows:

$$H_d = ab' + a'b = a \oplus b \tag{17-31}$$

$$C_{bo1} = a'b \tag{17-32}$$

The final difference is

$$S_d = h_d c'_{bo1} + h'_d b_{bo1} = h_d \oplus c_{bi1} \tag{17-33}$$

$$C_{bo2} = h'_d c_b \tag{17-34}$$

The borrow signal leaving this subtracting section is

$$\begin{aligned} C_{bo} &= C_{bo1} + C_{bo2} = a'b + h'_d b_b \\ &= b(a'b + ab') + h'_d c_b = h_d b + h'_d c_b \end{aligned} \tag{17-35}$$

The resulting relay circuit with two differential relays is shown in Fig. 17-32. It only differs from the corresponding adding circuit shown in Fig. 17-6 in that the position of the h contact in the carry or borrow circuit has been reversed.

Most types of adding circuits can be adapted in a rather simple way to the subtracting function. There is one difference that should be mentioned here: whereas the inputs of addend, augend, and carry signals in an adder may be interchanged, this cannot be done in a subtracter with minuend, subtrahend, and borrow signals because the subtraction does not follow the commutative law.

THE REPRESENTATION OF NEGATIVE BINARY NUMBERS

A special problem arises in subtracters when the subtrahend is larger than the minuend or in adders when addend or augend are negative. This problem can be met by a suitable representation of negative numbers in the binary system.

In binary arithmetic, two different systems are used for the representation of negative numbers, both systems having their advantages and disadvantages. Both systems have in common the fact that the first bit of

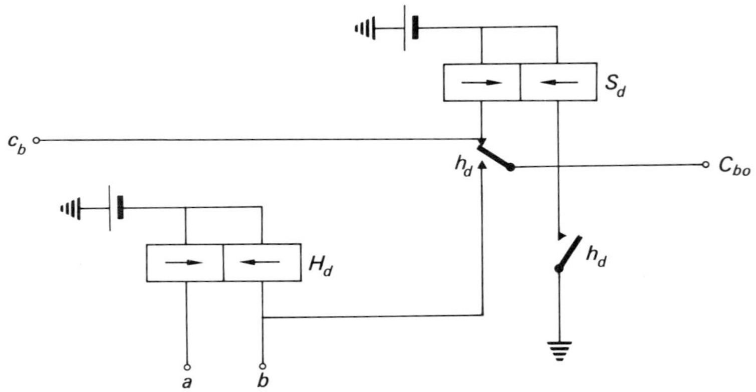

Fig. 17–32

an n-bit number represents the sign and the following $n-1$ bits are significant for the value of that number. A 0 as first bit indicates a positive number; a 1 as first bit makes it a negative number. In both systems, the positive numbers are coded in the ordinary binary way.

The ones-complement system Any number N $(-2^{n-1} < N < 2^{n-1})$ can be represented by a binary number N_b of n bits as follows:

$$N_b \equiv N(\text{mod } 2^n - 1) \tag{17-36}$$

with $0 \leqslant N_b < 2^n$.

If $N > 0$,

$$N_b = N \tag{17-37}$$

If $N = 0$,

$$N_b = 0 \qquad \text{or} \qquad N_b = 2^n - 1 \tag{17-38}$$

If $N < 0$,

$$N_b = N + (2^n - 1) \tag{17-39}$$

Example

$N = +11 \qquad N_b = 01011 \qquad n = 5$

$N = -11 \qquad N_b = -11 + 31 = 20 = 10100$

In this number system, the representation of the negative numbers is obtained by subtracting the absolute value of the number from $2^n - 1$ or, as can easily be seen in the example, by inverting all bits in the positive representation of the number. This explains the name of the ones-complement number system,

The twos-complement system Any number N $(-2^{n-1} \leqslant N < 2^{n-1})$ can uniquely be represented by a binary number N_b of n bits as follows:

$$N_b \equiv N(\text{mod}\, 2^n) \tag{17-40}$$

with $0 \leqslant N_b < 2^n$.

If $N \geqslant 0$,

$$N_b = N \tag{17-41}$$

If $N < 0$,

$$N_b = N + 2^n \tag{17-42}$$

Example

$N = +11 \qquad N_b = 01011 \qquad n = 5$

$N = -11 \qquad N_b = -11 + 32 = 21 = 010101$

In the twos-complement system, the representation of the negative numbers is obtained by subtracting the absolute value of the number from 2^n or, as can be seen in the example, by inverting all digits followed by the addition of 1.

The relation between the negative numbers of both systems is such that a twos-complement negative number can be formed from the corresponding ones-complement number by adding 1.

As an example, the numbers –7 to +7 are shown coded in both systems in Table 17-11.

It should be noted that in the ones-complement system there are two representations of 0, namely, +0 = 0000 and –0 = 1111. In the twos-complement system, however, there is an unambiguous representation of 0. In the positive numbers of both systems, all 0's left of the one with the highest weight are nonsignificant with respect to the value of that number. In the negative numbers, the 1's left of the most significant 0 are nonsignificant with respect to the value of the binary number.

Because the negative numbers in the ones-complement system are obtained from the corresponding positive numbers by inverting all digits, a multiplication of a negative number by 2^n can be performed simply by shifting that number over n places to the left and filling these places with nonsignificant 1's. However, in the twos-complement system these places

Table 17-11

Decimal number	*Binary number*	*Decimal number*	*Negative number ones complement*	*Negative number twos complement*
0	0 0 0 0	0	1 1 1 1	—
+1	0 0 0 1	−1	1 1 1 0	1 1 1 1
+2	0 0 1 0	−2	1 1 0 1	1 1 1 0
+3	0 0 1 1	−3	1 1 0 0	1 1 0 1
+4	0 1 0 0	−4	1 0 1 1	1 1 0 0
+5	0 1 0 1	−5	1 0 1 0	1 0 1 1
+6	0 1 1 0	−6	1 0 0 1	1 0 1 0
+7	0 1 1 1	−7	1 0 0 0	1 0 0 1
		−8	—	1 0 0 0

have to be filled with 0's because in the multiplying shifting process of a negative number in that system the already mentioned surplus value of 1 increases in n steps to 2^n. This is $2^n - 1$ too much. Filling the n empty shifting places with 0's instead of 1's compensates for this undesired increase in value when multiplying.

ADDITION AND SUBTRACTION OF POSITIVE AND NEGATIVE NUMBERS IN THE ONES-COMPLEMENT SYSTEM

The subtraction of two positive numbers P and Q will yield the same result as the addition $P + (-Q)$. Thus $P - Q = P + (-Q)$, $(-Q)$ being a negative number, further to be indicated by N. Using Eq. (17-39), the desired difference or sum S_b can be determined via an intermediate sum S_p, possibly one bit larger in representation than S_b:

$$S_p = P_b + N_b = P + N + 2^n - 1 \tag{17-43}$$

There are two possibilities to be investigated:

1. $P + N > 0$: $\quad 2^n + 2^{n-1} - 2 \geqslant S_p \geqslant 2^n$ (17-44)

2. $P + N \leqslant 0$: $\quad 2^{n-1} < S_p \leqslant 2^n - 1$ (17-45)

In case 1 sum S_p will fall between the following limits in binary for $n=5$:

$$1\,\underline{0}\,1\,1\,1\,0 \geqslant S_p \geqslant 1\,\underline{0}\,0\,0\,0\,0 \tag{17-46}$$

In case 2, sum S_p will show the following limits for $n=5$:

$$\underline{1}\,0\,0\,0\,0 < S_p \leqslant \underline{1}\,1\,1\,1\,1 \tag{17-47}$$

In all possible cases $P+N>0$, an overflow bit with value 2^n will be generated, and a 0 digit will show up in the place of the sign digit. This is in agreement with the fact that a positive result is required. In all possible cases $P+N \leqslant 0$, no overflow bit with value 2^n is generated, and a 1 digit will show up in the place of the sign digit, which is in agreement with the fact that a negative result is required.

The desired sum S_b can be determined from S_p as follows:

$$S_b \equiv S_p (\operatorname{mod} 2^n - 1) \tag{17-48}$$

This equation means that the $(n+1)$st overflow bit in Eq. (17-46) must be used as an end-around carry and added to the least significant bit of the binary number. Equation (17-46) then changes into

$$\underline{0}\,1\,1\,1\,1 \geqslant S_p \geqslant \underline{0}\,0\,0\,0\,1 \tag{17-49}$$

No further elucidation is necessary other than the statement that in the case of $P+N>0$ and $n=5$, S_p will be at least 1 and a maximum of 15, whereas in the case of $P+N \leqslant 0$, S_p will be at least 0 and a maximum of -15, which result follows from the above.

It also follows from the above that in the case of the addition $P+N>0$, an end-around carry will always be generated. It can be shown in an analogous way that in the case of the subtraction $P-Q<0$, an end-around borrow will always be generated, which has to be treated in the same way as the end-around carry.

ADDITION AND SUBTRACTION OF POSITIVE AND NEGATIVE NUMBERS IN THE TWOS-COMPLEMENT SYSTEM

In this system, the negative numbers are represented mod 2^n, so that an overflow carry or borrow appearing in the addition or subtraction may be omitted from the result.

Example

Ones complement				*Twos complement*			
−11	10100	+11	(1)01011	−11	10101	+11	(1)01011
+15	01111	+15	01111	+15	01111	+15	01111
	+		−		+		−
	100011		11100	+4	~~1~~00100	−4	11100
	00001		1 00001				
	+		−				
+4	00100	−4	11011				

It should be noted that an end-around borrow can never go beyond the digit place where it was generated.

17-8 THE ADDITION OF BINARY NUMBERS BY MEANS OF CORE CIRCUITS

The marginal forming of switching functions on cores leads to two interesting examples. The first, shown in Fig. 17-33, is the equivalent of a code converter

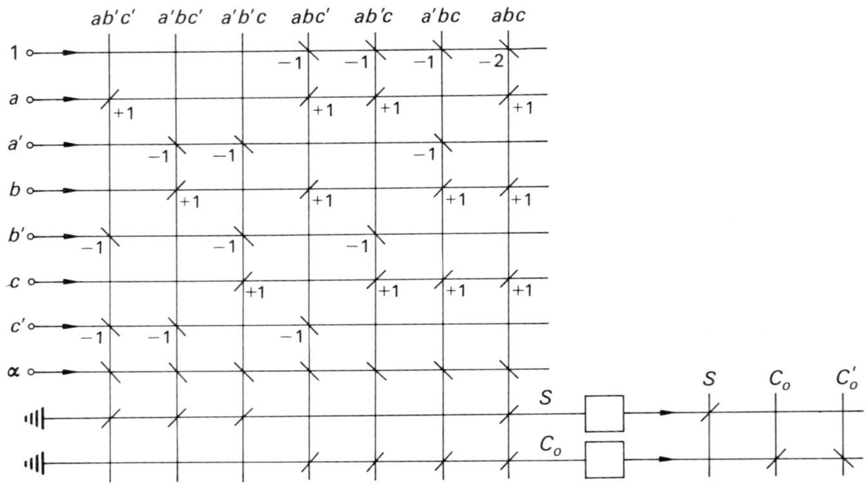

Fig. 17–33

translating a three-digit binary code into an eight-position decimal code. Seven cores are required to form the various terms of the switching functions involved:

$$S = ab'c' + a'bc' + a'b'c + abc \tag{17-50}$$

$$C_o = abc' + ab'c + a'bc + abc \tag{17-51}$$

These are shown in the circuit of Fig. 17-33. This circuit operates in such a way that one of the cores is always put into the 1 state by the various possible combinations of input variables, except for $a = b = c = 0$. The α clock pulse reverses the magnetic flux of that core, so that a pulse will be transmitted either to core S or to core C_o (and C_o') or to both simultaneously. The carry storage is shown as a pair of cores C_o and C_o'.

The C_o core has to be reset to 0, which normally occurs upon reading out, and the C_o' core has to be reset to 1 after the read-out process. All types of transfer loops can be used between these core stages, but by using a balanced split-winding transfer loop for the read-in of the C_o cores, special reset circuits can be avoided. The output signals of the core pair C_o which are obtained, for example, on a β clock pulse can be used as c and c' input signals of the same adder if it is a part of a serial adder.

It should be noted that the core which produces the term abc is the most critical element from the point of view of the marginal operation. The input variables a, a', b, and b' are supposed to originate from other core circuits, e.g., as outputs of a core shift register.

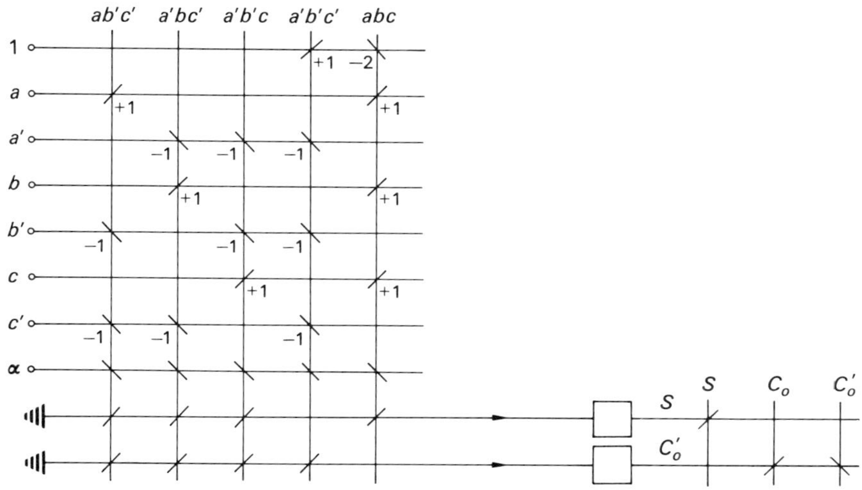

Fig. 17–34

The number of cores in an adder of the type shown in Fig. 17-33 can be reduced to five if C_o' is generated instead of C_o. There is no objection to doing this since C_o' has to be generated too. The resulting circuit is shown in Fig. 17-34.

A more intelligent use of the possibilities of the marginal core logic would be to start with the following equations:

$$S = (a + b + c) \cdot (ab + ac + bc)' + abc = xy' + z \tag{17-52}$$

$$C_o = ab + ac + bc \tag{17-53}$$

in which $x = a + b + c$, $y = ab + ac + bc$, and $z = abc$. (17-54)

Only one core is required for implementing functions x, y, and z. This is shown in the circuit of Fig. 17-35. Function $f(x)$ controls the C_o core. The S core is controlled by x, y, and z. The output of the C_o core on a β clock pulse can be used immediately for the read-in of the three cores representing x, y, and z.

Equations (17-50) and (17-51) can also be put in their maxterm form:

$$S = (a + b + c)(a + b' + c')(a' + b + c')(a' + b' + c) \tag{17-55}$$

$$C_o' = (a + b' + c')(a' + b + c')(a' + b' + c)(a' + b' + c') \tag{17-56}$$

This maxterm representation allows the circuit to be implemented with five cores plus one without using marginal effects on the cores. In this type of core circuit, the marginal effects occur in the read-out circuit as an addition and subtraction of voltages. The resulting core circuit is shown in Fig. 17-36. With certain types of integrating circuits, it is possible to combine different read-out circuits on the cores.

Negated input variables can be avoided in this type of circuit by using Heyn's mixed algebraic-Boolean representation of the switching formulas. Equation (17-50) can be transformed as follows:

$$\begin{aligned}
S &= ab'c' + a'bc' + a'b'c + abc \\
&= a(1 - b)(1 - c) + (1 - a)b(1 - c) + (1 - a)(1 - b)c + abc \\
&= a(1 - b - c + bc) + b(1 - a - c + ac) + c(1 - a - b + ab) + abc \\
&= a - ab - ac + b - ab - bc + c - ac - bc + 4abc \\
&= a + b + c - 2[a + b - (a \vee b)] - 2[a + c - (a \vee c)] \\
&\quad - 2[b + c - (b \vee c)] + 4[a + b + c - (a \vee b) - (a \vee c) \\
&\quad - (b \vee c) + (a \vee b \vee c)] \\
&= a + b + c - 2(a \vee b) - 2(a \vee c) - 2(b \vee c) + 4(a \vee b \vee c)
\end{aligned} \tag{17-57}$$

Equation (17-51) can be transformed in the same way:

$$\begin{aligned}
C_o &= abc' + ab'c + a'bc + abc \\
&= ab(1 - c) + ac(1 - b) + bc(1 - a) + abc \\
&= ab + ac + bc - 2abc \\
&= a + b - (a \vee b) + a + c - (a \vee c) + b + c - (b \vee c) \\
&\quad - 2(a + b + c) + 2(a \vee b) + 2(a \vee c) + 2(b \vee c) - 2(a \vee b \vee c) \\
&= (a \vee b) + (a \vee c) + (b \vee c) - 2(a \vee b \vee c)
\end{aligned} \tag{17-58}$$

The expression for abc in the mixed algebraic-Boolean representation can easily be derived from the expression given for $ab = a + b - (a \vee b)$. The resulting core circuit is shown in Fig. 17-37. It contains seven cores and the read-out is more marginal than in the circuit of Fig. 17-36.

The various adding principles can be used as a basis for the design of serial adders. However, the implementation of parallel adders by means of

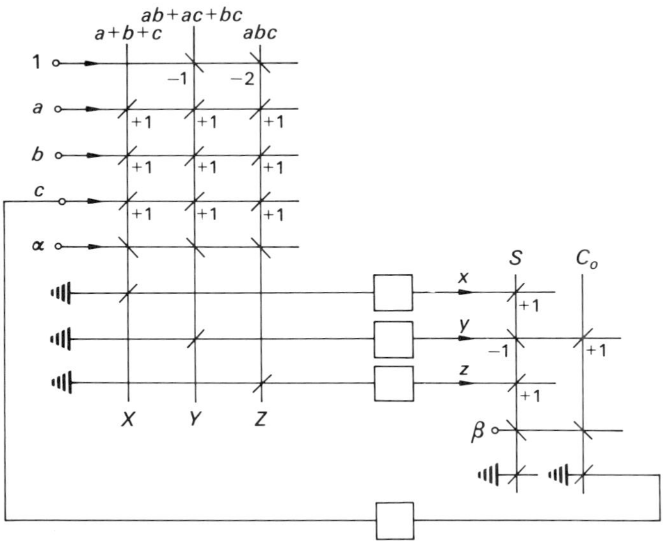

Fig. 17–35

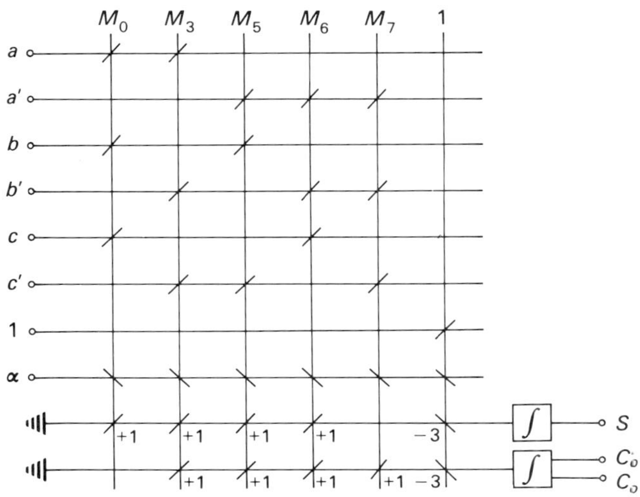

Fig. 17–36

core circuits requires some attention because the rapid carry transfer through a number of adding sections is a special problem in this type of circuit.

The "ripple through" of a carry signal cannot be solved in a rigid clock-pulse-driven circuit in which the operation of all sections is performed simultaneously, because the carry propagation requires some time, this time varying with the information processed.

As was already shown in the previous sections, a section of an adder for two binary numbers cannot generate a carry to a higher-order section and simultaneously transfer a carry signal from a lower-order section to that higher-order section. This fact can readily be seen in Table 17-5.

The ripple-through action can be obtained essentially in the following way by means of magnetic cores. A section of a parallel adder with the ripple-through feature is shown in Fig. 17-38. The circuit is in certain respects the equivalent of the relay circuit of Fig. 17-6, which has an immediate carry transfer. The ripple-through feature, which is required in order to obtain a high-speed parallel adder with magnetic cores, cannot be achieved by means of pure clock-pulse circuits. The circuit shown in Fig. 17-38 makes it possible to design such an adder.

If a core is preconditioned by a variable αh and if a variable c is subsequently used to read out that core, this core will generate the switching function $S = hc$. This is shown in the core circuit of Fig. 17-39.

The circuit shown in Fig. 17-38 operates as follows. Addend a and augend b, under the control of a clock pulse β, generate the functions ab', $a'b$, and ab to be stored on cores AB', $A'B$, and AB. On a clock pulse α, the function $H = \alpha(ab' + a'b)$ is obtained which preconditions core HC and conditions core H. At the end of clock pulse α, that is, on a clock pulse α', core AB is read out so that the outgoing carry signal $c_o = ab$ is generated in the read-out winding of core AB. This signal opens transistor T_1 so that a carry pulse is fed to the next section of the parallel adder. This pulse will be received on the c_i terminal of that section. If core HC is in the 1 state, the incoming carry pulse $c_i = 1$ will reverse the state of that core, so that transistor T_1 of that section is opened and a carry pulse $c_o = ch$ is transferred to the next higher-order section. This carry transfer terminates in the first section of the adder in which core HC is in the 0 state following the α clock pulse.

The pulse $c_o = 1$ entering a section of the parallel adder conditions core C so that after the ripple-through time, cores H and C in all sections are conditioned. Simultaneously with the reading in of a new addend and augend into cores AB', $A'B$, and AB, cores HC' and $H'C$ can be conditioned on the β clock pulse, so that the final sum $S = hc' + h'c$ will be available on the following α pulse.

In the circuit of Fig. 17-38, a pulse amplifier in the form of a transistor T_1 has been connected in the wire of the outgoing carry $C_o = ab + ch$, because

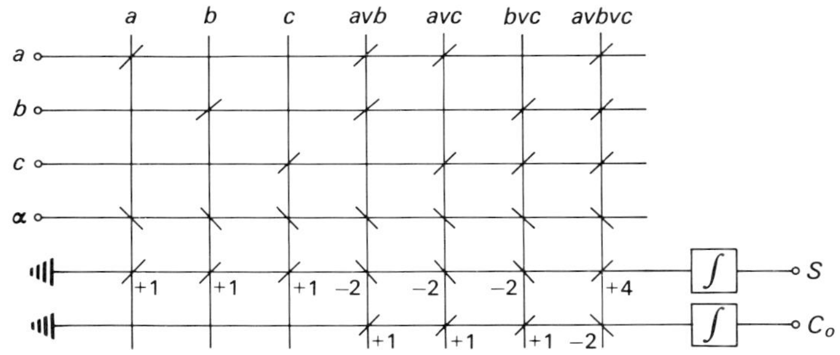

Fig. 17–37

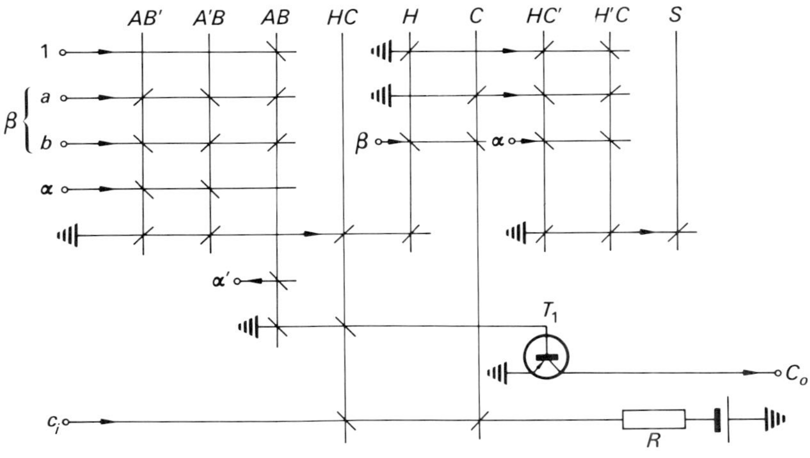

Fig. 17–38

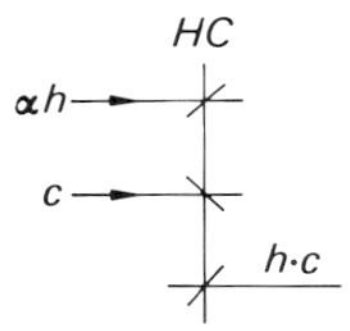

Fig. 17–39

a carry pulse can at most be transferred through all sections of a parallel adder.

Cores H, C, HC', and $H'C$ can be eliminated from the circuit as is shown in Fig. 17-40 by introducing a second pulse amplifier, for example, in the form of a transistor T_2 and a special circuit for core S

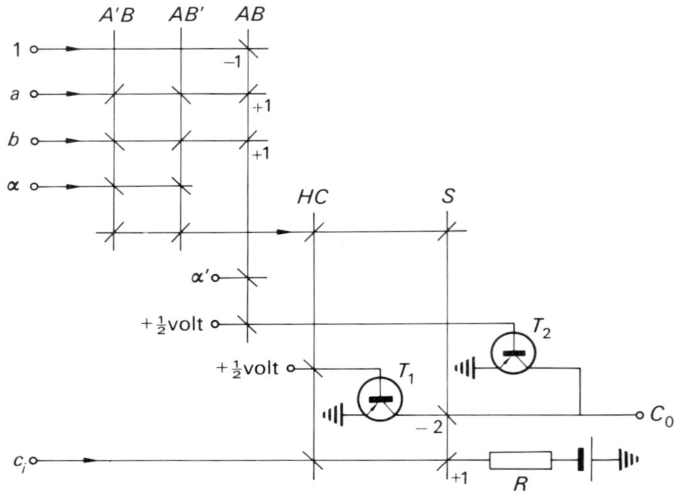

Fig. 17–40

The carry pulse $c_{o1} = ab$ is now amplified by transistor T_2, while the carry pulse $c_{o2} = ch$, which is transferred through the section, is amplified by transistor T_1. Core S is preconditioned by $\alpha(ab' + a'b) = h$. An incoming carry pulse, which generates the outgoing carry $c_{o2} = ch$, reverses the state of core S from 1 to 0 via the read-in winding -2. If, however, $\alpha(ab' + a'b) = 0$, the incoming carry c_i is not transferred through the section, but it terminates in that section and has to bring the S core in the 1 state via the $+1$ winding of Fig. 17-40.

The circuits shown in Figs. 17-38 and 17-40 are examples of a much larger group of core circuits. The carry-transfer problem of a parallel adder can also be solved by means of the core logic explained in Sec. 7-6.

17-9 CONDITIONAL-SUM LOGIC

The conditional-sum logic devised by Sklansky is not a special switching trick designed to overcome the varying time delay in carry transfer; it should be regarded rather as a refined method of organizing the addition, thereby making use of the fact that the addition of two binary digits can have only two different results depending on the incoming carry. The principle that any binary operation can have only two results was already successfully applied in the serial conversion of reflected to ordinary binary numbers.

The principle of the conditional-sum logic can best be explained by means of the addition of the following two binary numbers y and z, as shown in Table 17-12.

Table 17-12

	7	6	5	4	3	2	1	0	
153 =	1	0	0	1	1	0	0	1	y
92 =	0	1	0	1	1	1	0	0	z
S_0	1	1	0	0	0	1	0	1	t_1
C_0	0	0	0	1	1	0	0	0	
S_1	0	0	1	1	1	0	1	x	
C_1	1	1	0	1	1	1	0	x	
S_0	1	1	1	0	0	1	0	1	t_2
C_0	0		0		1		0		
S_1	0	0	1	1	1	0			
C_1	1		0		1				
S_0	1	1	1	0	0	1	0	1	t_3
C_0	0				1				
S_1	1	1	1	1					
C_1	0								
S = 245 =	1	1	1	1	0	1	0	1	t_4

During the first time interval t_1, a sum S_0 and an outgoing carry C_0 are determined in all columns under the assumption that the incoming carry is 0. In addition, a sum S_1 and an outgoing carry C_1 are arrived at under the assumption that the incoming carry is 1. In column 0, no sum S_1 and no carry C_1 are to be found because for the digits of the first column there cannot possibly be an incoming carry of 1. Rows S_0 and S_1 (t_1) contain all digits of the correct sum S. The first step in the process of choosing the correct sum digits is performed in pairs of adjacent columns. The fat lines enclose the columns that are grouped together as pairs.

A carry $C_0 = 0$ in column 0 indicates that sum S_0 (t_1) in column 1 has been formed under the correct assumption. A carry $C_0 = 0$ in column 2 indicates that the sum S_0 (t_1) in column 3 has also been formed under the correct assumption, if preceding columns are not taken into account.

It will be clear that in an analogous manner a carry C_1 (t_1) in column 2 indicates that sum S_1 in column 3 has been formed correctly, provided that the pair of digits in columns 2 and 3 receive a carry 1 from the pair of columns 0 and 1.

A carry $C_0 = 1$ (t_1) in column 4 indicates that sum S_0 in column 5 has been determined under the wrong condition if the incoming carry in the pair of columns 4 and 5 is 0.

In an analogous way, it can be shown that sum S_1 (t_1) in column 5 has been determined under a correct condition provided that the corresponding incoming carry in the pair of columns 4 and 5 is 1.

Finally, it can be seen that sums S_0 and S_1 (t_1) in column 7 have been determined correctly provided that the incoming carries in the pair of columns 6 and 7 are 0 and 1, respectively.

In a second operation (t_2), two pairs of sum and carry rows are again obtained, leaving out all carries which have already been used to check and correct the situation in the second column of a pair.

All pairs of S and C in the second column of a pair, which were found to have been incorrectly determined in operation t_1, are replaced by the sum and carry from the other pair of rows. In this example, only S_0 and C_0 in column 5 have been replaced by S_1 and C_1, as is indicated by enclosures in Table 17-6.

Since carry C_0 in column 0 (operation t_1) is always 0, sum S_1 in columns 0 and 1 of operation t_2 can be omitted from the table.

Carry $C_0 = 0$ in column 1 of operation t_2 indicates that sums S_0 in columns 2 and 3 have been formed under the correct assumption. The same happens for sums S_0 in columns 6 and 7, because C_0 in column 5 of operation t_2 is 0.

However, $C_1 = 0$ in column 5 of operation t_2 indicates that in this phase of the adding process, sums S_1 in columns 6 and 7 have been formed under an erroneous assumption, so that these sums must be replaced in a third operation (t_3) by the corresponding sums S_0. Carry $C_1 = 1$ in column 7 of operation t_2 must also be replaced in this third operation by the corresponding carry $C_0 = 0$.

Since sums S_0 in columns 0 to 3 (operation t_3) are already final, no sums S_1 are indicated in these columns. All carries used have also been omitted in this part of Table 17-12.

Finally, $C_0 = 1$ in column 3 of operation t_3 indicates that sums S_0 in columns 4 to 7 have been formed under the wrong assumption $(C_0 = 0)$ and that the corresponding sums S_1 have to be chosen instead; this is shown in operation t_4. Carry $C_1 = 0$ in column 7 indicates that no overflow occurs, so that the final sum $S = 245$.

The substitution performed in the various steps of the adding process have all been indicated by means of self-explanatory enclosures.

In this conditional-sum logic, a final sum can be formed in $({}^2\log n) + 1$ steps, in which $({}^2\log n)$ represents the next larger integer.

It is good to know the principle of the conditional-sum logic; however, its operating speed is no improvement compared with the high-speed carry transfer and with some special numerical systems. Circuits will not be discussed in this text.

17-10 GROUP-CARRY LOGIC

At the cost of a relatively large amount of additional hardware, a great reduction in carry-propagation time can be achieved by means of a group-carry logic. Only the principle of this method will be discussed here. The outgoing carry C of a section n of a parallel adder can be written in the following form:

$$C = ch + ab \tag{17-59}$$

or

$$C = ch + bh' = ch + ah' \tag{17-60}$$

where a and b represent the addend and augend digits and $h = a \oplus b = a'b + ab'$. Term ch represents the propagated carry; terms ab or bh' or ah' give the generated carry. The outgoing carry is related to all preceding sections of the adder by the incoming carry signal c, which is in turn the outgoing carry C_0 of section $n - 1$.

Equation (17-60) will now be applied as follows to define the signals coming from the various sections of the adder.

The first section of a group of sections has an incoming carry ${}^0c$ and generates an outgoing carry

$$\begin{aligned} {}^0C_o &= {}^0(h'a) + {}^0(hc) = {}^0(h'a) + {}^0h \cdot {}^0c \\ {}^0C'_o &= \qquad\qquad\qquad\quad = {}^0(h'a') + {}^0h \cdot {}^0c' \end{aligned} \tag{17-61}$$

The second section of this group of sections produces an outgoing carry

$$\begin{aligned} {}^1C_o &= {}^1(h'a) + {}^1(hc) = {}^1(h'a) + {}^1h \cdot {}^1c \\ &= {}^1(h'a) + {}^1h \cdot {}^0C_o \\ &= {}^1(h'a) + {}^1h \cdot {}^0(h'a) + {}^1h \cdot {}^0h \cdot {}^0c \\ {}^1C'_o &= {}^1(h'a') + {}^1h \cdot {}^0(h'a) + {}^1h \cdot {}^0h \cdot {}^0c' \end{aligned} \tag{17-62}$$

The third section of this group yields an outgoing carry

$$
\begin{aligned}
{}^2C_o &= {}^2(h'\,a) + {}^2(h'\,c) = {}^2(h'\,a) + {}^2h\cdot{}^2c \\
&= {}^2(h'\,a) + {}^2h\cdot{}^1C_o \\
&= {}^2(h'\,a) + {}^2h\cdot{}^1(h'\,a) + {}^2h\cdot{}^1h\cdot{}^0(h'\,a) + {}^2h\cdot{}^1h\cdot{}^0h\cdot{}^0c \qquad (17\text{-}63) \\
{}^2C_o' &= {}^2(h'\,a') + {}^2h\cdot{}^1(h'\,a') + {}^2h\cdot{}^1h\cdot{}^0(h'\,a') + {}^2h\cdot{}^1h\cdot{}^0h\cdot{}^0c'
\end{aligned}
$$

etc.

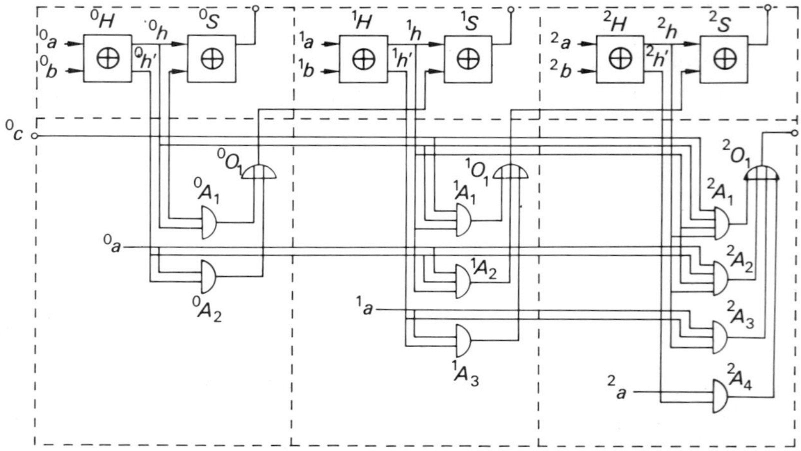

Fig. 17–41

It is possible to provide each section of a parallel adder with an instantaneous carry generating circuit as defined in Eqs. (17-61) to (17-63). As the number of sections in a group increases, these circuits require an ever-increasing number of components, so that this method as such ultimately requires prohibitive amounts of hardware. The block diagram of a group of sections of a parallel adder with instantaneous group-carry generation based on Eqs. (17-61) to (17-63) is given in Fig. 17-41.

The circuit of Fig. 17-41 has more or less the character of a flow chart. In the top of the figure, the actual adder circuits of three sections are indicated in half-adder symbols. Any one of the half-adder circuits given in this chapter can be used. Each of the sections of this group of three is provided with a group of gates practically instantaneously generating the carry for the next section of the group.

The speed of carry propagation can best be expressed in terms of the maximum number of switching orders through which the carry signal has to be propagated. In a group of adder sections as shown in Fig. 17-41, this is the number of switching orders in a half-adder H and two switching orders in the logical carry networks. If a parallel adder is subdivided into m groups of n sections, the next group starting again with an incoming carry generated by the last section of the preceding group, then for each group of sections two switching orders are added in the carry path. Finally, a carry path ends in a half-adder S.

The carry-propagation time of a parallel adder operating in accordance with the block diagram of Fig. 17-41, will amount, as a maximum, to twice the operating time of a half-adder plus $2m$ times the operating time of a gate.

For adders with a very large number of sections, the operating time can be further reduced by instantaneous group-carry generation. For example, a 100-digit adder can be subdivided into 20 groups of 5, the circuit being as shown in Fig. 17-41 and the group-carry generating circuits consisting of 5 adder sections. The principle is the same as above, so that no further illustration is necessary.

It should be noted that if the group-carry-propagation time equals the propagation time of a carry in a group of sections without any special means of boosting the carry-propagation speed, the special carry-generating means of each section, except those which instantaneously generate the outgoing carry of a group, can be omitted. In that case, the number of sections in a group can best be chosen as the square root of the total number of sections of the adder.

The last two methods for boosting the carry-propagation speed by means of special circuit layouts and special logical circuits must be regarded —and as such they have their place in this textbook—as special disciplines, the use of which need not be limited only to adders. The relatively large amount of hardware required in both methods makes them suitable for use only in adders to which the high-speed carry transfer path with primed transistors cannot be applied for some reason or other.

17-11 SPECIAL NUMERICAL SYSTEMS

Additions of numbers coded in the ordinary numerical systems, such as the denary or binary scale, have a variable processing time which is related to the transfer of the carry digits from lower-order columns to higher-order columns. In the extreme case, a transfer is even possible from the lowest-order column to the highest-order column. This phenomenon severely limits the operating speed of those types of parallel adders for which no special means are provided for rapid carry transfer.

It has proved possible to devise types of numerical systems of such a nature that in the addition of two numbers no carry digits are found to occur or that the transfer of the carry digit is limited to the next-higher-order column. The addition thus acquires a "totally parallel" character, and this is also true for the multiplication. The following discussion will be limited to the principles of the numerical systems of Valach and Avizienis. These systems will be explained by means of some illustrative examples without going too far into the mathematical background which, since it forms no part of the switching techniques, falls more or less outside the scope of this textbook.

THE SYSTEM OF RESIDUAL CLASSES

Valach has indicated that high-speed parallel addition is possible without any carry transfer when the algebra of residual classes is used as a basic system. In this system, the numbers are represented by means of a number of positive residues taking mod p_i $(i = 0, 1, 2, \ldots, N)$ from the decimal number to be represented, these residues being relatively prime. A number $X \geqslant 0$ can then be expressed in a series of integers x_i $(i = 0, 1, 2, \ldots, N)$ that are defined by the following equation:

$$x_i \equiv X(\mathrm{mod}\, p_i) \qquad 0 \leqslant x_i < p \tag{17-64}$$

The various residues x_i will be called the digits of the numbers in the new numerical system. The index i indicates the order of the digit and p_i represents the basis of the digit of order i. The digits x_i $(i = 0, 1, 2, \ldots, N)$ can be used to represent all integers up to a certain maximum if the numbers are expressed in a constant number of digits.

Example

Order i =	0	1	2	3	4
Basis p_i =	3	4	5	7	11
7 =	(1	3	2	0	7)
15 =	(0	3	0	1	4)
31 =	(1	3	1	3	9)
63 =	(0	3	3	0	8)

This representation of numbers in the system of residual classes implies that many decimal numbers will be represented by the same figures and that these decimal numbers will have the same distance, called period P. This period is determined by the product of the basis numbers:

$$P = p_0 \cdot p_1 \cdot p_2 \cdots p_N \tag{17-65}$$

In order to obtain an unambiguous system of number representation, it is a rule that a number in the system of residual classes represents the smallest possible positive number X_0 of all possible numbers X with the same representation. This can be expressed by means of the following equation:

$$X_0 \equiv X(\mathrm{mod}\ P) \qquad \text{with } 0 \leqslant X_0 < P \tag{17-66}$$

Period P in the example just given is $P = 3 \cdot 4 \cdot 5 \cdot 7 \cdot 11 = 4{,}620$, so that the numbers in the example can be expressed from 0 up to 4,619 inclusive.

It should be noted that numbers in the system of residual classes are put between brackets, with the lowest basis number on the left-hand side.

Addition The addition of two numbers $X + Y = Z$ follows simply from the addition of the corresponding modulus rests:

$$z_i = x_i + y_i (\text{mod}\, p_i) \tag{17-67}$$

A carry that must be transferred from lower-order columns to higher-order columns does not exist in this numerical system. No difficulties will arise in the addition of two such numbers as long as the sum does not exceed the period P. All additions can be performed in the same time.

Example

$$\begin{array}{rl} \text{Basis} & 3\ 4\ 5 \qquad P = 60 \\ 8 = & (2\ 0\ 3) \\ 17 = & (2\ 1\ 2) \\ \hline 25 = & (1\ 1\ 0) \end{array} +$$

The subtraction is defined by $z_i = x_i - y_i$ (mod p_i).

Multiplication The multiplication $X \cdot Y = Z$ follows simply from the multiplication of the corresponding modulus rests:

$$z_i = x_i \cdot y_i (\text{mod}\, p_i) \tag{17-68}$$

The multiplication can also be performed in a totally parallel way, because no carries can originate from any column.

Example

$$\begin{array}{rl} \text{Basis} & 3\ 4\ 5 \\ 8 = & (2\ 0\ 3) \\ 7 = & (1\ 3\ 2) \\ \hline 56 = & (2\ 0\ 1) \end{array} \times$$

These examples were chosen in such a way that sum and product were smaller than period P. Of course, it is not always possible when programming a problem to predict whether a certain sum or product will or will not exceed period P. Numbers larger than P can always be expressed by means of an extension of the number of basis numbers.

For example, 40 can be expressed by means of basis numbers 3, 4, and 5 as (1 0 0). On this same basis this is also the representation of 40^2. Extension of the basis with numbers 7 and 11 allows an unambiguous representation of both numbers 40 and 40^2 as (1 0 0 5 7) and (1 0 0 4 5), respectively.

Division The system of residual classes allows a simple division of a dividend by a divisor when the divisor is one of the basis numbers or a product of these basis numbers. A number $X = (x_0, x_1, \ldots, x_k, \ldots, x_j, \ldots, x_N)$ is divisible by p_i if $p_k = 0$ and by $p_k \cdot p_j$ if x_k and x_j are 0. The division is defined by the following formula:

$$q_i \cdot p_{k_i} (\text{mod}\, p_i) \equiv x_i \qquad \text{where } 0 \leqslant q_i < p_i \tag{17-69}$$

The quotient number q in the order i of the division when multiplied by the divisor p_{k_i}—one of the basis numbers—must equal dividend x_i on the basis mod p_i. This division gives a well-defined solution if $i \neq k$. The difficulty arising from the undefined division for $i = k$ can be circumvented by reducing the basis, leaving out p_k.

Example

Basis	3	4	5	7
55 =	(1	3	0	6)
5 =	(2	1	0	5) :
11 =	(2	3	—	4)

This type of division is rather difficult to perform. The division operation can be replaced by a specially defined multiplication, which will not be discussed here.

Decimal value of a number The decimal value of a number in this system of residual classes can be determined by means of a mixed subtraction-division process which can best be explained by means of an example.

Example

```
Basis p_i        3  4  5  7 11
Given number    (1  3  2  1  6)

                (1  1  1  1  1) ≡ 1
                ─────────────── – Subtract
                (0  2  1  0  5)
                (0  3  3  3  3) ≡ 3
                ─────────────── : Divide
                   (2  2  0  9)
                   (2  2  2  2) ≡ 2
                   ──────────── – Subtract
                   (0  0  5  7)
                   (0  4  4  4) ≡ 4
                   ──────────── : Divide
                      (0  3 10)
                      (0  5  5) ≡ 5
                      ───────── : Divide
                         (2  2)
                         (2  2) ≡ 2
                         ────── – Subtract
                         (0  0)
```

Thus, the decimal value of the given number (1 3 2 1 6) will be

$$[(0 + 2 \cdot 4 \cdot 5) + 2] \cdot 3 + 1 = 127$$

The decimal-conversion process consists of a succession of subtractions and divisions. In each subtracting phase of the conversion process, the lowest basis number is made 0, so that the resulting number can be divided by the basis p_i of that column. The decimal value of the given number can then be calculated easily from the subtracted numbers and the basis number p_i. The whole process, however, is so complicated that in most cases it will be not be worthwhile to use it solely for the purpose of obtaining very-high-speed arithmetic operations.

SIGNED-DIGIT NUMBER REPRESENTATIONS

Avizienis designed the so-called signed-digit number representations in which the carry propagation during the operations of addition and subtraction is always limited to one position to the left, thus eliminating carry-propagation chains, or special carry-generating networks covering groups of digits. This can be achieved by means of a redundancy in the number representations for the operands. This redundancy allows the addition to be split into two operations in which each sum (or difference) digit is a function only of the two corresponding digits to be added and of a carry (or borrow) generated by the two corresponding digits in the next lower order of the numbers. The result is an addition time that is independent of the number of digits and of their value.

In a first operation, a provisional sum digit w_i and a carry digit t_i are formed from the addend and augend digits y_i and z_i; in a second operation, the final sum digit s_i is obtained by adding the provisional sum digit w_i and the carry digit t_{i-1} originating from the next lower order. These two operations are defined as follows:

$$y_i + z_i = rt_i + w_i \tag{17-70}$$

and

$$s_i = w_i + t_{i-1} \tag{17-71}$$

in which r is the radix of the numerical system. It can be shown that this type of addition (or subtraction) of two numbers implies that

$$t_i = -1, 0, \text{ or } +1 \tag{17-72}$$

and

$$r > 2 \tag{17-73}$$

whereas in a numerical system with radix $r = 10$ the ranges of numbers become for w_i:

$$\bar{5}, \bar{4}, \bar{3}, \bar{2}, \bar{1}, 0, 1, 2, 3, 4, 5 \tag{17-74}$$

and for y_i, z_i, and s_i:

$$\bar{6}, \bar{5}, \bar{4}, \bar{3}, \bar{2}; \bar{1}, 0, 1, 2, 3, 4, 5, 6 \tag{17-75}$$

in which $\bar{x}$ means $-x$.

Digits 0 to 6 remain unchanged for addend and augend, but digits 7 to 9 must be changed into $1\bar{3}$, $1\bar{2}$, and $1\bar{1}$, respectively ($10-3$, $10-2$, $10-1$) in a code-conversion process which involves possibly a carry through all digits of a number.

The addition of two numbers in signed-digit representation will be illustrated by means of the following example:

$$0.79470 + 0.37636 = 1.17106$$

$$y_i = 0.79470 = 00.1\bar{3}\ 1\bar{1}\ 04\ 1\bar{3}\ 00 = 1.\ \bar{2}\ \ \bar{1}\ \ 5\ \ \bar{3}\ \ 0$$

$$z_i = 0.37636 = 00.03\ 1\bar{3}\ 06\ 03\ 06 = 0.\ 4\ \ \bar{3}\ \ 6\ \ 3\ \ 6$$

$$\overline{\hspace{20em}} +$$

$$y_i + z_i = rt_i + w_i \qquad = 01.02\ 0\bar{4}\ 11\ 00\ 1\bar{4}$$

$$s_i = t_{i-1} + w_i \qquad = 1.\ 2\ \ \bar{3}\ \ 1\ \ 1\ \ \bar{4}$$

The resulting sum, which is of course in the signed-digit representation, can be converted into the ordinary decimal representation by a simple subtraction of a subtrahend consisting of all negative digits from a minuend consisting of all positive digits of the number to be converted. The resulting sum S (decimal representation) will be

$$1.20110 - 0.03004 = 1.17106$$

The carry-propagation problem reappears in this code conversion from signed-digit representation into the ordinary decimal representation.

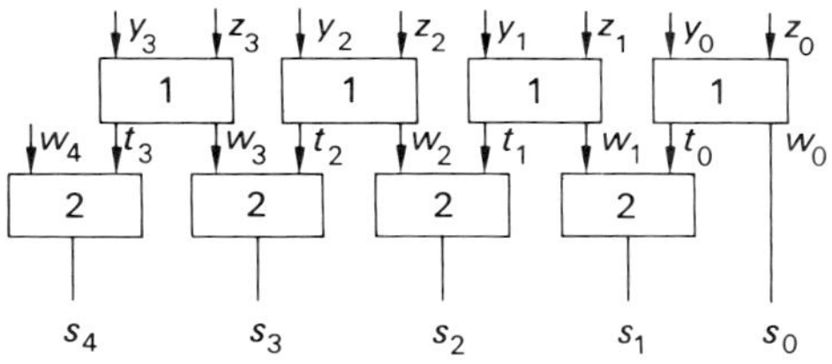

Fig. 17–42

In the signed-digit representation, the addition of two numbers, which is not totally parallel as was the case with the system of rest coding, can be performed by means of an adding circuit as symbolically shown in Fig. 17-42.

It should be noted that Avizienis's signed-digit representation cannot be applied to the binary number system with radix $r = 2$. This limits the applicability of this interesting type of numerical system to decimal systems.

Subtraction, multiplication, and division of numbers in the signed-digit representation can be readily derived from the above discussion of the addition problem.

The signed-digit representation apparently requires a code conversion of all numbers at the input to the calculating system. This is not true, however, if the allowed range of digits for s_i, y_i, and z_i is taken to be as follows:

$$s_i\colon \bar{1}\ 0\ 1\ 2\ 3\ 4\ 5\ 6\ 7\ 8\ 9 \tag{17-76}$$

With this choice of digits, no code conversion of decimal numbers is required at the input to the calculating system. As a result of the first operation

$$y_i + z_i = x_i = rt_i + w_i$$

digits x_i can have a value ranging from -2 to $+18$ inclusive. These values can be represented as follows:

$$
\begin{array}{lll}
x_i = t_i\,w_i & x_i = t_i\,w_i & x_i = t_i\,w_i \\
-2 = \bar{1}\;8 & +5 = 0\;5 & +12 = 1\;2 \\
-1 = \bar{1}\;9 \text{ or } 0\;\bar{1} & +6 = 0\;6 & +13 = 1\;3 \\
0 = 0\;0 & +7 = 0\;7 & +14 = 1\;4 \\
+1 = 0\;1 & +8 = 0\;8 & +15 = 1\;5 \\
+2 = 0\;2 & +9 = 0\;9 \text{ or } 1\;\bar{1} & +16 = 1\;6 \\
+3 = 0\;3 & +10 = 1\;0 & +17 = 1\;7 \\
+4 = 0\;4 & +11 = 1\;1 & +18 = 1\;8
\end{array}
\qquad (17\text{-}77)
$$

This table of possible digits x_i, t_i, and w_i shows that Eq. (17-72), which limits t_i to -1, 0, and $+1$ in the addition of two numbers, is also valid for the numerical representation of Eq. (17-70). Furthermore, it shows that there are two possible representations for $x_i = -1$ and $x_i = +9$.

In accordance with Eq. (17-71), $s_i = w_i + t_{i-1}$, the second operation is possible when the sum of digits w_i and t_{i-1} falls within the range -1 to $+9$ inclusive. Thus no difficulties can arise for all digits w_i falling within the range 0 to 8 inclusive. Only if $w_i = -1$ or $+9$ can digit s_i eventually fall outside the given range. This can only happen for $x_i = -1$ or $x_i = +9$. For both these x_i there are accordingly two representations. The question now arises to see whether it is possible or not, on the basis of the carry digit t_i generated by the next-lower order pair of digits, to choose one of these representations in such a way that s_i will fall within the range given by Eq. (17-76).

Thus, the coincidence of $x_i = -1$ and $t_{i-1} = -1$ and the coincidence of $x_i = +9$ and $t_{i-1} = +1$ must be avoided. This leads to the following criteria for the choice of the representation of digits $x_i = -1$ and $x_i = +9$:

1. $x_i = -0$ must be represented by

   $\bar{1}\;9$, if $x_{i-1} = -2$ to $+8$ inclusive

   and by

   $0\;\bar{1}$, if $x_{i-1} = +9$ to $+18$ inclusive

2. $x_i = +9$ must be represented by

   $0\;9$, if $x_{i-1} = -2$ to $+8$ inclusive

   and by

   $1\;\bar{1}$, if $x_{i-1} = +9$ to $+18$ inclusive

That the coincidence mentioned above is avoided if the criteria applied can easily be checked in (17-77).

It should be noted that the group of numbers $x_{i-1} = 0$ to $+8$ inclusive can be assigned arbitrarily to either of the representations $x_i = -1$ and $x_i = +9$ in both criteria, because the carry, being $t_{i-1} = 0$, cannot be of any influence on the range limits of s_i. Preferably both criteria should be chosen to be identical.

The above will be illustrated by means of the following example in which addend and augend already contain some digits $\bar{1}$, which will be the general case in the arithmetic unit of a computer.

Example

$$
\begin{array}{r|ccccccccccccc}
 & 8 & 7 & 0 & \bar{1} & 0 & \bar{1} & 6 & 0 & 5 & 4 & 3 & 2 & 1 \\
 & 1 & 2 & \bar{1} & 0 & \bar{1} & \bar{1} & 3 & \bar{1} & 3 & 5 & 6 & 7 & 9 \\
\hline
t_i w_i = & 1\bar{1} & 09 & \bar{1}9 & \bar{1}9 & \bar{1}9 & \bar{1}8 & 09 & \bar{1}9 & 08 & 1\bar{1} & 1\bar{1} & 1\bar{1} & 10 \\
s_i = 1 & \bar{1} & 8 & 8 & 8 & 8 & 8 & 8 & 9 & 9 & 0 & 0 & 0 & 0
\end{array}
\quad +
$$

The signed-digit representation of the sum corresponds to the decimal number 9888888990000.

It should be noted that the advantage of not having to perform a conversion at the input of the system can also be obtained in the signed-digit representation as devised by Avizienis when the asymmetrical range $-1 \cdots +8$ is chosen for w_i and the asymmetrical range $-2 \cdots +9$ for y_i, z_i, and s_i, in which case the addition becomes even simpler than in the previous example.

The principle of the previous numerical system, however, can also be applied in a system with radix $r = 2$. In such a case, only the following possibilities will appear in the addition:

$x_i = t_i w_i$

$-2 = \bar{1}\ 0$

$-1 = \bar{1}\ 1$ or $0\ \bar{1}$

$0 = 0\ 0$

$+1 = 0\ 1$ or $1\ \bar{1}$

$+2 = 1\ 0$

The criteria for the choice of the matching representation of digits $x_i = -1$ and $x_i = +1$ are as follows:

1. $x_i = -1$ must be represented by

   $\bar{1}\ 1$, if $x_{i-1} = -2$ to 0 inclusive

   $0\ \bar{1}$, if $x_{i-1} = +1$ or $+2$

2. $x_i = +1$ must be represented by

$0\ 1$, if $x_{i-1} = -2$ to 0 inclusive

$1\ \bar{1}$, if $x_{i-1} = +1$ or $+2$

This may be illustrated by means of the following example:

$$
\begin{array}{rl}
y_i = & 1\ \ 0\ \ 0\ \ \bar{1}\ \ 0\ \ \bar{1}\ \ 1\ \ 0\ \ 0\ \ 0\ \ 1\ \ 0\ \ 1 \\
z_i = & 0\ \ 1\ \ \bar{1}\ \ 0\ \ \bar{1}\ \ \bar{1}\ \ 0\ \ \bar{1}\ \ 0\ \ 1\ \ 0\ \ 1\ \ 1 \\
\hline
 & \qquad\qquad\qquad\qquad + \\
t_i w_i = & 1\bar{1}\ 01\ \bar{1}1\ \bar{1}1\ \bar{1}1\ \bar{1}0\ 01\ \bar{1}1\ 00\ 1\bar{1}\ 1\bar{1}\ 1\bar{1}\ 10 \\
s_i = & 1\ \ \bar{1}\ \ 0\ \ 0\ \ 0\ \ 0\ \ 0\ \ 0\ \ 1\ \ 1\ \ 0\ \ 0\ \ 0\ \ 0
\end{array}
$$

The result of this addition, the binary number 1000000110000, can easily be checked.

PROBLEMS

1. Explain the operation of a binary adder as a decoding process of the decimal sum of a number of binary digits. Show that for certain numbers of binary symbols, some adding-circuit arrangements are to be preferred from the point of view of economical design.

2. Explain by means of an example how the carry transfer can cause a variable operating time within a binary adder (parallel addition of binary numbers); and show by means of a circuit that it is possible to design parallel relay adders with an immediate carry transfer.

3. Determine the truth table and switching formulas of one section of a

(*a*) Full adder
(*b*) Half-adder

Design the circuit of a full adder by carefully factoring the switching formulas, so that 20 diodes at most are involved in the logical gates and that a maximum of 2 inverters is required in order to obtain the correct output signals.

Design the circuit of a full adder consisting of 2 half-adders, with 18 diodes at most used in the logical diode gates and a maximum of 4 inverters used in order to obtain the correct output signals.

Note that both phases are available for all input signals.

4. Factor the switching formulas of a full adder so that no negated input variables are required. Draw the circuit of one section of this adder.

5. Determine the switching formulas for a half-adder so that the circuit can be implemented by means of three transistors. Draw the circuit of a full adder consisting of two half-adders of this type.

6. Show that a pulse-halving circuit can be converted by means of logical gates into an adding circuit. Give the switching formulas and discuss the carry-propagation circuit.

7. Show that an *RC* delay circuit (slave memory element) and a trigger (master memory element) can be arranged for the addition of two binary digits. Draw the circuit.

8. Show that a serial adder can be designed by means of MS memory elements.

9. Explain the principle of a Kirchhoff adder and give the circuit of a transistorized full adder.

10. Determine the switching functions of a full adder having no inverted input variables that can be implemented by means of five magnetic cores.

11. Determine the switching functions of a full adder in their maxterm representation and give a circuit using six magnetic cores.

12. Determine the switching functions of a full adder in the mixed algebraic-Boolean representation without using inverted input variables. Draw the circuit with seven magnetic cores.

13. Draw a magnetic core circuit for a full-adder circuit with amplification of the generated and propagated carry pulse.

14. Explain the subtraction of two binary numbers, and especially the "borrow" transport problem in a parallel subtracter.

15. Design the circuit of a section of a parallel subtracter for binary numbers containing two differential relays.

16. Explain the purpose of the "end-around borrow" in the subtraction of two binary numbers, the subtrahend being larger than the minuend. Give a formula for the representation of negative binary numbers by means of their ones complement.

17. Show that the subtraction of two binary numbers can be performed in an adder by means of the ones complement of the addend. Add +6 and −9. Add −6 and +9.

18. Show that the addition of two binary numbers can be performed in a subtracter by means of the ones complement of the subtrahend. Subtract +6 and −9. Subtract −6 and +9.

19. Draw a fast carry-line circuit with transistors in the path of the propagated carry. Explain its operation and why it is a fast circuit.

20. Determine the switching functions of a carry circuit of an adder circuit operating with group-carry logic. Draw a symbolic circuit of a group of adder circuits.

SELECTIVE BIBLIOGRAPHY

1. Allen, M. W.: A Decimal Addition-subtraction Unit, *Proc. IRE*, vol. 103, pp. 138–145, 1956.
2. Avizienis, Algirdas: Signed-digit Representations for Fast Parallel Arithmetic, *IRE Trans.*, vol. EC-10, pp. 389–400, 1961.
3. Baker, G. T.: Computation Problems in Circuit Design, *Post Office Elec. Eng. J.*, vol. 39, pp. 58–63, 1946.
4. Barnes, R. C. M., G. A. Howells, and G. H. Cooke-Yarborough: Transistor Arithmetic Circuits for an Interleaved Digit Computer, *Proc. Inst. Elec. Engrs.* (*London*), pt. B, suppl. 3, vol. 103, pp. 371–381, 1956.
5. Bental, L. J., and J. K. Wood: An Economical Relay Operated Accumulator, *Electron. Eng.*, vol. 28, p. 409, 1956.
6. Chao, S. C.: A Generalized Resistor-transistor Logic Circuit and Some Applications, *IRE Trans.*, vol. EC-8, pp. 8–12, 1959.
7. Chen, Ma-Chao: A Magnetic Core Parallel Adder, *IRE Trans.*, vol. EC-7, pp. 262–264, 1958.
8. Florida, C. D.: Digital Computer Adding and Complementing Circuit, *Electron. Eng.*, vol. 30, pp. 429–435, 1958.
9. Garner, H. L.: The Residue Number System, *IRE Trans.*, vol. EC-8, pp. 25–31, 1959.

10. Gilchrist, B., J. H. Pomerene, and S. Y. Wong: Fast Carry Logic for Digital Computers, *IRE Trans.*, vol. EC-4, pp. 133–135, 1955.
11. Harris, J. R.: A Transistor Shift Register and Serial Adder, *Proc. IRE*, vol. 40, pp. 1597–1602, 1952.
12. Horton, J. W.: A Full Binary Adder Employing Two Negative Resistance Diodes, *IBM J. Res. Develop.*, vol. 2, pp. 223–231, 1958.
13. Keen, A. J.: Relay Adding Circuits, *Electron. Eng.*, vol. 29, p. 143, 1957.
14. Kilburn, T., D. B. G. Edwards, and D. Aspinall: A Parallel Arithmetic Unit Using a Saturated-transistor Fast-carry Circuit, *Proc. Inst. Elec. Engrs.* (*London*), vol. 48, pp. 573–584, 1960.
15. MacSorley, O. L.: High-speed Arithmetic in Binary Computers, *Proc. IRE*, vol. 49, pp. 67–91, 1961.
16. Maynard, F. B.: Half Adders Drive Simultaneous Computer, *Electronics*, vol. 31, pp. 80–82, 1958.
17. Morgan, L. P., and D. B. Jarvis: Transistor Logic Using Current Switching and Routing Techniques and Its Application to a Fast-carry Propagation Adder, *Proc. Inst. Elec. Engrs.* (*London*), pt. B, vol. 106, pp. 467–469, 1959.
18. Oberman, R. M. M.: Lower Limits in the Invention of Switching Circuits, *De Ingenieur*, vol. 67, pp. E 161–164, 1953.
19. Poel, W. van der: Some Special Subjects in Switching Algebra (Enige Byzondere Onderwerpen uit de Schakelalgebra), *De Ingenieur*, pp. E 9–14, 1955 (in Dutch).
20. Reijns, G. L.: High-speed Binary Parallel Adder, *Tijdschrift Nederlands Radiogenootschap*, vol. 27, pp. 1–16, 1962.
21. Richards, R. K.: "Arithmetic Operations in Digital Computers," chap. 4, D. Van Nostrand Company, Inc., Princeton, N.J., 1955.
22. Scarrot, C. G., W. J. Harwood, and K. C. Johnson: Logical Devices Using Saturable Magnetic Cores, *Proc. Inst. Elec. Engrs.* (*London*), vol. 103, pp. 302–312, 1956.
23. Sklansky, J.: Conditional-sum Addition Logic, *IRE Trans.*, vol. EC-9 pp. 226–231, 1960.
24. Sklansky, J.: An Evaluation of Several Two-summand Binary Adders, *IRE Trans.*, vol. EC-9, pp. 213–226, 1960.
25. The Staff of the Computation Laboratory: "Synthesis of Electronic Computing and Control Circuits," chap. 12, Harvard University Press, Cambridge, Mass., 1951.
26. Stock, J. R.: An Arithmetic Unit for Automatic Digital Computers, *Z. Angew. Math. Phys.*, vol. 5, pp. 168–172, 1954.
27. Svoboda, A.: "The Numerical System of Residual Classes in Mathematical Machines," pp. 388–397, Congresso International de Automatica, Instituto de Electricidad y Automatica, Madrid, 1958.
28. Svoboda, A.: "The Numerical System of Residual Classes in Mathematical Machines, Information Processing," pp. 419–422, R. Oldenbourg Verlag, Munich, 1959.
29. Tocher, K. D., and M. D. Lehman: A Fast Parallel Arithmetic Unit, *Proc. Inst. Elec. Engrs.* (*London*), pt. B, suppl. 3, vol. 103, pp. 371–381, 1956.
30. Valach, M.: Vznik Kodu a Císelné Soustavy Zbytkových Trid. Stroje na Zpracování Informací. *Sb. III*, *Nakl CSAV* (*Prague*), 1955.
31. Weinberger, A., and J. L. Smith: A One Micro-second Adder Using One-megacycle Circuitry, *IRE Trans.*, vol. EC-5, pp. 65–72, 1956.
32. Williams, F. C., A. A. Robinson, and T. Kilburn: Universal High Speed Computers; Serial Computing Circuits, *Proc. Inst. Elec. Engrs.* (*London*), part II, vol. 99, pp. 107–123, 1952.
33. Wood, J. K.: An Economical Relay Operated Accumulator, *Electron. Eng.*, vol. 28, pp. 250–253, 1956.
34. Woods-Hill, W.: An Electronic Arithmetic Unit, *Electron. Eng.*, vol. 27, pp. 212–217, 1955.

18
Multiplication and Division of Two Binary Numbers

18-1 MULTIPLICATION METHODS

The multiplication of two binary digits, i.e., two digits of a number system with radix 2, is defined by truth table 18-1, in which a is the multiplicand, b the multiplier, and P the product.

Table 18-1

a	b	P
0	0	0
1	0	0
0	1	0
1	1	1

The multiplication of two binary numbers containing a number of binary digits each consists mainly of the formation of all partial products P_{mn}

resulting from the binary multiplication of each digit of the multiplicand with each digit of the multiplier and the addition of these partial products in the correct combinations. The multiplication of two binary numbers is illustrated in Fig. 18-1 where $A = a_3 \cdot 2^3 + a_2 \cdot 2^2 + a_1 \cdot 2^1 + a_0 \cdot 2^0$ is the multiplicand and $B = b_3 \cdot 2^3 + b_2 \cdot 2^2 + b_1 \cdot 2^1 + b_0 \cdot 2^0$ the multiplier. The product P resulting

			a_3	a_2	a_1	a_0	Multiplicand
			b_3	b_2	b_1	b_0	Multiplier
							×
			a_3b_0	a_2b_0	a_1b_0	a_0b_0	Partial products
		a_3b_1	a_2b_1	a_1b_1	a_0b_1		
	a_3b_2	a_2b_2	a_1b_2	a_0b_2			
a_3b_3	a_2b_3	a_1b_3	a_0b_3				
							+

$$P = K_6 2^6 + K_5 2^5 + K_4 2^4 + K_3 2^3 + K_2 2^2 + K_1 2^1 + K_0 2^0 \quad \text{Product}$$

Fig. 18–1

from the multiplication shown in Fig. 18-1 is not a binary number. Functions K_0, K_1, and so on are not binary; function K_0 can be 0 or 1 and function K_1 can be 0, 1, or 2; the function K_i can have a decimal value equal to the number of corresponding partial products.

Since the final product P is wanted in a binary form, it is necessary to replace the K functions by binary functions. The splitting of the K functions into binary functions can be performed during the adding process of the partial products. In the same way as was already explained in Chap. 17, carry digits can be transferred to higher-order additions of partial products. In the example shown in Fig. 18-1, i.e., the multiplication of 2 four-digit binary numbers, only first- and second-order carries are possible. The index number N of the binary digit with the highest weight in the final product is determined by

$$N = m + n + 1 \tag{18-1}$$

where m and n represent the indices of the digits with the highest weight in the multiplicand and in the multiplier, respectively. The introduction of carries between successive orders of the addition of partial products causes the multiplication shown in Fig. 18-1 to change into the one illustrated in Fig. 18-2.

Functions S_0, S_1, etc., are binary functions. The carries c are provided with two index numbers indicating the originating and terminating order of the addition of the partial products. The double-headed arrows between the various carries in Fig. 18-2 indicate mutually exclusive carries.

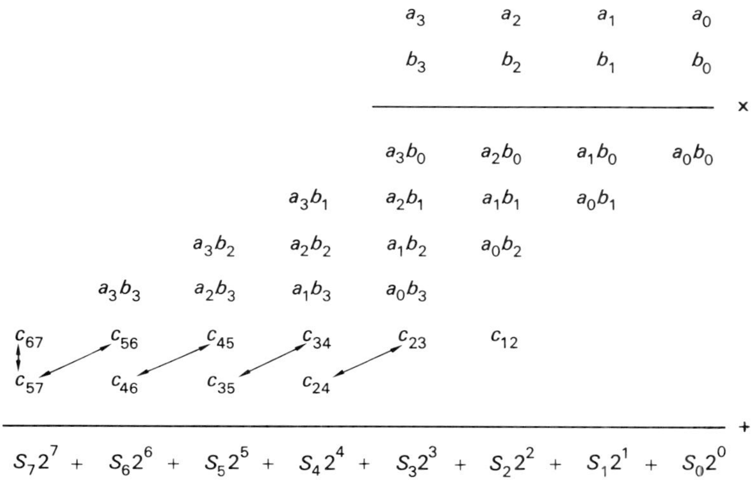

Fig. 18–2

The partial products of the multiplication can be easily obtained by means of the multiplication of two binary digits in accordance with truth table 18-1. Because this truth table is the same as the table for the logical or Boolean multiplication of two switching variables, this multiplication can be implemented by means of AND gates.

The difficulty in the multiplication of two binary numbers does not lie in the multiplication of all combinations of the binary digits of multiplicand and multiplier to obtain the partial products but in the adding of these partial products, in the formation of the carries, and in the transfer of these carries.

This binary multiplication is complicated even further if multiplicand and multiplier are given as decimal digits in a binary-coded form. This makes it necessary to generate a binary-coded-decimal carry between successive binary-coded-decimal sections of the adders of the partial products.

The formation of the partial products, the addition of these partial products, and the carries originating from the addition of the partial products of lower-order sections can be performed basically in three different ways:

1. All the operations can be performed simultaneously, so that a parallel multiplier is obtained. Multiplicand and multiplier are given on a parallel basis.

2. The operations can be performed on a serial-parallel basis, the multiplicand being supplied in parallel and the multiplier in serial form.
3. The multiplication can be split up completely into a series of operations and performed by means of a one-digit serial adder combined with shift registers. Multiplicand and multiplier both appear in serial form only.

If the adding part of a parallel multiplier is to consist exclusively of half-adders and full adders of two binary digits and a carry, the diagram shown in Fig. 18-2 can be rearranged as shown in Fig. 18-3 or 18-4. A parallel multiplier is already described in U.S. Patent 2,428,812. Both these diagrams can be applied to serial-parallel multipliers; however, the diagram of Fig. 18-4 is best suited for a serial multiplier.

The diagram of Fig. 18-3 shows eight steps, which can be performed simultaneously. In the first step, the provisional sums ${}^1S_3 = a_3b_0$, ${}^1S_2 = a_2b_0$, ${}^1S_1 = a_1b_0$, and the final sum $S_0 = a_0b_0$ are obtained. No carries are generated during this phase of the multiplying process.

In the second step, partial product a_3b_1 leads to provisional sum ${}^2S_4$, partial product a_2b_1 and ${}^1S_3$ yield provisional sum ${}^2S_3$ and carry ${}^2C_3$, partial product a_1b_1 and ${}^1S_2$ yield provisional sum ${}^2S_2$ and carry ${}^2C_2$, and partial product a_0b_1 and ${}^1S_1$ yield sum $S_1 = a_0b_1 + a_1b_0$ and carry ${}^2C_1$.

In the flow diagram of Fig. 18-3, a carry ${}^nC_m$ generated in a row of fundamental operations is processed in the next row of operations. This creates no difficulty in a parallel multiplier, but in a serial multiplier operating in the mode of the flow chart of Fig. 18-3, all these carries have to be stored for a number of steps equal to the number of digits of the multiplicand. In the flow diagram of Fig. 18-4, it can be seen that these carries generated in a certain operation can be processed in the next operation, except the carries (${}^1C_4$, ${}^2C_5$, and ${}^3C_6$) that are generated in the last processing step of a row.

During each of the eight steps of operation shown in Fig. 18-3, a sum digit S is generated. This generation of sum digits is carried on in the same way even after all partial products have been converted into provisional sums and carries until a sum digit S_N has been formed with an index $N = m + n + 1$ [Eq. (18-1)].

The flow chart of Fig. 18-3 describing the multiplication of two binary numbers can be implemented by means of a number of adders performing the operations as indicated within the rectangles in dotted lines. The diagram of Fig. 18-3 can also be realized by means of a device combining $m - 1$ adders and $m - 1$ carry memory elements through which the provisional sums are shifted. This latter case corresponds to a serial-parallel multiplier.

It can easily be shown that the flow chart of Fig. 18-3 leads to the same product as the type of multiplication performed in Fig. 18-2.

A variation on the flow chart of Fig. 18-3 is shown in Fig. 18-4. In this flow chart, the carries are processed in the adding operation following that

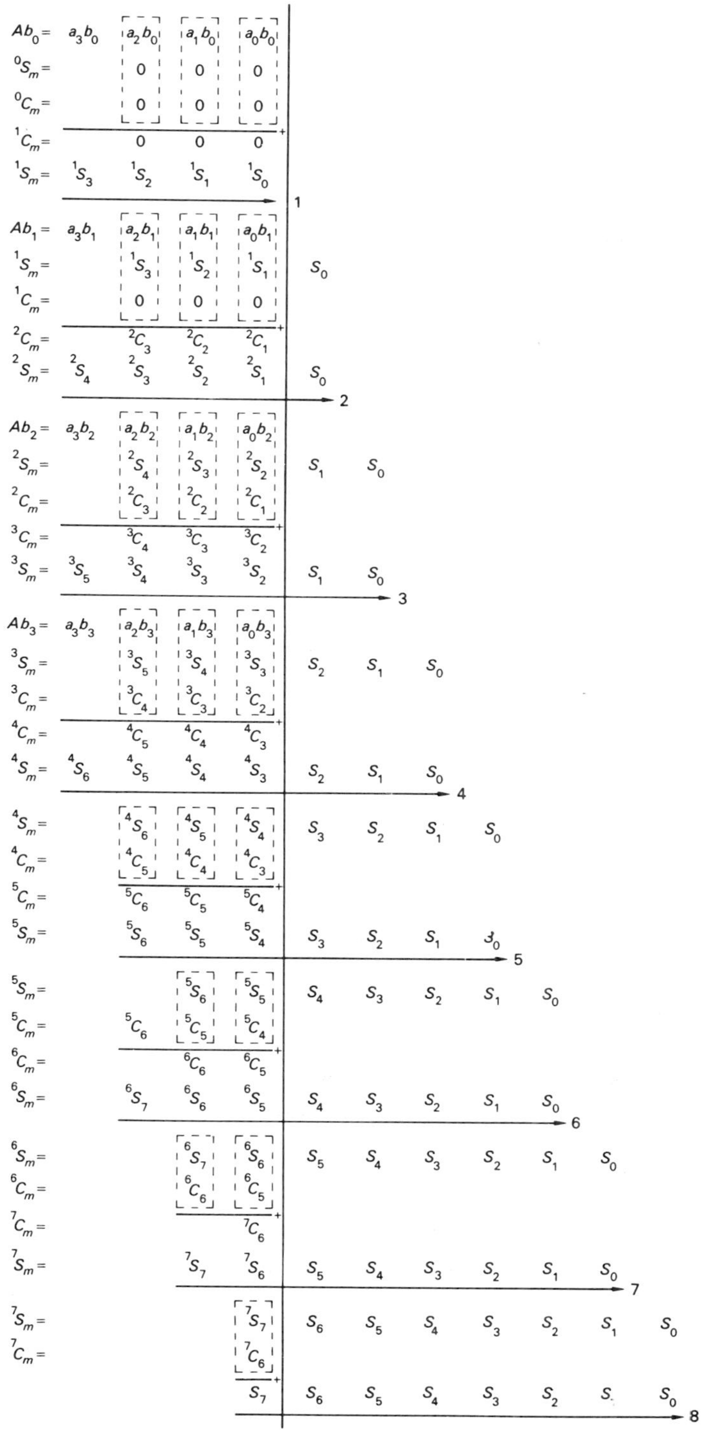

Fig. 18–3

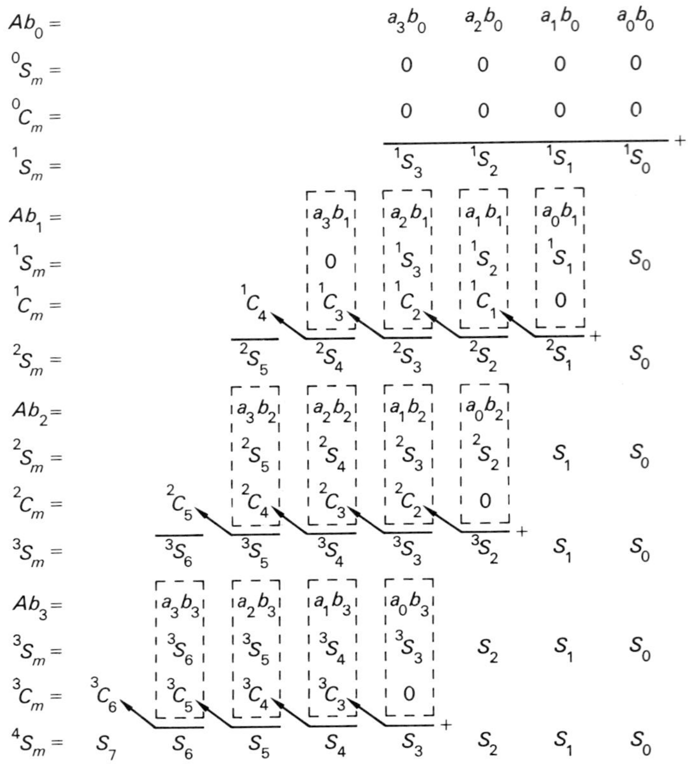

Fig. 18–4

in which they are generated, except for the carries generated in the last operation of a row.

In the first row of the diagram of Fig. 18-4, the group of partial products formed by the multiplicand and the first digit of the multiplier are indicated. Partial product a_0b_0 is at the same time the first product digit S_0. In the second-row operations, this group of partial products is added to the over-one-place shifted group of partial products formed by the multiplicand and the second digit of the multiplier, thus forming the second product digit S_1, the three intermediate sum digits ${}^2S_2$ to ${}^2S_4$, and the four carry digits ${}^1C_1$ to ${}^1C_4$. These carry digits (except carry ${}^1C_4$, for which no next-higher-adder column is available) are processed in the next-higher-adder column of the diagram. Carry ${}^1C_4$ can also be considered as an intermediate sum ${}^2S_5$, which will be processed in the next row of the diagram.

The diagram of Fig. 18-4 needs no further explanation. It is slightly simpler than the diagram of Fig. 18-3.

18.2 PARALLEL MULTIPLIERS

Parallel multipliers can be designed in a number of ways. The block diagram of a Kirchhoff adder for several binary numbers as shown in Fig 17-28 is very suitable as a basis for the adding part of a parallel multiplier operating in accordance with the multiplication diagram shown in Fig. 18-2. The portion of such a parallel adder in which the partial products are formed can consist of a matrix of AND gates as shown in Fig. 18-5. The Kirchhoff parallel adder must be adapted so as to accept four binary numbers of seven digits from which a number of digits are permanently 0, as can be seen in the diagram of Fig. 18-2. The input resistors of these 0 inputs of the Kirchhoff networks have been omitted from the block diagram of Fig. 18-5. For each of the Kirchhoff networks, they can be replaced by one resistor connected to the 0 signal voltage. The 0 inputs make it possible to simplify the carry circuits, as can be deduced from the diagram of Fig. 18-2. The complete block diagram of the parallel multiplier based on the Kirchhoff parallel adder is shown in Fig. 18-5.

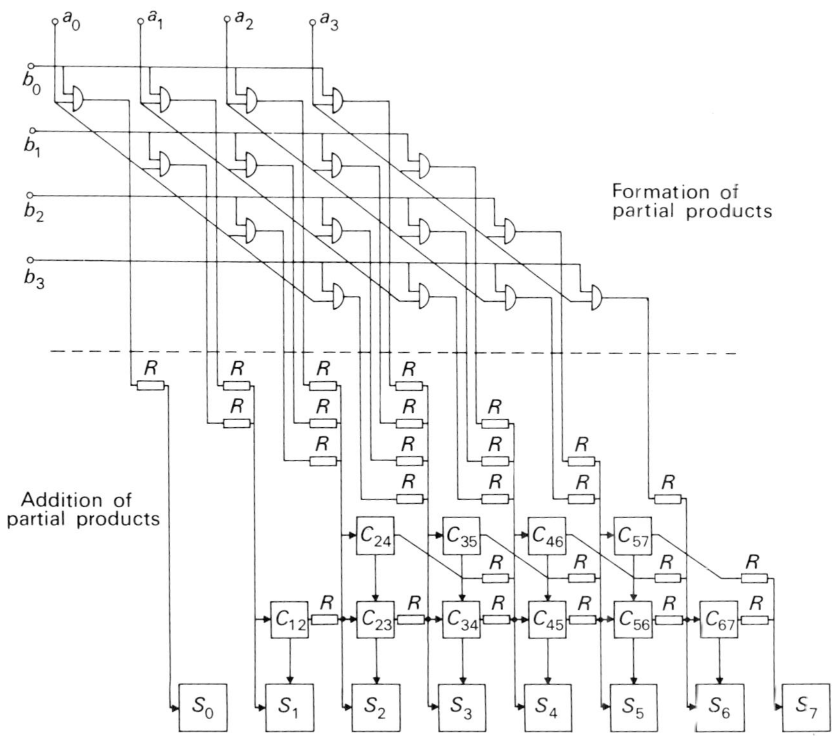

Fig. 18–5

The block diagram of the parallel multiplier shown in Fig. 18-5 can be converted into a practical network by means of transistor decoders of a type essentially similar to the circuit shown in Fig. 17-26. The operating speed of this parallel adder depends mainly on the speed of carry propagation in the addition of the partial products.

The marginal operation of the parallel adder of the parallel multiplier can be remedied altogether by performing the adding process of the partial products by means of a number of adders, half-adders, and some logical gates. The diagram of Fig. 18-3 shows the organization of such a parallel multiplication process for two four-digit binary numbers. This diagram can be implemented by means of the circuit shown in Fig. 18-6.

The speed of the parallel multiplication in this type of circuit is again limited by the speed of the carry propagation in the device used for adding the partial products. The resulting diagram of the parallel multiplier is completely systematic. It consists of a matrix of AND gates where the partial products are obtained and of an adding part which contains in this case six full adders $F_{1/6}$, nine half-adders $H_{1/9}$, and two OR gates $O_{1/2}$. The two OR gates can be combined into one.

It should be noted that there are known to exist parallel adders of the type shown in Fig. 18-6 that contain a slightly smaller number of half-adders (one full adder equals two half-adders) with a slightly larger maximum carry-propagation time. The diagram of Fig. 18-2 shows that regardless of the number of digits of the multiplicand and multiplier, the sum S_0 is formed from a single partial product; and the last digit (highest weight) of the product can never be the result of an addition, because the possible carries that can form the last digit of the product (highest-weight digit) are mutually exclusive. This means that the number of full and half-adders in the right-hand column of Fig. 18-6 is

$$N_a = m + n \tag{18-2}$$

where m is the index of the multiplicand digit with the highest weight and n is the multiplier digit with the highest weight. If the carry-propagation time of full and half-adders is the same, the maximum carry propagation time will be

$$t_p = (m + n)\, t_a \tag{18-3}$$

in which t_a represents the carry-propagation time of one adding section.

If full adders are thought of as consisting of two half-adders, the carry-propagation time given by Eq. (18-3) must be increased by a number corresponding to the number of full adders in the right-hand column of the diagram of Fig. 18-6. Since output $a_0 b_0$ of the first AND gate in the right-hand column of the AND-gate matrix always leads to output terminal S_0, and since

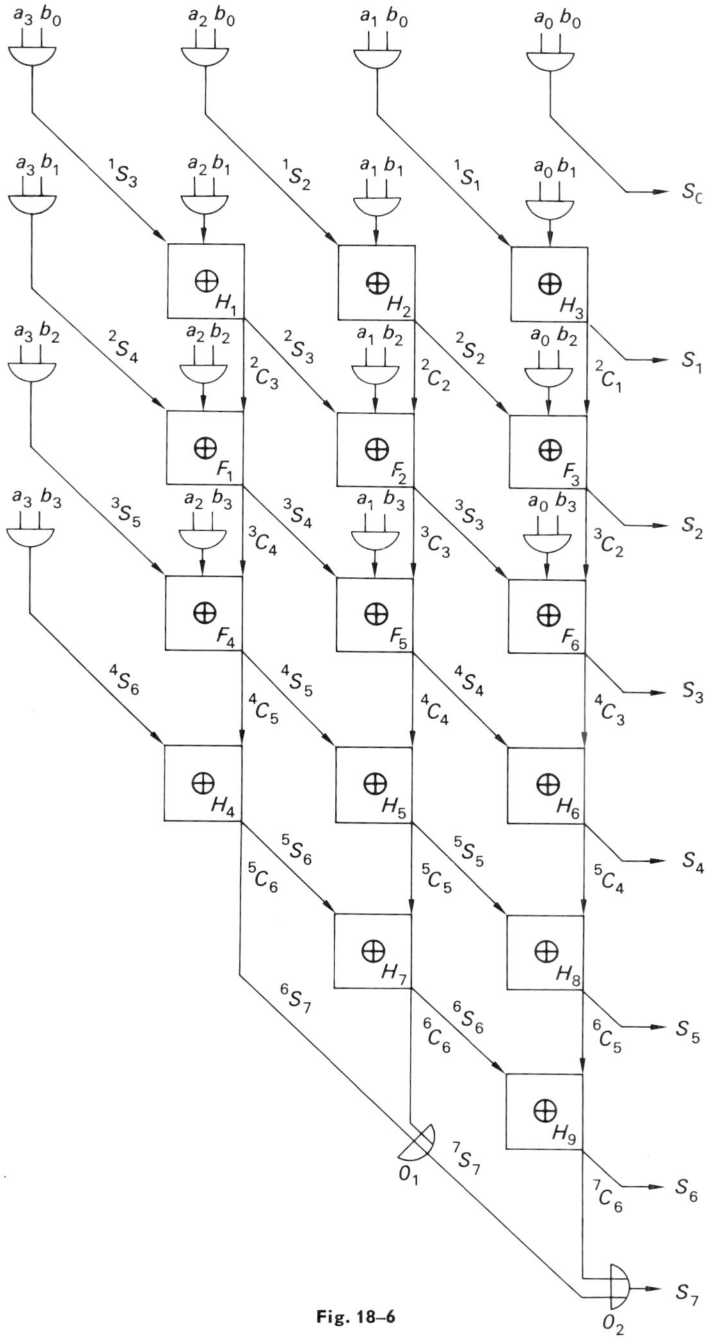

a3 b0
a2 b0
a1 b0
a0 b0
a3 b1
a2 b1
a1 b1
a0 b1
a3 b2
a2 b2
a1 b2
a0 b2
a3 b3
a2 b3
a1 b3
a0 b3
1S3
1S2
1S1
S0
H1
H2
H3
2S4
2S3
2S2
2C3
2C2
2C1
S1
F1
F2
F3
3S5
3S4
3S3
3C4
3C3
3C2
S2
F4
F5
F6
4S6
4S5
4S4
4C5
4C4
4C3
S3
H4
H5
H6
5S6
5S5
5C6
5C5
5C4
S4
H7
H8
6S7
6S6
6C6
6C5
S5
H9
O1
7S7
7C6
S6
S7
O2

Fig. 18–6

output a_0b_1 of the second AND gate in that column always leads to the first adder of the right-hand column of adders, which as a half-adder yields digit S_1, $m-1$ outputs of the AND gates of the right-hand matrix column will always lead to $m-1$ full adders in the right-hand column of the adding devices. The maximum carry-propagation time will thus be increased to

$$t_p = (2m + n - 1) \cdot t_a \tag{18-4}$$

The diagram shown in Fig. 18-6 is given in the form of a block diagram. The various blocks can be implemented by means of all types of switching components, provided that they can be used to arrive at the desired switching functions. The various adding circuits need not differ from those already mentioned in the discussion of the addition and subtraction of binary numbers in Chap. 17.

The discussion above of the maximum carry-propagation time in parallel multipliers is not applicable if adding circuits are used with high-speed carry wires in accordance with Sec. 17-5.

The flow diagram of Fig. 18-4 leads to a type of parallel multiplier differing from that shown in Fig. 18-6. A block circuit of a 4×4 bit parallel multiplier of this type is shown in Fig. 18-7. The circuits of Figs. 18-6 and 18-7 are arranged in an analogous way. The carry propagation from adder to adder forms the main difference between both parallel multiplier circuits. In the diagram of Fig. 18-6, the carry paths run vertically between the various half- and full adders; in the diagram of Fig. 18-7, these connections run mainly in a horizontal way.

Taking the amount of hardware of a full adder twice that of a half-adder, the number of half-adders in the circuit of Fig. 18-7 is one less than that in the circuit of Fig. 18-6. The maximum number of carry-propagation steps is the same in both diagrams.

The parallel multipliers shown in Figs. 18-6 and 18-7 do not differ very much from the practical point of view.

18-3 SERIAL-PARALLEL MULTIPLIERS

The subdivision of the addition of partial products of multiplicand and multiplier as shown in the diagram of Fig. 18-3 can also be used as a basis for designing a serial-parallel multiplier. This diagram shows $m + n + 2$ main operations which can be performed serially instead of in parallel, as in the block diagrams of Figs. 18-6 and 18-7.

If the eight main operations shown in Fig. 18-3 are performed serially, multiplicand A must be fed to the multiplying circuit on a parallel basis, while multiplier B must be fed serially. The block diagram of Fig. 18-8 can perform one main operation at a time.

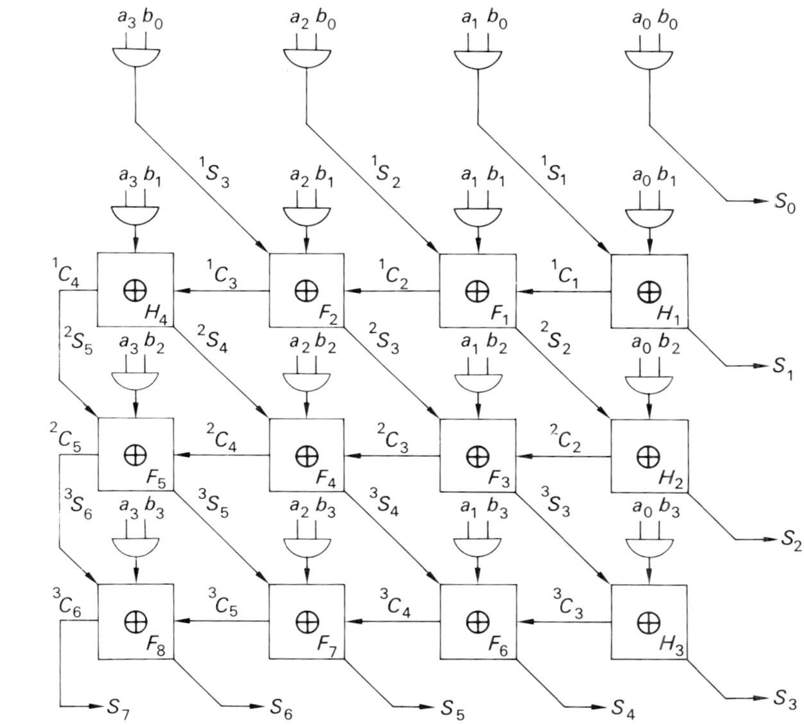

Fig. 18–7

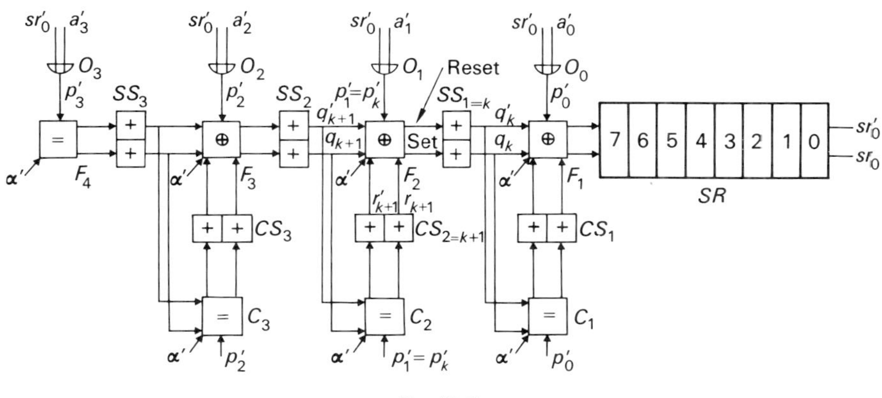

Fig. 18–8

The adding process in Fig. 18-8 is performed under the control of one α' clock pulse only. The diagram of Fig. 18-8 contains a shift register SR in which multiplier B is stored and in which the final product can be stored if so desired. In the latter case, shift register SR should consist of $m+n+2$ sections. The digits of multiplicand A apply permanent signals to one input of $m+1$ OR gates, while the other inputs of all these OR gates $O_{0/m}$ are connected to the output of shift register SR. Gates $O_{0/m}$ form the partial products $a_i b_j$. The multiplier furthermore consists of three carry stores CS, three sum stores SS, three full adders $F_{1/3}$, an auxiliary circuit F_4, and three carry-forming networks C. In each of these three adding sections $F_{1/3}$, the operations will be performed as indicated in the corresponding sections enclosed in dotted lines in the flow chart of Fig. 18-3.

The block diagram of Fig. 18-8 clearly shows that three of the provisional sum digits are stored for a brief moment and presented for the next main operation to the adder of the following section, while any carry digit produced during the adding process of the partial products is also stored for a moment but is again applied to its generating adder for the next main operation.

The final digits of the product, which are shown in Fig. 18-3 on the right of the vertical line, are generated by the F_1 adder and transferred to shift register SR.

The set and reset signals of the sum store SS_k are as follows, with signals p, q, and r being, respectively, a partial product, a provisional sum from the preceding adder section, and a carry signal from the preceding adding operation:

$$\begin{aligned}\text{Set } SS_k &= p_k' q_{k+1}' r_{k+1} + p_k' q_{k+1} r_{k+1}' + p_k q_{k+1}' r_{k+1}' + p_k q_{k+1} r_{k+1} \\ &= r_{k+1}'(p_k' \oplus q_{k+1}') + r_{k+1}(p_k' \oplus q_{k+1}) \qquad (18\text{-}5)\end{aligned}$$

$$\begin{aligned}\text{Reset } SS_k &= p_k' q_{k+1}' r_{k+1}' + p_k' q_{k+1} r_{k+1} + p_k q_{k+1}' r_{k+1} + p_k q_{k+1} r_{k+1}' \\ &= r_{k+1}'(p_k' \oplus q_{k+1}) + r_{k+1}(p_k' \oplus q_{k+1}') \qquad (18\text{-}6)\end{aligned}$$

The set and reset signals of the carry store CS_{k+1} are as follows:

$$\begin{aligned}\text{Set } CS_{k+1} &= p_k' q_{k+1} r_{k+1} + p_k q_{k+1}' r_{k+1} + p_k q_{k+1} r_{k+1}' + p_k q_{k+1} r_{k+1} \\ &= (p_k \oplus q_{k+1}) r_{k+1} + p_k q_{k+1} \qquad (18\text{-}7)\end{aligned}$$

If $r_{k+1} = 1$, there is no need to set CS_{k+1}, because this carry store is then already in the set condition. The set formula for this carry store can then be simplified to

$$\text{Set } CS_{k+1} = p_k q_{k+1} \qquad (18\text{-}8)$$

In an analogous way, it is found that

$$\text{Reset } CS_{k+1} = p'_k q'_{k+1} \tag{18-9}$$

The set and reset signals for sum store SS_m are p_m and p'_m, respectively. The resulting circuits for sum store SS_k and its full adder F_{k+1} are shown in Fig. 18-9a, the corresponding circuits for carry store CS_{k+1} in Fig. 18-9b, and circuit F_4 in Fig. 18-9c.

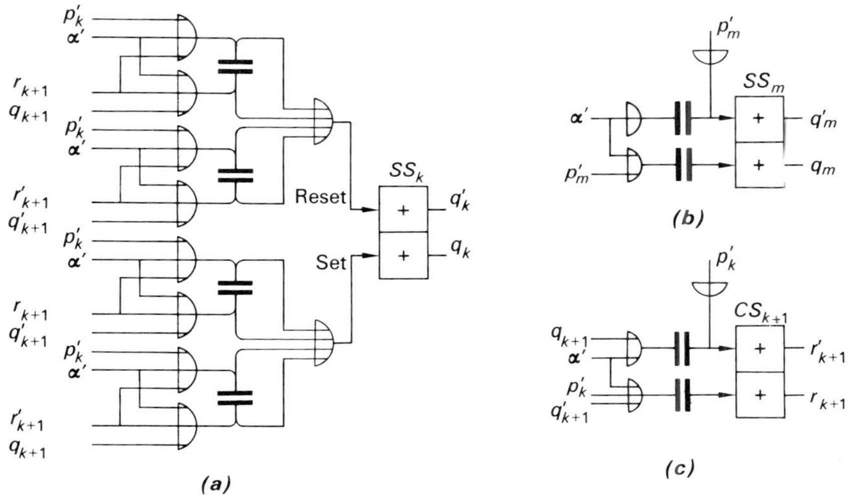

Fig. 18–9

Two variants of the serial-parallel multiplication process can be obtained, using the diagram of Fig. 18-4. In the construction of the diagram of Fig. 18-3, it was assumed that all carries generated in each of the adding operations are stored separately and are used again in the next operation performed in the corresponding adder, the result being $m + n + 2$ main operations of constant length.

The addition of partial products can also be performed by means of parallel adders with immediate carry transfer to higher-order sections. Once the carry transfer through the parallel adder is completed, the provisional sum stored in the adding sections must be shifted over one digit to lower-order sections, or the multiplicand must be shifted over one digit to higher-order sections, after which the next main operation can be performed, and so on. The carry transfer introduces the problem of carry propagation in this type of multiplier. The problem of immediate carry transfer is sufficiently discussed in the preceding chapter. Serial-parallel multipliers using this idea are based on the diagram of Fig. 18-4.

A block circuit corresponding to the first variant of the series-parallel multiplier is shown in Fig. 18-10 for the multiplication of two four-digit binary numbers, a four-section shift register $SR_{0/3}$, four full adders $F_{0/3}$, four intermediate sum stores $SS_{1/4}$, four carry-forming networks $CP_{1/4}$, four partial-product OR gates $O_{0/3}$, and four multiplicand-digit stores $SA_{0/3}$.

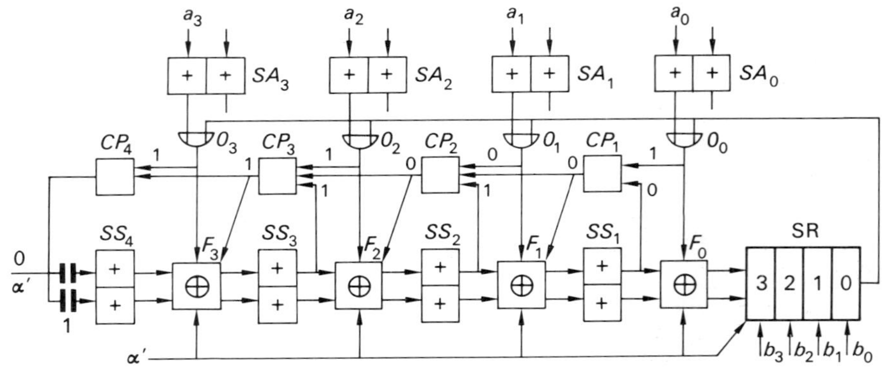

Fig. 18–10

Memory elements $SA_{0/3}$ contain multiplicand A. During the multiplication operation, no change occurs in the state of these memory elements. At the start of the multiplication, multiplier B is stored in shift register $SR_{0/3}$, as indicated in Fig. 18-10. One input terminal of each of the four OR gates $O_{0/3}$ is connected to a corresponding output terminal of a memory element of the multiplicand store. The other input terminals of the O gates are linked together and connected to the output terminal of the lowest-order section of multiplier shift register $SR_{0/3}$.

The multiplier digit stored in section SR_0, combined with the multiplicand digits a_0 to a_3, yields the first group of partial products a_3b_0, a_2b_0, a_1b_0, and a_0b_0, which indicate corresponding signals to the full adder circuits $F_{0/3}$.

On the first clock pulse α', the first group of partial products forms the only input to the full adders $F_{0/3}$. Memory elements $SS_{1/4}$ are then in the 0 state. Partial product a_0b_0 is then transferred to SR_3 and partial products a_1b_0, a_2b_0, and a_3b_0 are stored in the sum stores $SS_{1/3}$. The multiplier is shifted over one position in shift register SR so that digit b_1 will control the formation of the second group of partial products.

In the pause between the first and second clock pulse, the second group of partial products is formed by gates $O_{0/3}$, so that the outputs of these gates indicate signals corresponding to these partial products, to the full adders $F_{0/3}$, and to the carry-forming networks $CP_{1/4}$. As a result of the operation of the carry-forming networks $CP_{1/3}$, a carry signal will eventually be indicated to full adders $F_{1/3}$, so that the possible carry signal is directly taken into account in the addition of the higher-order digits. In this carry-forming process, a fourth network CP_4 is required to take account of a possible overflow carry. This carry signal is stored on the next (second) clock pulse in memory element SS_4. The other provisional sums (including possible carries) are stored in memory elements $SS_{1/3}$. On each clock pulse, a final-product digit is produced by full adder F_0 and transferred to shift register SR.

		1 (SS_3)	1 (SS_2)	0 (SS_1)	1 (SR_3)
	1 (O_3)	1 (O_2)	0 (O_1)	1 (O_0)	
1 (SS_4)	0 (SS_3)	0 (SS_2)	1 (SS_1)	1 (SR_3)	1 (SR_2)

Fig. 18-11

The operation of the carry-forming network is illustrated in Fig. 18-10 with a number of 0's and 1's which could occur during a multiplication. If, for example, after the first clock-pulse number 1 1 0 1 is stored in SS_3, SS_2, SS_1, and SR_3, while digit $b_1 = 1$, the outputs of OR gates $O_{3/0}$ indicate also number 1 1 0 1. Both numbers are shown in Fig. 18-11 in their correct relative positions. The resulting sum and the memory elements in which the digits of this sum are stored are also shown in Fig. 18-11.

A multiplier of this type with immediate carry transfer requires a number of operating steps equal to the number of digits of the multiplier. With a serial transport of the numbers, it requires more operating steps to clear the multiplier for the next multiplication.

The two serial-parallel multipliers of Figs. 18-8 and 18-10 differ very little. The latter type requires perhaps slightly less hardware than the former, but with serial transport of the numbers, it is probable that with the first type a higher operating speed can be achieved when in the second type the carry-propagation time between clock pulses is not negligible.

A second variant of the serial-parallel multiplier can be obtained by using a parallel adder (or accumulator) with $m + n + 2$ sections ($S_{0/7}$), the last of which is exclusively used as a carry store, a multiplicand shift register SA containing $m + n + 1$ sections, a multiplier shift register SB with $n + 1$ sections, and $m + n + 1$ AND gates A that yield the partial products.

Figure 18-12 shows the initial state of the multiplier. The multiplicand fills the first $m + 1$ sections of the SA shift register. The other places are filled with 0's. The last section of multiplier shift register SB controls one input terminal of all partial product gates A_0 to A_6. The second input terminal of all these gates is connected to the output terminal of the corresponding section of multiplicand shift register SA. AND gate G_1 connects the multiplier shift register and the partial-product gates A. The second input terminal of this gate is connected to clock pulse α.

On the first clock pulse α, the first group of partial products a_3b_0, a_2b_0, a_1b_0, and a_0b_0 are added to the 0 state in sections S_3 to S_0 of the parallel adder. Clock pulse α', which is applied after clock pulse α, causes both

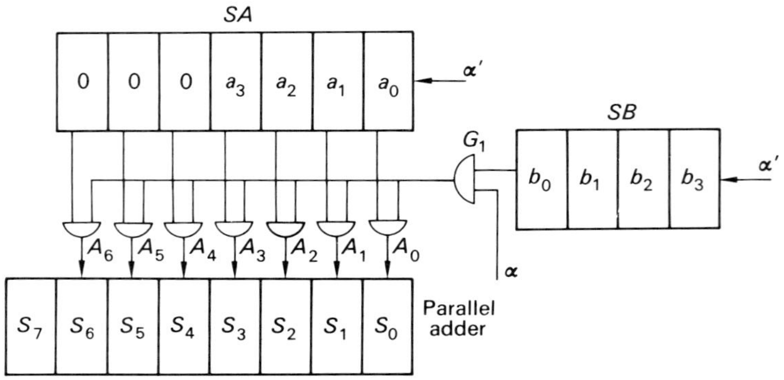

Fig. 18–12

multiplicand and multiplier to shift one section in their respective shift registers. The first section of multiplicand shift register SA is thereby put in the 0 state. The second group of partial products a_3b_1, a_2b_1, a_1b_1, and a_0b_1 is obtained on the second clock pulse α and added to sections S_4 to S_1, respectively. During this addition, a carry can be generated which can occupy section S_5.

The second α' clock pulse brings the b_2 digit in the left-hand section of the SB shift register and the multiplicand in sections 2, 3, 4, and 5 of the SA shift register. The third clock pulse α yields the third group of partial products, which are added in the correct places to the parallel adder.

The third clock pulse α' brings the $(n + 1)$st (fourth) digit of the multiplier in the last section of the shift register SB and brings the multiplicand in sections 4, 5, 6, and 7 of the SA shift register. The fourth clock pulse α opens the A gates so that the $(n + 1)$st (fourth) group of partial products is obtained and added to sections S_3 to S_6, respectively, of the parallel adder.

A carry digit can fill the eighth section of the parallel adder. The desired product is now obtained.

The operation of a serial-parallel multiplier conforming to the block diagram of Fig. 18-12 can easily be illustrated by means of the example given in Fig. 18-13. The multiplicand is 1 0 1 1 and the multiplier is 1 1 0 1.

The block diagram of Fig. 18-12 can also be operated in a single clock-pulse system, but this would complicate the diagram. With respect to the speed of multiplication, there is no basic difference between the series-parallel multipliers of Figs. 18-10 and 18-12. The carry-propagation problem is the same. There is, however, one difference between these two multipliers. If the

Start	0	0	0	0	0	0	0	0	
(α) add 1·(1101)					1	1	0	1	+
	0	0	0	0	1	1	0	1	
(α′) shift and (α) add 1·(1101)				1	1	0	1		−
	0	0	1	0	0	1	1	1	
(α′) shift and (α) add 0·(1101)			0	0	0	0			+
	0	0	1	0	0	1	1	1	
(α′) shift and (α) add 1·(1101)		1	1	0	1				+
Multiplicand	1	0	0	0	1	1	1	1	

Shifting multiplier

Fig. 18-13

information transfer in the computer is performed entirely on a serial basis, then it will take $m + n + 2$ steps to transfer the result from the multiplier of Fig. 18-12 to another memory; whereas in the multiplier of Fig. 18-10, the first $m + 1$ digits can, if so desired, be transferred directly to another memory, so that only $n + 1$ shift steps have to be performed after the end of the multiplication to clear the multiplier.

It should be noted that there are several ways of applying the multiplicand to the partial-product gates A. A shift register has been chosen in the block diagram of Fig. 18-12; but it is also possible to use for the multiplicand a nonshifting register in combination with a number of diode gates connecting at the proper moments the various digits to the correct partial-product gates A. The preferable method falls under the category of general switching problems and is not specifically related to multipliers. This question and the problem of the use of floating-point multipliers will not be gone into in this book.

18-4 SERIAL MULTIPLIERS

The serial multiplier consists essentially of one AND gate for the formation of the partial products of multiplicand and multiplier digits, and of one full adder for the adding of these partial products. The rest is much more a matter of circuit layout and of available memory type for storing the sums of the partial products than a switching problem. This intermediate storing of the

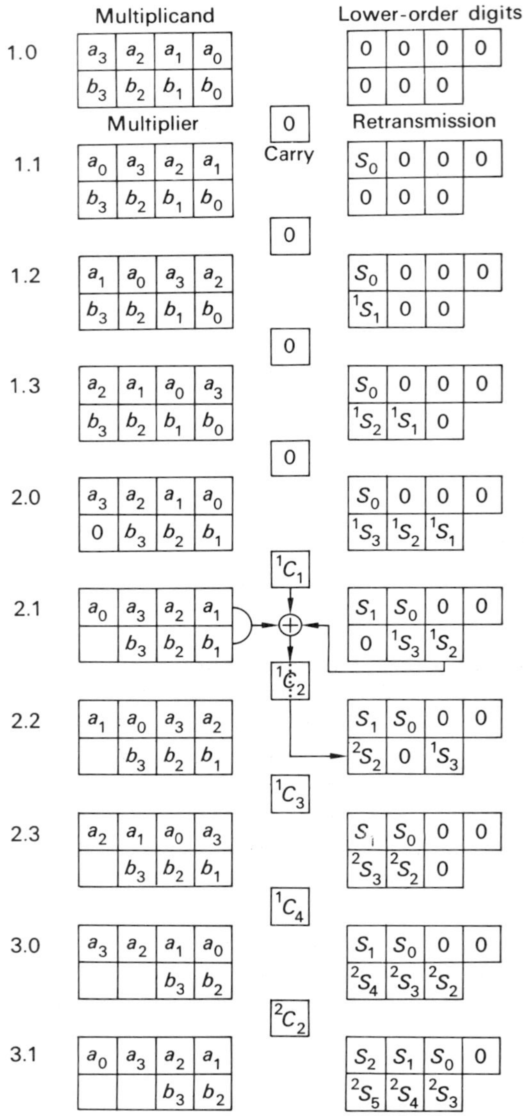

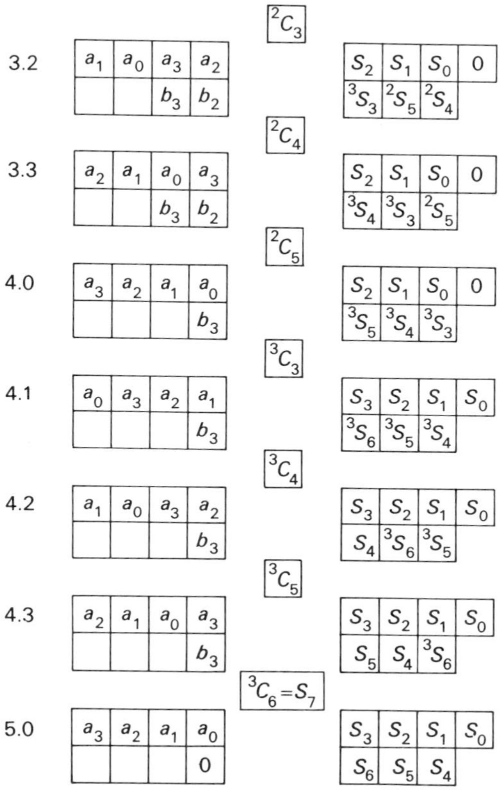

Fig. 18–14

sums of partial products can be performed in a number of ways, e.g., delay lines, tracks on magnetic drums, core memories, or shift registers. The last method will be used here because it allows a rather simple explanation of the operating principles of a serial multiplier—although essentially the same principles hold for all these types of switching components.

Figure 18-14 gives an example of a possible organization diagram for the multiplication of a multiplicand and a multiplier with, respectively, m and n as indices of the digits with the highest weight. This diagram comprises $(m + 1) \cdot (n + 1)$ steps. The various grids in the diagram represent shift registers. Multiplicand and multiplier are stored in the left-hand grids. The multiplicand is shifted one digit to the right on each clock pulse, and the multiplier is shifted to the right over one digit on each fourth clock pulse. This offers the possibility of feeding all the required partial products successively to an AND gate, which is symbolically shown in grid 2.1.

The right-hand pair of grids is filled successively with final and intermediate sums, in accordance with the diagram of Fig. 18-4. The grids of Fig. 18-14 are constructed in such a way that on a clock pulse the situation in one group of grids changes in the situation shown in the next group. The basic mutual relations between these grids are shown in grid 2.1. With the aid of diagram 18-4, the various steps in the diagram of Fig. 18-14 can be easily followed. However, one step—the step from 3.0 to 3.1, for example—in each

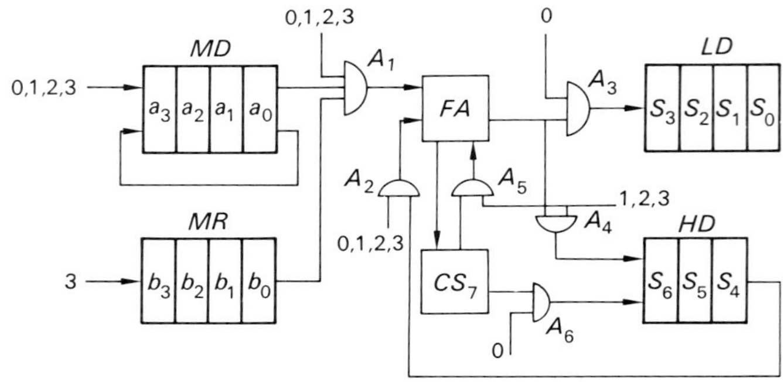

Fig. 18–15

group of $m + 1$ steps needs some explanation. In that step, partial product a_0b_2 is added to the provisional sum ${}^2S_2$ and stored as the final sum S_2 in the upper-right-hand grid 3.1. Also, carry ${}^1C_4$ is stored as provisional sum ${}^2S_5$ in the lower-right-hand grid 3.1.

In the final step, final sums S_0 to S_6 are stored in the right-hand grid, while sum S_7 is still registered in the carry store. The product with digits S_0 to S_7 is now ready for serial transport to another register.

A block diagram of a serial multiplier conforming to the diagram of Fig. 18-14 is shown in Fig. 18-15. This diagram shows the four shift registers for multiplicand (MD), multiplier (MR), low-order digits (LD), and retransmission or high-order digits (HD) in the same relative position as the grids in Fig. 18-14. The full adder FA and its carry store CS are controlled by gates A_1 (partial products) and A_2 (retransmission of provisional sums). The output digits of adder FA are routed to the low-order-digit store LD via gate A_3, or to the high-order-digit store via gate A_4. The digits of the carry store CS are routed either via AND gate A_5 to adder FA or via AND gate A_6 to the high-order-digit store HD. The various gates are controlled by a series of clock pulses, as indicated near the inputs.

It is possible to speed up a serial multiplier based on the block diagram of Fig. 18-15. On the average, half the digits of binary numbers are 0's. In case of a 0 digit of the multiplier, gate A_1 remains closed during a complete operating cycle. The high-order-digit store receives in the first step of such a cycle the provisional sum stored in the far-right place of the low-order-digit store. The effect of the rest of the steps of such an operating cycle is that the provisional sums stored in the low-order-digit store are shifted one place to the right. In this case, it is possible to make a one-clock-pulse shortcut on an operating cycle. This requires a special clock-pulse generating circuit in addition to that already shown in Fig. 18-15 and an extra gate between the output of the HD shift register and the input of the LD shift register.

It should be noted that computers have in principle no need for a special multiplier. It is possible to introduce a programmed multiplication, thereby using the adding device, which is always an integral part of any computer.

With 50- and 100-bit shift registers in one integrated circuit, serial multipliers are quite feasible. Their only disadvantage is a long operating time.

18-5 DIVISION OF BINARY NUMBERS

In a number of general-purpose computers, the process of division of the dividend by a divisor, resulting in a quotient and a remainder, is merely a question of programming. These machines have not been provided with special circuits for division because there are certain iterative methods by means of which the rather difficult division process can be performed through addition or subtraction. The programming methods for division and the way of operation of a special circuit for the division are of course analogous.

The division of a dividend D by a divisor d giving a quotient q and a remainder r is defined by the following alogrithm:

$$D = qd + r \tag{18-10}$$

Just as in the ordinary division of decimal numbers, the difficult operation is the check of whether or not the divisor can be subtracted by a number not more than $b - 1$ times from a certain group of digits of the dividend, b being the radix of the number system used. Computers generally do not have the ability to perform this inspection of two numbers unless one is willing to pay a penalty in time and hardware for this feature.

Fortunately the division of two binary numbers is simple enough that the divisor can be subtracted or not subtracted from a certain group of digits of the dividend, because in this number system $b = 2$. Furthermore, the binary number system has a unique feature which allows division to be performed in a straightforward operation.

In the so-called restoring division process, the following typical steps can be found to be used:

Step 1	$D - d \cdot 2^{m-n} = +E$	(quotient digit 1, value 2^{m-n})
Step 2	$E - d \cdot 2^{m-n-1}$	(results in a negative partial remainder) (quotient digit 0, value 2^{m-n-1})
Step 3	$E - d \cdot 2^{m-n-2} = +G_1$	(quotient digit 1, value 2^{m-n-2})

After each step of the division process, the divisor is shifted over one digit place. In the first and third step of this example, the subtraction yields a positive partial remainder. In the second step of this example, however, it is assumed that the subtraction will yield a negative partial remainder. There are three ways to deal with this difficulty.

1. The subtraction of divisor and partial remainder is always performed; and after the test of a negative partial remainder, the divisor can be added in the same position to that negative partial remainder, thus restoring the last positive partial remainder. After that, the divisor is shifted over one digit place and the process is tried again.
2. The restoring process can be replaced by a testing process performed by means of a comparing circuit. When the comparing circuit test reveals that the subtraction will yield a negative partial remainder, the divisor is shifted over one digit place without performing the subtraction. After that, the comparing circuit tests again, and so on. (In method 1 one is paying a penalty in time; in method 2, in the hardware of the comparing circuit.)
3. The division process can also be performed in the following way. The divisor is subtracted from the dividend without a preceding test of whether or not this will yield a positive partial remainder. When this partial remainder proves to be positive, a quotient digit 1 has to be registered and the divisor is shifted over one digit place. After that, the process is repeated. However, when the partial remainder proves to be negative, a quotient digit 0 is registered and the divisor is also shifted over one digit place. In the next operation, the divisor is added to the negative partial remainder just obtained. The resulting partial remainder in this so-called "nonrestoring" division process is of course the same as that obtained on the same place in the restoring division process. The fact that these two different division processes yield the same result depends on the property of the binary number system that

$$2^p - 2^{p-1} = 2^{p-1} \tag{18-11}$$

Using the same general type of example, the following three steps are successively performed in the nonrestoring division.

Step 1 $D - d \cdot 2^{m-n} = +E$ (quotient digit 1, value 2^{m-n})

Step 2 $E - d \cdot 2^{m-n-1} = -F$ (quotient digit 0, value 2^{m-n-1})

Step 3 $-F + d \cdot 2^{m-n-2} = +G_2$ (quotient digit 1, value 2^{m-n-2})

The division process performed so far can be represented in the following formulas:

$$D - d \cdot 2^{m-n} - d \cdot 2^{m-n-1} + d \cdot 2^{m-n-2} = +G_2 \tag{18-12}$$

or

$$D - d \cdot 2^{m-n} - d \cdot 2^{m-n-2} = G_2 = G_1 \tag{18-13}$$

There are a number of aims to be kept in mind when designing a circuit for the division of binary numbers.

1. When designing a serial-parallel division circuit, the number of digits of the parallel adder or subtracter should preferably equal the number of digits of the divisor.
2. Circuits for comparing two binary numbers should preferably be avoided. These types of circuits are rather expensive with respect to the amount of hardware they require.

The division can be performed basically in three different ways:

1. All the operations can be performed simultaneously so that a parallel division circuit is obtained. Dividend and divisor are given on a parallel basis.
2. The operations can be performed on a serial-parallel basis, with the divisor being supplied in parallel and a number of digits of the dividend equal to the number of digits of the divisor but the rest given in serial form.
3. The division can be split up completely into a series of operations and performed by means of a one-digit serial adder or subtracter combined with a number of shift registers. Dividend and divisor, being available in shift registers, appear for purposes of processing in serial form only.

The general form of the division process in accordance with point 2 is as shown in Fig. 18-16. In this figure, it is assumed that the third processing step would yield a negative partial remainder. Divisor d is shifted there over two places. The indices of the partial remainder jump there from 2 to 4. Moreover, in the example of Fig. 18-16, no difficulties can arise from the representation of negative numbers because these numbers are avoided.

In the example shown in Fig. 18-17, the general way of division in accordance with point 3 is demonstrated. In this type of division, negative

Divisor	Dividend	Quotient
$d_3 \; d_2 \; d_1 \; d_0$	$D_7 \; D_6 \; D_5 \; D_4 \; D_3 \; D_2 \; D_1 \; D_0$	$q_4 2^4 + q_3 2^3 + q_1 2^1 + q_0$

$$
\begin{array}{cccccccc}
D_7 & D_6 & D_5 & D_4 & D_3 & D_2 & D_1 & D_0 \\
d_3 & d_2 & d_1 & d_0 & & & & \\
\hline
{}^1D_7 & {}^1D_6 & {}^1D_5 & {}^1D_4 & D_3 & & & \\
 & d_3 & d_2 & d_1 & d_0 & & & \\
\hline
 & {}^2D_6 & {}^2D_5 & {}^2D_4 & {}^2D_3 & D_2 & D_1 & \\
 & & & d_3 & d_2 & d_1 & d_0 & \\
\hline
 & & & {}^4D_4 & {}^4D_3 & {}^4D_2 & {}^4D_1 & D_0 \\
 & & & & d_3 & d_2 & d_1 & d_0 \\
\hline
 & & & & {}^5D_3 & {}^5D_2 & {}^5D_1 & {}^5D_0
\end{array}
\quad \text{Remainder}
$$

Fig. 18–16

numbers cannot be avoided. In order to avoid difficulties with possible end-around carries or borrows, the twos-complement representation of negative numbers is used.

The only important difference between the division process shown in Figs. 18-16 and 18-17 is that in Fig. 18-17 a straightforward continuous process is shown using subtraction and addition after a negative partial remainder is obtained. If so desired, the subtraction can be avoided by using the complement representation of the divisor.

The four examples that follow illustrate the division processes given in Figs. 18-16 and 18-17 and their difficulties. In all cases, the division of 248 by 9 is used.

$$
\begin{array}{cccc|cccccccc|l}
\multicolumn{4}{c|}{\text{Divisor}} & \multicolumn{8}{c|}{\text{Dividend}} & \text{Quotient} \\
d_3 & d_2 & d_1 & d_0 & D_7 & D_6 & D_5 & D_4 & D_3 & D_2 & D_1 & D_0 & q_4 2^4 + q_3 2^3 + q_1 2^1 - q_0 \\
 & & & & d_3 & d_2 & d_1 & d_0 & & & & & \\
\cline{5-8}
 & & & & {}^1D_7 & {}^1D_6 & {}^1D_5 & {}^1D_4 & D_3 & & & & - \\
 & & & & & d_3 & d_2 & d_1 & d_0 & & & & \\
\cline{5-9}
 & & & & & {}^2D_6 & {}^2D_5 & {}^2D_4 & {}^2D_3 & D_2 & & & - \\
 & & & & & & d_3 & d_2 & d_1 & d_0 & & & \\
\cline{6-10}
 & & & & & & {}^3D_5 & {}^3D_4 & {}^3D_3 & {}^3D_2 & D_1 & & - \\
 & & & & & & & d_3 & d_2 & d_1 & d_0 & & \\
\cline{7-11}
 & & & & & & & {}^4D_4 & {}^4D_3 & {}^4D_2 & {}^4D_1 & D_0 & + \\
 & & & & & & & & d_3 & d_2 & d_1 & d_0 & \\
\cline{8-12}
 & & & & & & & & {}^5D_3 & {}^5D_2 & {}^5D_1 & {}^5D_0 & - \quad \text{Remainder}
\end{array}
$$

Fig. 18–17

Example 1

$$
\begin{array}{rcccc}
 & \text{Divisor} & \text{Dividend} & \text{Quotient} & \\
 & 1001\,| & 11111000\,| & 11011 & \\
 & & 1001 & & \\
 & & - \overline{} & & \\
248 - 9\cdot 2^4 = 104 & & 1101 & & \\
 & & 1001 & & \\
 & & - \overline{} & & \\
104 - 9\cdot 2^3 = 32 & & 10000 & & \\
 & & 1001 & & \\
 & & - \overline{} & & \\
32 - 9\cdot 2^1 = 14 & & 1110 & & \\
 & & 1001 & & \\
 & & - \overline{} & & \\
14 - 9\cdot 2^0 = 5 & & 0101 & \text{Remainder} &
\end{array}
$$

In this example, no negative numbers can come into being during the division process. It corresponds to the division process given in Fig. 18-16.

In the next example, however, a negative partial remainder will be obtained that has to be considered as a negative number in the twos-complement representation so that the overflow borrow may be omitted without any more. Special sign digits are not required in this example. The encircled overflow digits may be considered as such.

Example 2

	Divisor	Dividend	Quotient
	1 0 0 1	1 1 1 1 1 0 0 0	1 1 0 1 1
		1 0 0 1	
		−	
$248 - 9 \cdot 2^4 = 104$		(0)1 1 0 1	
		1 0 0 1	
		−	
$104 - 9 \cdot 2^3 = 32$		(0)1 0 0 0	
		1 0 0 1	
		−	
$32 - 9 \cdot 2^2 = -4$		(1)1 1 1 0	
		1 0 0 1	
		+	
$-4 + 9 \cdot 2^1 = +14$		(0)1 1 1 0	
		1 0 0 1	
		−	
$14 - 9 \cdot 2^0 = +5$		(0)1 0 1	Remainder

At first sight, the binary form of the partial remainder -4 might not be recognized. In full this partial remainder is 1 1 1 1 0 0 (or -4 in twos-complement representation). In the short-hand notation used in the division process the 0's are indicated in the partial remainders only insofar as they are required in the corresponding step of the operation.

It is of course possible to use only addition or subtraction in the division process. In using addition only, the addition of a positive and a negative number will always yield an overflow digit 1 when their sum is greater than 0 and an overflow digit 0 when their sum is smaller than or equal to 0 [see Eqs. (17-44) to (17-47)]. Thus in all cases in which a positive partial remainder will be obtained, this overflow digit can be used immediately as the quotient digit, which can be shifted into the quotient register.

In all cases in which the partial remainder is positive, the twos complement of the divisor has to be used. This will not lead to difficulties because in the twos-complement representation of negative numbers a number can be multiplied by a power n of 2 by shifting this number over n digit places to the left and filling these places with 0's. These 0's play no further role in the division process. The adding in the division process can thus be limited to a number of digits equal to the number of digits in the divisor. The example of 248:9 is now repeated here, using addition only.

Example 3

```
              Divisor        Dividend
             0 1 0 0 1 | 0 1 1 1 1 1 0 0 0 |
                         1 0 1 1 1
                         ---------- +
                      [1]0 0 1 1 0 1
                           1 0 1 1 1
                           ---------- +
                      [1][1]0 0 1 0 0 0
                             1 0 1 1 1
                             ---------- +
                      [1][1][0]1 1 1 1 1 0
                               0 1 0 0 1
                               ---------- +
                      [1][1][0][1]0 0 1 1 1 0
                                 1 0 1 1 1
                                 ---------- +
Quotient              [1][1][0][1][1]0 0 1 0 1      Remainder
```

In the example above the division was performed in a number of steps equal to the number of digits of the quotient. However, it can happen in nonrestoring division that a negative remainder is generated. This will be found to occur in cases in which the last quotient digit becomes 0. The correct positive remainder can then be obtained by adding the divisor to the negative remainder in an extra processing step. This will be illustrated in Example 4 by the division 240 : 9.

Example 4

```
                   Divisor        Dividend

                  0 1 0 0 1 | 0 1 1 1 1 0 0 0 0 |
                              1 0 1 1 1
                              ---------- +
                           [1]0 0 1 1 0 0
                                1 0 1 1 1
                                ---------- +
                           [1][1]0 0 0 1 1 0
                                  1 0 1 1 1
                                  ---------- +
                           [1][1][0]1 1 1 0 1 0
                                    0 1 0 0 1
                                    ---------- +
                           [1][1][0][1]0 0 0 1 1 0
                                      1 0 1 1 1
                                      ---------- +
Quotient = 26 =            [1][1][0][1][0]1 1 1 0 1      Negative remainder, −3
                                      0 1 0 0 1
                                      ---------- +
                                      0 0 1 1 0      Positive remainder, +6
```

It is of course also possible in this type of division to look one step back for the correct remainder and to omit the extra step used in the example above.

18-6 PARALLEL DIVIDERS

In Fig. 18-18, a block diagram of a parallel divider is given for the division of a dividend with six significant and one sign digit by a divisor with three significant and one sign digit. Quotient and remainder have the same length as the divisor. For shorter divisor numbers, the number of digits of the quotient will increase and that of the remainder will decrease. In the block

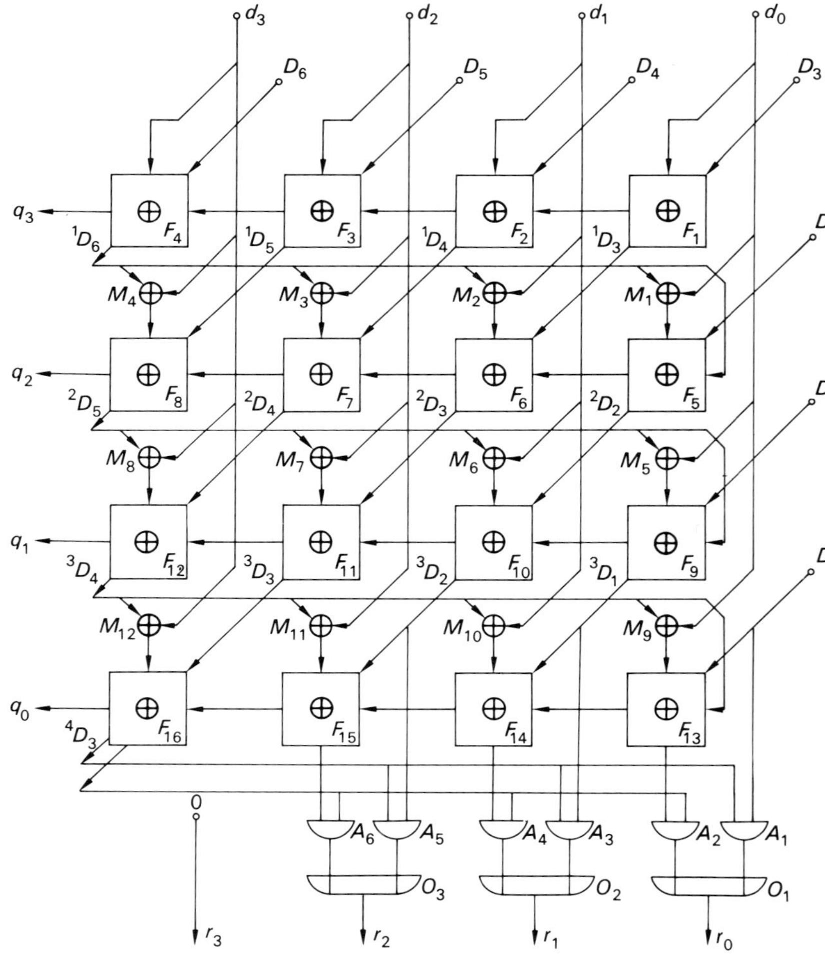

Fig. 18–18

diagram of Fig. 18-18, there is no self-adaptation to variations in length of dividend and divisor.

Dividend D_0 to D_6 in its positive form is fed into the block diagram of Fig. 18-18 via terminals with the same indications. Divisor d_0 to d_3 in its twos-complement representation is fed to the terminals with corresponding indications. In the first row of adders F_1 to F_4, dividend digits D_3 to D_6 and divisor digits d_0 to d_3 are added, resulting in quotient digit q_3 and ${}^1D_6$ which is the sign digit of the first partial remainder. When this sign ${}^1D_6$ is 0, a positive partial remainder is obtained, so that in the operation of the adders in the second row (adders F_5 to F_8) the divisor has again to be used in its twos-complement representation. However, when this digit ${}^1D_6$ is 1, indicating that a negative partial remainder has been obtained, the positive form of the divisor has to be used.

It should be noted that the twos-complement form TC of a number N is obtained by adding 1 to the ones-complement form OC. Number N can be obtained from its twos-complement form in the following way.

$$N' = OC$$

$$TC = OC + 1 = N' + 1$$

or

$$TC - 1 = N'$$

or

$$TC' + 1 = N \tag{18-14}$$

Thus, the positive form of a number N can be obtained from the twos-complement representation by inverting all digits and adding 1. In the second row of the block diagram of Fig. 18-18, this property is used to change the twos-complement representation of the divisor into the actual positive number. When a 1 results from the addition in F_4, mod 2 adders M_1 to M_4 invert digits d_0 to d_3 of the divisor; this 1 is also fed as a carry into adder F_5 of the second row of adders.

In the second row of adders in block diagram 18-18, digit D_2 and digits ${}^1D_3$ to ${}^1D_5$ form the partial remainder to be processed with the divisor or its twos-complement representation, depending on the result of the addition performed in the first row of adders.

The relations between any pair of successive rows in the block diagram of Fig. 18-18 are the same. When the last quotient digit q_0 is 1, a positive remainder will result; however, when q_0 is 0, a negative remainder has been obtained. In that case, digits ${}^3D_2$, ${}^3D_1$, D_0 are the correct significant digits. Gates A_1 to A_6 and O_1 to O_3 connect the correct remainder to outputs r_0 to r_2. There is no need for any switching in the output of the sign digit r_3 of the remainder. That has to be a 0 because the remainder must always be positive.

18-7 SERIAL-PARALLEL DIVIDERS

Many variations are possible in serial-parallel dividers. A couple of examples will be given here. When the use of a comparing circuit is allowed, the diagram of Fig. 18-19 can be used to accomplish the division process illustrated in Fig. 18-16.

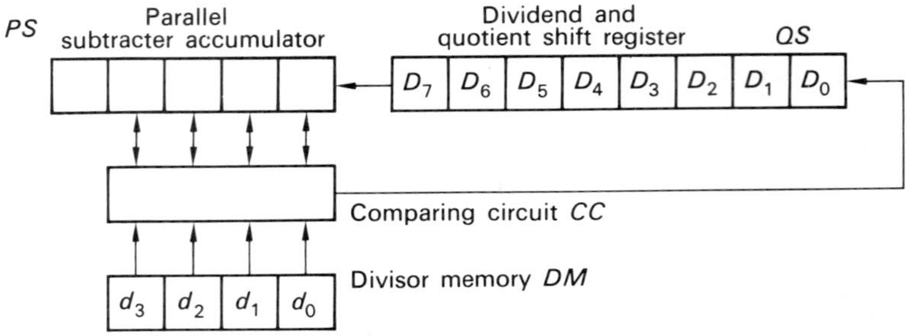

Fig. 18–19

In the block diagram of Fig. 18-19, the divisor is presented by register *DM*. The dividend is initially placed in the quotient shift register *QS* and during the division process is shifted into and through the parallel subtracting accumulator *PS*. On each shifting step, comparing circuit *CC* checks whether or not it is possible to subtract the divisor from the dividend digits so far shifted into *PS*. Each time this subtraction proves not possible, a 0 is shifted into *QS* and one next dividend digit is shifted into *PS*.

Each time the check circuit *CC* finds the subtraction possible, this subtraction is performed, a 1 is fed into *QS*, and the next dividend digit is shifted into *PS*.

At the end of the division process, the quotient is found correctly placed in *QS* and the digits in the parallel subtracter form the remainder.

It should be noted that the parallel subtracter contains an extra section for storing one digit. This section is necessary in cases where the divisor starting with a 1 is larger than the corresponding digits of the dividend or partial remainder, which also has a 1 as highest-valued digit. In such a case, the subtraction proves not possible, so that this 1 is shifted into the extra section. The subtraction is then always possible.

In this type of block circuit, no difficulties arise with variations in length of dividend and divisor. The finding of the correct relative position of divisor and dividend is automatically performed by feeding 0's into *QS* as long as the comparing circuit finds the subtraction not possible. When the divisor and dividend have reached their correct relative position for the beginning of the actual division, comparing circuit *CC* generates its first 1 automatically.

It should be noted that the subtraction in the parallel subtracter accumulator in which use is made of MS memory elements is performed in such a way that on a clock pulse the result is stored one digit position to the left, while the next D digit is fed in at the same time.

The serial-parallel divider shown in Fig. 18-20 as a block diagram does not differ very much from the one given in Fig. 18-19; this circuit operates basically in accordance with the dividing scheme given in Fig. 18-19.

In this serial-parallel adder, the highest-valued digits of dividend and divisor are supposed to be sign digits. Use is made of addition only. By means

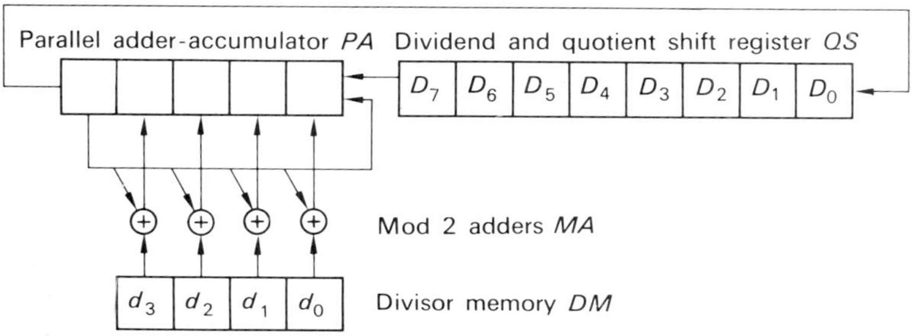

Fig. 18–20

of mod 2 adders MA, the positive representation of the divisor can be changed in the twos complement or the reverse. The required extra 1 is entered in the parallel adder as a carry into the first section.

The parallel addition in PA is performed in such a way that the result is shifted one digit position to the left so that the next dividend digit can be fed in simultaneously.

The block circuit of Fig. 18-20 presents no difficulties in finding the correct relative position of divisor and dividend. This feature has already been described for the circuit of Fig. 18-19.

At the end of the division process, the remainder is found in the parallel adder PA. If this remainder is negative, the divisor in its positive representation has still to be added to find the correct positive remainder, which presents some difficulties in the organization of the dividing process.

18-8 SERIAL DIVIDERS

A serial divider is in certain respects analogous to a serial multiplier. The main circuit is formed by four n-bit shift registers, a full adder, a carry store, and a number of gates controlling the sequence of all operations. In Fig. 18-21, a block circuit of a serial divider is shown for the division of a dividend of $2n$ bits by a divisor of n bits. In this diagram the various clock-pulse circuits are not shown; only the clock inputs of the gates are indicated in this figure. The operating moments of the gates are indicated by the numbers of the corresponding steps in the flow diagram.

It should be noted that the highest-valued digits of dividend and divisor are sign digits, which are very usful in the nonrestoring division process to be used in the serial circuit.

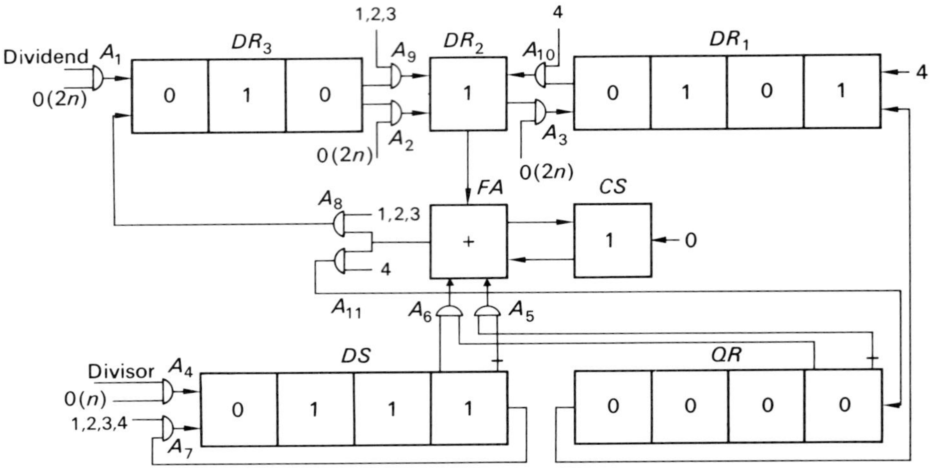

Fig. 18–21

Before the actual division process starts, divisor and dividend have to be lined up; or, in other words, the most significant 1 of the divisor in its positive representation has to be placed under the most significant 1 of the dividend also in its positive form. The lining-up procedure will not be discussed in the following description of the serial divider, nor will the difficulties arising from dividend and divisor numbers of other lengths than $2n$ and n, because they are no part of the actual division process.

In Fig. 18-21, shift registers DR_1, DR_2, and DR_3 contain the $2n$-bit dividend, the lowest-valued bit in the most-right-hand section of shift register DR_1. The most-left-hand section of shift register DR_3 contains the sign digit of the dividend. Via gates A_1, A_2, and A_3, the dividend can be fed into these shift registers during a series of $2n$ clock pulses indicated by $0(2n)$. Shift register DS contains the n-bit divisor with the lowest-valued bit in the most-right-hand place. The divisor can be fed into this shift register by a series of n clock pulses indicated by $0(n)$ on one of the inputs of gate A_4. It is assumed that both dividend and divisor are given in their positive representation. This means that when using the addition as main operation in the division process, the negative representation of the dividend must be used in the first group of operations. It follows from the discussion in the preceding paragraphs that the twos-complement representation can best be used for the negative representation of the numbers.

The quotient digits produced in the division process show in fact whether or not the subtraction of the divisor from the dividend or the partial remainder was possible, which means that the quotient digits determine whether or not the positive representation of the divisor must be used in the next group of operations, i.e., the next subtraction of the divisor from the partial remainder. The choice between the positive and negative form of the divisor thus can be controlled by the lowest-order section of quotient register QR. This control of the representation of the divisor can be limited in the case of a complete serial operation to the digit in the lowest-order section of register DS. For this purpose, gates A_5 and A_6 have been provided in the connections between this section of DS and full adder FA.

It follows from the preceding discussion that the carry store CS must be primed with a 1 when the twos complement of the divisor has to be used for a group of operations (subtraction of the divisor). This 1 is necessary in the first step only of the group of operations. When starting the division process, carry store CS can be primed into the 1 state by a clock pulse 0. During the division process, no priming is required because when in the last step of a group of operations (subtraction of the last digit of the divisor) carry store CS is set to 1, the twos complement of the divisor must be used in the next group of operations.

In the block circuit of Fig. 18-21, the division process is completed in $n^2 + n$ clock pulses after the various shift registers have been primed with dividend and divisor. In this example, shift register DS recycles in four steps, so that a new group of operations can be started for each four clock pulses. On clock pulses 1, 2, and 3 of a group, the digits of shift register DR_3 are shifted via gate A_9 into the single shift section DR_2. This shift section presents its contents for processing to full adder FA. The circuit produces a sum digit of the partial remainder, which is shifted into register DR_3 via gate A_8, and possibly a carry digit, which is temporarily stored in the carry store CS.

A group of processing operations ends with the fourth clock pulse. On this clock pulse, the next digit of shift register DR_1 is shifted into DR_2 as a preparation for the next group of operations. The sum digit produced in the last operation of a group is shifted via gate A_{11} into quotient register QR. It should be noted that this sum is the complement of the desired quotient digit. This presents no difficulty when using shift registers with single inputs.

In this way, the serial division process can be completed in $n^2 + n$ clock pulses. It should be noted that when using divisors with less digits, the quotient will get more digits. These extra digits can be stored in DR_1. When a negative remainder is obtained, an extra group of operations is required in order to obtain the correct positive remainder, thus increasing the number of operating steps from $n^2 + n$ to $n^2 + 2n$. The gating requirements for this extra group of operations are not shown in the block diagram of Fig. 18-21.

Example

Dividend: 0 1 0 1 0 1 0 1 = 85 Divisor: 0 1 1 1 = 7

In the following flow chart, shift registers *DR*, *DS*, *QR*, and carry store *CS* only will be shown. The result of the processing of the digits in one situation will be given in the next situation.

1.0	*DR*	[0 1 0 1]		[0 1 0 1]				
			[1]					
	DS	[1 0 0 0]	*CS*	[0 0 0 0] *QR*				
1.1		0 0 1 0		0 1 0 1	2.1	1 1 1 0		1 0 1 –
			1				0	
		0 1 0 0		0 0 0 0		1 0 1 1		0 0 0 $\underline{0}$
1.2		1 0 0 1		0 1 0 1	2.2	1 1 1 1		1 0 1 –
			0				0	
		0 0 1 0		0 0 0 0		1 1 0 1		0 0 0 $\underline{0}$
1.3		1 1 0 0		0 1 0 1	2.3	0 1 1 1		1 0 1 –
			0				1	
		0 0 0 1		0 0 0 0		1 1 1 0		0 0 0 $\underline{0}$
1.4		1 1 0 0		1 0 1 –	2.4	0 1 1 1		0 1 – –
			0				1	
		0 1 1 1		0 0 0 $\underline{0}$		1 0 0 0		0 0 $\underline{0}$ 1
3.1		0 0 1 1		0 1 – –	4.1	1 0 0 0		1 – – –
			1				0	
		0 1 0 0		0 0 $\underline{0}$ 1		0 1 0 0		0 $\underline{0}$ 1 1
3.2		0 0 0 1		0 1 – –	4.2	0 1 0 0		1 – – –
			1				0	
		0 0 1 0		0 0 $\underline{0}$ 1		0 0 1 0		0 $\underline{0}$ 1 1
3.3		0 0 0 0		0 1 – –	4.3	0 0 1 0		1 – – –
			1				0	
		0 0 0 1		0 0 $\underline{0}$ 1		0 0 0 1		0 $\underline{0}$ 1 1
3.4		0 0 0 0		1 – – –	4.4	0 0 1 1		– – – –
			1				0	
		1 0 0 0		0 $\underline{0}$ 1 1		0 1 1 1		$\underline{0}$ 1 1 0
5.1		0 0 0 1		– – – –	6.1	1 1 0 1		– – – $\underline{0}$
			1				0	
		1 0 1 1		$\underline{0}$ 1 1 0		1 0 1 1		1 1 0 0
5.2		1 0 0 0		– – – –	6.2	0 1 1 0		– – – $\underline{0}$
			1				1	
		1 1 0 1		$\underline{0}$ 1 1 0		1 1 0 1		1 1 0 0
5.3		0 1 0 0		– – – –	6.3	0 0 1 1		– – – $\underline{0}$
			1				1	
		1 1 1 0		$\underline{0}$ 1 1 0		1 1 1 0		1 1 0 0 Quotient
5.4		1 0 1 0		– – – $\underline{0}$	6.4	0 0 0 1		Remainder
			0					
		0 1 1 1		1 1 0 0				

18-9 HIGH-SPEED ARITHMETIC IN MULTIPLIERS

The serial-parallel multipliers discussed in the preceding paragraphs show in their repetitive addition a combined adding and shifting operation requiring n operation steps (n being the number of digits of the multiplier) to complete the multiplication. In serial-parallel multipliers in which adding and shifting are separate operations, there is a possibility of decreasing the operation time required by taking advantage of some special properties of the binary number system so that less than $2n$ operation steps are required to perform a multiplication.

The simplest property of the binary number system that can lead to an increased operating speed is the fact that on the average, half of the digits of a binary number will be 0. It is possible to shift over these 0 digits in the multiplier. However, this will not decrease operating time by 50 percent because the shifting process will also take some time.

Another possible improvement is of a more complex nature, as was demonstrated by MacSorley. The ordinary binary representation of a number can be transformed in a signed-digit minimal binary representation. Any group of $p + 1$ successive 1's in a binary number can be replaced by two 1's, as defined by the following basic equation:

$$2^n + \cdots + 2^{n-p} = 2^{n+1} - 2^{n-p} \tag{18-15}$$

For $p = 1$, this equation reduces to

$$2^n = 2^{n+1} - 2^n \tag{18-16}$$

Equation (18-15) states that each group of $p + 1$ successive 1's (preceded and followed by a pair of 0's at least) in a binary number can be replaced by a group of p 0's on digit places n to $n - p + 1$, while the lowest-valued 1 has to be replaced by a -1 and the 0 preceding the group of $p + 1$ 1's has to be replaced by a $+1$.

Example

0 0 1 1 1 1 0 0	Binary representation
0 +1 0 0 0 −1 0 0	Signed-digit representation using Eq. (18-15)

When the digits of the ordinary binary representation are indicated by B and those of the signed-digit representation by b, the following equation, which can be easily verified, gives b as a function of B:

$$b_n = (-1)^{B_{n+1}}(B_n \oplus B_{n-1}) \tag{18-17}$$

However, when the group of successive 1's in the binary representation is followed or preceded by another group of 1's separated by a single 0, the following transformation of digits will be obtained when applying Eq. (18-15).

$\ldots 1 \quad 1 \quad 0 \quad 1 \quad 1 \ldots$ Binary representation

$\ldots 0 \; -1 \; +1 \quad 0 \quad 0 \ldots$ Signed-digit representation

Using Eq. (18-16) in the reverse direction, this last representation can be improved to the following minimal representation:

$\ldots 0 \quad 0 \; -1 \quad 0 \quad 0 \ldots$ Minimal signed-digit representation

Thus, in the minimal signed-digit representation no two adjacent digits are 1. Moreover, application of Eq. (18-16) makes the higher-valued digits in the signed-digit representation dependent on the value of the next-lower digit already obtained. If this next-lower digit produced by Eq. (18-15) is already a 1, the corresponding higher-valued digit in the signed-digit representation must be 0. Equation (18-17) can be adapted to that requirement by multiplication with b'_{n-1}, as can be easily verified. The resulting transformation equation is then as follows:

$$b_n = (-1)^{B_{n+1}} (B_n \oplus B_{n-1}) \cdot b'_{n-1} \tag{18-18}$$

Factor b'_{n-1} leaves all digits produced by Eq. (18-17) untouched, except the -1 digits generated by Eq. (18-15), on a place one higher in value than the 1 digit determined by that same equation as the highest-valued digit of the lower-order group of 1's, because all the other 1's are generated on places where the neighboring lower-valued digit b is a 0.

Example

$118 = 2^6 + 2^5 + 2^4 + 2^2 + 2^1 = \quad 1 \quad 1 \quad 1 \quad 0 \quad 1 \quad 1 \quad 0$
Binary representation

Equation (18-15): $118 = 1 \quad 0 \quad 0 \; -1 \quad 1 \quad 0 \; -1 \quad 0$
Signed-digit representation

Equation (18-16): $118 = 1 \quad 0 \quad 0 \quad 0 \; -1 \quad 0 \; -1 \quad 0$
Minimal representation

The number of 1's in the minimal representation of 118 has decreased from 5 to 3; however, the number of digits has increased by 1—but this is not objectionable.

It should be noted that in a number of cases a signed-digit representation can be obtained by application of Eq. (18-15) only, this representation having already the minimum number of 1's. Furthermore, application of Eq. (18-15) may be started either from the high-order end or from the low-order end of the binary number. The resulting powers of 2 and their signs will differ in a number of cases: however, the resulting number of powers of 2 will be the same in both cases.

Example

$$58 = 1\ 1\ 1\ 0\ 1\ 0 = 2^6 - 2^3 + 2^2 - 2^1$$

Equation (18-15) from left to right :: $2^6 - 2^2 - 2^1 = 64 - 4 - 2 = 58$

Equation (18-15) from right to left :: $2^6 - 2^3 + 2^1 = 64 - 8 + 2 = 58$

The first form can be transformed in the second form by application of Eq. (18-15) for a second time.

It should be noted that application of Eq. (18-15) on all groups of 1's and single 1's of the binary representation of a number changes its notation into the reflected binary notation of the double of the original number.

Example $29 = 1\ 1\ 1\ 0\ 1 = 2^4 + 2^3 + 2^2 + 2^0 = 2^5 - 2^2 + 2^1 - 2^0 = 1\ 0\ 0\ {-1}\ 1\ {-1}$

which is the reflected binary notation for $63 - 7 + 3 - 1 = 58$.

The question to ask now is: How can the above properties of the binary number system be used to increase the operating speed or to reduce the number of operations of a multiplier?

It is not necessary to convert both multiplicand and multiplier from the ordinary binary representation into the signed-digit representation. The conversion of the multiplier is sufficient for the speeding up of the multiplication because the number of 1's in the multiplier determines the number of operations to be performed. However, the marking of the multiplier digits with + and − signs is not very practicable, because the signs would have to be stored in addition to each 1. A ternary registration can be used for this purpose, but this also is not a very attractive plan.

When the conversion of the digits of the binary number is limited to the application of Eq. (18-15) only, the successive 1 digits, eventually spaced by one or more 0's, will have alternating signs, so that there is no need for ternary registration. Moreover, the number of cases to which Eq. (18-16) can be applied for a further reduction of the number of 1's is not very large.

MacSorley has shown that there is no need for a conversion of the multiplier from the ordinary representation into the minimal signed-digit representation if the operating rules given below are followed. In the application of these rules, the multiplier must be shifted through an indicator circuit that determines the kind of operation to be performed on the multiplicand. The number of shift steps of the multiplier through the indicator circuit is the same as the number of steps that the multiplicand has to be shifted in the formation of the partial products. If the lowest-order digit of the multiplier is a 1, the indicator circuit will treat it as 1 0, the 0 being an imaginary nonsignificant digit.

OPERATING RULES

1. When the indicator circuit tests 001 or 010 (which means tests shifting across 1's from right to left digit pair 00 or shifting across 0's digit pair 01), add the multiplicand and shift (jump) across all following 0's.
2. When the indicator circuit tests 101 or 110, subtract the multiplicand and shift across all following 1's.

It should be noted that after each operation of the indicator circuit at least two shift steps must be performed, because there are no adjacent 1 digits in the minimal representation. The operating rules above represent the exact translation of Eq. (18-18). The rules will be illustrated by means of an example.

Example

```
                  441 × 1365 = 601965
                  1 0 1 0 1 0 1 0 1 0 1   = 1365
                  0 0 1 1 0 1 1 1 0 0 1   = 441
                   +     -     -     +
                  _____________________ ×
  0 0 0 0 0 0 0 0 1 0 1 0 1 0 1 0 1 0 1
  1 1 1 1 1 1 0 1 0 1 0 1 0 1 0 1 0 1 1
  1 1 1 0 1 0 1 0 1 0 1 0 1 0 1 1 1 1 1
  1 0 1 0 1 0 1 0 1 0 1 0 0 0 0 0 0 0 0
  _____________________________________ +
1 0 1 0 0 1 0 0 1 0 1 1 1 1 0 1 1 0 1 0 1 1
  1 0 0 1 0 0 1 0 1 1 1 1 0 1 1 0 1 1 0 1   = 601695
```

Two of the additions in the example above are in fact subtractions that have been transformed into additions by using the ones-complement representation. This causes an end-around carry in the final sum of the example. The example contains all the possibilities inherent in the rules given above. The number of adding operations is reduced from six to four; three elementary shift steps can be performed each time as a jump; the cost in hardware is rather great; a double-length accumulator and an indicator circuit are required; and the possibility of taking a variable number of shift steps has to be taken into account. The example was so chosen that a gain in speed was reached; however, in many multiplications no, or almost no, gain will be obtained, making it questionable whether the amount of hardware and complication in the operation are worthwhile.

PROBLEMS

1. Explain the multiplication of two binary numbers by means of a generalized example and show that the number of digits of the product is at most equal to the sum of the number of digits of multiplicand and multiplier.

2. Give a general organization scheme of a multiplier for two binary numbers which could be used as a basis for a parallel multiplier.

3. Transform the organization diagram resulting from Prob. 2 into a symbolic circuit of a parallel multiplier. Determine an equation for the maximum propagation time of the carry in such a circuit as a function of the number of digits of multiplicand and multiplier.

4. Explain the principle of a serial-parallel multiplier by means of a block diagram. Illustrate the operation of the circuit by means of an example.

5. Determine the switching equations of the sum memory element and the carry memory element of the kth section of a serial-parallel multiplier and draw a symbolic circuit of this section using MS-memory-element techniques.

6. Explain the principle of a serial-parallel multiplier operating by means of a parallel adder. Illustrate the operation of the circuit by means of an example.

7. Draw the circuit of a serial multiplier and explain its operation.

8. Give the general diagram of a serial-parallel divider.

9. Draw a block diagram of a circuit operating in accordance with the diagram of Prob. 8.

10. Show that the number of comparing sections of the circuit of Prob. 9 need not be higher than the number of digits of the divisor if the dividend instead of the divisor is shifted during the operation.

11. Show by means of an example that the division of two binary numbers can be organized as a repeated straightforward operation.

12. Show that the division process resulting from Prob. 11 can be implemented in a circuit containing a parallel adder with the number of sections equal to the number of digits of the divisor. Determine the correct remainder.

13. Show that when using signed digits in the binary representation of numbers the number of operations in a multiplier can be decreased. Multiply 29 (multiplier) and 23 (multiplicand).

SELECTIVE BIBLIOGRAPHY

1. Garner, H. L.: A Ring Model for the Study of Multiplication for Complement Codes, *IRE Trans.*, vol. EC-8, pp. 25–30, 1959.
2. Lehman, M.: Short-cut Multiplication and Division in Automatic Binary Digital Computers, *Inst. Elec. Engrs.* (*London*), pt. B, vol. 105, pp. 496–504, 1958.
3. MacSorley, O. L.: High-speed Arithmetic in Binary Computers, *Proc. IRE*, vol. 49, pp. 67–91, 1961.
4. Richards, R. K.: "Decimal Multiplication and Division," chaps. 5 and 9, D. Van Nostrand, Company, Inc., Princeton, N.J., 1955.
5. Robertson, J. E.: Two's Complement Multiplication in Binary Parallel Digital Computers, *IRE Trans.*, vol. EC-4, pp. 118–119, 1955.
6. Robinson, A. A.: Multiplication in the Manchester University High Speed Digital Computer, *Electron. Eng.*, vol. 25, pp. 6–10, 1953.
7. Shimson, M.: An Improved Technique for Fast Multiplication on Serial Digital Computers, *Electron. Eng.*, vol. 30, pp. 504–505, 1958.
8. The Staff of the Computation Laboratory: "Synthesis of Electronic Computing and Control Circuits," Harvard University Press, Cambridge, Mass., 1951.

19
Transmitters and Receivers of Binary-coded Information

19-1 OSCILLATORS

In information-processing techniques, binary-coded information must be sent from one place to another, e.g., from a machine to a printing mechanism, or from a tape to a machine or printer. Some of the problems inherent to the transmission of coded information will be discussed in this chapter. This topic has come to cover such a wide field that it is necessary to limit this chapter mainly to some problems of the start-stop transmission of binary information. Transmitters and receivers that are readily adaptable to the requirements of synchronous transmission can be designed essentially on the basis of the fundamental switching circuits described in the previous chapters. The square-wave oscillators used in these transmitters and receivers fall more or less outside the scope of the logical treatment of switching circuits that we have been following in the preceding chapters. However, it is essential to have a fairly good idea of such square-wave oscillators operating mostly in a start-stop mode before one can attempt to design the matching transmitting and receiving circuitry.

SQUARE-WAVE OSCILLATORS WITH TRANSISTORS

The time chart of a transistor start-stop square-wave oscillator is given in Fig. 19-1, and the corresponding circuit is shown in Fig. 19-2. The oscillator consists of two transistors ($T_{1/2}$) with a common emitter resistor R_5, collector resistors R_3 and R_4, and the frequency-determining circuits R_1/CC_1 and R_2/CC_2. The resistance of R_1 and R_2 can be of the order of $\frac{1}{2}\cdot 10^6$ ohms. The fact that the voltage drop between the emitter and collector of a transistor is very small makes it difficult to draw the various curves in the time chart

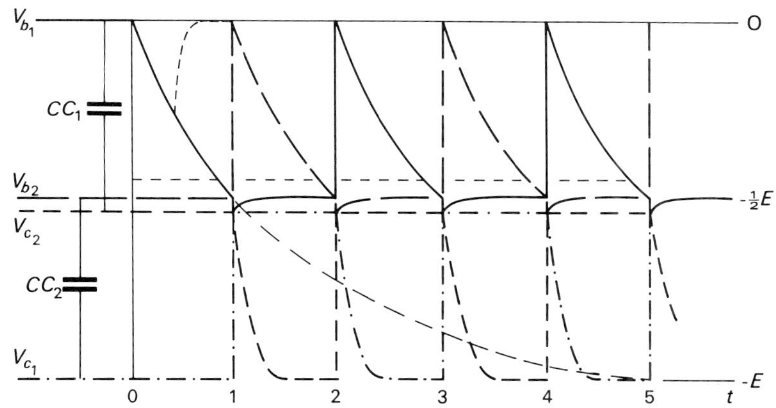

Fig. 19–1

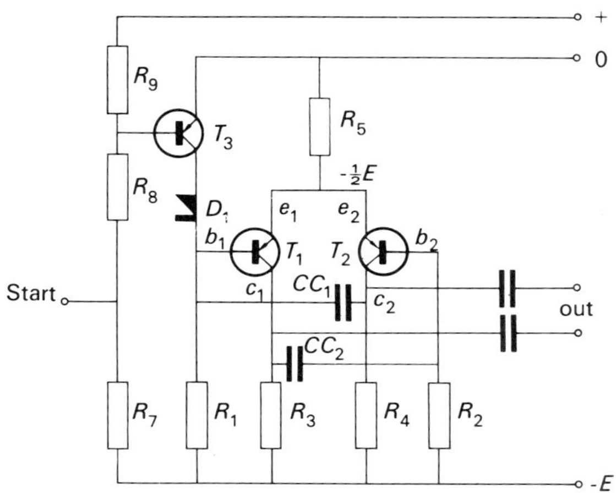

Fig. 19–2

of Fig. 19-1. They are shown in the middle of the diagram in their true proportional spacing: the distances between the horizontal parts of the characteristic voltages of bases and collectors in the middle of Fig. 19-1, however, have been blown up for the sake of clarity.

The starting circuit of the transistor oscillator is limited in Fig. 19-2 to a single transistor T_3 (which can be part of a trigger); it is conducting in the rest condition of the oscillator, which keeps the base voltage V_{b1} of transistor T_1 before moment $t = 0$ at 0 potential. This is the voltage level that base b_1 will have after each complete generated cycle. The duration of the first half-cycle will then be the same as that of the following half-cycles.

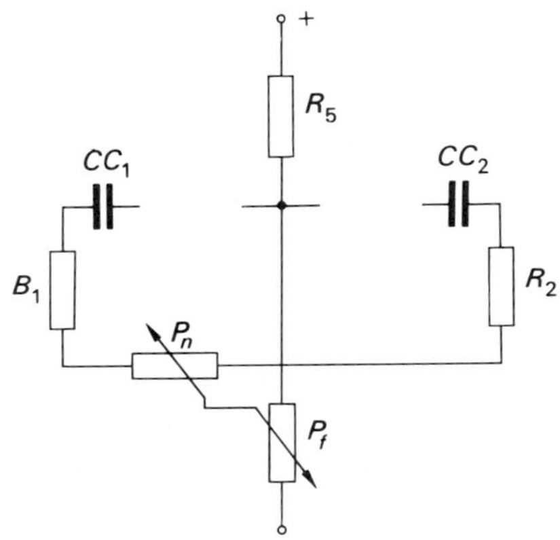

Fig. 19–3

The oscillator is started by feeding a start signal to the input terminal "start"; this makes transistor T_3 nonconducting. It also causes timing capacitor CC_1 to be charged via timing resistor R_1 in accordance with the e curve V_{b1} of Fig. 19-2. This curve is cut off at moment 1 when voltage V_{b1} equals $-\frac{1}{2}E$, which is the voltage of the bases of transistors T_1 and T_2 of the oscillator.

Transistor T_1, which was nonconducting in the rest condition of the oscillator, becomes conducting. A triggering action causes the collector voltage V_{c1} of this transistor to change rapidly from $-E$ to about $-\frac{1}{2}E$. The base voltage V_{b2} of transistor T_2 consequently changes from about $-\frac{1}{2}E$ to 0. (Collector c_1 of transistor T_1 and base b_2 of transistor T_2 are interconnected via capacitor CC_2.) This makes transistor T_2 nonconducting. Capacitor CC_1 is now rapidly charged via collector resistor R_4 and conducting transistor T_1, while the duration of the next half-cycle will be determined by the charging of timing capacitor CC_2 via timing resistor R_2. The base voltage level V_{b2} at moment 2 will be $-\frac{1}{2}E$. This again starts a triggering action which reverses the condition of the oscillator and puts it back into its initial state at moment 0.

The speed and neutrality of the oscillator of Fig. 19-2 can be adjusted by means of the arrangement of two voltage dividers P_f and P_n as shown in Fig. 19-3. The voltage level to which the capacitors will be charged can be adjusted by means of voltage divider P_f. Voltage divider P_f determines the voltage to which a capacitor would be charged if no triggering action of transistors T_1 and T_2 were to interfere. Voltage determines the frequency to be generated. By means of voltage divider P_n, the neutrality of the oscillator can be adjusted and the value of resistor R_1 can be increased slightly. (The value of resistor R_2 is decreased accordingly, and vice versa.) The adjustments of frequency and neutrality are practically independent of each other.

Two problems inherent in transistor oscillators should be mentioned here. When using germanium transistors as active elements in the oscillator, leak currents may flow in the collector path. This small collector current tends to reduce the capacitor voltages, which, in turn, affects the timing. This disadvantage is of little practical significance with respect to silicon transistors, for the leak currents in them are very small.

Another difficulty with respect to the oscillator of Fig. 19-2 is the rather high voltage difference between the emitters and collectors of transistors T_1 and T_2. This difference amounts to a little less than E volts, but it can be clipped at a little more than $\frac{1}{2}E$, as is shown in the circuit of Fig. 19-4.

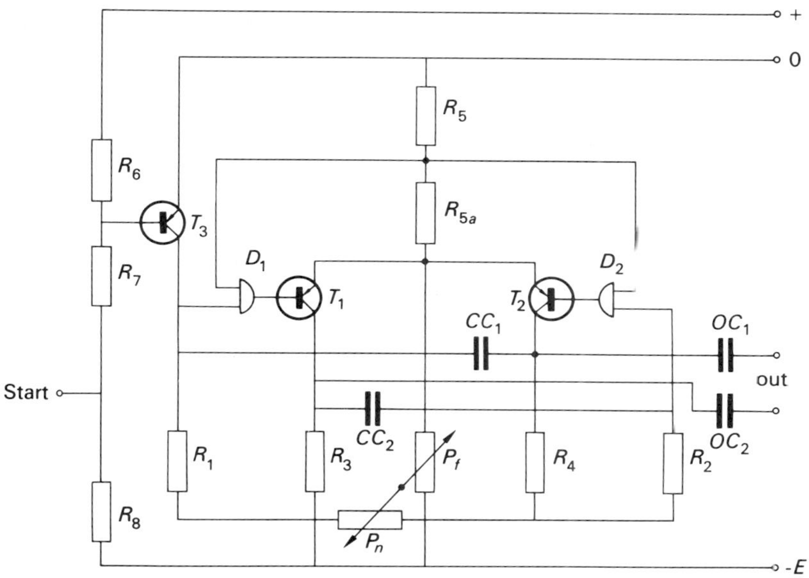

Fig. 19–4

The output terminals of AND gates D_1 and D_2 are connected to the bases of transistors T_1 and T_2. One of the input terminals of these AND gates is connected to a voltage slightly more positive than the emitter voltage, which is about $-\frac{1}{2}E$. The other input terminals of AND gates D_1 and D_2 are connected to the corresponding timing circuits R_1CC_1 and R_2CC_2. These gates ensure the clipping of base voltages V_{b1} and V_{b2} in accordance with the horizontal dotted lines shown in Fig. 19-1.

The clipping gates D_1 and D_2 tend to reduce the effect of the leak current of the transistors.

It should be noted that the oscillators have a typical feature; i.e., when the start is interrupted before the first state change of the trigger, the voltage of capacitor CC_1 is rapidly restored to its initial value without any further effect on the oscillator. Such a feature is essential in the case of an oscillator of a receiver in a start-stop system. A false start signal can thus be suppressed without having to let the receiver run through a complete receiving cycle.

Oscillators of the type described above cannot be loaded. The collectors of transistors T_1 and T_2 are connected via capacitors OC_1 and OC_2 to a trigger which can be loaded.

19-2 TRANSMITTERS OF BINARY START-STOP SIGNALS

A transmitter of binary information signals consists of a memory in which the information to be transmitted is stored and of the transmitting circuitry which performs the parallel-serial transformation of the information to be transmitted. According to their basic behavior, transmitting circuits can be classified as follows:

1. Continuously running transmitters
2. Transmitters operating in a start-stop mode

The continuously running transmitters produce a series of signals under the control of one start signal, whereas the start-stop transmitters have to be started for each signal to be transmitted. A transmitter of group 2 can be used in the mode of those in group 1, but a transmitter of group 1 cannot be used under start-stop control.

One could use as an oscillator for a transmitter of group 1 a high-quality standard generator, for example, that serves several other purposes; however, transmitters of group 2 have to be equipped individually with a start-stop oscillator.

Transmitters can also be subdivided into two groups according to the features of the memory:

a. Transmitters controlled by a memory in which the information signals are continuously available *b.* Transmitters controlled by a memory in which the information is available only for a short time interval during the transmission of a signal

The information to be transmitted can be stored in groups of triggers or in core matrices, in the form of the predetermined wiring of a sequence switch, or on magnetic or perforated paper tapes. If the memory produces a group of continuous signals as do contacts and triggers, there is no need for an auxiliary memory for the code combination to be transmitted in the transmitting circuit during the parallel-serial transformation; however, in transmitters controlled, say, by tape readers, which have a rather long stepping time, or by core matrices, such an auxiliary memory is necessary.

The first example of the design of a transmitting circuit will be of the type 2*b*. This transmitter is to be so designed that it can be used for the start-stop transmission of signals with five information elements preceded by a start element of so-called start polarity and followed by a stop element of so-called stop polarity. The duration of an information element and of a start element is normally the same; that of the stop element is generally taken to be about $1\frac{1}{2}$ times the duration of an ordinary signal element.

The most important parts of a type 2*b* transmitter are a start-stop oscillator, a distributor, and a memory. Experience alone will show that this type of transmitter can best be made by means of a seven-section shift register as shown in Fig. 19-6; these sections act at the same time as distributor and memory. A transmitter comprising a distributor and a separate memory would be uneconomical.

The symbolic circuit of this transmitter is shown in Fig. 19-6. The basic shift register of Fig. 16-9 has been used here. For the transmission of a start-stop signal with five information elements, the memory requires only six sections because the polarities of the start and stop element of the signal are always the same for each signal. If this six-section memory of a shift register is extended with one section, it can be used as a counting shift register, so that the functions of the distributor and memory of the transmitter can be combined into one circuit.

It should be noted that any of the other types of shift registers treated in Chap. 16 could rather easily be incorporated in the symbolic diagram of Fig. 19-6.

The transmission of signals by means of the transmitter shown in Fig. 19-6 can be started by means of start contact S_c. The operation of this contact is represented as line *a* in the time chart of Fig. 19-5. If this contact is closed permanently, a series of start-stop signals will be transmitted. If it is closed for a short moment, or opened during the transmission of a signal, the signal being transmitted will always be completed. The transmitter stops only at the end of this signal and waits for the next operation of start contact S_c. It should be noted that instead of a start contact S_c, one might also use an

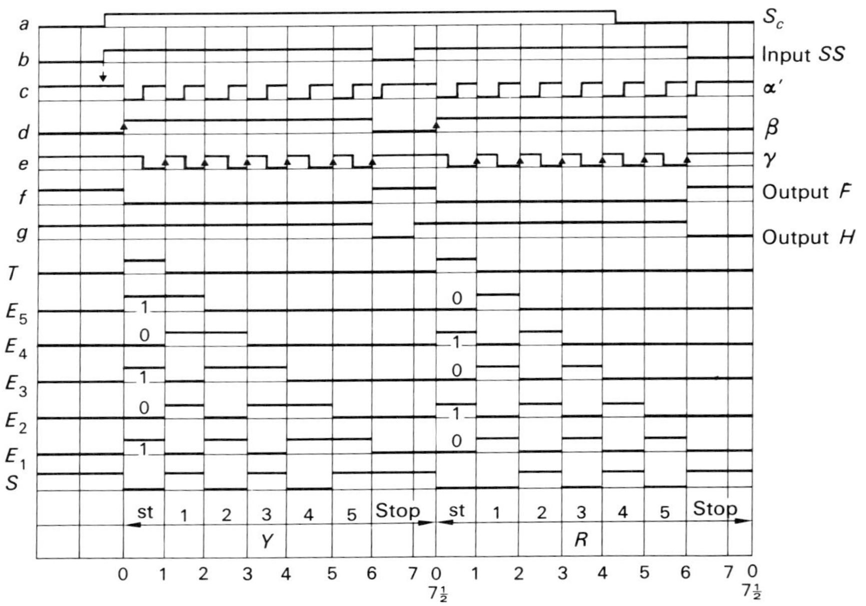

Fig. 19–5

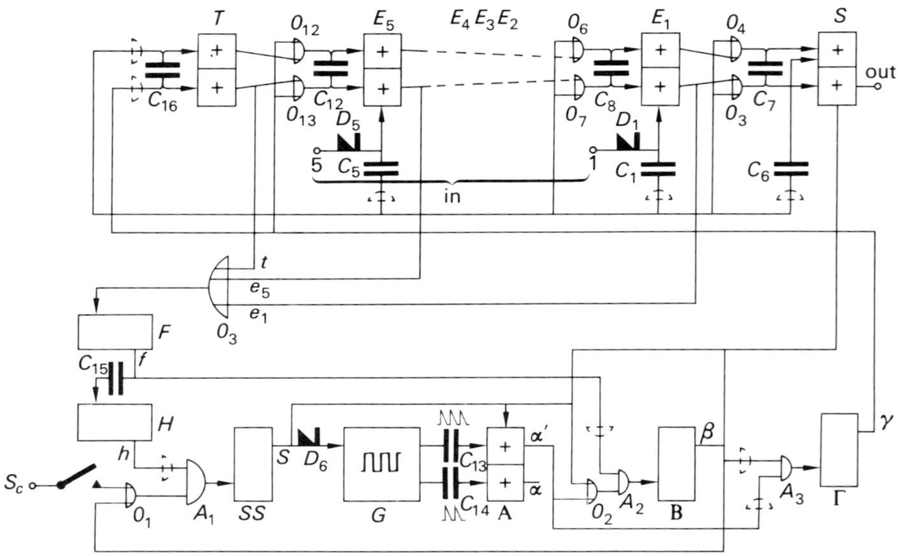

Fig. 19–6

electronic signal; in the present example, however, which is specially designed for operation with a tape reader, contact S_c is a mechanical contact in the tape reader.

If an oscillator of the type shown in Fig. 19-2 is used, the first change of state will occur after a time interval equal to the duration of a half-signal element. In the time chart of Fig. 19-5, contact S_c is closed half an element before moment $t = 0$. The first change of state of the oscillator occurs at $t = 0$, as is shown on line c.

Start contact S_c starts oscillator G in the symbolic circuit of Fig. 19-6 via an OR gate O_1, an AND gate A_1, and a start-stop inverter SS. The operation of this inverter is shown on line b of the time chart. The output terminals of oscillator G are capacitively connected to an auxiliary trigger A in order to prevent a variable loading of this circuit, which would have an unfavorable influence on stability of the generated frequency.

Signal α', generated by the square-wave oscillator through auxiliary trigger A, is shown on line c of the time chart of Fig. 19-5. The α' signal, together with other signals, controls two auxiliary pulse sources B and Γ. Inverter B produces a signal $\beta = 1$ for all moments $t = 0$ of a series of signals, as is shown on line d of the time chart. The β signal is restored to 0 at moment $t = 6$ of each operating cycle.

The following operations must be performed on signal $\beta = 1$ at moment $t = 0$:

1. There must be simultaneous positioning of sections E_1 to E_5 of the shift register in accordance with the code combination supplied by the memory—in this case, the tape reader.
2. At the end of each operating cycle, all sections of the shift register, except section S which provides the output signal, remain in the 0 state. Section S is then in the 1 state, so that a signal of stop polarity is produced. A signal $\beta = 1$ applied at moment $t = 0$ via the capacitor of the RC delay memory element C_6 of Fig. 19-6 brings section S of the shift register in the 0 state, so that a signal of start polarity is then transmitted.
3. A signal $\beta = 1$ at moment $t = 0$ applied via capacitor C_{16}—which is part of an RC delay memory element—puts section T of the shift register into the 1 state. Six 0 pulses are then shifted into the T section, so that at moment $t = 6$ the initial state of the T section is shifted into section S. At that moment, the sections E_1 to E_5 and T remain in the 0 state. This marks the end of the operating cycle. Section S, which is now in the 1 state, begins to transmit a signal of stop polarity.

These three operations occur simultaneously at moment $t = 0$, as can be seen from the time chart. In the present example, the code combination of the letter Y of the International Telegraph Alphabet 2 has been used to set the five memory sections E_1 to E_5 of the shift register for the transmission of the first signal as indicated in the time chart of Fig. 19-5. The five signals from the tape reader must be connected to the five corresponding input terminals (1 up to 5) of the circuit, as shown in Fig. 19-6. Diode D_1 and capacitor C_1, etc., together form a differentiating RC delay circuit of the type shown in Fig. 10-16. A signal $\beta = 1$ (line d) given at moment $t = 0$ sets the sections of the shift register in accordance with the signal being fed in at that moment at the corresponding input terminals 1 to 5.

Number of shift sections

	7	6	5	4	3	2	1
t	T	E_5	E_4	E_3	E_2	E_1	S
	0	0	0	0	0	0	1
0 →							
	1	x_5	x_4	x_3	x_2	x_1	0
1 →							
	0	1	x_5	x_4	x_3	x_2	x_1
2 →							
	0	0	1	x_5	x_4	x_3	x_2
3 →							
	0	0	0	1	x_5	x_4	x_3
4 →							
	0	0	0	0	1	x_5	x_4
5 →							
	0	0	0	0	0	1	x_5
6 →							
	0	0	0	0	0	0	1

Fig. 19–7

The further operation of the transmitter of Fig. 19-6 and the time chart of Fig. 19-5 can perhaps be better explained by means of the simplified diagram of Fig. 19-7. This diagram shows in the horizontal direction the sections of the shift register and their contents, which can be 0, 1, or X. The indication X means that the section contains one of the information elements to be transmitted. The moments $t = 0$ to $t = 6$ are seen to occur between the horizontal lines.

The transmitted signal leaves the diagram of Fig. 19-7 at the right-hand side, i.e., out of section S.

The following operations are controlled by the γ signal produced by inverter Γ. The operation of this inverter is shown on line e of the time chart. Whenever signal γ changes from 0 to 1, the following operations are performed in the circuit.

1. Six shifting operations—so that the information contents of all the sections arrive sequentially in section S through which this information is transmitted.
2. Six shifting operations, by means of which a 0 is shifted into section T. These 0's fill up the empty sections of the shift register, so that at moment $t = 6$ the situation is as shown on the bottom line of the diagram of Fig. 19-7. Sections E_1 to E_5 and T then contain a 0. During a transmission cycle, there can be no other moment with the same state of these shift sections. This marks the end of this cycle. The OR gate O_3 operates inverter F, as is shown on line f of the time chart.

It should be pointed out that the situation in which the last six sections of the shift register contain a 0 cannot occur with the 32nd combination of the International Telegraph Alphabet 2 (0 0 0 0 0) because section T will contain a 1 between moments 0 and 1.

The actual end of the signal to be transmitted is reached at moment $7\frac{1}{2}$. A number of operations still have to be performed between moment $t = 6$ and $t = 7\frac{1}{2}$:

1. At moment $t = 6$, inverter F interrupts the operation of inverter B via the AND gate A_2, as indicated on line d of the time chart.
2. The B inverter stops via the AND gate A_3 the operation of the Γ inverter, as is shown on line e of the time chart. No further shift pulse can be transmitted to the shift register.
3. In the case of a transmission at the standard telegraph speed of 50 baud, the F inverter operates inverter H for 20 msec, as is indicated on line g of the time chart.
4. Inverter H—via AND gate A_1 and inverter SS—stops the operation of oscillator G for at least 20 msec. After moment $t = 6$, oscillator G still generates a pulse of short duration, as can be seen on line c of the time chart, but this cannot have any further effect on the circuit because the operation of inverters B and Γ has already been stopped.
5. Inverter H reverts to normal at moment $t = 7$ (line g). If at that moment contact S_c is still closed, a new operating cycle is started, as is the case in the time chart of Fig. 19-5. If at that moment contact S_c is open, as is the case at the end of the second operating cycle in the time chart, the transmitter stops and waits for contact S_c to be closed again.

Some additional remarks should be made on the circuit of Fig. 19-6.

1. The output terminal of inverter SS is connected to an extra input terminal of trigger S. This is a clamping connection which prevents trigger

S from being reset erroneously as a result of which it would start to transmit a signal of permanent start polarity.

2. A number of OR gates with a single input have been shown in dotted lines in the circuit of Fig. 19-6. These single input gates are necessary in the case of a transistor circuit in order to provide the collector current for the controlling (*pnp*) transistor.
3. The feedback connection linking the output of inverter B to OR gate O ensures the completion of the transmitted signals, whatever the state of contact S_c may be.

All these operating requirements lead to the symbolic circuit shown in Fig. 19-6. It is possible to translate some of these requirements into switching equations, but the main body of the circuit is determined by the standard organization of the shift register.

Another type of transmitter is obtained when the information stores make the code combination to be transmitted permanently available to the transmitting circuit during its cycle of operation. If, for example, a tape reader is used as information store, the sensing contacts or other types of devices remain in the operated condition during the transmitting cycle. The transmitting circuit can then consist of a distributor performing the parallel-serial transformation.

The time chart of Fig. 19-8 shows the operation of a seven-step distributor in which three of the sections are always in the operated condition (lines 1 to 7). During each transmission cycle consisting of a start element, five information elements, and a stop element, the distributor receives seven driving pulses (line d) at moments 0 to 6. The block diagram of this transmitter is shown in Fig. 19-9.

In the rest condition of the distributor, sections 5, 6, and 7 are in the operated or 1 condition. It is sufficient to sense the first- and last-operated section of a group of three to indicate a certain step of the distributor. The output of AND gate A_1 indicates the seventh step of the distributor (line e of the time chart of Fig. 19-8).

The distributor can be started for one or more cycles of operation by closing start contact sc of the tape reader (line a of the time chart) once a monostable trigger MT has finished its 20-msec start-blocking pulse (line f in the time chart). In this case, the start-stop unit SS is brought into the operated condition (line b of the time chart) and the oscillator is started (lines c and d of the time chart). The oscillator starts its operation after a delay lasting half an element; produces seven pulses that drive the seven-step distributor; is driven through these seven steps because AND gate A_2 produces via the OR gate O_1 a holding signal for start-stop unit SS; and stops at moment 6 when the distributor again reaches the seventh position, because

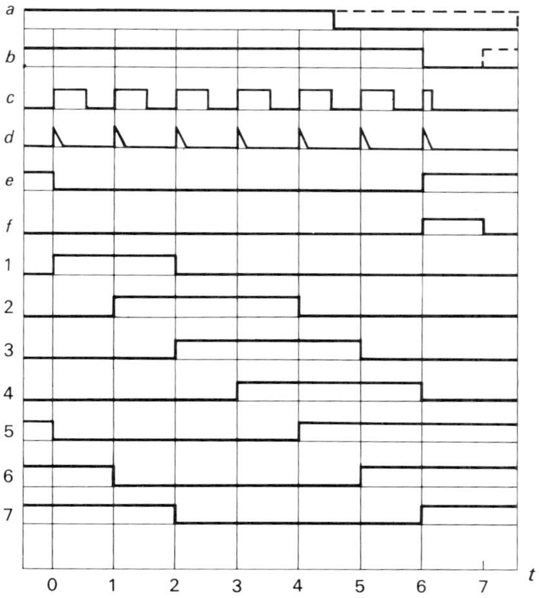

Fig. 19–8

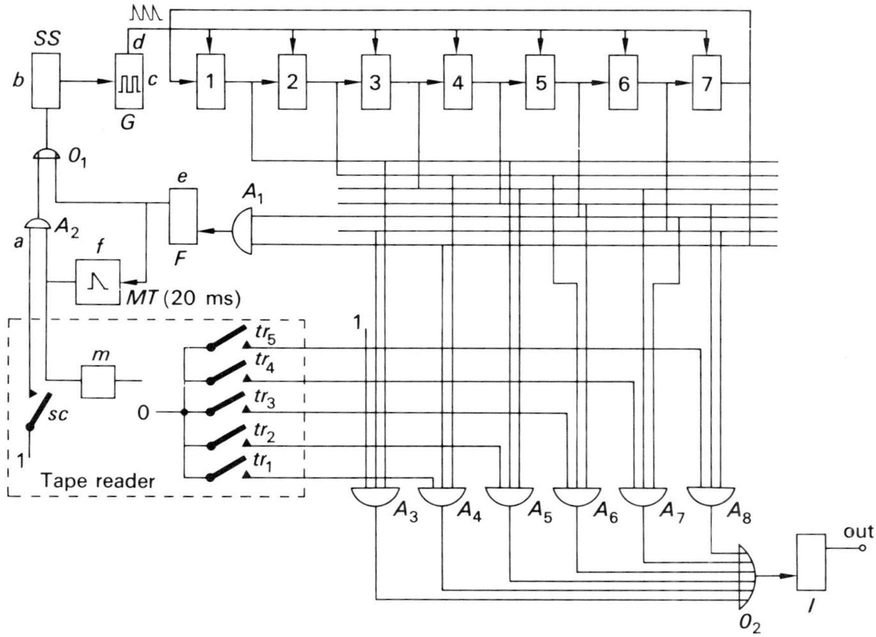

Fig. 19–9

AND gate A_1 interrupts the holding signal and starts a 20-msec start-blocking pulse in the monostable trigger MT that blocks AND gate A_2. The 20-msec delay is correct for a 50-baud transmission speed. This delay should be about 22 msec for $45\frac{1}{2}$ telegraph systems.

If starting contact sc remains closed (dotted line a in Fig. 19-8), start-stop unit SS can come into the operated condition at moment 7. The transmission of the next signal starts then at moment $7\frac{1}{2}$ because of the half-element starting delay of the oscillator. In this way, the minimum length of a signal in an uninterrupted series of signals is adjusted to its correct value.

The five contacts of the tape reader that sense the perforated tape, or any other suitable kind of device for sensing the information store, present the information to be transmitted in parallel form to AND gates A_4 to A_8. A sixth AND gate A_3 controls the polarity of the start element of the signal to be transmitted. Each of these AND gates has three input terminals. Two of these terminals are connected to the outputs of the first and last section of the group of conducting sections characterizing a certain step of the distributor; the third input terminal is connected to the corresponding memory element of the information store.

The information input terminal of gate A_3 is connected permanently to 1. During the first step of the distributor, the other two input terminals of gate A_3 are also characterized by 1 signals, so that start polarity (0) is transmitted via OR gate O_2 and inverter I. Two of the input terminals of gate A_4 are characterized with 1 signals by the distributor during its second step. The second step of the distributor corresponds to the transmission of the first information element of the signal. If the information store indicates a 0 (open contact tr_1), the transmission of start polarity will be continued; however, if a 0 is indicated (closed contact tr_1), the transmitted signal element will be of stop polarity.

The other gates (A_5 up to A_8) will be opened by the distributor in the sequence of their numbering for the transmission of the other information elements of the signal. Finally, a stop element of stop polarity must be transmitted before another signal can be started. During the seventh step of the distributor, none of the gates A_3 to A_8 is characterized. This means that the output terminals of all these gates are in the 0 condition, and this results in the transmission of stop polarity (1).

It is not absolutely necessary to use a seven-step distributor in the transmitter of the type shown in Fig. 19-9; one can use instead a counter consisting of three pulse halvers which has to be reset at a suitable moment.

A transmitter for start-stop signals may have a continuously running oscillator provided that it is not started by a signal with an inaccurate $7\frac{1}{2}$-element rhythm independent of the functioning of the transmitter itself. An elegant application of pulse halvers combined to form a distributor is possible in this case. The time chart of such a distributor driven by a

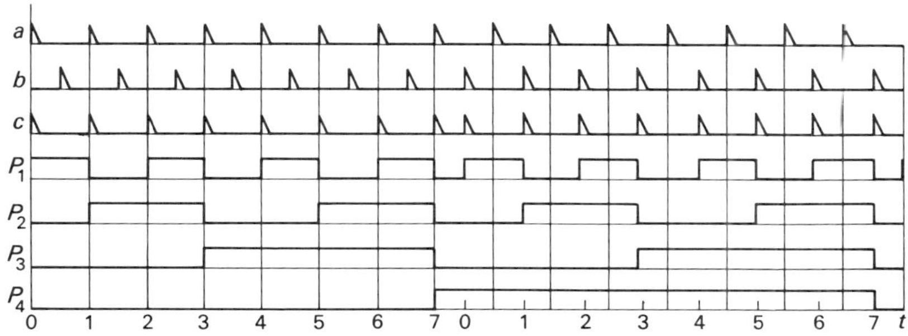

Fig. 19–10

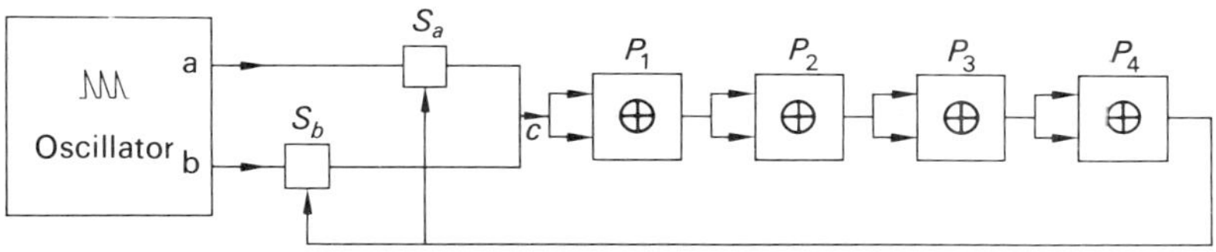

Fig. 19–11

continuously running oscillator is shown in Fig. 19-10; the block diagram of the distributor is given in Fig. 19-11.

The oscillator is started at the beginning of the transmission of a series of signals. The square-wave oscillator produces two series of driving pulses on wires a and b (Figs. 19-10 and 19-11), which are shifted one half element with respect to each other. Pulse halver P_4 determines which one of the two series of pulses is going to be used for driving the counter consisting of four pulse halvers (P_1 to P_4). This pulse halver sets switch S so that, for example, S_a is conducting and S_b is nonconducting if they remain in the 0 or in the 1 condition, respectively, and vice versa.

The pulse series that drives the distributor is shown on line c of the time chart. At moment 7 of the first cycle, pulse halver P_4 is put into the 1 condition, while the other pulse halvers P_1 to P_3 are reset in the 0 condition. Switch S_a becomes nonconducting and S_b becomes conducting.

This connects line b to the distributor, so that the next pulse which arrives after an interval of one half-element starts a new transmission cycle. The pulse on line b given at moment 7 of the second transmission cycle resets pulse halver P_4. The third transmission cycle, which is now again under control of the pulses on line a, starts after an interval of one half-element after moment 7 of the second cycle. The shifting back and forth of the control of the distributor limits the transmission cycle to $7\frac{1}{2}$ cycles per signal or to 15 cycles for two signals.

19-3 RECEIVERS FOR BINARY START-STOP SIGNALS

A receiver for binary start-stop signals must perform the serial-parallel transformation of signals which may—or may not—come in as an uninterrupted series.

The design problems for a receiver are

1. The possible difference in speed between the oscillator of the transmitter and that of the receiver
2. The distortion of the start-stop transition between two signals

The first problem makes it necessary for the oscillators to remain constant within certain limits. The limits depend on the number of information elements in the signals. For a five-unit transmission, an accuracy of $+$ or $-\frac{3}{4}$ percent is sufficient [recommendations of the Comité Consultatif International Télégraphique et Téléphonique (CCITT)].

The difference in speed between a transmitter and a receiver in a start-stop transmission system can cause a slight variation in the moment of the stop-start transition of two consecutive signals. This is of minor significance. More important, however, is the speed requirement for the sensing moment of the last information element, as well as the loss of margin (safety of operation) if this sensing moment is not correctly timed at the receiving side.

Furthermore, it is necessary that the operating cycle of a receiver be shorter than that of the associated transmitter, because the receiver must be started from a stop position for each signal to be received so that possible differences in speed between transmitter and receiver are eliminated at the beginning of each operating cycle.

The second problem is more serious from the point of view of the receiver design. All elements of a signal will be more or less distorted during transmission, which means that the transitions between signal elements of different polarity experience a shift from their theoretical place during transmission. The stop-start transition, which is one of the possible transitions, is also vulnerable to distortion; thus, in an uninterrupted series of transmitted signals, the stop-start transition can shift back and forth away from its theoretical place. This causes the start element of the next signal to be too long or too short; consequently, there is a shift of the sensing moments of the following signal away from the theoretical place owing to the distortion or displacement of the stop-start transition. A loss of margin is found to occur with respect to a signal the information elements of which should not be affected at all by distortion. This possible loss of margin is illustrated in Fig. 19-12.

Figure 19-12 shows a signal as part of an uninterrupted sequence of transmitted signals. The stop-start transition of the signals on line *a* of this figure is not affected by distortion. The sensing moment of the first information element remains at its theoretical place, i.e., the middle of that information element. This place ensures the maximum receiving margin, because this margin is determined by the shortest allowable displacement of the sensing moments. The signals shown on lines b and c of Fig. 19-12 are affected by distortion of the stop-start transition, which results, respectively,

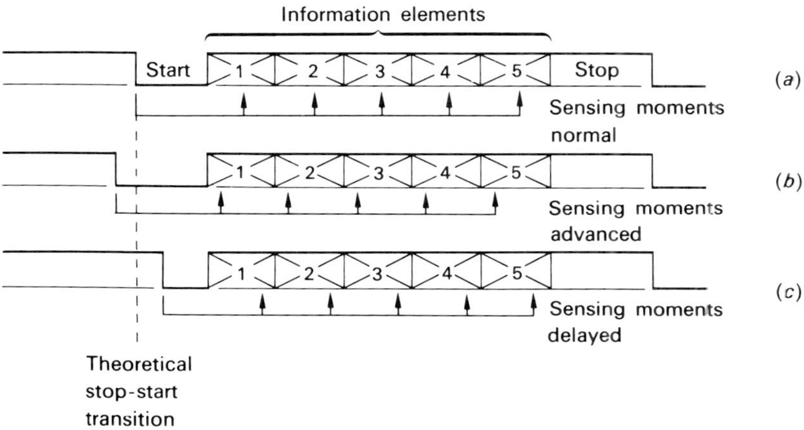

Fig. 19–12

in an increase and a decrease in the duration of the start element. Whereas the sensing moment of the first information element normally occurs $1\frac{1}{2}$ elements after the stop-start transition, this sensing moment in the example shown on line b is seen to occur before, and in the example of line c after, the theoretical moment. In both cases, a loss of margin is unavoidable.

This loss of margin can be limited for displacements of the stop-start transition that increase the length of the start element—or alternatively decrease the length of the stop element of the previous signal—by setting the minimum receiving cycle at, say, about $6\frac{3}{4}$ signal elements for a 7-unit transmission or at about $7\frac{1}{4}$ elements for a $7\frac{1}{2}$-unit transmission.

The following two different types of receivers can be designed depending on the main distributing and receiving elements used:

1. A shift register operating as a receiving memory
2. A distributor and an n-element memory

The general block diagrams of these two types of receivers are shown in Figs. 19-13 and 19-14, where the sensing of the incoming information elements is illustrated. The block diagram of the shift-register receiver has

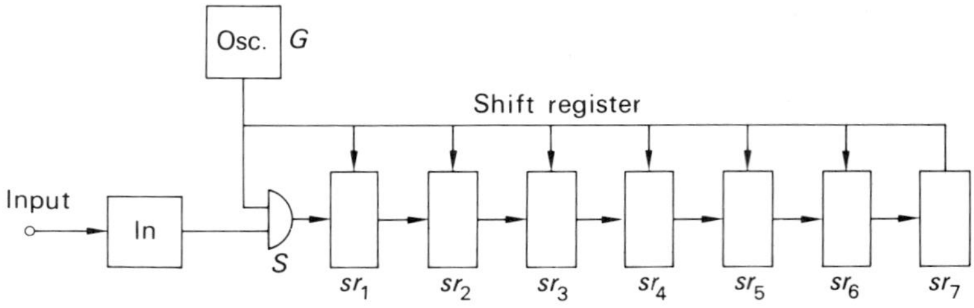

Fig. 19–13

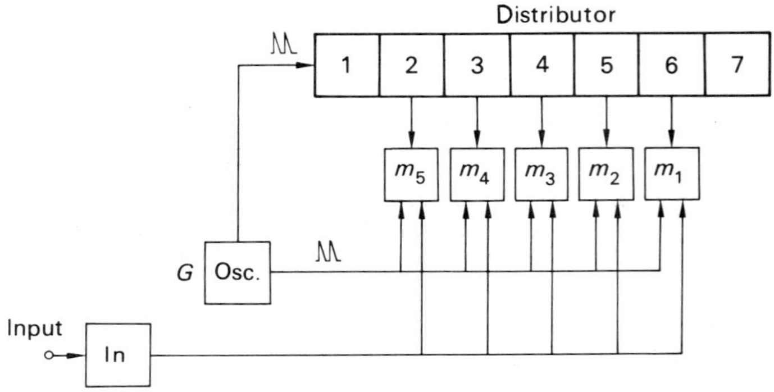

Fig. 19–14

only one sensing gate S controlled by the incoming signals and by the oscillator of the receiver. The sensed signals control the input to a shift register. After $n + 2$ clock pulses of the oscillator (n information elements and one start and one stop element), the reception of the signal is completed. The $n = 5$ information elements are registered temporarily in the shift sections sr_2 to sr_6.

In the receiver using a distributor, the incoming signal controls n sensing gates, which are shown in Fig. 19-14 as part of the five memory elements m_1 to m_5. These sensing gates can be opened successively under the control of their associated distributor sections at moments determined by the oscillator clock pulse.

The time chart of a receiver based on the block diagram of Fig. 19-13 is given in Fig. 19-15. The incoming signal is shown on line a of that time chart. The stop-start transition is assumed to occur at the theoretical moment 0. The following signal consisting of a start, five information elements, and a stop element is shown undistorted. The stop-start unit SS (Fig. 19-16) of oscillator G is brought immediately into the operated condition. This starts

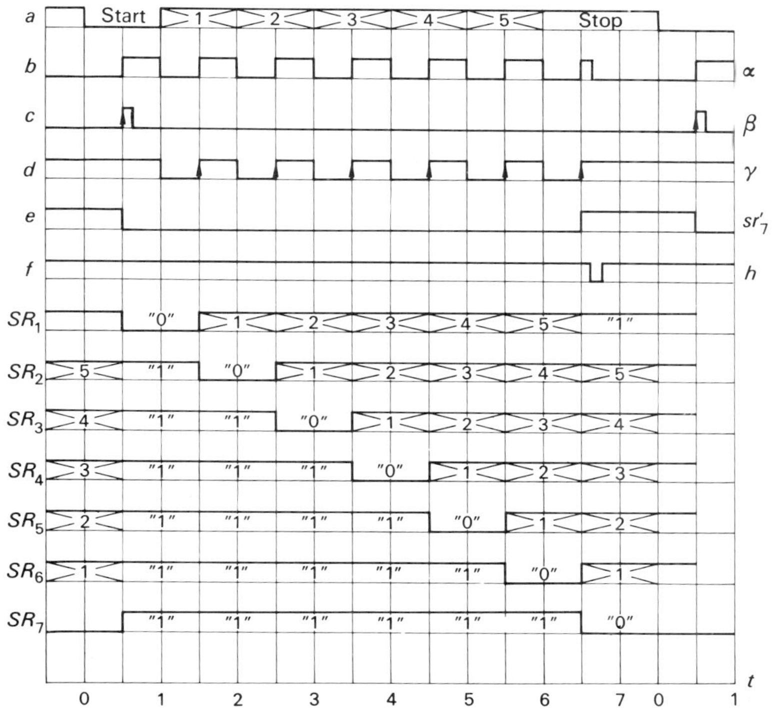

Fig. 19–15

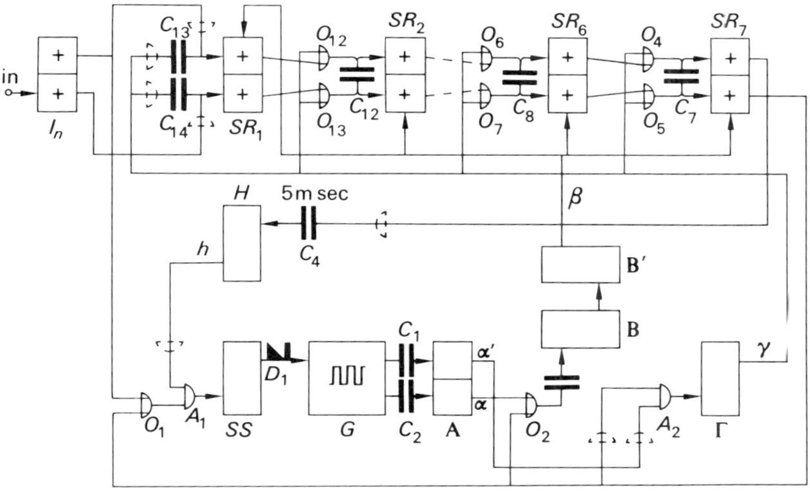

Fig. 19–16

oscillator G after a delay of one half-signal element. During this time interval, false start signals can be suppressed. Signals of start polarity received after a stop element of a previous signal are identified as start elements if their duration exceeds one half-element. The first change of condition of square-wave oscillator G occurs at moment $\frac{1}{2}$ (line b). As a result of this first change of condition, a pulse $\beta = 1$ is produced on line c. This pulse preconditions the sections SR_1 to SR_7 of the shift register as follows: 0 1 1 1 1 1 1. The information of the last signal received is destroyed at moment $\frac{1}{2}$. The γ signal produced by the Γ inverter under the control of oscillator G changes six times from 0 to 1, namely, at moments $1\frac{1}{2}$ to $6\frac{1}{2}$, as is shown on line d of the time chart. The RC delay circuits with capacitors C_{13} and C_{14} at the input of shift section SR_1 sense at these moments the polarity of the incoming signal, whereas the information contained in sections SR_1 to SR_6 is shifted over one place in the register, so that the sensed incoming information can be stored for the duration of one element in section SR_1. It should be noted that trigger In, which is connected to the input terminal and which acts as an adaptor between the transmission circuit and the receiver, is free to follow the change of polarity of the incoming signals, while section SR_1 of the shift register can follow the polarity of these signals at predetermined moments, as shown in the time chart.

At each of the moments $1\frac{1}{2}$ to $6\frac{1}{2}$, a new information element is transferred into the shift register, and the stored information is pushed one section further with the 0 of the start element in front. At moment $6\frac{1}{2}$, this 0 of the start element reaches section SR_7. The state of section SR_7 is then changed from 1 to 0, and this indicates that the stop element of the received signal has been sensed by the input gates of section SR_1 and that the receiving cycle is now being completed.

The shifting of the 0 information of the start element of the signal can be seen clearly in the time chart of Fig. 19-15. The information elements of the received signals have been numbered from 1 to 5 in the time chart, while all standard information, which is the same for each receiving cycle, has been marked with a 0 or a 1.

At moment $6\frac{1}{2}$, once the γ signal has changed the state of section SR_7 from 1 to 0 (see line SR_7 in the time chart), an sr_7 signal closes gate O_1, so that oscillator G is stopped. The oscillator still produces a short pulse after moment $6\frac{1}{2}$, as is shown on line b of the time chart. The sr_7' signal starts the RC delay circuit with capacitor C_4 and inverter H, which produces a pulse of about 5 msec, so that oscillator G is blocked via gate A_1 against a possible restart by an early stop-start transition of a following signal that is due to distortion in the transmission circuit.

The 5 msec blocking pulse is suitable for telegraph systems with the standard 50-baud speed using signals with a duration of 7 elements. For $7\frac{1}{2}$-element signals, this blocking time may be increased to 12 or even 15 msec.

In general, oscillator G has to be blocked up to about a quarter element before the theoretical moment of the stop-start transition. This is more than sufficient to avoid difficulties due to differences in speed between the oscillators of the transmitter and of the receiver. Moreover, this blocking can improve the receiving margin of the next signal.

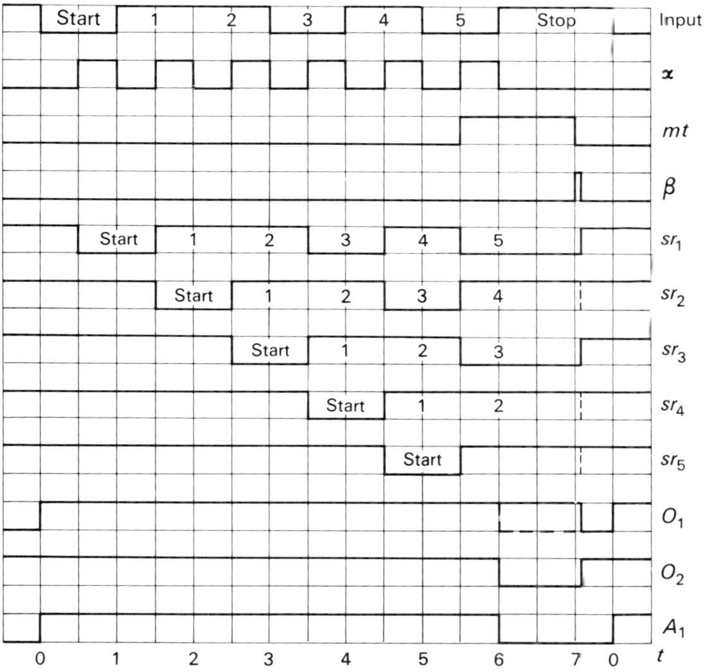

Fig. 19–17

The received information can be transferred from sections SR_2 to SR_6 into other memories or used for printing, for punching tapes, etc. This can be performed between moments $6\frac{1}{2}$ and $7\frac{1}{2}$ in the time chart.

Finally, it should be noted that transmitters and receivers based on the symbolic circuits of Figs. 19-6 and 19-16 can be designed to operate at speeds of a couple of thousand baud if so desired.

In the time chart of Fig. 19-15, the signal received during the preceding cycle is wiped out by a β pulse at moment $\frac{1}{2}$, while the complete information is already available at moment $6\frac{1}{2}$ of the previous cycle. In a number of cases, it is not necessary to keep the received information for such a long time for the purpose of ensuring its processing. However, from the point of view of maintenance, it is a desirable to do so, because it permits a check of the operation of the receiver by allowing a look at the received code combination after the end of a receiving cycle if no further signal follows.

The circuit of a receiver for start-stop signals can be simplified further when the received information is already wiped out during the stop element, e.g., at moment 7 of the receiving cycle, as is shown in Fig. 19-17. This figure shows the time chart of such a simplified receiving circuit.

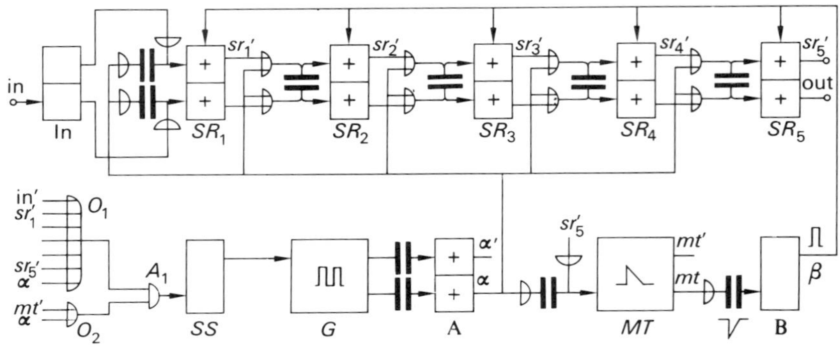

Fig. 19–18

At the sensing moment of the fifth information element, a one-shot trigger MT is operated for the duration of about $1\frac{1}{2}$ signal elements. The main purpose of this one-shot trigger MT is to limit the contraction of the stop element which may be caused by a shift of the stop-start transition under the influence of distortion. The return to 0 of MT can be used to generate a β pulse for the reset of all the memory elements of the receiving shift register. In the time chart of Fig. 19-17, this will occur shortly after moment 7 of the receiving cycle.

In this case, use can be made of the fact that the complete received information is available after the sensing moment of the fifth information element, as can be seen in the time chart of Fig. 19-17 between moments $5\frac{1}{2}$ and 7. The receiving shift register, which contains seven sections in Fig. 19-16, can now be reduced to five sections, and the auxiliary circuits can also be reduced to a certain extent, as is shown in the circuit of Fig. 19-18.

Gate O_1 controls the start of a receiving cycle. A start may occur only if all memory elements SR_1 to SR_5 are in the 1 state and oscillator G is in the rest condition. In that case $\alpha = 0$. When a stop-start transition is received, signal in$'$ changes from 0 to 1, and this starts oscillator G. The operation of oscillator G is secured during the further operation cycle by the fact that the start element of the received signal is shifted through all sections SR_1 to SR_5, so that one of the signals sr_1' to sr_5' is always a 1.

At the sensing moment of the fifth information element, the control of oscillator G is taken over by gate O_2. This gate allows oscillator G to finish its last reversal between moments $5\frac{1}{2}$ and 6. Thereafter, the oscillator is stopped and blocked by one-shot trigger MT until shortly after moment 7.

The reception of the actual signal need not be explained for the circuit of Fig. 19-18 after what was said about the circuit of Fig. 19-16.

PROBLEMS

1. Draw the circuit of a square-wave transistor oscillator and explain the operation of this circuit by means of a time chart.

(*a*) How can the influence of leak currents in the transistors be reduced?

(*b*) How can the oscillator be started and stopped in such a way that the first half-period of each operating cycle is equal to all following half-periods?

2. Draw the time chart of a transmitter for five-unit start-stop signals in which a shift register is used as a time base for the elements of the transmitted signal and as an auxiliary store for the information elements of these signals.

(*a*) Explain how a shift register can be used to count the number of elements transmitted.

(*b*) Explain how the information of a tape reader can be transferred from the punched tape into the corresponding sections of the shift register.

(*c*) Explain how the transmitter can be made to operate only in complete transmitting cycles.

3. Draw the time chart of a start-stop transmitter controlled by a continuously running oscillator, and draw a symbolic circuit.

4. Explain the combined operation of a receiver of start-stop signals with a transmitter of such signals. Show by means of a time chart how the sensing moments of the incoming information elements can be delayed or advanced depending on the distortion of the stop-start transitions in the series of incoming signals.

5. Draw a block diagram of

(*a*) A receiver for start-stop signals using a counting shift register as time base.

(*b*) A receiver for start-stop signals using a distributor as time base and explain the differences between the two types of receiver.

SELECTIVE BIBLIOGRAPHY

1. Aschmoneit, E. K.: Elektronik in der Fernschreibtechnik, *Elektronik*, vol. 6, pp. 127–130, 199–203, 244–245, 1957, and vol. 7, pp. 103–108, 232–241, 1958.
2. Biggam, F. D.: A Transistorized Time Division Multiplex Telegraph Set, *Trans. AIEE*, pt. 1, vol. 75, pp. 780–785, 1956.
3. Carter, R. O., L. K. Wheeler, and A. C. Frost: An Electronic Regenerative Repeater for $7\frac{1}{2}$ Unit Start-Stop Telegraph Signals, *Post Office Elec. Eng. J.*, vol. 41, pp. 222–227, 1949.
4. Hearn, R. B.: Electronic Regeneration of Telegraph Signals, *Bell Lab. Record*, vol. 27, pp. 173–177, 1948.
5. Lommel, H. L. van, and P. J. Clemens: Automatic 7E Rotary Teleprinter Switching System, *Elec. Commun.*, vol. 34, pp. 38–55, 1957.

6. Oberman, R. M. M.: Circuit for the Reception and Retransmission of Telegraph Signals with a Vibrating Relay as Oscillator (in Dutch: Schakeling voor het ontvangen en eventueel weer doorzenden van telegraafteekens met een vibreerend relais als taktgever), Netherlands Patent 58,845, priority Oct. 6, 1942.
7. Oberman, R. M. M.: Entzerrender Uebertrager mit Doppelspeicherung für nach dem Start-Stop-Verfahren Uebertragene Telegraphiersignale, German Patent 959,558, priority Feb. 6, 1955.
8. Oberman, R. M. M.: Regenerative Telegraph Repeaters, *Het P.T.T.-Bedrijf*, vol. 5, pp. 164–172, 1953.
9. Oberman, R. M. M., and A. Snijders: Regenerative Repeater, U.S. Patent 2,599,345, priority May 11, 1948.
10. Ostendorf, B.: A Completely Electronic Regenerative Repeater, *Bell Lab. Record*, vol. 27, pp. 436–441, 1949.
11. Snijders, A.: Electronic Transmitter and Receiver for Signals in a Start-Stop Code, U.S. Patent 2,879,334, priority March 20, 1952.
12. Snijders, A.: Electronic Transmitter, Receiver and Regenerative Repeater for Telegraph Signals in a Start-Stop Code, U.S. Patent 2,842,616, Nov. 1952.
13. Snijders, A.: Regenerative Repeater System, U.S. Patent 2,850,567, priority March 24, 1952.
14. Snijders, A.: Apparatus for Automatically Determining at a Distance the Largest Distortion of Telegraph Signals in a Start-Stop Code Transmitted by a Subscriber's Station, British Patent 800,308, priority Dec. 30, 1954.
15. Wilder, H. F.: Electronic Regeneration of Teleprinter Signals, *Trans. AIEE*, vol. 65, pp. 34–40, 1946.

20
Codes

20-1 NUMERICAL SYSTEMS

Automatic devices operate on coded signals, which can be externally fed or internally generated. These coded signals or codes, i.e., the language of automatic devices, can be considered as corresponding to the numbers of a numerical system with a certain radix. In small automatic devices, the machine language can be so simple that it can hardly be considered a coding system; but in large equipments, such as computers and in information-transmission systems, the choice of the correct coding system is of primary importance. For computing devices, the coding systems used should allow easy addition; and for the transmission of coded signals in the presence of noise, error-detecting and/or error-correcting codes will be a likely feature. There is a large number of such possible coding systems, since each numerical system, whatever its radix, can be used as such. The more practical systems can be divided into two categories:

1. Digital codes, or numerical systems with a small radix
2. Analog codes, or numerical systems with a very high radix

The following discussion will be limited to category 1 because the switching circuits discussed in this textbook all belong to this group.

When the choice of the coding system is governed only by economic considerations with regard to the storage of numbers, the merit of storage devices can be assessed as follows.

A counter with counting wheels, such as the message register of an automatic-telephone subscriber or the mileage counter of an automobile, consists of a number of counting wheels, each having the 10 decimal digits 0 to 9 on its periphery. The counters can count up to, say, 999 on 3 of these counting wheels and can be read in the numerical system with radix 10.

It would also be possible to provide these counters with wheels having only the numbers 0 and 1 on their periphery. By means of 10 of the wheels, numbers up to 1023 could be indicated. This device would have approximately the same storing capacity. These wheels provide a coding system with radix 2, commonly known as the binary number system.

The total number of digits used on the periphery of all counting wheels of comparable counters can also serve to assess the efficiency of the numerical system. The decimal counter given as an example here uses $3 \times 10 = 30$ digits, while the binary counter uses only $10 \times 2 = 20$ digits. From this point of view, the binary number system can be considered more efficient, because approximately the same group of numbers can be represented by means of a much smaller total number of digits.

The relation between a numerical system with radix 10, which will be used as reference system, and a system with radix b having the same capacity is

$$10^n = b^m \tag{20-1}$$

where n and m are the number of digits required for representing identical numbers in both systems.

It follows from Eq. (20-1) that

$$\frac{m}{n} = \frac{1}{\log b} \tag{20-2}$$

which relates the number of digits in a numerical system with radix b to the decimal system.

The relative efficiency of a storage device operating in a system with radix b can be expressed in terms of the ratio between the total number of digits at the periphery of the counting wheels in both systems for equivalent counters. This figure of merit can be represented by the following equation:

$$M = \frac{b \cdot m}{10 \cdot n} = \frac{b}{10 \cdot \log b} \tag{20-3}$$

The values obtained for M for numerical systems with a radix up to 12 and the values of ratio m/n are given in Table 20-1. A maximum storage efficiency is obtained for a system with radix e.

Table 20-1

b	m/n	M	b	m/n	M
1			7	1.18	0.828
2	3.32	0.664	8	1.11	0.885
3	2.10	0.629	9	1.05	0.943
4	1.66	0.664	10	1.00	1.000
5	1.43	0.715	11	0.96	1.056
6	1.29	0.771	12	0.93	1.112

The figures given in Table 20-1 are interesting from a theoretical point of view. There are a number of practical factors, however, which play such an important part that most usable codes are based on a numerical system with radix 2 for the complete code symbols or radix 2 per decimal figure to be coded.

20-2 THE TERNARY NUMBER SYSTEM

From the theoretical point of view developed above, the ternary number system is the most efficient because $M = 0.629$ and $m/n = 2.10$. There are two ways to indicate the three digits of the ternary number system, viz., by using the symbols 0, 1, and 2 or -1, 0, and $+1$. The last choice has some advantages, because the use of the figures 0 and 1 suggests some relation to the binary number system. The ternary code combinations and their decimal equivalents according to the second system (-1, 0, $+1$) are given in Table 20-2.

The ternary code combinations of Table 20-2 are given in the sequence of the decimal equivalents obtained by addition or subtraction of the weights 3^n of the digits indicated by $+1$ or -1. This ternary number system has the advantage of giving positive as well as negative numbers. Three ternary digits are sufficient to represent the decimal numbers from -13 up to $+13$. Four binary digits are required for coding the 10 decimal digits 0 up to 9, and an extra digit must be used to indicate the sign. From the point of view of efficiency of registration, especially with respect to the decimal system in common use, the ternary number system seems quite satisfactory. However, in most available techniques the use of $+1$, 0, and -1 signals is not very reliable, and this is especially true for the 0 signal. The storage of $+1$, 0, and -1 signals on magnetic tapes and drums and other magnetic devices requires a selective demagnetization at high speeds. This demagnetization is difficult to achieve and must be performed by means which are entirely different from

Table 20-2

Decimal number	9	3	1	*Decimal number*	9	3	1
0	0	0	0	0	0	0	0
1	0	0	+1	−1	0	0	−1
2	0	+1	−1	−2	0	−1	+1
3	0	+1	0	−3	0	−1	0
4	0	+1	+1	−4	0	−1	−1
5	+1	−1	−1	−5	−1	+1	+1
6	+1	−1	0	−6	−1	+1	0
7	+1	−1	+1	−7	−1	+1	−1
8	+1	0	−1	−8	−1	0	+1
9	+1	0	0	−9	−1	0	0
10	+1	0	+1	−10	−1	0	−1
11	+1	+1	−1	−11	−1	−1	+1
12	+1	+1	0	−12	−1	−1	0
13	+1	+1	+1	−13	−1	−1	−1

those needed for the registration of +1 and −1 signals. Moreover, an unmagnetized registration area is vulnerable to all kinds of magnetic interference. A rather high noise level can be expected around the 0-digit areas. Even for the binary registration on magnetic means, the actual 0 state of the magnetic medium has been carefully avoided.

There is, however, one practical technical application of the ternary system—the so-called "cable code" used on long-distance submarine cables. In this code, a +1 signal is used for a dot, a −1 signal for a dash, and a 0 signal for a space between two symbols. But for all other applications in telegraphy for which economic considerations were not so constraining as with the very expensive long-distance submarine cable, circuits, and some radio circuits, the use of the ternary code system has been avoided.

Before leaving this survey of the practicability of the ternary number system, it is necessary to examine in some detail the fundamental circuits that could be used in such a system, i.e., the three-state trigger and the adding circuits.

There are two possible ways to form a three-state trigger:

1. A binary trigger can be given a condition in which both tubes or transistors are equally conducting. The output signals appear marginally on one terminal.
2. A three-state trigger (three inverters) can be used as the fundamental circuit of a ternary coded system. The output signals are obtained as binary signals on three output terminals.

This type of classification also holds for the fundamental circuits of code systems based on any radix b.

It is not difficult to design a ternary trigger. A circuit for such a trigger is shown in Fig. 20-1. It is more difficult, however, to process the marginal output signals necessary for the operation of a ternary adder.

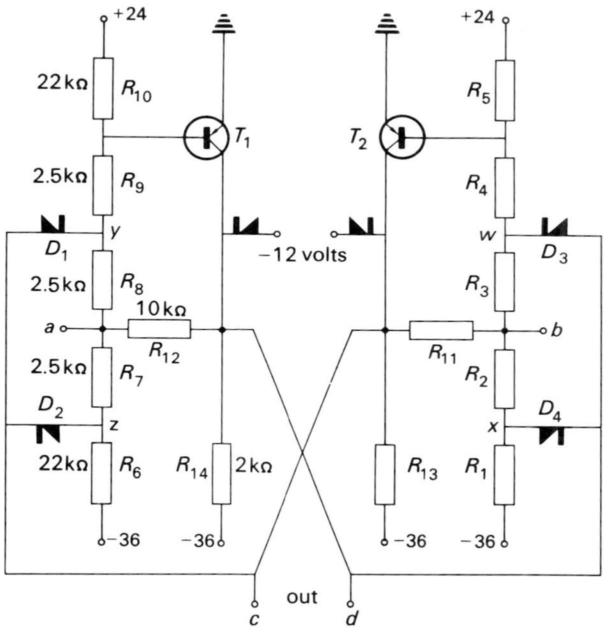

Fig. 20–1

The circuit design of the ternary trigger shown in Fig. 20-1 falls outside the scope of the logical methods described in the previous chapters. This ternary trigger operates as follows. The idealized voltages at the output terminals c and d in the circuit of Fig. 20-1 are given in Table 20-3 for the

Table 20-3

State	*Voltage* c	d
+1	0	−12
0	−6	−6
−1	−12	0

three possible states +1, 0, and −1. The ternary trigger is in fact the result of superimposing a normal two-state trigger on a circuit in which the current of the two active elements (tubes or transistors) is automatically balanced. These two component circuits are shown in Fig. 20-2*a* and *b*.

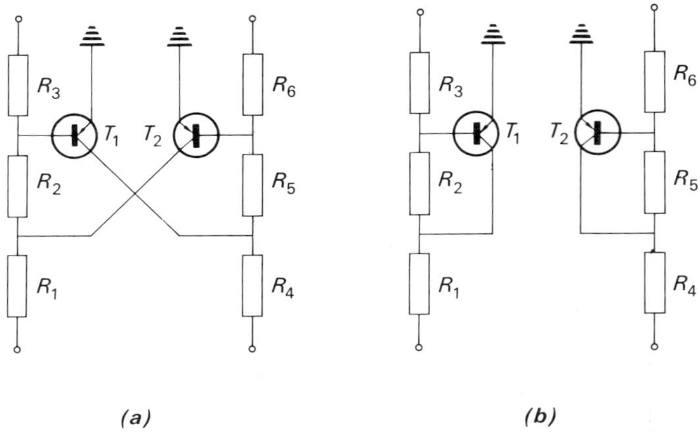

Fig. 20–2

The two-state trigger shown in Fig. 20-2*a* needs no further explanation. If the collector current of transistor T_1 in the circuit of Fig. 20-2*b* tends to increase, the voltage drop over the collector resistor R_1 also tends to increase, so that the voltage over the voltage divider $R_{2/3}$ tends to decrease. The basis current of transistor T_1 tends to become more positive relative to the emitter, and this effect balances out the assumed increase in collector current.

If the circuit of Fig. 20-1 is in the 0 state, points y and z have a voltage of, say, −3 and −9 volts, and points w and x have the same voltages. This means that the pairs of diodes D_1/D_2 and D_3/D_4 are blocking—so that the situation of the circuit of Fig. 20-2*b* arises. Both bases are connected now to the same controlling voltage, so that the trigger remains in its midposition. The voltages of points a and b can be shifted over a rather large interval before the circuit triggers into one of the other two stable states. This happens in one direction when, under the control of point a, point y becomes negative with respect to point c, and in the other direction when point z becomes positive with respect to point c. The start of this triggering can be described also in terms of the voltage difference between point d and points w and x.

In the first case, diodes D_1 and D_4 become conducting; in the second case, the conducting diodes are D_2 and D_3. Moreover, in the first case, the resulting state of the trigger is −1; in the second case, this state is +1.

It should be noted that the input gates have been omitted from the circuit shown in Fig. 20-1. The resulting circuit is slightly more complex than a binary two-state trigger, so that it is satisfactory in terms of efficiency of storing numbers in the ternary system.

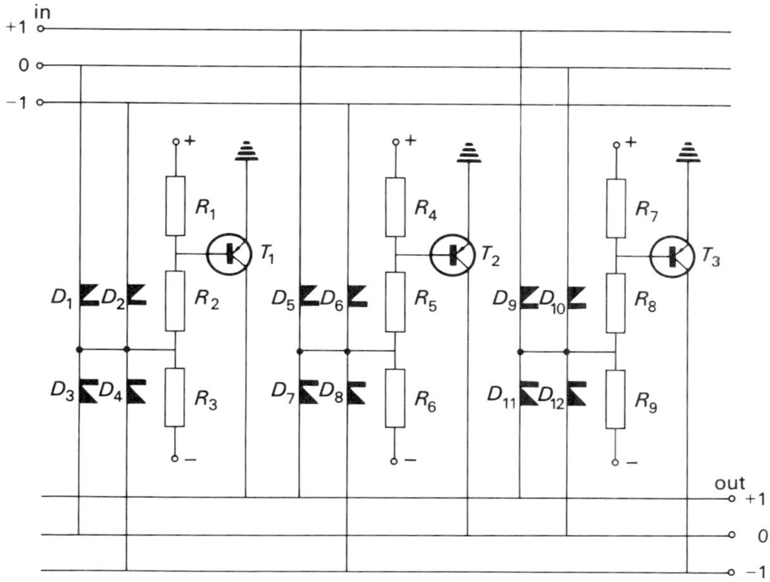

Fig. 20–3

One of the possible solutions of a three-state three-section trigger containing the minimum number of diodes is illustrated in Fig. 20-3. In this example, it is shown as a transistor circuit, but no special difficulties would be encountered in the design of the corresponding tube circuit.

The circuit is provided with three input terminals −1, 0, and +1 (or 0, 1, and 2) and three output terminals with the same designation. One of the three transistors is always conducting; the other two are always blocking. An analogous circuit with the reverse condition would be easy to design but would not result in an improvement in economy with respect to hardware.

The type of three-state three-section trigger shown in Fig. 20-3 is proportionally more expensive than the two-state two-section binary trigger. The blocking and controlling circuits of multistate multisection triggers require an ever larger number of diodes per section as radix b increases. From this point of view, the efficiency figures given in Table 20-1 do not represent the actual efficiency that can be reached under practical circumstances.

The type of circuit shown in Fig. 20-3 has an important advantage over the circuit shown in Fig. 20-1. One of the output terminals is always marked with a standard 1 signal, while the other two are always 0. Moreover, only one of the input terminals has to be marked with a standard 1 signal in order to obtain the corresponding state of the ternary trigger. The associated switching circuits can be based on the methods described in the previous chapters. This is not possible for the marginal type of trigger shown in Fig. 20-1.

As far as the logical (diode) circuit is concerned, a ternary switching system has some disadvantage compared with the binary system. This may be demonstrated by means of a comparison of the ternary and binary additions. This example is of a more or less general nature. The ternary addition is defined by truth table 20-4, which shows the ternary addition of two digits a and b, the resulting sum S, and the resulting carry C. The same

Table 20-4

a	b	S	C
0	0	0	0
+1	0	+1	0
0	+1	+1	0
+1	+1	−1	+1
−1	0	−1	0
0	−1	−1	0
−1	−1	+1	−1
−1	+1	0	0
+1	−1	0	0

type of truth table could be drawn up for the 0, 1, 2 ternary system. The notation used does not affect the issue.

The desired ternary adder can be described by means of the following equations:

$$S_0 = a'b' + a_{-1}b_{+1} + a_{+1}b_{-1} \tag{20-4}$$

$$S_{+1} = a_{+1}b' + a'b_{+1} + a_{-1}b_{-1} \tag{20-5}$$

$$S_{-1} = a_{+1}b_{+1} + a_{-1}b' + a'b_{-1} \tag{20-6}$$

$$C_0 = a'b' + a_{+1}b' + a'b_{+1} + a_{-1}b' + a'b_{-1} + a_{-1}b_{+1} + a_{+1}b_{-1} \tag{20-7}$$

$$C_{+1} = a_{+1}b_{+1} \tag{20-8}$$

$$C_{-1} = a_{-1}b_{-1} \tag{20-9}$$

This circuit can be implemented by means of 52 diodes; 23 diodes are required for the adding circuit of two binary digits. (It is true that these circuits can be simplified to a certain extent, but the improvement is proportionally the same for both types of circuits.) The amounts 52 and 23 cannot be compared on the basis of their appearance alone, because of the better efficiency grading of the ternary system. An insight into the magnitude can roughly be achieved by noting the fact that $3^5 \approx 2^8$—that is, six sections of a ternary adder are equivalent to about nine sections of a binary adder; thus, the ternary switching circuits need about $1\frac{1}{2}$ times as many components as do the equivalent binary circuits.

A ternary full adder of two numbers can be designed by means of Kirchhoff networks with about the same amount of hardware as a binary full adder. In a numerical system with radix n, the sum of two numbers can be as a maximum $(n-1)+(n-1)=(n-2)+n=(n-2)+C_1$. This means that the first section of an adder can generate a carry $C=1$. The sum of the two numbers and carry $C=1$ from the first section can amount in the second section to, at most, $(n-1)+(n-1)+1=(n-1)+n=(n-1)+C_1$. This means that in the addition of two numbers in a numerical system with radix n, the maximum carry between two sections of an adder can be $C=1$. Table 20-5 gives the truth table of a section of a ternary full adder. The addend and augend are a and b, the incoming carry is c, the resulting sum is S, and the outgoing carry is C_o.

Table 20-5

a	b	c	S	C_o
0	0	0	0	0
1	0	0	1	0
0	1	0	1	0
1	1	0	2	0
2	0	0	2	0
0	2	0	2	0
2	1	0	0	1
1	2	0	0	1
2	2	0	1	1
0	0	1	1	0
1	0	1	2	0
0	1	1	2	0
1	1	1	0	1
2	0	1	0	1
0	2	1	0	1
2	1	1	1	1
1	2	1	1	1
2	2	1	2	1

The outgoing carry can be generated easily by means of a Kirchhoff network and an associated transistor circuit of the type shown in Fig. 20-4. The signal voltages corresponding to the digit values 0, 1, and 2 are assumed to be 0, 10, and 20 volts. The Kirchhoff network shown in Fig. 20-4 consists of three equal-value resistors combined in a Y circuit. The signal voltages

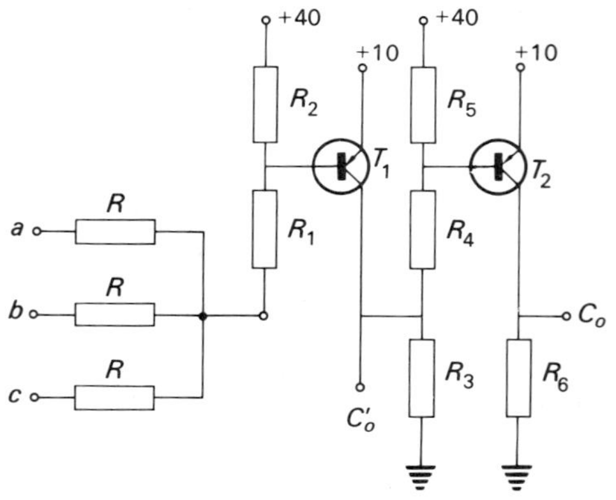

Fig. 20–4

of addend, augend, and carry-in are connected to points a, b, and c, respectively. Since the signal voltages can be 0, 10, or 20 volts, the star point of the Kirchhoff network can assume six voltage levels above 0, as is indicated in Table 20-6. The number of the voltage level in Table 20-6 corresponds to the sum of $a + b + c$ of Table 20-5.

The outgoing carry C_o can be obtained simply by a discrimination between the second and third switching levels. This discrimination can be performed by means of transistors with a rather sharp transition from the

Table 20-6

Switching level	*Volts*
6	20
5	16.6
4	13.3
3	10.0
2	6.6
1	3.3
0	0

open to the closed state. In this case, however, the voltage intervals obtained by means of the Kirchhoff network are only one-third of the assumed standard interval of the control voltage of 10 volts. It is necessary to restore the 0-to-6.6-volt output voltages of the Kirchhoff network to the standard 0 level and the 10-to-20-volt output voltages to the standard 10-volt level.

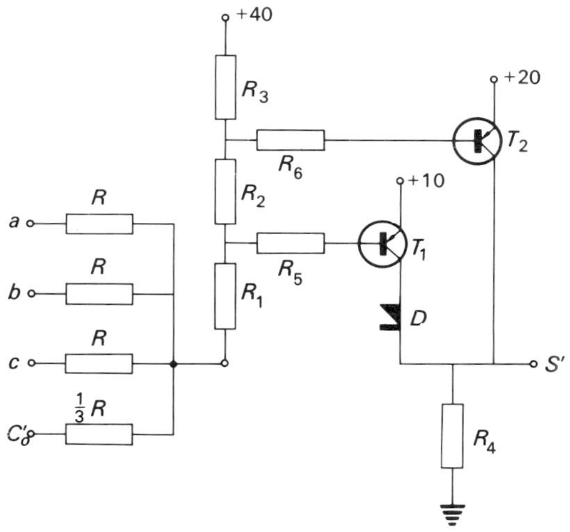

Fig. 20–5

Voltage divider $R_{1/2}$, which has a high value compared with the resistors of the Kirchhoff network, shifts the resulting voltage level of the Kirchhoff network so that transistor T_1 is blocking from level 3 upwards.

By means of a second transistor circuit (T_2), the outgoing carry C_o can be obtained in the correct phase.

Sum S cannot be obtained in such an easy way, because S can be either 0, 1, or 2 and, moreover, the sum S has to be 0 for switching levels 0 and 3, 1 for levels 1 and 4, and 2 for levels 3 and 5, as can be seen in Table 20-5. This operation can be performed by means of a Kirchhoff network of the type shown in Fig. 20-5, provided with a fourth resistor $\frac{1}{3}R$ connected to input terminal C_o'. Input terminal C_o' must be connected to the output terminal with the same designation as in the circuit shown in Fig. 20-4. As can be seen in Table 20-5, the shifting down of output signal S can be performed under the control of outgoing carry signal C_o'. The resulting output voltages of the Kirchhoff network of Fig. 20-5 are shown in Table 20-7.

A discrimination must be made between 5 and 6.66 volts, and another discrimination has to be made between 6.66 and 8.33 volts in order to distinguish S_0, S_1, and S_2. This can be performed by means of the transistor circuit connected to the Kirchhoff network as shown in Fig. 20-5. Voltage

Table 20-7

Switching level	C_o'	*Volts*
6		
5	0	8.33
4	0	6.66
3	0	5.00
2	+10	8.33
1	+10	6.66
0	+10	5.00

divider $R_{1/2/3}$ with two taps shifts the output voltages of the Kirchhoff network, so that for a switching level 0 of input signals a, b, and c and $C_o'=1$ transistor T_2 is made conducting. Output signal S' is then 20 volts. This blocks diode D, so that transistor T_1 is ineffective. Transistor T_2 is blocked for a combination of input signals a, b, and c in switching level 1 and $C_o'=1$, and transistor T_1 then becomes conducting. The output signal S' is thus 10 volts. Both transistors T_1 and T_2 are blocking for every combination of input signals of level 2 and $C_o'=1$. The output signal voltage S' in this case is 0. The same sequence of output voltages S' is repeated for input switching levels 3, 4, and 5 of a, b, and c and $C_o'=0$. The output terminal of the circuit of Fig. 20-5 produces S', which can be restored to normal by connecting this signal to the ternary trigger shown in Fig. 20-1.

In terms of hardware, the ternary addition performed in the manner described above is equivalent to the binary addition by means of Kirchhoff networks. Both have a rather large number of switching and voltage levels in common between which a discrimination process is required. From the point of view of discrimination of voltage levels, transistors may be better than high-vacuum tubes, because transistors have a very rapid cutoff of the collector current. However, in view of the rather large number of switching levels with rather narrow voltage intervals, some stabilization of the power supply is required. It is possible to extend this marginal switching principle to coding systems with a radix larger than 3, but this would be too complex from a practical point of view.

It is possible to design a computer operating in the ternary system, but the communication of that computer with the outside world would always have to be performed via a code converter, because all practical communication means are based on binary methods (punched tapes, magnetic tapes, indicator lamps). Such code converters can be designed by electronic means, but the introduction of these elements would merely complicate the circuitry. A ternary switching algebra has been developed by Mühldorf.

20-3 SYSTEMS WITH A RADIX LARGER THAN 3

From the point of view of storage efficiency the quaternary system ($b = 4$) is qualitatively similar to the binary system. It is possible to design trigger circuits with four or even more stable states. From the point of view of code conversion, the objections against a quaternary system are about as serious as those against the binary system. The code conversion from a quaternary system into a binary system and the reverse is easy. Each group of two binary digits corresponds to one quaternary digit.

The three possible designs for a quaternary storage element are

1. A four-state trigger producing a marginal output signal on one terminal (0, 1, 2, 3)
2. A four-section four-state trigger producing a one-out-of-four output signal on one of four channels
3. A four-state trigger built up of two binary triggers, which is equivalent to two ordinary binary storage elements for two binary digits

The last design needs no further explanation. That under 2 is not practical compared with the one under 3 because the information flow is divided over four channels in the second case, as opposed to two channels in the third case. From the practical point of view, solution 1 is also less suitable for implementation.

In principle, however, it is quite possible to design triggers with any number of stable states. There are various possible solutions. Figure 20-6 shows a circuit for a four-state trigger using a diode-triode technique. It can easily be converted into a diode-transistor circuit with a larger number of stable states.

The circuit shown in Fig. 20-6 consists of the following elements:

1. A trigger with tubes T_1 and T_2, anode resistors $R_{1/2}$ and $R_{5/6}$, and cross-connected voltage dividers $R_{9/10}$ and $R_{12/13}$. This part of the circuit has to be dimensioned in such a way that the amplification of the tube circuits is 1, so that this trigger alone can remain in all stable states.
2. A circuit to provide extra control voltages to the grids of tubes T_1 and T_2. This network consists of voltage dividers R_{15-19} and R_{20-24} and diode gates D_{1-12} and D_{13-24}. The output voltages at points X and Y of this network are connected via resistors R_{11} and R_{14} to the grids of tubes T_1 and T_2.

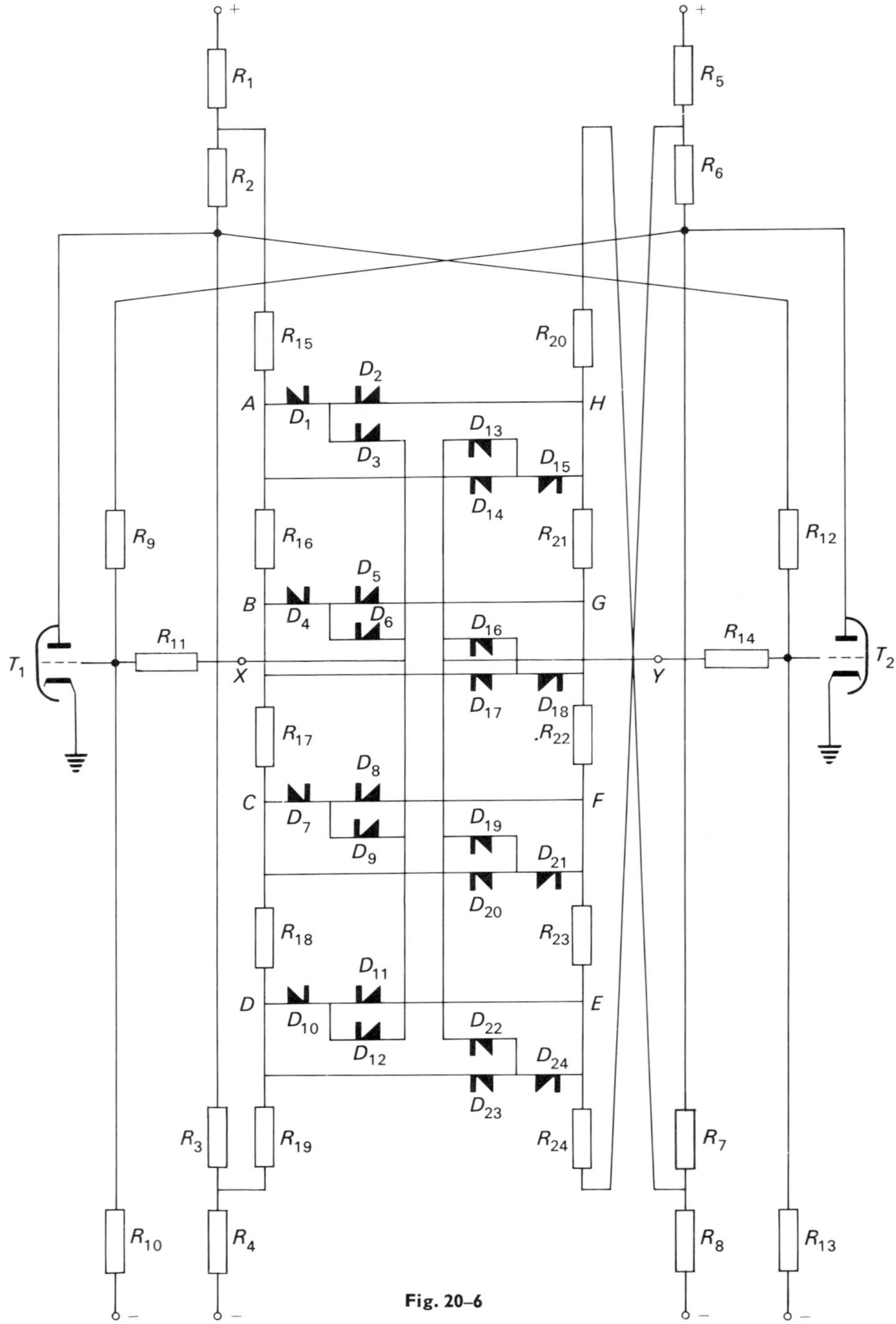
R1
R2
R3
R4
R5
R6
R7
R8
R9
R10
R11
R12
R13
R14
R15
R16
R17
R18
R19
R20
R21
R22
R23
R24
D1
D2
D3
D4
D5
D6
D7
D8
D9
D10
D11
D12
D13
D14
D15
D16
D17
D18
D19
D20
D21
D22
D23
D24
A
B
C
D
E
F
G
H
X
Y
T1
T2
+
−

Fig. 20–6

Voltage dividers $R_{1/2}$, $R_{3/4}$, $R_{5/6}$, and $R_{7/8}$ are so dimensioned that the voltage drop of points A to D and E to H as a function of the grid voltage have parallel characteristics.

Diode gates $D_{1/2}$, $D_{4/5}$, $D_{7/8}$, and $D_{10/11}$ select the most positive, and diode gates $D_{14/15}$, $D_{17/18}$, $D_{20/21}$, and $D_{23/24}$ the most negative of all voltages of the pairs of points A/H, B/G, C/F, and D/E, while diode gate $D_{3/6/9/12}$

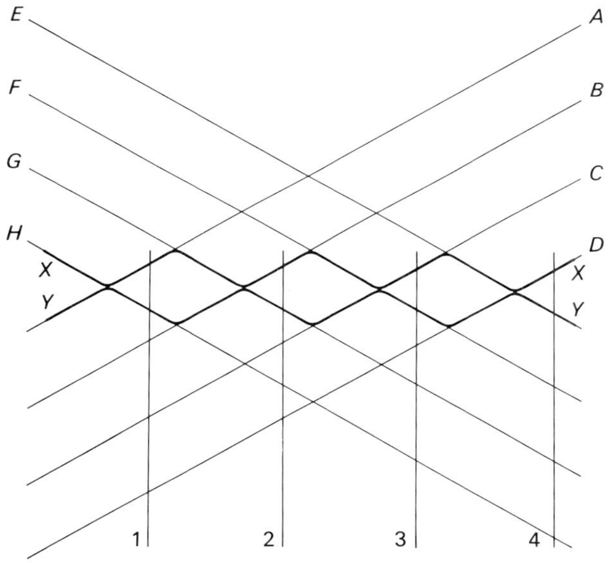

Fig. 20–7

selects the most negative voltage in the first group and diode gate $D_{13/16/19/22}$ selects the most positive voltage in the second group of voltages. The output voltages of both parts of the diode network are shown as heavy lines X and Y in the diagram of Fig. 20-7.

The effect of the diode network on the grids of both tubes is such that the grid of tube T_1 receives a control voltage from its own voltage divider R_{15-19} when the X voltage line increases, while the grid of tube T_2 receives a control voltage from its own voltage divider R_{20-24} when the Y voltage line decreases. The circuit will be in a stable condition in the parts indicated by the four vertical lines in Fig. 20-7.

During the decreasing portions of the X voltage line and the increasing portions of the Y voltage line, the control grid of tube T_1 is cross-connected to voltage divider R_{20-24}, and the control grid of tube T_2 is cross-connected to voltage divider R_{15-19}, so that the triggering effect is increased. There will be no stable state during these parts of the characteristics.

It is not strictly necessary to use a double-diode network as shown in Fig. 20-6. One-half of that circuit would be adequate. The complexity of this type of circuit for a 10-state trigger is already such that it requires about as much hardware as 4 binary triggers.

The use of multistate triggers is theoretically possible, but involves too many marginal levels. The ternary system seems to be the limit of what is practicable with respect to marginal levels under the present conditions.

There are some arguments in favor of a code with radix 8. Three binary memory elements are sufficient for the storing of one octonary digit. The possibility of errors due to the similarity between octonary and the decimal numbers is so great, however, that the use of this system should be avoided for practical reasons.

The radix 10 gives the widely used and well-known decimal system, which is very well adapted to the needs of practically every type of computation work. Nevertheless, since it involves the use of 10-state switching circuits, it is a rather unsuitable system from the point of view of technical feasibility.

The best way of implementing switching circuits for the decimal number system is to use a system with radix 16 from which the numbers 10 up to 15 are omitted. In this system, one digit can be stored in four binary elements. This type of coding is frequently referred to as the "binary-decimal" system. This type of coding for the machine language gives the least number of difficulties as regards the conversion of the decimal into the binary-decimal machine code.

The fact that 6 out of the 16 numbers of the system with radix 16 are not used in the binary-decimal coding system allows for a great number of variations in the binary coding of decimal digits. However, most of the practical applications of binary-decimal codes are limited to systematically weighted codes.

20-4 THE BINARY-DECIMAL SYSTEM

POSITIVELY WEIGHTED FOUR-BIT CODES

The first group of binary-decimal codes covers codes with systematically weighted elements with positive values only. There are some rules governing the systematic weighting of the code elements. The sum of the weights of the 1 elements gives the decimal number of that code symbol. This requirement results in the two following rules:

1. The weight of a digit or element of a code symbol is at most equal to the sum of the weights of the preceding digits or elements plus 1.
2. The sum of the weights of all (four) digits or elements must be at least 9.

These rules result in the 17 possible 4-bit binary-decimal codes given in Table 20-8. All these codes are equivalent with respect to the amount of hardware required for translating from the decimal into the binary system, and vice versa.

The 8421 code is the most practical code from the point of view of arithmetic operations involving decimal digits. The meaningless symbols 10 to 15 have to be eliminated after each processing step. Inversion of the elements of the code symbols yields the so-called fifteens complement. For some types of operation, the nines complement can be useful; it can be obtained by subtracting 6 from the result after inversion, or by adding 6 to the digit before inversion. There is one code given in Table 20-8 which is self-complementary with respect to 9, that is, the 4221 code, which is given in Table 20-9 as a 2421 code. The self-complementing features of this code are better illustrated by means of this table.

Table 20-8

8421	7321	6221	6311	5211
7421	6321	5221	5311	
6421	5321	4221	4311	
5421	4321			
4421	3321			

Table 20-9

No.	2	4	2	1	*No.*	2	4	2	1
0	0	0	0	0	9	1	1	1	1
1	0	0	0	1	8	1	1	1	0
2	0	0	1	0	7	1	1	0	1
3	0	0	1	1	6	1	1	0	0
4	0	1	0	0	5	1	0	1	1

A nonweighted four-bit code can be derived from the binary 8421 code. It is the so-called "excess-three" code. This code can be obtained by adding a binary 3 to each of the code symbols of the conventional 8421 code. This code is given in Table 20-10.

Table 20-10

Excess three

0	0	0	1	1	9	1	1	0	0
1	0	1	0	0	8	1	0	1	1
2	0	1	0	1	7	1	0	1	0
3	0	1	1	0	6	1	0	0	1
4	0	1	1	1	5	1	0	0	0

The elementary arithmetical operations are rather easy to perform with this systematic code, although it is in fact the nonweighted type. A binary 3 must be subtracted from each sum of two excess-three code symbols. The manipulation is completely binary. The decimal carry is readily obtained; it is in fact the carry generated by the highest-order binary digits because the code symbol resulting from each sum is in fact excess-six, whereby the six unwanted binary code symbols are eliminated.

POSITIVELY AND NEGATIVELY WEIGHTED FOUR-BIT CODES

Codes in which one or more of the elements have a constant negative weight are not very much in use. Some reflected codes in which successive code symbols differ in one element only have found practical applications. The reflected binary code described in Chap. 15 and shown here as Table 20-11 leads to some interesting self-complementary reflected binary-decimal codes.

These codes have found important applications in the field of analog-to-digital conversion, where they are used for the code-converting discs that produce a reflected binary-decimal code. A $9 \pm 7 \pm 3 \pm 1$ reflected binary code can be easily derived from the $15 \pm 7 \pm 3 \pm 1$ reflected binary code. This code is shown in Table 20-12. The self-complementary effect is obtained by inverting only the element with the value 9. It is also possible to derive a nonweighted excess-three reflected binary code from the one shown in Table 20-12. This excess-three code is shown in Table 20-13.

It is rather difficult, however, to perform arithmetical operations with these codes without converting them into the ordinary binary code.

Table 20-11

No.	15	±7	±3	±1
0	0	0	0	0
1	0	0	0	1
2	0	0	1	1
3	0	0	1	0
4	0	1	1	0
5	0	1	1	1
6	0	1	0	1
7	0	1	0	0
8	1	1	0	0
9	1	1	0	1
10	1	1	1	1
11	1	1	1	0
12	1	0	1	0
13	1	0	1	1
14	1	0	0	1
15	1	0	0	0

Table 20-12

No.	9	±7	±3	±1
0	0	0	0	0
1	0	0	0	1
2	0	0	1	1
3	0	0	1	0
4	0	1	1	0
5	1	1	1	0
6	1	0	1	0
7	1	0	1	1
8	1	0	0	1
9	1	0	0	0

Table 20-13

No.	Excess three			
0	0	0	1	0
1	0	1	1	0
2	0	1	1	1
3	0	1	0	1
4	0	1	0	0
5	1	1	0	0
6	1	1	0	1
7	1	1	1	1
8	1	1	1	0
9	1	0	1	0

20-5 ERROR-DETECTING CODES

In all systems in which digital information has to be transmitted from one place to another, whether remote or near, or that processes such information, there is a risk that this information might be garbled by faulty operation of the circuits owing to noise or any other interference or failure of components. For example, in nationwide telephone dialing systems and in computers in which very many operations are performed in a rather short time interval, it is essential not only that errors leading to wrong numbers or faulty computation results be detected, but also that the part of the system in which the fault occurred be automatically traceable.

It is true that much can be done by preventive maintenance, such as marginal tests and routine tests, but intermittent faults can still remain undetected during these tests and cause faulty operation of the machines during the actual work. There are three main means of transmission of digital information which should be considered:

1. The dialing in automatic telephone and telegraph networks. Many subscribers dial simultaneously and use parallel paths throughout the system. A faulty path can affect one subscriber at a time and does not lead to a complete blocking of the system. The transmission channels, provided that they do not comprise radio circuits, are not very vulnerable to interference from the outside.
2. The transmission of digital information in long-distance telegraph channels consisting of many links. Radio links can be vulnerable to interference from the outside, i.e., radio interference.
3. The transmission of information in computers. The information is passed through the same paths very many times before the desired result is obtained. These paths are to a very high degree invulnerable to outside interference, but the number of times that the information passes through these same pieces of equipment can be so great that even the slightest intermittent component failure can disturb the correct operation.

The transmission of digital information can be protected by means of error-detecting and/or error-correcting codes. This will be explained by means of codes using symbols with a constant number of elements.

In a code without the slightest redundancy, i.e., a code in which all possible combinations or symbols have a meaning, every error changes a code symbol into another possible code symbol. There is a so-called "distance" of at least one between the symbols of such codes. Error-detecting and/or error-correcting features can be obtained by adding redundancy in a systematic way.

The introduction of a single-parity-check bit, that is, the addition of an extra element to the code symbols by which the number of 1's of each symbol is, for example, made odd, enlarges the minimum distance of the code symbols to two. This distance is the number of elements in which two code symbols differ. The parity check gives in an encrypted way one additional information item about all elements of the protected symbol. A single error mutilates such a code symbol into a recognizable nonexistent code symbol. One-half of the possible code symbols of a code with n elements contains an odd number of 1's—these are the checked symbols—while the other half with an even number of 1's are used as nonexistent symbols. This is the odd-parity check. It is also possible to use an even-parity check.

If the total number of elements or digits of a code is n and the number of information elements is m, the redundancy can be defined as

$$R = \frac{n}{m} \tag{20-10}$$

An example of a code with an odd-parity check is given in Table 20-14. A four-digit binary code (8421 code) is used in this example, but the system is applicable to all other types of codes.

Table 20-14

		Information						*Information*			
No.	*Check*	8	4	2	1	*No.*	*Check*	8	4	2	1
0	1	0	0	0	0	8	0	1	0	0	0
1	0	0	0	0	1	9	1	1	0	0	1
2	0	0	0	1	0	10	1	1	0	1	0
3	1	0	0	1	1	11	0	1	0	1	1
4	0	0	1	0	0	12	1	1	1	0	0
5	1	0	1	0	1	13	0	1	1	0	1
6	1	0	1	1	0	14	0	1	1	1	0
7	0	0	1	1	1	15	1	1	1	1	1

This simple one-bit odd-parity check detecting one single error only is in certain respects a weak test. Depending on the origin of the error (intermittent component faults, radio interference, or other sources), it can easily happen that a hit or a miss affects two or more adjacent digits of a code symbol, so that there is a good chance that a good signal will be mutilated and converted into one of the other good symbols.

For this purpose, a more thorough parity check will be obtained by limiting the code symbols to those symbols with the same number of 1's. The 10 decimal digits can be represented by a two-out-of-five code, the standard teleprinter alphabet by means of a three-out-of-seven code. The number of code symbols in such types of codes is given by the equation

$$P_m^n = \frac{n!}{(n-m)!\,m!} \tag{20-11}$$

Some of the numbers of the groups of code symbols resulting from Eq. (20-11) are given in Table 20-15.

Table 20-15

m	n 5	6	7	8	9	10
2	10	15	21	28	36	45
3	10	20	35	56	84	120
4	5	15	35	70	126	210

The resulting codes used for representing the 10 decimal digits are now no longer systematic. Two codes that are near-systematic are given as Tables 20-16 and 20-17. In Table 20-16 the 0 symbol is $7 + 4$. This is the only exception in that code table to the systematic weight system. In Table 20-17, the elements of the right-hand columns have varying weights.

Table 20-16

No.	7	4	2	1	0
0	1	1	0	0	0
1	0	0	0	1	1
2	0	0	1	0	1
3	0	0	1	1	0
4	0	1	0	0	1
5	0	1	0	1	0
6	0	1	1	0	0
7	1	0	0	0	1
8	1	0	0	1	0
9	1	0	1	0	0

Table 20-17

No.	4	3	2	1/6	0/5
0	1	0	0	1	0
1	0	0	0	1	1
2	0	0	1	0	1
3	0	0	1	1	0
4	0	1	0	1	0
5	0	1	1	0	0
6	1	0	1	0	0
7	1	1	0	0	0
8	0	1	0	0	1
9	1	0	0	0	1

The advantage of the m-out-of-n codes is that the check fails only if within the time interval of one complete code symbol one or more pairs of opposing errors occur. Experience gained in radio telegraphy has shown that, in general, radio interference does not lead to hits and misses within the time period of one telegraph signal. Radio interference has the character of a burst.

It should be noted that the simple parity check detecting in fact an odd number of errors in a code symbol can be made to perform almost as well as the check in the m-out-of-n code if the detection device is provided with an extra checking circuit for the unallowable code combinations.

An m-out-of-n code that has come into widespread use for radio transmission purposes is the three-out-of-seven code which provides for thirty-five possible code symbols and which is recommended by the Comité Consultatif International Télégraphique et Téléphonique (CCITT).

20-6 ERROR-CORRECTING CODES

SINGLE-ERROR CORRECTION

The single-error-detecting features of a code can be extended to at least single-error-correcting features by adding more redundancy, that is, more superfluous information about the actual message content of the code symbol. The simplest and at the same time the most expensive way to form an error-correcting code would be to transmit all information three times. The number n of elements contained in the code symbols of such a code is always three times greater than the number of information elements m which are actually required. Hence [see Eq. (20-10)]

$$R = \frac{n}{m}$$

In such codes, the majority of analogous elements will decide on the resulting value of a symbol element. An error in two analogous elements results in a fault. On the other hand, m errors in any group of m successive elements in a code symbol do not lead to an erroneous symbol.

In some types of transmission systems, the signal elements are sensed three times at different moments. This can yield good results in those systems in which the duration of a fault is short compared with the duration of the elements of the code symbols.

Before going into a discussion of the various methods of generating error-correcting codes, an attempt will be made to limit this consideration to a single form of error correction.

Error correction can be wanted for two different types of errors:

1. Arbitrarily occurring errors
2. Burst errors

In radio-transmission systems, errors are generally burst errors, that is, errors of such a nature that during a short interval all 1 elements of a group are changed into 0 or the reverse. It practically never happens that in one and the same burst the fault character reverses.

In cable-transmission systems, many errors stem from short interruptions in the transmission. The shorter the interruptions, the more frequent they are. With a very high occurrence of these interruptions in transmission, one should be inclined to seek a remedy in the use of error-correcting codes with the power to correct arbitrarily occurring errors. However, the power of any error-correcting code is limited to a certain number of errors within a code symbol. The use of an error-correcting code is no guarantee against faultless transmission. Even the use of an error-correcting code can increase the number of errors when the probability of error occurrence is high. In such a case, a high redundancy and consequently a large number of check elements are required. This problem will be discussed later.

The following discussion will be mainly limited to the type of error correction that is the most likely in radio and cable transmission. It can be applied successfully in cases in which the occurrence of two burst errors within the same code symbol is improbable. The required redundancy and consequently the required number of check elements are not very great.

The power of a group of code symbols to correct independent errors is related to the so-called distance d of the symbols. The distance d is the number of elements in which the symbols differ. In a group of code symbols with the distance $d = 1$, each error in an element of a symbol causes a change into another symbol of that group. A group of code symbols of this type has no redundancy for error detection and error correction. A single parity check on all the elements of the symbols widens the distance d of the symbols to $d = 2$. In fact, the parity check doubles the number of possible symbols; but for actual purposes, only half of that group is used. A single error changes a symbol of the used group into one of the unused group, and the change can be detected. However, it is impossible to correct the erroneous symbol.

Widening the distance to $d = 3$, so that the symbols of the used group differ in at least three elements, allows a correction of one element per code symbol. A single error changes a used symbol into an unused symbol, but that unused symbol cannot originate by a single error from any other of the used symbols of that group. For larger distances, the situation becomes more complicated; and this situation will be discussed later. However, it can be stated here that for the correction of two arbitrary errors in a code symbol, a distance $d = 5$—and for three arbitrary errors, a distance $d = 7$, etc.—is required.

The distance d of the symbols of a code can be extended to any desired value by means of an extra number of elements representing parity checks that must be applied as evenly as possible on the information elements. For a given distance d, the check elements of the symbols must contain at least $d - 1$ independent messages about each of the corresponding information elements. The check elements contain these messages in an encrypted form.

The power of a group of code symbols to correct burst errors is not directly related to the concept of distance between the code symbols of the group. This statement can be illustrated by the fact that, for example, code symbols consisting of 15 elements, that is, 3 times the same group of 5 information elements, has the power to correct any combination of up to 5 errors in its elements provided that they occur in a group of 5 successive elements. The minimum distance in this group of code symbols is three, so that when using the concept of distance, the power of error correction of this of code is limited to a single error.

Check								8				4		2	1
Information	15	14	13	12	11	10	9		7	6	5		3		
1	x		x		x		x		x		x		x		X
2	x	x			x	x			x	x			x	X	
4	x	x	x	x					x	x	x	X			
8	x	x	x	x	x	x	x	X							

Fig. 20–8

It should be noted that for the purposes of design of practical error-correcting equipment, the elements of the code symbols are divided in a group of information elements and a group of check elements. Both groups of elements together form the code symbols with the required error-correcting power. This excludes certain groups of symbols from application in the error-correcting systems.

Hamming discovered error-correcting codes and gave a solution for the single-error-correcting case. His method will be explained by means of Fig. 20-8. In this figure, code symbols with 15 elements are used. These elements are numbered 1 to 15 and subdivided into a row of information elements and a row of check elements. The check elements are numbered 1, 2, 4, 8, etc., which are the numbers of the fundamental weights of the binary number system. These check elements are defined by the following equations:

$$
\begin{aligned}
y_1 &= y_3 \oplus y_5 \oplus y_7 \oplus y_9 \oplus y_{11} \oplus y_{13} \oplus y_{15} \\
y_2 &= y_3 \oplus y_6 \oplus y_7 \oplus y_{10} \oplus y_{11} \oplus y_{14} \oplus y_{15} \\
y_4 &= y_5 \oplus y_6 \oplus y_7 \oplus y_{12} \oplus y_{13} \oplus y_{14} \oplus y_{15} \\
y_8 &= y_9 \oplus y_{10} \oplus y_{11} \oplus y_{12} \oplus y_{13} \oplus y_{14} \oplus y_{15}
\end{aligned}
\tag{20-12}
$$

By this definition, check elements y_1, y_2, y_4, y_8, etc., contain at least two independent messages about each of the information elements, as can easily be verified in the above equations. The x's in Fig. 20-8 show from right to left an arrangement in accordance with the binary system. If the rows are given the weights 1, 2, 4, and 8 from top to bottom, the checks concerning a specific element of the symbol can be numbered as indicated in the top row.

Code symbols of this construction may be garbled during transmission in one of the elements, information or check. At the receiving end of the connection, this error may be detected and corrected by a recheck of all elements, as given by the following equations:

$$\begin{aligned}
Y_1 &= y_1 \oplus y_3 \oplus y_5 \oplus y_7 \oplus y_9 \oplus y_{11} \oplus y_{13} \oplus y_{15} \\
Y_2 &= y_2 \oplus y_3 \oplus y_6 \oplus y_7 \oplus y_{10} \oplus y_{11} \oplus y_{14} \oplus y_{15} \\
Y_4 &= y_4 \oplus y_5 \oplus y_6 \oplus y_7 \oplus y_{12} \oplus y_{13} \oplus y_{14} \oplus y_{15} \\
Y_8 &= y_8 \oplus y_9 \oplus y_{10} \oplus y_{11} \oplus y_{12} \oplus y_{13} \oplus y_{14} \oplus y_{15}
\end{aligned} \tag{20-13}$$

When no error has garbled the transmission, the recheck Y_8, Y_4, Y_2, $Y_1 = 0\,0\,0\,0$. When an error has occurred, the recheck will show the binary number of the erroneous element because only those receiving checks will result in a 1 which are involved in the encrypted message about that element. The erroneous element can be corrected by inversion.

From a practical point of view, Hamming's ingenious method shows a disadvantage in that the checking means are rather complicated and the method is applicable only to code symbols that allow the correction of a single erroneous element.

After Hamming, codes with greater error-correcting capabilities have been devised by Abramson, Melas, Fire, Green and San Soucie, and Bose and Ray-Chauduri. Finally, Meggitt has shown the common base of these codes and has devised practical means for transmitting and receiving error-correcting code symbols.

The possibilities of the use of cyclic burst-error-correcting codes will be explained by means of the properties of mod 2 feedback shift registers, as discussed in Chap. 16. Before the construction of transmitters and receivers for these codes can be treated, some further properties of mod 2 feedback shift registers will be considered.

As a consequence of the discussion in Chap. 16, output sequence P of the mod 2 adding feedback network of the shift register shown in Fig. 20-9 can be represented as a function of the feedback constellation $f(x) = \sum_{j=1}^{k} c_j x^j$, the initial-state polynomial $m(x)$ determined by the initial state $y_{-1}, y_{-2}, \ldots, y_{-k}$ of the shift register and input sequence T.

$$P(x) = T(x) \sum_{j=1}^{k} c_j x^j \oplus m(x) \tag{20-14}$$

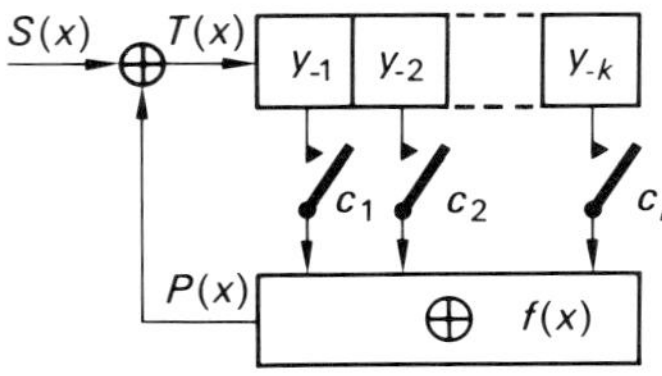

Fig. 20–9

Furthermore, the constellation of the feedback network was expressed by characteristic polynomial

$$\phi(x) = 1 \oplus f(x) \tag{20-15}$$

so that

$$P(x) = T(x)\,[1 \oplus \phi(x)] \oplus m(x) \tag{20-16}$$

In the circuit of Fig. 20-9, input sequence T of the shift register is formed by sequences P and S:

$$T(x) = P(x) \oplus S(x) \tag{20-17}$$

or

$$T(x) \oplus S(x) = T(x)\,[1 \oplus \phi(x)] \oplus m(x)$$

$$S(x) = T(x) \cdot \phi(x) \oplus m(x)$$

and

$$T(x) = \frac{S(x)}{\phi(x)} \oplus \frac{m(x)}{\phi(x)} \tag{20-18}$$

If the initial state is 0, 0, ..., 0, then $m(x) = 0$ so that

$$T(x) = \frac{S(x)}{\phi(x)} \tag{20-19}$$

Substitution of Eq. (20-18) into Eq. (20-17) gives

$$\frac{S(x)}{\phi(x)} \oplus \frac{m(x)}{\phi(x)} = S(x) \oplus P(x)$$

or

$$P(x) = S(x) \oplus \frac{S(x)}{\phi(x)} \oplus \frac{m(x)}{\phi(x)}$$

or

$$P(x) = S(x)\left[1 \oplus \frac{1}{\phi(x)}\right] \oplus \frac{m(x)}{\phi(x)} \tag{20-20}$$

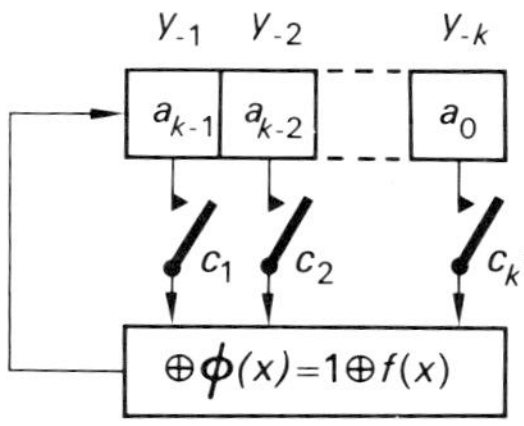

Fig. 20–10

Let us now consider a shift register that runs at a certain moment through the state determined by a_{k-1}, a_{k-2}, ..., a_o, as is shown in Fig. 20-10—a state to be considered further as initial state $(y_{-1}, y_{-2}, \ldots, y_{-k})$ of the next process step.

$$A(x) = a_0 \oplus a_1 x \oplus a_2 x^2 \oplus \cdots \oplus a_{k-1} x^{k-1} \tag{20-21}$$

The relation between $A(x)$ and $m(x)$ will now be determined. It follows from Fig. 20-10 that, in general, for all $i = 0, 1, 2, \ldots, k-1$,

$$y_{i-k} = a_i \tag{20-22}$$

Let characteristic polynomial $\phi(x)$ have period n; then S is also cyclic over n, so that

$$y_{i+n} = y_i \tag{20-23}$$

for all $i = -k, -k+1, -k+2, \ldots$.

Hence,

$$y_{i+n-k} = y_{i-k} = a_i \tag{20-24}$$

It follows from Chap. 16 that

$$m_{n-k}(x) = \frac{m(x) \oplus \phi(x) \sum_{i=0}^{n-k-1} y_i x^i}{x^{n-k}}$$

$$= x^{-(n-k)}\left[m(x) \oplus \phi(x)\left(\sum_{i=0}^{n-1} y_i x^i \oplus \sum_{i=n-k}^{n-1} y_i x^i\right)\right]$$

$$= x^{-(n-k)}\left\{m(x) \oplus \phi(x)\left[S(x) \oplus x^n S_n(x) \oplus \sum_{i=0}^{k-1} y_{i+n-k} x^{i+n-k}\right]\right\} \tag{20-25}$$

Because S is periodic over n, $S_n(x) = S(x)$; hence,

$$m_{n-k}(x) = x^{-(n-k)}\left\{m(x) \oplus \phi(x)\left[S(x) \oplus x^n S(x) \oplus x^{n-k} \sum_{i=0}^{k-1} y_{i+n-k} x^i\right]\right\} \tag{20-26}$$

Substitution of Eq. (20-24) into Eq. (20-26) gives

$$m_{n-k}(x) = x^{-(n-k)}\left\{m(x) \oplus \phi(x)\left[S(x) \oplus x^n S(x) \oplus x^{n-k} \sum_{i=0}^{k-1} a_i x^i\right]\right\}$$

or using Eq. (20-21),

$$m_{n-k}(x) = x^{-(n-k)}\{m(x) \oplus \phi(x)\,[S(x) \oplus x^n S(x) \oplus x^{n-k} A(x)]\} \tag{20-27}$$

This equation can be simplified, when substituting $S(x) = m(x)/\phi(x)$, to

$$m_{n-k}(x) = x^k m(x) \oplus A(x)\,\phi(x) \tag{20-28}$$

Hence

$$A(x)\,\phi(x) = m_{n-k}(x) \oplus x^k m(x) \tag{20-29}$$

Because the degree of initial-state polynomial $m_{n-k}(x)$ is at most $k-1$, $[m_{n-k}(x)]_k$ will be zero, so that

$$[A(x)\,\phi(x)]_k = m(x) = m^{(A)}(x) \tag{20-30}$$

in which $A(x)\phi(x)$ represents the sequence $A(x)\phi(x)$ shifted over k places—in the same way that $S_k(x)$ was obtained from $S(x)$.

In the construction of transmitters and receivers for error-correcting codes, the problem of connecting a shift register for a certain period of time with an incoming sequence of information elements will be encountered. This situation is shown in principle in Fig. 20-11. In this figure, an information channel is shown with sequence U and a switch SW connecting at moment t_1 the output to the input and disconnecting output and input at moment

t_2. Thus switch SW starts sequence W at moment t_1 and stops this sequence at moment t_2.

$$W(x) = \sum_{i=t_1}^{t_2-1} u_i x^i$$

or

$$W(x) = x^{t_1} U_{t_1}(x) \oplus x^{t_2} U_{t_2}(x) \tag{20-31}$$

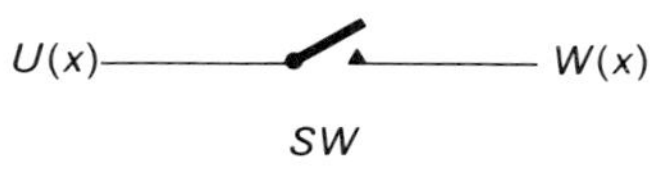

Fig. 20–11

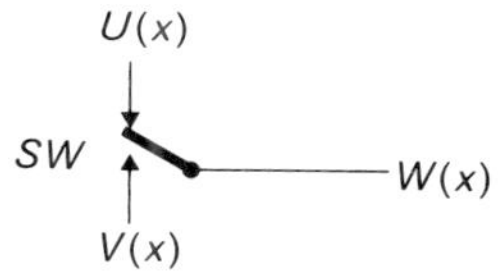

Fig. 20–12

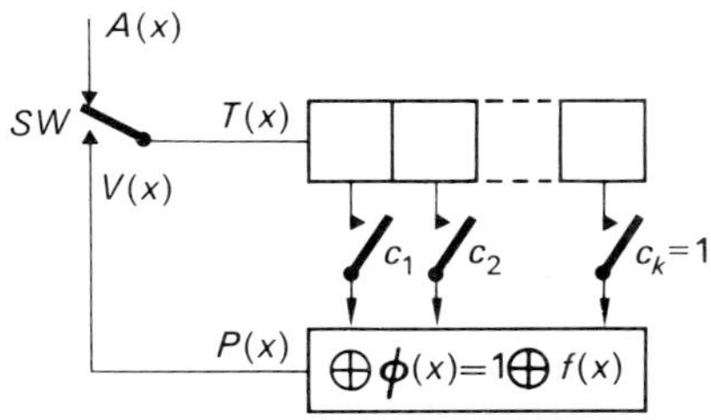

Fig. 20–13

Another situation to be considered, following from the last one, is shown in Fig. 20-12. In this circuit, switch SW stops the transmission of sequence U at moment t_1 and starts that of sequence V.

$$W(x) = \sum_{i=0}^{t-1} u_i x^i \oplus \sum_{i=t}^{\infty} v_i x^i$$

or

$$W(x) = U(x) \oplus x^t U_t(x) \oplus x^t V_t(x) \tag{20-32}$$

The above results will now be applied to the shift-register circuit shown in Fig. 20-13. Let sequence A be represented by the following polynomial:

$$A(x) = a_0 \oplus a_1 x \oplus \cdots \oplus a_{k-1} x^{k-1} \tag{20-33}$$

Then $A_k(x) = 0$. Switch SW will be reversed after k pulses. Sequence T can then be represented by the following equation:

$$T(x) = A(x) \oplus x^k A_k(x) \oplus x^k V_k(x) = A(x) \oplus x^k V_k(x) \tag{20-34}$$

Substitution of Eq. (20-34) into Eq. (20-16) gives

$$P(x) = A(x) \oplus x^k V_k(x) \oplus A(x)\,\phi(x) \oplus x^k V_k(x)\,\phi(x) \oplus m(x) \qquad (20\text{-}35)$$

Sequence P_k will follow from sequence P by shifting this sequence over k places. Because the degree of $m(x)$ and $A(x)$ is at most $k - 1$, the following result will be obtained:

$$\begin{aligned} P_k(x) &= A_k(x) \oplus [x^k V_k(x)]_k \oplus [A(x)\,\phi(x)]_k \oplus [x^k V_k(x)\,\phi(x)]_k \oplus [m(x)]_k \\ &= V_k(x) \oplus [A(x)\,\phi(x)]_k \oplus V_k(x)\,\phi(x) \end{aligned} \qquad (20\text{-}36)$$

Substitution of Eq. (20-30) into Eq. (20-36) leads to

$$P_k(x) = V_k(x)\,[1 \oplus \phi(x)] \oplus m^{(A)}(x) \qquad (20\text{-}37)$$

Switch SW will be reversed at moment n to the position shown in Fig. 20-13. Input sequence T of the shift register can then be described by the following equation:

$$T(x) = A(x) \oplus x^k A_k(x) \oplus x^k V_k(x) \oplus x^n V_n(x) \oplus x^n A_n(x) \qquad (20\text{-}38)$$

It follows from Eq. (20-33) that $A_k = 0$ and from $n > k$ that $A_n = 0$. This reduces Eq. (20-38) to

$$T(x) = A(x) \oplus x^k V_k(x) \oplus x^n V_n(x) \qquad (20\text{-}39)$$

This equation can now be substituted into Eq. (20-16), which leads to the following result:

$$P_k(x) = [V_k(x) \oplus x^{n-k} V_n(x)]\cdot[1 \oplus \phi(x)] \oplus m^{(A)}(x) \qquad (20\text{-}40)$$

In the circuit of Fig. 20-13,

$$V(x) = P(x) \qquad \text{and} \qquad V_k(x) = P_k(x) \qquad (20\text{-}41)$$

Substitution of Eq. (20-41) into Eq. (20-40) leads to the following equation for $V_k(x)$:

$$V_k(x) = \frac{m^{(A)}(x)}{\phi(x)} \oplus x^{n-k}\cdot\frac{V_n(x)}{\phi(x)}[1 \oplus \phi(x)] \qquad (20\text{-}42)$$

An expression for $V_n(x)$ can easily be obtained from Eq. (20-42) because

$$\begin{aligned} V_n(x) &= [V_k(x)]_{n-k} \\ &= \left[\frac{m^{(A)}(x)}{\phi(x)}\right]_{n-k} \oplus \frac{V_n(x)}{\phi(x)}[1 \oplus \phi(x)] \end{aligned} \qquad (20\text{-}43)$$

Finally, substitution of Eqs. (20-42) and (20-43) into Eq. (20-39) gives the final equation for the input sequence of the shift register of Fig. 20-13:

$$T(x) = A(x) \oplus x^k\cdot\frac{m^{(A)}(x)}{\phi(x)} \oplus x^n\left[\frac{m^{(A)}(x)}{\phi(x)}\right]_{n-k} \qquad (20\text{-}44)$$

TRANSMITTERS

In the above part of this section, the necessary preparations were given for the process of determining a code symbol of n bits, the first n bits of which are information bits and the last k bits are check bits. The degree of checking depends on the number of check bits in relation to the number of information bits in a code symbol and the characteristic polynomial of the feedback network of the shift register involved in the transmitting circuit. In the following, it is assumed that the check bits can be determined in the circuit of Fig. 20-14.

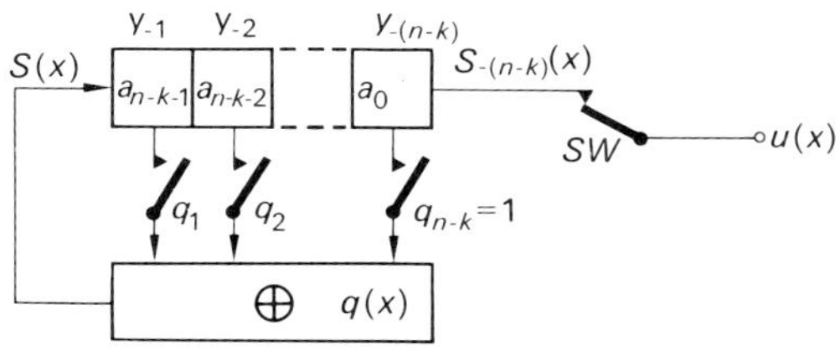

Fig. 20–14

In this figure, a mod 2 feedback shift register is shown with a number of $n - k$ sections, that is, the length of the information portion of the code symbol to be transmitted. The information bits are $a_{n-k-1}, a_{n-k-2}, \ldots, a_0$ corresponding to the initial state $y_{-1}, y_{-2}, \ldots, y_{-(n-k)}$. The switches connecting the shift register with the mod 2 adding feedback network are now called $q_1, q_2, \ldots, q_{n-k}$, in order to get a clear difference with respect to the transmitter circuit of Fig. 20-16. Characteristic polynomial $q(x)$ is chosen in such a way that its period is n. The generated sequence S (and thus also $S_{-(n-k)}$) is periodic over n. The transmitted sequence U consisting of n bits can be obtained by giving n shift pulses to the shift register preset with $n - k$ information bits. The k check bits are generated by the mod 2 adding feedback network.

The finite sequence u of n bits can be obtained from an infinite sequence by means of switch SW. Sequence u is then just one period of sequence $S_{-(n-k)}(x)$, $S(x)$ being the sequence fed into the shift register.

$$u(x) = S_{-(n-k)}(x) \oplus x^n[S_{-(n-k)}(x)]_n = S_{-(n-k)}(x) \oplus x^n S_k(x) \qquad (20\text{-}45)$$

Polynomial $S_{-(n-k)}(x)$ can easily be determined from $S(x)$ by the following equation:

$$S_{-(n-k)}(x) = x^{n-k}S(x) \oplus \sum_{i=0}^{n-k-1} y_{-(n-k)+i}\, x^i = x^{n-k}S(x) \oplus \sum_{i=0}^{n-k-1} a_i x^i$$

$$= x^{n-k}S(x) \oplus A(x) \qquad (20\text{-}46)$$

Substitution of Eq. (20-46) into Eq. (20-45) leads to

$$u(x) = A(x) \oplus x^{n-k}S(x) \oplus x^n S_k(x) \tag{20-47}$$

which is a rather obvious result.

Another formulation of $u(x)$ can be made using the fact that S and $S_{-(n-k)}$ have a period n which fact can be expressed by

$$S_{-(n-k)}(x) = S_k(x) \tag{20-48}$$

Substitution of Eq. (20-48) into Eq. (20-45) gives the following expression for $u(x)$:

$$u(x) = S_k(x) \oplus x^n S_k(x) = (1 \oplus x^n) S_k \tag{20-49}$$

Furthermore, if $m^{(A)}(x)$ is the initial-state polynomial corresponding to initial state $(a_{n-k-1}, a_{n-k-2}, \ldots, a_0)$, then

$$S(x) = \frac{m^{(A)}(x)}{q(x)} \tag{20-50}$$

and

$$S_k(x) = \frac{m_k{}^{(A)}(x)}{q(x)} \tag{20-51}$$

with

$$m_k{}^{(A)}(x) = \frac{m^{(A)}(x) \oplus q(x) \sum_{i=0}^{k-1} y_i x^i}{x^k} \tag{20-52}$$

Substitution of Eq. (20-51) into Eq. (20-49) leads to the following result:

$$u(x) = \phi(x) \cdot m_k{}^{(A)}(x) \qquad \text{with } \phi(x) = \frac{1 \oplus x^n}{q(x)} \tag{20-53}$$

In the above, polynomial $u(x)$ was generated by a mod 2 feedback shift register with $n - k$ sections and a feedback polynomial $q(x)$. Equation (20-53) shows that polynomial $u(x)$ can also be generated by another mod 2 feedback shift register with k sections and a characteristic polynomial $\phi(x)$ with $\phi(x) = (1 \oplus x^n)/q(x)$, which equation is the same as Eq. (16-97) giving the relation of $\phi(x)$ and the first cycle of the Q sequence $[m(x) = 1]$. From this, it may be clear why $q(x)$ was chosen as characteristic polynomial of the feedback network of the transmitting shift register of Fig. 20-14.

In the transmitting circuit of Fig. 20-14, the shift register was primed with the information bits of the code symbol to be transmitted. In many practical cases, it will be necessary to insert serially the k check bits after a presentation of the $n - k$ information bits in serial form. This can be performed in the transmitting circuit of Fig. 20-15.

In this case, polynomial $u(x)$ follows directly from Eq. (20-44) by replacing k by $n-k$:

$$u(x) = A(x) \oplus x^{n-k} \cdot \frac{m^{(A)}(x)}{q(x)} \oplus x^n \left[\frac{m^{(A)}(x)}{q(x)}\right]_k \tag{20-54}$$

Sequence $u(x)$ [(20-54)] is equivalent to sequence u, which is obtained from substitution of Eq. (20-50) into Eq. (20-47). It should be noted that switch SW has to be reversed at moments $n-k$ and n.

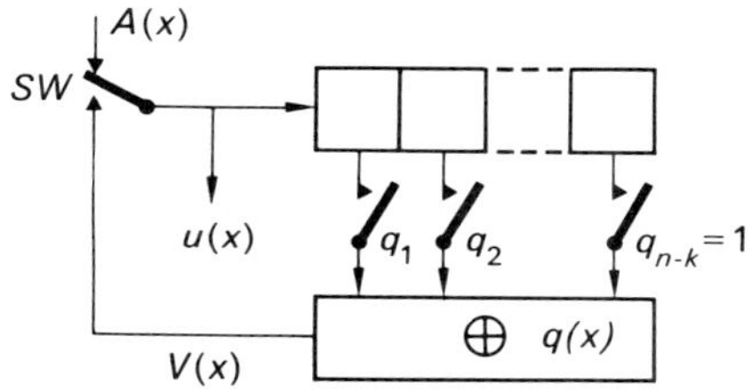

Fig. 20–15

In the two types of transmitting shift registers discussed above, the number of sections in the shift register is $n-k$, this being the degree of $q(x)$. In most practical cases of error-correcting codes, the number $n-k$ of information elements in a code symbol is much greater than the number of check elements k. The question arises whether or not it is possible to design a transmitter with a shift register of k sections generating error-correcting code symbols of n elements by means of a feedback network with characteristic polynomial $\phi(x)$. It is obvious that this shift register of k sections cannot be primed with the $n-k$ information elements $A(x)$ when $k < n-k$.

The transmitter circuits described above show that a code symbol can be represented as follows:

$$u(x) = A(x) \oplus x^{n-k} u_{n-k}(x) \tag{20-55}$$

Substitution of Eq. (20-53) into Eq. (20-55) leads to

$$u(x) = A(x) \oplus x^{n-k}[\phi(x) m_k^{(A)}(x)]_{n-k} \tag{20-56}$$

This equation shows that sequence $u(x)$ can be formed from a sequence $A(x)$ and a sequence $P(x)$ with the property that $P_{n-k}(x) = [\phi(x) m_k^{(A)}(x)]_{n-k}$ fed to both sides of a switch SW which is reversed at moment $n-k$. A possible circuit configuration is shown in Fig. 20-16, having switch SW, a shift register with characteristic polynomial $\phi(x)$, and an input sequence $T(x)$ of shift register SR. The questions now are which sequence $T(x)$ of shift register SR has to be fed in this shift register and what priming has to be applied in order to generate a suitable sequence $P(x)$.

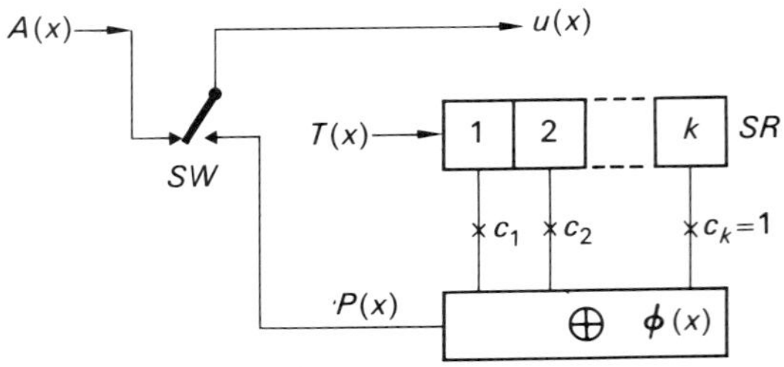

Fig. 20-16

Phase $n - k$ of this sequence represents the check bits of the code symbol to be transmitted, which are placed behind the information bits when reversing switch SW at moment $n - k$. In this case, there is no other sensible choice for priming shift register SR than that with an all-zero combination. The output sequence of the mod 2 adding feedback network is then as follows:

$$P(x) = [1 \oplus \phi(x)]\, T(x) = T(x) \oplus T(x) \cdot \phi(x)$$

Hence,

$$[\phi(x) \cdot m_k{}^{(A)}(x)]_{n-k} = T_{n-k}(x) \oplus [\phi(x) \cdot T(x)]_{n-k}$$

This relation is satisfied by

$$T(x) = m_k{}^{(A)}(x) \tag{20-57}$$

because

$$T_{n-k}(x) = [m_k{}^{(A)}(x)]_{n-k} = 0$$

Substitution of Eq. (20-57) into Eq. (20-53) leads to the following result:

$$m_k{}^{(A)}(x) = T(x) = \frac{u(x)}{\phi(x)} \tag{20-58}$$

This equation incorporates the result obtained for the circuit of Fig. 20-9 in Eq. (20-19). The resulting transmitting circuit with a mod 2 feedback shift register of k sections is shown in Fig. 20-17 (the transmitter circuit

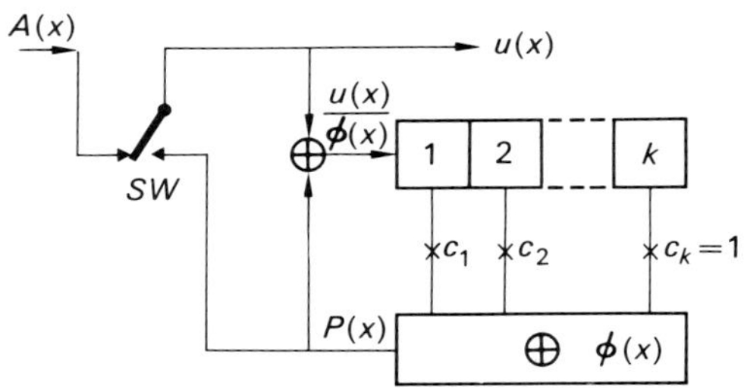

Fig. 20-17

designed by Meggitt). Using Eq. (20-20), the following expression can be obtained:

$$P(x) = u(x)\left[1 \oplus \frac{1}{\phi(x)}\right] \oplus \frac{m(x)}{\phi(x)} \tag{20-59}$$

This equation also holds for moment $n - k$:

$$P_{n-k}(x) = u_{n-k}(x) \cdot \left[1 \oplus \frac{1}{\phi(x)}\right] \oplus \frac{m_{n-k}(x)}{\phi(x)} \tag{20-60}$$

It should be noted that P_{n-k} and u_{n-k} are phases of a sequence but m_{n-k} is not. It follows from Eq. (20-60) that

$$m_{n-k}(x) = \phi(x)\, P_{n-k}(x) \oplus u_{n-k}(x) \cdot [1 \oplus \phi(x)] \tag{20-61}$$

It follows from the circuit of Fig. 20-17 that

$$u_{n-k}(x) = P_{n-k}(x)$$

Hence,

$$\begin{aligned} m_{n-k}(x) &= \phi(x)\, u_{n-k}(x) \oplus \phi(x)\, u_{n-k}(x) \oplus u_{n-k}(x) \\ &= u_{n-k}(x) \end{aligned} \tag{20-62}$$

Equation (20-62) shows that after $n - k$ pulses, that is, the moment that all $n - k$ information elements of $A(x)$ have been received, the initial-state polynomial of the actual content of shift register SR represents the check elements. These check elements can also be expressed in $A(x)$ and $\phi(x)$. It follows from Eqs. (20-53) and (20-55) that

$$\phi(x) \cdot m_k{}^{(A)}(x) = A(x) \oplus x^{n-k}\, u_{n-k}(x)$$

$$m_k{}^{(A)}(x) = \frac{A(x)}{\phi(x)} \oplus x^{n-k} \frac{u_{n-k}(x)}{\phi(x)}$$

$$[m_k{}^{(A)}(x)]_{n-k} = \left[\frac{A(x)}{\phi(x)}\right]_{n-k} \oplus \frac{u_{n-k}(x)}{\phi(x)} = 0$$

Hence,

$$u_{n-k}(x) = \phi(x)\left[\frac{A(x)}{\phi(x)}\right]_{n-k} \tag{20-63}$$

The complete code symbol can now be represented by the following:

$$u(x) = A(x) \oplus x^{n-k}\, \phi(x)\left[\frac{A(x)}{\phi(x)}\right]_{n-k} \tag{20-64}$$

The shift register SR in the transmitting circuit of Fig. 20-17 will now be replaced by a memory circuit of the type of Fig. 16-49. This type of memory circuit must be provided with an external input, as shown in Fig. 20-18.

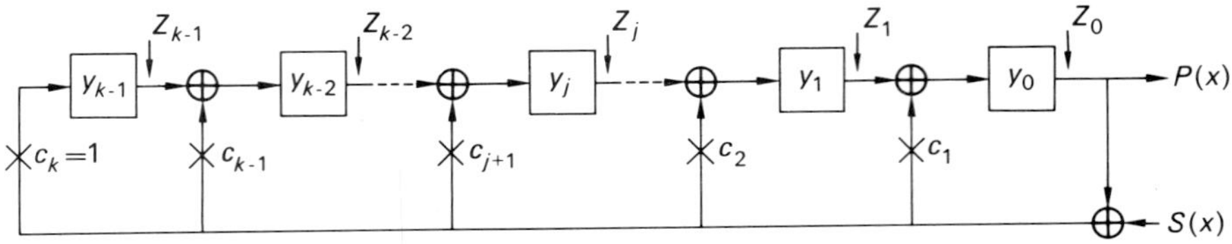

Fig. 20–18

It follows from this circuit that

$$Z_0(x) = y_0 \oplus xZ_1(x) \oplus c_1 x[Z_0(x) \oplus S(x)]$$

Hence,

$$xZ_1(x) = y_0 \oplus (1 \oplus c_1 x)\, Z_0(x) \oplus c_1 xS(x) \tag{20-65}$$

In the next section,

$$Z_1(x) = y_1 \oplus xZ_2(x) \oplus c_2 x[Z_0(x) \oplus S(x)]$$

Hence,

$$x^2 Z_2(x) = y_1 x \oplus xZ_1(x) \oplus c_2 x^2 Z_0(x) \oplus c_2 x^2 S(x) \tag{20-66}$$

Substitution of Eq. (20-65) into Eq. (20-66) leads to

$$x^2 Z_2(x) = (y_0 \oplus y_1 x) \oplus (1 \oplus c_1 x \oplus c_2 x^2)\, Z_0(x) \oplus (c_1 x \oplus c_2 x^2)\, S(x) \tag{20-67}$$

The following general form of Eq. (20-67) can now be obtained:

$$x^j Z_j(x) = \sum_{i=0}^{j-1} y_i x^i \oplus \left[1 \oplus \sum_{i=1}^{j} c_i x^i\right] Z_0(x) \oplus \sum_{i=1}^{j} c_i x^i S(x) \tag{20-68}$$

Hence, for $j = k - 1$,

$$x^{k-1} Z_{k-1}(x) = \sum_{i=0}^{k-2} y_i x^i \oplus \left[1 \oplus \sum_{i=1}^{k-1} c_i x^i\right] Z_0(x) \oplus \sum_{i=1}^{k-1} c_i x^i S(x) \tag{20-69}$$

It follows from the circuit of Fig. 20-18 that

$$\begin{aligned} Z_{k-1}(x) &= y_{k-1} \oplus c_k x[Z_0(x) \oplus S(x)] \\ &= y_{k-1} \oplus c_k xZ_0(x) \oplus c_k xS(x) \end{aligned} \tag{20-70}$$

Substitution of Eq. (20-70) into Eq. (20-69) leads to the following result:

$$\sum_{i=0}^{k-1} y_i x^i \oplus \left[1 \oplus \sum_{i=1}^{k} c_i x^i\right] Z_0(x) \oplus \sum_{i=1}^{k} c_i x^i S(x) = 0 \tag{20-71}$$

Equation (20-71) can also be written as follows:

$$Y(x) \oplus \phi(x)\, Z_0(x) \oplus [1 \oplus \phi(x)]S(x) = 0$$

Because $P(x) = Z_0(x)$, this leads to

$$P(x) = S(x) \oplus \frac{S(x)}{\phi(x)} \oplus \frac{Y(x)}{\phi(x)} \tag{20-72}$$

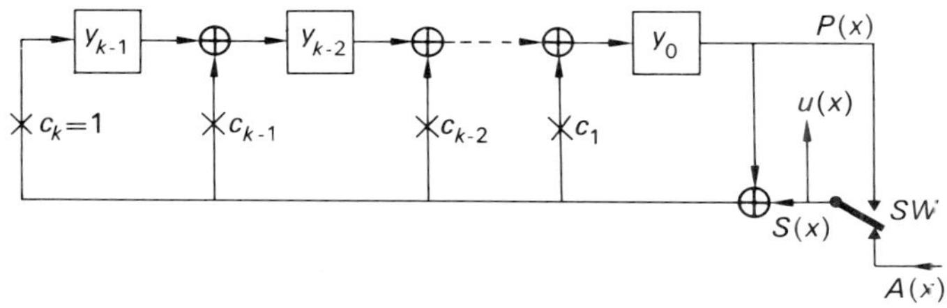

Fig. 20–19

Using the memory circuit of Fig. 20-18, the transmitter circuit of Fig. 20-19 can now be designed. The transmitting circuit of Fig. 20-19 also has two modes of operation. During the $n-k$ pulses of the first mode of operation

$$A(x) = S(x) = u(x)$$

Hence, it follows from Eq. (20-72) that

$$P(x) = A(x) \oplus \frac{A(x)}{\phi(x)} \tag{20-73}$$

because the check-digit generating process is started from the zero state of the circuit in which $Y(x) = 0$. Hence,

$$\begin{aligned} P_{n-k}(x) &= A_{n-k}(x) \oplus \left[\frac{A(x)}{\phi(x)}\right]_{n-k} \\ &= \left[\frac{A(x)}{\phi(x)}\right]_{n-k} \end{aligned} \tag{20-74}$$

Equation (20-72) applied to moment $n-k$ in mode 1 leads to

$$P_{n-k}(x) = S_{n-k}(x) \oplus \frac{S_{n-k}(x)}{\phi(x)} \oplus \frac{Y^{(n-k)}(x)}{\phi(x)}$$

Hence,

$$\left[\frac{A(x)}{\phi(x)}\right]_{n-k} = A_{n-k}(x) \oplus \frac{A_{n-k}(x)}{\phi(x)} \oplus \frac{Y^{(n-k)}(x)}{\phi(x)}$$

$A_{n-k}(x) = 0$, $A(x)$ being a polynomial of a degree less than $n - k$.

$$\left[\frac{A(x)}{\phi(x)}\right]_{n-k} = \frac{Y^{(n-k)}(x)}{\phi(x)}$$

Hence,

$$Y^{(n-k)}(x) = \phi(x)\left[\frac{A(x)}{\phi(x)}\right]_{n-k} \tag{20-75}$$

Equation (20-75) shows that the content $Y^{(n-k)}(x)$ of the memory elements in the circuit of Fig. 20-19 is equal to the check digits described by Eq. (20-63). In the transmitting circuit of Fig. 20-19, these check digits are directly visible as a content of the memory elements.

It should be noted that switch SW in the transmitting circuit of Fig. 20-19 is placed in such a way that during the series of pulses from $n - k + 1$ to n the memory elements form an ordinary shift register. The check elements are shifted from this shift register as the trailing end of the generated code symbol. At the end of this operation it is empty.

Before discussing the circuits and their capabilities in the receiving end of an error-correcting data transmission link, it will be useful to discuss the problem of whether or not the symbols generated in one of the equivalent transmitting circuits of Figs. 20-14, 20-17, or 20-19 fulfill the requirement of producing code symbols (information elements and check elements) that have a minimum distance d of at least 3, this depending on the desired error-correcting capabilities of the code.

In the above transmitting circuits, characteristic polynomial $\phi(x)$ or $q(x)$ is chosen in such a way that its period equals the length of the code-symbol information and check elements. The code symbols are cyclic over n. They can be generated in the transmitter of Fig. 20-14 by a polynomial

$$q(x) = \frac{1 \oplus x^n}{\phi(x)} \tag{20-76}$$

The problem in error-correcting transmission systems is determining what can be reached in distances d between the code symbols. However, determining the distance of the code symbols of a group is rather time-consuming, especially when the number of information elements is not small. In the special case of cyclic code symbols generated by mod 2 feedback shift registers, as shown in Figs. 20-14, 20-17, and 20-19, there is a way around this problem.

In each group of check symbols, an encrypted message is given about each of the information elements of the code symbol. For example, for a distance of 3, at least three independent messages about each of the information elements has to be contained in the complete symbol, because this is the minimum that can give a majority decision (2:1) in case of a single error. Two of these messages are contained in the check, and one is the information element itself. In the same way, it can be easily seen that in general for an error-correcting capability of a code symbol of r errors at least $2r + 1$ messages about each element must be contained in the code symbol. Because the symbols are of cyclic construction, each group of $n - k$ successive elements can function as check group, so that the errors may be as well in the information elements as in the check elements of the code symbol. Because the distances d for single, double, etc., error correction are 3, 5, ..., and the minimum number of messages about each element of the code symbol for the same error-correcting capability are also 3, 5, ..., these two concepts can be considered to be equivalent in our case. It should be noted that the messages about the information elements of a code symbol must be linearly independent; otherwise the information encrypted in the check is ambiguous.

The number of messages about each of the information elements of a code symbol contained in the check can be determined from the all-zero combination of information elements, except a 1 on the specific place, by applying the regular check-construction procedure. As an example, the error-correcting capabilities of a transmitting circuit with a shift register of nine sections as shown in Fig. 20-14 and a feedback polynomial

$$\begin{aligned} q(x) &= (1 \oplus x)(1 \oplus x \oplus x^4)(1 \oplus x^3 \oplus x^4) \\ &= 1 \oplus x^2 \oplus x^3 \oplus x^6 \oplus x^7 \oplus x^9 \end{aligned} \tag{20-77}$$

are determined in Fig. 20-20. The corresponding polynomial is

$$\begin{aligned} \phi(x) &= (1 \oplus x \oplus x^2)(1 \oplus x \oplus x^2 \oplus x^3 \oplus x^4) \\ &= 1 \oplus x^2 \oplus x^3 \oplus x^4 \oplus x^6 \end{aligned} \tag{20-78}$$

Both polynomials have a period $n = 15$. The resulting nine six-bit check groups show, as is indicated in the rightmost column of Fig. 20-20, that the minimum number of messages about a specific information element is four. The $2^9 = 512$ possible code symbols will have a maximum minimum distance 4. With this code, it is possible to correct one error in a symbol and to detect one error more.

The fact that the function of $q(x)$ and $\phi(x)$ may be interchanged needs no further elucidation. A transmitting circuit as given in Fig. 20-14 with feedback polynomial (20-78) can produce $2^6 = 64$ different code symbols each with nine check bits. The generation of the check bits of the all-zero-except-one groups of information elements is shown in the six left-hand columns of Fig. 20-20. The resulting number of messages about each of the information elements of these code symbols is shown on top of these six columns. The resulting distance of this group of code symbols is 6.

Number of messages	6	6	8	8	6	6										
Information x^6	0	0	0	0	0	1										
Information	0	0	0	0	1	0										
Information x^4	0	0	0	1	0	0										
Information x^3	0	0	1	0	0	0										
Information x^2	0	1	0	0	0	0										
Information	1	0	0	0	0	0	Information									Number of messages
Check	0	1	1	1	0	1	0	0	0	0	0	0	0	0	1	5
Check	1	1	1	0	1	0	0	0	0	0	0	0	0	1	0	5
Check	1	0	1	0	0	1	0	0	0	0	0	0	1	0	0	4
Check	0	0	1	1	1	1	0	0	0	0	0	1	0	0	0	5
Check	0	1	1	1	1	0	0	0	0	0	1	0	0	0	0	5
Check	1	1	1	1	0	0	0	0	0	1	0	0	0	0	0	5
Check	1	0	0	1	0	1	0	0	1	0	0	0	0	0	0	4
Check	0	1	0	1	1	1	0	1	0	0	0	0	0	0	0	5
Check	1	0	1	1	1	0	1	0	0	0	0	0	0	0	0	5

$q(x) = 1 \oplus x^2 \oplus x^3 \oplus x^6 \oplus x^7 \oplus x^9$

Fig. 20–20

Table 20-18 shows what can be reached in distances for various lengths of the groups of information elements in code symbols with 15 elements, which can be described by

$$q(x)\,\phi(x) = 1 \oplus x^{15} = (1 \oplus x)\,(1 \oplus x \oplus x^2)\,(1 \oplus x \oplus x^4)$$
$$\times\,(1 \oplus x^3 \oplus x^4)\,(1 \oplus x \oplus x^2 \oplus x^3 \oplus x^4)$$

Table 20-18

$n-k$	$q(x)$	n	N	n of $\phi(x)$	d
4	$1 \oplus x \oplus x^4$	15	16	15	8
5	$(1 \oplus x)(1 \oplus x \oplus x^4) = 1 \oplus x^2 \oplus x^4 \oplus x^5$	15	32	15	7
6	$(1 \oplus x \oplus x^2)(1 \oplus x \oplus x^4) = 1 \oplus x^3 \oplus x^4 \oplus x^5 \oplus x^6$	15	64	15	6
7	$(1 \oplus x)(1 \oplus x \oplus x^2)(1 \oplus x \oplus x^4) = 1 \oplus x \oplus x^3 \oplus x^7$	15	128	15	5
8	$(1 \oplus x \oplus x^4)(1 \oplus x^3 \oplus x^4)$ $= 1 \oplus x \oplus x^3 \oplus x^4 \oplus x^5 \oplus x^7 \oplus x^8$	15	256	15	4
9	$(1 \oplus x)(1 \oplus x \oplus x^4)(1 \oplus x^3 \oplus x^4)$ $= 1 \oplus x^2 \oplus x^3 \oplus x^6 \oplus x^7 \oplus x^9$	15	512	15	4
9	$(1 \oplus x)(1 \oplus x \oplus x^4)(1 \oplus x \oplus x^2 \oplus x^3 \oplus x^4)$ $= 1 \oplus x \oplus x^4 \oplus x^5 \oplus x^6 \oplus x^9$	15	512	15	3
10	$(1 \oplus x \oplus x^2)(1 \oplus x \oplus x^4)(1 \oplus x \oplus x^2 \oplus x^3 \oplus x^4)$ $= 1 \oplus x \oplus x^2 \oplus x^4 \oplus x^5 \oplus x^8 \oplus x^{10}$	15	1024	15	4
11	$(1 \oplus x)(1 \oplus x \oplus x^2)(1 \oplus x \oplus x^4)(1 \oplus x \oplus x^2 \oplus x^3 \oplus x^4)$ $= 1 \oplus x^3 \oplus x^4 \oplus x^6 \oplus x^8 \oplus x^9 \oplus x^{10} \oplus x^{11}$	15	2048	15	3

This table gives a choice of possible combinations. From it, one can see that the checking capabilities of a group of code symbols can increase with the number of check elements—but this is not a regular function.

When $\phi(x)q(x) = 1 \oplus x^n$, the checks of the all-zero-except-one information symbols of both groups of code symbols show an interrelation, as expressed in Fig. 20-20. It is obvious from this figure that when $\phi(x)$ has a period dividing that of $q(x)$, a repetition of the lines of check elements will show up, automatically reducing the distance of that group of code symbols to 2. Then the check is no longer unambiguous.

An example can be found in $q(x) = 1 \oplus x^5 \oplus x^{10}$ and $\phi(x) = 1 \oplus x^5$, having periods of 15 and 5, respectively.

The next problem is to reconstruct the original code symbol at the receiving end of a transmission link when this symbol has been garbled during transmission. The check indicators as constructed by means of a suitable characteristic polynomial from the all-zero-except-one groups of information elements form the basis of the following error-detection and -correction process.

Let us first consider the case of a single-error-correcting code with distance 3. As an example, a code with 15 element symbols consisting of 11 information elements and 4 check elements will be taken. Code symbols of this construction can be generated by a shift register with a characteristic polynomial $1 \oplus x^3 \oplus x^4 \oplus x^6 \oplus x^8 \oplus x^9 \oplus x^{10} \oplus x^{11}$ (see Table 20-18). The corresponding grid of check indicators is shown in Fig. 20-21. There are two possibilities when a received code symbol is garbled, viz., (1) the error is in one of the information elements or (2) the error is in one of the check elements.

For both cases, it is sufficient to form at the receiving end of the transmission link a new or reconstructed check from the information elements and to compare that reconstructed check with the check received by adding mod 2 to the corresponding elements.

It is obvious that when no error has occurred, both checks will agree so that their mod 2 sum will be 0.

When a single error has occurred in one of the information elements of any of the possible code symbols, it is obvious that the mod 2 sum of both checks will be the check indicator of the all-zero-except-one information combination of the erroneous information element. In Fig. 20-22, a transmitted symbol of the 15-element code used in the example is given with its check. On the second line of this figure, the received symbol is shown with an error in the seventh information element. The reconstructed check is added. On the third line, the mod 2 sum of both checks is shown, resulting in the check indicator of the seventh (erroneous) information element.

It should be noted that the transmitted symbol is of course not known at the receiving end. However, the elements of this symbol have been shown in Fig. 20-22 and used in the mod 2 addition of all the elements of both symbols, in order to show that the seventh element was garbled.

It is now a question of circuit design to find the relation between the mod 2 sum of both checks and the number or place of the erroneous information element.

In case of an error in one of the elements of the check, the correct check will be generated in the receiving circuit, so that the mod 2 sum of both checks indicates directly the place of the single error. An example is shown in Fig. 20-23.

The error-detecting and -correcting process becomes slightly more complicated for two independent errors for which code symbols are required having a distance d of at least 5. A group of code symbols with 15 elements (7 information and 8 check elements) having a distance 5 can be obtained by means of a shift-register transmitter with characteristic polynomial $1 \oplus x \oplus x^3 \oplus x^7$. The corresponding check-indicator grid is shown in Fig. 20-24.

The place of a single error in the 15-element code with 7 information and 8 check elements as generated by characteristic polynomial $1 \oplus x \oplus x^3 \oplus x^7$ is determined in the same way as shown in the previous example.

In Fig. 20-25, the transmitted and corresponding received code symbols are shown garbled in two arbitrary information elements.

As follows from the mod 2 sum of the transmitted and the received-code-symbol-information elements, numbers 4 and 7 have been garbled. However, this conclusion cannot be drawn that simply in the receiving end of a transmission link, because the actual transmitted symbol is not available. The mod 2 sum of the received and the reconstructed check contains that

Check indicator				Information elements											Number
1	0	0	1	0	0	0	0	0	0	0	0	0	0	1	1
1	0	1	1	0	0	0	0	0	0	0	0	0	1	0	2
1	1	1	1	0	0	0	0	0	0	0	0	1	0	0	3
0	1	1	1	0	0	0	0	0	0	0	1	0	0	0	4
1	1	1	0	0	0	0	0	0	0	1	0	0	0	0	5
0	1	0	1	0	0	0	0	0	1	0	0	0	0	0	6
1	0	1	0	0	0	0	0	1	0	0	0	0	0	0	7
1	1	0	1	0	0	0	1	0	0	0	0	0	0	0	8
0	0	1	1	0	0	1	0	0	0	0	0	0	0	0	9
0	1	1	0	0	1	0	0	0	0	0	0	0	0	0	10
1	1	0	0	1	0	0	0	0	0	0	0	0	0	0	11

Fig. 20–21

	Check				Information				↓							
Transmitted symbol	1	1	0	1	1	1	0	1	1	1	0	1	0	1	1	
Received symbol	0	1	1	1	1	1	0	1	0	1	0	1	0	1	1	⊕
Check indicator	1	0	1	0	0	0	0	0	1	0	0	0	0	0	0	= 7

Fig. 20–22

	Check		↓		Information											
Transmitted symbol	1	1	0	1	1	1	0	1	1	1	0	1	0	1	1	
Received symbol	1	1	1	1	1	1	0	1	1	1	0	1	0	1	1	⊕
mod 2 sum	0	0	1	0	0	0	0	0	0	0	0	0	0	0	0	

Fig. 20–23

Check indicator	Information element	Number
0 0 0 1 0 1 1 1	0 0 0 0 0 0 1	1
0 0 1 0 1 1 1 0	0 0 0 0 0 1 0	2
0 1 0 1 1 1 0 0	0 0 0 0 1 0 0	3
1 0 1 1 1 0 0 0	0 0 0 1 0 0 0	4
0 1 1 0 0 1 1 1	0 0 1 0 0 0 0	5
1 1 0 0 1 1 1 0	0 1 0 0 0 0 0	6
1 0 0 0 1 0 1 1	1 0 0 0 0 0 0	7

Fig. 20–24

Check	↓ ↓	
1 1 1 1 0 1 1 1	0 1 0 0 0 1 1	Transmitted symbol
1 1 0 0 0 1 0 0	1 1 0 1 0 1 1 ⊕	Received symbol
	⊕	
0 0 1 1 0 0 1 1	1 0 0 1 0 0 0	

Fig. 20–25

Check indicator information element number		
	4	1 0 1 1 1 0 0 0
	7	1 0 0 0 1 0 1 1
		⊕
		0 0 1 1 0 0 1 1

Fig. 20–26

	↓ (position 6), ↓ (position 12)	
Received check and information elements	1 1 1 1 0 0 1 1 0 1 0 1 0 1 1	
Reconstructed check elements	0 1 0 0 1 1 1 1	⊕
	1 0 1 1 1 1 0 0	
Check indicator of element 4	1 0 1 1 1 0 0 0	⊕
Place of error element in check	0 0 0 0 0 1 0 0	

Fig. 20–27

information in an encrypted form. This mod 2 sum does not correspond now to one of the check indicators of Fig. 20-24; rather, it corresponds to the mod 2 sum of the check indicators of both erroneous elements, as may be seen from Fig. 20-26.

The correct pair of erroneous elements could be determined in this case by systematically adding mod 2, starting with that of information element number 1, the check indicators to the mod 2 sum of received and reconstructed check, and stopping when one of the other check indicators results. The pair of check indicators involved indicates the numbers of the erroneous information elements.

When both errors in question are in the check elements, the sum mod 2 of the received and reconstructed check will show two 1's on the places of the erroneous elements.

Figure 20-27 gives a case in which one error is in the information elements and the other is in the check elements. The same transmitted code symbol as used in Fig. 20-25 is taken in this example. The check indicators of the various information elements have to be added mod 2 to the resulting sum mod 2 of the received and the reconstructed check elements. In this case, no sum 0 or check indicator will result but rather a sum with a single 1, indicating the place of the erroneous element in the check.

In most transmission systems, the character of the errors is such that a number of successive elements is garbled either into 1's or into 0's (hits and misses, respectively), both types of error not happening simultaneously within a symbol. This means that there is a geometric relation between the erroneous elements in these so-called bursts of errors. The errors in these bursts show patterns depending on the information covered by the burst. In Table 20-19,

Table 20-19

4	3	2	1
			1
		1	1
	1	0	1
	1	1	1
1	0	0	1
1	0	1	1
1	1	0	1
1	1	1	1

the error patterns for burst errors up to four bits have been indicated. A 1 indicates an erroneous symbol element; a 0 indicates an element without error. The error pattern of a burst will always start with a 1 and end with a 1. In between, the elements can be in all combinations with or without error. If the error is a "hit" burst, all symbol elements in the burst will be changed in 1; however, the elements in the burst already being a 1 will show up in the burst pattern as 0. It will follow from Table 20-19 without any difficulty that the number of different burst patterns is 2^{b-1}, b being the length of the burst pattern in number of bits.

It should be noted that there is a maximum length of burst errors that can be corrected in a code symbol of n elements, k of which are check elements. This can be shown as follows.

Let A_1 and A_2 be two code symbols differing in the check elements and in the last information element.

$$\begin{aligned} A_1 &= c_{k-1}, \ldots, c_0, a_{m-1}, a_{m-2}, \ldots, a_0 \\ A_2 &= c^*_{k-1}, \ldots, c^*_0, a'_{m-1}, a_{m-2}, \ldots, a_0 \end{aligned} \tag{20-79}$$

Both code symbols have a length of $n = m + k$ elements, m being the number of information elements and k being the number of check elements. The first $m - 1$ information elements of code symbols A_1 and A_2 are equal. Thus, these code symbols can differ as a maximum in $k + 1$ elements.

Let these code symbols now be affected by burst errors of, as a maximum, b elements, b being the smallest integer larger than $\frac{1}{2}k$ (if k is even, then $b = \frac{1}{2}k + 1$; if k is odd, then $b = \frac{1}{2}k + \frac{1}{2}$).

Symbol A_1 is now supposed to be affected in check elements $c_{k-1}, \ldots, c_{k-b}$ in such a way that these check elements change in the correspondingly numbered check elements of code symbol A_2.

$$A^*_1 = c^*_{k-1}, \ldots, c^*_{k-b}, c_{k-b-1}, \ldots, c_0, a_{m-1}, a_{m-2}, \ldots, a_0 \tag{20-80}$$

Symbol A_2 is now supposed to be affected in its check elements $c^*_{k-b-1}, \ldots, c^*_0$ and information elements a'_{m-1}, so that they change in the correspondingly numbered elements of code symbol A_1.

$$A^*_2 = c^*_{k-1}, \ldots, c^*_{k-b}, c_{k-b-1}, \ldots, c_0, a_{m-1}, a_{m-2}, \ldots, a_0 \tag{20-81}$$

With these two different bursts of length b,

$$A^*_1 = A^*_2 \tag{20-82}$$

so that in this case there is no possibility for correction. Hence, the maximum burst length r that can be corrected must be equal to or smaller than $\frac{1}{2}k$:

$$r \leqslant \tfrac{1}{2}k \tag{20-83}$$

It can easily be shown that $r = \frac{1}{2}k$ is a possible case.

Example Let us now consider a code with seven symbols consisting of $m = 3$ information elements and $k = 4$ check elements. The check elements can be formed in a maximum-length sequence-generating shift register with $q(x) = 1 \oplus x \oplus x^3$.

	Check				*Information*		
A_1	1	1	1	0	0	1	0
A_2	0	1	0	1	1	1	0

A burst of length $b = \frac{1}{2}k + 1 = 3$ with pattern 1 0 1 will be applied to elements 5 to 7 of code symbol A_1 and a burst with length $t = 2$ with pattern 1 1 will hit elements 3 and 4 of code symbol A_2. This gives the following result:

	Check				*Information*		
A_1^*	0	1	0	0	0	1	0
A_2^*	0	1	0	0	0	1	0

It can easily be checked that bursts of length $b = 2$ hitting both code symbols in any place never can change these symbols into the same garbled symbol.

A practical way of performing the decoding of the check sum mod 2 can now be discussed. From the examples given above, it follows that:

1. In the case of a single error, the mod 2 sum of the received and reconstructed check digits is again a check indicator if the error is in the information elements; and if the error is in the group of check elements, their mod 2 sum is a single 1 on the place of the error.
2. In the case of a multiple error, the mod 2 sum of the received and reconstructed check digits will be the mod 2 sum of the check indicators of the erroneous information elements if all errors are in the information elements only. This mod 2 sum can also be a code combination differing in one or more elements from a check indicator or a mod 2 sum of a number of check indicators, depending on the distribution of the errors in the group of information elements and the group of check elements and provided that the check-generating characteristic polynomial has sufficient error-correcting power.

In case 2 there are three different types of error patterns to be considered:

1. Burst errors.
2. Arbitrary errors with an error pattern not exceeding the length of the group of check digits.

3. Arbitrary errors with an error pattern exceeding the length of the group of check digits. This case was not shown in the examples given above.

RECEIVERS

That a receiver for error-correcting code symbols will contain some of the other circuit analogous to that used in the transmitter for the generation of the check elements needs no further explanation.

Equation (20-72) represents the operation of the register circuit of Fig. 20-18 starting from an initial state $Y(x)$ under the influence of sequence $S(x)$ externally fed into that register. For the receiver, it is useful to know the register content $Y(x)$ at an arbitrary moment t. For an arbitrary moment t, Eq. (20-72) changes as follows:

$$P_t(x) = S_t(x) \oplus \frac{S_t(x)}{\phi(x)} \oplus \frac{Y^{(t)}(x)}{\phi(x)} \tag{20-84}$$

Sequence $P(x)$ at moment t can be represented by $P^{(t)}(x)$—but because this is also phase t of that sequence $P^{(t)}(x) = P_t(x)$. The same holds for all polynomials representing sequences. In general, $Y^{(t)}(x) \neq Y_t(x)$ because $Y(x)$ represents a register content in polynomial form.

It follows from Eq. (20-84) that

$$Y^{(t)}(x) = S_t(x) \oplus \phi(x) S_t(x) \oplus \phi(x)\, P_t(x) \tag{20-85}$$

Using Eq. (20-72),

$$\begin{aligned} Y^{(t)}(x) &= S_t(x) \oplus \phi(x) S_t(x) \oplus \phi(x)\left[S(x) \oplus \frac{S(x)}{\phi(x)} \oplus \frac{Y(x)}{\phi(x)}\right]_t \\ &= S_t(x) \oplus \phi(x)\left[\frac{S(x) \oplus Y(x)}{\phi(x)}\right]_t \end{aligned} \tag{20-86}$$

Hence, if $S_t(x) = 0$,

$$Y^{(t)}(x) = \phi(x)\left[\frac{S(x) \oplus Y(x)}{\phi(x)}\right]_t \tag{20-87}$$

Now consider a receiving circuit as shown in Fig. 20-28. This circuit contains a shift register SR_1 with at least n sections in which the complete code symbol can be stored during the receiving cycle of n pulses; a register circuit SR_2 of the type shown in Fig. 20-18; a detection circuit DC for detecting the moments of correction; and mod 2 adders M_1 and M_3 interconnecting these portions of the receiving circuit.

The transmitted code symbol can be represented by the following equation [as shown in (20-64)]:

$$u(x) = A(x) \oplus x^{n-k}C(x) = A(x) \oplus x^{n-k}\phi(x)\left[\frac{A(x)}{\phi(x)}\right]_{n-k}$$

$A(x)$ represents the information elements and $C(x)$ the check elements corresponding to these information elements.

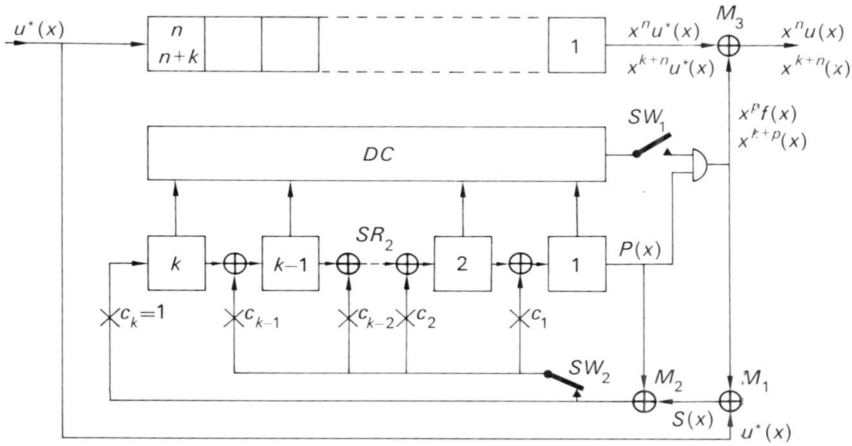

Fig. 20–28

A code symbol can be garbled during transmission in such a way that one or more symbol elements may be erroneous. This can be represented by an error polynomial $F(x)$, so that the received code symbol can be represented by the following:

$$u^*(x) = u(x) \oplus F(x) \tag{20-88}$$

Error polynomial $F(x)$ can be split up into an error $a(x)$ in the information elements and an error $c(x)$ in the check elements:

$$F(x) = a(x) \oplus x^{n-k}c(x) \tag{20-89}$$

Hence, the received code symbol can be written as follows:

$$u^*(x) = [A(x) \oplus a(x)] \oplus x^{n-k}[C(x) \oplus c(x)] \tag{20-90}$$

The information and check elements actually received are then

$$A^*(x) = A(x) \oplus a(x) \tag{20-91}$$

and

$$C^*(x) = C(x) \oplus c(x) \tag{20-92}$$

or

$$u^*(x) = A^*(x) \oplus x^{n-k}C^*(x) \tag{20-93}$$

In accordance with Eq. (20-75), the reconstructed check $C^{(r)}(x)$ will be available in register circuit SR_2 of Fig. 20-28 at moment $n-k$ of the receiving cycle:

$$Y^{(n-k)}(x) = C^{(r)}(x) = \phi(x)\left[\frac{A^*(x)}{\phi(x)}\right]_{n-k} \tag{20-94}$$

Equation (20-94) also follows from Eq. (20-87) for $t = n-k$, $Y(x) = 0$, and $S(x) = A^*(x)$.

It should be noted that mod 2 adder M_1 is not operating during the receiving cycle. Switch SW_1 is then open.

Using Eqs. (20-75) and (20-91), Eq. (20-94) can be given the following form:

$$C^{(r)}(x) = C(x) \oplus \phi(x)\left[\frac{a(x)}{\phi(x)}\right]_{n-k} \tag{20-95}$$

Switch SW_2 in the receiving circuit of Fig. 20-28 was closed during the first $n-k$ pulses of the receiving cycle. Opening of this switch changes the operation of register circuit SR_2 into that of a straight shift register. The reconstructed check elements $C^{(r)}(x)$ are now fed to mod 2 adder M_2. The other input of M_2 is then fed with the received check elements, so that the following mod 2 sum is formed:

$$C^{(s)}(x) = C^*(x) \oplus C^{(r)}(x) \tag{20-96}$$

Substitution of Eqs. (20-92) and (20-95) into Eq. (20-96) gives

$$\begin{aligned} Y^{(n)}(x) = C^{(s)}(x) &= C(x) \oplus c(x) \oplus C(x) \oplus \phi(x)\left[\frac{a(x)}{\phi(x)}\right]_{n-k} \\ &= c(x) \oplus \phi(x)\left[\frac{a(x)}{\phi(x)}\right]_{n-k} \\ &= \phi(x)\left[\frac{a(x) \oplus x^{n-k}c(x)}{\phi(x)}\right]_{n-k} \\ &= \phi(x)\left[\frac{F(x)}{\phi(x)}\right]_{n-k} \end{aligned} \tag{20-97}$$

Error polynomial $F(x)$ can now be transformed into an error pattern $f(x)$ starting with the first erroneous symbol element p:

$$F(x) = x^p f(x) \tag{20-98}$$

Polynomial $f(x)$ then always starts with a 1. In the following, the degree of $f(x)$ will be supposed to be smaller than k; that is, the maximum length of the error pattern equals the length of the check.

Substitution of Eq. (20-98) into Eq. (20-97) gives

$$\begin{aligned} Y^{(n)}(x) &= \phi(x)\left[\frac{x^p f(x)}{\phi(x)}\right]_{n-k} \\ &= \phi(x)\left[\frac{f(x)}{\phi(x)}\right]_{n-k-p} \end{aligned} \tag{20-99}$$

By substitution of Eq. (20-99) into Eq. (20-87) with $t = k + p$, $S(x) \to S_n(x) = 0$, and $Y(x) \to Y^{(n)}(x)$—switch SW_2 in its rest position after moment n—it is found that

$$\begin{aligned} Y^{(n+k+p)}(x) &= \phi(x)\left\{\left[\frac{f(x)}{\phi(x)}\right]_{n-k-p}\right\}_{k+p} \\ &= \phi(x)\left[\frac{f(x)}{\phi(x)}\right]_n \\ &= f(x) \end{aligned} \tag{20-100}$$

Equation (20-100) shows that $n + k + p$ pulses after the beginning of the receiving cycle or $k + p$ pulses after the beginning of the correction cycle, error pattern $f(x)$ forms the content of register circuit SR_2 in the receiver of Fig. 20-28. If shift register SR_1 of this receiver is provided with $n + k$ sections, then the first erroneous element of the code symbol will be in section $n + k$ of that shift register at moment $k + p$ of the correction cycle. The actual correction is now a question of recognition of error pattern $f(x)$ in register circuit SR_2.

It is a practical nuisance in the receiver of Fig. 20-28 that the correction cannot be performed starting with moment p of the correction cycle but has to start at moment $p + k$.

Let us now investigate what will happen in the receiving circuit of Fig. 20-28 when switch SW_2 is not opened at moment $n - k$. What will be the content of register circuit SR_2 at moment n? This content follows from Eq. (20-87) with $S(x) = u^*(x)$ and $Y(x) = 0$:

$$\begin{aligned} Y^{(n)}(x) &= \phi(x)\left[\frac{u^*(x)}{\phi(x)}\right]_n \\ &= \phi(x)\left[\frac{u(x) \oplus x^p f(x)}{\phi(x)}\right]_n \\ &= \phi(x)\left[\frac{u(x)}{\phi(x)}\right]_n + \phi(x)\left[\frac{x^p f(x)}{\phi(x)}\right]_n \end{aligned}$$

Code symbol $u(x)$ has been formed in such a way that it can be divided by $\phi(x)$. Quotient $u(x)/\phi(x)$ will be of lower degree than n so that its phase n will be zero. Hence,

$$Y^{(n)}(x) = \phi(x)\left[\frac{f(x)}{\phi(x)}\right]_{n-p} \tag{20-101}$$

In a way analogous to Eq. (20-100), the following equation can be obtained:

$$Y^{(n+p)}(x) = f(x) \tag{20-102}$$

Thus when leaving switch SW_2 in the circuit of Fig. 20-28 in its closed position, k pulses can be spared in the correction cycle, as was desired.

Example An example will be given of the operation of the receiver of Fig. 20-28. The example of Fig. 20-25 will be used. The correction process represented by Fig. 20-29 starts on the first line with the mod 2 sum $C^{(s)}(x)$ of the received and reconstructed checks, which is contained in register circuit SR_2 at the end of the receiving cycle when switch SW_2 is open after moment $n - k$.

The arrows above the first line of Fig. 20-29 and pointing to it indicate the places of the mod 2 adders between the sections of register circuit SR_2 in the circuit of Fig. 20-28. The x's in the second column indicate the (unknown) places of the garbled symbol elements in shift register SR_1. On each clock pulse, these symbol elements are shifted one place to the right in SR_1. After $k + p = 11$ of these clock pulses, the first garbled element is in section $n + k$ of SR_1. At this moment (reading from right to left) the content of register circuit SR_2 shows $f(x) = 1 \oplus x^q$.

An analogous picture of the correction process will be obtained when one error is in the information elements and one error is in the check elements with a distance not exceeding k.

As can easily be checked, in the event that switch SW_2 remains closed during the receiving cycle, the correction process will then start on line $k + 1$ of Fig. 20-29.

When at moment $k+p$ of the correction cycle register circuit SR_2 shows error pattern $f(x)$, a parallel correction is possible which can be performed between the correspondingly numbered sections of register circuits SR_1 and SR_2.

A serial correction is also possible starting at moment $k+p$ via a mod 2 adder connected to the rightmost section of SR_1. During that correction process, the position of switch SW_2 is arbitrary. Register circuit SR_2 operates again as an ordinary shift register now shifting out $f(x)$, indicating the garbled symbol elements via mod 2 adder M_3 of Fig. 20-28.

In both cases, detection circuit DC has to test for a 1 in the first section of register circuit SR_2 and no 1 or one other 1 in one of the other sections, because in our example a double error can be corrected.

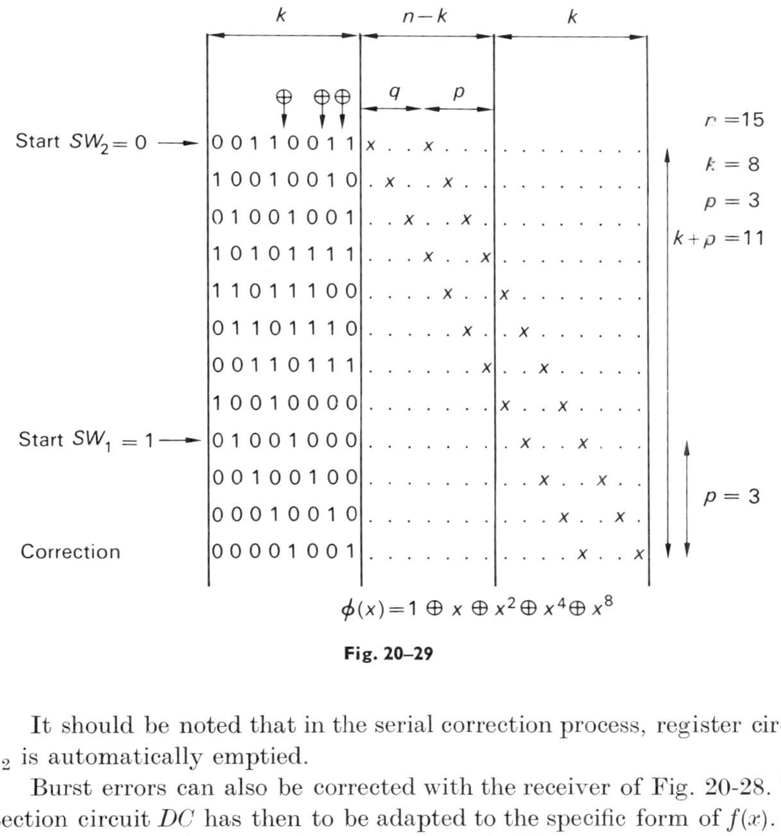

Fig. 20-29

It should be noted that in the serial correction process, register circuit SR_2 is automatically emptied.

Burst errors can also be corrected with the receiver of Fig. 20-28. The detection circuit DC has then to be adapted to the specific form of $f(x)$. For a maximum burst length r, detection circuit DC has to test for a 1 in the right-hand section of memory circuit SR_2 and $k-r$ successive 0's in sections $k-r+1$ to k.

The choice of characteristic polynomial $\phi(x)$ determines the error-correcting power of the code symbols. The first requirement to be made by this choice is that different error-pattern polynomials $f(x)$ used as a priming of shift register SR_2 lead to nonequivalent sequences $f(x)/\phi(x)$.

The second requirement in the choice of $\phi(x)$ is that all error patterns $f(x)$ leads to sequences $f(x)/\phi(x)$ with cycle length n.

The burst-error-correcting power of a group of code symbols determined by characteristic polynomial $\phi(x)$ can easily be determined by testing whether or not the various error-pattern polynomials $f(x)$ lead to different sequences. For example, characteristic polynomial $\phi(x) = 1 \oplus x^3 \oplus x^4 \oplus x^5 \oplus x^6$ for a group of code symbols with $n = 15$ and $k = 6$ has a maximum minimum distance $d = 3$, giving it an error-correcting power of a single error. However, notwithstanding the fact that $d = 3$, it has also the power to correct burst errors with $r = 3$. The error-correcting power of this group of code symbols can easily be investigated by using the various burst-error patterns as priming of a register circuit of the type of SR_2. The result of this is given in Table 20-20.

Table 20-20 shows all the possible states of register circuit SR_2 for burst errors with a maximum length of $r = 3$. In this case, there are four different error patterns $f(x)$ which form the last lines of the four columns of this table. In an arbitrary correction process, register circuit SR_2 starts on one of the lines of this table and ends at moment p on the bottom line of the same column. The detection circuit DC then detects $f(x)$.

Table 20-20

No.	1	11	101	111
1	1 1 1 1 0 0	1 1 1 1 0 1	1 1 1 1 1 0	1 1 1 1 1 1
2	0 1 1 1 1 0	1 0 0 0 1 0	0 1 1 1 1 1	1 0 0 0 1 1
3	0 0 1 1 1 1	0 1 0 0 0 1	1 1 0 0 1 1	1 0 1 1 0 1
4	1 1 1 0 1 1	1 1 0 1 0 0	1 0 0 1 0 1	1 0 1 0 1 0
5	1 0 0 0 0 1	0 1 1 0 1 0	1 0 1 1 1 0	0 1 0 1 0 1
6	1 0 1 1 0 0	0 0 1 1 0 1	0 1 0 1 1 1	1 1 0 1 1 0
7	0 1 0 1 1 0	1 1 1 0 1 0	1 1 0 1 1 1	0 1 1 0 1 1
8	0 0 1 0 1 1	0 1 1 1 0 1	1 0 0 1 1 1	1 1 0 0 0 1
9	1 1 1 0 0 1	1 1 0 0 1 0	1 0 1 1 1 1	1 0 0 1 0 0
10	1 0 0 0 0 0	0 1 1 0 0 1	1 0 1 0 1 1	0 1 0 0 1 0
11	0 1 0 0 0 0	1 1 0 0 0 0	1 0 1 0 0 1	0 0 1 0 0 1
12	0 0 1 0 0 0	0 1 1 0 0 0	1 0 1 0 0 0	1 1 1 0 0 0
13	0 0 0 1 0 0	0 0 1 1 0 0	0 1 0 1 0 0	0 1 1 1 0 0
14	0 0 0 0 1 0	0 0 0 1 1 0	0 0 1 0 1 0	0 0 1 1 1 0
$f(x) = 15$	0 0 0 0 0 1	0 0 0 0 1 1	0 0 0 1 0 1	0 0 0 1 1 1

It should be noted that $f(x)$ forms the lowest-valued binary combination in a column. The various sequences through which SR_2 can run can easily be ordered in accordance with these lowest-valued binary combinations. The sequences of Table 20-20 having all $n = 15$ can be arranged in shorthand notation as indicated in Table 20-21.

Table 20-21

$f(x)$	n
0 0 0 0 0 1	15
0 0 0 0 1 1	15
0 0 0 1 0 1	15
0 0 0 1 1 1	15

$$\phi(x) = 1 \oplus x^3 \oplus x^4 \oplus x^5 \oplus x^6$$

Not many characteristic polynomials $\phi(x)$ have an efficiency in useful sequences so great as that of the example of Table 20-20.

The operation of the receiver for error-correcting code symbols of Fig. 20-28 can now be fully understood. The problem of how this type of receiver will operate when register circuit SR_2 is replaced by an ordinary shift register has still to be investigated. A block circuit of this type of receiver is shown in Fig. 20-30. It is a Meggitt-type receiver.

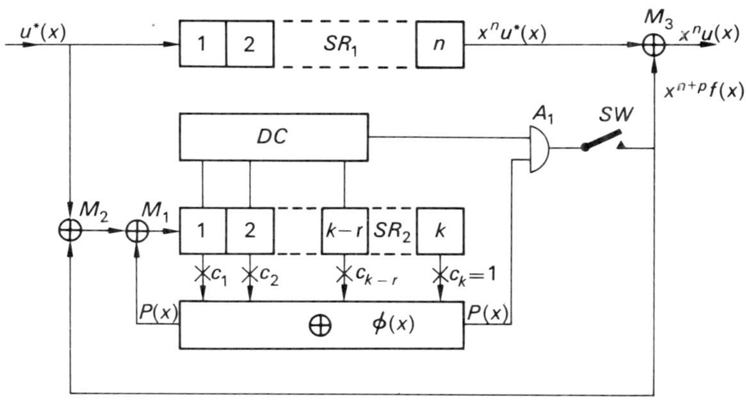

Fig. 20-30

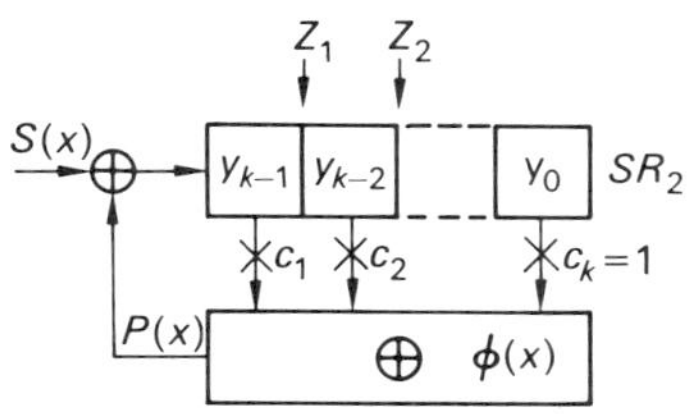

Fig. 20-31

In a way completely analogous to that used for Eq. (20-72), the following equation can be determined for the operation of the shift register given in Fig. 20-31. The following formula follows from Eq. (20-20):

$$P(x) = S(x) \oplus \frac{S(x)}{\phi(x)} \oplus \frac{m(x)}{\phi(x)} \tag{20-103}$$

The only difference between Eq. (20-72) and Eq. (20-103) is that the actual memory content $Y(x)$ has been replaced by the corresponding initial-state polynomial $m(x)$.

Moment $n - k$ of the receiving cycle was important in the receiver of Fig. 20-28. At that moment, the reconstructed check $C^{(r)}(x)$ was known. In the receiver configuration of Fig. 20-30, the following equation will be found analogous to Eq. (20-94):

$$m_{n-k}(x) = \phi(x)\left[\frac{A^*(x)}{\phi(x)}\right]_{n-k} = C^{(r)}(x) \tag{20-104}$$

For moment n of the receiving cycle, Eq. (20-104) changes into

$$m_n(x) = \phi(x)\left[\frac{u^*(x)}{\phi(x)}\right]_n \tag{20-105}$$

$$= \phi(x)\left[\frac{u(x)}{\phi(x)}\right]_n \oplus \phi(x)\left[\frac{F(x)}{\phi(x)}\right]_n \tag{20-106}$$

In Eq. (20-106),

$$\left[\frac{u(x)}{\phi(x)}\right]_n = \left[\frac{m_k{}^{(A)}(x)\,\phi(x)}{\phi(x)}\right]_n = [m_k{}^{(A)}(x)]_n = 0 \tag{20-107}$$

Hence,

$$m_n(x) = \phi(x)\left[\frac{F(x)}{\phi(x)}\right]_n \tag{20-108}$$

Substitution of $F(x) = x^p f(x)$ into Eq. (20-108) leads to the following result:

$$m_{n+p}(x) = f(x) \tag{20-109}$$

Equation (20-109) is analogous to Eq. (20-102). In both types of receivers, $Y(x)$ and $m(x)$ are interchangeable. However, the operation of the receiver of Fig. 20-30 is more difficult to explain because error-pattern polynomial $f(x)$ has been obtained in the form of an initial-state polynomial that is not directly visible as the content of a group of memory elements.

Now consider the shift-register circuit of Fig. 20-32, which is primed with initial-state polynomial $m_{n+p}(x)$.

In the circuit of Fig. 20-32,

$$P(x) = [1 \oplus \phi(x)]\, T(x) \oplus m_{n+p}(x) \tag{20-110}$$

If $T(x) = 0$, then

$$P(x) = m_{n+p}(x) = f(x) \tag{20-111}$$

Equation (20-111) shows that if starting at moment $n + p$ shift register SR_2 of Fig. 20-30 is fed with 0's, a sequence $P(x)$ will be generated which is identical to error polynomial $f(x)$. When this sequence is fed to mod 2 adder M_3 of the circuit of Fig. 20-30, correction of the garbled symbol elements will be performed.

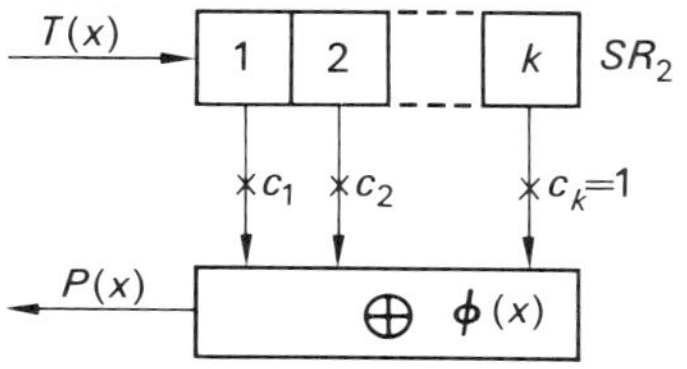

Fig. 20–32

It is not difficult to see in the circuit of Fig. 20-30 that when sequence $P(x) = f(x)$ is fed to mod 2 adder M_2, shift register SR_2 will be fed with 0's, so that at the end of the correction cycle shift register SR_2 will be empty.

The final problem in the receiver of Fig. 20-30 is how to detect moment p of the correction cycle, i.e., the moment that the first garbled symbol element has reached the right-hand section of shift register SR_1 in which the complete code symbol has been stored during the receiving cycle.

For burst-error correction, detection circuit DC has to be arranged in such a way that initial-state polynomial $m_{n+p}(x)$ having a degree of $r - 1$ as a maximum can be detected from the actual constant $Y_{(n+p)}(x)$ of shift register SR_2.

It follows from Eq. (20-30) that

$$m_{n+p}(x) = [\phi(x)\ Y^{(n+p)}(x)]_k \tag{20-112}$$

If $Y^{(n+p)}(x)$ is of degree g and $\phi(x)$ is of degree k, than $m^{(n+p)}(x)$ and $f(x)$ are of degree

$$(k + g) - k = g \tag{20-113}$$

Equation (20-113) states that $Y^{(n+p)}(x)$, $m_{n+p}(x)$, and $f(x)$ have the same degree. Hence, the degree of $m_{n+p}(x) = f(x)$ being $r - 1$ as a maximum, sections Z_1 to Z_{k-r} of shift register SR_2 will contain 0's when the first erroneous symbol element is fed to mod 2 adder M_3. From this it follows that detection circuit DC must test for an all-zero content of sections Z_1 to Z_{k-r}, and when testing this content, feed a 1 to AND gate A_1 for the rest of the correction cycle.

Sequence $P(x) = f(x)$ starting at moment $n + p$ fed to the other input of AND gate A_1 provides for the various correction pulses. These correction pulses are fed to mod 2 adder M_3 for the actual correction of the symbol elements shifted out of shift register SR_1 and to mod 2 adder M_2 to provide for the condition $T^{(n+p)}(x) = 0$.

It should be noted that switch SW has to be closed at moment n, that is, at the beginning of the correction cycle.

In the Meggitt-type of receiver, it is not so easy to introduce a detection of arbitrary errors with an error-pattern polynomial having a degree not exceeding k. This problem will not be discussed here.

Example The design required is that of an error-correcting group of code symbols consisting of $m = 9$ information elements and a number of check elements so that all burst errors of length $r = 3$ can be corrected. There are four possible error patterns, so that $\phi(x)$ has to be chosen in such a way that the check-generating shift register can generate at least four different sequences with a cycle length $n = m + k$. It follows from $r = 3$ that $k = 2r = 6$ as a minimum. If this should prove possible, $n = 15$ and the check-generating shift register must have at least six sections.

Characteristic polynomial $\phi(x)$ can be formed as follows:

$\phi_1(x) = 1 \oplus x \oplus x^2$ and $\phi_2(x) = 1 \oplus x \oplus x^4$

Hence,

$$\phi(x) = \phi_1(x) \cdot \phi_2(x) = 1 \oplus x^3 \oplus x^4 \oplus x^5 \oplus x^6$$

generating sequences 1(1), 1(3), 4(15), which is required.

The distance $d = 3$ follows from the diagram below.

	Check	*Information*
5	1 1 1 0 0 1	0 0 0 0 0 0 0 0 1
4	0 0 1 0 1 1	0 0 0 0 0 0 0 1 0
4	0 1 0 1 1 0	0 0 0 0 0 0 1 0 0
4	1 0 1 1 0 0	0 0 0 0 0 1 0 0 0
Distance $d = 3$	1 0 0 0 0 1	0 0 0 0 1 0 0 0 0
6	1 1 1 0 1 1	0 0 0 1 0 0 0 0 0
5	0 0 1 1 1 1	0 0 1 0 0 0 0 0 0
5	0 1 1 1 1 0	0 1 0 0 0 0 0 0 0
5	1 1 1 1 0 0	1 0 0 0 0 0 0 0 0

It should be noted that solely from the point of view of the concept of distance between code symbols, the group of 512 code symbols as determined by $\phi(x) = 1 \oplus x^3 \oplus x^4 \oplus x^5 \oplus x^6$ has an error-correcting power of a single error only. The geometry of this group of code symbols, however, provides for the burst-error correcting power of $r = 3$.

In the transmitter of Fig. 20-16, the possibility of generating different sequences has no consequences. Only in the receiver of Fig. 20-30 is this possibility of vital importance in the correction cycle following the receiving cycle. In this receiver, shift register SR_2 is filled with 0's if a code symbol is received not affected by errors. However, if a code symbol is received garbled in one or more elements, shift register SR_2 is filled at the end of the receiving cycle with a code combination belonging to one of the four possible cycles. How these four possible cycles can run through shift register SR_2 during the correction cycle is shown in Table 20-22.

The four code combinations in the four cycles of shift register SR_2 corresponding to the possible initial-state polynomials $f(x)$, the error pattern, are shown on top of the various columns. The code combinations are the lowest-valued code combinations in the corresponding sequence.

Table 20-22

Error pattern $f(x)$	1	11	111	101
1	000001	000010	000100	000111
2	100000	100001	100010	100011
3	010000	110000	110001	010001
4	001000	011000	111000	101000
5	100100	101100	111100	110100
6	110010	010110	011110	111010
7	111001	001011	101111	011101
8	011100	100101	010111	101110
9	001110	010010	101011	110111
10	100111	101001	110101	111011
11	110011	010100	011010	111101
12	011001	101010	001101	111110
13	001100	010101	100110	111111
14	000110	001010	010011	011111
15	000011	000101	001001	001111

For a burst error with the 101-pattern, shift register SR_2 is in the beginning of the correction cycle filled with one of the combinations shown in the corresponding column of Table 20-22. When the first garbled element reaches section n of shift register SR_1, shift register SR_2 contains code combination 000111, the error indicator. When performing the actual correction of the first garbled element, the content of shift register SR_2 is changed in 000011—which is a code combination shown in column 1—because in this case, after correction of the first element of the burst error, the second element needs no correction. After the next step of shift register SR_2, the third and last element of the burst error reaches the last section of shift register SR_1. Shift register SR_2 now contains code combination 000001. During the correction of this element of the code symbol, the content of shift register SR_2 is changed into the all-zero combination indicating that all errors have been corrected.

In Table 20-22, the correction process of a burst error with pattern 111 is also shown.

It should be noted that the code used in this example has a rather high efficiency, not reached in many cases.

In the synopsis following, some groups of error-correcting codes as devised by various authors will be briefly discussed. These codes are generated by means of shift registers with characteristic polynomials from which it can be predicted that they will generate at least a number of b cycles each containing one of the b required error indicators corresponding to the b error-pattern polynomials. However, in many cases these codes are not optimal; i.e., the number of check elements is too large or the possible number of information elements for a required error-correcting power is too small. It is better to investigate the behavior of various characteristic polynomials with respect to the lowest-valued combinations (the error indicators) in the various cycles by means of a computer.

SYNOPSIS

The error-correcting codes of Hamming, Abramson, Fire, and Bose and Ray-Chaudhuri can best be considered as special applications of Meggitt's method. The error-correcting power of a group of code symbols is determined by the characteristic polynomial describing the construction of the mod 2 adding feedback network of shift register SR_2 in the receiver (Fig. 20-31). The characteristics of the polynomials generating the error-correcting codes of Hamming, Abramson, Fire, and Bose and Ray-Chaudhuri can be summarized as follows.

Hamming The starting point for a Hamming code is a primitive polynomial of degree k. This polynomial ${}^k\phi(x)$ generates a maximum-length sequence of length $2^k - 1$. Because in this case the primitive polynomial can generate only one cycle, the error-correcting power will be limited to a single error. The number of elements in the code symbols must be equal to the length $2^k - 1$ of the shift-register cycle. The number of check elements is equal to the length of the shift register or, in other words, equal to the degree k of the characteristic polynomial of the shift register used. From this it follows that the number of information elements in the code symbols will be $2^k - k - 1$.

Example Polynomial $\phi(x) = 1 \oplus x \oplus x^3$ generates a sequence of length $2^3 - 1 = 7$. The code symbols contain $k = 3$ check elements and $2^k - k - 1 = 8 - 3 - 1 = 4$ information elements. In this code, there are 15 possible symbols and, as an extra, the all-zero combination, so that 2^4 code symbols in total may be transmitted.

Abramson A code with the power to correct burst-of-two errors will be obtained when a polynomial of the form ${}^k\phi(x) \cdot (1 \oplus x)$ is used as a starting point, in which ${}^{k}\phi(x)$ is a primitive polynomial of degree k. This form of characteristic polynomial yields two sequences of length $2^k - 1$, so that the information about $2\cdot(2^k - 1)$ error situations can be given. One sequence performs the correction of the single errors, the other that of the first element of a burst-of-two error.

The total number of elements in the code symbols is again equal to the length of the sequences that can be generated by the used polynomial. The number of check elements is again equal to the degree of the characteristic polynomial which in this case is $k + 1$. Thus, the number of information elements will be $2^k - k - 2$.

Example Polynomial $(1 \oplus x \oplus x^3)\cdot(1 \oplus x) = 1 \oplus x^2 \oplus x^3 \oplus x^4$ yields two sequences of length $2^k - 1 = 2^3 - 1 = 7$. The number of check elements is $k + 1 = 4$, so that the number of information elements in the symbols of this code will be $2^k - k - 2 = 2^3 - 3 - 2 = 3$.

Fire In this code, the characteristic polynomial has the form

$$\phi(x) = {}^{k}\phi(x)\cdot(1 \oplus x^p)$$

in which ${}^{k}\phi(x)$ is again a primitive polynomial of degree k. Furthermore, it is required that p and $2^k - 1$ are relatively prime and that $k \geqslant \frac{1}{2}p$. Characteristic polynomials of this type yield a large number of sequences of length $p\cdot(2^k - 1)$. Up to a burst length of $\frac{1}{2}p$ when p is even and up to a burst length of $\frac{1}{2}(p + 1)$ when p is odd, these sequences provide for the correct error indicators. Moreover, some longer patterns can be corrected.

The number of elements in the code symbol is again equal to the length of the generated sequences, in this case $p\cdot(2^k - 1)$. The number of check elements is equal to the degree of the characteristic polynomial used, that is, $k + p$. From this it follows that the number of information elements in the code symbols will be $p\cdot(2^k - 1) - (k + p) = p\cdot(2^k - 2) - k$.

Example Polynomial $\phi(x) = (1 \oplus x^2 \oplus x^5)\cdot(1 \oplus x^5) = 1 \oplus x^2 \oplus x^7 \oplus x^{10}$ yields sequences of length $5\cdot(2^5 - 1) = 155$. Thus, the number of elements of the code symbols will be 155. The number of check elements, which is equal to the degree of the given polynomial, is 10, so that the number of information elements will be 145.

According to Fire, burst errors to a maximum length of $\frac{1}{2}(p + 1) = \frac{1}{2}(5 + 1) = 3$ can be corrected in this system. It can easily be checked that this code can correct the following burst-error patterns: 1, 11, 101, 111, 1101, and 1111, which include all the error patterns for bursts up to a length of 3, and as an extra two of the four error patterns of burst length 4.

It should be noted that the extra error patterns for bursts of length 4 that are incomplete decrease the efficiency of the code in this example. With the same length of shift register, a far more efficient code can be obtained when using the following characteristic polynomial:

$$\phi(x) = (1 \oplus x \oplus x^2)\cdot(1 \oplus x \oplus x^6 \oplus x^7 \oplus x^8)$$

This polynomial yields four sequences of length 255 and covers all burst errors of length 3. This polynomial increases the number of information elements in the symbols from 145 to 245.

Example Polynomial $(1 \oplus x \oplus x^3 \oplus x^4 \oplus x^6)\cdot(1 \oplus x^{11})$ can correct all burst errors up to a maximum length of 6. The total number of elements in the code symbols is $11\cdot(2^6 - 1) = 693$; the number of check elements is $11 + 6 = 17$.

The importance of the Fire codes is that a suitable polynomial can easily be determined when a maximum-burst length is given. However, it is a disadvantage that a number of longer error patterns not forming a complete group can also be corrected. This decreases the efficiency of the Fire code, as was demonstrated above.

Bose–Ray-Chaudhuri The symbols of these codes can be constructed in such a way that up to p arbitrary errors can be corrected. The construction of the characteristic polynomial of the code-symbol-generating shift register is as follows

The fundamental polynomial of Bose–Ray-Chaudhuri codes is the following:

$$\phi(x) = {}^k\phi(x)\cdot\phi_3(x)\cdot\phi_5(x)\cdots\phi_{2p-1}(x) \tag{20-114}$$

in which ${}^k\phi(x)$ is again a primitive polynomial of degree k, while polynomials $\phi_3(x)$, $\phi_5(x)$, ..., $\phi_{2p-1}(x)$ are chosen in such a way that $\phi_3(x^3)$, $\phi_5(x^5)$, ..., $\phi_{2p-1}(x^{2p-1})$ can be divided by ${}^k\phi(x)$. Polynomials $\phi_3(x)$, $\phi_5(x)$, ..., $\phi_{2p-1}(x)$ can be chosen in such a way that the length of the sequence generated by a shift register with a mod 2 adding feedback circuit in accordance with Eq. (20-114) is $2^k - 1$. Thus, the number of elements in the code symbols will also be $2^k - 1$. The number of check elements is again equal to the degree of characteristic polynomial (20-114). However, the degree of polynomials $\phi_3(x)$, $\phi_5(x)$,..., $\phi_{2p-1}(x)$ is generally not known beforehand, so that the number of check elements has to be determined in each example. It can be proved that polynomials $\phi_3(x)$, $\phi_5(x)$, ..., $\phi_{2p-1}(x)$ can be chosen so that their degree is at most k. The number of check elements is then at most $p\cdot k$.

Example 1 Determining a group of code symbols with 15 elements having the power to correct all single and double errors is required. For the solution of this problem, polynomial

$${}^k\phi(x) = 1 \oplus x^3 \oplus x^4$$

can be used as a starting point. Only polynomial $\phi_3(x)$ has to be determined because $p = 2$ and $2p - 1 = 3$. For this purpose, we try to find a multiple of ${}^k\phi(x)$ in which only the following powers of x occur:

$$x^3, x^6, x^9, x^{12}, \ldots$$

After some trials, we find that

$$(1 \oplus x^3 \oplus x^4)\cdot(1 \oplus x^4 \oplus x^6 \oplus x^7 \oplus x^8) = 1 \oplus x^3 \oplus x^6 \oplus x^9 \oplus x^{12} \qquad (20\text{-}115)$$

It follows from Eq. (20-115) that

$$\phi_3(x^3) = 1 \oplus x^3 \oplus x^6 \oplus x^9 \oplus x^{12}$$

Hence,

$$\phi_3(x) = 1 \oplus x \oplus x^2 \oplus x^3 \oplus x^4$$

The characteristic polynomial having the power to correct all single and double errors in code symbols of length 15 can now be chosen as follows:

$$\begin{aligned}\phi(x) &= {}^k\phi(x)\cdot\phi_3(x) = (1 \oplus x^3 \oplus x^4)\cdot(1 \oplus x \oplus x^2 \oplus x^3 \oplus x^4)\\ &= 1 \oplus x \oplus x^2 \oplus x^4 \oplus x^8 \qquad (20\text{-}116)\end{aligned}$$

The number of check elements, being equal to the degree of polynomial (20-116), is in this case eight, so that the number of information elements in the code symbols will be seven.

It should be noted that polynomial (20-116) is not the most effective code-symbol-generating polynomial of degree 8 with single- and double-error-correcting power. Both irreducible nonprimitive polynomials

$$\phi(x) = 1 \oplus x^3 \oplus x^4 \oplus x^5 \oplus x^8$$

and

$$\phi(x) = 1 \oplus x \oplus x^2 \oplus x^4 \oplus x^6 \oplus x^7 \oplus x^8$$

have the same error-correcting power in the form of a distance $d = 5$, but the number of elements in the code symbols can be larger (17) so that the number of information elements in these symbols is increased from 7 to 9.

Example 2 Extending the error-correcting power of the 15-element symbols of Example 1 to all triple errors is now required. Thus $p = 3$ and $2p - 1 = 5$, so that we have to find $\phi_5(x)$ such that $\phi_5(x^5)$ can be divided by ${}^k\phi(x)$. For this purpose, it is necessary to find a multiple of ${}^k\phi(x)$ in which appear only $x^5, x^{10}, x^{15}, \ldots$.

After some trials, it is easily found that

$$(1 \oplus x^3 \oplus x^4)\cdot(1 \oplus x^3 \oplus x^4 \oplus x^5 \oplus x^6) = 1 \oplus x^5 \oplus x^{10}$$

so that the following choice for $\phi_5(x)$ can be made:

$$\phi_5(x^5) = 1 \oplus x^5 \oplus x^{10}$$

Hence,

$$\phi_5(x) = 1 \oplus x \oplus x^2$$

Thus, the complete characteristic polynomial $\phi(x)$ can be chosen as follows:

$$\begin{aligned}\phi(x) &= (1 \oplus x^3 \oplus x^4)\cdot(1 \oplus x \oplus x^2 \oplus x^3 \oplus x^4)\cdot(1 \oplus x \oplus x^2)\\ &= 1 \oplus x^2 \oplus x^5 \oplus x^6 \oplus x^8 \oplus x^9 \oplus x^{10}\end{aligned}$$

The number of check elements is in this case 10, so that the number of information elements is decreased from 7 to 5 by the increase of the error-correcting power from all single and double errors to all single, double, and triple errors.

The fact that the ratio of the number of information elements and check elements is rather unfavorable and that in most actual cases burst errors will occur make this group of codes, which are very interesting from a mathematical point of view, less attractive.

The question that still has to be investigated is whether an error-correcting code gives any improvement in a transmission system. The number of elements in the code symbols has been increased; concurrently, the code symbols have become more vulnerable to transmission irregularities.

Let p be the probability of an erroneous element; then the probability of a correct element will be $1-p$. The probability of finding no error in a code symbol consisting of n elements will be $(1-p)^n$. The probability of finding one error in a code symbol consisting of n elements will be $np(1-p)^{n-1}$.

In a single-error-correcting code, the symbols that have no error or a single error are classified as good. The probability of such a type of symbol occurring is

$$np(1-p)^{n-1}+(1-p)^n \tag{20-117}$$

In a single-error-correcting code, the probability of having an erroneous symbol is then

$$1-[np(1-p)^{n-1}+(1-p)^n] \tag{20-118}$$

The probability of having a correct m-element symbol is

$$(1-p)^m \tag{20-119}$$

The probability of having an error in a symbol consisting of m elements is

$$1-(1-p)^m \tag{20-120}$$

The ratio between the probabilities (20-118) and (20-120) shows the reduction in the error probability P resulting from the introduction of the single-error-correcting code:

$$P=\frac{1-[np(1-p)^{n-1}+(1-p)^n]}{1-(1-p)^m} \tag{20-121}$$

For small values of p, the function $(1-p)^n$ may be taken as $1-np$, so that

$$P\approx\frac{1-np[1-(n-1)\,p]-(1-np)}{1-(1-mp)} \tag{20-122}$$

$$P\approx\frac{m}{n}\cdot(n-1)\,p=pR(n-1) \tag{20-123}$$

with $n/m=R$.

Equation (20-122) shows that an important reduction of the probability of an erroneous signal can be obtained by the use of a single-error-correcting code, provided that the probability of an erroneous element is already small. However, there is a value of p for which P becomes larger than 1. In such a case, the use of a single-error-correcting code merely increases the number or errors in the transmission link.

PROBLEMS

1. What is the storage efficiency for a hypothetical number system with radix $e = 2.71828$?

2. Give the truth table for the addition of two ternary numbers (sum and output carry) and derive the switching equations.

3. Design a section of an adder for two ternary numbers having a carry input. Use Kirchhoff networks.

4. State the rules for positively weighted four-bit binary-decimal codes and give all possible groups of four weights.

5. Explain the different ways that a four-bit binary-decimal group of code symbols can be error-detecting.

6. Explain the concept of distance between binary code symbols of a group. What is the minimum distance for single-error- and double-error-correcting power of a group of code symbols.

7. Explain the operation of a transmitter for error-correcting code symbols with a number of shift sections equal to the number of information elements of the code symbol. Give an equation for the transmitted code symbol.

8. Explain the operation of a transmitter for error-correcting code symbols with a number of shift sections equal to the number of check elements of the code symbol. Derive an equation for the transmitted code symbol using polynomial $A(x)$ representing the information elements and characteristic polynomial $\phi(x)$.

(*a*) The transmitter consists of an ordinary shift register with an external mod 2 adding feedback network.

(*b*) The transmitter consists of a register circuit with mod 2 adders between the sections.

9. What is the relation between the concept of distance between binary code symbols and the number of 1's in those code symbols of that group having a single 1 in the information elements. Show this with an example [such as, $q(x) = 1 \oplus x \oplus x^3 \oplus x^7$].

10. Show by means of an example that the code symbols of Prob. 9 have an error-correcting power of two independent errors, either in the information elements or in the check elements, or a single error in both.

11. What are burst errors? How many burst-error patterns can be distinguished for a maximum-burst length of r? What is the relation between r and k (the number of check elements in the code symbols)?

12. Draw a receiver circuit for error-correcting code symbols and show that error-pattern polynomial $f(x)$ can be traced back either in parallel form or in serial form during the correction cycle of n pulses following the receiving cycle of n pulses ($n =$ the number of elements in the code symbols). Show by means of an example [$\phi(x) = 1 \oplus x \oplus x^3 \oplus x^7$] the operation of the receiver during the correction cycle.

SELECTIVE BIBLIOGRAPHY

1. Abramson, N. M.: A Class of Systematic Codes for Non-independent Errors, *IRE Trans.*, vol. IT-5, pp. 150–157, 1959.
2. Abramson, N.: A Note on Single-error Correcting Binary Codes, *IRE Trans.*, vol. IT-6, pp. 502–503, 1960.
3. Abramson, N. M., and B. Elspas: Double-error-correcting Codes and Decoders for Non-independent Binary Errors, Unesco Information Processing Conference, Paris, 1959.
4. Albert, A. A.: "Fundamental Concepts of Higher Algebra," The University of Chicago Press, Chicago, Ill., 1956.
5. Birkhoff, G., and S. Mac Lane: "A Survey of Modern Algebra," The Macmillan Company, New York, 1941.
6. Bose, R. C., and D. K. Ray-Chaudhuri: On a Class of Error-correcting Binary Group Codes, *Inform. Control*, vol. 3, pp. 68–79, 1960.
7. Bose, R. C., and D. K. Ray-Chaudhuri: Further Results on Error-correcting Binary Group Codes, *Inform. Control*, vol. 3, pp. 279–290, 1960.
8. Brown, D. T.: Error Detecting and Correcting Binary Codes for Arithmetic Operations, *IRE Trans.*, vol. EC-9, pp. 333–337, 1960.
9. Diamond, J. M.: Checking Codes for Digital Computers, *Proc. IRE*, vol. 43, pp. 487–488, 1955.
10. Elspas, B.: The Theory of Autonomous Linear Sequential Networks, *IRE Trans.*, vol. CT-6, pp. 45–60, 1959.
11. Fire, P.: A Class of Multiple-error-correcting Binary Codes for Non-independent Errors, *Sylvania Lab. Rep.* RSL-E-2, 1959.
12. Green, J. H., Jr., and R. L. San Soucie: An Error-correcting Encoder and Decoder of High Efficiency, *Proc. IRE*, vol. 46, pp. 1741–1744, 1958.
13. Hagelbarger, D. W.: Recurrent Codes: Easily Mechanized Burst-correcting Binary Codes, *Bell System Tech. J.*, vol. 38, pp. 969–984, 1959.
14. Hamming, R. W.: Error-detecting and Error-correcting Codes, *Bell System Tech. J.*, vol. 29, pp. 147–160, 1950.
15. Meggitt, J. E.: Error Correcting Codes for Correcting Bursts of Errors, *IBM J. Res. Develop.*, vol. 4, pp. 329–334, 1960.
16. Melas, C. M.: A New Group of Codes for Correction of Dependent-errors in Data Transmission, *IBM J. Res. Develop.*, vol. 4, pp. 58–65, 1960.
17. Peterson, W. W.: "Error-correcting Codes," John Wiley & Sons, Inc., New York. 1961.

21
The Organization of Sequential Circuits

21-1 STATE DIAGRAMS AND RACE CONDITIONS

In the preceding chapters, a number of stringent laws were given regarding the use of logical elements as contacts and diodes in dipole and multipole circuits, as well as some laws concerning the use of memory elements as relays, triggers, and capacitors in switching circuits. However, the time charts of the various types of circuits which illustrate graphically the operation of the memory elements were given without any further comment about the way in which they could be achieved starting from the fundamental requirements governing the circuit.

In this chapter, we shall describe a method by means of which an internal organization of the operation of a circuit can be designed.

Generally speaking, circuits with gates and triggers (or relays or any other bistable switching element) and circuits with gates and MS memory elements can be considered as belonging to two different main classes. In general, the difference between these two main classes depends on the type of internal operation of the circuit.

Circuits in which one external change of state can be followed by the change of state of a single memory element only. The circuit remains in that state until the following external change of state occurs.

Ib. Circuits in which one external change of state can be followed by the change of state of a number of memory elements. In such circuits, a number of unstable states are introduced, which are known as transient states or race conditions, through which the circuit has to run before the next external change of state occurs.

II. Circuits in which an external change of state can produce any internal change of state only under the control of, say, a clock pulse, which is given at predetermined intervals. Circuits of this class contain MS memory elements, thereby eliminating the occurrence of race conditions.

It is possible, however, to design circuits which belong to class I*a* but are in fact circuits of class II. The pulse halver is an example of a circuit of class I*a* if all possible external states are considered as in the time chart of Fig. 11-1; but if the α pulse is considered as a clock pulse as shown in the time chart of Fig. 11-19, this pulse halver can be considered as belonging to class II. The circuit of Fig. 11-6 can then be simplified to that of Fig. 11-20 or 11-21.

The circuits of class I are sometimes called asynchronous circuits, while the circuits of class II are often referred to as synchronous circuits. The distinction between these two groups of circuits, however, is not very sharp.

In this chapter, the operation of these classes of switching circuits will be more readily explained by means of a type of diagram that is different from the time chart describing the final form of operation of a circuit.

The time chart of a pulse halver consisting of two triggers (see Fig. 11-6) describes four distinct combinations of the states of the individual memory elements. These four states of the individual memory elements will be replaced here by four states of the complete circuit. These four states may be numbered arbitrarily 0, 1, 2, 3 or 1, 2, 3, 4. From this number of states, it may be concluded that if all operating combinations of the memory elements to be used are allowed, this circuit with four different states can be implemented by means of two triggers. Two triggers can produce at most $2^2 = 4$ different states. The states of the circuit can then be coded in, say, the binary notation. The individual digits of the numbers of the code combinations represent the states of the memory elements involved.

The coding process of the circuit states will be discussed in a following section. It is one of the most important parts of the design of a circuit.

The four states of the pulse halver as indicated in the time chart of Fig. 11-1 can be represented by means of the state diagram of Fig. 21-1. Each state of the circuit is indicated by a numbered circle. The circles are inter-

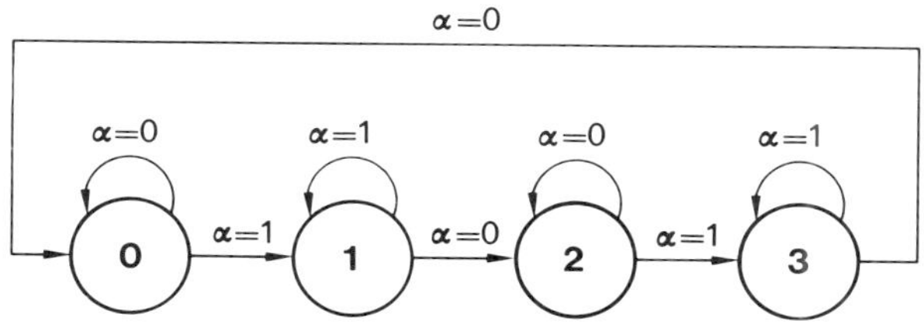

Fig. 21-1

connected by means of arrows. These interconnecting arrows indicate the time period between the change of the external state of the circuit and the following adaptation of the internal state of the circuit. The markings above the arrows show the external circuit condition leading to the next state. The arrows originating from a state circle and terminating in the same circle indicate a stable state that can be left only after a change in the external state of the circuit.

State 0 coincides with the time interval in the time chart of Fig. 11-1 before moment t_0; state 1 corresponds to the time interval between moments t_0 and t_1, etc. The transition between these two successive states has been obtained by applying an external control signal $\alpha = 1$, as is indicated by the arrow between states 0 and 1.

This time chart shows more clearly than a state diagram that state changes affect only one memory element at a time and that all these changes are controlled directly by the external-control signal α. If two memory elements are used, the states being coded in binary notation are as shown in the state diagram of Fig. 21-2.

However, if the pulse halver is considered as a circuit containing only one MS memory element, the time chart reduces to the state of the master part of this memory element, as is illustrated in Fig. 11-19. The state of this part of the MS memory element will normally be sufficient because it generates the output signals. The state diagram corresponding to this time chart is shown in Fig. 21-3.

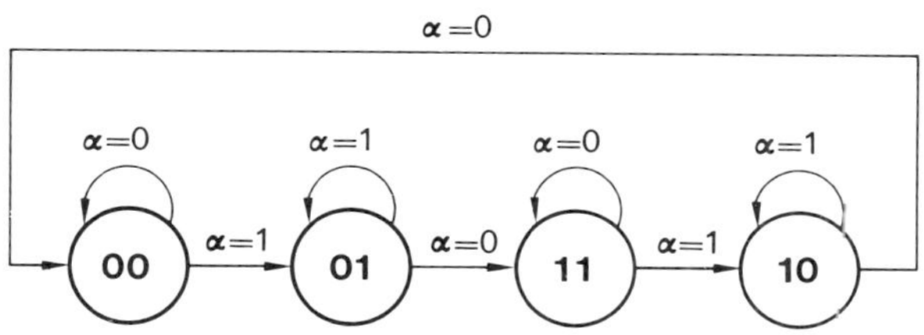

Fig. 21-2

This state diagram shows only two circuit states (0 and 1) because the state of the RC delay memory element(s) has been left out of consideration. The curved arrows $\alpha' = (1,0)$ indicate that during the successive intervals that the α' signal is 1 respectively 0, the state of the master part of the MS memory element does not change. The change of state of this master part occurs only on the trailing edge of the α' counting pulses, which is indicated in Fig. 21-3 by $\alpha = 0 \rightarrow 1$ (straight arrows). The meaning of the curved arrows

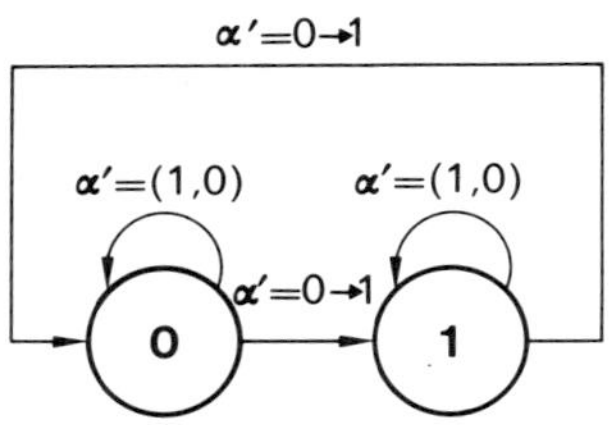

Fig. 21–3

originating from and terminating in the same state circle is slightly different in the state diagrams of Figs. 21-2 and 21-3. The curved arrows in the last diagram indicate in which phase of the counting (or clock) pulse the state of the slave parts of the MS memory elements is changed. During this phase, the master parts of the MS memory elements are in their stable state.

A pulse halver based on the state diagram of Fig. 21-3 belongs to class II as defined above. The only difference between a pulse halver belonging to class I and another belonging to class II is that the circuit of the first class uses two master memory elements both of which can produce output signals respectively on the front and on the back of the control pulse, whereas a pulse halver of the second class produces an output signal only by means of its single master memory element. This limitation of the pulse-halving circuit of the second class can generally be considered to be of minor importance.

An example of a more sophisticated relay circuit belonging to class Ia is shown in Fig. 15-13. This circuit, which represents a reflected binary counting chain, can be so designed as to belong to class II, as is shown in Fig. 15-8. The variant of the circuit of the reflected binary counting chain shown in Fig. 15-6 belongs to class Ib.

The circuits of class Ib require some further explanation. An example of a state diagram of a cyclic circuit belonging to class Ib is given in Fig. 21-4. Under the control of a signal $\alpha = 1$, state 0 1 1 changes into state 1 1 0 via an intermediary state 0 1 0, which can be called transient state. Subsequently, state 1 1 0 is changed under the control of a signal $\alpha = 0$ to 1 0 1 via transient state 1 1 1.

The state diagram of Fig. 21-4 shows that all states of the circuit are stable except states 010 and 111, which have terminating and originating arrows with the same marking and no curved arrows. The diagram of Fig. 21-4 shows that the circuit runs in a predetermined way through these transient states. It is a serial operation. In this type of organization of the intermediary states between two stable states, the time required to reach the second stable state is directly proportional to the number of memory elements changing their state under the control of one external change of state.

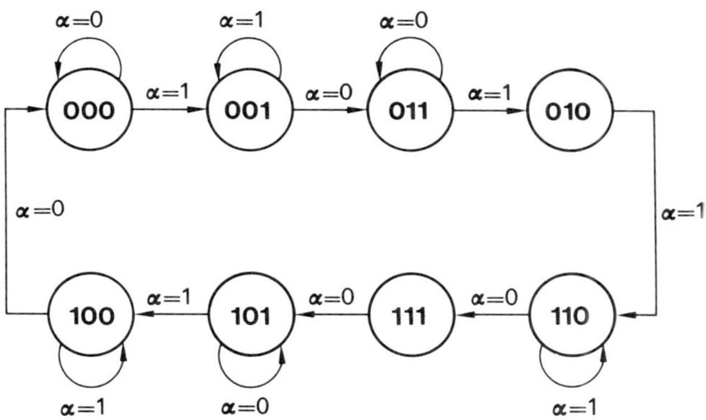

Fig. 21–4

In the example of the state diagram of Fig. 21-4, a cyclic circuit has been chosen—but this is not essential.

If a large number of memory elements (relays or triggers) have to change their states under the control of one external change of state, this can be achieved in two different ways:

1. The circuit can run through a predetermined series of intermediary states that will in this case be indicated by "transient states." The state diagram of Fig. 21-4 gives a simple example of this type of operation. It has a disadvantage in that it may take too long a time.
2. The circuit can run from one stable state of the circuit to the next, following an arbitrary way in a more or less parallel manner, depending on the operating speed of the memory elements. This enables high-speed operation at the cost of a relatively large number of states that cannot be used as stable states of the circuit. They are called race conditions.

In Figs. 21-5 and 21-6, two parts of a larger state diagram have been drawn with the (parallel) transition of a stable circuit state to the next, where these states differ in three variables.

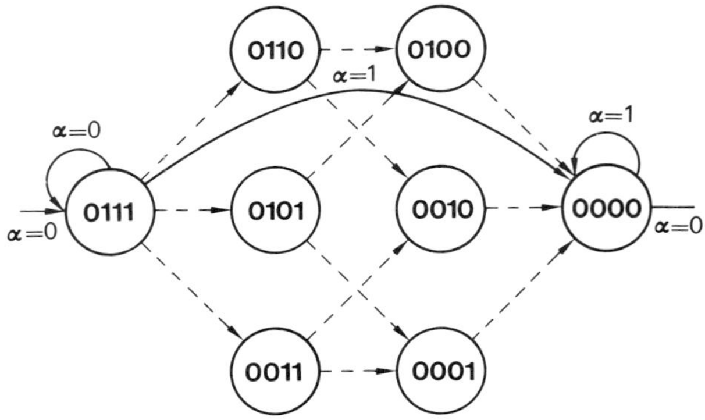

Fig. 21–5

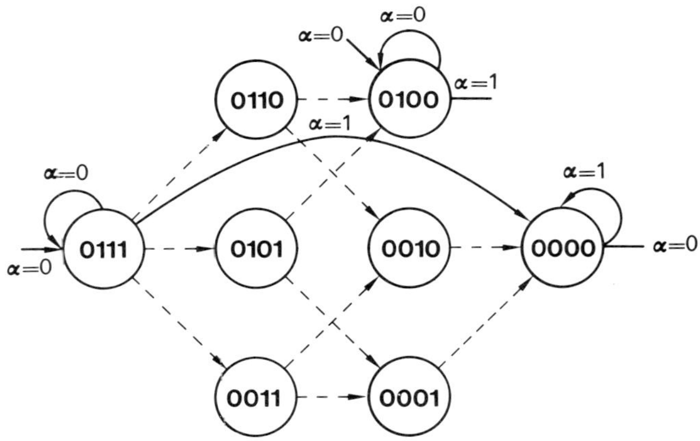

Fig. 21–6

The intermediary states, in this case called "race conditions," through which the circuit can run are shown in Fig. 21-5 in successive levels. The level that is reached first shows all possible race conditions that differ in one variable from the original state. The second level comprises those race conditions in which two variables have already changed their state. It could be thought that the transition time of one stable state to the next in the state diagram of Fig. 21-5 is directly proportional to the number of variables involved. However, one of the possible race paths will occur only during a very short interval of time. The transition time equals the operating time of the slowest memory element. If none of the possible race paths contains a state that is in another part of the diagram also used as stable state, all race conditions are of the noncritical type. If, however, a race path contains a state that is also used as stable state in another part of the diagram, the way of operation of the circuit will depend on the operating speeds of the memory elements. An example of a state diagram containing such a critical race condition is shown in Fig. 21-6. In this figure, it is assumed that state 0 1 0 0 is also used as a stable state in a part of the diagram not shown in the figure.

It follows from Fig. 21-5 that the number of race conditions and stable states involved in the change of state of n variables is

$$N = 2^n \tag{21-1}$$

so that the number of race conditions involved in such a state change will be

$$M = 2^n - 2 \tag{21-2}$$

A state change of a circuit in which more than one variable changes at a time should be avoided if at all possible because of the disproportionate increase in the number of race conditions involved that cannot be used as stable states. This type of circuit (class I*b*) in which master memory elements are normally used exclusively becomes uneconomical if the introduction of race conditions leads to a large increase in the number of memory elements, as can easily happen.

The operation of the circuits of class I can become very complex if too much emphasis is put on the economical use of the memory elements. On the other hand, there is complete freedom in the design of the circuits of class II. In the construction of the circuits of this group, the designer is hindered neither by race conditions nor by a complication of the circuit design as a consequence of the utmost combinational use of the MS memory elements. The problem of race conditions does not exist in circuits of class II.

The examples of state diagrams given above all relate to counting circuits in which a single input variable is applied in a regular sequence. Most sequential circuits will be of a different type. In a clock-pulse system, for instance, the clock pulse does not contribute to the information to be processed. Only the information signals are important in such cases. In the indications of state changes, clock pulses need not be mentioned. Consequently, they will be omitted in all further state diagrams.

In state diagrams of circuits of class II, the curved arrows indicating the stable states may be omitted because there are no race conditions in these types of circuits, so that all states are stable.

If the desired circuit is required to produce certain combinations of output signals in a circuit state resulting from a certain combination of input variables, this can be indicated in combination with these input signals.

21-2 REDUNDANT STATES IN SEQUENTIAL CIRCUITS

A sequential circuit of a completely general nature is provided with k input lines, l output lines, and m memory elements. The signals on the output lines as well as in the memory elements are supposed to have a binary character. A number of output lines may be connected to the input lines of the same circuit. A circuit of this general character is shown in Fig. 21-7.

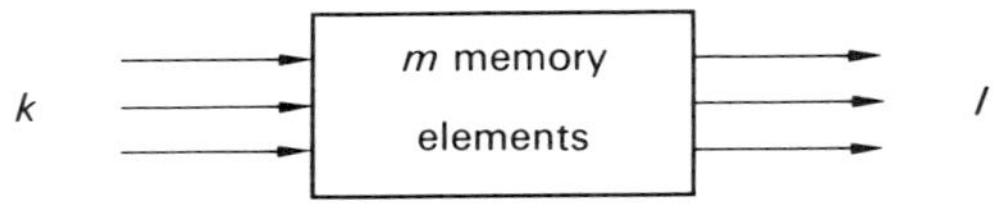

Fig. 21-7

The present input combination of signals on the k input lines and the present state of the m memory elements determine uniquely the next state of these memory elements and the present combinations of the signals on l output lines.

It is possible that two circuits controlled by identical sequences of combinations of input signals show identical sequences of combinations of output signals, even though their internal structure is different. These circuits may be considered to be equivalent.

The state diagram of one of these circuits may have a redundant number of states; consequently, the circuit may have too large a number of memory elements to represent all the circuit states including the redundant ones. If the number of memory elements of a circuit is larger than the minimum possible number, this does not necessarily stem from a state diagram with a redundant number of states; it can also result from the relatively ineffective coding of the states of that circuit. For example, a section of a decimal counter can be designed by means of four MS memory elements in a binary operation. It is also possible to assign the states of the same state diagram in such a way that a linear decimal counter with 10 MS memory elements is obtained.

With respect to the synthesis of sequential circuits, the above considerations lead to the four following stages of design:

1. The construction of the state diagram or equivalent truth table to be derived from the original set of requirements.

2. The detection of equivalent circuit states and the removing of all but one of these equivalent circuit states from the state diagram or truth table, so that the circuit obtained in this way is equivalent to the one resulting from the original state diagram.
3. The coding of the circuit states or state assignment. The coding process can be performed by

   a. Use of a minimum number of memory elements
   b. Implementation of a special mode of operation of the memory elements.

   Before the coding process of the states offering a wide variety of possibilities can be started, a choice has to be made between a circuit of class I or class II as mentioned in Sec. 21-1.
4. The construction of switching equations that describe the desired logical circuits between the memory elements, as derived from the final truth table with coded circuit states.

Circuits that show identical sequences of combinations of output signals on identical sequences of combinations of input signals will contain equivalent states which can be classified as follows.

1. *Two states of a sequential circuit are equivalent* if they show the same combination of output signals and lead under identical combinations of input signals to the same next states, regardless of the combination of input signals that have led to these present states.

In Figs. 21-8 and 21-9, two parts of a larger state diagram have been shown to illustrate this type of circuit equivalence. The arrows between the various circuit states and stable states are provided with an indication of the input combination—the state number before the oblique stroke and the generated output of the circuit behind the stroke, respectively. Output indications on the arrows between circuit states are not of direct importance

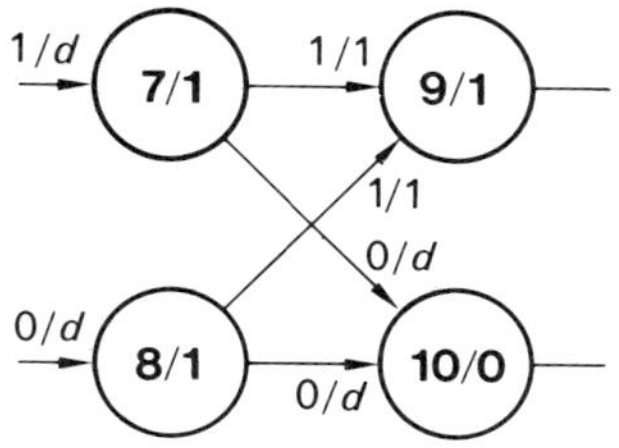

Fig. 21–8

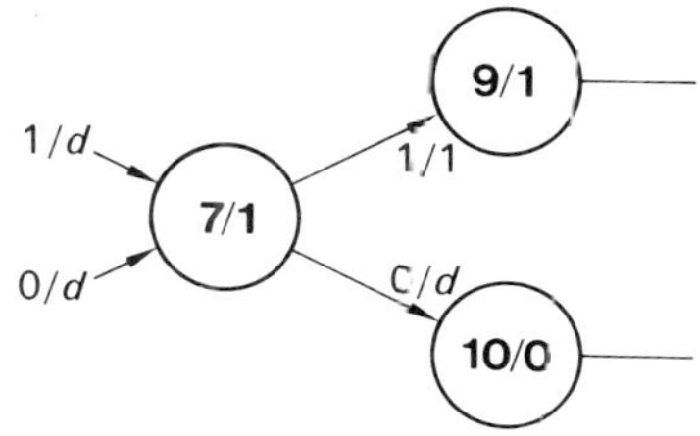

Fig. 21–9

in the problem of circuit-state equivalence. These indications show the desired output combination during the transient condition of the circuit. Generally, the simplest circuit will be obtained when these indications are chosen in accordance with the output combination of the stable state on the originating end of the arrow. In our example, the desired output combination during the transient state of the circuit will be indicated by d (don't care) if the output changes.

The indications given in Fig. 21-8 can be arranged in a state table as shown in Table 21-1. The diagram of Fig. 21-8 relates to a synchronous circuit with MS memory elements. However, the following simplification process can also be applied to asynchronous circuits.

Table 21-1

States			*Outputs*		
Present	*Next*		*Present*	*Next*	
	0	1		0	1
7	10	9	1	0	1
8	10	9	1	0	1
9	x	x	1	x	x
10	x	x	0	x	x

In state table 21-1, the lines relating to the present states 7 and 8 are important. They show what is required by the definition. The next states and all output entries of lines 7 and 8 are identical, so that states 7 and 8 may be considered as equivalent states. In the complete state table that describes the operation of the full circuit, all entries 8 can thus be replaced by 7, or vice versa. This process, devised by Huffman, is called merging. The effects of transient conditions can be considered by adding extra output columns showing the wanted outputs during these conditions.

As a consequence of the merging process, the circuit part concerning states 7, 8, 9, and 10 of Fig. 21-8 can be simplified as shown in Fig. 21-9 in which state 8 has been eliminated.

As may be illustrated by the following example, the merging process will not always be completed in one step. It should be pointed out, however, that the example has been constructed for the purpose of demonstration and therefore may very well never be found in actuality, because a circuit designer will generally try to avoid such cases by careful formulation of his requirements. The state diagram to be used as a starting point is shown in Fig. 21-10. The corresponding state table is given as Table 21-2. This table shows that

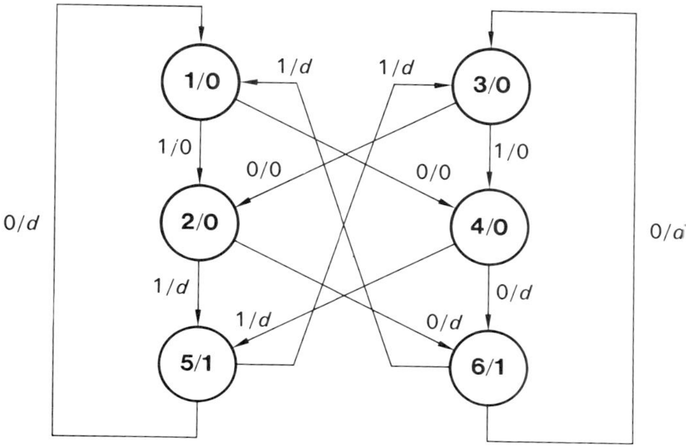

Fig. 21–10

as a first step only two states (2 and 4) can be merged. State 4 will be eliminated in favor of state 2, so that all entries 4 will be replaced by 2 in Table 21-2. The resulting state table without line 4 is shown as Table 21-3.

Table 21-2

States			*Outputs*		
Present	*Next*		*Present*	*Next*	
	0	1		0	1
1	4	2	0	0	0
2	6	5	0	1	1
3	2	4	0	0	0
4	6	5	0	1	1
5	1	3	1	0	0
6	3	1	1	0	0

States 1 and 3 of Table 21-3 can now be merged. All entries 3 are replaced by 1, as shown in the resulting Table 21-4. State 3 has been omitted from Table 21-4. As a result of the second merging process, a merging of states 5 and 6 proves possible in Table 21-4. This reduces state table 21-4 to Table 21-5, which contains only three states. The resulting state diagram is shown in Fig. 21-11.

Table 21-3

States			*Outputs*		
Present	*Next*		*Present*	*Next*	
	0	1		0	1
1	2	2	0	0	0
2	6	5	0	1	1
3	2	2	0	0	0
5	1	3	1	0	0
6	3	1	0	0	0

Table 21-4

States			*Outputs*		
Present	*Next*		*Present*	*Next*	
	0	1		0	1
1	2	2	0	0	0
2	6	5	0	1	1
5	1	1	1	0	0
6	1	1	1	0	0

Table 21-5

States			*Outputs*		
Present	*Next*		*Present*	*Next*	
	0	1		0	1
1	2	2	0	0	0
2	5	5	0	1	1
5	1	1	1	0	0

However, as will be seen in the next section dealing with the second class of equivalent circuit states, the fact that all possibilities for merging circuit states have been exhausted does not necessarily mean that all redundant circuit states have been eliminated.

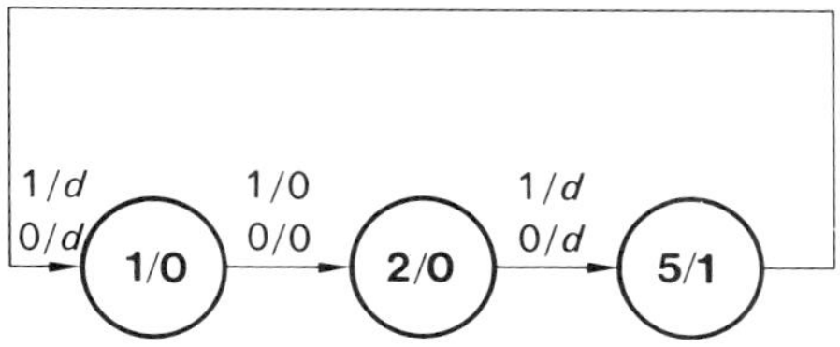

Fig. 21-11

2. *Two sequences of states of a sequential circuit are equivalent* if they comprise corresponding states that lead for identical combinations of input and output signals to next states that also have identical combinations of input and output signals.

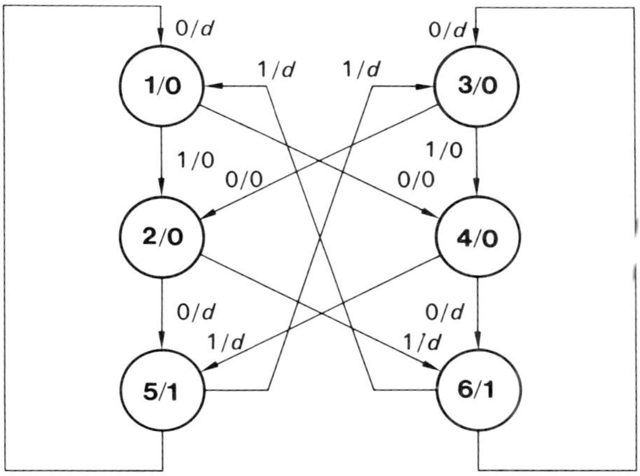

Fig. 21-12

An example of a state diagram containing such sequences is given in Fig. 21-12, the corresponding state table being Table 21-6. State table 21-6 does not contain states that can be merged. The state diagram of Figs. 21-10 and 21-12 differ in the indications of two transitions. Nevertheless, the merging procedure cannot be applied. Mealy has developed a general way to solve this problem, which can be applied in both cases of equivalent circuit states as defined above.

In the first step of this simplification procedure, the present circuit states as specified in the original state table are subdivided into a number of classes equal to the number of different combinations of output signals of the present circuit states. This number of classes may be considered as the number of circuit states that can be reached under the most favorable conditions in the simplification procedure.

Table 21-6

States			*Outputs*		
Present	*Next*		*Present*	*Next*	
	0	1		0	1
1	4	2	0	0	0
2	5	6	0	1	1
3	2	4	0	0	0
4	6	5	0	1	1
5	1	3	1	0	0
6	3	1	1	0	0

By way of an example, state table 21-6 has been provided with these class indications and is now shown as Table 21-7.

Table 21-7

States			*Outputs*			
Present	*Next*		*Present*	*Next*		
	0	1		0	1	*Class*
1	4	2	0	0	0	*A*
2	5	6	0	1	1	*A*
3	2	4	0	0	0	*A*
4	6	5	0	1	1	*A*
5	1	3	1	0	0	*B*
6	3	1	1	0	0	*B*

If there are as many class indications as there are states in the state table, no further simplification is possible, because a circuit state cannot produce two different combinations of output signals.

If this is not the case—and it is not in state table 21-7—the simplification procedure can be carried one step further. In this second step, a group of "next-class" columns is added to the state table. If all identical class indications of the present state then lead to identical groups of next-class indications, the simplification process is ready. The classes then indicate a state of the simplified state table or state diagram. The differing elements lead, under the control of identical combinations of input signals, to identical next states.

State table 21-7 has been extended to Table 21-8 with next-state-class indications.

Table 21-8

States			*Outputs*			*Classes*		
Present	*Next*		*Present*	*Next*		*Present*	*Next*	
	0	1		0	1		0	1
1	4	2	0	0	0	*A*	*A*	*A*
2	5	6	0	1	1	*A*	*B*	*B*
3	2	4	0	0	0	*A*	*A*	*A*
4	6	5	0	1	1	*A*	*B*	*B*
5	1	3	1	0	0	*B*	*A*	*A*
6	3	1	1	0	0	*B*	*A*	*A*

Because the class indications of the present states in the state table 21-8 do not lead to identical groups of next-state-class indications, the simplification process of the example is not yet completed. It affects states 1 and 3 on the one hand, and states 2 and 4 on the other. Their class indications are AAA and ABB, respectively.

In the third step of the simplification process, the identical present-state-class indications, which are followed by different pairs of next-state-class indications, are replaced by new-present-state-class indications, while all next-state indications are adjusted accordingly. This process is carried on until all identical present-state-class indications are followed by identical pairs of next-state-class indications. The simplification procedure is then

Table 21-9

States			*Outputs*			*Classes*		
Present	*Next*		*Present*	*Next*		*Present*	*Next*	
	0	1		0	1		0	1
1	4	2	0	0	0	*A*	*C*	*C*
2	5	6	0	1	1	*C*	*B*	*B*
3	2	4	0	0	0	*A*	*C*	*C*
4	6	5	0	1	1	*C*	*B*	*B*
5	1	3	1	0	0	*B*	*A*	*A*
6	3	1	1	0	0	*B*	*A*	*A*

ready. The classes of circuit states are now reduced to classes that contain exclusively equivalent circuit states.

The third step in the simplification process of the state table of this example is shown in Table 21-9.

The simplification process now proves to be ready. Each of the three classes of states proves to contain two equivalent circuit states. The resulting state table with only three circuit states is shown as Table 21-10. The state diagram corresponding to this table is shown in Fig. 21-13.

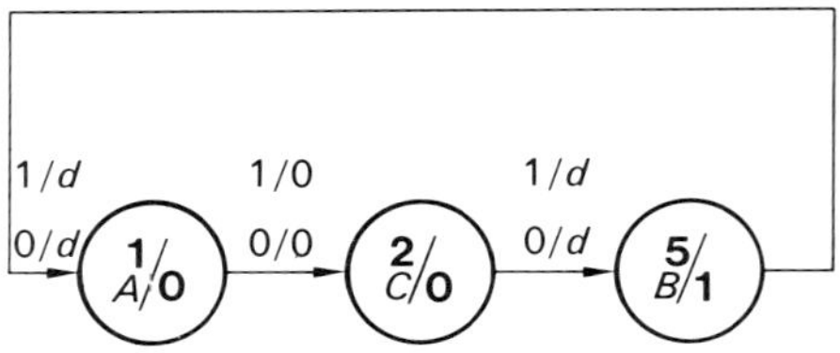

Fig. 21–13

Table 21-10

States			*Outputs*			*Classes*		
Present	*Next*		*Present*	*Next*		*Present*	*Next*	
	0	1		0	1		0	1
1	2	2	0	0	0	*A*	*C*	*C*
2	5	5	0	1	1	*C*	*B*	*B*
5	1	1	1	0	0	*B*	*A*	*A*

In order to avoid difficulties with the indications of the circuit states numbered as they are in arabic numerals, the classes of circuit states are indicated by means of capitals.

The examples of the two types of equivalent circuit states given above result in the same circuit. The first example can also be dealt with by means of Mealy's method, which is rationalized by Humphrey. This method leads for both types of state diagram to the same number of states. The resulting state diagram is in both cases the same.

The examples above have two input variables, one of which is the clock (not indicated) and one the actual input variable. The merging of states and Mealy's method can however be extended to the simplification of state diagrams of circuits with an arbitrary number of input and output variables.

In addition to the simplification of a state diagram by means of its equivalent states, there are two possibilities for further reduction in the number of states if

1. A number of combinations or sequences of inputs do not occur
2. A number of combinations or sequences of outputs do not occur

Both possibilities, which can occur in combination, can be treated in the same way. They result in blank entries in the truth table. These are the so-called don't-care conditions. It is possible to merge any two lines of circuit states of a state table which do not show contradictory entries. The simplification resulting from this possibility need not be unambiguous. Sometimes there are several merging possibilities, all of which have to be investigated in order to find the largest possible state reduction.

21-3 STATE ASSIGNMENT

After the completion of the state diagram and state table, eventually followed by the elimination of redundant states from the original circuit requirements, a number of circuit states are obtained which are the minimum possible number with which these requirements can be fulfilled. The circuit states have now to be implemented by means of combinations of operated and non-operated memory elements. At this point of the circuit design, the final choice has to be made between the two classes of circuits mentioned in Sec. 21-1.

If an asynchronous circuit is desired, the state assignment, sometimes called coding of states, has to be performed in such a way that adjacent states in the diagram do not differ in general by more than the state of one memory element. The memory elements to be used can be relays or triggers or similar devices.

If a synchronous circuit is required, the assignment of the necessary (MS) memory elements to the various states may be completely arbitrary, but such an arbitrary assignment will in general not lead to a circuit with a simple logic system in the connections between the various memory elements. The gate circuits in those logical connections will contain too many diodes or components performing the gating function in the input circuits of the memory elements.

The state assignment can also be considered as the actual organization of the circuit.

It is not very difficult to devise a feasible state assignment: but the number of possible state assignments is so large, even for circuits with a very small number of memory elements, that it is not possible to try them all out empirically. For example, in a circuit with 16 states, which requires a minimum of 4 memory elements, these 4 memory elements can be assigned in $2^n! = 2^4! \approx 20 \cdot 10^{12}$ ways to the 16 states.

The number of state assignments to be investigated in order to obtain a minimum of logical connections between the memory elements is not so great, since all assignments which can be derived from each other by means of a process of inverting and/or interchanging the binary variables will lead to circuits with the same logical complexity, and these need not therefore be investigated. If the number of required states is 2^n, the number N of assignments to be investigated can be limited to

$$N = \frac{(2^n - 1)!}{n!} \tag{21-3}$$

It is obvious that this number of assignments to be investigated is far too large for a systematic empirical investigation, such a process being both impractical and uneconomical in the design of switching circuits. Moreover, a large circuit containing the minimum possible number of memory elements will generally be a very difficult circuit from the point of view of maintenance. One cannot even be quite sure that this circuit will be the most economical. Adaptation of these types of circuits to changed operating requirements can lead to very complicated situations.

Fortunately, the design of sequential circuits with large numbers of memory elements is so complicated if all possible states of the memory elements of the circuit have to be used that a subdivision of the memory elements into groups is necessary. These groups perform specific functions such as the counting of pulses, the shifting of information, the adding of numbers, the transmitting or receiving of binary information, and so forth. For these purposes, specially organized circuits or disciplines have been designed, as shown in the former chapters of this book. Most of these circuits show a rather systematic and simple organization, which allows for a rather good survey of their operation and thus enables maintenance to be performed by a staff with normal skills.

The problem of state assignment, i.e., finding a good way out of the multitude of possibilities, is not tantamount to finding a needle in a haystack in a haphazard way. It is the task of the circuit designer to detect the fundamental organization of his switching problem. A circuit that involves this problem is very often the best circuit from the practical and economical points of view.

This principle, which is in fact nothing else than the principle of subdividing a circuit in functional groups of associated memory elements, can best be explained by means of examples that at first sight do not even seem to belong, even partially, to one of the groups of standard circuits described in the former chapters.

These examples are also useful in demonstrating the usefulness of the methods for eliminating redundant states from state diagrams. Both examples belong to the so-called group of electric lock circuits. These circuits receive sequences of input combinations and produce output combinations only after a predetermined set of requirements regarding the sequence of the input combinations has been fulfilled. The lock circuits can be subdivided into two groups:

1. Lock circuits that detect every predetermined sequence of input combinations
2. Lock circuits that detect fixed-cycle groups of input combinations

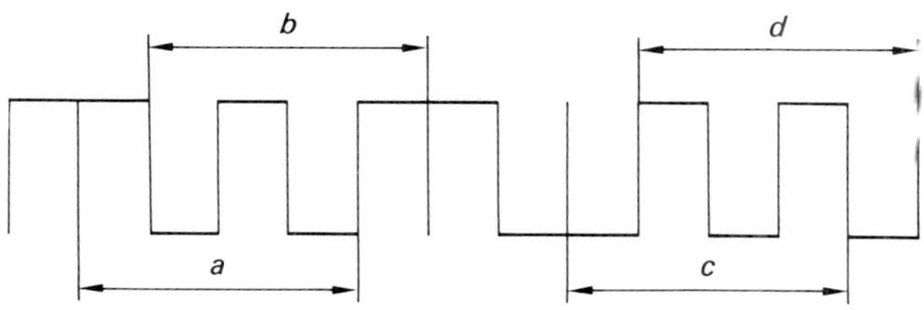

Fig. 21–14

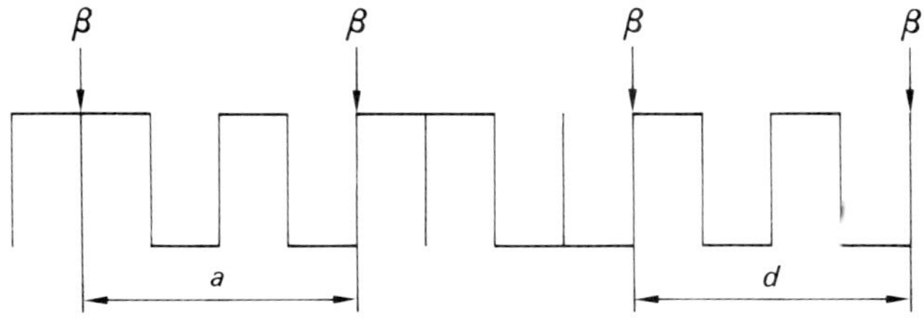

Fig. 21–15

The difference between these two groups of circuits will be demonstrated by means of Figs. 21-14 and 21-15. Both figures show the same sequence of input combinations. In the case of Fig. 21-14, the problem consists of designing a detection device to operate on every sequence of four input pulses 1 0 1 0 and 0 1 0 1 regardless of the position of these sequences with respect to a starting point or a group clock pulse. As shown in this figure, the detection device will operate on four sequences *a*, *b*, *c*, and *d*, notwithstanding the fact that sequences *b* and *d* are only shifted by one pulse with respect to sequences *a* and *c*.

A detection device operating on predetermined groups of input sequences marked by a group clock pulse β operates only twice on the signal sequence shown in Fig. 21-15. Input sequences *b* and *c* as indicated in Fig.

21-14 both belong to different clock-pulse groups, as is shown in Fig. 21-15. The group-detection device required for detecting the input combinations 1 0 1 0 and 0 1 0 1 will operate only on groups a and d, as is indicated in Fig. 21-15.

The truth table and state diagram will now be constructed for both examples. A circuit operating on clock pulses is required for this purpose.

The detection of all sequences of input signals 1 0 1 0 and 0 1 0 1 in a continuous series of pulses will be dealt with first. The detection device can stay at any moment in a situation which is the result of any of the 16 possible combinations of binary input signals. Any of these combinations can be the starting combination. Only the last three signals received are important in the construction of the truth table.

These last three signals together with the next input signal determine what the next state of the circuit will be and whether an output signal will be generated or not. In a clock-pulse system with MS memory elements, this output signal will be a pulse generated during the transient condition between two states. This pulse is used to set a trigger. The conditions of the last three input signals can be considered as representing all the possible states of the detection device. Thus, there are eight present states which may all be the starting state. The eight possible series of the last three input signals are numbered from 1 to 8 in Table 21-11.

Table 21-11

Last three input signals	*Number present state*
0 0 0	1
0 0 1	2
0 1 0	3
0 1 1	4
1 0 0	5
1 0 1	6
1 1 0	7
1 1 1	8

Table 21-12

States			*Transient outputs*	
Present	*Next*			
	0	1	0	1
1	1	5	0	0
2	1	5	0	0
3	2	6	0	1
4	2	6	0	0
5	3	7	0	0
6	3	7	1	0
7	4	8	0	0
8	4	8	0	0

The resulting state table is given as Table 21-12. The construction of state table 21-12 is in fact tantamount to determining the next states of the detection device on the basis of the input signal to be received and of the two previous input signals. If the circuit remains, say, in state 1, the last three signals received are 0 0 0. With a next input signal 1, the circuit will go to a state corresponding with that input signal and the two previous signals, that is, 0 0. In this case the next state will be 1 0 0, that is, state 5.

Output signals for the setting of an MS memory element not belonging to the actual detection device are generated in two cases under the control of the clock pulse. The actual detection device plays no further role in the output signals, so that no present- and next-state outputs have to be mentioned in state table 21-12.

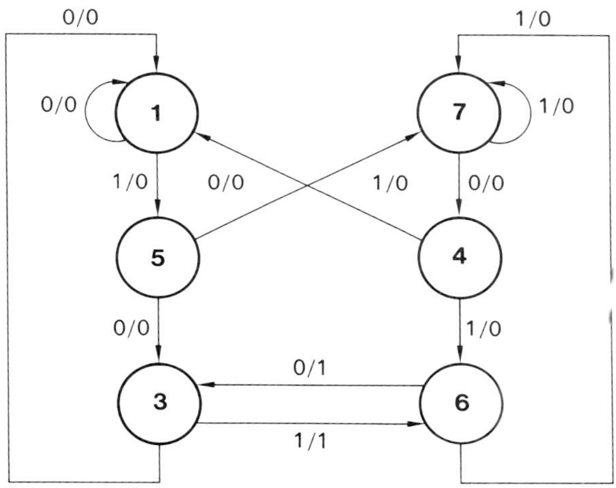

Fig. 21–16

State table 21-12 shows some obvious merging possibilities. States 1 and 2 can be merged, and the same holds true for states 7 and 8. The resulting state table is shown as Table 21-13, where the states have not been renumbered. The corresponding state diagram is given in Fig. 21-16. If a state assignment is chosen for the necessary three MS memory elements as indicated in Table 21-11, an efficient circuit will be obtained at the cost of some rather complex gating circuits that require an elaborate and time-consuming simplification of the switching formulas.

No specific solution of the state assignment problem will be given for this example because this type of problem can generally be solved more satisfactorily by means of shift registers with $n-1$ sections, n representing the number of pulses in the sequence to be checked. The specific function of the checking device is then obtained by means of the output gates. A circuit performing the checking process of Fig. 21-14 is shown in Fig. 21-17. Without using formulas, the operation of this type of checking circuit can be changed by means of a simple rewiring of the output gates. It is always better to organize the operation of a circuit in a general way than in a specific

Table 21-13

States			*Transient outputs*	
Present	*Next*			
	0	1	0	1
1	1	5	0	0
3	1	6	0	1
4	1	6	0	0
5	3	7	0	0
6	3	7	1	0
7	4	7	0	0

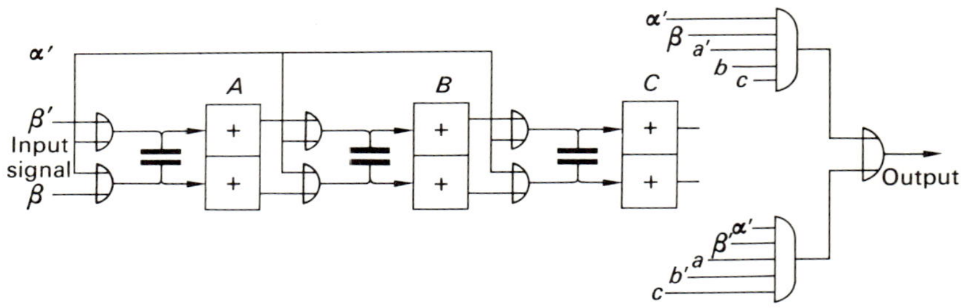

Fig. 21-17

way. The specific function of the circuit must then be incorporated in the output gate circuit. If the circuit of Fig. 21-17 is to be used for checking more than two input sequences, the required number of output gates can be added without any difficulty.

It can happen in cases with more complicated detection requirements of non-fixed-cycle sequences of signals, such as the detection of all sequences starting with a prescribed combination of two successive input signals followed by a series of four containing, e.g., at least two-out-of-four 1's, that the solution obtained via the state-diagram method results in a circuit with one memory element less than the shift-register solution. However, the more-simple gates of the shift-register solution make up for the additional memory element.

The fixed cycle detection device which detects the input-signal sequences 1 0 1 0 and 0 1 0 1 has to be provided with some means for counting the groups of four signals to be investigated if no group clock pulse is available. It will be assumed in the following discussion of the fixed-cycle detection device that such a group clock pulse is not available. It is of course possible to realize a detection device that meets the requirements by means of a three-section shift register and a two-section binary counter which counts the sequences of four input signals. It will be shown by means of a truth table defining the desired operation and of the corresponding state diagram that this detection device can be constructed by means of four MS memory elements.

In the case of a fixed-cycle detection device or electric lock circuit, the circuit must always start on the first signal of a group of four from a well-determined starting state on the fourth signal of each cycle. The state table showing the 15 possible states of this fixed-cycle detection device is given in Table 21-14.

Table 21-14

States			*Transient outputs*	
Present	*Next*			
	0	1	0	1
1	2	3	0	0
2	4	5	0	0
3	6	7	0	0
4	8	9	0	0
5	10	11	0	0
6	12	13	0	0
7	14	15	0	0
8	1	1	0	0
9	1	1	0	0
10	1	1	0	1
11	1	1	0	0
12	1	1	0	0
13	1	1	1	0
14	1	1	0	0
15	1	1	0	0

State table 21-14 shows four sections of entries corresponding to the possible states of the circuit as a result of the corresponding input signals of the circuit. It is not possible to apply the merging process on states that belong to different sections of state table 21-14, because this would involve a merging of states which belong to different steps of the fixed-cycle detection. For some input sequences, the cycle of the detection device would be reduced from four to three steps if two states of adjacent sections of the state table were to be merged. The state diagram corresponding to Table 21-14 is shown in Fig. 21-18.

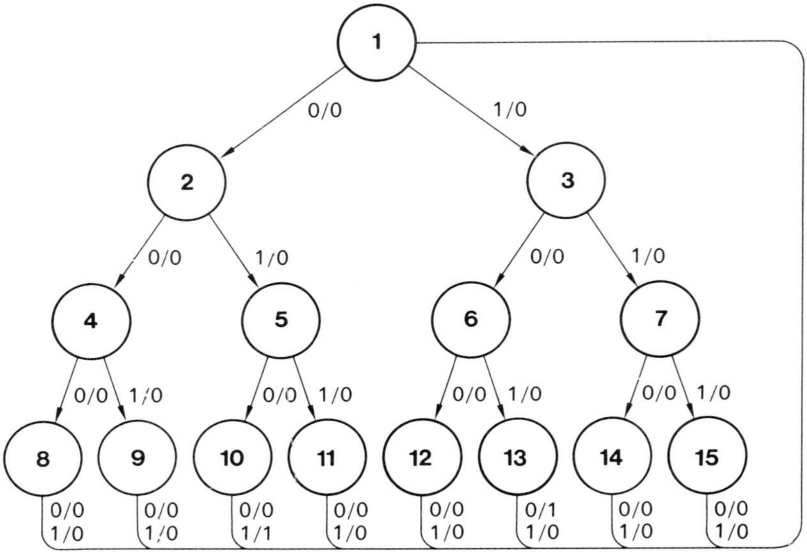

Fig. 21-18

State table 21-14 can be simplified by means of a merging process to yield the nine-line state table 21-15. The corresponding state diagram with the minimum number of circuit states is shown in Fig. 21-19.

It can be proved, using Mealy's rule, that state table 21-15 gives the minimum number of states. It should be noted that an experienced circuit designer will not draw up the state diagram of Fig. 21-18 but will immediately arrive at the diagram of Fig. 21-19.

The difficulties in the design of such a detection device begin to arise when one tries to assign the circuit states to combinations of MS memory elements, especially for states 5 and 8 of the state diagram of Fig. 21-19, which are part of the no-output cycle. These difficulties cannot be solved adequately before the operation of the circuit is fully understood by its designer.

In the fixed-cycle detection device for signal sequences, the length of the cycle has to be determined in some way. For the detection device described in the example, two of the four necessary memory elements can thus be assigned the special task of counting series of four clock pulses and

Table 21-15

States			*Transient outputs*	
Present	*Next*			
	0	1	0	1
1	2	3	0	0
2	5	4	0	0
3	6	5	0	0
4	7	8	0	0
5	8	8	0	0
6	8	9	0	0
7	1	1	0	1
8	1	1	0	0
9	1	1	1	0

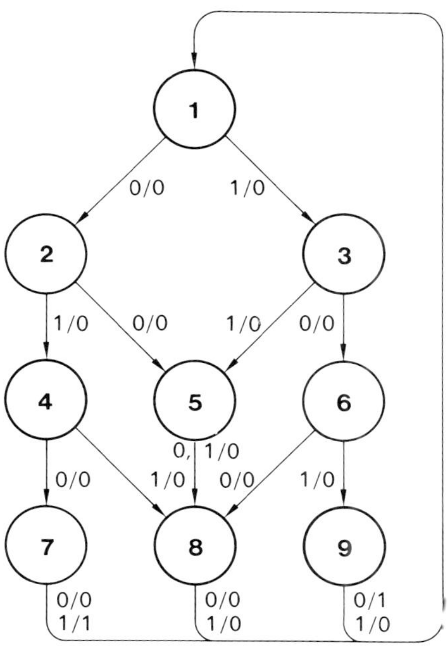

Fig. 21–19

then returning to zero. The resulting code assignment of the MS memory elements A and B is shown in the state diagram of Fig. 21-20.

The D memory element can be used to distinguish between a cycle starting with a 0 or a 1, as is shown in Fig. 21-21. The difficulty in this coding

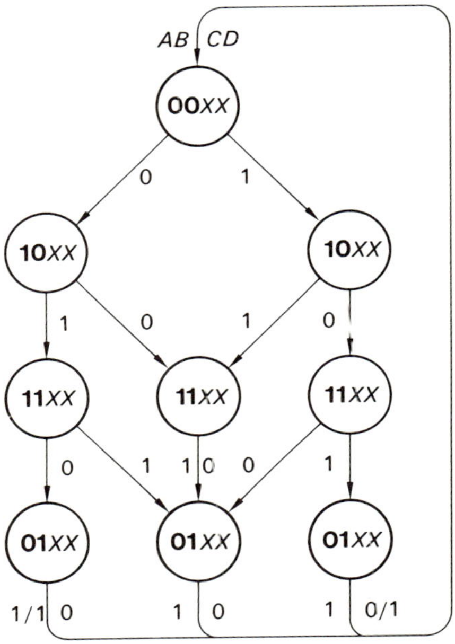

Fig. 21-20

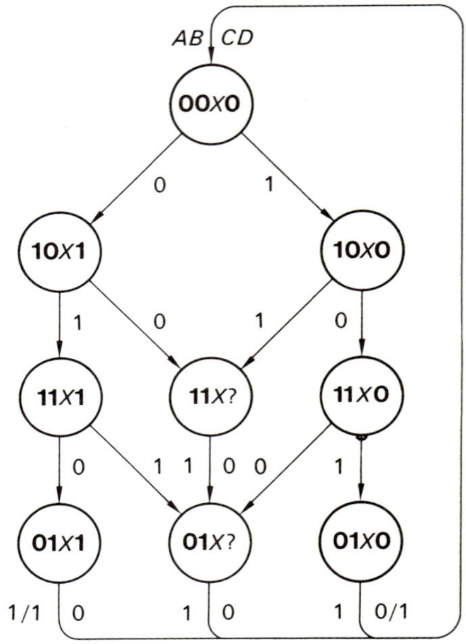

Fig. 21-21

process arises in the assignment of states 5 and 8. Because these states can be reached from the left- as well as from the right-hand sequence as shown in Fig. 21-19, a choice has to be made for the state of memory element D. If the code assignment of states 5 and 8 is made in such a way that memory element D is not operated in these states, this element must be reset if the states 5 and/or 8 are reached via the left-hand cycle of the state diagram. Consequently, the gates of the D memory element will become more complex than in the case in which the state of the D memory element is not affected by the way that states 5 and/or 8 are reached.

This situation can be solved by splitting states 5 and 8 into states $5a$, $5b$, $8a$, and $8b$, respectively. States $5a$ and $8a$ can then be used for the no-output cycles stemming from the left-hand sequence, while states $5b$ and $8b$ are used for the no-output cycle stemming from the right-hand sequence in the state diagram of Fig. 21-20. The C MS memory element is kept in the 0 state as long as the sequence of input signals forms a part of the cycles to be detected. The resulting state diagram is shown in Fig. 21-22. All-zero-output indications have been omitted.

The actual switching equation can finally be derived from the state diagram as well as from the coded truth table. The form of the truth table best suited for this purpose shows the present states and the next states in their assigned codes. This truth table that corresponds to the state diagram of Fig. 21-22 is given below as Table 21-16.

The simplest form of the resulting switching formulas can be obtained through manipulation with Karnaugh maps. The switching formulas for the A and B MS memory elements, which together form a binary counter consisting of two sections, need not be gone into any further. The fundamental circuit shown in Fig. 14-4 can be used here for this purpose.

The necessary terms of a set or reset function of a memory element can be derived most easily from the state diagram. All arrows involving a state change of the memory element concerned correspond to a necessary term respectively in the set or reset function.

Table 21-16

	Present state	*Next states*		*Transient outputs*	
	A B C D	0	1	0	1
1	0 0 0 0	1 0 0 1	1 0 0 0	0	0
2	1 0 0 1	1 1 1 1	1 1 0 1	0	0
3	1 0 0 0	1 1 0 0	1 1 1 0	0	0
4	1 1 0 1	0 1 0 1	0 1 1 1	0	0
5*a*	1 1 1 1	0 1 1 1	0 1 1 1	0	0
5*b*	1 1 1 0	0 1 1 0	0 1 1 0	0	0
6	1 1 0 0	0 1 1 0	0 1 0 0	0	0
7	0 1 0 1	0 0 0 0	0 0 0 0	0	1
8*a*	0 1 1 1	0 0 0 0	0 0 0 0	0	0
8*b*	0 1 1 0	0 0 0 0	0 0 0 0	0	0
9	0 1 0 0	0 0 0 0	0 0 0 0	1	0

The don't-care conditions can be determined in two different ways, as follows:

1. Each arrow in the state diagram for which no state change of the memory element concerned is found to occur can be used as a don't-care condition.
2. Each nonexisting present state can be used as don't-care condition.

This will be demonstrated for the construction of the set function of the C memory element of the state diagram of Fig. 21-22. This memory element must be set during state changes $2 \rightarrow 5a$, $6 \rightarrow 8b$, $4 \rightarrow 8a$, and $3 \rightarrow 5b$. The corresponding set function is

$$S_c = \alpha\beta' ab'c'd + \alpha\beta' abc'd' + \alpha\beta abc'd + \alpha\beta ab'c'd' \tag{21-4}$$

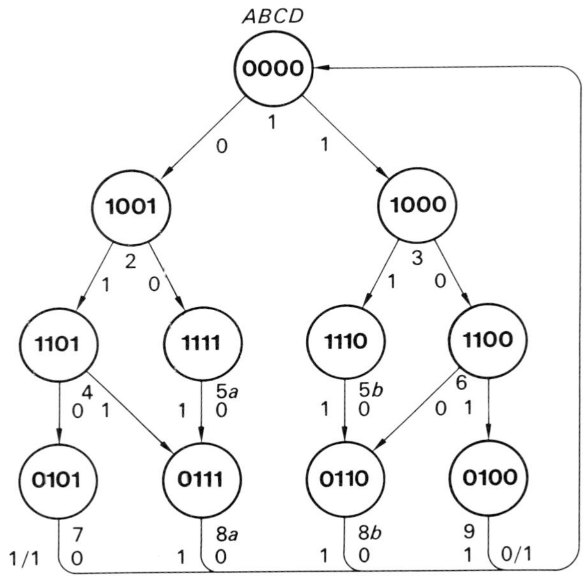

Fig. 21–22

in which α is the clock pulse and β the information signal to be checked. The don't-care conditions which belong to class 1 as mentioned above and which can be used for simplifying Eq. (21-4) are

$$S_{cd_a} = \alpha\beta abcd + \alpha\beta' abcd' \tag{21-5}$$

The don't-care conditions belonging to class 2 which can be used in this case are

$$S_{cd_b} = \alpha\beta' ab'cd + \alpha\beta ab'cd' \tag{21-6}$$

The resulting set function of the trigger in the C MS memory element then becomes

$$S_c = \alpha a[\beta'(b'd + bd') + \beta(b'd' + bd)] \tag{21-7}$$

The same equation can be obtained via factoring by means of a Karnaugh map as shown in Fig. 21-23.

The Karnaugh map in this case is a five-variable map with variables a, b, c, d, and β. The clock pulse α can be left out of consideration in this connection. The four essential terms of the set function of the trigger of the C MS memory element have been indicated as 1 in Fig. 21-23. All don't-care

	$c'd'\beta'$	$cd'\beta'$	$cd\beta'$	$cd'\beta'$	$c'd\beta$	$cd\beta$	$cd'\beta$	$c'd'\beta$
$a'b$	0	0	0	0	0	0	0	0
ab	1	d	d	0	1	d	d	0
ab'	0	d	d	1	0	d	d	1
$a'b'$	0	d	d	d	d	d	d	0

Fig. 21–23

conditions, regardless of their origin, have been marked with a d. The symmetry of the map shows that it will be better not to factor term $\beta'ab'c'd$ with the three adjacent don't-care conditions but only with the one indicated in the map.

Equation (21-7) shows that the c variable itself is not involved in the final form of the set function of the trigger in the C MS memory element.

The other formulas for the set and reset signals of the trigger in the various memory elements can be determined in the same way. They are as follows:

$$S_a = \alpha b' \qquad R_a = \alpha b \tag{21-8}$$

$$S_b = \alpha a \qquad R_b = \alpha a' \tag{21-9}$$

$$S_c = \alpha a[\beta'(b'd + bd') + \beta(b'd' + bd)] \qquad R_c = \alpha a' \tag{21-10}$$

$$S_d = \alpha\beta' a'b' \qquad R_d = \alpha a'b \tag{21-11}$$

$$S_0 = \alpha a'bc'(\beta d + \beta' d') \tag{21-12}$$

If Eqs. (21-8) to (21-12) are considered as the set and reset equations of the triggers in the MS memory elements, then the equations of the input gates can be derived by negation. If the other type of trigger is used, the equations may be left unchanged.

The resulting circuit will not be given here because an even simpler circuit can be obtained by means of a rather intriguing expansion and rearrangement of the state diagram of Fig. 21-22, as shown in Fig. 21-24. In this state diagram, two different starting conditions have been assumed. Which of the two starting conditions will be used depends on the polarity of the last signal of the preceding detection cycle.

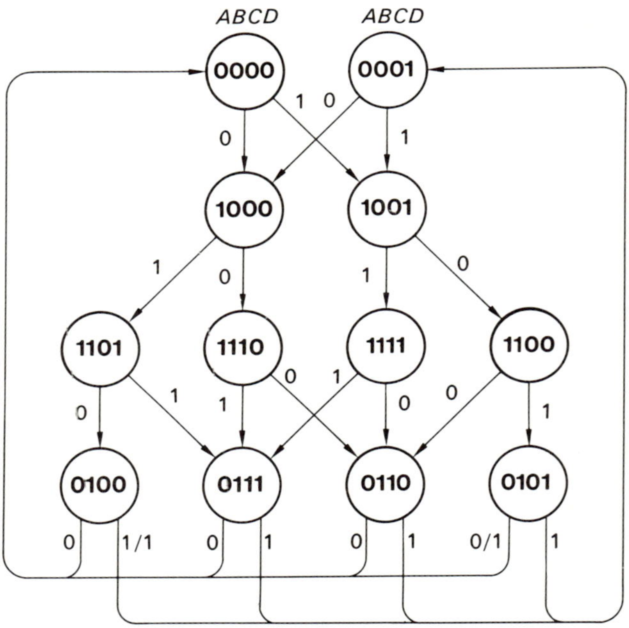

Fig. 21–24

The operation of the A and B MS memory elements follows the same pattern as in the state diagram of Fig. 21-22. The C and D MS memory elements follow a pattern which leads to a simpler gating circuit for the C MS memory element and the output circuit. The switching equations then become

$$S_a = \alpha b' \qquad R_a = \alpha b \tag{21-13}$$

$$S_b = \alpha a \qquad R_b = \alpha a' \tag{21-14}$$

$$S_c = \alpha a(\beta' d' + \beta d) \qquad R_c = \alpha a' \tag{21-15}$$

$$S_d = \alpha\beta \text{ (or } \alpha\beta d') \qquad R_d = \alpha\beta' \text{ (or } \alpha\beta' d) \tag{21-16}$$

$$S_o = \alpha a' bc'(\beta' d + \beta d') \tag{21-17}$$

The resulting circuit is shown in Fig. 21-25. This circuit with four MS memory elements requires 28 diodes in the various input gates of the MS memory elements and the output circuit. A suitable combination of the reset gates of MS memory elements B and C is used. No orders of diode gates with alternating polarity are required for implementing this circuit. This example shows how important a careful organization of the state diagram of the circuit can be.

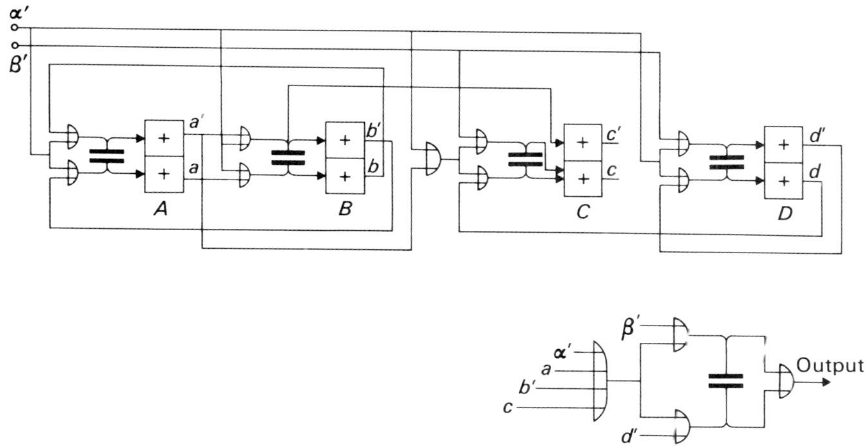

Fig. 21–25

The problem of detecting a predetermined series of input signals in a fixed cycle can also be solved in a general way by means of a counting shift register, as indicated by Snijders. Shift registers with a counting function have already been described in Chap. 19 on transmitters and receivers of binary-coded information.

The problem stated in the example, in which groups 1 0 1 0 and 0 1 0 1 are to be detected in a fixed-cycle device, will also be solved by means of a counting shift register. The state diagram describing the operation of this type of detection device is the same as the diagram already given in Fig. 21-18. In this case, the shift register must consist of four sections. The output is determined by the input signal and by the state of the sections of the shift register.

At the beginning of the first step of each operating cycle, the four sections of the shift register are preconditioned in the state 1 0 0 0. As a result of the third clock pulse of a cycle, the fourth section of the shift register will be brought into the 1 state, so that on the fourth clock pulse of a cycle, in combination with the state of the fourth section of the shift register, the state of the input and of the first three sections can be determined. The operation of this kind of circuit can be demonstrated further by means of Table 21-17, which shows in arabic numerals the shifting of the primed state through the four sections of the shift register. The shifting of the input signals is marked with + or − signs.

In step 4 of the first full cycle shown in Table 21-17, section D remains in the 1 state, while the input and the first three sections of the counting shift register show the input sequence − + − + (0 1 0 1). The detection device will then produce an output signal as required.

Table 21-17

Next input	*Present state* A	B	C	D	*Step No.*
−	0	0	0	1	
+	1	0	0	0	1
−	+	1	0	0	2
+	−	+	1	0	3
−	+	−	+	1	4
−	1	0	0	0	1
−	−	1	0	0	2
+	−	−	1	0	3
−	+	−	−	1	4

The switching functions for the triggers of the MS memory elements which form the sections of this shift register can be derived rather easily from the state diagram of Fig. 21-18 or from Table 21-17. The greatest economy in the use of diodes in gate circuits is reached when single-input MS memory elements are used. In this case the set and reset functions at the input gates of the MS memory elements become as follows:

$$S_a = \alpha' + a \qquad R_a = \alpha' + \beta + d \tag{21-18}$$

$$S_b = \alpha' + a' + d \qquad R_b = \alpha' + b' \tag{21-19}$$

$$S_c = \alpha' + b' + d \qquad R_c = \alpha' + c \tag{21-20}$$

$$S_d = \alpha' + c' + d \qquad R_d = \alpha' + d' \tag{21-21}$$

$$S_0 = d(\beta a' bc' + \beta' ab' c) \tag{21-22}$$

The resulting shift register is shown in Fig. 21-26. The output circuit is shown as a slave memory element controlling a trigger which is not shown.

This part of the circuit operates under control of a d' signal as clock pulse, so that the α' clock pulse could be omitted. Because one of the diodes of the output gate is already contained in the trigger that it controls, only 31 diodes are involved in the logic of this counting shift register. The gates used have not more than one output, whereas a maximum of two orders of gate circuits (of the same kind) are used in series. This result is equivalent to the best circuit given above with already reasonably good internal organization.

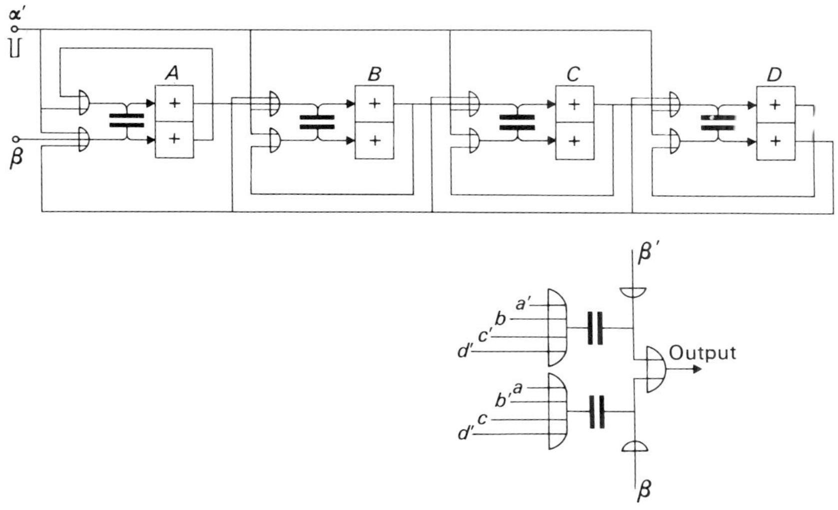

Fig. 21–26

The fact that the circuit with the minimum number of memory elements and the simplest gating circuits need not correspond to the state diagram with the minimum number of states will again be illustrated by means of the following example. This rather complicated example will serve to emphasize that the circuit designer must first develop a good organization chart for the operation of the circuit and in so doing he will automatically arrive at simple gating circuits.

A design of a sequential circuit that is to detect series of input signals of two 1's followed by at least two 1's out of the next four signals is required. The circuit must start the detection process of a sequence of six input signals whenever two 1's occur outside an operating cycle which is already under way. The state diagram with the minimum number of states resulting from these requirements is given in Fig. 21-27.

The state assignment in this mixed non-fixed- and fixed-cycle detection problem can be performed by means of a subdivision of the problem into a

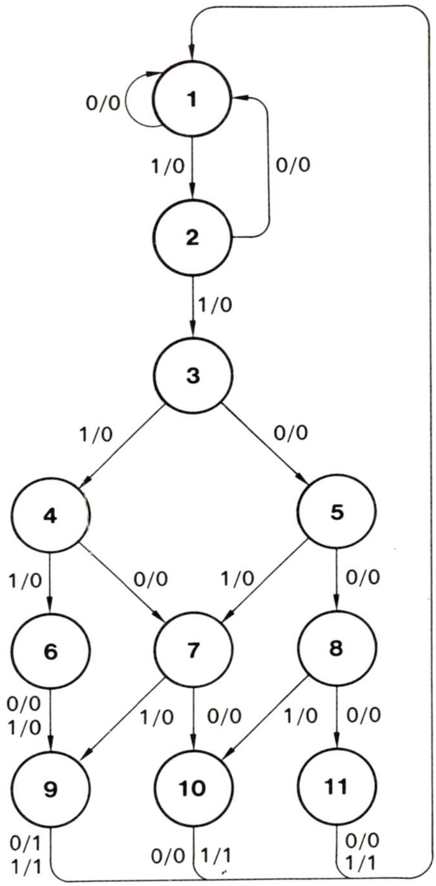

Fig. 21–27

counting problem and an information-shifting problem. A relatively uncomplicated operating layout of this circuit can be obtained by means of four MS memory elements. MS memory elements A and B perform a shifting function, and elements C and D a counting function, as is shown in the state diagram of Fig. 21-28.

The A and B MS memory elements are used first as a shift register which shifts a 1 signal into the next section but destroys the first 1 stored in the A MS memory element if the second signal is a 0. After receiving two successive 1 signals, the state of MS memory elements A and B is 1 and the fixed four-step counting cycle is started under the control of a two-section binary counter that consists of the C and D MS memory elements. During

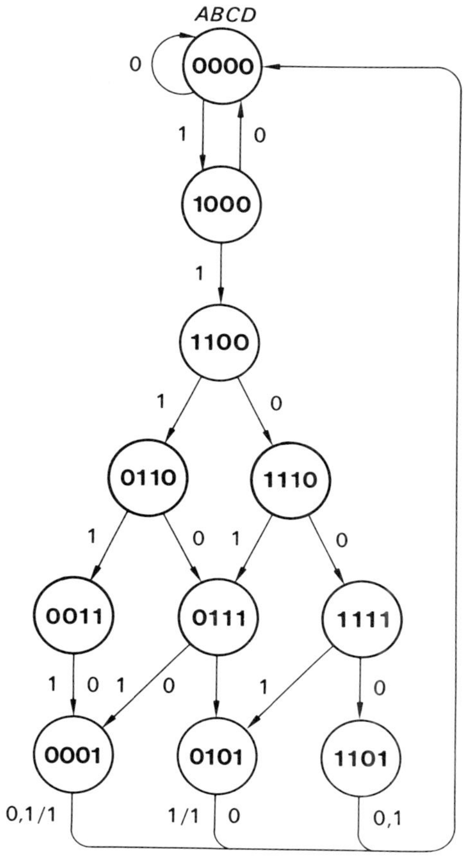

Fig. 21–28

that counting cycle, the operation of the A and B MS memory elements is changed in such a way that a 1 signal is shifted as a 0 into shift section A. The information in this section is not affected by the reception of a 0 signal.

The following set and reset functions of the triggers of the MS memory elements can be derived from the state diagram of Fig. 21-28:

$$S_a = \alpha\beta b' d' \qquad R_a = \alpha\beta b + \alpha c' d + \alpha\beta' b' \tag{21-23}$$

$$S_b = \alpha\beta a d' \qquad R_b = \alpha c' d + \alpha\beta a' \tag{21-24}$$

$$S_c = \alpha b d' \qquad R_c = \alpha d \tag{21-25}$$

$$S_d = \alpha c \qquad R_d = \alpha c' \text{ (or } \alpha c' d) \tag{21-26}$$

$$S_o = \alpha c' d(\beta a' + b') \tag{21-27}$$

The circuit corresponding to the switching functions (21-23) to (21-27) is shown in Fig. 21-29. It consists of 4 MS memory elements, 31 diodes in the logic gate circuits, and 3 diodes to provide extra inputs on triggers A and B.

The author hesitates to guarantee that the circuits given above are the best possible solutions. However, the examples may have served to show that a good circuit can only be obtained once an adequate organization of the

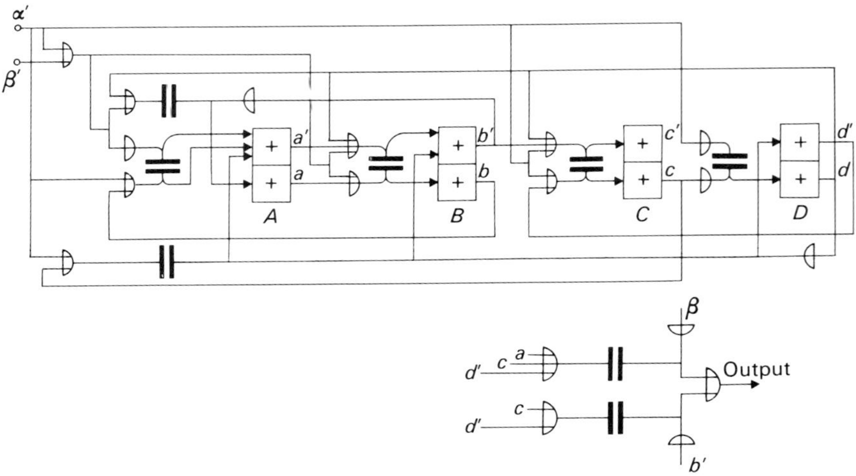

Fig. 21–29

operation of the circuit has been arrived at. The best results will be achieved when the binary counting and shifting operations are used to the largest possible extent, because the circuits required for these standard operations are the simplest circuits possible. The circuit with the minimum hardware need not be the circuit corresponding to the state diagram with the minimum number of states; nor has it been shown conclusively that a circuit with the minimum number of MS memory elements will necessarily be the most economical. It is not always easy to arrive at an adequate operating organization, but the well-organized circuit will generally require substantially less hardware than would be necessary with a rather arbitrary code assignment of the memory elements.

Finally, it should be pointed out that a number of gates in some detection circuits are used for several purposes. This decreases the number of diodes involved considerably. Such a combined use of gates also follows from good organization of the state diagram.

The principles of state equivalence in circuits apply equally well to the circuits of class I in which normally only one memory element changes its state as a result of one change in the external conditions of the circuit. The difference in treatment of both groups of circuits becomes most apparent with the state assignment. The assignment of the various successive states in the diagram of a circuit belonging to class I must be performed in such a way that these states generally do not differ by more than the state of one memory element (relay or trigger). If such is not possible, race conditions that cannot be used as stable states of the circuit must be introduced.

The design of a class I circuit will be illustrated here by means of a multiple-input example. Circuits with multiple inputs can in principle be treated in the same way as circuits with single inputs. The problem has been raised by Washburn, Keister, and Ritchie and is used as an example by Mealy.

A rotating shaft carries a single-grounded brush which makes contact with three stationary commutator segments arranged symmetrically around the shaft. A relay circuit is required which will indicate the direction of rotation of the shaft by lighting a lamp when the shaft is rotating clockwise. The shaft may reverse its direction at any time. Let us assume that the shaft is driven in such a way that a brush contact closure lasts for 0.25 sec and that the open time between the brush leaving one segment and reaching another is also 0.25 sec. When the direction of rotation of the shaft is changed, the output indication of the detection circuit must change as quickly as possible, or at most within 2 sec.

Three noncoinciding inputs are given as grounded contacts α, β, and γ. These inputs produce combinations of input signals which can be represented by the following binary numbers: 0 0 0, 1 0 0, 0 1 0, and 0 0 1. The first input combination is produced when the brush does not make contact with any of the segments. The detection device will have to distinguish six states when the shaft runs clockwise, i.e., three different states when the brush is moving on the three contact segments and three states corresponding to the movement of the brush between the segments. These last three intermediary states can best be defined as circuit states that occur when a brush leaves a certain segment.

The detection device must also discriminate between the same number of states when the shaft is rotating counterclockwise. Since the direction of rotation of the shaft can be reversed at any moment, the brush can leave any of the segments in both directions. Because there are no indications available to detect in which direction the brush leaves a contact segment, the state of the detection circuit will be the same in both cases. The detection of the reversal of the direction of rotation of the shaft can happen when the brush reaches a following segment.

If the direction of rotation of the shaft is reversed when the brush is moving between two contact segments, nothing can happen in the detection device because it is already impossible to detect in which direction the brush leaves a contact segment. Thus, it will not be possible to detect anything about the direction of rotation of the shaft when the brush returns to a certain contact segment. The detection of the direction can be performed only when the brush reaches another contact segment.

Figure 21-30 shows the shaft and the contact brush moving in the clockwise direction, together with the state indications of the detection circuit. Figure 21-31 shows the same for the shaft and the contact brush moving in the counterclockwise direction.

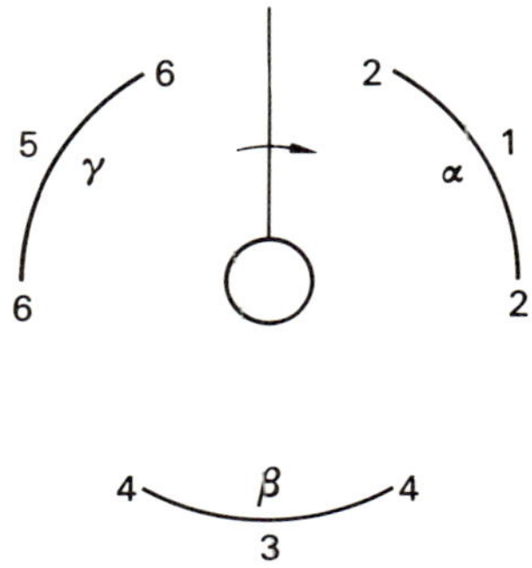

Fig. 21–30

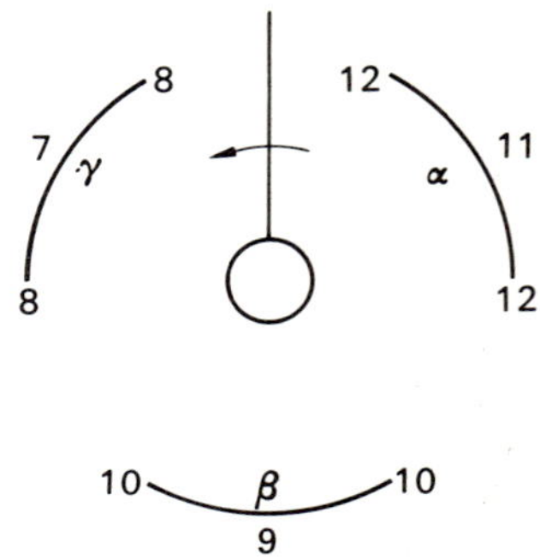

Fig. 21–31

The desired circuit states and the desired outputs of the detection device can be combined in a state table as shown in Table 21-18. The first six lines are related to the clockwise rotation of the shaft, while the second group of six lines describe the circuit states when the shaft is moving counterclockwise. Lines 2, 4, and 6 show what happens when the direction of rotation is reversed to counterclockwise, and lines 8, 10, and 12 indicate what happens when the counterclockwise movement of the shaft is reversed to clockwise.

If the detection circuit is in state 1 on segment α, then the next state of the circuit will always be 2, irrespective of whether the clockwise rotation of the shaft is continued or reversed during the movement of the brush over that contact segment. If the brush continues its clockwise movement, the circuit will assume state 3 when the brush reaches segment β. If the brush stops on segment α, the detection circuit will remain in state 1; and if the rotation of the shaft is reversed, the detection circuit will reach state 7 via state 2. The output columns show the desired output for each state of the detection circuit. The other lines of Table 21-18 can be interpreted in an analogous manner.

The nonexisting states in the various lines of the truth table 21-18 may be considered as don't-care conditions for which any other entry may be filled in. It is then obvious that states 1 and 2, 3 and 4, 5 and 6, 7 and 8,

Table 21-18

Present state	*Next state*				*Output*			
	000	100	010	001	000	100	010	001
1	2	1	-	-	1	1	-	-
2	2	1	3	7	1	1	1	0
3	4	-	3	-	1	-	1	-
4	4	11	3	5	1	0	1	1
5	6	-	-	5	1	-	-	1
6	7	1	9	5	1	1	0	1
7	8	-	-	7	0	-	-	0
8	8	1	9	7	0	1	0	0
9	10	-	9	-	0	-	0	-
10	10	11	9	5	0	0	0	1
11	12	11	-	-	0	0	-	-
12	12	11	3	7	0	0	1	0

9 and 10, and finally 11 and 12 may be merged. Table 21-19 is the outcome of this merging process.

It can be easily demonstrated by grouping the states in classes that Table 21-19 contains the minimum possible number of states necessary for the detection circuit to be designed. The corresponding state diagram is shown in Fig. 21-32.

It should be noted that only those arrows have been provided with a d for the output which involve a reversal in the direction of rotation of the shaft. For the other arrows, the output is no problem.

Table 21-19

Present state	*Next state*				*Output*			
	000	100	010	001	000	100	010	001
1	1	1	3	7	1	1	1	0
3	3	11	3	5	1	0	1	1
5	5	1	9	5	1	1	0	1
7	7	1	9	7	0	1	0	0
9	9	11	9	5	0	0	0	1
11	11	11	3	7	0	0	1	0

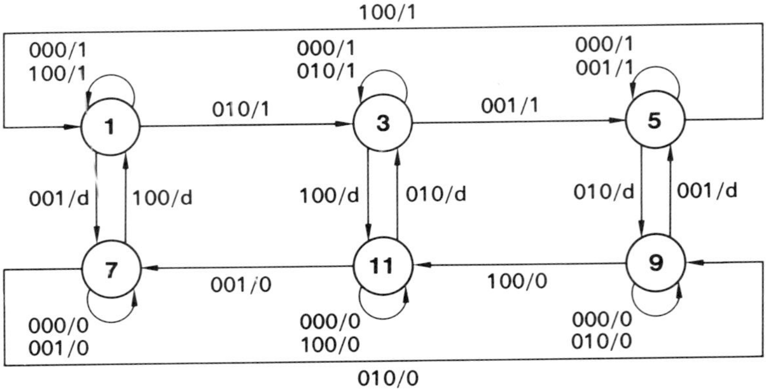

Fig. 21–32

It does not prove possible to assign states in this diagram in such a way for the three relay case that all successive states differ in state by not more than one relay. A possible state assignment with three relays, with the introduction of two transient states, is shown in the state diagram of Fig. 21-33. The three relays (or memory elements) involved have eight different states, so that only two states are available to cover the transient states. This makes it necessary to run in series through a predetermined transient state in the clockwise and in the counterclockwise detection cycle of the circuit.

The state equations for the various memory elements (relays) of the detection device can easily be derived from the state diagram. The terms of these state equations concern the actual state changes of the relays. The state equations are given here in the form

$$S_z = (s_z + z_0)\, r_z' \tag{21-28}$$

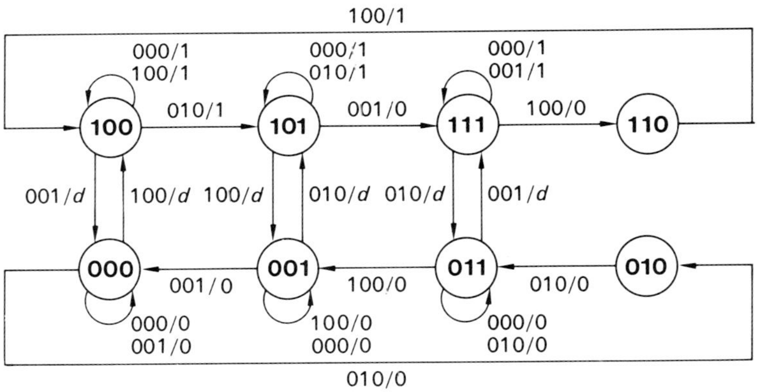

Fig. 21–33

This form of the state equation describes a relay in shunt-down operation. A solution of this type is not objectionable because the detection device operates at very slow speeds. The release time of relays in shunt-down operation is rather high.

The state equations for the three relays A, B, and C of the state diagram of Fig. 21-33 become as follows:

$$A = (\alpha b'c' + \beta b'c + \gamma bc + a)\,(\alpha b'c + \beta bc + \gamma b'c')' \tag{21-29}$$

$$B = (\beta a'c' + \gamma ac + b)\,(\alpha a'c + \alpha ac')' \tag{21-30}$$

$$C = (\beta ab' + \beta a'b + c)\,(\alpha ab + \gamma a'b')' \tag{21-31}$$

These equations can be represented by means of the circuit shown in Fig. 21-34. No efforts were made to find other possible forms of these state equations which might perhaps have resulted in a saving of a couple of contacts. If the don't-care outputs on the arrows between the states of both cycles of Fig. 21-33 are chosen in accordance with those of the states at the originating end of these arrows, then the output circuit can be formed by means of a contact of relay A only.

This example calls for a number of comments as regards the design of circuits, the application of state diagrams, and the state assignment in these diagrams.

First, with respect to the state assignment in the circuit of Fig. 21-34, and especially the assignment of transient states, it should be recalled that the coding problem of states was solved in the state diagram of Fig. 21-33 by means of the introduction of two transient states (0 1 0 and 1 1 0) which make the operation of the circuit unsymmetrical. The pure three-phase system is thus converted into a concealed four-phase system, and this suggests that it will not be possible to achieve a symmetrical three-phase detection system on the reversal of the shaft rotation.

It is possible, however, to design a perfectly symmetrical circuit by means of the introduction of a transient state between each successive state of the clockwise and counterclockwise ring of states in the diagram of Fig. 21-32. This does not require the introduction of more memory elements (relays in the present case), because all three transient states between the stable states of the clockwise rotation may be the same, and all three transient states between the stable states of the counterclockwise rotation may also be identical, although different from that used for the other direction of rotation. The resulting state diagram is shown in Fig. 21-35. The 0 0 0 state of the circuit has been chosen as the general transient state of the detection circuit during the clockwise rotation of the shaft and the 1 1 1 state as the general transient state between the stable states of the detection circuit during the counterclockwise rotation of the shaft.

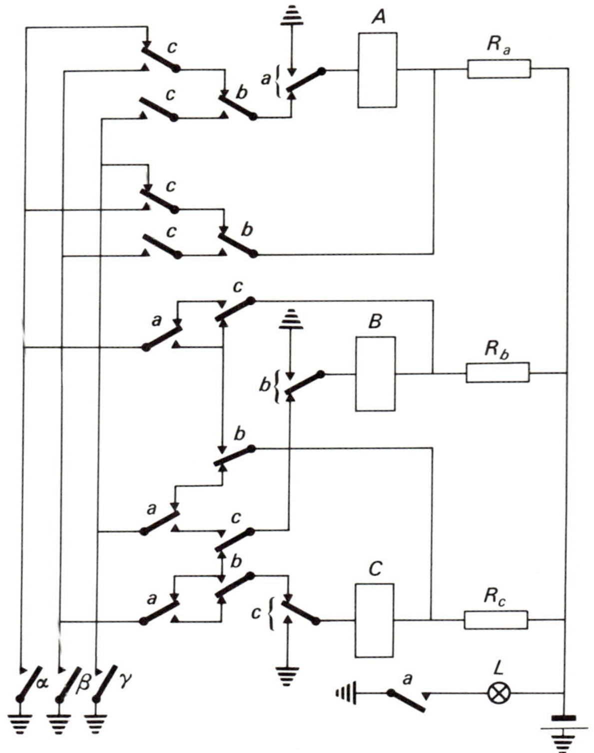

Fig. 21–34

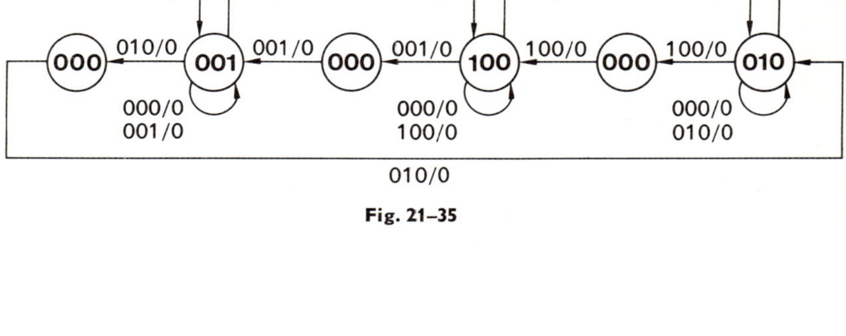

Fig. 21–35

The set and reset functions that can be derived from this state diagram are as follows:

$$S_a = \alpha b'c + \alpha b'c' + \alpha bc = \alpha(b' + c) \tag{21-32}$$

$$R_a = \gamma b'c + \gamma b'c' + \gamma bc = \gamma(b' + c) \tag{21-33}$$

$$S_b = \beta ac' + \beta a'c' + \beta ac = \beta(c' + a) \tag{21-34}$$

$$R_b = \alpha ac' + \alpha a'c' + \alpha ac = \alpha(c' + a) \tag{21-35}$$

$$S_c = \gamma a'b + \gamma a'b' + \gamma ab = \gamma(a' + b) \tag{21-36}$$

$$R_c = \beta a'b + \beta a'b' + \beta ab = \beta(a' + b) \tag{21-37}$$

$$S_o = abc' + ab'c + a'bc + abc = ab + bc + ab'c \tag{21-38}$$

The circuit corresponding to these equations is shown in Fig. 21-36. The circuit is perfectly symmetrical. When the shaft is rotating in the clockwise direction, at least two out of three relays are always operated; and when the shaft is rotating in the counterclockwise direction, one out of three relays at most is operated. The symmetrical operation results in an important saving in the number of contacts on the relays. The output circuit, however, is more complex than the circuit required in the original solution of the problem as given in Fig. 21-34. Two winding relays have been used in the circuit of Fig. 21-36 in order to avoid sneak paths.

It should be noted that the circuit shown in Fig. 21-36 can be simplified even further if the grounded control contacts α, β, and γ are replaced by separate pairs of contacts.

The state diagram of Fig. 21-35 can be changed slightly in order to introduce a useful feature in the detection device which is not included in the original problem. Moreover, the resulting detection circuit is even simpler. The modified state diagram is shown in Fig. 21-37.

The state cycle for the rotation of the shaft in the clockwise direction is the same in the state diagram of Fig. 21-37 as in Fig. 21-35. However, during the rotation of the shaft in the counterclockwise direction, the detection circuit runs via a transient state from one stable state to the other through the stable states of the clockwise direction. During the clockwise rotation of the shaft, detection lamp L will not light while running through the transient states; and during the counterclockwise rotation of the shaft, lamp L will light when passing through the transient states. These periods of the transient states are short compared with the other operating periods of the circuit, so that no ambiguity can arise between the two directions of rotation. This feature makes it possible to detect whether the shaft is still running or not. The other detection circuits described above are not provided with such a useful detection tool; however, the circuit of the detection lamp of Fig. 21-38 will provide this facility in the circuit of Fig. 21-36.

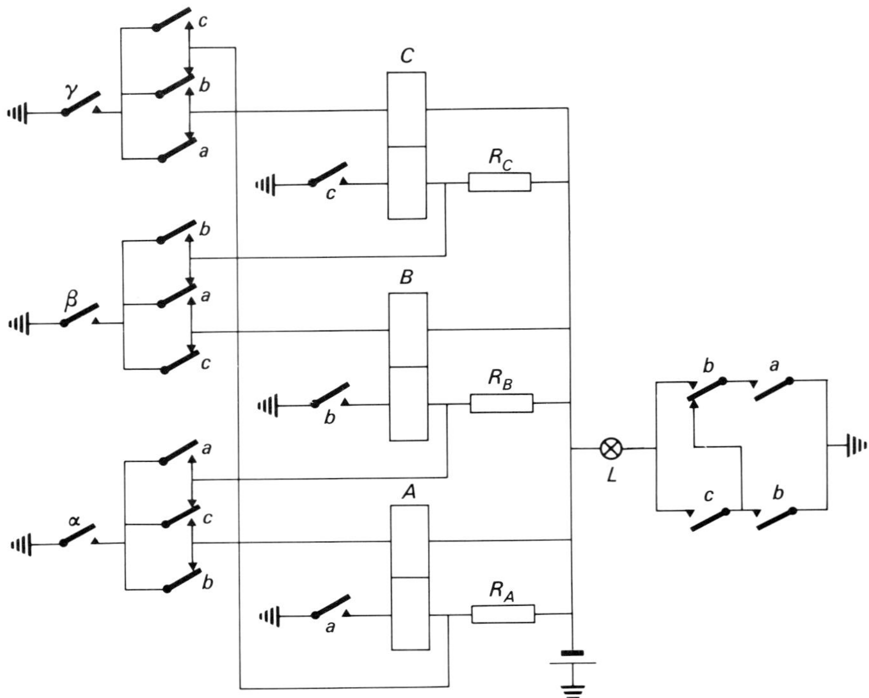

Fig. 21–36

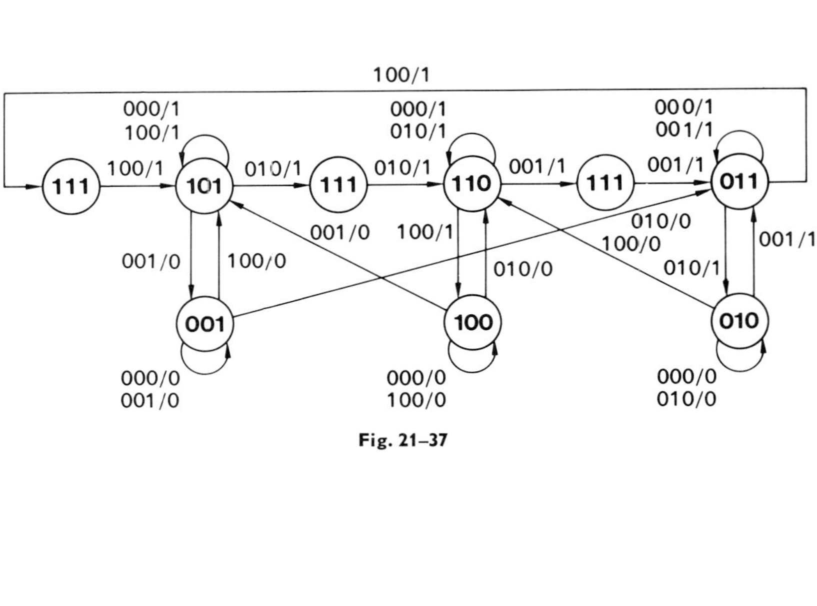

Fig. 21–37

The set and reset functions of the relays of the detection device operating in accordance with the state diagram of Fig. 21-37 reduce to

$$S_a = \alpha \qquad R_a = \gamma c \tag{21-39}$$

$$S_b = \beta \qquad R_b = \alpha a \tag{21-40}$$

$$S_c = \gamma \qquad R_c = \beta b \tag{21-41}$$

$$S_o = abc' + ab'c + a'bc \tag{21-42}$$

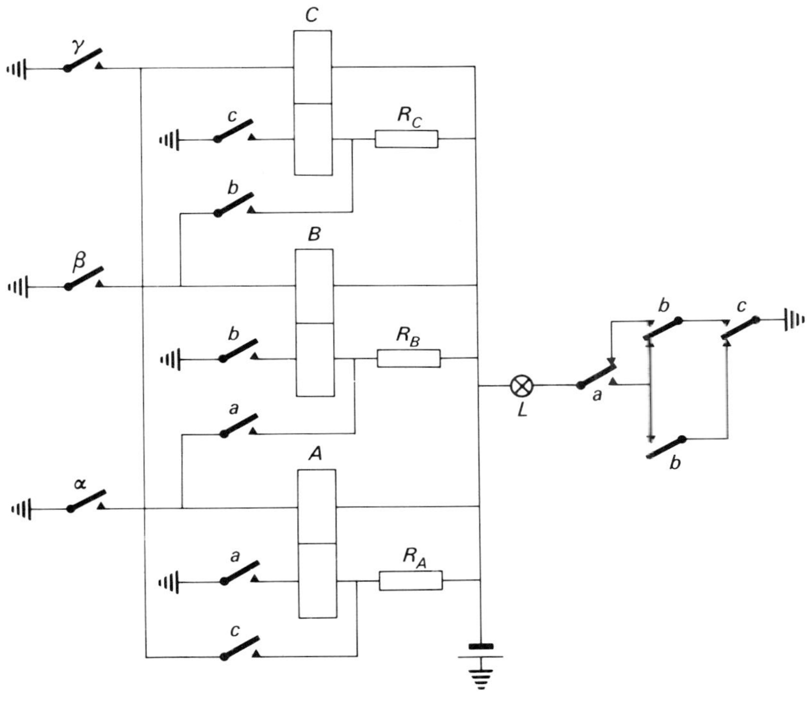

Fig. 21-38

The resulting circuit is given in Fig. 21-38. The improvement obtained in this circuit as compared with the original solution given in Fig. 21-34 is quite obvious, but this last solution is still not the simplest way to solve the problem as stated. This remarkable result is the outcome of careful consideration and organization of the operation of the desired circuit.

A second example of a circuit of class I should be dealt with because it also raises a number of problems that are essential to a practical solution of the state assignment in sequential circuits.

This problem concerns the design of a circuit with two inputs as given by Huffman. The inputs receive α and β signals respectively in an arbitrary sequence, by means of the grounding of the corresponding input wires. No two signals will be given simultaneously, so that the input combination 11 is excluded. The required circuit will have four output leads feeding four lamps numbered I to IV. One of these lamps at a time is lit. When an α signal is received, the circuit advances the position of the illuminated lamp; when a β signal is received, the change in position of the illuminated lamp occurs backward. The state diagram derived from these general requirements is shown in Fig. 21-39. It consists of two cycles of states, a clockwise cycle and a counterclockwise cycle, and the possible transitions between the two cycles have been indicated. These transitions have to be effected in such a way that the lamp which lights up must be next to the lamp which was last lit. All arrows in the state diagram that relate to a change in the output are provided with the lamp indication of the last stable state.

The state table resulting from the state diagram of Fig. 21-40 is given as Table 21-20.

Table 21-20

Present state	*Next state*			*Output*		
	0 0	1 0	0 1	0 0	1 0	0 1
1	1	2	16	I	II	IV
2	3	2	-	II	II	-
3	3	4	14	II	III	I
4	5	4	-	III	III	-
5	5	6	12	III	IV	II
6	7	6	-	IV	IV	-
7	7	8	10	IV	I	III
8	1	8	-	I	I	-
9	9	8	10	IV	I	III
10	11	-	10	III	-	III
11	11	6	12	III	IV	II
12	13	-	12	II	-	II
13	13	4	14	II	III	I
14	15	-	14	I	-	I
15	15	2	16	I	II	IV
16	9	-	16	IV	-	IV

A double-merging process can be performed on the state table of this problem. It is immediately apparent from this table that states 1 and 15, 3 and 13, 5 and 11, and also 7 and 9 can be merged. The open places in the

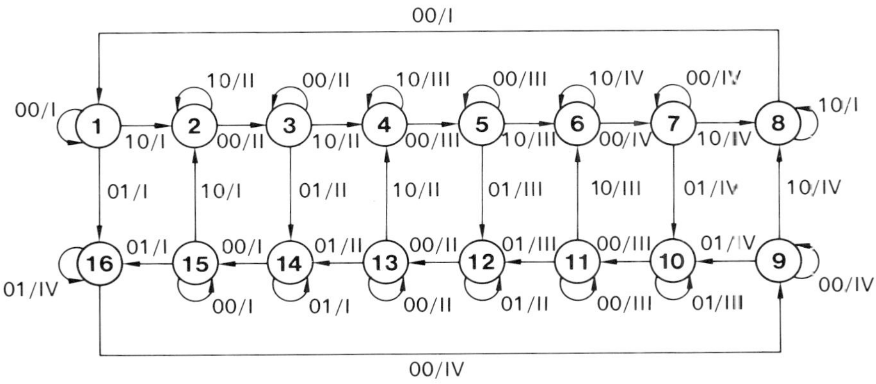

Fig. 21–39

columns of the next states may be considered as don't-care conditions, so that a second merging operation can be performed on the lines of states 2 and 12, 4 and 10, 6 and 16, and finally 8 and 14. The resulting state table is given as Table 21-21. The corresponding state diagram is shown in Fig. 21-40.

State table 21-21 shows eight lines. These eight different states can be realized by means of three relays, provided that the various successive states can be coded in such a way that they differ in the state of one memory element only. In principle, the only possible way of making this state assignment with three relays is shown in the state diagram of Fig 21-40. The switching formulas that can be derived from this state diagram are

$$A = s_a + a(r_a)' = (\alpha b'c' + \beta bc) + a(\alpha bc' + \beta b' c)' \qquad (21\text{-}43)$$

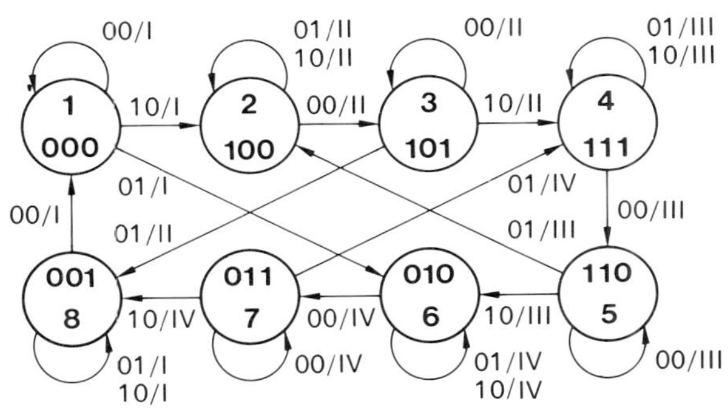

Fig. 21–40

Table 21-21

Present state	*Next state*			*Output*		
	0 0	1 0	0 1	0 0	1 0	0 1
1	1	2	6	I	II	IV
2	3	2	2	II	II	II
3	3	4	8	II	III	I
4	5	4	4	III	III	III
5	5	6	2	III	IV	II
6	7	6	6	IV	IV	IV
7	7	8	4	IV	I	III
8	1	8	8	I	I	I

If it is considered that the hold and reset circuit of a relay need not be closed during the setting of that relay, Eq. (21-43) can be simplified by adding the set terms to the reset terms to

$$A = (\alpha b'c' + \beta bc) + a(\alpha bc' + \alpha b'c' + \beta b'c + \beta bc)'$$

$$= (\alpha b'c' + \beta bc) + a(\alpha'c' + \beta'c) \qquad (21\text{-}44)$$

The state equation for the B relay can be obtained in the same way:

$$B = (\alpha ac + \beta a'c') + b(\alpha a'c + \beta ac')'$$

$$= (\alpha ac + \beta a'c') + b(\alpha'c + \beta'c') \qquad (21\text{-}45)$$

Similarly, the state equation for the C relay is as follows:

$$C = \alpha'\beta'(ab' + a'b) + c[\alpha'\beta'(a'b' + ab)]'$$

$$= \alpha'\beta'(ab' + a'b) + c(\alpha + \beta) \qquad (21\text{-}46)$$

State equations (21-43) to (21-47) make it necessary to introduce two auxiliary relays in order to provide for the necessary α and β contacts. The resulting circuit is shown in Fig. 21-41. The output equations are

$$\text{I} = a'b' \qquad \text{II} = ab' \qquad \text{III} = ab \qquad \text{IV} = a'b \qquad (21\text{-}47)$$

In spite of the fact that only three memory elements—i.e., three relays—are required for the desired circuit according to the state minimizing theory, five relays are necessary from a practical point of view. The two extra relays provide for the necessary α and β contacts. The actual automatic circuit contains 23 contacts; the lamp circuit requires six extra contacts. A discussion of the state assignment and the construction of this circuit from the switching point of view will be given in the next section.

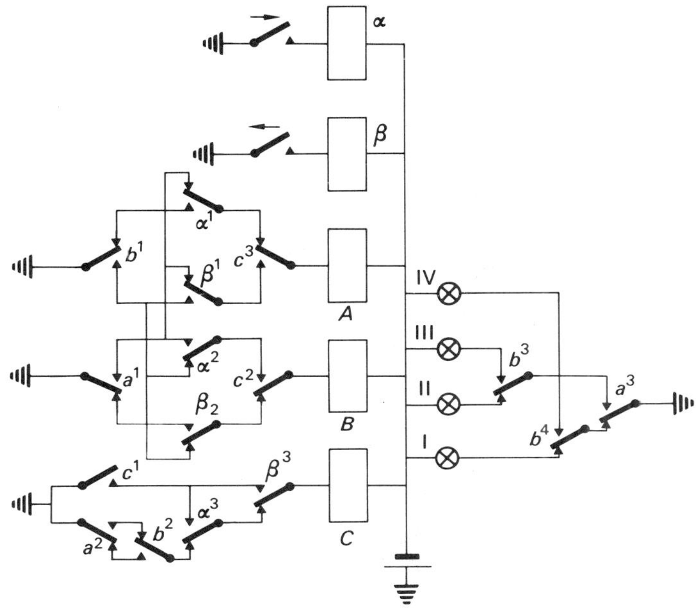

Fig. 21-41

21-4 SOME OBSERVATIONS ON STATE TABLES, STATE ASSIGNMENT, AND CONSTRUCTION OF CIRCUITS

The minimum number of memory elements required to satisfy a given set of switching requirements can be accurately determined by means of the methods described in the preceding sections. However, the circuit that comprises the minimum number of states is not always the best solution because of the economy of the circuit, nor is it necessarily the most desirable from the point of view of maintenance. Some important design steps will be described in this section which have to be carried out before, during, or after the state minimizing process described in the former sections.

The design steps concern the following points:

1. In the state table, it is necessary to specify no more transitions between the various states than are strictly required by the problem to be solved. This results in a maximum number of merging possibilities between the various states, and thus in a minimum number of memory elements. These are the *internal requirements* of the circuit.
2. After the state table has been reduced to a minimum number of lines, it may be necessary to review the *external requirements*, i.e., the terminating and originating control signals that are normally given in the problem

specifications. It may happen, for instance, that the resulting minimum state diagram can be controlled better by another set of terminating control signals or that the required control signals originate from too complex a group of memory elements. The circuit then has to be adapted to these practical requirements.

3. The basic circuits of the memory elements used must be adapted to the switching equations resulting from the state diagram, so that auxiliary switching elements which have no other function than a repetition (contact multiplication) of the terminating-control signals can be omitted. This is especially important in relay circuits.
4. Once all the preceding design steps have been carried out carefully, it may happen that a reassignment of the states of the circuit, eventually with more than the minimum number of memory elements, results in a circuit with fundamentally simpler gating circuits between memory elements. It may happen that this number of memory elements is even smaller than the total number of relays (auxiliary relays included).

In this section, these four points will be clarified by means of a discussion of some of the examples given in the preceding section. The examples discussed here deal with relay circuits, but it should be borne in mind that, generally speaking, points 1 to 4 apply equally well to electronic circuits, as will follow from the examples of electronic circuits given in the preceding section.

The state diagram of the rotating-shaft problem is repeated here as Fig. 21-42. The diagram shows that the reversal of the rotation of the shaft can be detected on each of the contact segments. This fulfills the stated requirement that this reversal should be detected as soon as possible. On the other hand, however, it has also been specified that this reversal should be detected

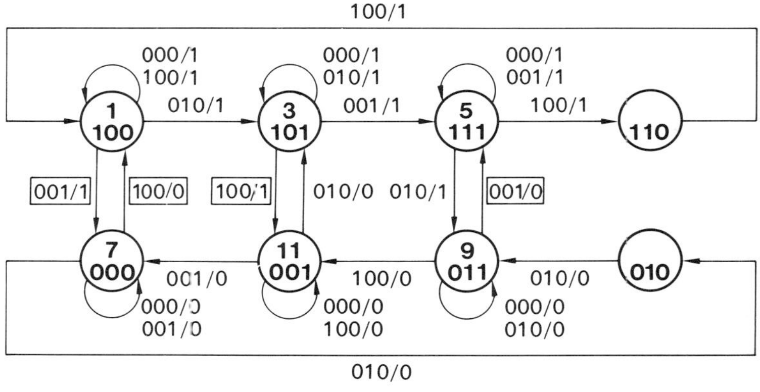

Fig. 21–42

at least within 2 sec, that is, within $\frac{4}{3}$ revolution of the shaft. From an engineering point of view, it is necessary to investigate the consequences and possibilities of this allowed margin of 2 sec in the detection time. This margin can be used with a view to reducing the number of state changes allowing the detection of the reversal of the shaft rotation. The detection of the reversal can be limited, say, to the β segment. The other segments are then used for preparing the detection circuit. The complete state table 21-19 will then reduce to Table 21-22. Limiting the detection to the β segment means that all state-change arrows between the upper and lower row of states in the diagram of Fig. 21-42 which are enclosed have to be omitted as shown in state table 21-22. The pairs of states 1 and 3 and 7 and 9 can obviously be merged so that state table 21-23 is obtained after a renumbering of the states.

Table 21-22

Present state	*Next state*				*Output*			
	000	100	010	001	000	100	010	001
1	1	1	3	-	1	1	1	-
3	3	-	3	5	1	-	1	1
5	5	1	9	5	1	1	0	1
7	7	-	9	7	0	-	0	0
9	9	11	9	-	0	0	0	-
11	11	11	3	7	0	0	1	0

Table 21-23

Present state	*Next state*				*Output*			
	000	100	010	001	000	100	010	001
1	1	1	1	2	1	1	1	1
2	2	1	3	2	1	1	0	1
3	3	4	3	3	0	0	0	0
4	4	4	1	3	0	0	1	0

It can easily be proved by means of Mealy's rule that state table 21-23 contains a minimum number of states. The greatly simplified state diagram corresponding to this table is shown in Fig. 21-43.

The only way that is in principle available for coding the four necessary states is also indicated in the state diagram of Fig. 21-43. The following state equations for shunt-down operated relays result from this state diagram:

$$A = (s_a + a)\,r_a' = (\gamma b + \gamma b' + a)\cdot(\alpha b + \alpha b')' = (\gamma + a)\cdot(\alpha)' \tag{21-48}$$

$$B = (s_b + b)\,r_b' = (\beta a' + b)\cdot(\beta a)' \tag{21-49}$$

The very simple resulting circuit is shown in Fig. 21-44. It comprises only two relays.

The longest detection time in case of a reversal of the shaft rotation will occur if the shaft has nearly reached the γ segment when rotating clockwise. In that state of the detection circuit, relay A is not operated and relay B is operated, as is shown in Fig. 21-45. The first segment to be reached then in the counterclockwise rotation of the shaft is β. The state of the circuit remains unchanged, after which segment α is reached, again without a state change of the circuit. This is described in the state diagram of Fig. 21-43 by the arrow originating from and terminating in state 0 1. The next segment in the counterclockwise direction is now γ. The state of the circuit is now changed from 0 1 to 1 1. Relay A is operated. The contact brush will subsequently reach segment β, so that the state of the circuit will change from 11 into 10. Relay B releases its armature, so that the circuit of the detection lamp of the clockwise rotation of the shaft in Fig. 21-44 is interrupted, while the circuit of the detection lamp of the counterclockwise rotation is closed. All these actions are illustrated by Fig. 21-45.

This figure shows the contact segments and the brush at the moment when the rotation of the shaft is reversed just before the γ segment is reached, and the brush at the moment when the β segment is reached, where the reversal of the shaft rotation is detected. The state of relays A and B is indicated above the segments. The detection takes at most three contact segment times and four pauses so that the maximum detection time is in total 7×250 msec $= 1{,}750$ msec. This time is well below the stated upper limit of 2,000 msec.

It should be noted that the solution obtained in terms of relays and contacts can be easily adapted to a circuit with electronic elements.

When the simplicity of the solution given in Fig. 21-44 is compared with the circuit shown in Fig. 21-34, there will be no doubt that the considerations stated under 1 of this section are very important with respect to the practicability of the result. This even holds when one compares the result of Fig. 21-44 with the circuit of Fig. 21-38.

It should also be recalled here that one disadvantage of this example is that the final solution is of such striking simplicity that it does not offer a state-assignment problem. In many other circumstances, this type of circuit in which all possible operating combinations of the memory elements are used

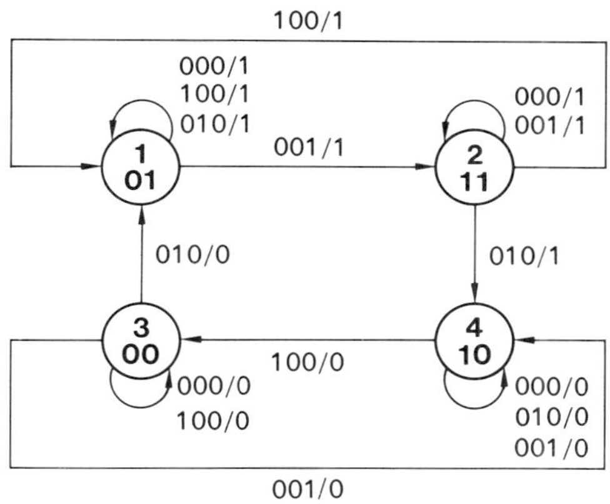

Fig. 21–43

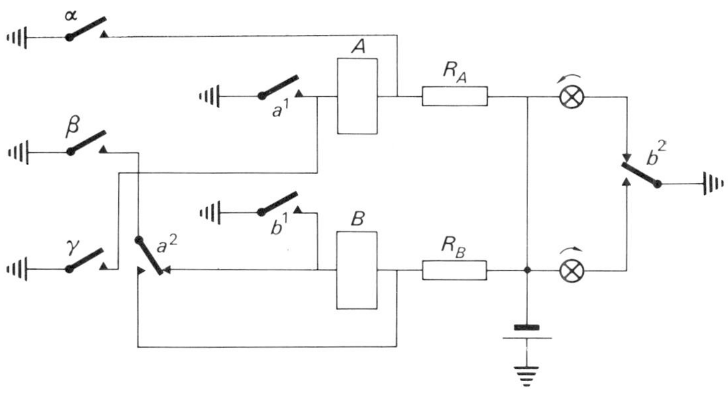

Fig. 21–44

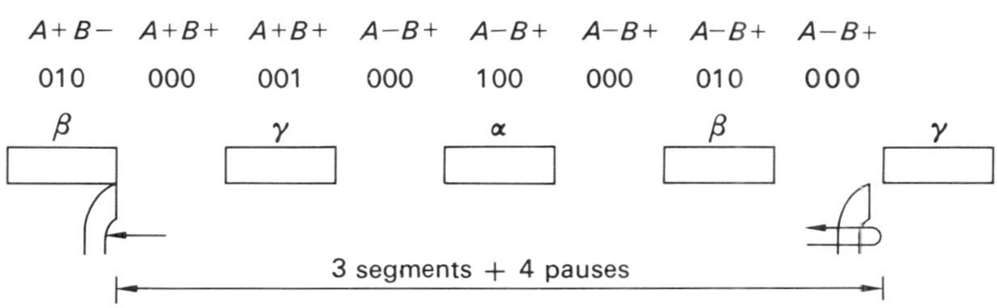

Fig. 21–45

will have a rather complicated internal organization. This is mostly a disadvantage from the point of view of easy maintenance. The operation of those types of circuits is difficult to understand.

The state diagram shown in Fig. 21-33, which represents the original set of requirements of the detection problem, shows a certain lack of organization and symmetry. This was already remedied to a certain extent by the state diagram of Fig. 21-35, but there is still another solution that must be considered. The state diagram of Fig. 21-33 raises the question of whether it would not have been preferable to choose four contact segments instead of three for the input signals. This would reduce the detection time, at the cost of a greater complexity of the circuit, because the number of the required states would be increased. By the introduction of a fourth contact segment, the two transient states required in the original solution are changed into stable states of the circuit.

The state diagram of a detection device operating on four contact segments, but detecting the reversal of the rotation of the shaft at two segments only, is shown in Table 21-24. The corresponding state diagram

Table 21-24

Present state	*Next state*					*Output*				
	0000	1000	0100	0010	0001	0000	1000	0100	0010	0001
1	1	1	1	2	4	1	1	1	1	0
2	2	1	3	2	2	1	1	0	1	1
3	3	4	3	3	2	0	0	0	0	1
4	4	4	1	3	4	0	0	1	0	0

with coded states is shown in Fig. 21-46. The state diagram is entirely systematic in form.

The state equations that result from the state diagram of Fig. 21-46 then become

$$A = (\gamma b + \gamma b' + a)\cdot(\alpha b + \alpha b')' = (\gamma + a)\cdot(\alpha)' \tag{21-50}$$

$$B = (\beta a' + \delta a + b)\cdot(\beta a + \delta a')' \tag{21-51}$$

State equations (21-50) and (21-51) can be represented by means of the circuit shown in Fig. 21-47. If contact δ and contact a^3 are omitted from the circuit shown in this figure, it reduces to the circuit shown in Fig. 21-44.

If it is assumed in the case of four contact segments that these contacts are opened and closed during equally long periods and that the rotation speed of the shaft is the same, the maximum detection time will be about

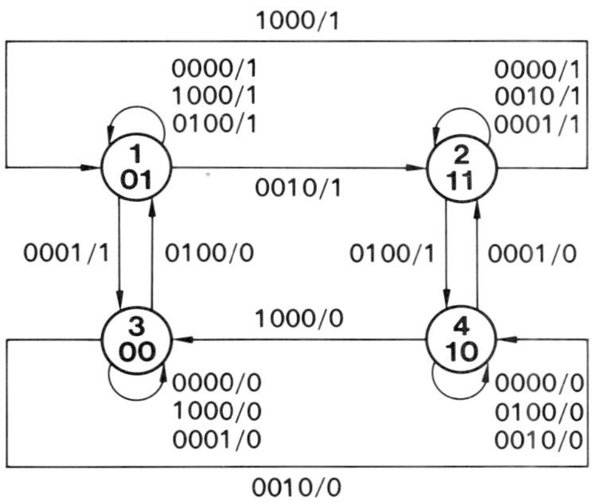

Fig. 21-46

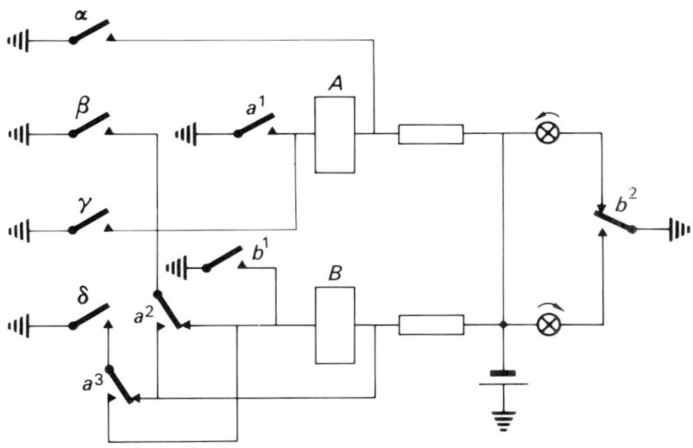

Fig. 21-47

$\frac{5}{8} \times 1{,}500 = 940$ msec, as can be derived easily from a diagram analogous to that given in Fig. 21-45. This detection time of 940 msec is hardly more than the maximum detection time in the original solution, which was 750 msec.

One problem has still to be solved before we leave the discussion of the detection of the direction of rotation of a shaft. In terms of circuits with memory elements, there seems to be no reason why there should not be a single-memory-element solution, because there is only one bit of information to remember, viz., the direction of rotation of the shaft. Could it not be that the possibility for this solution of the problem has already been cut off by the definition of the inputs as three 60° segments? This is indeed the case. The circuit of the single-relay solution (with pip) having a maximum detection time of 1.5 sec is shown in Fig. 21-48.

In the clockwise detection cycle, relay A is operated except for the small gap between segments β and α. Segment α is long enough to operate relay A; the gap between segments α and β is so small that the operating circuit changes make-before-break into the holding circuit via contact a^1. In the counterclockwise detection cycle, relay A is not operated except for the period of segment α.

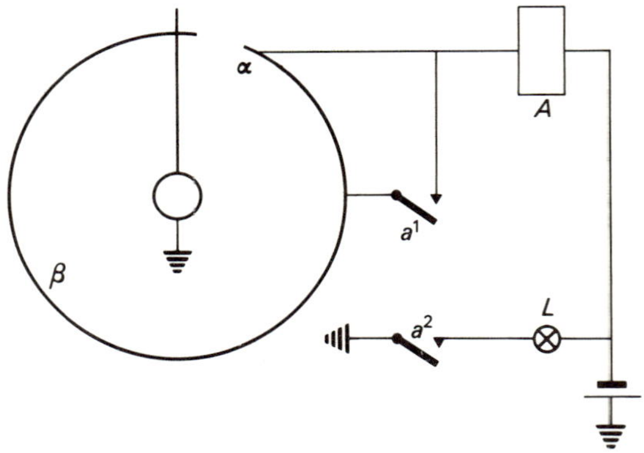

Fig. 21–48

Even after having given this solution, the question still remains of whether the detection of the direction of rotation of a shaft should be performed by means of a relay circuit and not by means of a small friction clutch or a small permanent magnet stuck on the shaft and a single contact.

The problem of counting in the forward direction on signals from one source and counting in the reverse direction on signals from another source is very suitable for illustrating point 4 given at the beginning of this section.

An interesting variant of the forward and backward counting problem is possible when four memory elements are used. When more than the necessary minimum number of memory elements is used in a circuit, the logical organization of the operation of the memory elements can lead to great difficulties if the designer does not know beforehand how he wants to organize his circuit. The problem of state assignment and possibly the redesigning of the state diagram are very important steps in the design of a circuit, as the reader may have gathered from the examples in the preceding paragraph.

The state diagram of a four-step forward and backward counting circuit is shown in Fig. 21-49, where four memory elements are involved. The intermediary steps of the circuit during the periods when controlling contacts α and β are closed are different for both directions of counting. It is pretty difficult to recognize an organization pattern in the states as given in Fig. 21-49. Moreover, the state diagram shows more than one state change per step.

A rearrangement of the code pattern and a grouping in pairs as illustrated in Fig. 21-50 shows that the circuit can be implemented by means of two pulse-halving circuits that count in the forward direction under the control of the α pulse and in the reverse direction under control of the β pulse.

To some extent it is better to illustrate this straightforward type of circuit organization by means of time charts as used in the preceding chapters.

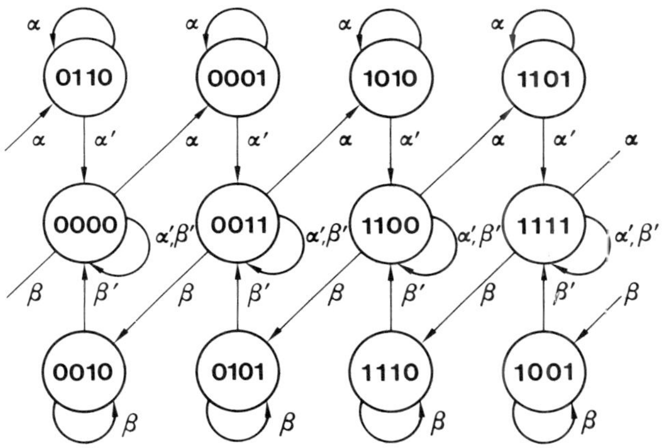

Fig. 21–49

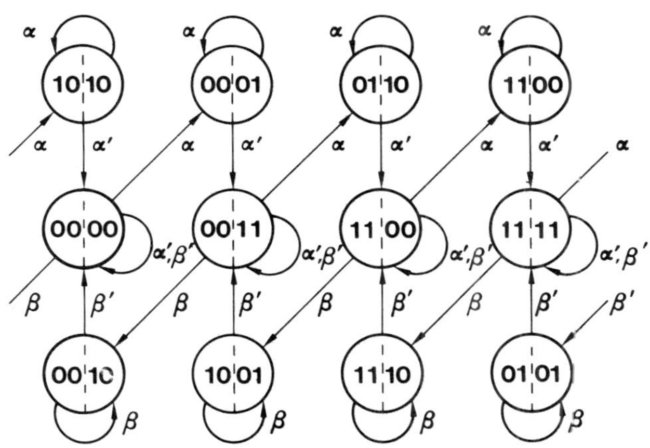

Fig. 21–50

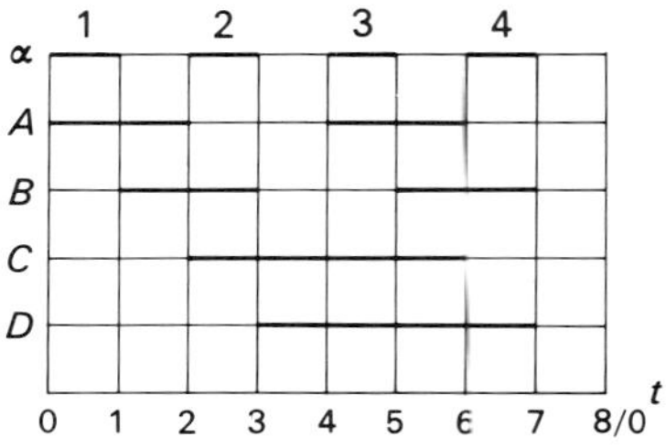

Fig. 21–51

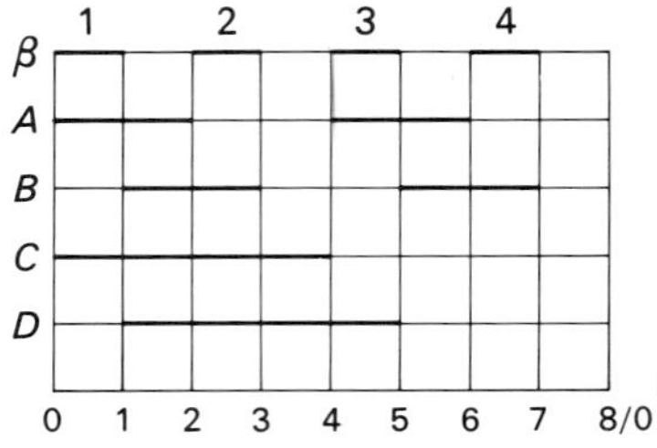

Fig. 21–52

In this example, two time charts are necessary, one to show the operation of the circuit in the forward direction and another to show its operation in the reverse direction. These two time charts are shown in Figs. 21-51 and 21-52.

The time charts show clearly that the operation of the pulse halver A/B is not affected by the direction of counting. The second pulse halver C/D causes the circuit to count in the forward direction when controlled by αb signals, while the circuit counts in the reverse direction when the second pulse halver is controlled by $\beta b'$ signals.

It should be noted that the number of pulses which can be counted in either direction can be increased by the introduction of additional pulse halvers. The resulting circuit for two sections (four pulses) is shown in Fig. 21-53.

If the pulse-halving circuit of Fig. 12-19 is used, the resulting circuit will contain 2 pulse contacts α and β, 4 relays and 10 contacts for logical purposes, and 2 diodes and 6 contacts for controlling the lamps. The first solution given in this chapter uses 2 pulse contacts, 5 relays, and 25 contacts for logical purposes, no diodes, and 6 contacts for switching the indicator

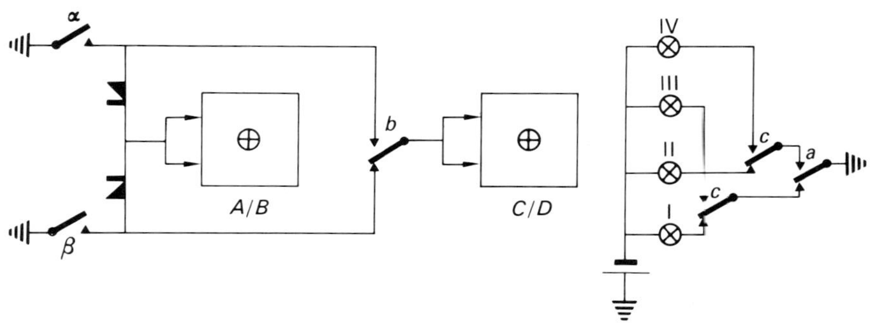

Fig. 21–53

lamps. It is easy to see which circuit is preferable from the economical point of view.

The forward and backward relay counting chain can be transformed without fundamental difficulties into an electronic circuit with n pulse-halving circuits, or n MS memory elements if the number of pulses to be counted in either direction does not exceed 2^n pulses.

The examples discussed in this section and in the previous section may have shown that the state minimizing procedures are very good per se but that a careful examination of the initial requirements is necessary in order to arrive at the most practical circuit. Moreover, the final state diagram coded in a rather arbitrary way leads perhaps to a circuit with the minimum number of memory elements but a maximum number of logical interconnections between those memory elements, and thus certainly not to the most practical circuit. It is necessary to incorporate a systematic organization in the state diagram of the circuit, possibly with an extension of the number of states. The better the circuit can be organized using the fundamental operations of binary counting and shifting, the simpler the final circuit will be, because these fundamental operations can be performed with the simplest means. This is especially true for electronic circuits that operate under the control of a clock pulse.

It is the task of the designer of a switching circuit to consider very carefully the switching problem as a whole and not in its component parts as can easily happen. In many cases, the problem can be solved in a rather easy and successful way if the circuit designer has a thorough knowledge of a number of standard solutions of standard problems. A number of the standard problems such as counting, shifting, adding, transmitting, etc., have been dealt with in the various chapters of this book. Without knowledge about them, or with the knowledge only of the handling of switching formulas, state equations, and the minimizing of state tables, it will be very difficult for the circuit designer to find solutions that show a practical and simple internal organization.

PROBLEMS

1. A rotating shaft carries two perpendicular grounded brushes making contact with four stationary commutator segments arranged symmetrically around the shaft as shown in Fig. 21-54. There are five possible situations:

1. The brushes make contact with segments a and b
2. The brushes make contact with segments b and c
3. The brushes make contact with segments c and d
4. The brushes make contact with segments d and a
5. The brushes make no contact with any of the segments

The commutator segments form the input of a circuit in black box S, the output of which is a circuit with lamp L connected to battery. The requirement of the design of the circuit in black box S is such that the output circuit via lamp L is closed in situations 1, 2, and 3 and open in situations 4 and 5. Design the circuit in the black box.

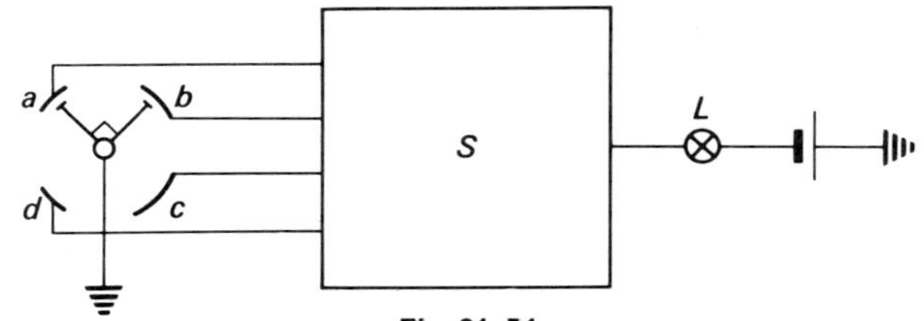

Fig. 21–54

2. A switch S with four positions as shown in Fig. 21-55 is given. Contacts a, b, and c of this switch form the input of a circuit in black box X. Contact T forms the fourth input of black box X. The requirement of the design of the circuit in the black box is such that outputs y and z indicate the numbers of the positions of switch S in the natural binary code when T is open and in the reflected binary code when T is closed. This is summarized in the table below.

S	$T = 0$		$T = 1$	
	y	z	y	z
0	0	0	0	0
1	0	1	0	1
2	1	0	1	1
3	1	1	1	0

The complete black-box circuit is shown in Fig. 21-56.

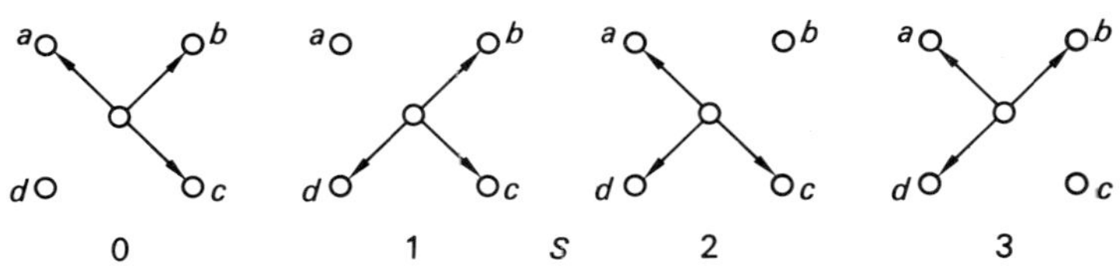

Fig. 21–55

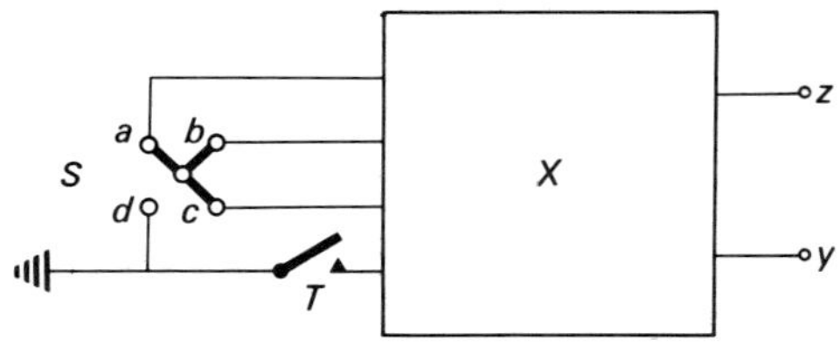

Fig. 21–56

3. A rotating shaft carries a single grounded brush which makes contact with three stationary commutator segments arranged as is shown in Fig. 21-57. A relay circuit S is required which will indicate the direction of rotation of the shaft by lighting map L when the shaft is rotating clockwise. Commutator segments a and b are sufficiently long to allow the operation or release of a relay. The brush and commutator segments are ideally shaped so that during the rotation of the shaft there is no contact interruption of the brush and no bridging of successive segments by the brush.

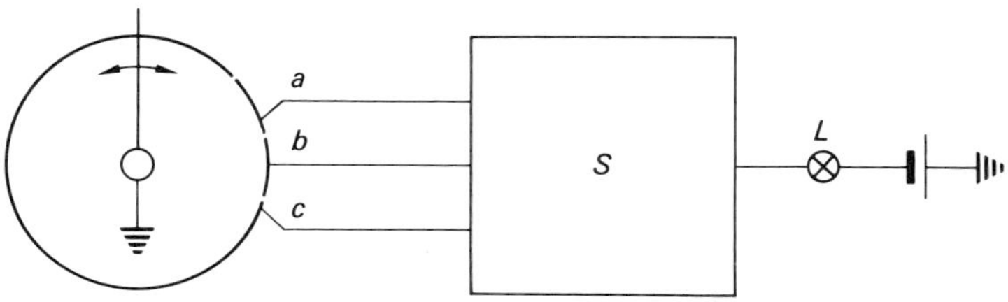

Fig. 21-57

4. In Prob. 3, a relay circuit is required which will indicate the direction of rotation of the shaft by lighting lamp L for a long period with a short interruption when the shaft is rotating clockwise and by lighting lamp L for a short period with a long interruption when the shaft is rotating counterclockwise.

5. The black box S in the circuit of Fig. 21-58 has three inputs a, b, and c, and one output. The output is connected to lamp L and to input c. Devise a relay circuit in black box S such that when signal $b = 1$, the circuit of lamp L will be closed except $a = b = c' = 1$ and will remain closed when $b = 0$ except when $a' = b = 0$. Signals a and b can be given in any sequence.

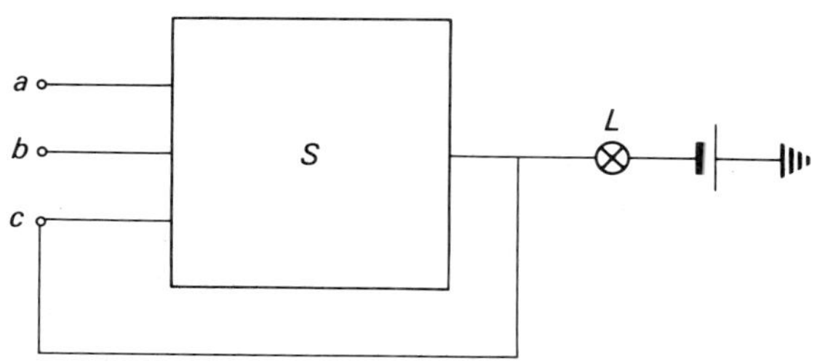

Fig. 21-58

6. The design of an electronic circuit with two inputs α and β and six outputs is required. This configuration is shown in Fig. 21-59. The inputs receive α and β signals respectively in an arbitrary sequence. No two signals $\alpha = 1$ and $\beta = 1$ will be given simultaneously, so that input combination 11 is excluded. The required circuit has six outputs feeding six lamps L_1 to L_6. One of these lamps at a time is lit. When an α signal is received, the circuit advances the position of the illuminated lamp; when a β signal is received, the change in position of the illuminated lamp occurs backward.

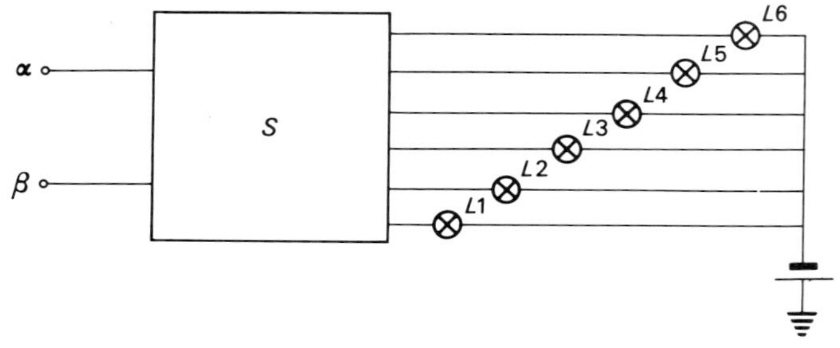

Fig. 21–59

7. Design an electronic circuit having one input and one output only (Fig. 21-60). The input signals are clock pulses at regular intervals. The output signals must be 11 element cycles of the following character:

1 0 1 1 0 1 1 1 0 0 0

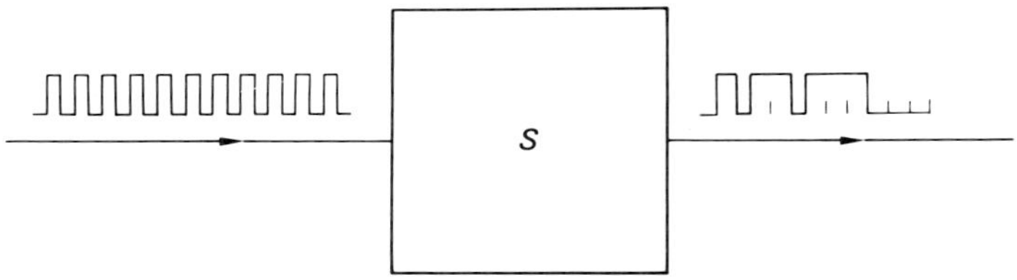

Fig. 21–60

8. Devise an electronic circuit with three inputs and one output. This circuit receives 0 and 1 signals of constant duration in an arbitrary sequence on a first input, each 1 signal having a space to the next 1 signal of at least one 0 signal, a clock signal with the same repetition rate on a second input, whereas a switch SW is connected to a third input. Switch SW when closed blocks the output. Switch SW is closed for 10 clock pulses and then open for 10 clock pulses, and so on. When switch SW is closed, the incoming 1 signals are stored in such a way that when switch SW is opened again they are fed to the output spaced with a single 0; and when this is no longer possible, the incoming 1 signals are retransmitted with their original spacing.

9. Design an electric lock circuit that generates an output signal when detecting in a series of six successive input signals three 1 signals or four 0 signals.

10. Design an electronic counting circuit operating in cycles of 10 such that an output signal is generated each tenth pulse. The code of counting must be such that successive code combinations of the counter differ in one element only and code combinations 0 to 4 and the corresponding complements to nine (5 to 9) also differ in one element only.

SELECTIVE BIBLIOGRAPHY

1. Aufenkamp, D. D.: Analysis of Sequential Machines, *IRE Trans.*, vol. EC-7, pp. 299–306, 1958.
2. Aufenkamp, D. D., and F. E. Hohn: Analysis of Sequential Machines, *IRE Trans.*, vol. EC-6, pp. 276–285, 1957.
3. Ginsburg, S.: A Synthesis Technique for Minimal State Sequential Machines, *IRE Trans.*, vol. EC-8, pp. 441–449, 1959.
4. Ginsburg, S.: "An Introduction to Mathematical Machine Theory," Addison-Wesley Publishing Company, Inc., 1962.
5. Huffman, D. A.: The Synthesis of Sequential Circuits, *J. Franklin Inst.*, vol. 257, pp. 161–190 and 275–303, 1954.
6. Humphrey, W. S., Jr.: Sequential Circuits, chap. 10 of "Switching Circuits with Computer Applications," McGraw-Hill Book Company, 1958.
7. Mealy, G. H.: A Method for Synthesizing Sequential Circuits, *Bell System Tech. J.*, vol. 34, pp. 1045–1079, 1955.
8. Moore, E. F.: Gedanken Experiments on Sequential Machines, *Automata Studies*, pp. 129–153, 1956.
9. Seshu, S., R. E. Miller, and G. Metze: Transition Matrices of Sequential Machines, *IRE Trans.*, vol. CT-6, pp. 5–12, 1959.

22
Drawing Symbols and Methods

For carefree operation of the circuit, the drawing of a switching circuit is nearly as important as the correct design itself. In the preceding chapters, switching circuits were designed and expressed in switching formulas. These switching formulas have to be converted into circuit diagrams. Depending on the kind of circuit to be designed, the final diagram will be shown by means of symbols representing the actual circuit elements or groups of those circuit elements. The circuit elements are, for example, a diode, a transistor, a relay, a contact, and even a resistor and a capacitor.

A list of the symbols used in this book will be given in this chapter with some remarks about their use. The list is not at all a reference list of all possible symbols in the field of switching. So far, no generally accepted list of drawing symbols for switching purposes has been agreed on by the International Electrical Committee (IEC). Most of the symbols given are more or less in general use in the specific form or as a variant thereof. The choice of the symbols for this book has been governed by the criterion that a symbol has to express something of the function which it fulfills and that it should be easy to draw. This last consideration, which has been neglected far too often, is necessary to keep drawing costs low.

The various circuits described in the preceding chapters all have a more or less systematic build-up, which is reflected in the drawing. Detailed switching circuits are mostly drawn in such a way that the positive pole of the power source is on the top and the negative pole on the bottom of the figure. This allows for an easy means of description of the operation of the circuit if voltage levels are involved.

The circuits are also mostly drawn in such a way that the operation starts at the left-hand side and ends at the right-hand side. The relay circuits are all drawn in the so-called detached-contact method, i.e., a method in which the contacts of a relay are not drawn in the immediate vicinity of the operating winding but are scattered all over the drawing of the circuit and placed in the functionally correct position, so that the circuit lines in the drawing are as short as possible and can be placed from left to right in the correct sequence of operation of the circuit.

All relay and contact circuits are always shown in the situation of disconnected power supply. The starting point of a circuit description will therefore always be the same. The advantages of the detached-contact drawing method may seem negative at first. This may be true for very simple circuits; but it is definitely not the case for complicated circuits. The detached-contact drawing method can simplify the task of explaining complicated and very complicated contact circuits in combination with the systematic arrangement of the various circuits from left to right in the sequence of their operation.

The aspect of the drawing of a complicated switching circuit will be influenced largely by the possibility of subdividing the circuit into smaller units performing a specific function such as counting, adding, shifting, or the like. The most suitable type of diagram can then be chosen for the unit circuits. From the point of view of easy maintenance, a circuit that can be drawn in a clear and systematic way is generally preferable to a circuit in which all or nearly all possible combinations of memory elements are used. This last type of circuit generally has a fairly complicated logic that inhibits easy fault-finding by means of the types of simple mental processes required for the majority of the maintenance personnel.

No.	*Symbol*	*Description*	*Remarks*
1	General symbols		
1.1		Battery	
1.2	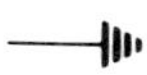	Earth	

No.	*Symbol*	*Description*	*Remarks*
1	General symbols		
1.3		Voltage source	
1.4		Current source	
1.5	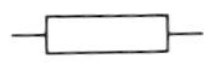	Resistor	Long rectangle.
1.6		Self-inductor	
1.7	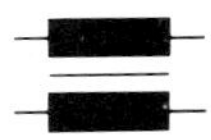	Transformer	
1.8		Capacitor	
1.9	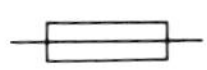	Fuse	
2	Relay windings		
2.1	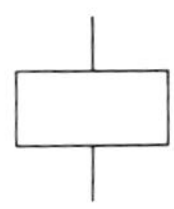	Relay winding	This symbol must be drawn for each winding of a relay. The symbols may be placed apart or side by side.
2.2	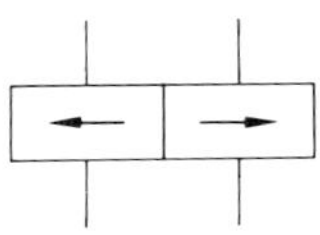	Differential relay windings	If necessary, the direction of magnetization of the relay windings can be indicated by an arrow.
2.3	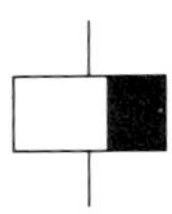	Slow-releasing relay	
2.4	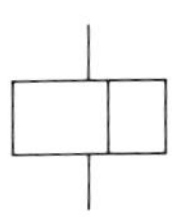	Slow-operating relay	

No.	*Symbol*	*Description*	*Remarks*
2	Relay windings		
2.5	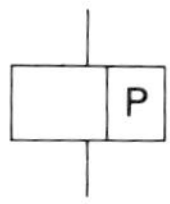	Polarized relay	
2.6	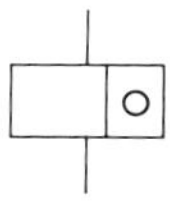	Message register	
3	Relay contacts		
3.1		Make contact	
3.2		Break contact	
3.3	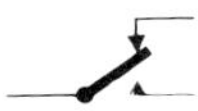	Transfer contact	
3.4	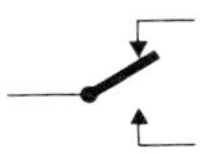	Symmetrical transfer contact	For polar relays.
3.5	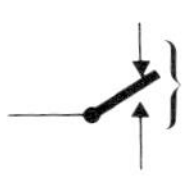	Make-before-break contact	
3.6		Symbol for contact *z*	
4	Tubes		
4.1		Gas-filled diode	Indicator tube.
4.2		High-vacuum diode	

No.	*Symbol*	*Description*	*Remarks*
4	Tubes		
4.3		Triode	
4.4		Double triode	The two halves of a double triode may also be drawn separately.
4.5		Pentode	
5	Semiconductors		
5.1*a*		Diode	
5.1*b*			
5.2	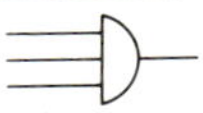	AND gate	
5.3	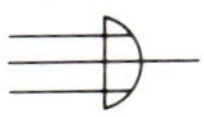	OR gate	
5.4		*pnp* transistor	
5.5		*npn* transistor	
6	Magnetic cores		
6.1	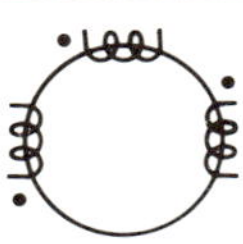	Magnetic core	Physical symbol with dot notation for polarity indication.
6.2	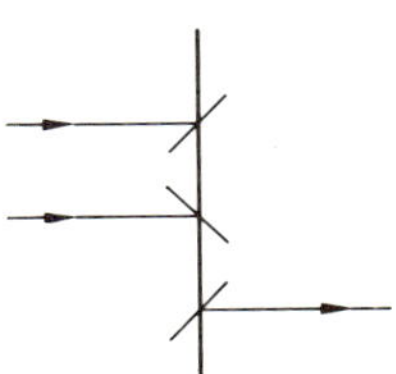	Magnetic core	Mirror method showing polarities of induced currents more easily.

No.	*Symbol*	*Description*	*Remarks*
7	Composite symbols		
7.1	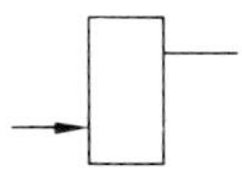	Inverter	Corresponding generally to a triode or transistor inverter as shown below. 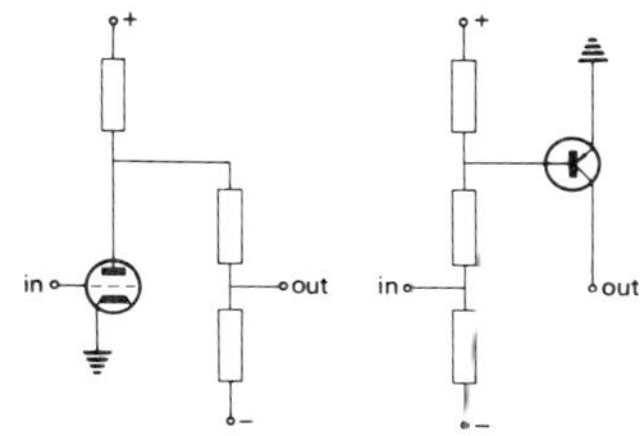 This symbol can also be used for other types of inverters, such as pulse transformers.
7.2*a*	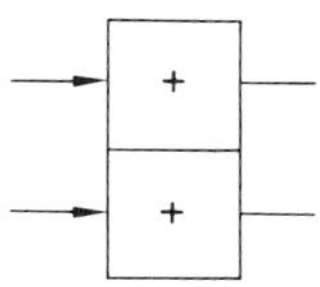	Trigger	A trigger is formed by two inverters. The output of one inverter is connected to the input of the other, and vice versa. The external control is applied via an AND gate or an OR gate depending on the polarity of the control pulse. These gates are sometimes degenerated by replacing the diodes by resistors. The indication + or · can then be omitted. If an input lead of the trigger of symbol 7.2*a* receives a control signal changing from 0 to 1, the corresponding output lead also changes from 0 to 1, and the other from 1 to 0. The reverse is true for the trigger of symbol 7.2*b*.
7.2*b*	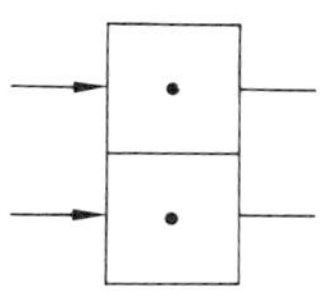		
7.3	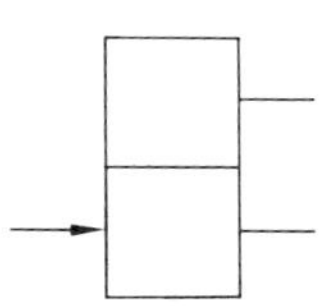	Single-input trigger	Trigger with a one-wire control (set and reset on the same wire). Remains in its position without external control.

No.	*Symbol*	*Description*	*Remarks*
7	Composite symbols		
7.4	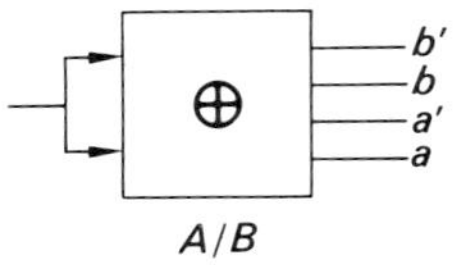	Pulse halver	Circuit comprising two triggers or one trigger and a delay storage element. Performs the addition mod 2 on incoming control pulse. Output wires may be omitted if not available or required.
7.5	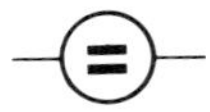	Integrating delay storage element	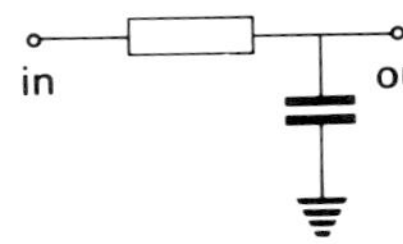
7.6*a*	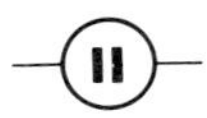	Differentiating delay storage element	In many cases, the delay storage element shown below is used in a more complicated form.
7.6*b*			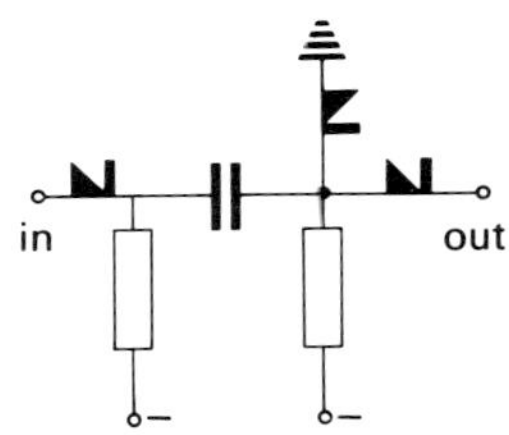
7.7	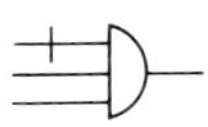	Gate with inhibiting input	If an inverter is connected to one or more inputs of a gate, this can be indicated with a cross line as shown in the gate symbol.
7.8	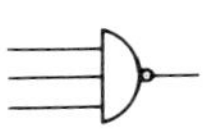	Gate with inhibiting output	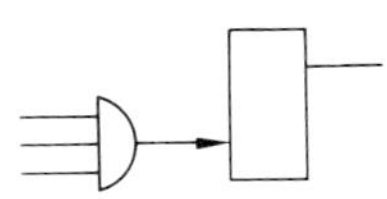

Index